# SYSTEM RELIABILITY

Edited by **Constantin Volosencu**

**System Reliability**
http://dx.doi.org/10.5772/66993
Edited by Constantin Volosencu

**Contributors**

Atanda Raji, Filippo De Carlo, Maria Antonietta Arleo, Vinicius Jacques Jacques Garcia, Magdiel Schmitz, Maria Clara Ferreira Almeida Da Silva, Daniel P Bernardon, Lynceo Falavigna Braghirolli, Júlio Schenato Fonini, Marco Boaski, Mauricio Sperandio, Caio Santos, Maicon Ramos, Daniel Porto, Sébastien Jacques, Seongwoo Woo, Bernard Tonderayi Mangara, Xiaojian Yi, Jian Shi, Peng Hou, Nataša Kontrec, Stefan Panić, Ivanka Stanimirovic, Zdravko Stanimirovic, Denis Migov, Alexey Rodionov, Stephan Neumann, Lothar Wöll, Georg Jacobs, Katharina Schick, Achim Kramer, Christophe Letot, Lucas Equeter, Clément Dutoit, Pierre Dehombreux, Carlos Barriquello, Joseph Bernstein, Rabah Medjoudj, Djamil Aissani, Hassiba Bediaf, Jon T. Selvik, Eric P. Ford, Michał Szewczyk, Ibrahima Dit Bouran Sidibe, Djelloul Imene, Giovanna Adinolfi, Giorgio Graditi

**Published by InTech**
Janeza Trdine 9, 51000 Rijeka, Croatia



**Publishing Process Manager** Romina Skomersic
**Technical Editor** SPi Global
**Cover** InTech Design team

First published December, 2017
Printed in Croatia
Legal deposit, Croatia: National and University Library in Zagreb

Additional hard copies can be obtained from orders@intechopen.com

System Reliability, Edited by Constantin Volosencu
p. cm.
Print ISBN 978-953-51-3705-4
Online ISBN 978-953-51-3706-1

For EU product safety concerns:
IN TECH d.o.o., Prolaz Marije Krucifikse Kozulić 3, 51000 Rijeka, Croatia,
info@intechopen.com or visit our website at intechopen.com.

# Meet the editor

Constantin Volosencu is a professor at "Politehnica" University from Timisoara, Department of Automation. He is the author and editor of 14 books, author of over 200 scientific papers published in journals and conference proceedings, author of 27 patents and manager of research grants. He is a member of editorial boards of international journals, former plenary speaker, a member in scientific committees and the chair at international conferences. He has researches in the field of control systems, electrical drives, power ultrasonic, fuzzy logic, neural networks, fault detection and diagnosis, sensor networks and distributed parameter systems. He developed electrical equipment for machine tools, spooling machines, high-power ultrasonic processes and others, with homologation of 30 prototypes and series zero.

# Contents

# Preface

System reliability has its importance in engineering, to identify the least reliable components in the system in order to improve the design. In the complex system, this becomes a difficult task. In this domain, the analysts need a mathematical approach to identify and to quantify the importance of each component in the system. For keeping a system in working conditions, activities of maintenance must be involved.

Researchers from the entire world write to figure out their newest results and to contribute new ideas or ways in the field of system reliability. They were inventive and see the field as inspiration for research. The kinds of applications they treat are indicative of each author's interest and engagement. The book shows the degree of information from authors in the field and their dedication, appreciation and enthusiasm of the field they care about.

The authors choose to publish their research project results to explain reliability patterns in different applications, to explain variations of parameter in different industrial processes and to test the consequences of parameter variations on processes; they conducted experimental tests of hypothesis, developed mathematical versions of reliability models and used fundamental ideas to address applied problems in novel ways.

The authors were motivated to write the chapters to keep them focused, energized and inspired, using the great opportunity for publishing in an open access way through a well-known publishing house. The preface must be seen as a good thought of wisdom about their hopes and their motivation and achieving their goals.

In a brief description, the book has four sections: reliability, reliability of electronic devices, power system reliability and feasibility and maintenance.

The book can become a valuable tool for professors, students and professionals, with its presentation of some issues that may be taken as examples applicable to new practical situations. Based on the authors' experience, the book may be popular among specialists.

The book presents cases that illustrate the application in a variety of domains. The first section begins with an interesting system reliability analysis based on goal-oriented methodology, followed by some applications as reliability design of water-dispensing systems, based on field data and parametric accelerated life tests; reliability evaluation of drivetrains for off-highway machines, taken into consideration the load cycles and components load, with examples of application of the power shift transmission and the offshore winch; a predictive analysis on equipment downtime in the oil and gas industry to improve maintenance and to prevent hazardous event that could harm people and the environment; a study for extending the useful life of asset, based on an adaptive maintenance methodology that uses the reliability-centred maintenance approach to define a degradation-based adaptive maintenance model; a study in the area of network reliability analysis to speed up obtaining of the cumulative bounds for faster feasibility decision; analysis of standard reliability parameters of technical systems' parts based on spare parts forecasting using Rayleigh and Weibull's models; and a study on cannibalisation for improving system reliability, with reduced adverse effects on maintenance costs and personnel morale. The second section begins with an interesting mathematical study on the multiple temperature operational life testing procedure, for electronic industry, to replace the standard high-temperature operational life, by predicting the system failure rate by testing a small number of components, calculating the mean time to failure. Some applications follow: reliability prediction of smart maximum PowerPoint converter in photovoltaic applications; the reliability of die interconnections used in plastic discrete power packages, dedicated to on-board electronic systems used in applications such as automotive industry; and the analysis of the effects of mechanical and electrical straining on performances of conventional thick-film resistors, from micro- and macro-structural, charge transport and low-frequency noise aspects. The third section presents the following: analysis of software and hardware development in phasor measurement unit technology to ensure the secure operation and stability of transmission systems in the electric power system; a study on electric interruptions and loss of supply in power systems, based on interruption modelling using Weibull-Markov process with degradation; a short overview of electricity distribution networks with its protection devices for assessment of protection and reliability; and a technical-economic feasibility study of autonomous hybrid AC/DC microgrid system. The fourth section presents a predictive modelling of emergency services in electric power distribution systems for resource planning during extreme weather events when considering the geographic dispersion of such services and also the time windows that comprise the amount of service time demanded; a web-based decision-support system aimed to aid distribution system operators for planning and executing customer and maintenance services in the electric power distribution system; a short review of some maintenance models, with examples of imperfect maintenance models in literature, to identify suitable models for real cases; and a mathematical model–based study of preventive maintenance of a repairable equipment whose lifetime distribution depends on the operating environment severity.

The editor thanks the authors for their excellent contributions in the field and understanding during the process of editing. Also, the editor thanks to all the editorial personnel involved in this book publication. The publishing provided a set of editorial standards, which ensured the quality of the scientific level of relevance of accepted chapters.

**Prof. Constantin Volosencu**
Politehnica University of Timişoara
Timisoara, Jud. Timis, Romania

Section 1

# Reliability

Chapter 1

# Complex System Reliability Analysis Method: Goal-Oriented Methodology

Yi Xiao-Jian, Shi Jian and Hou Peng

Additional information is available at the end of the chapter

http://dx.doi.org/10.5772/intechopen.69610

**Abstract**

Goal-oriented (GO) methodology is a success-oriented method for complex system reliability analysis based on modeling the normal operating sequence of a system and all possible system states. Recently, GO method has been applied in reliability and safety analysis of a number of systems, spanning defense, transportation, and power systems. This chapter provides a new approach for reliability analysis of complex systems, first, by providing its development history, its engineering applications, and the future directions. Then, the basic theory of GO method is expounded. Finally, the comparison of GO method, fault tree analysis and Monte-Carlo simulation is discussed.

**Keywords:** system reliability analysis, complex system, GO method, reliability model

## 1. Introduction

Quality and reliability are key attributes of economic success of a system because they result in an increase in productivity at low cost and vital for business growth and enhanced competitive position. The recent advances in electronics, computing, communication, control, and networking have resulted in integrated systems that are: (i) complex in structure, (ii) large in scope and scale, (iii) characterized by multimode operation, (iii) capable of working in varied working conditions, and (iv) hierarchically organized. The reliability of such complex systems is a critical factor of their fitness for their intended use and hence is vital for in design and manufacturing. Reliability analysis method of complex systems is conducive to prevent defects in the first place in all aspects when they do occur in operation, in order to improve their reliability and reduce their life-cycle cost.

Fault tree analysis (FTA) and Monte-Carlo simulation (MCS) are now the standard reliability and safety analysis methods. Different from them, goal-oriented (GO) methodology [1] is a success-oriented method for system reliability analysis based on modeling the normal operating sequence of a system and all possible system states. It is especially suitable for complex systems having time-sequence of operation, multiple states, and so on, and the quantitative analysis and the qualitative analysis of the GO method are conducted by the GO operation according to the GO model. The keys of the GO method are the GO model and the GO operation. Although the GO method was introduced in 1976 [1], it was largely unknown until recently. The GO method has become increasingly popular in recent years because of its advantages in terms of its ease of creating a model and of its representational and analysis power [2, 3]. This chapter is an attempt at providing a basic theory of GO method in terms of GO model, GO operation, and comparison with FTA and MCS.

### 1.1. Development of GO method

The major application of GO method is in establishing a system reliability model, and its quantitative and qualitative analysis. The chronological development of the GO method can be broadly divided into two periods of growth: (1) 1970s~the early 1990s, the basic model and theory of the GO method, its comparison with FTA method, and its GO operator type and function were explained in the research reports by the Electric Power Research Institute in US [4–8]; (2) After the late 1990s, the GO method has attracted more attention again, in particular in the People's Republic of China. Perhaps, the initial application of the GO method was in reliability and safety analysis of missile and weapons systems. Recently, the GO method has also been applied in the reliability analysis of defense systems, water, oil, and gas supply systems, manufacturing systems, transportation systems, power systems, and logistics management systems [2, 3]. Furthermore, the theory of GO method for a complex system with complex correlations (dependencies), closed-loop feedback, multiple functions, multiple fault modes, etc., has been developed from three aspects, which are GO model, basic GO algorithm, and GO method for complex systems with various characteristics [9–17]. Literature [2, 3] gave the overview of development of GO method in detail.

### 1.2. Further of GO method

The GO method has been widely used. There are a number of areas where the powerful advantages of the GO model and its reliability analysis methodology can be exploited.

Although there is a great deal of interest in the design of complex systems for reliability, research on deigning these systems for both functional and structural reliability and life-cycle cost is needed. To go further, there is still a substantial research gap in the optimal design of systems for reliability and life-cycle cost, taking into account the issues of structure, function, behavior, and other characteristics, such as active or cold standby redundancy, and fault-tolerant mechanisms. In terms of modeling, analysis, and software tools, there are a number of research issues to consider. These include: (i) How to integrate the product structure, behavior, and functions in a reliability model; (ii) How to conduct the reliability analysis of complex systems accurately, thoroughly, and quickly; (iii) How to optimally allocate reliability among subsystems, taking into consideration the structural and functional hierarchy, as well as redundancy management techniques under resource constraints; and (iv) How to develop software tools to support design of complex systems for reliability that are intuitive and support collaborative design. Considering the advantages of GO model and its reliability analysis method, the GO method not only can solve the existing problems above, but also it can further develop system reliability theory and application. Meanwhile, the corresponding software will also have a more extensive application prospects and important value.

Another important direction for research is the application of the GO methodology to related areas of quality control, fault diagnosis and prognosis, and condition-based maintenance.

## 2. GO model

The GO model is a key element of GO method. It is developed directly using product schematic diagrams, its structure, and its functional hierarchy. According to the GO model, the reliability analysis is conducted by GO operation. The GO model composed of GO operator and signal flow.

### 2.1. GO operator

GO operator contains function operator and logical operator, which represents the unit itself or logical relationship, respectively. Its data, type, and GO operation formula are the basic attributes of GO operator. There are 17 standard GO operators in basic GO theory, and their signs and description are shown in **Figure 1** and **Table 1**, respectively. In **Figure 1**, $S$, $C$, and $R$ are the input signal, GO operator itself, and output signal, respectively.

#### *2.1.1. Standard GO operator*

In this section, six frequently-used standard GO operators are illustrated from aspects of description, operation rule table, and GO operation formula, respectively. In operation rule table, $V_S$, $V_C$, and $V_R$ are the state of the input signal, GO operator itself, and output signal, respectively, and $0, \cdots, N$ represent their state values. In GO operation formula, $P_S(i)$, $P_R(i)$, and $P_C(i)$ are the state probability of the input signal, GO operator itself, and output signal, respectively. The state cumulative probability of input signal $A_S(i)$ and output signal $A_R(i)$ are defined as

$$\begin{cases} A_S(i) = \sum_{j=0}^{i} P_S(j) \quad i = 0, \cdots, N-1; A_S(N) = 1 \\ A_R(i) = \sum_{j=0}^{i} P_R(j) \quad i = 0, \cdots, N-1; A_R(N) = 1 \end{cases} \tag{1}$$

**1. Type 1 operator**

- **Description:** It describes the unit with two states, which are success state (enable signal flow pass) and failure state (stop signal flow passing). For example, electric resistance, switch, valve, and pipeline.

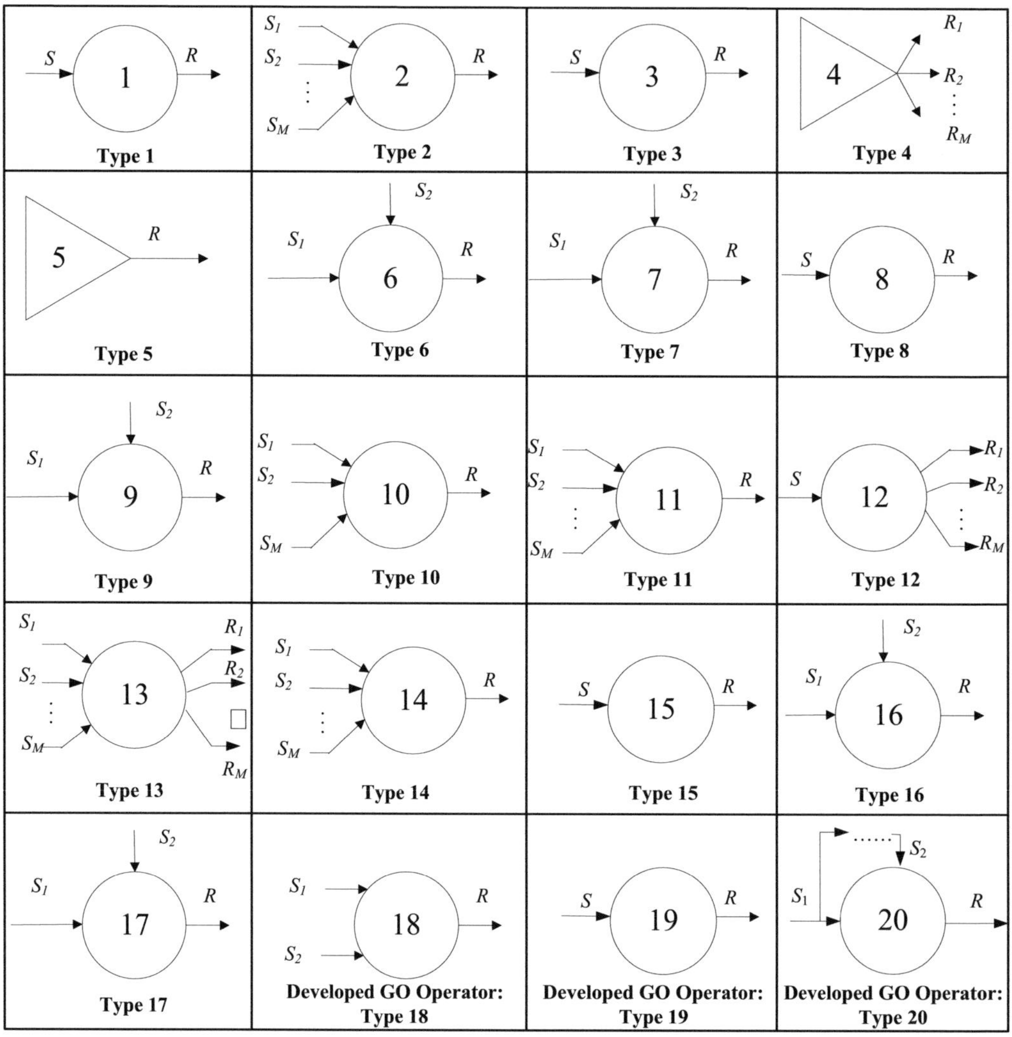


**Figure 1.** Signs of standard GO operators and three developed GO operators.

- **Operation rule table:**

| $V_S$ | $V_C$ | $V_R$ |
|---|---|---|
| 0, …, $N$–1 | 1 | 0, …, $N$–1 |
| $N$ | 1 | $N$ |
| 0, …, $N$ | 2 | $N$ |

- **GO operation formula:**

$$\begin{cases} P_R(i) = P_S(i) \cdot P_C(1), \quad i = 0, 1, \cdots, N-1 \\ P_R(N) = P_S(N) \cdot P_C(1) + P_C(2) \end{cases} \tag{2}$$

**2. Type 2 operator**

- **Description:** It describes the logical relationship OR among some inputs signal and one output signal.
- **Operation rule table:**

| $V_{S1}$ | $V_{S2}$ | … | $V_{SM}$ | $V_R$ |
|---|---|---|---|---|
| 0, …, $N$ | 0, …, $N$ | 0, …, $N$ | 0, …, $N$ | MIN{$V_{S1}$, $V_{S1}$,…, $V_{SM}$} |

- **GO operation formula:**

$$\begin{cases} P_R(0) = A_R(0) \\ P_R(i) = A_R(i) - A_R(i-1), i = 1, \cdots, N \end{cases} \tag{3}$$

**3. Type 3 operator**

- **Description:** It describes unit with failure state, operating state, and operating ahead state. For example, control system, contactor coil, and so on.
- **Operation rule table:**

| $V_S$ | $V_C$ | $V_R$ |
|---|---|---|
| 0, …, $N$ | 0 | 0 |
| 0, …, $N$–1 | 1 | 0, …, $N$–1 |
| $N$ | 1 | $N$ |
| 0, …, $N$ | 2 | $N$ |

- **GO operation formula:**

$$\begin{cases} P_R(0) = P_C(0) + P_S(0) \cdot P_C(1) \\ P_R(i) = P_S(i) \cdot P_C(1), i = 1, \cdots, N-1 \\ P_R(N) = P_S(N) \cdot P_C(1) + P_C(2) \end{cases} \tag{4}$$

**4. Type 5 operator**

- **Description**: It describes the single input unit, which as system input. For example, battery, water source, and so on.
- **GO operation formula:**

$$P_R(i) = \begin{cases} P_j, i = I_j \\ 0, i \neq I_j \end{cases}, j = 1, \cdots, L, i = 0, \cdots, N \tag{5}$$

**5. Type 6 operator**

- **Description:** It describes the unit receiving signal to turn on. For example, electric water pump, contactor, and so on.

| Operator type | Functional description |
|---|---|
| Type 1 | Unit with failure state and operating state |
| Type 2 | Logical relation "OR" |
| Type 3 | Unit with failure state, operating state, and operating ahead state |
| Type 4 | Multi-signal input unit |
| Type 5 | Single input unit |
| Type 6 | Unit receiving signal to turn on |
| Type 7 | Unit receiving signal to turn off |
| Type 8 | Unit with delayed response |
| Type 9 | Output signal decided by the state of two input signal |
| Type 10 | Logical relation "AND" |
| Type 11 | Logical relation "k-out-of-m" |
| Type 12 | Input signal can choose the output path |
| Type 13 | Unit with multiple input signals and output signals |
| Type 14 | Linear relation between multiple input signals and one output signal |
| Type 15 | Logical relation of output signal affected by the probability event of input signal |
| Type 16 | Unit requested to resume OFF-state |
| Type 17 | Unit requested to resume ON-state |

**Table 1.** Functional description of standard GO operators.

- **Operation rule table:**

| $V_{S1}$ | $V_{S2}$ | $V_C$ | $V_R$ |
|---|---|---|---|
| $I_1(0, \ldots, N)$ | $I_2(0, \ldots, N)$ | 0 | $I_1$ |
| $I_1(0, \ldots, N)$ | $I_2(0, \ldots, N)$ | 1 | $MAX\{I_1, I_2\}$ |
| $I_1(0, \ldots, N)$ | $I_2(0, \ldots, N)$ | 2 | $N$ |

- **GO operation formula:**

$$\begin{cases} A_R(i) = A_{S1}(i)[P_C(0) + A_{S2}(i) \cdot P_C(1), \ i = 0, \cdots, N-1 \\ P_R(N) = P_{S1}(N) + A_{S1}(N-1)[P_C(2) + P_{S2}(N) \cdot P_C(1)] \\ P_R(0) = A_R(0) \\ P_R(i) = A_R(i) - A_R(i-1) \ i = 1, \cdots, N-1 \end{cases} \quad (6)$$

**6. Type 10 operator**

- **Description**: It describes the logical relationship AND among some inputs signal and one output signal.

- **Operation rule table**:

| $V_{S1}$ | $V_{S2}$ | … | $V_{SM}$ | $V_R$ |
|---|---|---|---|---|
| 0, …, $N$ | 0, …, $N$ | 0, …, $N$ | 0, …, $N$ | MAX{$V_{S1}$, $V_{S1}$,…, $V_{SM}$} |

- **GO operation formula**:

$$\begin{cases} A_R(i) = \prod_{j=1}^{M} A_{Sj}(i), \ i = 0, \cdots, N-1 \\ P_R(N) = 1 - A_R(N-1) \\ P_R(0) = A_R(0) \\ P_R(i) = A_R(i) - A_R(i-1) \ i = 1, \cdots, N-1 \end{cases} \tag{7}$$

### 2.1.2. Developed GO operator

In this section, three developed GO operators are illustrated from aspects of description, operation rule table, and GO operation formula, respectively.

**1. Type 18 operator**

- **Description:** It describes the logical relation of standby mode, which is combination of primary equipment group *CG* and standby equipment group *CBG* working under condition of primary equipment group faulting. The input signals and output signal of Type 18 operator are denoted as $L_1$, $L_2$, and $R$, respectively. The signal flow $L_1$ represents primary equipment group working. The signal flow $L_2$ represents standby equipment group working under condition of primary equipment group faulting, and the signal flow $L_2$ is also the output signal of GO operator, which represents the standby equipment group. And the signal flow $R$ represents standby structure working. $L_1$, $L_2$, and $R$ have two states, which are state 1: success state and state 2: fault state.

- **Operation rule table:** The combination composed of a Type 18 operator and a Type 20 operator is often used to represent standby structure at any place. $S_1$, $S_2$, and $S_0$ represent the input signal flows and output signal flow of Type 20 operator, respectively, and $S_2$represents also the input signal flow of Type 18 operator, i.e., $L_1$.

| $V_{CG}$ | $V_{CBG}$ | $V_{S1}$ | $V_{S0}$ | $V_{S2}$ | $V_{L2}$ | $V_R$ |
|---|---|---|---|---|---|---|
| 1 | 1 | 1 | 2 | 1 | 2 | 1 |
| 1 | 1 | 2 | 2 | 2 | 2 | 2 |
| 1 | 2 | 1 | 2 | 1 | 2 | 1 |
| 1 | 2 | 2 | 2 | 2 | 2 | 2 |
| 2 | 1 | 1 | 1 | 2 | 1 | 1 |
| 2 | 1 | 2 | 2 | 2 | 2 | 2 |
| 2 | 2 | 1 | 1 | 2 | 2 | 2 |
| 2 | 2 | 2 | 2 | 2 | 2 | 2 |

- **GO operation formula:**

$$\begin{cases} P_R(1) = P_{L1}(1) + P_{L2}(1) \\ P_R(2) = 1 - P_R(1) \\ P_{L1}(1) = P_{S1}(1) \cdot P_{CG}(1) = P_{S1}(1) - P_{S0}(1) \\ P_{L2}(1) = P_{S0}(1) \cdot P_{CBG}(1) \end{cases} \tag{8}$$

where $P_{S0}(1)$ and $P_{CBG}(1)$ are the success probability of $S_0$ and *CBG*, respectively.

**2. Type 19 operator**

- **Description:** It describes the unit turning unstable operation into normal operating. The input signals and output signals of Type 19 operator are denoted as *S* and *R*, respectively. *S* is a multistate signal flow, which contains an operating state, a faulting state, and *m* unstable operation states. And *m* unstable operation states are divided into two kinds, which are q unstable operation states turned into operation state by Type 19 operator, and *m-q* unstable operation states. *R* is also a two-state signal flow, which contains operating state and faulting state. *C* is the unit itself, i.e., Type 19 operator, which contains operating state and faulting state.

- **Operation rule table:**

| $V_S$ | $V_C$ | $V_R$ |
|---|---|---|
| 0 | 0 | 0 |
| 0 | N | N |
| 1, …, q | 0 | 0 |
| 1, …, q | N | N |
| q+1,…, m | 0, N | N |
| N | 0, N | N |

- **GO operation formula:**

$$\begin{cases} P_R(0) = P_S(0) \cdot P_C(0) + \sum_{j=1}^{q} P_S(j) \cdot P_C(0) = \sum_{j=0}^{q} P_S(j) \cdot P_C(0) \\ P_R(N) = P_S(0) \cdot P_C(N) + \sum_{j=1}^{q} P_S(j) \cdot P_C(N) + \sum_{j=q+1}^{m} P_S(j) \cdot P_C(0) + \sum_{j=q+1}^{m} P_S(j) \cdot P_C(N) \\ \quad + P_S(N) \cdot P_C(0) + P_S(N) \cdot P_C(N) \\ \quad = \sum_{j=0}^{q} P_S(j) \cdot P_C(N) + \sum_{j=q+1}^{m} P_S(j) + P_S(N) \end{cases} \tag{9}$$

where $P_S(j)$ is state probability of unstable operation states for $S$, $j = 1, 2, \cdots, m$, $P_S(0)$ and $P_S(N)$ are state probability of operating state and faulting state for $S$, respectively,

$P_S(0) + P_S(N) + \sum_{j=1}^{q} P_S(j) + \sum_{j=q+1}^{m} P_S(j) = 1$; $P_C(0)$ and $P_C(N)$ are state probability of operating state and faulting state for $C$, $P_C(0) + P_C(N) = 1$.

**3. Type 20 operator**

- **Description:** It describes the signal flow of conditional operating mode, which is one of its input signal faults under the condition of another input signal success. The input signals and output signal of Type 20 operator are denoted as $S_1$, $S_2$, and $R$, respectively. $S_1$, $S_2$, and $R$ have two states, which are state 1: success state and state 2: fault state. It is often used in standby structure.
- **Operation rule table:**

| $V_{S1}$ | $V_{CG}$ | $V_{S2}$ | $V_R$ |
|---|---|---|---|
| 1 | 1 | 1 | 2 |
| 1 | 2 | 2 | 1 |
| 2 | 1, 2 | 2 | 2 |

- **GO operation formula:**

$$\begin{cases} P_R(1) = P_{S1}(1) \cdot [1 - P_{CG}(1)] = P_{S1}(1) - P_{S2}(1) \\ P_R(2) = 1 - P_R(1) \end{cases} \tag{10}$$

where $P_{S\alpha}(1)$, $P_{CG}(1)$, and $P_R(1)$ are the success probability of $S_\alpha$, $CG$, and $R$, respectively, $\alpha = 1, 2$; $P_R(2)$ is the fault probability of $R$.

### 2.2. Signal flow

Signal flow represents specific fluid flow, such as oil, gas, electricity, and so on, or a logical process. It describes the relationships among the GO operator, its inputs, and outputs. And its attribution includes state value and state probability. The signal flow is used to connect the GO operator and as the direction of GO operation.

### 2.3. GO model

GO model is developed by using signal flows to connect GO operators according to the system principle diagram, engineering drawing, and function constitute directly. And it reflects the system characteristics visually. The proper GO model should be satisfied with:

- The operator in GO model must be labeled with type and number, especially the number which is unique. The first number in the GO operators represents the type of operator and the second number represents the numbering of the operators.
- There is at least one input GO operator (such as Type 4 or Type 5). Generally, the number of GO operator begins with the input GO operator.

- For each GO operator in GO model, its input signal flow must be the output signal flow of the other GO operators. Each signal flow must be labeled with the unique number, and the numbers on the signal line represent the signal flow numbering.
- The signal flow sequence must start with any input GO operator and end with the output signal flow of system. The GO model does not allow being a cyclic model. Generally, the number of signal flow should be labeled with the output signal flow of an input GO operator.

**Example:**

The structure diagram of pressurized water reactor (PWR) purification system and its GO model are, respectively, shown in **Figures 2** and **3**, and the GO operator type of component is presented in **Table 2**.

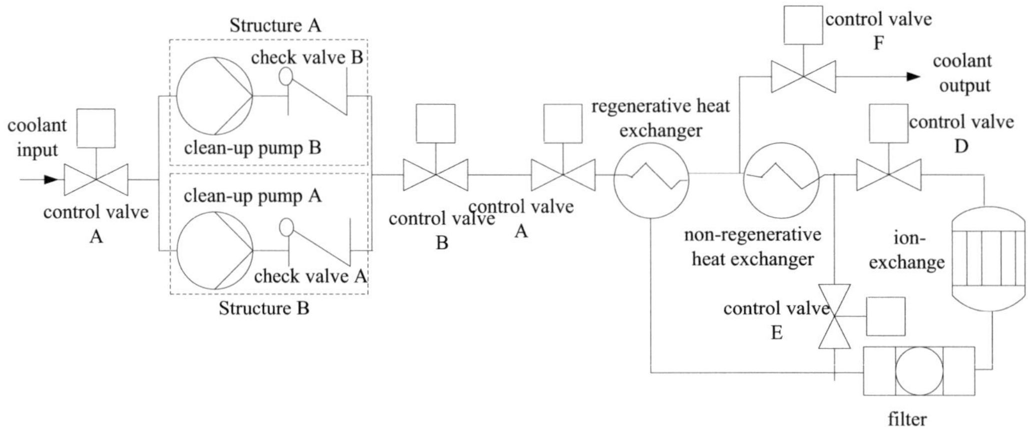


**Figure 2.** Structure diagram of PWR purification system.

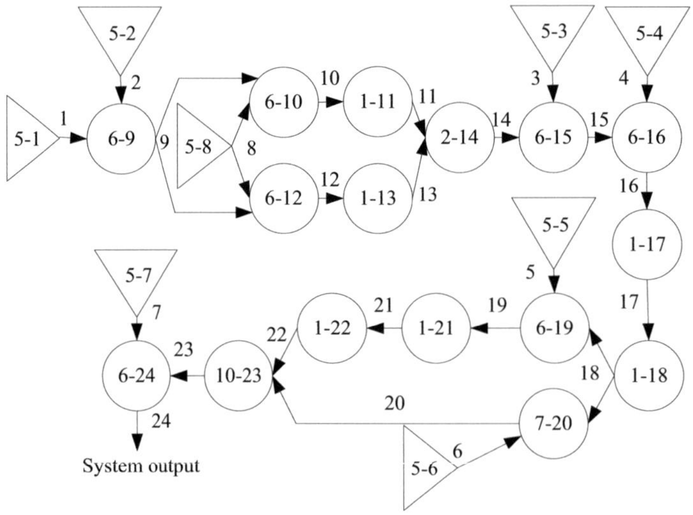


**Figure 3.** GO model of PWR purification system.

| Operator number | Component | Operator type |
|---|---|---|
| 1 | Coolant input | 5 |
| 2, 3, 4, 5, 6, 7 | Control signal | 5 |
| 8 | Control power | 5 |
| 9 | Control valve A | 6 |
| 10 | Clean-up pump A | 6 |
| 11 | Check valve A | 1 |
| 12 | Clean-up pump B | 6 |
| 13 | Check valve B | 1 |
| 15 | Control valve B | 6 |
| 16 | Control valve C | 6 |
| 17 | Regenerative heat exchanger | 1 |
| 18 | Nonregenerative heat exchanger | 1 |
| 19 | Control valve D | 6 |
| 20 | Control valve E | 7 |
| 21 | Ion exchange | 1 |
| 22 | Filter | 1 |
| 24 | Control valve F | 6 |

**Table 2.** GO operator type of component for **Figure 2**.

## 3. GO operation

GO operation is defined that it begins with output signal flow of input GO operators in GO model to calculate the state probability and state value of output signal flow for the next GO operator, and it will be finished until the system output signal flow is calculated after the sequence of signal flow. GO operation contains the quantitative and qualitative analyses. And the quantitative and qualitative analyses are conducted using the GO operation based on the GO algorithm, following the reliability analysis process of GO method. GO algorithm and GO analysis process are the key elements of GO operation.

### 3.1. GO algorithm

The operational efficiency and accuracy of analysis are thus affected by the GO algorithm. The GO algorithm is comprised of a state combination algorithm [4] and a probability formula algorithm [18–21]. The number of state combinations for a complex system is very large, and the probability of a combination state cannot be easily computed. The probability formula algorithm is faster and easier than the state combination algorithm, and it is the mainstream GO algorithm. Thus, this section illustrates two kinds of probability formula algorithms, which are direct GO algorithm and GO algorithm with shared signals.

### *3.1.1. Direct GO algorithm*

Direct GO algorithm is based on the calculation of state probability for signal flow, and the proper direct GO algorithm should be satisfied with:

- The state probability of input GO operator is the state probability of its output signal flow. The output signal flow of input GO operator is the input signal flow of the next GO operator.
- The state probability of output signal flow for the next GO operator is calculated based on its GO operation formula and data, and this output signal flow will be the next GO operator's input signal flow.
- Based on the above rules, following the signal flow sequence, the output signal flow of every GO operator in GO model can be obtained, and the GO operation will be finished until the state probability of output signal flow represented system output is calculated.
- When GO operator is executed with quantitative analysis, it is not necessary to list the corresponding state combination.

### *3.1.2. GO algorithm with shared signals*

In GO model, the output signal flow of a GO operator often connects multiple GO operators, and they are the input signal flows of more than one GO operators, such output signal flow is called shared signal. If the GO operation adopts the direct GO algorithm, the quantitative analysis results will have biases. Thus, GO algorithm with shared signals was proposed in order to obtain more accurate result.

**1. Shared signal**

There are two situations in GO model on shared signal, which are as follows:

- Completely contain. It means all items in formula of a signal flow's state probability have state probability of shared signal.
- Not completely contain. It means some items in formula of a signal flow's state probability have not state probability of shared signal.

The rules of the processing shared signals in GO algorithm with shared signals are as follows:

- Behind a shared signal, all of signal flows in sequence of GO model still have this shared signal, and there are two situations, which are completely contain and not completely contain.
- For multiple signal flows with the same shared signal, their joint state probability cannot be obtained by using the product of their state probabilities directly.
- For two signal flows completely contain the same one shared signal, their joint state probability can be obtained by using the product of their state probabilities to divide the state probability of shared signal.
- For multiple signal flows with the same one shared signal, their joint state probability formula can be obtained by turning a high stage term of shared signal in the formula into a one stage term of shared signal.

## 2. Probability formula

For a system with $L$ shared signals, the probability formula of GO algorithm with shared signals is given by Eq. (10).

$$P_R = \sum_{K_1=0}^{1}\sum_{K_2=0}^{1}\cdots\sum_{K_l=0}^{1} P_{RK_1K_2\cdots K_l}\prod_{l=1}^{L}[(1-P_{Sl})(1-K_l)+P_{Sl}K_l] \quad (11)$$

where $P_{RK_1K_2\cdots K_L}$ is the cumulative probability of system output under a combination of all shared signals, and $Sl$ is a shared signal in the system. $K_l = 0$ and $K_l = 1$ are the failure and success states of the shared signal $l$, respectively. $P_{Sl}$ and $P_R$ are success probability of shared signal $l$ and system output, respectively. The item of $\prod_{l=1}^{L}[(1-P_{Sl})(1-K_l)+P_{Sl}K_l]$ is the state probability for each combination of shared signals.

## 3. Calculating form

It is difficult and complex to derive mathematical formulae for a complex system with a large number of shared signals. A new form of Eq. (10) involves probabilistic weighting of shared signals. The probabilistic weighting improves operation efficiency greatly and avoids the need for complex mathematical formulae. The calculation process is shown in **Table 3**. In **Table 3**, numbers 1 and 0 represent success state and failure state of a shared signal $Sl$, respectively.

The success probability of system can be obtained by Eq. (11).

$$P_R = \sum_{j=1}^{2^L} A_j B_j \quad (12)$$

where $A_j$ is the state probability for each combination of shared signals, and $B_j$ can be obtained by the GO operation of the system, which sets success and failure probability of a shared signal to 1 and 0 depending on the state of shared signal for each combination of shared signals.

### 3.2. Reliability analysis process of GO method

The reliability analysis process of GO method is the criterion and prerequisite for conducting quantitative analysis and qualitative analysis. Generally, the steps of GO analysis process are as follows:

| State combination of shared signal | | | | State probability of combination | Success probability of system |
|---|---|---|---|---|---|
| $S_1$ | $S_2$ | … | $S_L$ | | |
| 0 | 0 | 0 | 0 | $A_1$ | $B_1$ |
| 0 | 0 | 0 | 1 | $A_2$ | $B_2$ |
| ⋮ | ⋮ | ⋮ | ⋮ | ⋮ | ⋮ |
| 1 | 1 | 1 | 1 | $A_2{}^L$ | $B_2{}^L$ |
| **Success probability of system** | | | | $P_R$ | |

**Table 3.** Calculation form of GO algorithm with shared signal.

**Step 1.** Conducting system analysis. The system analysis is the base of GO method, and it directly affects the developing GO model and conducting GO operation. First, to analyze system structure and system function constitutes according to the principle diagram, engineering drawing, or function flowchart of system. Second, to determine system characteristics, such as correlations, multistate, and so on. Then, to determine the interfaces, inputs, and output of system. Finally, to define the success rule of system according to system analysis result.

**Step 2.** Developing GO model. First, to select GO operator according to system analysis results, and then to establish GO model through the signal flow to connect GO operator.

**Step 3.** Processing data of GO operator. According to engineering practice, to obtain the state probabilities of GO operator.

**Step 4.** Operating quantitative analysis. If the GO model does not contain shared signal, the direct algorithm can be selected to conduct GO operation. If the GO model contains shared signal, the GO algorithm with shared signal should be selected to conduct GO operation. Otherwise, it will cause a big error.

**Step 5.** Operating qualitative analysis. Setting the reliability of a function GO operator in GO model is 0, and the reliabilities of other GO operators are kept constant; in this case, if system reliability is 0 by GO operation, this GO operator will be a one-order minimum cut set. Setting the reliabilities of two function GO operators in GO model is 0 except one-order minimum cut set, and the reliabilities of other GO operators are kept constant; in this case, if system reliability is 0 by GO operation, the two GO operators will be a two-order minimum cut set. In the same way, the higher-order minimum cut sets of system can be obtained.

**Step 6.** Evaluating system. The quantitative analysis result and qualitative analysis result can be used as a guidance and theoretical basis for improving system and fault diagnosis of system, and so on.

Above all, the reliability analysis of GO method is formulated, as shown in **Figure 4**.

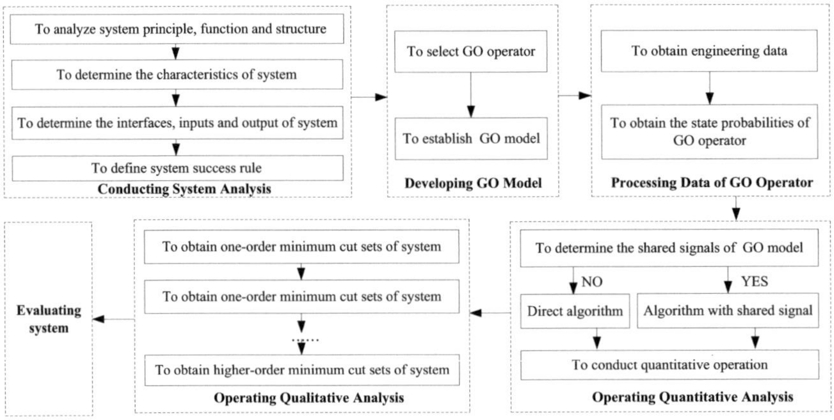


**Figure 4.** Reliability analysis process of GO method.

# 4. Example

In this section, taken a hydraulic oil supply system (HOSS) of an armored vehicle as a case, its reliability analysis is respectively conducted by GO method, FTA, and MCS in order to illustrate the usage and characteristics of GO method.

## 4.1. Reliability analysis of HOSS based on GO method, FTA, and MCS

### *4.1.1. Reliability analysis of HOSS based on GO method*

**1. Conducting system analysis**

- **To analyze system principle, function, and structure:** HOSS supplies oil for an armored vehicle, pump-motor system, pump-motor control system, hydraulic torque converter, and lubrication system. HOSS consists of pressure oil tank, pump P1, P2, P3, and P4, oil filter LF1, LF2, and LF3, pressure relay, bypass valve LF2B and LF3B, one-way valve CV1 and CV2, constant pressure valve RV1, RV2, and RV3, hydraulic torque converter TC, radiator HE, and so on, as shown in **Figure 5**. Oil is extracted by P1 from oil pan via LF1, and then oil is injected into pressure oil tank via LF2 and case inner passage. When LF2

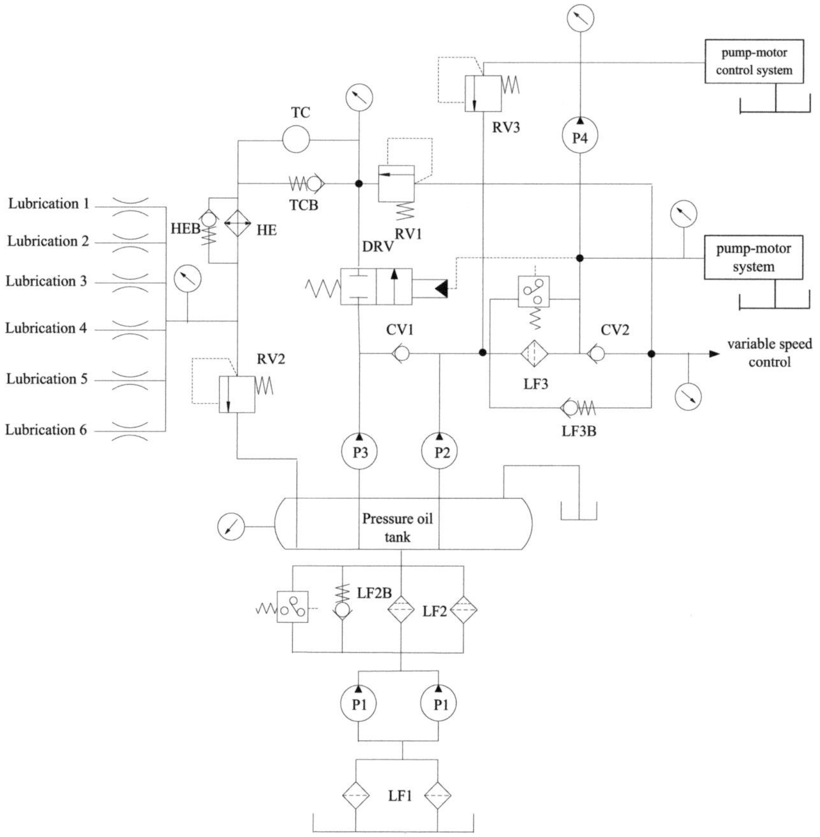


**Figure 5.** Diagram of HOSS.

group is obstructed and pressure between input and output becomes more than 0.5 mega Pascal, oil will be injected into pressure oil tank via LF2B. Oil is extracted by P2 from pressure oil tank; then oil is injected into CV2 via LF3 and then it is injected into hydraulic manifold block as the pressure oil provided for oil cylinder of variable speed control system and pump-motor control system by P4. LF3 and LF3B are another parallel structure, and they are same as LF2 group and LF2B. Because of the pressure of control oil which decreases a little at the situation of high speed, ingress oil of P2 can meet requirements of system. In addition, oil is extracted by P3 from pressure oil tank via DRV to TC. Then, ingress of oil is injected into lubrication system via HE. TC and TCB, and HE and HEB are same as LF2 group and LF2B. RV1, RV2, and RV3 are constant pressure valves of variable speed control and pump-motor system, lubrication system, and pump-motor control system, respectively.

- **To determine characteristics of system:** According to analysis of HOSS, the LF2 group and LF2B, LF3 and LF3B, TC and TCB, and HE and HEB are standby structures in HOSS. The standby equipments haven't the change-over switch.

- **To determine interfaces, input, and output of system:** According to analysis of HOSS, the oil from oil pan and pump group power are system inputs, and oil supply of variable speed control system, pump-motor system, pump-motor control system, hydraulic torque converter, and lubrication system are system outputs.

- **To define system success rule:** According to analysis of HOSS, success rule can be defined as a system that can provide oil to variable speed control system, pump-motor system, pump-motor control system, hydraulic torque converter, and lubrication system of an armored vehicle under high speed condition at steering situation without considering overload protection.

**2. Developing GO model**

- **To select GO operator:** According to the system analysis result and the types of GO operator, the functional GO operators and logical GO operators are selected to describe the units itself and logical relationships in HOSS, respectively, as presented in **Table 4**.

- **To establish GO model:** According to diagram of HOSS and analysis result of HOSS, the GO model of HOSS is developed from system input to system output, as shown in **Figure 6**.

**3. Processing data of GO operator**

According to statistical results of data from engineering, the success state probabilities of component in HOSS are presented in **Table 5**.

**4. Operating quantitative analysis**

It is shown in **Figure 6** that signal flow S1, S4, S7, S8, S15, S16, S17, and S22 are shared signals; the GO algorithm with shared signals should be adopted to conduct GO operation, and the calculation form is as presented in **Table 6**.

| No. | Component | Operator type | No. | Component | Operator type |
|---|---|---|---|---|---|
| 1 | Oil pan | 5 | 24 | RV3 | 1 |
| 2, 3 | LF1 | 1 | 25 | P3 | 6 |
| 5, 6 | P1 | 6 | 26 | DRV | 1 |
| 7 | Pump group power | 5 | 28 | TC | 1 |
| 9, 10 | LF2 | 1 | 30 | TCB | 1 |
| 13 | LF2B | 1 | 32 | HE | 1 |
| 15 | Pressure oil tank | 1 | 34 | HEB | 1 |
| 16 | P2 | 6 | 36 | RV2 | 1 |
| 17 | LF3 | 1 | 4, 8, 11, 27 | | 2 |
| 18 | CV2 | 1 | 37 | | 10 |
| 20 | LF3B | 1 | 14, 21,31, 35 | | 18A |
| 22 | RV1 | 1 | 12, 19, 29, 33 | | 20 |
| 23 | P4 | 6 | | | |

**Table 4.** GO operator in GO model of HOSS.

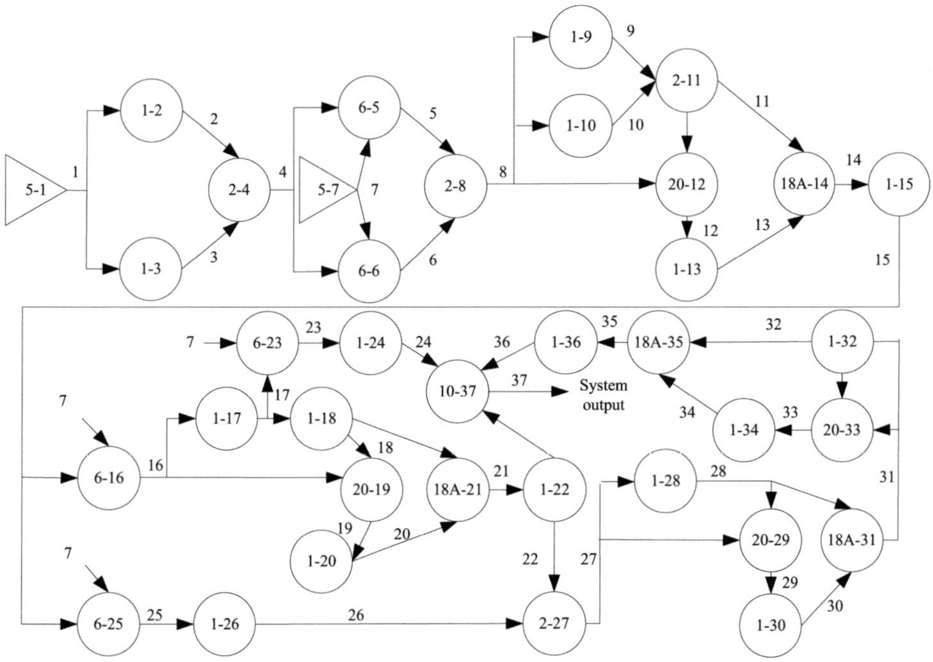


**Figure 6.** GO model of HOSS.

| No. | Success state probability | No. | Success state probability | No. | Success state probability |
|---|---|---|---|---|---|
| 1 | 0.99975006256 | 13 | 0.99867457268 | 24 | 0.99949819446 |
| 2 | 0.99413847887 | 15 | 0.99990821563 | 25 | 0.99867457268 |
| 3 | 0.99413847887 | 16 | 0.99950052524 | 26 | 0.99923302636 |
| 5 | 0.99950052524 | 17 | 0.99987532520 | 28 | 0.97865595028 |
| 6 | 0.99950052524 | 18 | 0.99891901058 | 30 | 0.99867457268 |
| 7 | 0.98676571976 | 20 | 0.99867457268 | 32 | 0.99548611722 |
| 9 | 0.99987532520 | 22 | 0.99949819446 | 34 | 0.99966858815 |
| 10 | 0.99987532520 | 23 | 0.99950052524 | 36 | 0.99949819446 |

**Table 5.** GO model of HOSS.

| State of shared signal | | | | | | | | State probability of combination | Success probability of system |
|---|---|---|---|---|---|---|---|---|---|
| S1 | S4 | S7 | S8 | S15 | S16 | S17 | S22 | | |
| 0 | 0 | 0 | 0 | 0 | 0 | 0 | 0 | 8.154e-32 | 0 |
| 0 | 0 | 0 | 0 | 0 | 0 | 0 | 1 | 1.620e-28 | 0 |
| 0 | 0 | 0 | 0 | 0 | 0 | 1 | 1 | 1.290e-24 | 0 |
| ⋮ | ⋮ | ⋮ | ⋮ | ⋮ | ⋮ | ⋮ | ⋮ | ⋮ | 0 |
| 1 | 1 | 1 | 1 | 1 | 1 | 1 | 1 | 0.98528272861 | 0.99846792597 |
| Success probability of HOSS ($P_{37}$) | | | | | | | | 0.9837732025 | |

**Table 6.** Quantitative analysis result by calculation form of GO algorithm with shared signal for **Figure 6**.

## 5. Operating qualitative analysis

According to the Step 5 in Section 3.2, all minimum cut sets of HOSS can be obtained by multiple GO operations, as presented in **Table 7**.

### *4.1.2. Reliability analysis of HOSS based on FTA and MCS*

## 1. Reliability analysis of HOSS based on FTA

The reliability analysis process of FTA mainly contains developing fault tree model of system, obtaining all minimum cut sets of system by using Fussell-Vesely method, and obtaining the system success probability according to the minimum cut sets of system. In this case, the brief fault tree model of HOSS is shown in **Figure 7**, and the brief analysis processes are presented in **Table 8**.

## 2. Reliability analysis of HOSS based on MCS

The reliability analysis process of MCS mainly contains generating random numbers of success probability of GO operators, establishing simulation model, and obtaining success probability of system by operating a specified number of simulation times according to simulation model.

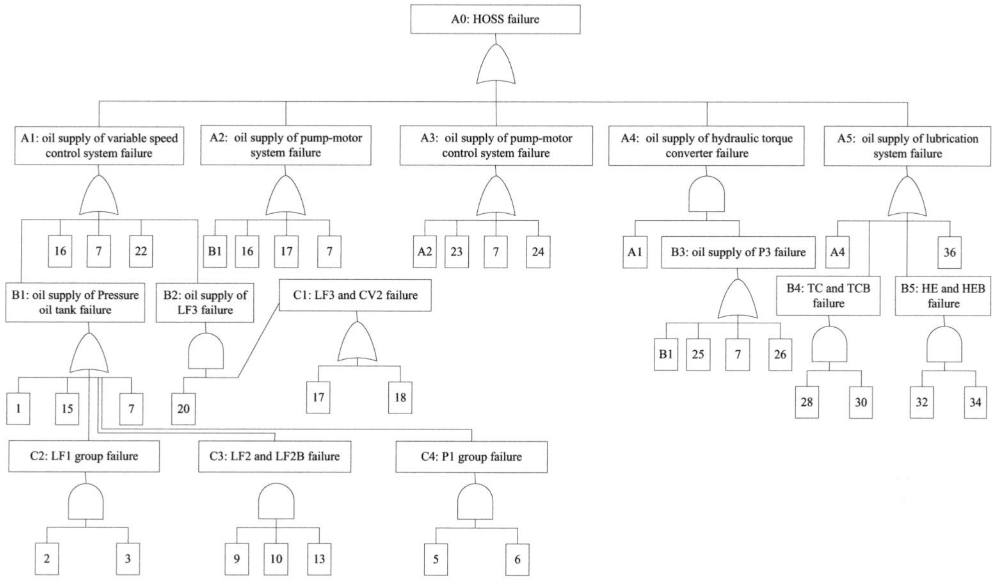


**Figure 7.** Brief fault tree model of HOSS.

| Order | No. | Minimum cut sets | Order | No. | Minimum cut sets |
|---|---|---|---|---|---|
| 1 | 1 | Oil pan | 2 | 2, 3 | LF1 group |
| | 7 | Pump group power | | 5, 6 | P1 group |
| | 15 | Pressure oil tank | | 18, 20 | CV2, LF3B |
| | 16 | P2 | | 28, 30 | TC, TCB |
| | 22 | RV1 | | 32, 34 | HE, HEB |
| | 23 | P4 | 3 | 9, 10, 13 | LF2 group, LF2B |
| | 24 | RV3 | | | |
| | 36 | RV2 | | | |
| | 17 | LF3 | | | |

**Table 7.** Qualitative analysis result by GO method for **Figure 6.**

In this case, the quantitative analysis result of HOSS by simulating 1 million times is 0.9838136000.

### 4.2. Comparison with FTA and MCS

According to reliability analysis process and analysis results of GO method, FTA, and MCS, the qualitative analysis results by the GO method are consistent with FTA, and the quantitative analysis result by the GO method is very close to the result by MCS. It illustrates that both the accurate quantitative analysis result and qualitative analysis result can be obtained by multiple GO operations based on GO method, and the comparisons of GO method, FTA, and MCS are presented in **Table 9.**

| Brief analysis processes | | | | | Minimum cut sets | Order |
|---|---|---|---|---|---|---|
| A0 | A1 | B1 | 1 | 1 | 1 | 1 |
| | | | C2 | 2, 3 | 2, 3 | 2 |
| | | | C3 | 9, 10, 13 | 9, 10, 13 | 3 |
| | | | 15 | 15 | 15 | 1 |
| | | | C4 | 5, 6 | 5, 6 | 2 |
| | | | 7 | 7 | 7 | 1 |
| | | 16 | 16 | 16 | 16 | 1 |
| | | 7 | 7 | 7 | 7 | 1 |
| | | B2 | C1, 20 | 17, 20 | - | - |
| | | | | 18, 20 | 18, 20 | 2 |
| | | 22 | 22 | 22 | 22 | 1 |
| | A2 | B1 | B1 | B1 | - | - |
| | | 16 | 16 | 16 | 16 | 1 |
| | | 7 | 7 | 7 | 7 | 1 |
| | | 17 | 17 | 17 | 17 | 1 |
| | A3 | A2 | A2 | **A2** | - | - |
| | | 23 | 23 | 23 | 23 | 1 |
| | | 7 | 7 | 7 | 7 | 1 |
| | | 24 | 24 | 24 | 24 | 1 |
| | A4 | A1, B3 | A1, B1 | A1, B1 | - | - |
| | | | A1, 25 | A1, 25 | - | - |
| | | | A1, 7 | A1, 7 | - | - |
| | | | A1, 26 | A1, 26 | - | - |
| | A5 | A4 | A4 | A4 | - | - |
| | | B4 | 28, 30 | 28, 30 | 28, 30 | 2 |
| | | B5 | 32, 34 | 32, 34 | 32, 34 | 2 |
| | | 36 | 36 | 36 | 36 | 1 |

**Table 8.** Brief analysis processes of FTA for HOSS.

| Feature | GO | FTA | MCS |
|---|---|---|---|
| Modeling oriented | Success | Failure | Logical |
| Modeling method | Decision-making tree | Fault tree | Logical language description |
| Model consistency | High | Poor | High |
| Model structure | Similar with schematic diagram | Hierarchy logical graph | Logical relationship |
| Model size | Small and compact | Hierarchy and large | Large |

| Feature | GO | FTA | MCS |
|---|---|---|---|
| Model element | Components and logical gate, characteristics | Failure event and logical gate | Components and logical language |
| Model description | Reflect system natural | Reflect cause and effect of failure | Reflect logical relation in system |
| Model sign | Various types and typical | Few types and not typical | Few types and not typical |
| Quantitative analysis | Accurate, stable | Approximative | Accurate, unstable |
| Qualitative analysis | Easy | Complex | - |

**Table 9.** Comparisons of GO method, FTA, and MCS.

## Author details

Yi Xiao-Jian[1,2]*, Shi Jian[3,4] and Hou Peng[2]

*Address all correspondence to: yixiaojianbit@sina.cn

1 Department of Overall Technology, China North Vehicle Research Institute, China

2 School of Mechatronical Engineering, Beijing Institute of Technology, China

3 Academy of Mathematics and Systems Science, Chinese Academy of Sciences, China

4 School of Mathematical Sciences, University of Chinese Academy of Sciences, China

## References

[1] Williams RL, Gateley WY. Use of the GO methodology to directly generate minimal cut sets. In: Fussell IB, editor. Nuclear System Reliability Engineering and Risk Assessment. Pennsylvania: Society for Industrial and Applied Mathematics; 1977. pp. 825–849

[2] Yi X-J, Dhillon BS, Mu H-N. Reliability analysis using GO methodology: A review. In: 22nd ISSAT International Conference Reliability and Quality in Design (RQD 2016); 4–6 August 2016; Los Angeles, USA. 2016. pp. 16–53

[3] Yi X-J, Dong HP, Wang QF, Zhang Z. A new system reliability analysis method: The current development of GO methodology in China. WIT Transactions on Engineering Sciences. 2015;**109**:222–229

[4] Chu BB. GO Methodology: Overview Manual. EPRI NP-3123. Vol. 1. Washington: Electric Power Research Institute; 1983

[5] Chu BB. GO Methodology: Application and Comparison of the GO Methodology and Fault Tree Analysis. EPRI NP-3123. Vol. 2. Washington: Electric Power Research Institute; 1983

[6] Chu BB. GO Methodology: GO Modeling Manual. EPRI NP-3123. Vol. 3. Washington: Electric Power Research Institute; 1983

[7] Chu BB. GO Methodology: GO User's Manual. EPRI NP-3123. Vol. 4. Washington: Electric Power Research Institute; 1983

[8] Chu BB. GO Methodology: Program and User's Manual (IBM Version). EPRI NP-3123. Vol. 5. Washington: Electric Power Research Institute; 1983

[9] Yi X-J, Shi J, Dong HP, et al. Reliability analysis of repairable system with multiple fault modes based on GO methodology. ASCE-ASME Journal of Risk and Uncertainty in Engineering Systems, Part B: Mechanical Engineering. 2016. DOI: 10.1115/1.4030971

[10] Yi X-J, Shi J, Dong HP, et al. Reliability analysis of repairable system with multiple failure modes based on GO methodology. In: ASME 2014 International Mechanical Engineering Congress & Exposition; 14–20 November 2014; Montreal, Canada. 2014

[11] Yi X-J, Dhillon BS, Shi J, et al. Quantitative reliability analysis of repairable systems with closed-loop feedback based on GO methodology. Journal of the Brazilian Society of Mechanical Sciences and Engineering. 2016. DOI: 10.1007/s40430-016-0665-9

[12] Yi X-J, Shi J, Mu HN, et al. Reliability analysis on repairable system with dual input closed-loop link considering shutdown correlation based on goal oriented methodology. Journal of Donghua University (English edition). 2016;**33**(2):25–29

[13] Yi X-J, Shi J, Mu HN, et al. Reliability analysis of repairable system with multiple-input and multi-function component based on GO methodology. In: ASME 2015 International Mechanical Engineering Congress & Exposition. Paper No. IMECE2015-51289; 13–19 November 2015; Houston, Texas. 2015

[14] Yi X-J, Shi J, Mu HN, et al. Reliability analysis of repairable system with multiple-input and multi-function component based on GO methodology. ASCE-ASME Journal of Risk and Uncertainty in Engineering Systems, Part B: Mechanical Engineering. 2016. DOI: 10.1115/1.4034744

[15] Yi X-J, Dhillon BS, Mu HN, et al. Reliability analysis method for multi-state repairable systems based on goal oriented methodology. In: Proceedings of ASME 2016 International Mechanical Engineering Congress & Exposition. Paper No. IMECE2016-65380; 11–17 November 2016; Phoenix, USA. 2016

[16] Yi X-J, Dhillon BS, Shi J, et al. Reliability analysis method on repairable system with standby structure based on goal oriented methodology. Quality and Reliability Engineering International. 2015. DOI: 10.1002/qre.1953

[17] Mu HN, Liu JW, Yi X-J, et al. Reliability analysis method of phased-mission nuclear power equipment based on goal oriented methodology. In: 2016 IEEE International Conference on Industrial Engineering and Engineering Management (IEEM); 4–7 December 2016; Bali, Indonesia. 2016

[18] Shen ZP, Gao J, Huang XR. A new quantification algorithm for the GO methodology. Reliability Engineering and System Safety. 2000;**67**(3):241–247

[19] Shen ZP, Gao J. GO methodology and improved quantification of system reliability. Journal of Tsinghua University. 1999;**39**(6):15–19

[20] Shen ZP, Gao J, Huang XR. An exact algorithm dealing with shared signals in the GO methodology. Reliability Engineering and System Safety. 2001;**73**(2):177–181

[21] Yi X-J, Dhillon BS, Shi J, et al. A new reliability analysis method for heavy vehicle systems based on goal oriented methodology. Proceedings of the Institution of Mechanical Engineers, Part D: Journal of Automobile Engineering. 2016. DOI: 10.1177/0954407016671276

# Reliability Design of Mechanical System-Like Water-Dispensing System in Refrigerator Subjected to Repetitive Impact Loading

Seong-woo Woo

Additional information is available at the end of the chapter

http://dx.doi.org/10.5772/intechopen.69255

**Abstract**

Based on field data and parametric accelerated life tests (ALT), the mechanical system-like water-dispensing system in a bottom-mounted freezer (BMF) was redesigned to find out the missing design parameter in the design phase. To carry out parametric ALTs using a force/momentum balance, the simple mechanical loads of the water-dispensing system were analyzed. At the first ALT, the hinge and front corner of dispenser lever were fractured. The failure shapes found experimentally were similar to those of the failed samples in field. Dispenser lever in water-dispensing system was modified as having its fillets and ribs. At the second ALT, the modified dispenser lever also was fractured because of not having enough strength for impact loading at its front corner. The missing design parameters of the dispenser lever were not enough to have corner fillet rounding and rib thickness. After parameter ALTs with corrective action plans, reliability of the newly designed water-dispensing system is assured to have B1 life 10 years with a failure rate of 0.1%/year.

**Keywords:** reliability design, water-dispensing system, parametric accelerated life testing, missing design parameter

## 1. Introduction

As customers want to have water-dispensing function, **Figure 1** shows the bottom-mounted freezer (BMF) refrigerator with the newly designed water-dispensing system. As shown in **Figure 1(b)**, it consists of the dispenser cover, spring, and dispenser lever. To dispense water for a product lifetime, the dispenser system needs to be designed to withstand the operating conditions subjected to it by the consumers who use the BMF refrigerator. Dispensing water in the BMF refrigerator has several operating steps: (1) press the lever and (2) dispense water.

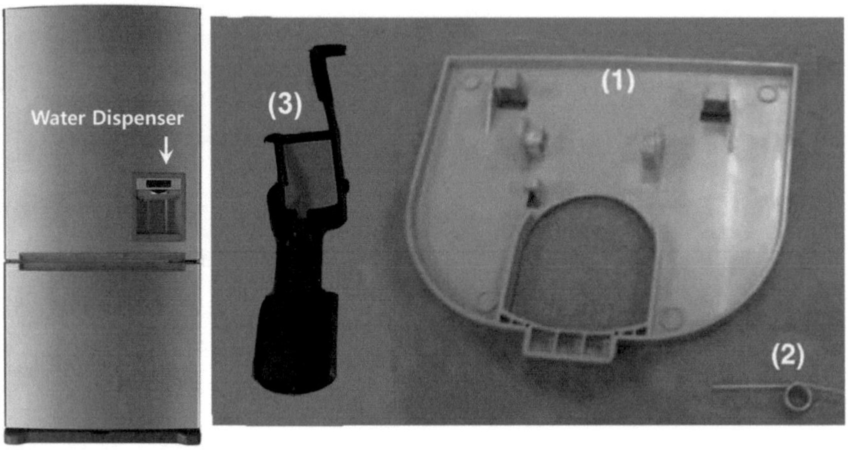


**Figure 1.** BMF refrigerator with the newly designed water-dispensing system. (a) BMF refrigerator. (b) mechanical parts of dispenser: (1) dispenser cover, (2) spring, and (3) dispenser lever.

He will push the dispenser lever if customer want to drink the cooling water. Consequently, the water-dispenser system will have a variety of repetitive mechanical loads when the consumer uses it, though depending on the consumer usage conditions.

The water-dispensing system in field had been fracturing, causing end-users to replace their product. When subjected to repetitive stresses in operating refrigerator, the failed water-dispensing system came from the design flaws. Market data also showed that the returned products had critical design flaws in structure, including stress risers—sharp corner angles and thin ribs—in the water-dispensing system. The design flaws that could not withstand the repetitive impact loads on the water-dispensing system could cause a crack to occur. Thus, the reliability design of the new water-dispensing system is required to robustly withstand repetitive loads under customer usage conditions (**Figure 2**).

A typical pattern of repeated load or overloading may cause structural failure in product lifetime under the customer usage conditions. Many engineers think such possibility can be assessed: (1) mathematical modeling like Newtonian method, (2) the time response of system simulation for (random) dynamic loads, (3) the rain-flow counting method, and (4) miner's rule that the system damage can be estimated [1–3]. However, because there are a lot of assumptions, this analytic methodology is exact but complex to reproduce the product failures due to the design flaws in field.

Robust design skills like Taguchi methods and statistical design experiments (SDE) [4–9] were studied by engineers and statisticians many years ago. Especially, Taguchi's robust design method utilizes parameter design to place it in a location where random noise factors do not cause effects and determines the optimal design parameters and their levels. As utilizing interactions between control factors and noise factors, the purpose of parameter design is to find the proper control factors that make the system's performance robust in relation to changes in the noise factors. In the robust design process, the noise factors are assigned to an outer array, and the control factors are assigned to an inner array in a complex matrix (**Figure 3**).

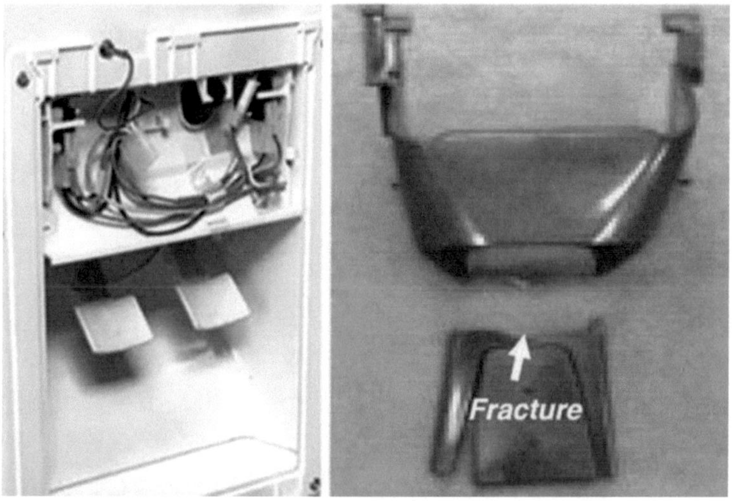


**Figure 2.** A damaged product after use.

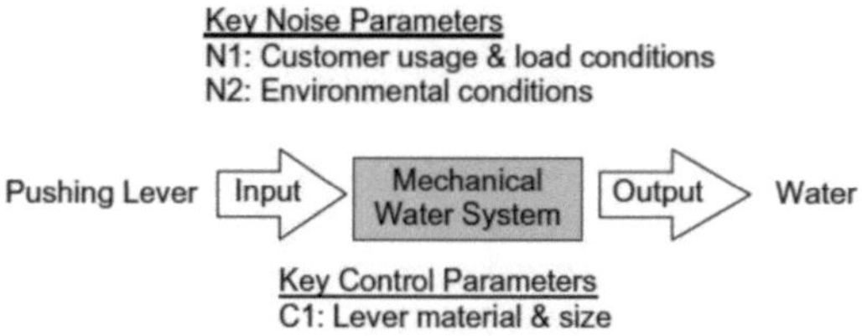


**Figure 3.** Parameter diagram of water-dispensing system.

However, as the noise array is calculated repeatedly for every row in the control array repetitive, experimental iterations in the Taguchi product array would require a lot of computing time. Because a mechanical structure has complex shape, Taguchi method would require infinite iterations. It is not easy to search out solutions in the robust design process, though a mechanical structure is simple. Newly designed refrigerators with the missing design parameters may result in reliability disasters and pay the huge quality costs.

In this study, we suggest a new parametric accelerated life testing (ALT) method that can enhance the reliability design of newly designed water-dispensing system. To confirm the design parameters, parametric ALT will utilize load analysis, new sample size equations with accelerated factor, and the modifications of dispenser system. It will benefit to confirm the final design of new product. Another case study of this new reliability testing methodology [10–23] will be suggested.

## 2. Load analysis

As seen in **Figure 4**, the mechanical dispensing system works like the functional design concept. To properly operate the water-dispensing system, its mechanical lever assembly consists of many mechanical structural parts. When a cup touches on the lever to release the

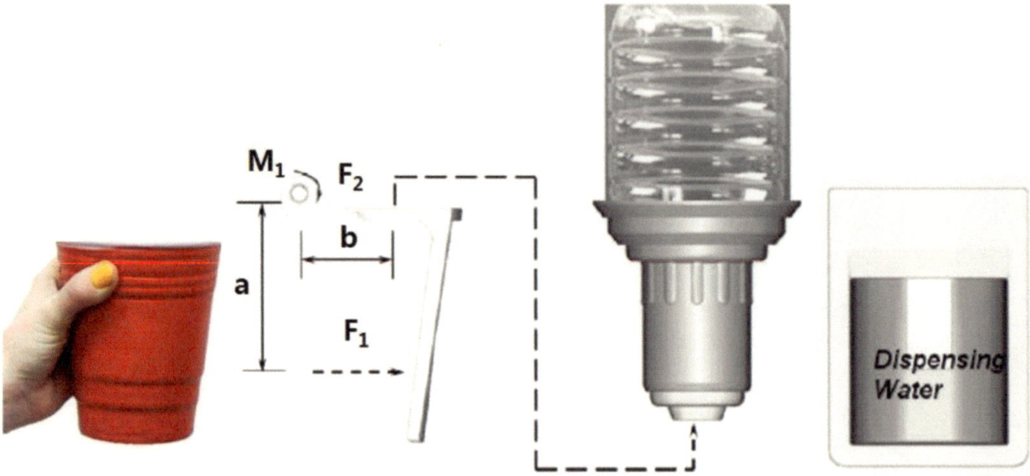


**Figure 4.** Functional design concept of the mechanical dispensing system.

water, water will dispense. Depending on the end-user usage conditions, the lever assembly goes through repetitive impact loads in the water-dispensing process. The concentrated stress of the pressing cup reveals sharp corner angles. As a result, the impact force applies on the hinge of lever. When designing the mechanical dispensing system, it is a critical design step to withstand these repetitive stresses. In the United States, the typical consumer claims the refrigerator to release water from 4 to 20 times a day.

From the free-body diagram of the simple lever system, the force and momentum at hinge can be represented as

$$F = F_1 \tag{1}$$

$$M = aF_1. \tag{2}$$

Because the stress of the dispenser lever depends on the applied force of cup, the life-stress model (LS model) [24] can be represented as

$$TF = A(S)^{-n} = A(F)^{-\lambda}. \tag{3}$$

The acceleration factor (AF) can be derived as

$$AF = \left(\frac{S_1}{S_0}\right)^n = \left(\frac{F_1}{F_0}\right)^\lambda. \tag{4}$$

## 3. Parametric accelerated life testing (ALT) in water-dispensing system

To derive the sample size equation and carry out parametric ALT, probability concepts in reliability engineering should be made out. First of all, the characteristic life $\eta_{\mathrm{MLE}}$ from the maximum likelihood estimation (MLE) can be derived as:

$$\eta_{\text{MLE}}^{\beta} = \sum_{i=1}^{n} \frac{t_i^{\beta}}{r}. \tag{5}$$

If the confidence level is 100(1 − $\alpha$) and the number of failure is $r \geq 1$, the characteristic life, $\eta_\alpha$, would be estimated from Eq. (4)

$$\eta_\alpha^{\beta} = \frac{2r}{\chi_\alpha^2(2r+2)} \cdot \eta_{\text{MLE}}^{\beta} = \frac{2}{\chi_\alpha^2(2r+2)} \cdot \sum_{i=1}^{n} t_i^{\beta}, \quad \text{for } r \geq 1. \tag{6}$$

If there are no failures, $p$-value is $\alpha$ and In $(1/\alpha)$ is mathematically equivalent to chi-squared value, $\frac{\chi_\alpha^2(2)}{2}$. The characteristic life, $\eta_\alpha$, would be expressed as:

$$\eta_\alpha^{\beta} == \frac{2}{\chi_\alpha^2(2)} \cdot \sum_{i=1}^{n} t_i^{\beta} = \frac{1}{\ln\frac{1}{\alpha}} \cdot \sum_{i=1}^{n} t_i^{\beta}, \quad \text{for } r = 0. \tag{7}$$

Because Eq. (6) is established for all cases $r \geq 0$, it can be redefined as:

$$\eta_\alpha^{\beta} = \frac{2}{\chi_\alpha^2(2r+2)} \cdot \sum_{i=1}^{n} t_i^{\beta}, \quad \text{for } r \geq 0. \tag{8}$$

From the Weibull reliability function, we can find out that the characteristic life can be converted into $L_B$ life as:

$$R(t) = e^{-\left(\frac{L_{BX}}{\eta}\right)^{\beta}} = 1 - x. \tag{9}$$

After logarithmic transformation, Eq. (9) can be expressed as:

$$L_{BX}^{\beta} = \left(\ln\frac{1}{1-x}\right) \cdot \eta^{\beta}. \tag{10}$$

If the estimated characteristic life of $p$-value $\alpha$, $\eta_\alpha$, in Eq. (8), is substituted into Eq. (10), we obtain the $BX$ life equation:

$$L_{BX}^{\beta} = \frac{2}{\chi_\alpha^2(2r+2)} \cdot \left(\ln\frac{1}{1-x}\right) \cdot \sum_{i=1}^{n} t_i^{\beta}. \tag{11}$$

For a 60% confidence level, the first term $\frac{\chi_\alpha^2(2r+2)}{2}$ in Eq. (11) can be approximated to $(r+1)$ [25]. By Taylor expansion, if the cumulative failure rate, $x$, is below 20%, the second term $\ln\frac{1}{1-x}$ approaches to $x$, and it can be represented as:

$$L_{BX}^{\beta} = \frac{x}{r+1} \cdot \sum_{i=1}^{n} t_i^{\beta}. \tag{12}$$

Most lifetime testing has insufficient samples. The allowed number of failures would not have as much as that of the sample size.

$$\sum_{i=1}^{n} t_i^{\beta} = \sum_{i=1}^{r} t_i^{\beta} + (n-r)h^{\beta} \geq (n-r)h^{\beta}. \tag{13}$$

If Eq. (13) is substituted into Eq. (12), the *BX* life equation can be modified as follows:

$$L_{BX}^{\beta} \geq \frac{x}{r+1} \cdot \cdot (n-r)\, h^{\beta} \geq L_{BX}^{*\beta}. \tag{14}$$

Then, sample size equation with the number of failures can also be modified as:

$$n \geq (r+1) \cdot \frac{1}{x} \cdot \left(\frac{L_{BX}^{*}}{h}\right)^{\beta} + r. \tag{15}$$

From the sample size Eq. (15), we can go on parametric ALT testing under any failure condition ($r \geq 0$). Consequently, it also confirms whether the failure mechanism and the test method are proper.

If the acceleration factors in Eq. (4) are replaced with the planned testing time $h$, Eq. (15) will be modified as:

$$n \geq (r+1) \cdot \frac{1}{x} \cdot \left(\frac{L_{BX}^{*}}{AF \cdot h_a}\right)^{\beta} + r. \tag{16}$$

Reliability target of the new water-dispensing system is over B1 life 10 years. The operating conditions and cycles of the dispenser system were examined, based on the customer usage conditions. Under B1 life 10 years, if the objective number of life cycles $L_{BX}$ and AF are given, the actual required test cycles $h_a$ can be obtained from Eq. (16). ALT equipment can then be conducted in accordance with the operation procedure of the dispenser system. In parameter ALT, we can obtain the missing design parameters.

## 4. Laboratory experiments

For the water-dispensing system in BMF refrigerator, the working conditions of customer are about 0–43°C with a relative humidity ranging from 0 to 95% and 0.2–0.24 g's of acceleration. The water dispensing happens approximately 4–20 times per day. With a product life cycle design point for 10 years, water-dispensing system approximately happens for 73,000 usage cycles.

The maximum force expected by the consumer in dispensing water was 19.6 N. For accelerated testing, the applied force makes from double to 39.2 N. With a quotient, $\lambda$, of 2, the total *AF* was approximately 4.0 using Eq. (4). To find out missing design parameter of the newly designed water-dispensing system, reliability target can be set to more than the B1 life 10 years. Presumed the shape parameter $\beta$ was 2.0 and $x$ was 0.01, the actual test cycles calculated in Eq. (16) were 65,000 cycles for sample size 8 units. If parametric ALT for water-dispensing system fails less than once during 65,000 cycles, it will be guaranteed to have a B1 life 10 years with about a 60% level of confidence (**Figures 5** and **6**).

**Figure 7(a)** and **(b)** shows the failed product from the field and the first accelerated life testing, respectively. The failure sites in the field and the first ALT occurred at the hinge and front corner of dispenser levers as a result of high-impact stress. **Figure 8** represents a graphical analysis if

**Figure 5.** Equipment used in accelerated life testing.

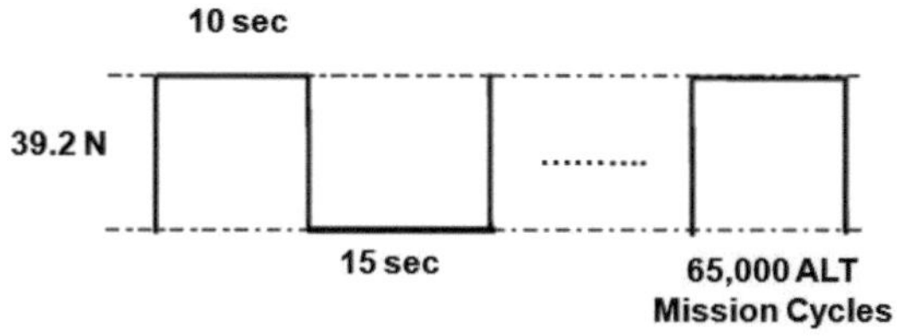


**Figure 6.** Duty cycles of repetitive impact load F on the water-dispensing system.

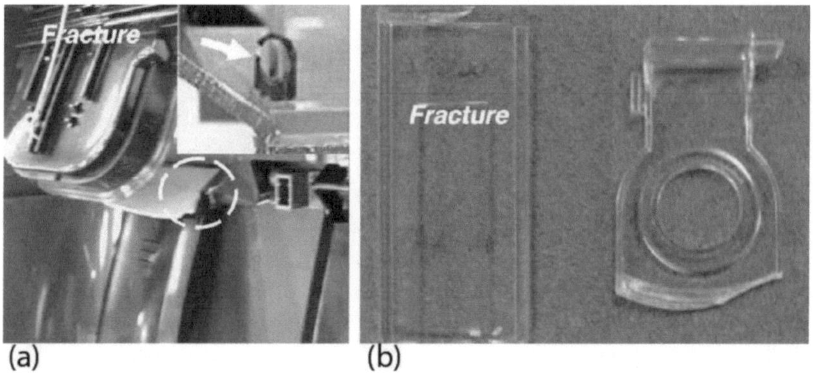


**Figure 7.** Failed products in field and crack after first ALT. (a) Failed products in field. (b) Crack after first ALT.

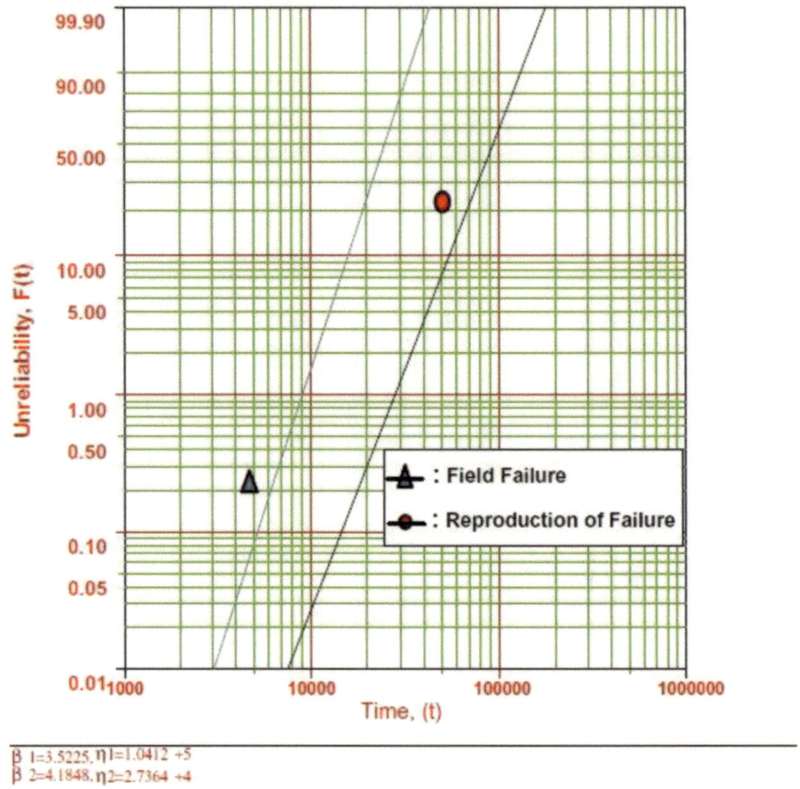


**Figure 8.** Field data and first ALT on Weibull chart.

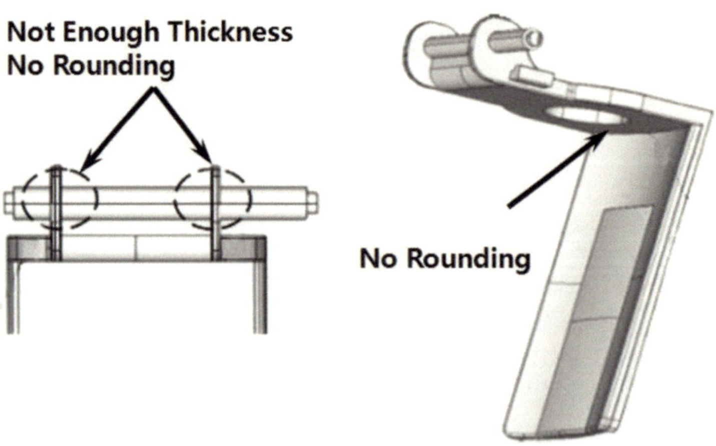


**Figure 9.** Structure of failing dispenser lever in field.

the ALT results and field data are plotted on Weibull distributions. For the shape parameter, the estimated value on the chart was 2.0. For the final design, the shape parameter was determined to be 3.5. The reduction factor $R$ was 0.01 from the experiment data—product lifetime $L_B$,

http://dx.doi.org/10.5772/intechopen.69255

acceleration factor *AF*, actual mission cycles $h_a$, and shape parameter $\beta$. It means that *R* is $\beta$ power of product lifetime versus testing cycle. Consequently, we know that this parameter ALT is effective to decrease the sample size because reduction factor is less than 0.1 from Eq. (16).

As seen in **Figures 7(b)** and **9**, the fracture of the dispenser lever in the first and second ALTs occurred in its hinge and front corner. As shown in **Table 1**, the missing design parameters of the dispenser lever can be listed. If dispenser lever is subjected to the repetitive impact load, we can conclude that its design flaws can result in a fracture.

To withstand the fracture of dispenser lever due to the repetitive food stresses, the dispenser lever was redesigned as follows: (1) increasing the rib rounding of hinge, C1, from R0 to R2.0 mm; (2) increasing the front corner rounding, C2, from R0 to R1.5 mm; (3) increasing the rib

| CTQ | Parameters | | | Unit |
|---|---|---|---|---|
| Crack | KNP | N1 | Impact loading | N |
| | KCP | C1 | Hinge rib rounding, fillet1 | mm |
| | | C2 | Front corner rounding, fillet2 | mm |
| | | C3 | Hinge rib thickness, rib1 | mm |
| | | C4 | Front side rounding, fillet3 | mm |
| | | C5 | Front lever thickness, rib2 | mm |

**Table 1.** Vital parameters based on ALTs.

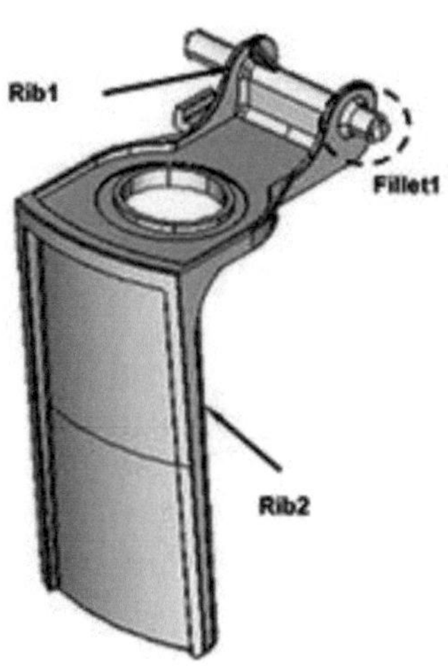


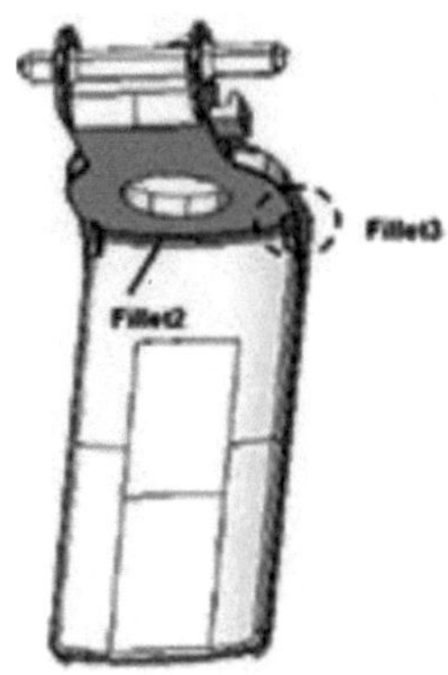


C1: Fillet1 R0 → R1.5 (1st ALT) → R2.0 (2nd ALT)

C2: Fillet2 R0 → R1.5 (1st ALT)

C3: Rib1 T1 → T1.8 (1st ALT)

C4: Fillet3 R0 → R8 (1st ALT) → R11 (2nd ALT)

C5: Rib2 T3 → T4 (3rd ALT)

**Table 2.** Redesigned dispenser lever.

thickness of hinge, C3, from T1 to T1.8; (4) increasing the front side rounding, C4, from R0 to R11 mm; and (5) thickening the front lever, C5, from T3 to T4 mm (**Table 2**).

As the design flaws make better, the parameter design criterion of the newly designed samples was secured to have more than the reliability target—B1 life 10 years. The confirmed value, $\beta$, on the Weibull chart was 3.5. For the second ALT with sample size 8 units, the actual test cycles in Eq. (16) were 38,000. In the third ALT, there were no design problems of the water-dispensing system until the test was carried out to 68,000 cycles. We therefore concluded that the modified design parameters found from first and second ALT were effective.

**Table 3** summarizes the ALT results. **Figure 10** also shows the results of the first ALT and third ALT plotted in the Weibull distribution. With the modified design parameters, final samples of the water-dispensing system were guaranteed to reliability target—B1 life 10 years.

| | **1st ALT** | **2nd ALT** | **3rd ALT** |
|---|---|---|---|
| | **Initial design** | **Second design** | **Final design** |
| In 38,000 cycles, lever has no crack | 25,000 cycles: 2/8 fracture | 32,000 cycles: 1/8 fracture | 38,000 cycles: 8/8 OK<br>56,000 cycles: 8/8 OK<br>68,000 cycles: 1/8 fracture |
| Freezer<br>Drawer<br>Structure |  |  | 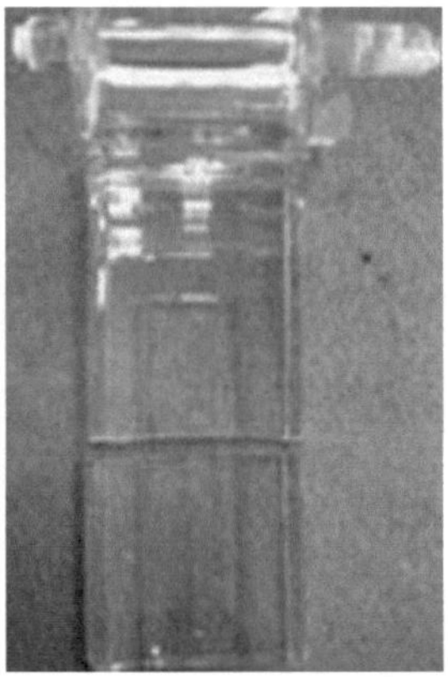 |
| Material and specification | C1: Fillet 1 R0 → R1.5<br>C2: Fillet2 R0 → R1.5<br>C3: Rib1 T1 → T1.8<br>C4: Fillet3 R0 → R8 | C1: Fillet 1 R1.5 → R2.0<br>C2: Fillet 2 R8.0 → R11.0 | C3: Rib2 T3.0 → T4.0 |

**Table 3.** Results of ALTs.

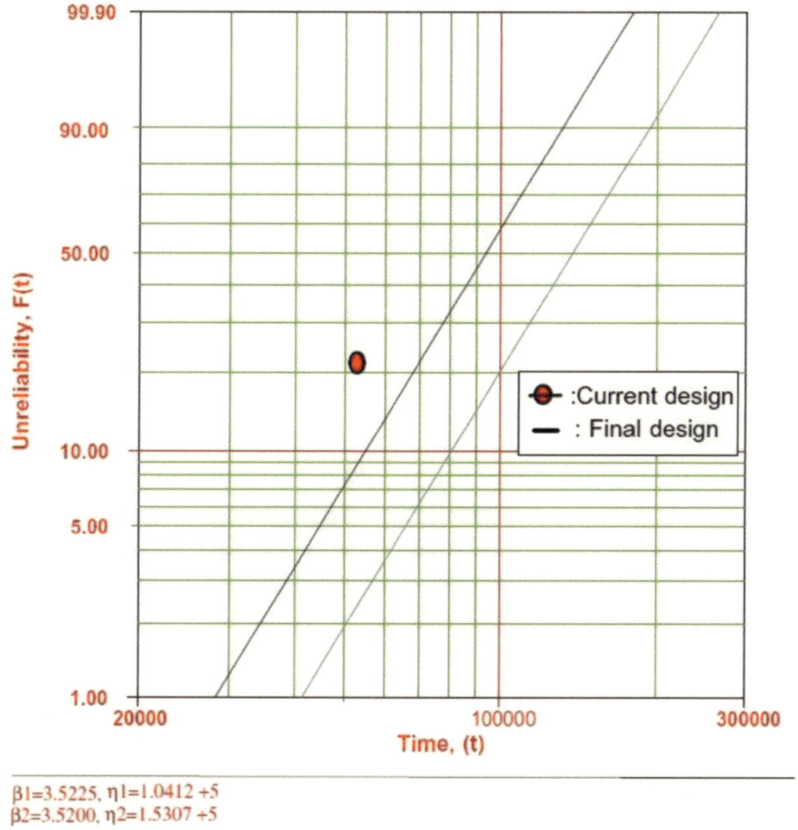


**Figure 10.** Results of ALT plotted in Weibull chart.

## 5. Conclusions

We suggested a new reliability design method for newly designed water-dispensing system in BMF refrigerators. The missing design parameters for the water-dispensing system were identified through parameter ALTs. At that time, the reliability target was B1 life 10 years.

As previously seen in the first ALT, the hinge and front corner of dispenser lever were fracturing because of design flaws due to repetitive impacting stress. As increasing the fillets and ribs of the dispenser lever, the water-dispensing lever system was corrected. During the second ALT, the front corner of dispenser lever also fractured because they did not have enough strength to withstand the repetitive impact loads. As additional reinforced ribs of dispenser lever were provided, we knew that the design of water-dispensing system was improved robustly.

As a result of these modified design parameters, there were no problems in the third ALT. Consequently, the modified design parameters were guaranteed to have the reliability requirement of water-dispensing system—B1 life 10 years. Through the inspection of returned products in field, load analysis, and parametric ALTs, we knew that the study of the missing design parameters of the water-dispensing system in the design phase was very effective in redesigning more reliable parts with significantly longer life.

## Nomenclature

| | |
|---|---|
| *AF* | Acceleration factor |
| *BX* | Durability index |
| *C1* | Hinge rib rounding, fillet1, mm |
| *C2* | Front corner rounding, fillet2, mm |
| *C3* | Hinge rib thickness, rib1, mm |
| *C4* | Front side rounding, fillet3, mm |
| *C5* | Front lever thickness, rib2, mm |
| $F(t)$ | Unreliability |
| $F$ | Force, $N$ |
| *F1* | Pushing force under accelerated stress conditions, $N$ |
| *F0* | Pushing force under normal conditions, $N$ |
| $h$ | Testing cycles (or cycles) |
| $h^*$ | Non-dimensional testing cycles, $h^* = h/L_B \geq 1$ |
| *KCP* | Key control parameter |
| *KNP* | Key noise parameter |
| $L_B$ | Target $B_X$ life and $x = 0.01X$, on the condition that $x \leq 0.2$ |
| $n$ | Number of test samples |
| *N1* | Consumer pushing force, $N$ |
| $R$ | Reduction factor, $R = (L_B/AF \cdot h)^{\beta} \geq 1$ |
| $r$ | Failed numbers |
| $S$ | Stress |
| $S_1$ | Mechanical stress under accelerated stress conditions |
| $S_0$ | Mechanical stress under normal conditions |
| $t_i$ | Test time for each sample |
| *TF* | Time to failure |
| X | Accumulated failure rate, % |
| $x$ | $x = 0.01 \cdot X$, on condition that $x \leq 0.2$ |
| *Greek symbols* | |
| $\eta$ | Characteristic life |

http://dx.doi.org/10.5772/intechopen.69255

| | |
|---|---|
| $\lambda$ | Cumulative damage exponent |
| $\mu$ | Friction coefficient |
| *Superscripts* | |
| $\beta$ | Shape parameter in a Weibull distribution |
| $n$ | Stress dependence, $n = -\left[\frac{\partial \ln(T_f)}{\partial \ln(S)}\right]_T$ |
| *Subscripts* | |
| 0 | Normal stress conditions |
| 1 | Accelerated stress conditions |

## Author details

Seong-woo Woo

Address all correspondence to: twinwoo@yahoo.com

Reliability Association of Korea, Seoul, Korea

## References

[1] Matsuishi M, Endo T. Fatigue of metals subjected to varying stress. The Japan Society of Mechanical Engineers. Fukuoka, Japan, March; 1968

[2] Mott RL. Machine Elements in Mechanical Design. 4th ed. Upper Saddle River: Pearson Prentice Hall; 2004, pp. 190–192

[3] Palmgren AG. Die Lebensdauer von Kugellagern. Zeitschrift des Vereines Deutscher Ingenieure. 1924;**68**(14):339–341

[4] Taguchi G. Off-line and on-line quality control systems. Proceedings of the International Conference on Quality Control. Tokyo, Japan; 1978

[5] Taguchi G, Shih-Chung T. Introduction to Quality Engineering: Bringing Quality Engineering Upstream. New York: American Society of Mechanical Engineering; 1992

[6] Ashley S. Applying Taguchi's quality engineering to technology development. Mechanical Engineering. July; 1992

[7] Wilkins J. Putting Taguchi methods to work to solve design flaws. Quality Progress. 2000;**33**(5):55–59

[8] Phadke M. Quality Engineering Using Robust Design. Englewood Cliffs: Prentice Hall; 1989

[9] Byrne D, Taguchi S. The Taguchi approach to parameter design. Quality Progress. 1987;**20**(12): 19–26

[10] Woo S, Pecht M. Failure analysis and redesign of a helix upper dispenser. Engineering Failure Analysis. 2008;**15**(4):642–653

[11] Woo S, O'Neal D, Pecht M. Improving the reliability of a water dispenser lever in a refrigerator subjected to repetitive stresses. Engineering Failure Analysis. 2009;**16**(5): 1597–1606

[12] Woo S, O'Neal D, Pecht M. Design of a hinge kit system in a kimchi refrigerator receiving repetitive stresses. Engineering Failure Analysis. 2009;**16**(5):1655–1665

[13] Woo S, Ryu D, Pecht M. Design evaluation of a French refrigerator drawer system subjected to repeated food storage loads. Engineering Failure Analysis. 2009;**16**(7):2224–2234

[14] Woo S, O'Neal D, Pecht M. Failure analysis and redesign of the evaporator tubing in a kimchi refrigerator. Engineering Failure Analysis. 2010;**17**(2):369–379

[15] Woo S, O'Neal D, Pecht M. Reliability design of a reciprocating compressor suction reed valve in a common refrigerator subjected to repetitive pressure loads. Engineering Failure Analysis. 2010;**17**(4):979–991

[16] Woo S, Pecht M, O'Neal D. Reliability design and case study of a refrigerator compressor subjected to repetitive loads. International Journal of Refrigeration. 2009;**32**(3): 478–486

[17] Woo S, O'Neal D, Pecht M. Reliability design of residential sized refrigerators subjected to repetitive random vibration loads during rail transport. Engineering Failure Analysis. 2011;**18**(5):1322–1332

[18] Woo S. Chapter 11: The reliability design of mechanical system and its Parametric ALT. In: Handbook of Materials Failure Analysis with Case Studies from the Chemicals, Concrete and Power Industries. Butterworth Heinemann: Elsevier; 2015. pp. 259–276

[19] Woo S, O'Neal D. Reliability design of mechanical systems subject to repetitive stresses. Recent Patents on Mechanical Engineering. 2015;**8**(4):222–234

[20] Woo S, O'Neal D. Improving the reliability of a domestic refrigerator compressor subjected to repetitive loading. Engineering. 2016;**8**(3):99–115

[21] Woo S, O'Neal D. Design of the hinge kit system subjected to repetitive loading in a commercial refrigerator. Challenge Journal of Structural Mechanics. 2016;**2**(2):75–84

[22] Woo S. Reliability design of ice-maker system subjected to repetitive loading. Engineering. 2016;**8**(9):618–632

[23] Woo S. Reliability Design of Mechanical Systems: A Guide for Mechanical and Civil Engineers. 1st ed. Switzerland: Springer; 2017

[24] McPherson J. Accelerated testing, packaging, electronic materials handbook. ASM International. 1989;**1**:887–894

[25] Ryu D, Chang S. Novel concept for reliability technology. Microelectronics Reliability. 2005;**45**(3):611–622

Chapter 3

# Down Time Terms and Information Used for Assessment of Equipment Reliability and Maintenance Performance

Jon T. Selvik and Eric P. Ford

Additional information is available at the end of the chapter

http://dx.doi.org/10.5772/intechopen.71503

**Abstract**

Reliability and maintenance data is important for predictive analysis related to equipment downtime in the oil and gas industry. For example, downtime data together with equipment reliability data is vital for improving system designs, for optimizing maintenance and in estimating the potential for hazardous events that could harm both people and the environment. The quality is largely influenced by the repair time taxonomy, such as the measures used to define downtime linked to equipment failures. However, although it is important to achieve high quality from maintenance operations as part of this picture, these often seem to receive less focus compared to reliability aspects. Literature and experiences from, e.g., the OREDA project suggest several challenging issues, which we discuss in this chapter, e.g., for the interpretation of "MTTR." Another challenge relates to the duration of maintenance activities. For example, while performing corrective maintenance on an item, one could also be working on several other items while being on site. This provides an opening for different ways of recording the mobilization time and repair time, which may then influence the data used for predictive analysis. Some relevant examples are included to illustrate some of the challenges posed, and some remedial actions are proposed.

**Keywords:** reliability, maintenance, data collection, downtime, MTTR, detection time

## 1. Introduction

Equipment reliability and maintenance (RM) data are widely collected in the oil and gas industry and are needed for predictive analysis, e.g., for oil and gas production systems, to achieve safe and cost-efficient solutions. An important benefit of such activity is optimized maintenance [1], for example, when finding the optimal inspection intervals for pipelines (see e.g., [2–4]) or when deciding upon the appropriate testing modes for safety instrumented systems (see e.g., [5]).

In the current chapter, we focus on the maintenance part of the RM data collection and mainly on the measures linked to maintenance activities associated with failed items, i.e., equipment with downtime. Thus, attention in this chapter is primarily on the time interval from when a failure of a reparable item occurs to the time when it is back in an upstate.

Updated information about equipment downtime is important and relevant for various types of analysis within the oil and gas industry to inform decision-making. In particular, the measure "mean time to repair," commonly abbreviated as "MTTR" is widely used within this industry for the purpose of reliability, availability and maintainability analysis (see e.g., [6–9]). It is also widely used in, for example, design planning (e.g., [10, 11]) and also in relation to safety integrity level verification for safety instrumented systems (e.g., [5, 12, 13]).

However, there are several obstacles related to the collection of such data. One of the challenges is simply to get sufficiently high quality in the data for downtime predictions, for example, for assessment of the expected repairing time, i.e., the "MTTR." The problem is typically twofold; the relevant population is too small (e.g., technology is changing, and old data, which has taken years to collect, may not be relevant anymore), and the taxonomy used for the data collection may not be sufficiently clear, which may give room for interpreting information differently.

This is the reason why the international standard ISO 14224 [14] is such an important document. The document is partly a result of industry field feedback implemented for more than 25 years within the OREDA project (see [1, 15]), which has led to the ISO 14224 standard [16]. It represents a main guidance document for collection of RM data within the oil and gas industry, applicable in par with the dependability standard IEC 60300-3-2 [17], and outlines principles and taxonomy for how to achieve quality data. It is a way of ensuring data quality, e.g., through a consistent taxonomy being used (see e.g., [18, 19]).

A much-ignored issue when collecting downtime data is the difficulty to measure the time to detect the failure, i.e., the exact time of failure. This time is normally referred to as the "detection time" (TD) (see e.g., [14]). It is a key value needed, for example, when assessing the expected time to achieve the repair of the failed item, i.e., the so-called mean time to repair, widely referred to by the abbreviation "MTTR." The main problem is that, in situations where failures are hidden and these are not evident before some demand occurs, it is most difficult for a data collector to assess or specify the exact time of the failures. Besides, it may not be possible to confirm whether in fact these values are true or not. Often, rather one attempts to ignore the issue by claiming that the precision of this value is negligible, as $TD \ll MTTR$, or $MTTR \ll MTTF$, where MTTF is the common abbreviation for the "mean time to failure." In other words, the numeric value of TD is assumed to be of low importance to the predictive analysis.

In addition, when using the data on repair times from a database, one normally mixes together situations where the failure is detected at once (being in continuous mode) and those situations where the failure is hidden (or dormant) for a while until a demand occurs (e.g., revealed from a functional test of the equipment). Consequently, to guide for more consistent data, the newly issued [14] has tried to limit the use of MTTR, despite the strong position and frequent use of this measure in the oil and gas industry.

Similarly, the time to mobilize or carry out the repair may also be subject to uncertainty, depending on how the data collector interprets these measures. Having a consistent interpretation of the key terms used to predict downtime is crucial in order to obtain high-quality maintenance data.

An objective of this chapter is to critically examine key terms and associated measures used in maintenance data collection, by studying the repair time taxonomy defined in Ref. [14], i.e., the different terms used in data collection to define downtime in relation to failed items (see Section 2).

The remaining part of the chapter is structured as follows. Section 2 gives a brief description and definitions of key measures, mostly based on the terminology in [14, 20], such as the MTTR. Section 3 provides two example cases illustrating the effects of using different interpretations of the MTTR. Section 4 links data collection to decision-making and presents findings from data collection experiences identified from published literature, addressing several challenging issues that could compromise the maintenance data quality and use. Then Section 5 provides a discussion on to what extent the industry is in line with ISO 14224 taxonomy and how to handle the issues identified. In Section 6 we give some conclusions.

## 2. Key downtime terms and measures in maintenance data collection

As indicated in the previous section, ISO 14224 [14] gives guidance both on how to collect and how to analyze downtime data and is a document strongly linked to ISO/TR 12489 [21] on reliability modeling and also the two dependability documents [17, 20].

The recently issued [14] specifies four different maintenance data categories; see **Table 1**, where details are given on what is the minimum data that must be collected. For example, any corrective maintenance action shall be paired with the associated failure event (i.e., failure record). Downtime data is also labeled as minimum, meaning that the data collector must specify the total length of the downtime interval.

It is common to split the downtime interval in three main parts, i.e., the active repair time, and the activities before and after (such as waiting, delay, start-up). Obtaining accurate information regarding these down time segments is challenging. One such example is already mentioned, i.e., the detection time.

Another example is the "active repair time," which could easily be confused with the "active maintenance time." [14] distinguishes between similar terms such as the "active maintenance time," "active repair time" and "overall repairing time" and uses these when addressing

| Data category | Examples | Minimum data |
| --- | --- | --- |
| Identification | Equipment tag number, failure reference | Unique maintenance identification, equipment identification/location, failure record |
| Maintenance data | Maintenance category = preventive or corrective | Date of maintenance, maintenance category |
| Maintenance resources | Type of main resource(s) and number of days used, e.g., drilling rig, diving vessel, service vessel | (No data specified as minimum) |
| Maintenance times | Time duration for active maintenance work being done on the equipment | Active maintenance time, downtime |

**Table 1.** Maintenance data categories (based on ([14], Table 8)).

measures, i.e., expected values, such as the "mean active repair time," "mean overall repairing time," "mean time to restoration" and "mean time to repair." We will further address and discuss the meaning of these terms and measures in the following sections.

## 2.1. Mobilization time

The mobilization time is normally a main part of the repair preparations. It includes mobilization of all types of resources required, such as vessels, personnel and ROVs. It includes all activities carried out to get the necessary resources available to execute the active repair of the failed items.

In Ref. [14], there is also a relevant note to the entry associated to the definition, which states that "time spent before starting the maintenance is dependent on access to resources, e.g., spare parts, tools, personnel, subsea intervention and support vessels." The mobilization time is therefore sometimes difficult to distinguish from delays caused by manufacturing time and transportation.

In practical terms, the mobilization of intervention vessels is often described using the term "opportunity maintenance," meaning that the intervention vessel is on site or called for in relation to other activities. For example, the vessel could already be on site when the failure is detected, or some critical failure is somehow making the mobilization more urgent and prioritized. To deal with such situations, it is important to have clear procedures for how to collect the mobilization data. Typically, the mobilization time $T_M$ is specified as $T_M = 0$, if other items are mainly responsible for the vessel order. However, for analysis purposes, it is important to be aware of which maintenance activities are included in the order and which are not.

Information about an intervention vessel can also be found in long-term schedules, where the time when the vessel is on site largely depends on the planned route. For example, the vessel could plan for 30 days at a "Site A," then 30 days at a "Site B" (which could then be operated by a different company), then at "Site C," etc. Mobilization will then depend on both whether the item is critical for production and safety and whether the intervention vessel by chance is at the site or soon is coming to this site.

Furthermore, it is not straightforward how to cope with the issue of multiple maintenance and mobilization activities. Often, several maintenance calls are issued, and the allocation of a maintenance vessel needs to be cost-efficient. Thus, one may find that there is a need to remobilize the vessel, as it was not able to complete the maintenance actions on site before moving to the next location.

### 2.2. Active maintenance time

The active maintenance time is, as defined in Ref. [20], "Duration of the maintenance action, excluding logistic delays." Per the definition, other delays such as technical delays are thus included in the active maintenance time. The measure is often referred to as "the effective time to repair" (see e.g., [12]). This is regardless of whether the repair is performed in one go.

However, there is nothing stating that the maintenance activity must be completed during the downtime. Hence, parts of the maintenance, or in some situations (and especially for preventive maintenance) most of the maintenance, relate to an upstate of the equipment. As mentioned, several separate activities may be performed before the maintenance is completed. For example, one may try different options and run checks on whether the performance is satisfying, or one might have to wait for equipment parts to arrive, etc. However, the time used to run-down or start-up the equipment is considered as part of the uptime and not the downtime.

In practice, the active maintenance time could include various testing of the equipment, to check its condition. Depending on the urgency of getting the item back into operation, the duration of this activity might be significant. This also relates to, e.g., wells that are alternating in production, which usually makes the maintenance activity less urgent and allows for more "experimentation." The most cost-efficient solution is then perhaps not the most time-efficient one. For example, one could opt to use time-demanding repairing tools with lower cost and lower efficiency.

The active maintenance time is, by including technical delay, interpreted in Ref. [14] as, "the calendar time during which maintenance work on the item is actually performed." Although normally not the case, it is possible for the active maintenance time interval to be larger than the total downtime. This would be true in situations where the equipment is running in operation while the maintenance activity is ongoing.

### 2.3. Active repair time and mean active repair time

The effective time needed to achieve the repair of an item is called the "active repair time" (see [14]) and is a key part of the downtime as shown in **Table 2**, in Phase No. 3. It consists of, per the IEC 60050-192 [20], three possible activities, i.e., the fault localization time, the fault correction time and the function checkout time. See also ([21], **Figure 5**), which compares repair time taxonomies provided in [20–22] (currently also available in the International Association of Drilling Contractors (IADC) Lexicon definition of "mean active repair time" [23].

The active repair time is different from the active maintenance time, as the preparations and delays are not included (Phase No. 2 and No. 4). But, it is not necessarily the same as the number

| Phase No. | Description | State of equipment |
|---|---|---|
| 1 | Time of failure and then run-down | Uptime |
| 2 | Preparations and delays | Downtime |
| 3 | Active repair time | Downtime |
| 4 | Waiting and delays | Downtime |
| 5 | Start-up | Uptime |

**Table 2.** Equipment state categorization.

of man-hours needed to achieve the repair. The number of man-hours also relates to the amount of personnel working with the repair and is thus not directly linked to the active repair time. Similarly, the time of resource use may only capture a part of, and give a misleading picture of, the overall active repair time. For example, an ROV is often used for subsea maintenance operations, but the use of the actual ROV time may be split on several simultaneous maintenance operations. There may be several other ROV activities carried out, while the ROV is subsea.

The expectation of the active repair time is a relevant key performance indicator (KPI) for downtime, i.e., the mean active repair time (MART). It is listed in ([14], Annex E), to have purpose and value through "indication of the productivity and work content of repair activities." It is noted that if one is also interested in the preparation and delay times, then the mean downtime (MDT), comprising Phase Nos. 2–4, is a relevant KPI or measure.

### 2.4. Overall repairing time and mean overall repairing time

ISO/TR 12489:2013 defines the mean overall repairing time (MRT) as the "expected time to achieve the following actions:

- The time spent before starting the repair; and,
- The effective time to repair; and,
- The time before the item is made available to be put back into operation."

[12] gives the same understanding of the elements included but instead refers to MRT as the "mean repair time," which although somewhat similar introduces a variation of the term that is found in practice but which is distinct from the "mean time to repair" (see Section 2.6).

By including the time spent before starting the repair and the time to prepare the item for operation, the measure is synonymous with the MDT when fault detection time is equal to zero. The MDT is simply the expectation of the downtime, i.e., the mean time the equipment is not in a standby or operating mode (upstate).

### 2.5. Detection time

Detection time is the period from when a fault occurs to the time when this is detected, where fault in this context refers to the equipment being unable to perform as required due to some internal state, in line with the definition in Ref. [20]. The term "fault detection time" is sometimes used for more specificity, as, e.g., in Ref. [21]. However, this term is not the same as the "fault localization time," i.e., the time taken to complete fault localization, although the two terms appear similar. Fault localization takes place after the fault is detected and during the period of corrective maintenance action. Fault localization often includes the activity of diagnosing at what time the fault occurred (see e.g., [24]).

Fault detection may be achieved through manual or automatic operations, depending on modes of operation and system characteristics. Faults of safety systems with possible long detection time, e.g., revealed through functional testing, often make it challenging to identify the exact detection time. Assumptions and estimates are then normally made based on the testing intervals.

The expectation of the fault detection time is called the "mean fault detection time" and is abbreviated as the "MFDT" (see e.g., [21]). For immediately revealed failures, the value of MFDT is equal to zero and negligible for failures with short detection time. Otherwise, this value strongly depends on the test policy for the equipment. For hidden failures, the detection time may represent the main part of the downtime.

Sometimes, for assessment of reliability and maintenance performance, the abbreviation "MFDT" is also used for the "mean fractional dead time." This term has a completely different meaning, i.e., a measure for the average unavailability expressing the expected fraction of time in a nonfunctional condition (refer to use in, e.g., ([25], p. 428, [26])). Obviously, the two terms should not be mixed.

### 2.6. Restoration time, mean time to restoration and mean time to repair

Restoration time (or time to restoration; see [14]) includes also the fault detection time in addition to the elements comprised by the overall repairing time (see above). The expectation of this value, i.e., the mean time to restoration (MTTRes), thus includes the full picture of:

- Fault detection time.
- Preparation and delays (administrative, logistic and technical delays).
- Active repair time.
- Delays after the item is repaired (mainly administrative).

The variation in the meaning of MTTR as the 'mean time to restoration' versus the 'mean time to repair' makes it unclear whether all the four elements are captured, i.e. the full picture above. It is often challenging to separate what definition is used in practice by only looking at the abbreviation "MTTR." The change of the meaning of MTTR in 1999 (see [27]) has the engineering population still divided between those using the present time definition, "mean time to restoration" and those keeping with the old definition, "mean time to repair" and a reluctance to change [28].

[14] also defines the mean time to repair (MTTR) as "expected time to achieve repair of a failed item." The problem with this definition is the fault detection time, which is either zero when the failure is revealed immediately, or it is unknown. If it is possible to include the detection time, the MTTR is equal to the MTTRes; otherwise MRT is considered a more appropriate

measure. When using the MTTR, the meaning could be all of the three measures above, as illustrated in **Figure 1**, depending on the length of fault detection, active repair time and preparations and delays, which is causing unnecessary confusion to data collection and analysis. [21] therefore avoids the use of the term MTTR.

## 2.7. Availability measures: intrinsic availability

The length of downtime and associated measures are important in computations of availability, where availability is often estimated (e.g., by Monte Carlo simulations) by the use of terms such as the MTTR and MTTF. This is the case when calculating the intrinsic (or inherent) availability for some component ($A_I$), where one considers the corrective maintenance downtime of the system:

$$A_I = MTTF/(MTTF + MTTR) \quad (1)$$

where MTTF is the mean time to failure. However, as already mentioned [14] notes that it is more appropriate to refer to the active maintenance time observed in the field, i.e., the MTTRes, which represents a more meaningful term compared with, e.g., the MTTR. The formula should therefore be instead.

$$A_I = MTTF/(MTTF + MTTRes) \quad (2)$$

The MTTRes here is not the same as the mean downtime, MDT, although the two measures may have the same value. Replacing MDT with MTTRes, and MTTF with mean uptime (MUT), would give the operational availability instead of the $A_I$ and is perhaps a more relevant measure from a maintenance perspective. Both are used to express proportion of time that the equipment or system is in an upstate, but the MDT is generally not considered an intrinsic property. The duration of MDT could in practice include a variety of delays (e.g., detection, isolation, spare parts, standby, repair duration, reinstatement, etc.; see ([14], Annex C)). The mean up- and downtime are measures that depend strongly on the system performance, i.e.,

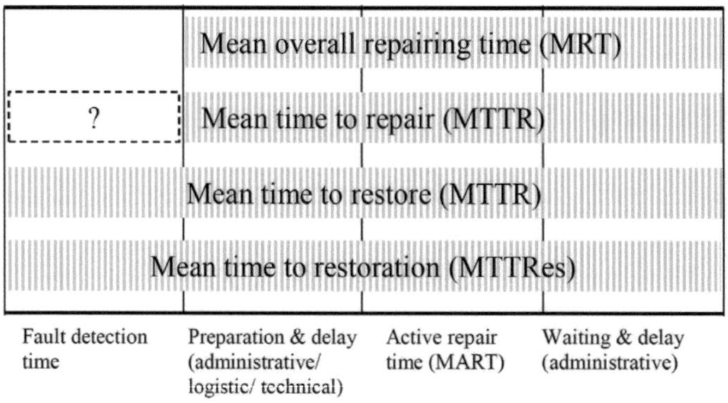


**Figure 1.** Different measures that could represent the mean time to repair (MTTR).

reliability, maintainability and maintenance support, and therefore dependent on the context in which they are used. The MTTRes, by focusing on the maintenance resources and disregarding external resources, thus allows for a more intrinsic analysis.

## 3. Significance of taxonomical differences

To illustrate the effects of different interpretations of maintenance times, two example cases are provided that illustrate the challenges related to situations where failures are detected with short-time versus long-time intervals. In the first case, we use a data set for subsea control modules (SCMs) obtained from the OREDA database. In the second case, we have constructed a data set for downhole safety valves (DHSVs) based on reliability data collected annually by the Petroleum Safety Authority Norway (PSA) to analyze and map the risk level on the Norwegian Continental Shelf, the so-called RNNP project (see e.g., [29]). Both data sets have been randomized and anonymized for confidentially reasons and are thus solely for illustration purposes.

### 3.1. Example I: short detection time

The data set consists of 375 SCMs, with a total time in service of $1.44 \times 10^7$ hour. During this period, a total of 255 failures have occurred (counting any type of failure severity, i.e., critical, degraded or incipient). An estimator for the mean time to failure, MTTF, is thus given as $1.44 \times 10^7/255 = 56{,}471$ hour. We now consider the implications of varying definitions of the repair time, in this context referred to simply as MTTR. Let us assume that there is uncertainty on all of the components that together form the MTTRes, i.e., fault detection time, administrative delay prior to repair, logistic delay, technical delay, active repair time and administrative delay post repair. For simplicity, we will assume that each of these parameters is represented by a triangular distribution denoted T1:

- T1 (1; 8; 20) hours, i.e., minimum = 1 hour, peak value = 8 hour and maximum value = 20 hour.
- All of the components of the MTTRes are assumed to be independent of each other.

This distribution is not based on any real data set, but delays and repair times on an SCM could realistically be at least 20 hour.

Recalling from the previous chapter we have:

- MART = active repair time (fault localization + correction + function checkout).
- MRT (overall repair time) = MART + administrative delay prior to repair + logistical delay + technical delay + administrative delay post repair.
- MTTRes (time to restoration) = MRT + fault detection time.

Thus, we can establish three definitions for the intrinsic (inherent, technical) system availability, A:

$$A_1 = MTTF/(MTTF + MTTRes) \quad (3)$$

$$A_2 = MTTF/(MTTF + MRT) \quad (4)$$

$$A_3 = MTTF/(MTTF + MART) \quad (5)$$

For now, we assume that the MTTF is deterministic, and thus all uncertainty is placed on the MTTR component. Running a Monte Carlo simulation using N = 10,000 gives the distributions for $A_1$, $A_2$ and $A_3$ shown in **Figure 2**.

As **Figure 2** shows, widening the definition of repair time to include delays and fault detection will lower the system availability and increase the standard deviation, as there is obviously more uncertainty. The deviations between the three definitions of repair time are not significant in this particular case, as the MTTF is relatively much larger. Consider however **Figure 3**, where a sensitivity analysis is run for the MTTF, showing the expected absolute and relative difference between $A_1$ and $A_3$ (blue line) and $A_2$ and $A_3$ (red line), where $A_3$ is a main element of both $A_2$ and $A_3$, i.e., the differences thus indicating the contribution of detection time. As the MTTF becomes lower, the significance in the varying interpretation of repair time becomes as expected greater.

**Figure 3** also shows the expected relative difference between the two measures, showing a decreasing value in the approximate MTTF interval = [500; 10,000]. Indicating that especially when the MTTF is higher than 10,000 hours, i.e., around 1 year, it is fully acceptable to ignore the contribution of detection time in such calculations.

While there is the obvious point that MTTR generally plays a more important part of availability the lower the MTTF is, there are also other reasons why the MTTR generally is skewed

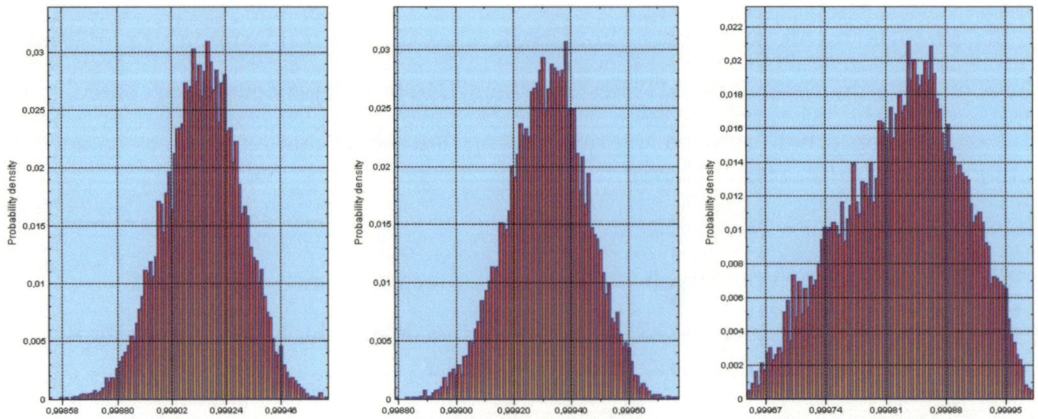


**Figure 2.** Intrinsic availability distribution for $A_1$, (left) using MTTR = MTTRes. Mean = 99.915% and stdev = $1.5 \times 10^{-4}$. $A_2$, (center) using MTTR = MRT. Mean = 99.932% and stdev = $1.4 \times 10^{-4}$. $A_3$, (right) using MTTR = MART. Mean = 99.983% and stdev = $7.0 \times 10^{-5}$.

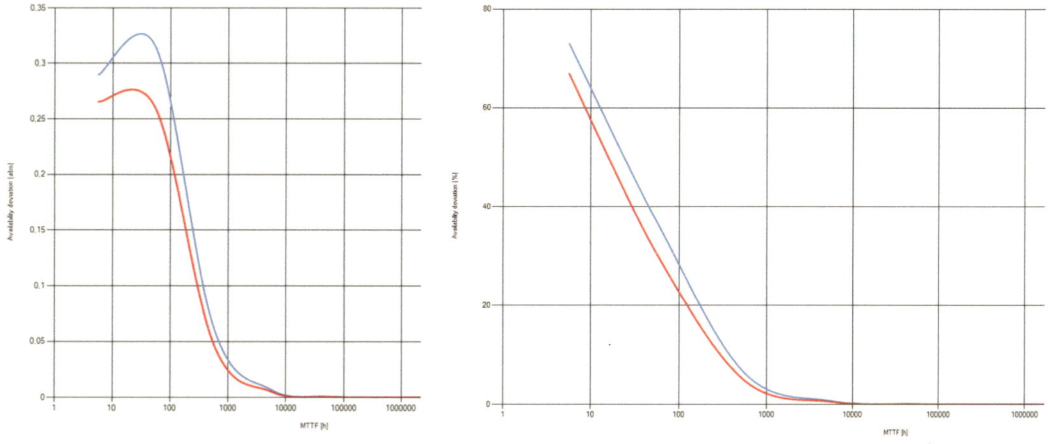


**Figure 3.** Expected absolute (left) and expected relative (right) difference between $A_1$ and $A_3$ (upper line) and $A_2$ and $A_3$ (lower line), as a function of varying MTTF [hours].

towards the right of **Figure 2**. When collecting subsea data, for example, for OREDA, this process is often time-demanding and costly; see e.g., [15] when done manually, which is often the case. For this reason, the priority is often to collect failures and any maintenance related directly to these. This comes at the cost of sometimes disregarding opportunity maintenance or other types of no-failure maintenance, where the equipment in question is actually in a downtime state, thus overestimating the total time in service and consequently also the MTTF. Essentially, the actual MTTF is bound to be lower than what is often used as an estimate, and thus the difference in repair time definition becomes greater. In addition to component-related maintenance, there are also at times planned shutdowns of wells or even fields, which are not always captured for the same reasons and which also emphasize the point made. Furthermore, as mentioned in Section 2.1, there is the difficulty of distinguishing mobilization time from delays caused by manufacturing time and transportation, meaning there is a possibility of potentially both longer or shorter times added to, or subtracted from, extended definitions of MTTR. According to Ref. [15], quality checks on various equipment items (not necessarily subsea items) yielded wrong interpretations or coding used during data collection in 39% of the cases. Such errors could also swing both the MTTF and the MTTR in either direction but will certainly give rise to variations in the expected system availability.

### 3.2. Example II: hidden failures

A main objective of this second example case is to address the important issue of hidden failures (also called dormant failures). These are failures that, according to Ref. [14], are not immediately evident to operations and maintenance personnel. This means that it may take some time before detection, as it is not possible to detect these failures unless some specific action, such as a periodic test, is performed.

For the second data set, we refer to a population of 8714 DHSV tests collected from 73 facilities operating at the Norwegian Continental Shelf during the time period in 2012–2016 from the

RNNP project [29], which in contrast to the SCM population is taken from a wide range of oil and gas operators. The availability requirements refer to an industry standard (see **Table 3**) based on the requirements set by the Norwegian oil and gas operator Statoil. These requirements are also referred to as the "failure fraction," FF, the ratio between number of safety critical failures revealed from periodic testing, $x$, and the corresponding number of tests performed, $n$:

$$FF = x/n \tag{6}$$

This measure may further be used to estimate the MTTF (see [30, 31]), which will then also depend on the time between the tests (the periodic testing interval).

For simplicity, we assume that the valves are tested at the maximum interval, i.e., twice a year (as defined in Ref. [32]), meaning that if a failure is detected from testing, the valve failure has occurred at some point in time within the interval = [0, 6] months. The data set then corresponds to an estimated 4357 DHSVs. These valves are associated with a total number of 200 failures recorded from this population.

Furthermore, we assume that, except for the detection time TD, each of the parameters is representable by the triangle distribution defined in Example I, T1 (1; 8; 20) hours. In addition, another triangular probability distribution is made for the time to detect the failure, denoted T2 (1; 2160; 4320) hours, i.e., a peak (mean) value of 3 months. This corresponds to a total downtime of $4.32 \times 10^5$ hours.

Although the various delay times may be correlated, we assume the parameters are independent of each other. For the time to failure, we calculate this based on the RNNP data in **Table 3**, showing the percentage of critical failures, i.e., exceeding the acceptance criteria of the barrier testing. The calculations based on PSA data [29] yield a total time in service of $7.49 \times 10^7$ hours. An estimator for the mean time to failure, MTTF, is thus given as $7.49 \times 10^7/200 = 3.74 \times 10^5$ hours. We now consider the implications of varying definitions of the repair time, in this context referred to simply as MTTR. We assume that there is uncertainty on all of the components which together form the MTTRes.

**Figure 4** shows, similarly to **Figure 2** for Example I, that widening the definition of repair time to include delays and fault detection will lower the system availability and increase the standard

| **Barrier element: DHSV** | |
|---|---|
| Number of facilities where tests were performed in 2016 | 73 |
| Average number of tests for facilities where tests were performed in 2016 | 119 |
| Number of facilities with percentage failures in 2016 greater than the industry standard | 25 |
| Total (mean) percentage of failures in 2016 | 0.023 (0.026) |
| Total (mean) percentage failures 2012–2016 | 0.021 (0.021) |
| Industry standard for availability (Statoil value) | 0.02 |

**Table 3.** General calculations and comparison with industry standards [29].

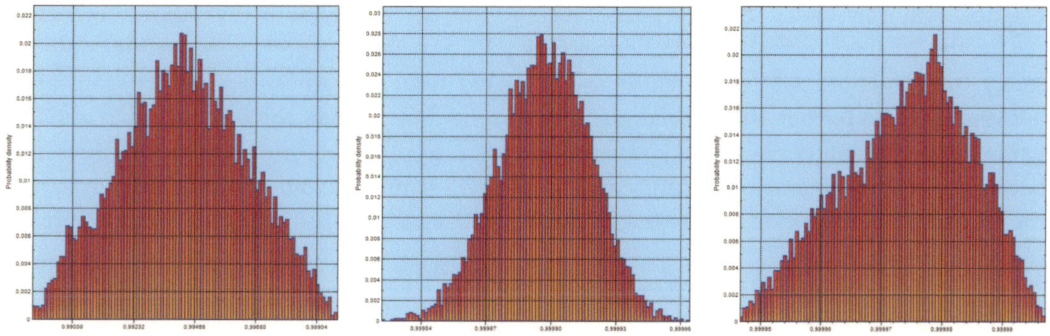

**Figure 4.** Intrinsic availability distribution for $A_1$, (left) using MTTR = MTTRes. Mean = 99.416 and stdev = $2.3 \times 10^{-3}$. $A_2$, (center) using MTTR = MRT. Mean = 99.990% and stdev = $0.2 \times 10^{-4}$. $A_3$, (right) using MTTR = MART. Mean = 99.997% and stdev = $0.1 \times 10^{-4}$.

deviation. For this example, in contrast to the previous, the detection time is significant. Also for this example, the MTTF is relatively high; however, now the differences between using MTTRes and MRT are much greater, where the deviations in the mean availability between the $A_1$ (i.e., MTTR = MTTRes) and $A_2$ (i.e., MTTR = MRT) is equal to 0.547% versus a deviation of 0.061% given for Example I. And although there is significant uncertainty related to the value of MTTF, the relevance of detection time is significant.

Consider also the value of the MTTF in **Figure 5**, where a sensitivity analysis is run, showing both the expected absolute and relative difference between $A_1$ and $A_3$ (blue line) and $A_2$ and $A_3$ (red line). As the MTTF becomes lower, the significance in the varying interpretation of repair time becomes as expected greater also in this example.

Example II indicates that the MTTF must be significantly higher before the availability deviations converge towards zero, at least a factor of 10 higher than the SCM situation (Example I). Only when the MTTF is in the region of 500,000 hours (60 years or more) can the contribution from detection time be considered negligible in this example.

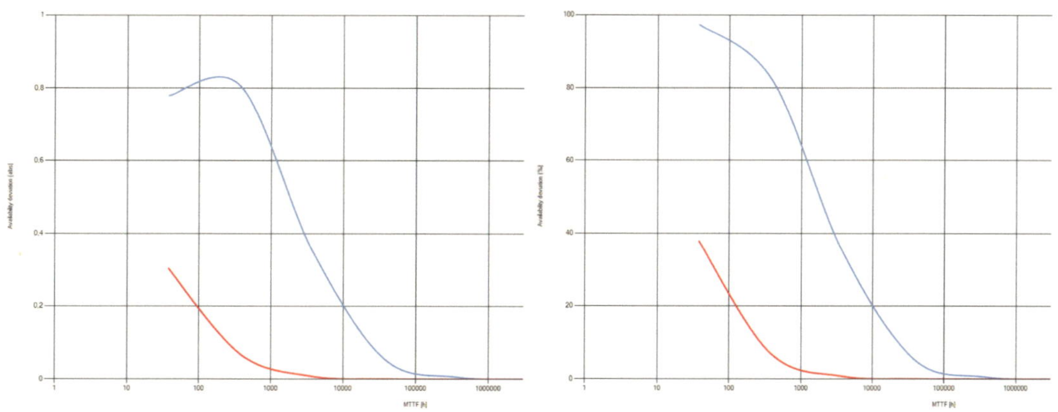

**Figure 5.** Expected absolute (figure to the left) and expected relative (figure to the right) difference between $A_1$ and $A_3$ (upper line) and $A_2$ and $A_3$ (lower line), as a function of varying MTTF [hours].

Besides, one may claim that one could see significant variations in this value. For example, in the 2010 edition of the PDS handbook [30], which presents recommended data for safety instrumented systems, DHSVs are assigned with overall failure rates in the range between 2.0 and 6.7 per $10^6$ hours depending on the data source, corresponding to a MTTF value inside the interval = [58, 17] years. Knowing also that there are significant differences between companies and operating conditions, the estimated value of detection time and the measure selected (i.e., MRT or MTTRes) could significantly influence the availability calculations.

## 4. Maintenance data collection link to decision-making and challenges

Information about downtime is highly important for decision-making in the oil and gas industry, including for subsea systems. For example, such data is needed to track maintenance KPIs and achieve the so-called maintenance excellence; refer to references [33, 34].

In general, having useful information about RM is important for high-quality decision-making, and it is one of the six key dimensions that are used to evaluate decision quality [35]:

1. Helpful frame (what is it that I am deciding?)
2. Creative alternatives (what are my choices?)
3. Useful information (what do I know?)
4. Clear values (what consequences do I care about?)
5. Sound reasoning (am I thinking straight about this?)
6. Commitment to follow through (will I really take action?)

The above six dimensions can be visualized as a chain of decision links (see also ([36], p. 55)), where the decision is not stronger than its weakest link, which simply means that poor information (i.e., the RM data in this context) deteriorates the decision quality. It is also pointed out in Ref. [35] that the information or data should be "useful" and hence should be compared with its area of use, in this case within the area of RM data applicability and how such data may create business value by contributing to good decisions.

Despite the link to business value, and the broad consensus that use of RM data strongly depends on its quality, collecting such data about RM performance in the industry has typically received considerably less attention compared with the use of the information.

To some extent it is inevitable that data is not always suited for the decision-making, as the system requirements, design and operations change over time. Neither are the databases sufficiently flexible to adopt for the data needs all the time. The data sources are, in many situations, unmanageable at the time when they are needed. For example, when analyzing and predicting downtime for some subsea safety valve, one typically uses the data already collected and at the time available from some database or source, such as

OREDA (see e.g., [1, 37]), WellMaster (see e.g., [38]) or some internal database. There may be at the time limited room for collecting new and more relevant information. Data collection requires time, personnel and software resources. It could take years to build a quality database, and strategic decisions should be taken about what equipment data and format are relevant in the future. At the time of analysis and decision-making, there may not be sufficient resources (e.g., time) to collect additional information.

RM data collection is often an issue of resources, such as cost. In Ref. [14], it is mentioned that "collecting RM data is costly and therefore it is necessary that this effort is balanced against the intended use and benefits." The RM data collection activity is considered an investment, but it is an issue that strongly depends on both how and why the data are collected. Besides, it is not always clear what time perspective one is considering. Several years of data collection could be required for the data to have significant population and value in decision-making.

A similar point is made in Ref. [39], with reference to the Well Integrity Management System (WIMS). This is a software application for data collection, and [39] claim that it is important that both accurate and reliable information are achieved, at least from a user perspective. Nowadays data is shared with, and should be compatible with, other systems used by the oil and gas companies, such as, the data management system SAP, to synchronize data quality. The taxonomy used in the RM database should to a large extent also be reflected in the sources where the data is first recorded, such that the transfer of failure data does not have to go through several interpretation steps where essential information is missing.

Another issue is cost and ownership, which is a main reason why the data collection process is often delayed. Experience shows that often the data collection is performed at a quite late stage compared to when the actual failure occurred. The main information is typically captured by, e.g., reports from the maintenance provider. Then it is later the task of some data collector to transfer the relevant information into a RM database, including identification of missing and low-quality information. For example, the time to mobilize is normally given with low accuracy in such reports. The same is the situation when studying data from typical replacements, for example, choke replacements. The number of hours to replace such an item is too often given as an estimate, for example, 24 hours, as the detailed information may not be available from the maintenance reports.

The information provided by the MTTR, which is a common and much used KPI in the oil and gas industry (see e.g., [40, 41]), is also challenging. The KPI has a strong position within this industry, especially for availability calculations, and it is highly difficult to avoid the use of this term by replacing it with MRT and MTTRes, as suggested in Ref. [21]. A quick search in "Google scholar" (January 2016), for the period 2013–2016, confirms that the term "mean time to repair" (MTTR) is used significantly and way more than the term "mean time to restoration" (MTTRes). The search results confirm that on several accounts, the term used is MTTR, while the actual basis for estimating this value is equivalent to "mean time to restoration," which is the specific term and abbreviation used in the two IEC documents [20, 22] but then with the meaning of MTTRes. The abbreviation is also used for the mean time to recover (or recovery), synonymous with the "mean time to repair."

## 5. Discussion

RM data is used for a vast array of different applications, both operational and engineering. Data use today is as relevant as it was 30 years ago, for example, according to Ref. [42], reliability data is used to design the operational phase in terms of evaluating the operational performance of equipment, adjusting maintenance intervals, optimizing test intervals, establishing failure probability distributions, optimizing spare parts and logistics and job priority scheduling. Of similar importance to the engineering of equipment components, reliability data provide input to analysis such as safety integrity level analyses, RAM studies, required maintainability, selection of equipment and parts based on reliability experience, choice of maintenance strategy and qualification testing.

The OREDA database, which has a taxonomy based directly on ISO 14224, contains more than 39,000 failures and 73,000 maintenance records, as well as over 2000 years of operating experience from subsea fields. Its current member list includes BP, Engie E&P Norge, Eni, Gassco, Petrobras, Shell, Statoil and Total. Expert members from the OREDA project are also involved in the development of ISO 14224, to make sure that the challenges experienced in practice are captured when revising the standard. When the latest edition was issued [14], it was possible to include several definitions of key measures relevant for downtime predictions, such as, the new definition of the mean time to restoration (MTTRes).

Considering the cost of subsea maintenance operations, downtime and mobilization of vessels, the range of affected decisions, operations and procedures and not least the extent to which safety systems are used in today's industry, the importance of RM data quality cannot be emphasized enough. We will next discuss some challenges relating to the practical use of the ISO 14224 taxonomy, what impacts this may have on decisions, operations and design and some suggestions to improvements.

### 5.1. Data separation

One of the key challenges when collecting RM data is related to the fact that the RM database in many cases exists as a separate entity which may only partially communicate or not communicate at all, with other relevant systems (e.g., SAP). It is not rare to observe that data stored in an RM system such as OREDA or WellMaster is essentially data extracted from another system or database, converted into an appropriate format and then re-entered. This creates several challenges relating to both data integrity and quality:

(1) Creates an unnecessary overhead; data could simply be stored in one place, possibly supplemented by automatic or manual conversion if a special format or interpretation was required, as is often the case for standards that use a specific taxonomy.

(2) Requiring access to multiple data systems, means spending more time interviewing and more time spent on data interpretation, since it is likely that the different systems have different data formats, requiring multiple "reformatting" efforts to reach taxonomical compliance. The data are not even necessarily found within the same company, but may be located with suppliers or sub-suppliers.

(3) Since crucial data, such as maintenance or vessel data, may not be stored directly into a RM system, then there is also the risk that original records do not contain all required fields, meaning that if the persons responsible for the operations cannot be contacted or do not recall specifics, then required data will be impaired, severely inaccurate or lost forever.

### 5.2. Mobilization time

The mobilization time, as previously mentioned, includes all resources, including personnel. The source for obtaining such data varies. Sometimes, such as in the event of unexpected, significant failures, specific detailed maintenance reports may outline the mobilization time of resources used. In other cases, such data are available through data stored in vessel log reports, and the data stored into the RM database is typically an interpretation of this. These reports however seldom store records of other resources than the vessels themselves, such as personnel. Maintenance records for which no records exist in neither vessel logs nor other systems have unknown mobilization times and must therefore be estimated. This is typically done based on expected duration to bring the vessel to its destination. Such estimates will thus be optimistic figures in cases where there were delays, where there were shortages in spare parts or where there were other delays that should have been considered, but were not. Conversely, in some cases, due to lack of accurate maintenance records or other source documents, the time allocated to mobilization becomes too large if the same maintenance campaign covers several maintenance jobs but where mobilization is not split across all maintenance records. There is also a fair chance that data collection will not necessarily capture any remobilization related to maintenance jobs, if these occur at a point after the maintenance data was collected or if this remobilization is not logically mapped to the original mobilization activity.

It may be challenging to use statistical data to estimate the mobilization time due to the issues discussed above. In many situations, one must rely on experts having insights into the planning of maintenance activities, to assess and recommend representative values of mobilization time. For example, the mobilization is often highly area and company sensitive, where the mobilization time could be significantly shorter within one specific geographic area, e.g., due to a higher number of cooperating vessels operating within the area.

### 5.3. Detection time

The abbreviation for "mean fault detection time" (i.e., MFDT) is, as already mentioned in Section 2.5, an abbreviation that may have two distinct meanings, both of which are relevant for assessment of reliability and maintenance performance. However, in practice, this discrepancy is not widely problematic, as the specific meaning of both terms is considered well known.

In many situations, it may be reasonable to ignore the contribution of detection time when making assessments involving downtime measures. However, as we may see from the two examples given in the previous section, the significance partly depends on the equipment dealt with and obviously the type of assessment. It is especially important to distinguish between the two types of "MTTR," i.e., "MRT" and "MTTRes," when the failure rate is high and the time of failure is uncertain.

The detection time may naturally be subject to high uncertainty when dealing with hidden failures. However, that is not always the situation. Detection time is closely linked to equipment degradation modeling, such as, the use of so-called P-F interval models (see e.g., ([25], p. 394)). A P-F interval is the time from when there is a potential for failure (TP) due to the equipment being in a condition where it is possible to reveal some fault (e.g., from periodic functional testing, condition monitoring or inspection) to the time when the failure occurs (TF); see **Figure 6**. If the item is subject to periodic testing, and the test interval is shorter than the P-F interval, then one may have a situation where a fault is always detected from testing before any failure occurs. By using such models, it is possible to assess the time from when the fault is revealed to the time of the fault by comparing the condition of the item and the P-F interval model.

Detection time should also be seen in relation to the probability of detection, which relates to both the quality of testing and inspection activities, and to the incentives of accepting versus failing tests. For functional testing of equipment exposed to hidden failures, there may be strong incentives of passing tests that are close to the acceptance criterion. For example, if failing a valve leakage test leads to a more frequent test schedule, documentation work, etc., then the test personnel might pass a test even though some initial result shows that the leakage is 2.1 bar and thus just over the acceptance criterion of 2.0 bar. One could find it convenient to extend the test or make some adjustments on site to achieve a time interval where the leakage result is found acceptable. The incentives of failing or accepting tests is likely to both influence testing schedules and the statistics concerning how many of the results in the area around the acceptance criterion are reported as "failures" and thus also the estimated MTTF value.

### 5.4. Mean time to repair (MTTR)

As with MFDT, MTTR is also an abbreviation that may have several different meanings. Within the oil and gas industry, the letters used in the abbreviation could refer to several terms which all make sense within the area of analysis, and may thus cause analytic ambiguity. The

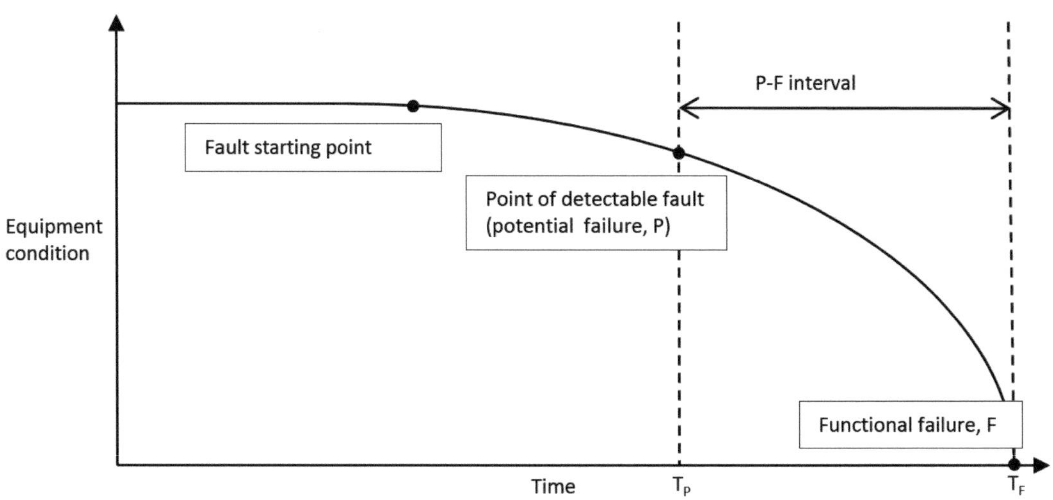


**Figure 6.** P-F interval model (equipment degradation model).

"TT" term consistently refers to "time to." However, the two other letters are found highly inconsistent in use. "M" could refer to "minimum," "mean" or "maximum" and likewise the "R" term could also refer to different meaningful words, such as "repair," "recovery," "respond" or "restore/restoration." It is therefore sometimes confusing when the abbreviation is not explicitly defined.

The use of the term "mean time to restoration," with the abbreviation MTTRes, as suggested in ISO 14224 [14] clearly reduces the chance of making wrong assumptions regarding use of MTTR.

The use of MTTR very much is an issue of quality. By comprising both MTTRes and MRT, it represents a measure that fails to give consistent information. In some situations, MTTR will provide the MTTRes information, and in some situations, it will provide a combination of the two. To use the MTTRes makes it clear that the intended meaning is captured. By keeping the MTTR as a useful term, one must ensure that the "R" refers to "restore" and not "repair." For decision-making purposes, the MTTR measure is better avoided, as also recommended in the international standard [14].

### 5.5. Failure fraction (FF)

The failure fraction, while being a measure that is simple to understand and use, also has some limitations with respect to time aspects. Using the FF as an availability measure, for example, makes it difficult to draw conclusions when the population varies significantly in detection time and number of demands. The value expressed from using this measure does not separate between equipment tested monthly or tested once a year, which makes it difficult to use the values for estimating the MTTF.

Failure fraction is typically used for equipment linked to hidden failures. If assuming that the fault does not occur during the testing, the measure provides relevant information about the number of tests that reveal hidden failures. However, such an assumption strongly depends on the testing interval, as there is a possibility that the fraction of test inducted faults may be high, and one could then argue that the FF, and consequently also the MTTF, would be far higher if the testing interval was reduced.

A main challenge is that the failure fraction ignores the number of demands or faults that occurred since the last test, which is important information regarding the availability of the system. Hence, it is possible that instead of one failure, there should have been recorded, e.g., two failures, where a shorter test interval could have detected the two distinct faults initiated at different points in time instead of only the one that was registered at the point of testing. Furthermore, some of the failures could be observed from a demand and maintained prior to the functional testing and thus not be included in the FF statistics.

### 5.6. Decision-making quality

Typically, the quality of data influences the quality of decision-making. The process also comprises other elements, as listed in Section 4. Part of this quality relates to consistency in the use of the downtime measures for, e.g., availability calculations, which we exemplify in the current chapter.

Based on the above discussion, the use of downtime measures is found ambiguous in the sense that they can be given different interpretations, such as the situation now appears to be with the term "mean time to repair," and, in particular, the abbreviation "MTTR," and may thus contribute to reduced data and decision-making quality. The attempt of [20] to reduce ambiguity by defining "MTTR" as "mean time to restore" and thereby include the detection time is welcome in that sense (see also [27]). However, as the "MTTR" is still being widely used with previous definitions, the "mean time to repair" and the "mean time to restore" could be difficult to separate. The ISO 14224 [14] term "mean time to restoration" with the abbreviation "MTTRes" is considered less ambiguous.

## 6. Conclusions

In this chapter, we have looked at different terms used to describe downtime. Different terms are used in data collection to define downtime, where some may be questioned to not provide adequate quality needed for associated analyzes and decision-making where the RM data is used.

The 'mean time to repair' is a term that is well-established and widely used for e.g. availability calculations. Although the meaning of this term has shifted over time, we find examples where MTTR refers to different meanings and thus is a challenging term to use. The solution proposed in Ref. [14], to avoid the use of this term, and instead use MTTRes and MRT, is considered an acceptable way to deal with this challenge. Data collection experience indicates that to complete a change of the MTTR meaning is difficult, as the term has such a strong position within the oil and gas industry.

Another challenge discussed in this chapter relates to the mobilization of maintenance resources. Experience shows that it is difficult to both interpret and quantify mobilization times in practice. Part of this problem is that resources may be linked to several maintenance activities on site, which provides an opening for different ways of recording the actual time used to mobilize and to repair the specific item. It becomes an interpretation issue, which may influence the values used for prediction of time needed to achieve repair of a failed item. Limited guidance, except for adequate definition, is given in Ref. [14] on this issue.

In general, we recommend that data collection is given higher attention compared with the situation today. Typically, investments are focused on building models and using the data rather than obtaining them and ensuring high quality. Especially, more focus should be on achieving high-quality data from maintenance operations.

## Acknowledgements

The current book chapter is based on the paper: "Maintenance data collection for subsea systems: A critical look at terms and information used for prediction of down time" [43], presented at the European Safety and Reliability (ESREL) conference in Portorož, Slovenia, June 18–22, 2017.

## Author details

Jon T. Selvik[1]* and Eric P. Ford[2]

* Address all correspondence to: jon.t.selvik@uis.no

1 University of Stavanger, Stavanger, Norway

2 International Research Institute of Stavanger (IRIS), Stavanger, Norway

## References

[1] Langseth H, Haugen K, Sandtorv H. Analysis of OREDA data for maintenance optimisation. Reliability Engineering and System Safety. 1998;**60**(2):103-110

[2] Dawotola AW, van Gelder PHAJM, Vrijling JK. Integrity maintenance of petroleum pipelines. Oil and Gas Facilities. SPE 162873. December 2012 https://www.onepetro.org/journal-paper/SPE-162873-PA

[3] Dawotola AW, Trafalis TB, Mustaffa Z, van Gelder PHAJM, Vrijling JK. Data-driven risk based maintenance optimization of petroleum pipelines subject to corrosion. In: Proceedings of the Twenty-first (2011) International Offshore and Polar Engineering Conference (ISOPE), Maui, Hawaii, USA, June 19-24, 2011

[4] Nezamian A, Rodriques C, Nezamian M. Asset integrity management and life extension of gas distribution facilities and pipeline network. SPE 180889-MS. In: Proceedings of the SPE Trinidad and Tobago Section Energy Resources Conference, Port of Spain, Trinidad and Tobago, 13-15 June 2016; https://www.onepetro.org/conference-paper/SPE-180889-MS

[5] Longhi AEB, Pessoa AA, de Almada Garcia PA. Multiobjective optimization of strategies for operation and testing of low-demand safety instrumented systems using a generic algorithm and fault trees. Reliability Engineering and System Safety. 2015;**142**:525-538

[6] Atwood CL, Engelhardt M. Bayesian estimation of unavailability. Reliability Engineering and System Safety. 2004;**84**:225-239

[7] Bevilacqua M, Ciarapica FE, Giacchetta G, Marchetti B. Evaluation of MTBF and MTTR trends for a selected equipment in an oil refinery. Article 18122. In: Proceedings of the 18th ISSAT International Conference on Reliability and Quality in Design. Boston, Massachusetts, U.S.A. International Society of Science and Applied Technologies. July 26-28, 2012

[8] Rajiv SK, Poja S. Computing ram indices for reliable operation of production systems. Advances in Production Engineering & Management. 2012;**7**(4):245-254

[9] Corvaro F, Giacchetta G, Marchetti B, Recanti M. Reliability, availability, maintainability (RAM) study, on reciprocating compressors API 618. Petroleum. 2017;**3**:266-272

[10] Gupta P, Gupta S, Gandhi OP. 2013. Modelling and evaluation of mean time to repair at product design stage based on contextual criteria. Journal of Engineering Design. 2013;**24**(7): 499-523

[11] Coulibaly A, Houssin R, Mutel B. Maintainability and safety indicators at design stage for mechanical products. Computers in Industry. 2008;**59**(5):438-449

[12] Ding L, Wang H, Kang K, Wang K. A novel method for SIL verification based on system degradation using reliability block diagram. Reliability Engineering and System Safety. 2014;**132**:36-45

[13] Jigar AA, Liu Y, Lundteigen MA. Spurious activation analysis of safety-instrumented systems. Reliability Engineering and System Safety. 2016;**156**:15-23

[14] Petroleum, Petrochemical and Natural Gas Industries—Collection and Exchange of Reli ability and Maintenance Data for Equipment. 3rd ed. Geneva, Switzerland: International Organization for Standardization (ISO); 2016

[15] Sandtorv HA, Hokstad P, Thompson DW. Practical experiences with a data collection project: The OREDA project. Reliability Engineering and System Safety. 1996;**51**:159-167

[16] Signoret J-P. Dependability & safety modeling and calculation: Petri nets. IFAC Proceedings Volumes. 2009;**42**(5):203-208

[17] IEC 60300-3-2: Dependability Management—Part 3-2: Application Guide—Collection of Dependability Data from the Field. Geneva, Switzerland: The International Electrotechnical Commission (IEC); 2004

[18] Brissaud F. Using field feedback to estimate failure rates of safety-related systems. Reliability Engineering and System Safety. 2017;**159**:206-213

[19] Selvik JT, Bellamy LJ. On the use of the international standard ISO 14224 on reliability data collection in the oil and gas industry: How to consider failure causes from a human error perspective. In: Walls L, Revie M, Bedford T, editors. Risk, Reliability and Safety: Innovating Theory and Practice. London: Taylor & Francis Group; 2017

[20] IEC 60050-192: International Electrotechnical Vocabulary – Part 192: Dependability. Geneva, Switzerland: The International Electrotechnical Commission (IEC); 2015

[21] ISO/TR 12489: Petroleum, Petrochemical and Natural Gas Industries—Reliability Model ling and Calculation of Safety Systems. Geneva Switzerland: The International Organi zation for Standardization (ISO); 2013

[22] IEC 61508: Functional Safety of Electrical/Electronic/Programmable Electronic Safety-Related Systems. Geneva, Switzerland: The International Electrotechnical Commission (IEC); 2010

[23] IADC—International Association of Drilling Contractors. Mean Active Repair Time. Lexicon. Available from: http://www.iadclexicon.org/mean-active-repair-time/: IADC; Accessed 01-09-2017

[24] Goot R, Levin I. Estimating the latent time of fault detection in finite automaton tested in real time. Automation and Remote Control. 2008;**69**(10):128-141

[25] Rausand M, Høyland A. System Reliability Theory – Models, Statistical Methods, and Applications. 2nd ed. Hoboken, New Jersey, USA: John Wiley & Sons, Inc.; 2004

[26] Ahmadi A, Ghodrati B, Garmabaki AHS, Kumar U. Optimum inspection interval for hidden functions during extended life. In: Proceedings of the 27th International Congress of Condition Monitoring and Diagnostic Engineering Management, 16-18 September 2014. Brisbane, Australia

[27] IEC 60050-191/ADM1: Amendment 1—International Electrotechnical Vocabulary. Chapter 191: Dependability and Quality of Service. Geneva Switzerland: The International Electrotechnical Commission (IEC); 1999

[28] Selvik JT, Signoret J-P. How to interpret safety critical failures in risk and reliability assessments. Reliability Engineering and System Safety. 2017;**161**:61-68

[29] PSA – Petroleum Safety Authority Norway: RNNP – Trends in risk level in the petroleum activity – Summary report 2016 – Norwegian Continental shelf. (2017). Available from: http://www.psa.no/getfile.php/1344338/PDF/RNNP%202016/ENG_summary_RNNP2016.pdf

[30] Hauge S, Onshus T. Reliability Data for Safety Instrumented Systems - PDS Data Handbook 2010 Edition. SINTEF Report no. A13502. SINTEF Technology and Society; 2010

[31] Selvik JT, Abrahamsen EB. A review of safety valve reliability using failure fraction information. In: Podofillini L et al., editors. Safety and Reliability of Complex Engineered Systems: ESREL 2015. London: Taylor & Francis Group; 2016

[32] NORSOK D-010: Well integrity in drilling and well operations. Edition 4:June 2013. Standards Norway. Available from: https://www.standard.no/en/sectors/energi-og-klima/petroleum/norsok-standard-categories/d-drilling/d-0104/

[33] Jansen M. Shell maintenance excellence. SPE 177964-MS: In: Proceedings of the Abu Dhabi International Petroleum Exhibition and Conference, 9-12 November 2015, Abu Dhabi, UAE; https://www.onepetro.org/conference-paper/SPE-177964-MS

[34] Smart K, Blakey K. 2014. Achieving maintenance excellence in Maersk Oil Qatar. International Petroleum Technology Conference; IPTC 17623; https://www.onepetro.org/conference-paper/IPTC-17623-MS

[35] Matheson D, Matheson J. The Smart Organization – Creating Value Through Strategic R&D, Boston. MA, USA: Harvard Business School Press; 1998

[36] Bratvold RB, Begg SH. Making Good Decisions. Richardson, TX, USA: Society of Petroleum Engineers; 2010

[37] Gjerstad T. OREDA—The reliability data reference for the offshore industry. SPE-14000-MS. In: Proceedings of Offshore Europe, 10-13 September 1985, Aberdeen, United Kingdom. https://www.onepetro.org/conference-paper/SPE-14000-MS

[38] Molnes E, Strand G-O. Application of a completion equipment database in decision-making. SPE 63112. In: Proceedings of the SPE Annual Technical Conference and Exhibition, 1-4 October 2000, Dallas, Texas. https://www.onepetro.org/conference-paper/SPE-63112-MS

[39] Corneliussen K, Sørli F, Brandanger Haga H, Tenold E, Menezes C, Grimbert B, Owren K. Well Integrity Management System (WIMS) – A systematic way of describing the actual and historic status of operational wells. SPE 110347. In: Proceedings of the SPE Annual Technical Conference and Exhibition, 11-14 November 2007, Anaheim, California, U.S.A. https://www.onepetro.org/conference-paper/SPE-110347-MS

[40] Reith CA, Lagstrom KB: Reliable subsea oil and gas transportation systems. In: Proceedings of the SNAME World Maritime Technology Conference 2015, Rhode Island, USA, 3-7 November 2015.

[41] Gustavsson F, Eisinger S, Kraggerud AG. Simulation of industrial systems. In: Proceedings of the ESREL 2003 Conference, Maastricht, The Netherlands, 15-18 June 2003.

[42] Lydersen S, Sandtorv H, Rausand M. Processing and Application of Reliability Data. Foundation for Scientific and Industrial Research at the Norwegian Institute of Technology. Trondheim, Norway: SINTEF Report; 1987

[43] Selvik JT, Ford EP. Maintenance data collection for subsea systems: a critical look at terms and information used for prediction of down time. In: Čepin M, Briš R, editors. Safety and Reliability – Theory and Applications. London: CRC Press/Taylor & Francis Group; 2017. p. 2989-2996

Chapter 4

# Updated Operational Reliability from Degradation Indicators and Adaptive Maintenance Strategy

Christophe Letot, Lucas Equeter,
Clément Dutoit and Pierre Dehombreux

Additional information is available at the end of the chapter

http://dx.doi.org/10.5772/intechopen.69281

**Abstract**

This chapter is dedicated to the reliability and maintenance of assets that are characterized by a degradation process. The item state is related to a degradation mechanism that represents the unit-to-unit variability and time-varying dynamics of systems. The maintenance scheduling has to be updated considering the degradation history of each item. The research method relies on the updating process of the reliability of a specific asset. Given a degradation process and costs for preventive/corrective maintenance actions, an optimal inspection time is obtained. At this time, the degradation level is measured and a prediction of the degradation is conducted to obtain the next inspection time. A decision criterion is established to decide whether the maintenance action should take place at the current time or postpone. Consequently, there is an optimal number of inspections that allows to extend the useful life of an asset before performing the preventive maintenance action. A numerical case study involving a non-stationary Wiener-based degradation process is proposed as an illustration of the methodology. The results showed that the expected cost per unit of time considering the adaptive maintenance strategy is lower than the expected cost per unit of time obtained for other maintenance policies.

**Keywords:** degradation-based reliability, degradation models, remaining useful life, reliability-based maintenance, predictive maintenance, numerical case study

## 1. Introduction

Maintenance is a keystone to ensure the competitiveness of any industry in terms of productivity, quality and availability. According to MIL-STD-3034, standard maintenance (preventive, corrective and inactive) is the action of performing tasks (time-directed, condition-directed, failure-finding, servicing and lubrication) at periodicities (periodic, situational and unscheduled) to ensure the item's functions (active, passive, evident and hidden) are available until the next scheduled maintenance period. Both preventive maintenance and corrective maintenance tasks are performed on industrial equipment through their operational lifetime. The balance between preventive and corrective maintenance actions is usually ruled by the long-term cost rate, the asset availability or safety criteria. Accordingly, different maintenance strategies are encountered in literature [1]. They concern the replacement of systems subject to random failures and whose states are identified at all time.

However, in some particular cases, the item state may be influenced by some factors especially for mechanical units that have to cope with variable mechanical stresses, a variable energy consumption, a modification of the operating conditions and the effect of the environment. Obviously, the reliability and remaining useful life (RUL) of such equipment will change accordingly. Consequently, the maintenance scheduling has to be updated considering the degradation history. This topic is covered by the degradation-based reliability approach that consists in monitoring degradation covariates with respect to a given threshold in order to trigger inspection or maintenance actions.

Several case studies highlighted that, usually, the failure of an item is to put in relation with a degradation process. Typical examples of such degradation processes are the crack-growth in a mechanical part due to fatigue loading, the wear of cutting tools in machining, the development of corrosion mechanism in reinforced concrete structures and the development of pitting on bearing race. Moreover, a large number of experiments and engineering phenomena show that items of the same category, even from one identical batch, degrade differently from one another in performance. As the failure of an item can lead to dramatic consequences, it is mandatory to assess the specific remaining useful life (RUL) accurately and to schedule the maintenance tasks accordingly for each item. The modelling of the degradation mechanism based on measurements and fitting procedures is a key element to achieve this objective.

Historically, the degradation was first considered at the design stage of an item, using empirical laws for the conception of mechanical parts for fatigue loading cycles (e.g., Palmgren fatigue life for bearings and Paris-Erdogan crack growth relationship). However, experience showed that these empirical degradation models were affected by a significant dispersion on the predicted life, thus enforcing the necessity to consider uncertainties for such models. Consequently, deterministic models were replaced by stochastic models to take into account the unit-to-unit variability and time-varying dynamics for the remaining useful life prediction. Thanks to the development of accurate real-time sensors and dedicated monitoring software, the tracking of the degradation is made possible by measuring related physical variables such as vibrations, temperatures, pressures and forces. The monitoring of those indicators allows to detect faulty behaviours and to forecast a degradation trend, thus allowing for a better remaining useful life prediction. To sum up, the reliability and the remaining useful life can be assessed at three different stages of an item life as illustrated in **Figure 1**:

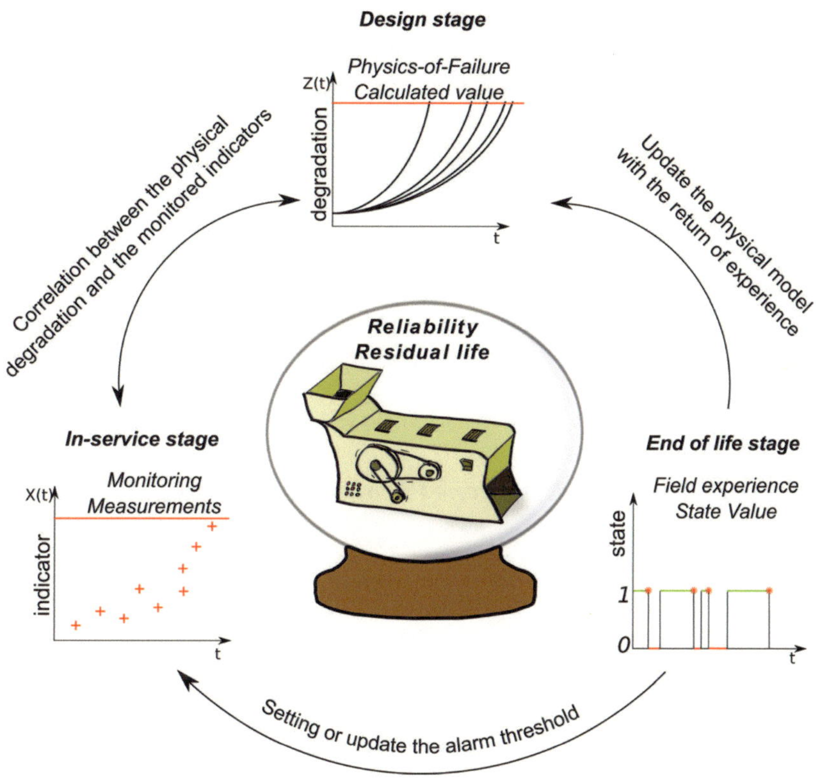


**Figure 1.** The three complementary approaches for the reliability and remaining useful life estimation.

1. In the design stage, the physical degradation mechanism is modelled taking into account the uncertainties in the parameters. This gives a nominal life expectancy of the item that depends on given conditions of usage.
2. The in-service stage during which the degradation indicators are monitored and alarm thresholds are set. Faulty behaviours due to a process perturbation or external cause can be detected.
3. The end of life stage from which failure data are used to update both the degradation models and the threshold values for the monitored indicators.

## 2. Degradation-based reliability

### 2.1. Reliability

The reliability of an item (a part, a machine or a system) is the probability that the item will perform its intended function throughout a specified time interval when operated in a normal (or stated) environment [2]. According to the standards, the term 'reliability' also refers to a reliability value and is considered as the probability for an item to be in a functional state. Given a random variable $T_f$ that represents the lifetime of an item, the reliability $R(t)$ is given by the following equation:

$$R(t) = P(T_f > t) \tag{1}$$

As previously mentioned, the reliability can be identified at different stages of an item life. The fitting step of reliability is usually performed using field failure data or simulated failure times from the design stage. The set of failure data is used to obtain the non-parametric failure function (also called unreliability) $F(t)$ that represents the distribution of the failure times:

$$F(t) = P(T_f \leq t) = 1 - R(t) \tag{2}$$

The probability density function $f(t)$ is derived from the failure function:

$$f(t) = \frac{dF(t)}{dt} \tag{3}$$

Finally, the failure rate (or hazard function) $h(t)$ is defined:

$$h(t) = \frac{f(t)}{R(t)} \tag{4}$$

The failure rate represents the conditional probability of failure of an item during $[t, t + \Delta t]$ given that this item has survived until time $t$. The failure rate is a first indicator on the evolution of an item state. An increasing failure rate indicates that the conditional probability of failure over time increases, thus implying a progressive degradation process.

Fitting a parametric reliability model on data is achieved using two methods: the regression method and the maximum likelihood method. For the regression method, the parametric reliability law is transformed into a linear form and a regression fit is performed. The latter is based on the likelihood function of the reliability model to identify the parameters that maximizes the probability of observing the failure data again.

The fitting procedure is illustrated on the two-parameter Weibull law for complete data as example.

### 2.1.1. The regression method

The failure function of a Weibull law is $F(t) = 1 - \exp\left(-\left(\frac{t}{\eta}\right)^{\beta}\right)$, $\eta$ being the scale parameter and $\beta$ the shape parameter. The linear form $y = Ax + B$ of this model for the regression fit is:

$$\ln\left(\ln\left(\frac{1}{1 - \hat{F}(t_i)}\right)\right) = \hat{\beta}\ln(t_i) - \hat{\beta}\ln(\hat{\eta}) \tag{5}$$

With $\hat{F}(t_i)$ a non-parametric estimator of the failure function assessed at the $i$th failure time considering that $n$ items were operational at the beginning of the survival study. Common non-parametric estimators of the failure function are [3]:

1. the Kaplan-Meier estimator $\hat{F}(t_i) = i/n$;
2. the mean rank estimator $\hat{F}(t_i) = i/(n+1)$;
3. the approached rank adjust estimator $\hat{F}(t_i) = (i - 0.3)/(n + 0.4)$.

From Eq. (5), the identification of the parameters for the linear regression gives:

$$\hat{\beta} = A \tag{6}$$

$$\hat{\eta} = \exp\left(-\frac{B}{\hat{\beta}}\right) \tag{7}$$

### 2.1.2. *The maximum likelihood method*

The likelihood function $L$ considers the product of the probability density function of the model governed by a set of parameters $\boldsymbol{\theta}$, each function being assessed at a failure time $t_i$:

$$L(t_i|\,\boldsymbol{\theta}) = \prod_{i=1}^{n} f(t_i) \tag{8}$$

For the Weibull case, using the log-transformation of the likelihood function, it follows:

$$\ln L(t_i|\,\beta,\,\eta) = \sum_{i=1}^{n} \left\{ \ln(\hat{\beta}) - \ln(\hat{\eta}) + (\hat{\beta} - 1)\left(\ln(t_i) - \ln(\hat{\eta})\right) - \left(\frac{t_i}{\hat{\eta}}\right)^{\hat{\beta}} \right\} \tag{9}$$

Taking the partial derivatives of the log-likelihood function, the estimators of the parameters are [3]:

$$\frac{\sum_{i=1}^{n}\left(t_i^{\hat{\beta}}\ln(t_i)\right)}{\sum_{i=1}^{n} t_i^{\hat{\beta}}} - \frac{1}{\hat{\beta}} - \frac{1}{n}\sum_{i=1}^{n} \ln\, t_i = 0 \tag{10}$$

$$\hat{\eta} = \left(\frac{1}{n}\sum_{i=1}^{n} t_i^{\hat{\beta}}\right)^{\frac{1}{\hat{\beta}}} \tag{11}$$

Sometimes, the produced fit does not match the experimental data. In this case, a third parameter has to be introduced, that is, the location parameter $\gamma$ that shifts the failure times accordingly. Then a convenient approach consists in testing different values of this location parameter, to apply the regression and to identify the best estimator $\hat{\gamma}$ for which the highest determination factor is obtained. The maximum likelihood method can also handle the case of the three-parameter Weibull estimation through numerical optimization of the likelihood function.

### 2.2. First hitting time process

First hitting (or passage) time processes are used in a wide area of applications including medicine, environmental sciences, engineering sciences, economy and sociology. Accordingly, such processes can describe either the sojourn duration of a patient in a hospital given the gravity of his illness, the time delay before a polluting product reaches an area, the lifetime of mechanical parts given a stochastic damage assessment, etc. Generally speaking, these processes aim at capturing the stochastic behaviour of a given diffusive mechanism to predict the hitting times of a critical threshold.

A first hitting time process has two components:

- A stochastic (degradation) process noted $\{Z(t),\ t \in T,\ z \in Z\}$, which describes the random evolution of a degradation process (e.g., physics-related processes in the areas of mechanics, chemistry and electricity or non-physics-related processes such as the evolution of quality or a performance indicator) with respect to elapsed time;
- A given state space boundary value noted $z_f$ that defines the failure level of the degradation process.

Given the initial degradation value $z_0$ at the starting time $t_0$, the first hitting time $T_f$ of reaching the critical threshold is [4]:

$$T_f = \inf\left(t \mid Z(t) - Z(t_0) \geq z_f - z_0\right) \tag{12}$$

Consequently, the first passage time for exceeding the degradation threshold is the first time $t$ for which the stochastic process $Z(t)$ has reached the threshold $z_f$ given that it started from the value $z_0$ at initial time $t_0$. Instead of considering the first hitting time, one may be interested in obtaining the probability of crossing the failure threshold:

$$F\left(t|Z(t), z_f, z_0, t_0\right) = P(T_f \leq t) = P\left(Z(t) - Z(t_0) \geq z_f - z_0\right) \tag{13}$$

The failure function $F(t)$ is now conditioned by the degradation process $Z(t)$ assessed over time, the failure threshold $z_f$, and the initial values $t_0$ and $z_0$.

### 2.3. Remaining useful life

Let $Z(t)$ be the evolution of the degradation over time, $z_f$ (a positive value) be the failure threshold and $X(t_j)$ a degradation measurement at inspection time $t_j$. It is supposed that the degradation process leads to a soft failure (at time $T_f$), which means that there are no other hard failure modes which compete for the failure time. Considering the first hitting time process of a given threshold, the RUL of an item given the conditional measurement $X(t_j)$ at inspection time $t_j$ and the preset threshold $z_f$ is:

$$\mathrm{RUL}(t_j) = \inf\{l : X(t_j + l) \geq z_f | l \geq 0,\ X(t_j) < z_f\} \tag{14}$$

In order to obtain an accurate estimation of the RUL, the degradation model $Z(t)$ should perfectly fit the degradation data $X(t)$ to minimize the error in the forecasted degradation value $Z(t_j + l)$. Practically, the RUL is obtained by the computation of the mean residual life MRL (i.e., the mean value of the RUL conditioned by the observations $X(t)$) using the following equation [5]:

$$\mathrm{MRL}(t_j) = E\Big(T_f - t | T_f > t, X(t_j) < z_f\Big) = \int_{t_j}^{\infty} R\Big(u \mid X(t_j)\Big) du \tag{15}$$

With $R(u \mid X | t_j))$ the conditional degradation-based reliability of the item at time $u > t_j$, given the degradation measurement $X(t_j)$.

### 2.4. Degradation models

According to Gorjian et al. [6], degradation models can be divided into two main families: normal degradation models and accelerated degradation models.

- Normal degradation model are dedicated to the estimation of reliability for asset operating at normal conditions. Examples of normal degradation models are the general degradation path model, the random process model, the (non-)linear regression models and the time series model. Normal degradation models can also consider some stress factors; such cases are the stress-strength interference model, the cumulative damage/shock model for which the degradation measure is a function of a defined stress.
- Accelerated degradation models make inference about the reliability at normal conditions given degradation data that were obtained at accelerated time/stress conditions. There exist two categories: the physics-based models and the statistics-based model. In physics-based models, the physical variables of the model (e.g., pressure, temperature and stress) are increased in order to obtain failure data under a reasonable timeframe. Examples of physics-based models are the Arrhenius model for temperature-related degradation mechanism and the inverse power model for non-thermal-related degradation mechanism (e.g., the fatigue damage in bearings). Statistics-based model uses data obtained in various operating conditions to establish a statistical model from a set of input explanatory variables. Example of statistics-based model is the Cox proportional hazards model that expresses the failure rate as the product of a baseline failure rate and a function of the covariates [7].

As previously mentioned, the RUL knowledge is a keystone to offer guidance for an optimal maintenance planning. It has been considered as a fundamental ingredient in the field of Prognostics and Health Management (PHM) [8]. The main challenge of RUL estimation lies in the presence of heterogeneity due to different inner states or external operating conditions of systems. The performance or degradation of a system is caused by interactions of both the inner deterioration and the working environment of the system, justifying the need to take into account the heterogeneity in the degradation model. In this way, it is affected by three kinds of heterogeneity that are the unit-to-unit variability for items from the same batch, the variability in the operating conditions over time and the diversity of tasks and workloads of systems during their life cycles. For each heterogeneity corresponds adequate degradation models. In this study, a focus on data-driven models with unit-to-unit variability and time-varying dynamics of systems is considered.

### 2.4.1. *Random coefficient regression models*

Random effects were first considered in random coefficient regression models [9]. At each inspection time $t_j$, a degradation value $X_j(t_j)$ is measured on an item $i$. The degradation model takes the form:

$$X_i(t_j) = Z(t_{ij}; \boldsymbol{\alpha}; \boldsymbol{\beta}_i) + \varepsilon_{ij} \tag{16}$$

With $\boldsymbol{\alpha} = (\alpha_1, \alpha_2, \ldots, \alpha_n)$ a vector of constant parameters that are characteristics of the tested population and $\boldsymbol{\beta}_i = (\beta_{i1}, \beta_{i2}, \ldots, \beta_{in})$ a vector of random parameters that are specific to each item $i$ (i.e., $\boldsymbol{\alpha}$ is the vector representing the common part of the degradation, while $\boldsymbol{\beta}_i$ represents the heterogeneity). The term $\varepsilon_{ij}$ represents the measurement error on the degradation value at time $t_j$ on the item $i$ and is supposed to follow a Gaussian distribution with a null mean and a standard deviation $\sigma_\varepsilon$. Common representations of this model are the linear form, the power form and the logarithmic form. However, this simple model has several drawbacks including the need for more historical degradation data from different items of the same category, the difficulty in capturing the time-varying dynamics of the items and the independency between random noise with time [10].

### 2.4.2. *Stochastic process-based models*

Stochastic process-based models with random coefficients are able to consider both time-varying dynamics and unit-to-unit variability. These processes may be represented by some specific models that are derived from the Levy Processes family [11]. A levy stochastic process has independent (non-)stationary increments which represent the sequence of successive random and independent displacements of a point in a space. Frequently used models from this family are the gamma process [12] and the Wiener process [13]. According to the results presented in the literature, it seems that stochastic process-based models with random effects can effectively improve the accuracy of RUL estimation in addition to extend the range of applications by considering both cases of monotonous and non-monotonous degradation processes, whether they are linear or non-linear [8]. However in industrial applications, the main drawback of stochastic process-based models with random effects is the computation issue that can be complex and highly dependent of the choice of random parameters and their distribution. Generally, the assumption of normally distributed parameters is chosen [14]. The next section is dedicated to the study of a non-stationary formulation of the Wiener process that is used in the illustrative example at the end of this chapter.

## 2.5. The Wiener process

The Wiener process has been widely applied to degradation modelling in various fields, for example, bearings, laser generators and milling machines [15]. The Wiener process is particularly a good candidate to represent the evolution of a degradation process that is made of an increasing trend over time with random Gaussian noise, both being proportional to elapsed time. It is characterized by continuous sample paths and independent, (non-)stationary and normally distributed increments [16].

### 2.5.1. *Definition and mathematical properties*

A Wiener process-based model has two kinds of parameters: one set related to the expected value of the degradation rate and one that represents the magnitude of the random noise. The generic formulation of a degradation process $Z_W(t)$ ruled by a Wiener process-based model is:

$$Z_W(t) = Z_W(t_0) + m(t; \boldsymbol{\theta}) + \sigma W(t) \tag{17}$$

With $Z_W(t)$ the initial degradation value at time $t_0$, $m(t; \boldsymbol{\theta})$ the trend function ruled by the set of parameters $\boldsymbol{\theta}$, $\sigma$ a parameter that represents the magnitude of the Gaussian noise perturbing the trend, $W(t)$ the standard Brownian motion that has the following characteristics:

- $W(0) = 0$;
- $W$ has independent increments, that is, for $0 \le t_1 \le t_2 \le t_3 \le t_4$, the increments $W(t_4) - W(t_3)$ and $W(t_2) - W(t_1)$ are independent random variables;
- $W$ is a continuous stochastic process, and for $0 \le t_1 \le t_2$, the increment $W(t_2) - W(t_1)$ has a normal distribution with mean equals to zero and standard deviation equals to $\sqrt{t_2 - t_1}$.

It follows that the Wiener process-based model can also be formulated as:

$$Z_W(t) = Z_W(t_0) + N\left(m(t; \boldsymbol{\theta}), \sigma\sqrt{t}\right) \tag{18}$$

With $N\left(m(t; \boldsymbol{\theta}), \sigma\sqrt{t}\right)$ the normal distribution with power density function $f_W(x)$

$$f_W(x) = \frac{1}{\sigma\sqrt{2\pi t}} \exp\left(-\frac{\left(x - m(t; \boldsymbol{\theta})\right)^2}{2\sigma^2 t}\right) \tag{19}$$

Therefore, the mathematical expectation and variance of a Wiener process-based degradation model are:

$$E\left(Z_W(t)\right) = Z_W(t_0) + m(t; \boldsymbol{\theta}) \tag{20}$$

$$V\left(Z_W(t)\right) = \sigma^2 t \tag{21}$$

### 2.5.2. *Fitting the Wiener process*

Given a set of $n + 1$ measurements of degradation data $z_0, z_1, z_2, \ldots, z_n$ at inspection times $t_0, t_1, t_2, \ldots, t_n$, the fitting procedure of a Wiener process-based degradation model is achieved mainly using the maximum likelihood method [17]. This method allows to obtain the value of the parameters $\boldsymbol{\theta}$ and $\sigma$ from the power density function (pdf) of the Wiener process-based model, each pdf function being assessed at the measurements points. The likelihood function is:

$$L(t, z|\boldsymbol{\theta},\ \sigma) = \prod_{i=0}^{n-1} \frac{1}{\sigma\sqrt{2\pi(t_{i+1}-t_i)}} \exp\left(-\frac{\left[(z_{i+1}-z_i) - \left(m(t_{i+1};\boldsymbol{\theta}) - m(t_i;\boldsymbol{\theta})\right)\right]^2}{2\sigma^2(t_{i+1}-t_i)}\right) \tag{22}$$

For the stationary Wiener process (i.e., $m(t;\ \boldsymbol{\theta})\ \mu.t$ is a linear function of time), the estimation of the parameters $\mu$, $\sigma$ is obtained by taking the partial derivative of the log-likelihood function and searching for the roots [17]:

$$\hat{\mu} = \frac{\sum_{i=0}^{n-1}(t_{i+1}-t_i)}{\sum_{i=0}^{n-1}(z_{i+1}-z_i)} = \frac{(z_n - z_0)}{(t_n - t_0)} \tag{23}$$

$$\hat{\sigma} = \sqrt{\frac{1}{n-1}\sum_{i=0}^{n-1}\frac{\left[(z_i - z_{i-1}) - \hat{\mu}(t_i - t_{i-1})\right]^2}{\Delta t_i}} \tag{24}$$

For non-stationary Wiener processes, the parameters are obtained using optimization techniques such as the Quasi Newton methods for instance [18].

*2.5.3. FHT and RUL distribution of a Wiener process*

Given a degradation threshold value $z_c$ and initial degradation value $z_0$, the hitting times of a Wiener process-based degradation model follow an Inverse Gaussian law with mean parameter equals to $m^{-1}(z_c - z_0 | \boldsymbol{\theta})$ (i.e., the inverse function of $m(t \mid \boldsymbol{\theta})$ and shape parameter equals to $(z_c-z_0)^2/\sigma^2$ that has the following power density function $f_{IG}$:

$$f_{IG}(t|z_c, z_0, \boldsymbol{\theta},\ \sigma) = \frac{z_c - z_0}{\sigma\sqrt{2\pi t^3}} \exp\left\{-\frac{(z_c - z_0)^2}{2t\sigma^2}\frac{\left[x - m^{-1}(z_c - z_0|\ \boldsymbol{\theta})\right]^2}{\left[m^{-1}(z_c - z_0|\ \boldsymbol{\theta})\right]^2}\right\} \tag{25}$$

The corresponding reliability considering the last measurement $z_i$ at inspection time $t_i$ is:

$$R(t|z_c, z_i, t_i, \boldsymbol{\theta},\ \sigma) = 1 - \int_{t_i}^{t} \frac{z_c - z_i}{\sigma\sqrt{2\pi x^3}} \exp\left\{-\frac{(z_c - z_i)^2}{2x\sigma^2}\frac{\left[x - m^{-1}(z_c - z_i|\ \boldsymbol{\theta})\right]^2}{\left[m^{-1}(z_c - z_i|\ \boldsymbol{\theta})\right]^2}\right\} dx \tag{26}$$

As the parameters of the Wiener process-based degradation model are updated for each new measurement, the reliability function given by Eq. (26) is a dynamic reliability, that is, the reliability is updated given the updated estimation of the parameters and the last degradation measurement. Consequently, it corresponds to the RUL distribution over time that is assessed at different inspection times.

## 3. Maintenance model

### 3.1. General assumptions

- The failure time of an item of equipment is ruled by a stochastic degradation process, that is, it corresponds to the hitting time of a degradation threshold.

- Whenever a failure occurred, a corrective maintenance action is performed immediately, that is, the degradation process cannot cross the threshold and there is no duration in failed state to consider.
- The degradation process itself is not altered by the maintenance actions, that is, the parameters of the degradation process remain unchanged. Maintenance actions only affect the recovery values of the degradation at inspection times.
- Both preventive maintenance and corrective maintenance are considered, that is, the failure of an item of equipment does not lead to dramatic consequences (only higher cost values). For preventive maintenance, an optimal inspection time $t_j = T_p$ is obtained considering the balance between preventive and corrective costs.
- For the predictive maintenance approach at each inspection time, two strategies are considered depending on the measured degradation level. If the degradation is lower than expected, then the maintenance action is postponed to another inspection time considering the predicted degradation distribution. Otherwise, the maintenance action is conducted at current inspection time. The decision to do or postpone the maintenance action is given by a cost criterion.
- In case of a corrective maintenance or a preventive replacement, the item is replaced by a new one (AGAN maintenance) so that $z_0 = 0$, that is, the expected life of the item corresponds to its mean time to failure (MTTF).
- Failure of the asset is tolerated, that is, it does not lead to catastrophic consequences.
- Downtimes due to unavailability of the maintenance staff or resources are not considered at this stage, that is, the duration of maintenance actions is considered as negligible compared to the life duration of the asset (i.e., MTTR < MUT).

### 3.2. Maintenance policies

Four types of maintenance policies are considered that are the corrective maintenance, the preventive systematic maintenance, the preventive condition-based maintenance (CBM) and the predictive maintenance [19].

#### *3.2.1. The corrective maintenance*

The maintenance task is carried out after failure of the asset to identify, isolate and rectify a fault in order to restore the failed equipment, machine or system in an operational condition. The timing for corrective maintenance can be immediate (the restoration process starts immediately after a failure) or deferred (the maintenance tasks are delayed given a set or maintenance rules). A corrective maintenance policy is mainly used for low value assets, equipment for which the failures do not lead to catastrophic consequences or item for which the RUL is hard to predict due to random failures.

#### *3.2.2. The preventive systematic maintenance*

Also known as calendar-based, clock-based or time-based maintenance, it is a maintenance action of an asset according to a scheduled timetable (i.e., a given periodicity between consecutives maintenance tasks). It is mainly applied for critical assets to prevent failures, or routine inspections that occur on a regular basis to control the state of safety equipment. The optimal periodicity is obtained given the reliability of the item, and the relative costs between a preventive maintenance and a corrective maintenance.

#### *3.2.3. The preventive condition-based maintenance (CBM)*

The preventive maintenance actions are based on the condition of the component being maintained. The condition of assets is tracked over time using statistical process control techniques, monitoring equipment performance through regular inspections. Measuring the variable of interest directly is usually difficult to achieve, and in this case, some other related variables are used to obtain the estimates of the variable of interest (e.g., bearings wear can be accessed through vibration, noise or temperatures analyses). Once the related indicators have crossed a given threshold, the preventive maintenance action is performed.

#### *3.2.4. The predictive maintenance*

An extension of the condition-based maintenance in that way that the state or degradation level of the asset is forecasted to predict the failure time and adapt the maintenance tasks accordingly. An alternative denomination is the adaptive maintenance as the maintenance scheduling is continuously adapted according to the updated actual degradation level and its forecasting.

### 3.3. The cost model

#### *3.3.1. Corrective and age replacement cost model*

In a context of a reliability-centred maintenance approach, the cost maintenance model is based on the reliability calculation that allows to obtain the most relevant time to perform the preventive maintenance in order to reach the optimum expected maintenance cost per unit of time. A generic age replacement model is used as preventive maintenance model [19]

$$\bar{c}_m(T_p) = \frac{F(T_p).C_c + R(T_p).C_p}{\text{MUT}|T_p} = \frac{F(T_p).C_c + R(T_p).C_p}{\int_0^{T_p} R(t)dt} \tag{27}$$

with $T_p$ is the time of preventive maintenance; $F(T_p)$ represents the probability of having a failure at time $T_p$ given the degradation-based reliability model; $C_c$ is the total corrective cost incurred when a failure occurs; $C_p$ is the total cost due to a preventive maintenance action; $\bar{c}_m$ is the average cost per unit of time that has to be optimized; $\text{MUT}\,|\,T_p$ is the mean up time under a preventive maintenance policy.

Considering the cost contributions, $C_c$ and $C_p$ are expressed as follows:

$$C_c = \text{MTTR}_c(\tau_{\text{sto}} + \tau_{\text{int}_c}) + C_{\text{cst}_c} + P_{\text{cst}_c} \tag{28}$$

$$C_p = \text{MTTR}_p(\tau_{\text{sto}} + \tau_{\text{int}_p}) + C_{\text{cst}_p} + P_{\text{cst}_p} \tag{29}$$

$MTTR_c$ and $MTTR_p$ are the mean times to restore, respectively, for a corrective maintenance and for a preventive maintenance, $\tau_{sto}$ the variable losses per unit of time due to unavailability of the asset, $\tau_{int}$ the variable costs per unit of time, $C_{cst}$ the fixed part of the costs, $P_{cst}$ the fixed part of the losses and the subscripts '$c$' and '$p$' standing for corrective and preventive maintenance, respectively.

From Eq. (27), considering an infinite time to perform a preventive maintenance leads to the pure corrective maintenance model, that is

$$\bar{c}_{mc}(\infty) = \frac{F(\infty).C_c + R(\infty).C_p}{\int_0^{\infty} R(t)dt} = \frac{C_c}{\text{MTTF}} \tag{30}$$

with MTTF the mean time to failure.

### 3.3.2. *Introducing the inspection cost*

Considering the condition-based and the predictive maintenance models, measurements are required to assess the current degradation level and to forecast its trend. It is considered that these measurements are performed through inspections that are considered as additional costs. In this case, the total maintenance cost over time is [20]:

$$C(t) = C_i N_i(t) + C_p N_p(t) + C_c N_c(t) \tag{31}$$

With $C_x$ and $N_x(t)$ the cost and the counter of inspection, preventive actions and corrective tasks. On the one hand, the inspections will increase the total cost, but on the other hand, it will allow to avoid failures as well as to increase the useful life of equipment. Consequently, there is an optimum number of inspections to consider.

Considering the condition-based maintenance scenario, the inspections should take place at a given periodicity that will reasonably decrease the probability of crossing the degradation threshold without being too frequent.

For the predictive maintenance scenario, it is considered that at least one inspection will take place at the time corresponding to the calendar-based preventive maintenance model (i.e., the inspection will guide the decision of performing the preventive maintenance action or postponing it at another inspection time). Practically, at the first inspection time $t_{j=1}$ corresponding to the time of preventive maintenance, the following criterion is assessed:

$$K_C(t_1 = T_p) = \frac{\frac{C_p}{t_1}}{\frac{F^*(t_2).C_c + R^*(t_2).C_p + C_i}{t_1 + \int_{t_1}^{t_2} R^*(t)dt}} \tag{32}$$

$F^*$ and $R^*$ being the updated failure and reliability function given the last degradation measurement $Z(t_1)$ and $t_2 = T_p^* = T_p\left(t_1|Z(t_1)\right)$ the next forecasted inspection time given the degradation level $Z(t_1)$ measured at the 1st inspection time $t_1$. This criterion represents the ratio between the strategy that replaces the equipment at time $t_1$ and the strategy to postpone the replacement at time $t_2 = T_p^*$. Actually, the numerator represents the cost rate obtained for a lifecycle $t_1$ and the denominator the cost rate for an expected lifecycle $t_2$ that is obtained given the last degradation measurement and considering the corrective and preventive cost values. If $K_C > 1$ it is cheaper to postpone the preventive maintenance action to the next inspection time predicted from the degradation-based reliability model. When $K_C \leq 1$, the maintenance is performed at the last inspection time reached.

At inspection time $t_{j+1}$, the criterion is assessed considering the total elapsed time and the number of inspections already performed, that is, the general form of the criterion is:

$$K_C(t_j) = \frac{\frac{C_p+(j-1).C_i}{t_j}}{\frac{F^*(t_{j+1}).C_c+R^*(t_{j+1}).C_p+j.C_i}{t_j+\int_{t_j}^{t_{j+1}} R^*(t)dt}} \tag{33}$$

As long as $K_C(t_j) > 1$, the maintenance action is delayed to next inspection time $t_{j+1}$.

## 4. Methodology

This section summarizes the methodology that consists of updating a degradation-based reliability model from data as well as the maintenance optimization for preventive replacement that leads to an adaptive maintenance model. Considering a completely new asset for which neither reliability nor degradation information is provided, the methodology focuses on four stages that correspond to the four maintenance policies related to the knowledge level of the reliability and degradation process of the asset.

### 4.1. Run-to-failure stage

As no information is available on the asset, the first stage consists to let the asset running until its failure before performing a corrective maintenance action to restore it in AGAN condition. This provides a set of failure times that is used to fit a parametric reliability model as presented in Section 2.1.

### 4.2. Systematic preventive maintenance stage

According to the parametric reliability of the asset and the corrective and preventive maintenance costs, an optimal periodicity $T_p$ is obtained using Eq. (27).

### 4.3. Monitoring the degradation and CBM stage

The third stage consists in monitoring the degradation process to fit a degradation model that will be used in the last stage. Consequently, the monitoring of the data should be tuned so that the measurements points are sufficient for the modelling. Two design variables are to be defined as the preventive degradation threshold beyond which the preventive task is performed and the degradation measurements periodicity. Generally, these variables are

http://dx.doi.org/10.5772/intechopen.69281

adjusted using experimental design, sensitivity analyses and return of experience. Given a penalty cost for the monitoring (e.g., inspection cost) and practical constraints, an optimal set of these parameters can be identified.

### 4.4. Forecasting the degradation and adaptive maintenance stage

Given the degradation measurements collected in the previous stage, a degradation model identification can be attempted. The selection of the most suitable model is complexed; it depends on the nature of the degradation data and the sample size. Goodness of fit criterion is used to give guidance on the most suitable degradation model. Once the degradation model has been identified, the adaptive maintenance stage defines the first inspection time as the time corresponding to the preventive systematic maintenance periodicity $t_1 = T_p$. Considering that the item has survived until this time, an inspection is performed and the degradation level is measured. Given the degradation model $Z(t)$, the distribution of the hitting times is updated so as the failure function density $F^*(t)$. A new reliability model is then fitted on this failure function from which we can deduce the mean residual life as well as a new optimized time $T_p^*$ for a preventive replacement. At this step, the cost criterion $K_c$ is assessed if $K_c > 1$, the maintenance is postponed to the next inspection time $t_{j+1} = T_p^*$; otherwise, the preventive maintenance action is performed at the current inspection $t_j$. **Figure 2** shows an illustration of the updating process of both the degradation and the threshold hitting times distribution.

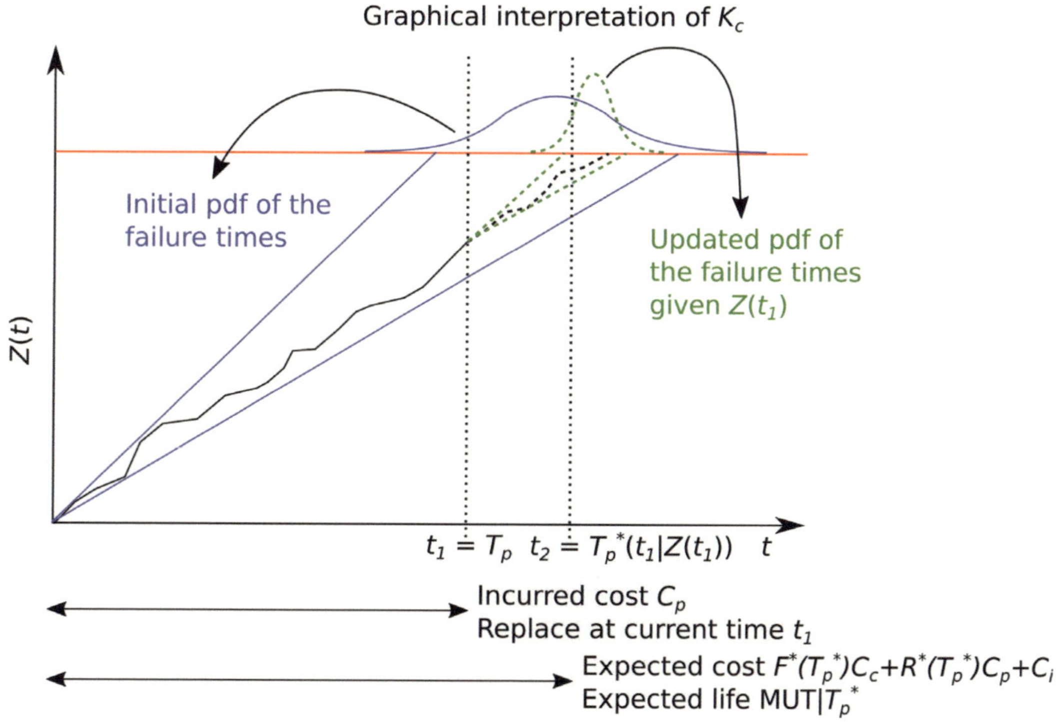


**Figure 2.** Illustration of the adaptive maintenance and graphical interpretation of the cost criterion $K_c$.

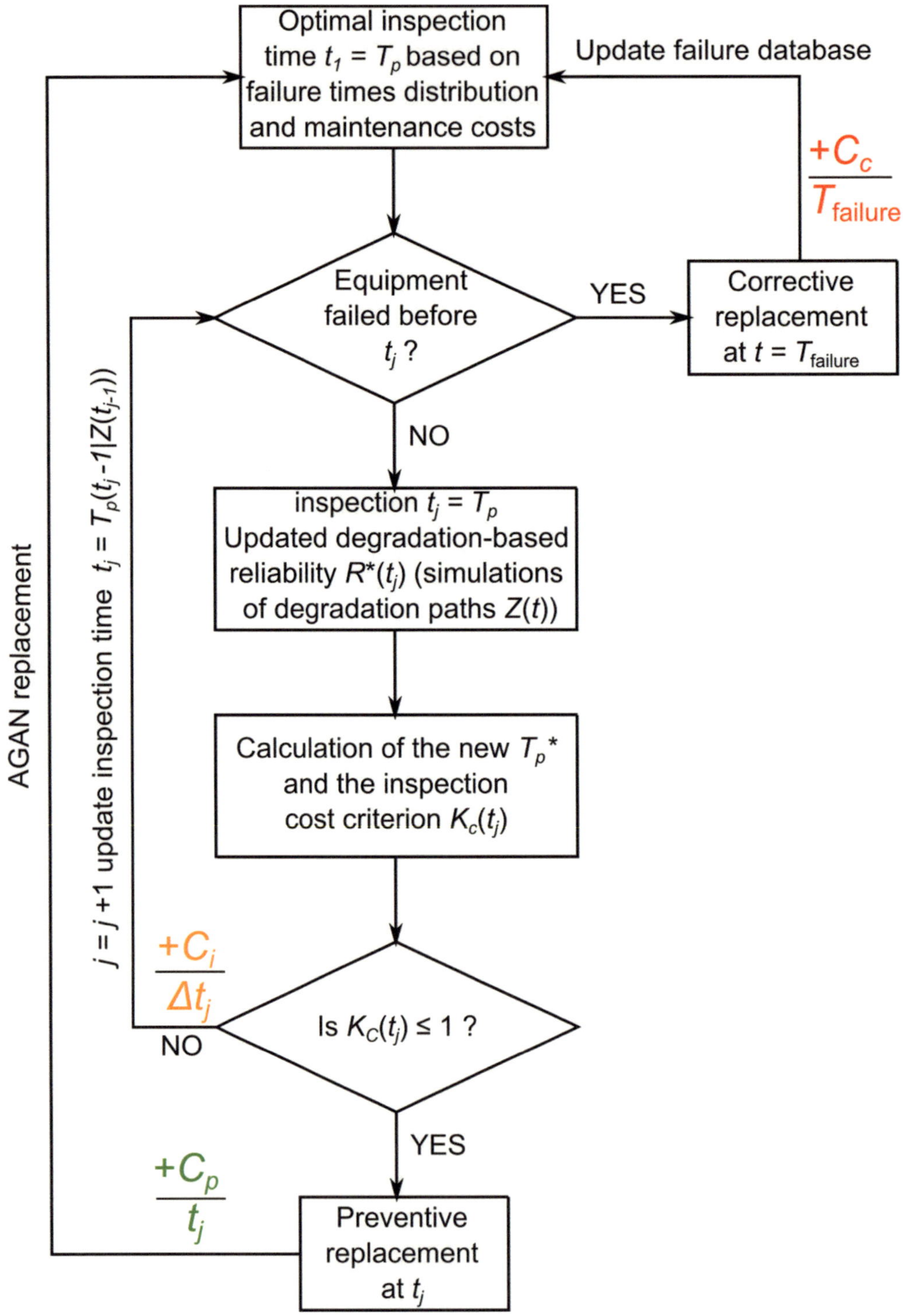


**Figure 3.** The simulation flowchart of the adaptive maintenance model.

The superscript '*' stands for any value or parametric law that is updated given the last degradation measurement. As the reliability model $R^*(t)$ is updated each time, a new degradation data is collected and the adaptive maintenance model is also updated, respectively. This methodology permits to increase the useful life of an item of equipment for which the specific degradation path is quite optimistic compared to the mean trend and to obtain a better estimation of the mean residual life in general. **Figure 3** presents the simulation flowchart of the adaptive maintenance model. Once the degradation model is identified, the first step consists in the simulation of random degradation paths and the computation of the hitting times of the degradation threshold corresponding to the failed state. From this collection of hitting times, a statistical generic reliability law is computed that determines the first inspection time $t_j = T_p$ given the costs of the different maintenance actions. At this time and considering that the item has not failed, an inspection is performed to measure the degradation level. From this degradation value and given the degradation process, new degradation trajectories are simulated in order to obtain a new set of hitting times of the threshold and the reliability law is updated accordingly. The next step is the assessment of the cost criterion $K_C(t_j)$ that decides whether or not a maintenance action should occur at the present inspection time. Obviously, the item may fail between consecutive inspection times, which leads to a corrective maintenance (AGAN replacement). In this case, the failure time is added on the failure times database in order to update the time of the first inspection.

## 5. An illustrative example

The methodology to obtain an adaptive maintenance model is applied on an illustrative example. The degradation model used in this example is a non-stationary Wiener process as presented in Section 2.5. The non-stationary Wiener process is as follows:

$$Z(t) = Z(t_0) + at^b + \sigma W(t) \tag{34}$$

With $a$, $b$ and $\sigma$ being random parameters for each degradation path. It is supposed that those parameters follow a uniform distribution with inferior and superior boundaries equalling to 0.8 and 1.2. The degradation failure threshold is set to $z_f = 100$, and each degradation path has an initial degradation $Z(t_0) = z_0 = 0$. This degradation model is supposed to be unknown for the first two stages of the study. The model is used to generate failure times during the first stage (i.e., the crossing times of the failure threshold). **Figure 4** shows three simulations of the degradation process.

The maintenance costs are as follows:

- Correction maintenance action, $C_c = 2500$ €
- Preventive maintenance action, $C_p = 500$ €
- Inspection cost, $C_i = 50$ €

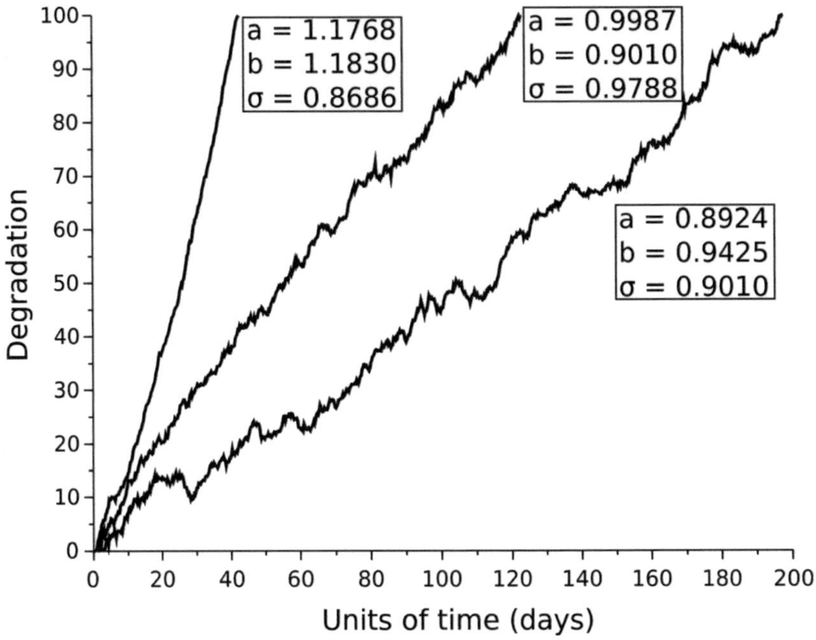


**Figure 4.** Three simulated paths of the Wiener-based degradation process. The corresponding failure times are, respectively, 42.53, 122.75 and 197.75 days.

### 5.1. Stage 1: run to failure

In the first stage, it is considered that the asset is running until failure. The distribution of the failure times is supposed to be unknown at first. From a collection of 5000 failure times, a three-parameter Weibull distribution is fitted using the maximum likelihood method. The estimated parameters are $\hat{\beta} = 1.2357$, $\hat{\eta} = 90.4343$, and $\hat{\gamma} = 39.0214$. The mean time to failure computed with the simulated failure times is MTTF = 126.76 days, and the expected value of the fitted Weibull distribution is 123.47 days. **Figure 5** shows a comparison of the pdf histogram obtained with failure data and the pdf of the fitted Weibull distribution. The expected corrective maintenance cost is $\overline{c}_{mc}(\infty) = 20.25$ €/day (see Eq. (30)).

### 5.2. Stage 2: systematic preventive maintenance

Considering the reliability obtained at the previous stage, the optimum systematic preventive maintenance periodicity is $T_p = 41.79$ days with an expected daily maintenance cost $\overline{c}_m(T_p) = 12.61$ €/day. **Figure 6** represents the evolution of the expected maintenance cost for different values of $T_p$ (see Eq. (27)).

### 5.3. Stage 3: condition-based maintenance (CBM)

In stage 3, the degradation is monitored. The purpose is to collect sufficient data for modelling the degradation process as well as performing the condition-based maintenance. In order to

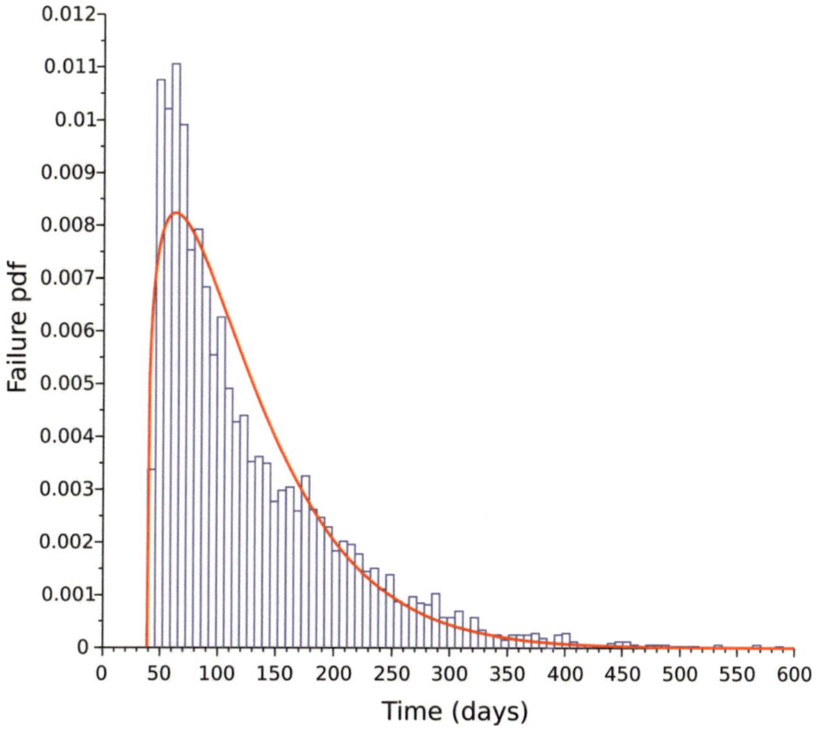


**Figure 5.** Probability density function from simulated failure times and estimated three-parameter Weibull pdf function.

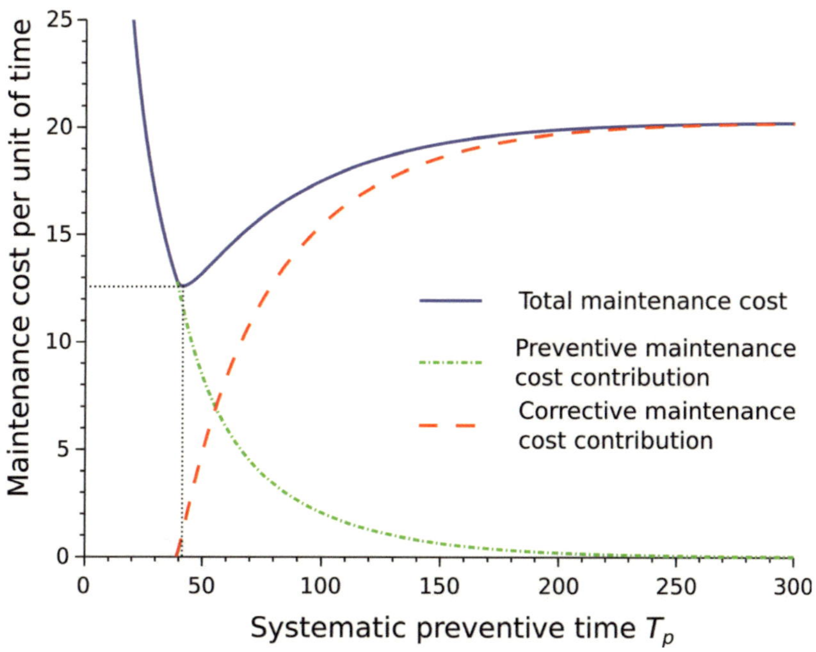


**Figure 6.** Optimum systematic preventive maintenance periodicity $T_p$.

find the optimal set of design parameters that are the periodicity of inspection $T_i$ and the preventive degradation threshold $z_{CM}$ for the condition monitoring, a Monte Carlo simulation approach was conducted. For each scenario run, 5000 simulations were performed to reach the stationary maintenance cost. A minimum condition-based maintenance cost value of 8.55 €/day was reached for the set of variables [$T_i$ = 22 days; $z_{CM}$ = 64]. **Figure 7** shows the surface plot of the related CBM cost per unit of time. The white sphere, located in the optimal region, represents the minimum cost value obtained. Conducting inspections too frequently leads to additional costs, and on the other hand, considering long duration between inspections increases the probability of failure and related corrective costs. Similarly, setting the condition monitoring degradation threshold close to the failure degradation level increases the likelihood of failure; and on the other hand, setting the condition monitoring degradation threshold to a very low level leads to precocious replacement of the asset thus shortening its useful life.

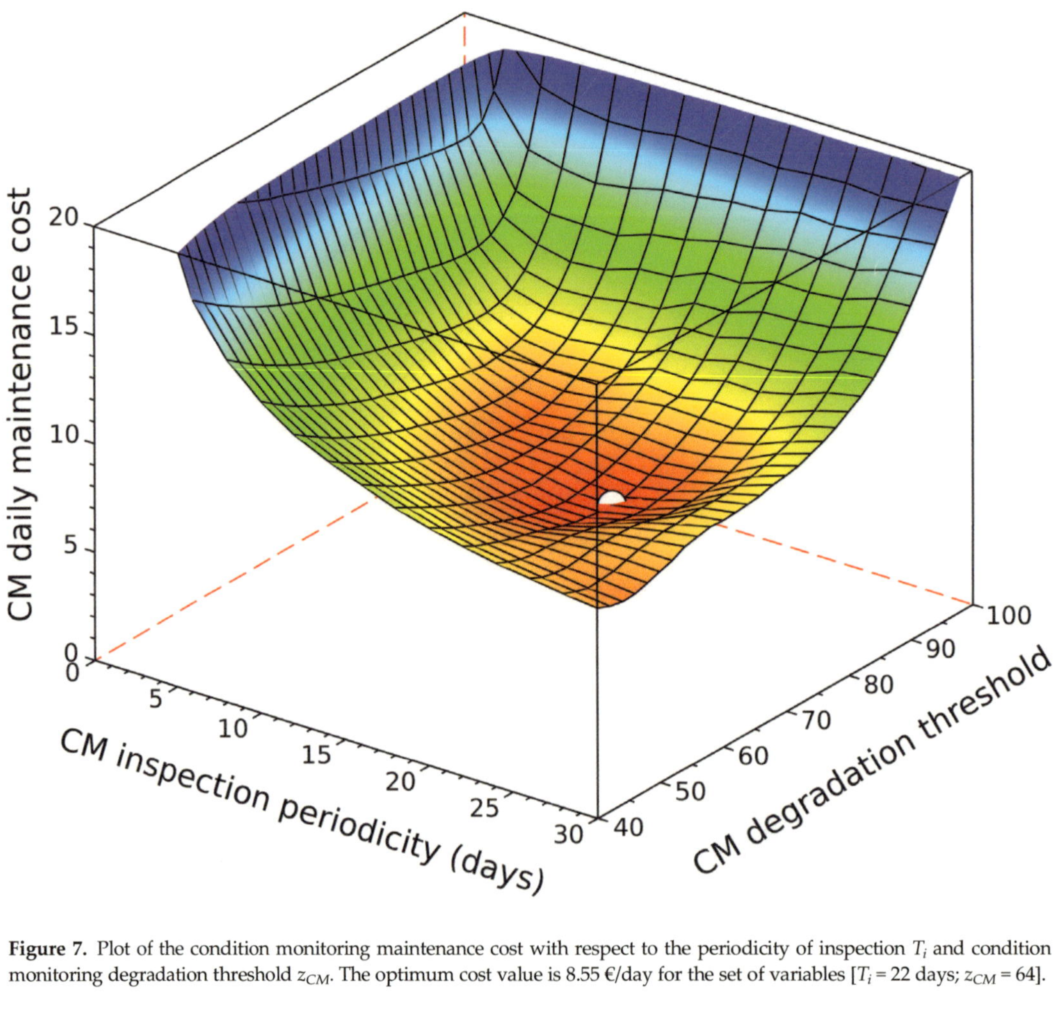


**Figure 7.** Plot of the condition monitoring maintenance cost with respect to the periodicity of inspection $T_i$ and condition monitoring degradation threshold $z_{CM}$. The optimum cost value is 8.55 €/day for the set of variables [$T_i$ = 22 days; $z_{CM}$ = 64].

http://dx.doi.org/10.5772/intechopen.69281

### 5.4. Stage 4: degradation-based adaptive maintenance

In this last stage, the adaptive maintenance methodology is set up. Given the monitoring of degradation data, the Wiener process-based degradation model $Z(t)$ can be identified for each degradation path using the maximum likelihood method. For each run, an inspection is conducted at $t_1 = 41.79$ days (i.e., the scheduled time for systematic preventive maintenance). The degradation level $Z(t_1)$ is measured and the cost criterion is assessed $K_c(t_1)$ according to Eq. (33). If $K_c(t_1) < 1$, the item is replaced at $t_1$ and only the cost of preventive maintenance is due; otherwise, the next inspection is scheduled to $t_2 = T_p(t_1|Z(t_1))$ given the updated RUL of the item. The same procedure is repeated for each inspection time $t_j$ until $K_c(t_j) < 1$. **Figure 8**

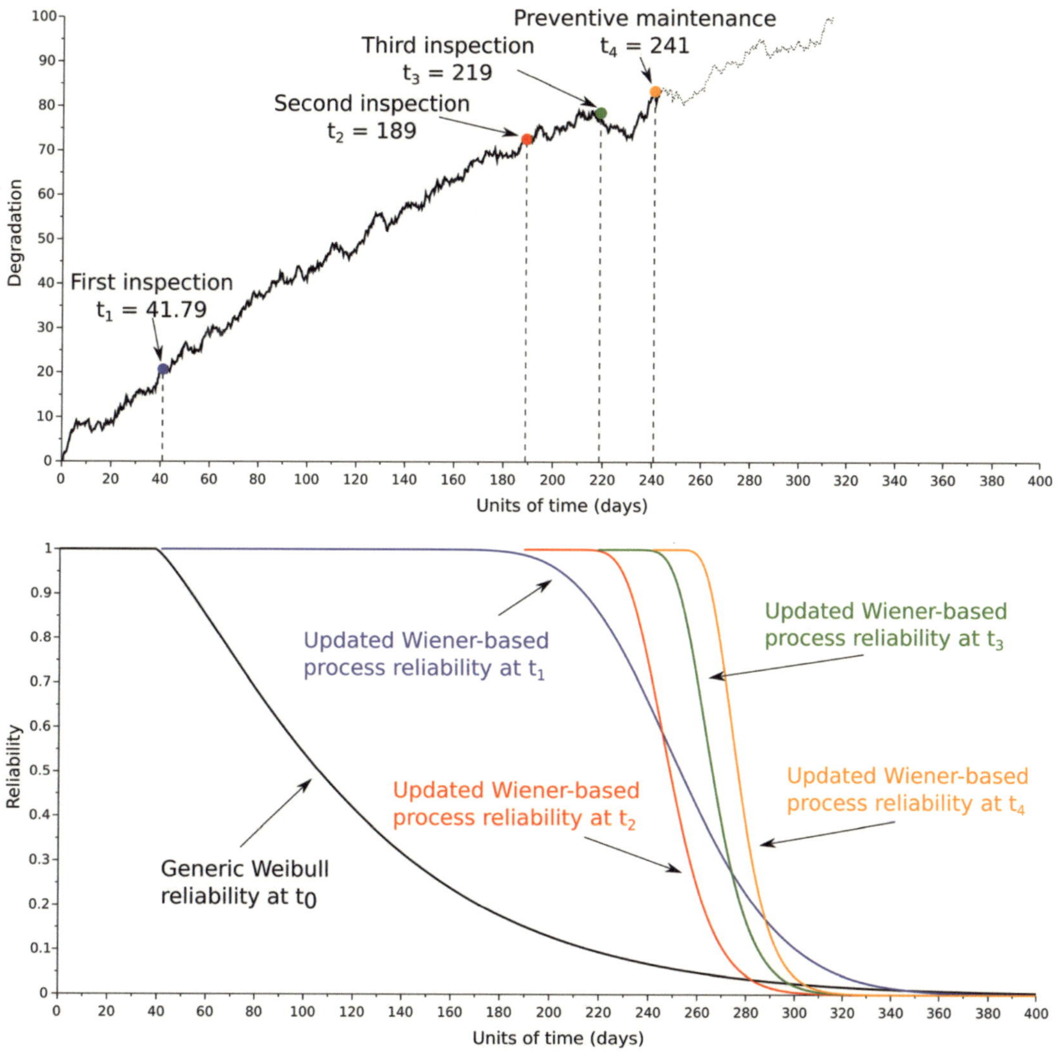


**Figure 8.** Illustration of the adaptive maintenance policy on a specific degradation path. The corresponding threshold hitting time being high, the adaptive maintenance allows to extend the usage of the asset, thus fully exploiting its useful life.

shows an example of the adaptive maintenance methodology. For each simulation, the degradation process is supposed to be known. For this specific degradation path, three inspections were performed and the item was preventively replaced at the fourth inspection time.

### 5.5. Maintenance policies comparison

**Figure 9** represents the maintenance cost per unit of time obtained over 5000 simulations for each maintenance policy. The expected theoretical maintenance costs for the corrective and preventive maintenance policies are also represented (dash lines).

At the end of the 5000 simulations, the number of inspections and failure events for each maintenance policy are the following:

- Corrective maintenance: 0 inspection and 5000 failures.
- Systematic preventive maintenance: 0 inspection and 40 failures.
- Condition-based maintenance: 20,093 inspections and 182 failures.
- Adaptive maintenance: 7397 inspections and 200 failures.

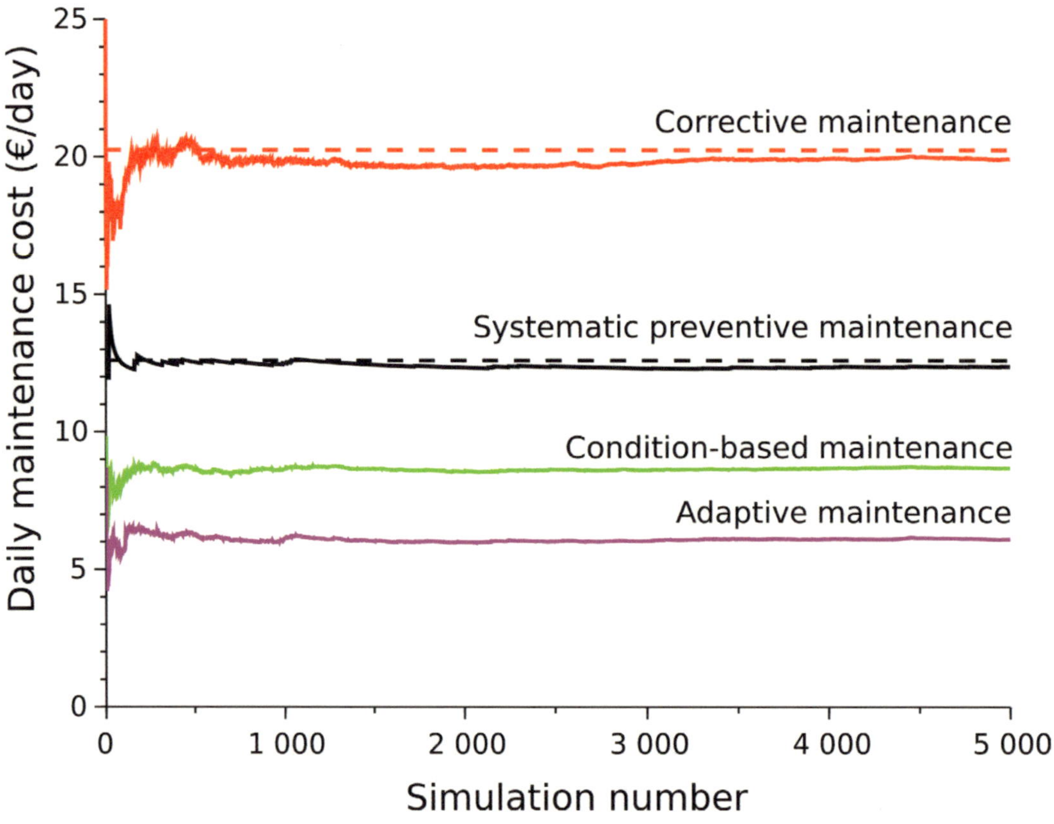


**Figure 9.** Maintenance cost per unit of time for the different maintenance policies.

Due to the distribution of the failure times, the calendar-based preventive maintenance had the minimum number of failure events; but on the other hand, it reduces the useful life of item since the replacement takes place at $t = 71.79$ days no matter the degradation level. Comparing the condition-based maintenance and the adaptive maintenance, the latter had slightly more failures events but required almost three times less inspections. The fact that both condition-based maintenance and adaptive maintenance had more failures than calendar-based maintenance makes sense: each time the preventive maintenance is postponed, there is a risk of a failure to occur between consecutive inspections.

## 6. Conclusion

This chapter was dedicated to the presentation of an adaptive maintenance methodology for extending the useful life of asset. The methodology uses the reliability-centred maintenance approach as well as the degradation-based reliability approach to define a degradation-based adaptive maintenance model. Background reliability information was presented in Section 1. Section 2 was devoted to the presentation of the degradation-based reliability approach with a focus on the stochastic processes. Section 3 detailed the age-based maintenance cost model and its extension to consider inspection costs, which lead to the definition of the cost criterion $K_C$ used to justify the best maintenance action to perform at each inspection time. Section 4 was a sum up of the methodology, highlighting the procedure of updating the reliability estimation step and degradation paths prediction given the last measurement. The methodology was applied on a numerical example using a non-stationary Wiener-based degradation process with random parameters. Four maintenance policies, from the run to failure to the adaptive maintenance stage, were compared. The results showed that the adaptive maintenance model had the minimum maintenance cost per unit of time. However, there are still challenges to cope with to improve the methodology, for example:

- The failure threshold definition that can be hard to set. An elegant solution would be to consider a probabilistic distribution of this threshold instead of a deterministic value or to combine expert judgement with fuzzy logic to take into account the uncertainties.
- The degradation modelling is also a tricky step, especially for degradation process with changing degradation rate and load dependant. While stochastic processes can consider both unit-to-unit variability and time-varying dynamics of systems, the fitting procedure of such process might lead to inaccurate model. Given the degradation data history, the fitting procedure should select the most relevant measurements to accurately predict the future behaviour especially for non-stationary and non-monotonous degradation mechanism.
- Finally, the adaptive maintenance methodology should be extended to the case of a system made of several components, each of them being ruled by its specific degradation mechanism.

This methodology can be applied for any asset given that degradation measurements and degradation modelling are possible. Examples of application are the replacement of cutting tools in machining process by monitoring the requested power, the replacement of bearings through vibration monitoring techniques and the maintenance scheduling of railway track sections given the assessment of railway track condition geometry.

## Author details

Christophe Letot*, Lucas Equeter, Clément Dutoit and Pierre Dehombreux

*Address all correspondence to: christophe.letot@umons.ac.be

Faculty of Engineering, Machine Design and Production Engineering Lab, Research Institute for the Science and Management of Risks, University of Mons, Mons, Belgium

## References

[1] Nakagawa T. Imperfect preventive maintenance. IEEE Transactions on Reliability. 1979;**28**(5): 402. DOI: 10.1109/TR.1979.5220657

[2] Blischke WR, Prabhakar Murty DN. Introduction and overview. In: Blischke WR, Prabhakar Murty DN, editors. Case Studies in Reliability and Maintenance. Hoboken, New Jersey: Wiley-Interscience; John Wiley and Sons; 2003. pp. 1–34. DOI: 10.1002/0471393002.ch1

[3] Prabhakar Murthy DN, Xie M, Jiang R. Parameter estimation. In: Shewhart WA, Wilks SS, editors. Weibull Models. Hoboken, New Jersey, USA:Wiley-Interscience, John Wiley & Sons; 2004. DOI: 10.1002/047147326X

[4] Bagdonavicius V, Nikulin M. Accelerated Life: Models Modeling and Statistical Analysis. In: Cox DR et al., editors. Monographs on Statistics and Applied Probability 94. Boca Raton, Florida, USA: Chapman and Hall/CRC; 2002. p. 334. DOI: 10.1201/9781420035872

[5] Huynh KT, Castro I, Barros A, Berenguer C. On the construction of mean residual life for maintenance decision-making. In: 8th IFAC Symposium on Fault Detection Supervision and Safety of Technical Processes; 29-31 August 2012; Mexico City, Mexico. 2012. pp. 654–659. DOI: 10.3182/20120829-3-MX-2028.00144

[6] Gorjian N, Ma L, Mittinty M, Yarlagadda P, Sun Y. A review on degradation models in reliability analysis. In: Kiritsis D, Emmanouilidis C, Koronios A, Mathew J, editors. Engineering Asset Lifecycle Management; 28-30 September 2009. Athens, Greece. London: Springer; 2010. pp. 369–384. DOI: 10.1007/978-0-85729-320-6_42

[7] Kleinbaum DG, Klein M. Survival Analysis: A Self-Learning Text. 3rd ed. London: Springer; 2012. p. 700. DOI: 10.1007/978-1-4419-6646-9

[8] Zhang Z, Si X, Hu C, Kong X. Degradation modeling-based remaining useful life estimation: A review on approaches for systems with heterogeneity. Proceedings of the Institution

of Mechanical Engineers Part O: Journal of Risk and Reliability. 2015;**1**:1–13. DOI: 10.1177/1748006X15579322

[9] Lu CJ, Meeker WQ. An accelerated life test model based on reliability kinetics. Technometrics. 1993;**37**(2):161–174. DOI: 10.2307/1269615

[10] Wang W, Christer A. Towards a general condition-based maintenance for a stochastic dynamic system. Journal of the Operational Research Society. 2000;**51**(4):145–155. DOI: 10.2307/254254

[11] Barndorff-Nielsen OE, Mikosch T, Resnick SI. Lévy Processes. 1st ed. Basel: Birkhäuser; 2001. p. 418. DOI: 10.1007/978-1-4612-0197-7

[12] Van Noortwijk JM. A survey of the application of gamma processes in maintenance. Reliability and System Safety. 2009;**94**(1):2–21. DOI: http://dx.doi.org/10.1016/j.ress.2007.03.019

[13] Wang X. Wiener processes with random effects for degradation data. Journal of Multivariate Analysis. 2010;**101**(1):340–351. DOI: 10.1016/j.jmva.2008.12.007

[14] Lu CJ, Meeker WQ. Using degradation measures to estimate a time-to-failure distribution. Technometrics. 1993;**35**(2):161–174. DOI: 10.2307/1269661

[15] Tang SJ, Guo XS, Yu CQ, Zhou ZJ, Zhou ZF, Zhang BC. Real time remaining useful life prediction based on nonlinear Wiener based degradation processes with measurement errors. Journal of Central South University. 2014;**21**(12):4509–4517. DOI: 10.1007/s11771-014-2455-9

[16] Si XS, Wang W, Hu CH, Chen MY, Zhou DH. A Wiener-process-based degradation model with a recursive filter algorithm for remaining useful life estimation. Mechanical Systems and Signal Processing. 2013;**35**(1-2):219–237. DOI: http://dx.doi.org/10.1016/j.ymssp.2012.08.016

[17] Kahle W, Lehmann A. The wiener process as a degradation model. In: Nikulin MS, Limnios N, Balakrishnan N, Kahle W, Huber-Carol C, editors. Advances in Degradation Modeling. 1st ed. Basel: Birkhäuser; 2010. p. 416. DOI: 10.1007/978-0-8176-4924-1

[18] Bonnans JF, Gilbert JC, Lemarechal C, Sagastizabal CA. Numerical Optimization: Theoretical and Practical Aspects. 2nd ed. Berlin: Springer-Verlag Berlin Heidelberg; 2006. p. 494. DOI: 10.1007/978-3-540-35447-5

[19] Hoang P, editor. Handbook of Reliability Engineering. 1st ed. London: Springer-Verlag; 2003. p. 663. DOI: 10.1007/b97414

[20] Huynh KT, Barros A, Berenguer C. Maintenance decision-making for systems operating under indirect condition monitoring: Value of online information and impact of measurement uncertainty. IEEE Transactions on Reliability. 2012;**61**(2):410–425. DOI: 10.1109/TR.2012.2194174

# Chapter 5

# Obtaining and Using Cumulative Bounds of Network Reliability

Alexey S. Rodionov and Denis A. Migov

Additional information is available at the end of the chapter

http://dx.doi.org/10.5772/intechopen.72182

**Abstract**

In this chapter, we study the task of obtaining and using the exact cumulative bounds of various network reliability indices. A network is modeled by a non-directed random graph with reliable nodes and unreliable edges that fail independently. The approach based on cumulative updating of the network reliability bounds was introduced by Won and Karray in 2010. Using this method, we can find out whether the network is reliable enough with respect to a given threshold. The cumulative updating continues until either the lower reliability bound becomes greater than the threshold or the threshold becomes greater than the upper reliability bound. In the first case, we decide that a network is reliable enough; in the second case, we decide that a network is unreliable. We show how to speed up cumulative bounds obtaining by using partial sums and how to update bounds when applying different methods of reduction and decomposition. Various reliability indices are considered: k-terminal probabilistic connectivity, diameter constrained reliability, average pairwise connectivity, and the expected size of a sub-network that contains a special node. Expected values can be used for unambiguous decision-making about network reliability, development of evolutionary algorithms for network topology optimization, and obtaining approximate reliability values.

**Keywords:** network reliability, factoring method, network connectivity, random graph, diameter constraint, probabilistic connectivity, pairwise connectivity, network topology optimization, estimation, cumulative updating

## 1. Introduction

The network reliability analysis is one of the primary tasks in the course of the network topology design and optimization. Usually, random graphs are used when modeling networks with unreliable elements [1–7]; therefore, the network reliability is defined as a connectivity measure in a random graph. There are many reliability indices of networks, for example, probabilistic connectivity [4], average pairwise reliability [5, 6], and diameter constrained reliability [7].

The most widespread network reliability index is the probabilistic connectivity of the corresponding random graph. Depending on the number of terminals (selected nodes that must be connected), there are three types of this measure such as two-terminal, all-terminal (ATR) and k-terminal reliabilities. The average pairwise reliability (APR) describes the network reliability from the point of connection between every two nodes, while a network may be still disconnected. The reliability of a network with a diameter constraint (DCR) is defined as the probability of each pair of nodes connectivity by paths that travel through a limited number of communication channels. This index is more useful, especially for P2P networks [8], wavelength division multiplexing networks, wireless sensor networks, and so on.

Another reliability index we consider in this chapter is the mathematical expectation of the size of a connected subgraph that contains a special node (MENC) [2], which describes the quality of monitoring area coverage.

Despite the fact that problems of exact reliability computations are NP-hard [9, 10], it is possible to conduct the exact calculation for networks of practical interest dimension, taking into account some special features of network topologies. For large-scale networks, various methods of reliability evaluation are widely used [11, 12]. Won and Karray suggested the cumulative updating of reliability bounds for uniqueness of decision on network feasibility [13]. Originally, this approach was presented for ATR. In this chapter, we present our results on cumulative updating of other reliability indices, and some improvements for ATR bounds cumulative updating. In addition, the main ideas of constructing the cumulative bounds for random graph characteristics are discussed along with examples of their usage in obtaining better evaluation of reliability indices and improving bionic optimization algorithms.

## 2. The basic notations

We simulate a network with perfectly reliable nodes and unreliable links by an undirected random graph $G = (V,E)$ with given presence probabilities $0 \le e_i \le 1$ of the edges. The notations which are used in this chapter are given below. Most of them coincide with the notations from [14].

$G$—Undirected probabilistic network;

$V$—Set of $n$ nodes;

$E$—Set of $m$ edges;

$e_i$; $e_{ij}$ – $i$-th edge or edge that connects $i$-th and $j$-th nodes, depending on context;

$p_j$ – Operating probability of $j$-th edge;

$w_i$ – Weight of node $i$, $WT = (w_1,\ldots,w_n)$;

$W(G)$ – Total weight of nodes of $G$;

$R(G)$ – All-terminal reliability of $G$, that is, probability that every two nodes are connected;

$K$—Set of terminal nodes;

$R_K^d(G)$—Diameter constrained reliability of $G$, that is, the probability that every two nodes from $K$ are connected by a path that travel through not more than $d$ edges;

$LB$, $UB$—Lower and upper bounds of $G$ reliability. In case of ATR, we use the original notation form [13]: $RL$, $RU$;

$R_0$—Predefined threshold for the network reliability value;

$N(G)$ ($M(G)$)—Mathematical expectation of the number of disconnected (or connected) pairs of nodes in $G$;

CS($G$; $s$)—Mathematical expectation of a number of nodes in a connected subgraph that contains a node $s$; if $s = 1$, then CS($G$);

$\overline{R}$—Average pairwise reliability of $G$;

$C$—Edge chain composed of edges $e1,\ldots,ek$;

$G/C$ ($G/e$)—Network $G$ with a contracted chain $C$ (edge $e$);

$G\backslash C$ ($G\backslash e$)—Network $G$ without chain $C$ (edge $e$); and

$G_e^*$—Network $G$ with an absolutely reliable edge $e$;

## 3. Factoring method and cumulative updating of network reliability bounds

The most common exact method for calculating various network reliability measures is the factoring method. According to this method, we divide the probability space, which consists of all particular network realizations, into two sets based on the presence or absence of a network element. Further on we refer to such a network element as a factored element, or a pivot element. As a result, for a given network $G$ and a factored element $e$, we will obtain the two networks: $G_e^*$, where $e$ is absolutely reliable, and $G\backslash e$, where $e$ is absolutely unreliable, so we could remove it. The probability of $G_e^*$ is equal to the pivot element reliability, and the probability of $G\backslash e$ is equal to the pivot element failure probability. For a reliability index $Rel$ of the initial network, the following expression holds (the total probability law):

$$Rel(G) = p_e Rel(G_e^*) + (1 - p_e) Rel(G\backslash e) \tag{1}$$

Then the obtained networks are subject to the same factoring procedure. Recursions continue until either an unreliable network is obtained (0 is returned) or an absolutely reliable network is obtained (1 is returned). For ATR (**Figure 1**), expression (1) turns to:

$$R(G) = p_e R(G/e) + (1 - p_e) R(G\backslash e). \tag{2}$$

We can speed up the factoring process by calculating the intermediate networks reliabilities directly, that is, without further factorization. For ATR calculation, the five-vertex graph reliability formula can be used [13]. Another way to accelerate the reliability computing by using

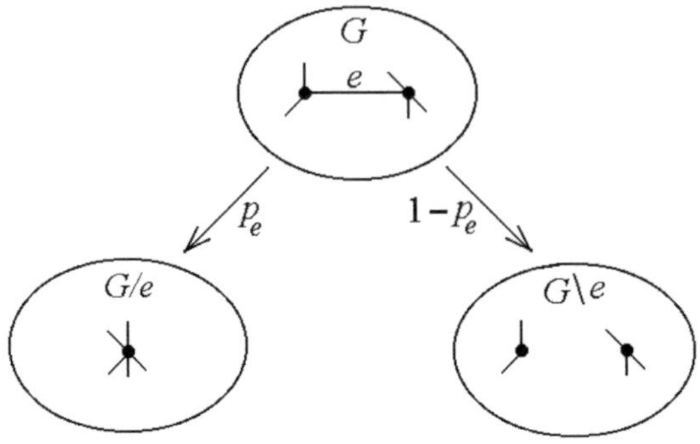


**Figure 1.** The factoring method.

various reduction [4] and decomposition [17] methods, such as serial-parallel transformation on each recursive call of the factoring procedure, biconnected decomposition, and other methods.

The idea of the cumulative updating method is an incremental updating of exact lower (*LB*) and upper (*UB*) reliability bounds and comparing them with a given reliability threshold $R_0$. If $LB > R_0$, then the network is reliable. If $UB < R_0$, then the network is unreliable. The *LB* must necessarily be non-decreasing, and the *UB* must necessarily be non-increasing. Another obligatory condition is the equality of *LB*, *UB* at the last step and exact reliability value *R*. One possible way of bounds updating is the usage of the factoring method. When a final graph or a disconnected graph is obtained, we can update bounds [13].

As a rule, any algorithm of calculation of mathematical expectation for a non-negative function μ of a random graph *G* practically comes to summarizing non-negative values. According to the definition of mathematical expectation in the discrete case [14]:

$$E[\mu(G)] = \sum_{H \in \Gamma} P(H) \cdot \mu(H), \tag{3}$$

where Γ is a set of all possible final realizations of *G* that are obtained during factorization.

If some realizations $\Gamma_0$ are obtained along with their probabilities, and a function μ is obtained for all these realizations, then the *LB* is greater or equal to the corresponding partial sum.

Now let us assume that for μ, its possible minimal $(\mu_m)$ and maximal $(\mu_M)$ values are known. In this case, we easily obtain the following bounds:

$$LB = \sum_{H \in \Gamma_0} P(H) \cdot \mu(H) + \mu_m \left(1 - \sum_{H \in \Gamma_0} P(H)\right); \quad UB = \sum_{H \in \Gamma_0} P(H) \cdot \mu(H) + \mu_M \left(1 - \sum_{H \in \Gamma_0} P(H)\right). \tag{4}$$

Finally, both bounds will obviously come to an exact value.

From (4), we easily obtain the equations for improving bounds when the new (*i*-th) realization $H_i$ is obtained and $\mu(H_i)$ is calculated:

$$LB_i = LB_{i-1} + P(H_i) \cdot [\mu(H_i) - \mu_m]; \quad UB_i = UB_{i-1} - P(H_i) \cdot [\mu_M - \mu(H)]. \tag{5}$$

For ATR, $\mu_m = 0$ and $\mu_M = 1$, so we arrive to the equations presented by Won and Karray in [13], while for MENC $\mu_m = 1$ and $\mu_M = n$, where $n$ is a total number of nodes in a graph. Initial values of bounds are equal to these values.

## 4. Improvements for ATR bounds cumulative updating

### 4.1. The chain branching

Our methods named chain branching (CB) and chain reduction (CR) [15] could be used [16] .as a basis for two kinds of the ATR bound updating (UAB) algorithms named UAB based on CB (UAB_CB), and UAB based on CR (UAB_CR). These methods can be used if a network contains a chain, that is, a sequence of adjacent two-degree nodes. The CB reduces the calculation of ATR for calculating the ATRs of networks obtained from $G$ by removing a pivot chain, or removing it and contracting its terminal nodes, while CR reduces it for calculating the ATR of the network $G_R$, in which, this chain is substituted by a single edge. Let a chain $C$ consist of $k$ edges, its terminal nodes are $s$ and $t$, and let an edge $e_{st}$ has a reliability $p_{st}$, then [16]

$$R(G) = \left[\prod_{i=1}^{k} p_i + p_{st} \sum_{i=1}^{k} (1 - p_i) \prod_{j \neq i} p_j\right] \cdot R(G/C) + (1 - p_{st}) \sum_{i=1}^{k} (1 - p_i) \prod_{j \neq i} p_j \cdot R(G \backslash C). \tag{6}$$

The best choice is the usage of the longest chain as a pivot. However, it requires finding of all the chains. We may accelerate the calculation by choosing a chain (or an edge) incidental to a node with a minimal degree for obtaining a new chain.

### 4.2. The usage of Cutnodes

Let a network $G$ have a cutnode $u$ that divides $G$ into two subnetworks $G_1$ and $G_2$, therefore, $R(G) = R(G_1)R(G_2)$. $RL^{(1)}$ and $RU^{(1)}$ are lower and upper bounds for $R(G_1)$, respectively. Like wise, $RL^{(2)}$ and $RU^{(2)}$ are the lower and upper bounds for $R(G_2)$, respectively. Then $RL^{(1)} \cdot RL^{(2)} \leq R(G) \leq RU^{(1)} \cdot RU^{(2)}$.

Let us estimate the ATR of subnetworks $G_1$ and $G_2$ in turn starting from $G_1$. While estimating the ATR of $G_1$, we have no information about the ATR of $G_2$, thus $RL^{(2)} = 0$ and $RU^{(2)} = 1$, and $RL$ for $R(G)$ is 0, while $RU$ for $R(G)$ is $RU(G_1)$. After some steps of estimating $R(G_1)$, we switch to estimating $RL^{(2)}$ and $RU^{(2)}$, thus improving $RL$ and $RU$ for $G$, and so on. Therefore, we calculate the bounds for $R(G_1)$ and $R(G_2)$ separately (possibly in parallel). Now let us continue with applying this scheme to calculating the bounds for $R(G_1)$ and $R(G_2)$ by the factoring method.

Similar to [15], we suppose that some networks $H_1, H_1, \ldots, H_I$ are obtained from $G_1$, and networks $K_1, K_2, \ldots, K_J$ from $G_2$ by factorization (1). Let us denote $x_i = P(H_i)$, $y_i = R(H_i)$, $1 \leq i \leq I$, and $s_i = P(K_i)$, $t_i = R(K_i)$, $1 \leq i \leq J$. Then [16]:

$$R(G_1) = \sum_{i=1}^{I} x_i y_i = 1 - \sum_{i=1}^{I} x_i (1 - y_i); \quad R(G_2) = \sum_{i=1}^{J} s_i t_i = 1 - \sum_{i=1}^{J} s_i (1 - t_i). \tag{7}$$

**Statement 1** *For all a,b such that* $1 \le a \le I$, $1 \le b \le J$, *the following inequalities hold:*

$$\sum_{i=1}^{a} x_i y_i \sum_{i=1}^{b} s_i t_i \le R(G) \le \left[1 - \sum_{i=1}^{a} x_i (1 - y_i)\right] \left[1 - \sum_{i=1}^{b} s_i (1 - t_i)\right]. \tag{8}$$

Proof of the statement can be found in [16], as well as the algorithm with the use of cutnodes (UAB_C).

### 4.3. The usage of two-node cuts

Now let us describe the UAB modification that uses two-node cuts. Let the network $G$ has a node cut $u,v$ that divides it into two subnetworks $G_1$ and $G_2$ (**Figure 2**). For any network $H$ that contains the nodes $u$ and $v$ we will denote a network that is obtained by contracting these nodes as $H'$.

For the first time, the approach based on two-node cut was presented by Kevin Wood [17] in 1989. It describes how to perform the reliability-preserving triconnected decomposition of a network for calculating its k-terminal reliability. Later in 2006, we have introduced a general method which allows using an arbitrary node cut for calculating all-terminal reliability [18, 19]. This method makes possible to reduce the reliability calculation to the calculation of reliabilities of a group of networks each one with a smaller dimension. The total number of these networks is $2^*B_W$, where $B_W$ is the Bell number and $W$ is the number of nodes in the cut. A particular case of two-node cut along with numerical experiments to demonstrate the efficiency for the ATR calculation was presented in [16, 20], and in [21]—for the k-terminal reliability calculation. Despite the fact that the method from [17] presents a graph transformation for further factoring, and the method from [20] presents an expression for the network reliability, both of them practically lead to similar calculations of the same complexity. Another result was with the ATR calculation using special kind of node cuts [22]: the so-called

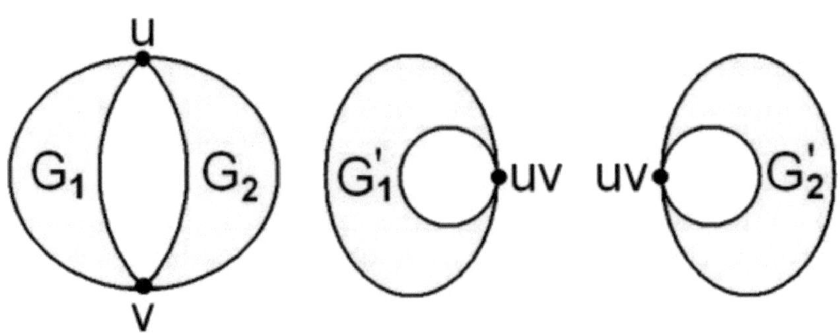


**Figure 2.** The network $G$ with two-node cut and auxiliary networks.

longitudinal and cycle cuts. All the listed results, which we have obtained, are described in detail in [19]. Following is the corresponding Eq. [20] for the ATR without proof:

**Theorem 1** *The following equation holds:*

$$R(G) = R(G_1)\left[R\left(G_2'(u,v)\right) - R(G_2)\right] + R(G_2)\left[R\left(G_1'(u,v)\right) - R(G_1)\right] + R(G_1)R(G_2). \tag{9}$$

The equation for three-node cut is sufficiently more complicated [18, 19]:

$$\begin{aligned} R(G) = \frac{1}{2}\Big[ & R\left(G_1^{1|23}\right)\left(R\left(G_2^{12|3}\right) + R\left(G_2^{13|2}\right) - R\left(G_2^{1|23}\right)\right) \\ & + R\left(G_1^{12|3}\right)\left(R\left(G_2^{1|23}\right) + R\left(G_2^{13|2}\right) - R\left(G_2^{12|3}\right)\right) + R\left(G_1^{13|2}\right)\left(R\left(G_2^{12|3}\right) + R\left(G_2^{1|23}\right) - R\left(G_2^{13|2}\right)\right) \\ & - R(G_1)\left(R\left(G_2^{12|3}\right) + R\left(G_2^{13|2}\right) + R\left(G_2^{1|23}\right)\right) - R(G_2)\left(R\left(G_1^{12|3}\right) + R\left(G_1^{13|2}\right) + R\left(G_1^{1|23}\right)\right) \\ & + R(G_1)R(G_2)\Big] + R(G_1)R\left(G_2^{123}\right) + R\left(G_1^{123}\right)R(G_2), \end{aligned} \tag{10}$$

where nodes 1, 2, and 3 composes the threee-node cut, $G_i^{x|yz}$ is a graph that is obtained by merging $y$ and $z$ nodes in $G_i$, and $G_i^{xyz}$ is obtained by merging $x$, $y$, and $z$ nodes. The expression for four-node cut has also been obtained [19], but we do not present it due to its huge size.

However, our results on using an arbitrary node cut for all-terminal reliability calculation were published only in Russian [18, 19], and only the two-node cut case was presented in English [16, 20]. Maybe, this is the reason that these results are not so widely known, and therefore, the same ideas were proposed again by Juan Manuel Burgos and Franco Robledo Amoza [23, 24] in 2016. As we have done it in [16–20], in [23, 24], the authors show how to calculate the all-terminal reliability of a network with a node cut via reliabilities of a group of smaller networks, but in a somewhat different way. As a result, Eqs. (9–10) were presented.

Let us describe how we can use expression (9) and update the ATR bounds separately for both subgraphs $G_1$ and $G_2$ (**Figure 2**) for obtaining the ATR bounds of the whole graph $G$.

Suppose that some networks $H_1, \ldots, H_I$ are obtained from $G_1$ during the factoring process. If this factoring does not involve edges incidental to $u$ or $v$, then parallel factoring in $G_1'(u,v)$ leads to obtain $H_1'(u,v), \ldots, H_I'(u,v)$. Let us introduce the following denotations: $x_i = P(H_i)$, $y_i = R(H_i)$, $x_i' = P\left(H_i'(u,v)\right)$, and $y_i' = R\left(H_i'(u,v)\right)$, for $1 \le i \le I$. As in both $G_1$ and $G_1'(u,v)$, the factoring is executed by the same pivot edges, we have $x_i' = x_i$. Obviously $y_i' \ge y_i$ because $H_1'(u,v)$ differs from $H_i$ by the edge $(u,v)$ that has reliability 1 in $H_1'(u,v)$. Similarly, by the factoring in $G_2$ by edges that are not incidental to $x$ or $y$, we obtain the networks $K_1, \ldots, K_j$, and by the factoring in $G_2'(u,v)$, we obtain $K_1', \ldots, K_J'$ . Let us denote $s_i P(K_i)$, $t_i = R(K_i)$, $s_i' = P\left(K_i'(u,v)\right)$, and $t_i' = R\left(K_i'(u,v)\right)$ for $1 \le i \le J$. Similar to the above case, we have $s_i' = s_i$, $t_i' \ge t_i$. Let $L = \max\{I, J\}$, and let $x_i = y_i = y_i' = 0$ for $I + 1 \le i \le L$, $s_i = t_i = t_i' = 0$ for $J + 1 \le i \le L$.

Let us obtain $RL_{1,i}$, $RU_{1,i}$, $RL_{1,i}'$, $RU_{1,i}'$, $RL_{2,i}$, $RU_{2,i}$, $RL_{2,i}'$, $RU_{2,i}'$ for $1 \le i \le L$:

$$RL_i^1 = \sum_{j=1}^{i} x_j y_j,\ RU_i^1 = 1 - \sum_{j=1}^{i} x_j\left(1 - y_j\right),\ RL'^1_i = \sum_{j=1}^{i} x_j y_{j'},\ RU'^1_i = 1 - \sum_{j=1}^{i} x_j\left(1 - y_{j'}\right),$$

$$RL_i^2 = \sum_{j=1}^{i} s_j t_j,\ RU_i^2 = 1 - \sum_{j=1}^{i} s_j(1 - t_j),\ RL'^{\,2}_i = \sum_{j=1}^{i} s_j t_{j'},\ RU'^2_i = 1 - \sum_{j=1}^{i} s_j\left(1 - t_{j'}\right).$$

As is shown in [13], these values are the upper and lower bounds for reliabilities of the corresponding networks; these bounds are exact for $i = L$. In the case $i = 0$, the lower bounds are 0 and the upper bounds are 1.

Note that $RL'_{1,i} \geq RL_{1,i}$, $RU'_{1,i} \geq RU_{1,i}$, as $y'_i \geq y_i$; $RL'_{2,i} \geq RL_{2,i}$, $RU'_{2,i} \geq RU_{2,i}$, as $t'_i \geq t_i$.

Now we obtain $RL_j$, and $RU_j$ for odd and even $j$ separately. For $j = 2i$, $0 \leq i \leq L$:

$$\begin{aligned} RU_{2i} &= RU_i^1 RU'^2_i + RU'^1_i RU_i^2 - RU_i^1 RU_i^2, \\ RL_{2i} &= RL_i^1\left(RL'^2_i - RL_i^2\right) + RL_i^2\left(RL'^1_i - RL_i^1\right) + RL_i^1 RL_i^2. \end{aligned} \tag{11}$$

For $j = 2i + 1, 0 \leq i \leq L - 1$:

$$\begin{aligned} RU_{2i+1} &= RU_{i+1}^1 RU'^2_i + RU'^1_{i+1} RU_i^2 - RU_{i+1}^1 RU_i^2,\ RL_{2i+1} \\ &= L_{i+1}^1\left(RL'^2_i - RL_i^2\right) + RL_i^2\left(RL'^1_{i+1} - RL_{i+1}^1\right) + RL_{i+1}^1 RL_i^2. \end{aligned} \tag{12}$$

**Statement 2** *For any $1 \leq j \leq 2\,L$, we have*

$$RU_{j-1} \geq RU_j \geq R(G) \geq RL_j \geq RL_{j-1}. \tag{13}$$

If $j = 2\,L$, then the second and the third inequalities become equalities. The proof of the statement can be found in [16], as well as the algorithm (UAB_2NC) with two-node cuts.

Note that both CB and CR can be used for speeding up the 2NC algorithm under the condition that the nodes $u$ and $v$ do not belong to a pivot chain. In the case of CR $RU_j^1 := RU_j^1 - x_j(1 - p)$, $RU'^1_j := RU'^1_j - x_j(1 - p)$, where $p$ is a multiplier by which the reliability of a reduced network must be multiplied in CR [4].

### 4.4. Case studies

We have compared the algorithms presented with node cuts UAB_C and UAB_2NC to the algorithm CR from [13] that does not take possible cutnodes or two-node cuts into account. The results [16] of the comparison are presented in **Tables 1–3**. The CR was used while factoring for better performance, and the five-node networks were considered as simple ones. The PC with Intel Core Duo 2.93 GHz was used for testing.

We have chosen the two networks $G_1$ and $G_2$ for testing UAB_C, and the two networks $G_3$ and $G_4$ for testing UAB_2NC.

| Network | $R_0$ | Algorithm | Time | Recursions |
|---|---|---|---|---|
| $G_1$ | 0.9 | UAB_CR | 39 s | 5,245,491 |
| $G_1$ | 0.9 | UAB_C | <1 ms | 2504 |
| $G_1$ | 0.94 | UAB_CR | 34 s | 639,562 |
| $G_1$ | 0.94 | UAB_C | <1 ms | 2643 |
| $G_2$ | 0.5 | UAB_CR | 2 s | 259,879 |
| $G_2$ | 0.5 | UAB_C | 60 ms | 4322 |
| $G_2$ | 0.901 | UAB_CR | 3 min 22 s | 24,765,569 |
| $G_2$ | 0.901 | UAB_C | 45 ms | 3233 |

**Table 1.** Comparison of UAB_CR and UAB_C.

| $R_0$ | *UAB_CR* | | *UAB_2NC* | |
|---|---|---|---|---|
| | Time (s) | Recursions | Time (s) | Recursions |
| 0.5 | 2.2 | 257,084 | 0.08 | 8696 |
| 0.6 | 43.2 | 5,055,553 | 0.18 | 19,399 |
| $R(G_3)$ | — | — | 0.28 | 27,180 |
| 0.92 | 13 | 1,532,293 | 0.08 | 8725 |
| 0.922 | 2.5 | 281,565 | 0.08 | 8696 |

**Table 2.** Results of the numerical experiments for the network $G_3$.

| $R_0$ | *UAB_CR* | | *UAB_2NC* | |
|---|---|---|---|---|
| | Time (s) | Recursions | Time (s) | Recursions |
| 0.2 | 0.12 | 16,814 | 0.1 | 10,963 |
| 0.3 | 27 | 3,630,494 | 0.87 | 80,756 |
| $R(G_4)$ | — | — | 8,8 | 805,358 |
| 0.996 | 113 | 14,878,088 | 0.02 | 1747 |
| 0.998 | 0.36 | 45,374 | < 1 ms | 35 |

**Table 3.** Results of the numerical experiments for the network $G_4$.

The network $G_1$ consists of two 9-vertex complete graphs which are connected by a cutnode. Thus, it contains 17 nodes and 72 edges.

The network $G_2$ consists of two $5 \times 5$ lattices which are connected by a cutnode. As a cutnode, we choose the corner. The resulting network contains 49 nodes and 80 edges. **Table 1** shows the results of numerical experiments.

The network $G_3$ (48 nodes and 79 edges) is composed of two $5 \times 5$ lattices which are connected by two nodes (**Figure 3**).

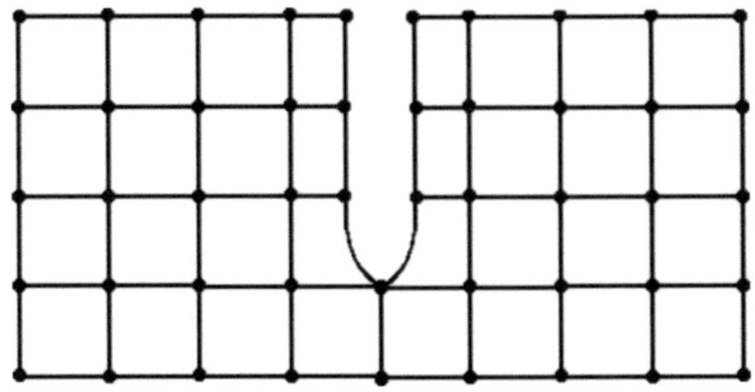

**Figure 3.** The union of two 5 × 5 lattices by two nodes.

The network $G_4$ (22 nodes and 108 edges) is composed of two 11-vertex complete graphs which are connected by two nodes without edge connecting them. A similar network that is composed of two 5-vertex complete graphs is presented in **Figure 4**.

The edges $G_1$ and $G_4$ are equally reliable with $p = 0.5$, while the edges $G_2$ and $G_3$ are equally reliable with $p = 0.9$. $R(G_1) = 0.9307194$, $R(G_2) = 0.883248$, $R(G_3) = 0.903168801959$, and $R(G_4) = 0.982472649148$.

**Tables 2** and **3** present the results for $G_3$ and $G_4$, respectively. The third row contains the results when the threshold value coincides with the exact reliability value. The results for the factoring algorithm with chain reduction are shown in the second column. The results for the UAB_2NC are shown in the third column.

As can be seen, the approach proposed has a great advantage. The efficiency of UAB as it is depends on the closeness of the threshold value $R_0$ to the reliability value $R(G)$.

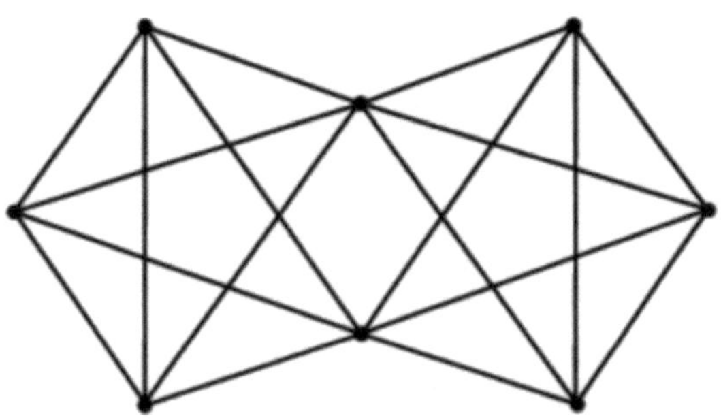

**Figure 4.** The union of two complete five-node networks by two nodes without connecting edge.

## 5. Cumulative updating of diameter constrained network reliability

For the DCR, expression (1) takes the following form:

$$R_K^d(G) = p_e R_K^d(G_e^*) + (1 - p_e) R_K^d(G \backslash e). \tag{14}$$

We denote this method as a simple factoring method (SFM). Cancela and Petingi have proposed a modified factoring method for calculating DCR [7], which is much faster than the already described classical factoring method in the diameter constrained case (14). The Cancela and Petingi factoring method (CPFM) operates with a list of paths instead of graphs. For any terminals *s*, *t*, the list $P_{st}(d)$ of all the paths with a limited length between *s*, *t* is generated. By $P_d$ the union of $P_{st}(d)$ for all the pairs of terminals is denoted. *P*(*e*) is composed of the paths of $P_d$ which include the edge *e*. The list of CPFM arguments are given below [7]:

- $np_{st}$: the number of paths of length at most *d* between *s* and *t* in *G*;
- $links_p$: the number of non-perfect edges (edges *e* such that $p(e) < 1$) in the path *p*, for every $p \in P_d$.
- $feasible_p$: this is a flag, which has the value False when the path is no longer feasible, that is, it includes an edge which failed; and True otherwise;
- $connected_{st}$: this is a flag, which has the value True when *s* and *t* are connected by a perfect path of length at most *d* and False otherwise;
- *connectedPairs*: the number of pairs of terminals which are connected by a perfect path of length at most d.

Pseudocode of the method proposed for the DCR bounds cumulative updating [27], which is based on CPFM, is presented below:

**Input**: *G* = (*V;E*), *d*, $P_d$, *P*(*e*), *np*(*s; t*), *links*(*p*), *feasible*(*p*), *connected*(*s,t*), *connectedPairs*, *RL* = 0, *RU* = 1, *Pl* = 1.
**Function** FACTO (*np*(*s,t*), *links*(*p*), *feasible*(*p*), *connected*(*s,t*),*connectedPairs*, *Pr*)
**if** $RL > R_0$ *or* $RU < R_0$ **then**
  *e* is an arbitrary edge: $0 < p_e < 1$
  contractEdge (*np*(*s; t*), *links*(*p*), *feasible*(*p*), *connected*(*s,t*),*connectedPairs*, *Pr*)
  deleteEdge (*np*(*s; t*), *links*(*p*), *feasible*(*p*), *connected*(*s,t*),*connectedPairs*, *Pr*)
**end** FACTO

**Function** contractEdge(*np*(*s; t*), *links*(*p*), *feasible*(*p*), *connected*(*s,t*),*connectedPairs*, *Pr*)
$Pr = Pr * p_e$
**for each** *p* = (*s,..,t*) *in P*(*e*) *such that feasible*(*p*) = *true* **do**
  *links*(*p*) = *links*(*p*) - 1
  **if** *connected*(*s,t*) = *false* **and** *links*(*p*) = 0 **then**
    *connected*(*s,t*) *true*
    *connectedPairs* = *connectedPairs* + 1
    **if** *connectedPairs* = |K|*(|K|-1)/2 **then** *RL* = *RL* + *Pr*
FACTO (*np*(*s,t*), *links*(*p*), *feasible*(*p*), *connected*(*s,t*), *connectedPairs*, *Pr*)
**end** contractEdge

**Function** deleteEdge(*np*(*s,t*), *links*(*p*), *feasible*(*p*), *connected*(*s,t*), *connectedPairs*, *Pr*)
$Pr = Pr *(1-p_e)$
**For each** *p* = (*s,…,t*) *in P*(*e*) *such that feasible*(*p*) = *true* **do**

*feasible(p)* = *false*
*np(s,t)* = *np(s,t)* - 1
**if** *np(s,t)* = 0 **then** *RU* = *RU* - *Pr*
FACTO (*np(s,t)*, *links(p)*, *feasible(p)*, *connected(s,t)*, *connectedPairs*, *Pr*)
**end** deleteEdge

For the speeding up of DCR calculation, it is possible to apply methods of reduction and decomposition. In our previous studies [25–27], we have obtained such methods which can calculate the DCR faster. These methods are the analogue of the well-known series-parallel transformation for CPFM, and the pivot edge selection strategy. Also, we have obtained decomposition methods for calculating DCR in the case with two terminals. The methods obtained allow us to significantly reduce the number of recursive calls in CPFM and the complexity of DCR computation. In the above algorithm, we assume that the series-parallel reduction has been already performed and the pivot edge selection strategy is used.

Below we show how *RL* and *RU* are changing during the procedure proposed for the Intellinet network topology (**Figure 5**). The edge reliability is equal to 0.7 for each edge and the diameter value is equal to 12. The threshold value was equal to the exact DCR value, which was previously

**Figure 5.** The Intellinet network.

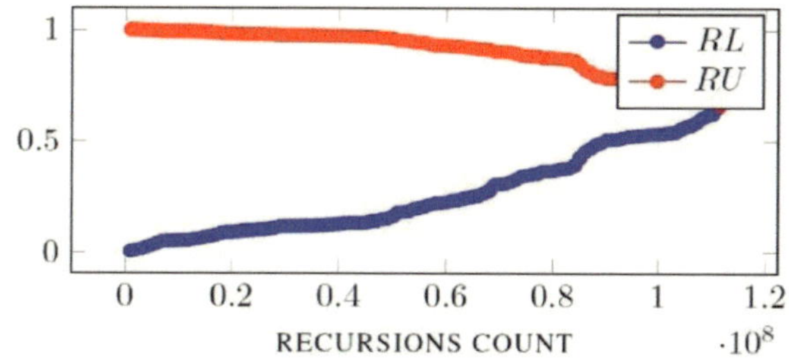


**Figure 6.** The behavior of DCR bounds for the Intellinet network.

calculated. The calculation time was about 12 min. Experiments were performed on Intel Xeon E31240 3.3 GHz, 8 cores (**Figure 6**).

## 6. Special problems of updating reliability bounds

When special methods of graph reduction or decomposition are applied for the speeding up of calculations, the corresponding equations for updating the bounds must be used. Derivation of such equations for different functions μ(G) is based on taking into account all "inner" changes of bounds that are concealed inside intermediate steps thus leading to final results of such a reduction or a decomposition. For example, when "branching by chain" [15] is applied, consequent factoring by edges of a chain is used for derivation, and each factoring may require a change of bounds.

Now let us turn to an example of a simple reduction, which is removing a dangling node. Let the graph $G$ have some attached trees in its structure is shown in **Figure 7**.

In the case of the ATR, we have the trivial equality $R(G) = p_e R(G \backslash e)$, where $e$ is the edge that connects the dangling node to the graph. As removing $e$ with probability $1 - p_e$ leads to graph

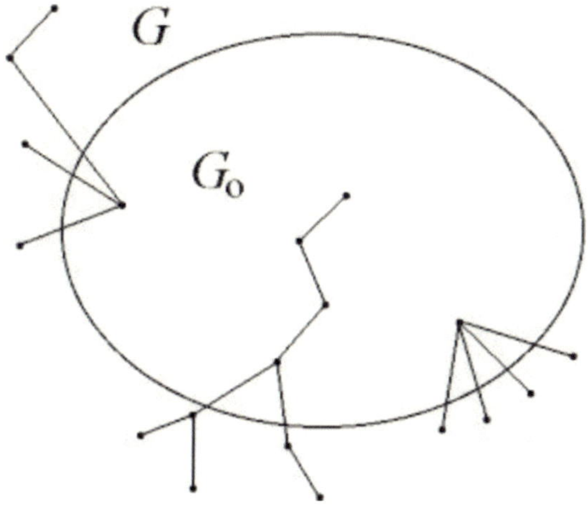


**Figure 7.** The graph with attached trees.

disconnection, the upper bound must be decreased by $1 - p_e$. The lower bound rests unchanged, but probabilities of all further realizations of $G\backslash e$ must be multiplied by $p_e$. If there are some attached trees as shown in **Figure 7**, then the ATR of the graph $G$ is equal to the ATR of this graph without all attached trees $(G_0)$ multiplied by the product of reliabilities of all the edges in these trees (let us denote it as *Pr*): $R(G) = \Pr \cdot R(G_0)$. Thus, we continue with the graph $G_0$, *LB* does not change, *Pr* must be included as a multiplier in probabilities of further realizations, and *UB* is reduced by probability of at least one edge of attached trees, that is, 1-*Pr*. If the reduction takes place at the initial step, then the upper bound starts from *Pr*.

The case of APR is not so simple. It is known that the task of obtaining this reliability index is equivalent to the one of obtaining mathematical expectation of the number of disconnected pairs of nodes in a random graph (EDP, see [28, 29]). Indeed, the following equations are valid:

$$\overline{R}(G) = \frac{C_n^2 - N(G)}{C_n^2}; N(G) = C_n^2\left(1 - \overline{R}(G)\right). \tag{15}$$

From these equations, bounds are easily obtained:

$$LB_{APR} = \frac{C_n^2 - UB_{EDP}}{C_n^2}, UB_{APR} = \frac{C_n^2 - LB_{EDP}}{C_n^2}. \tag{16}$$

In [14], the following equation was derived with deleting the dangling node $t$ that is incidental to the node $s$:

$$N(G) = N(G_0) + \left(1 - p_{st}\right) w_t W(G\backslash e_{st}). \tag{17}$$

Here $w_t$ is the weight of the node $t$, initially 1. This weight shows an expected number of nodes that are merged in this node during the graph transformations: when using the factoring method, for example, one should remember the number of nodes that are merged if we assume that an edge between them is reliable. This weight is used for calculating the number of disconnected pairs of nodes when the graph is divided. Let $W(G)$ be the total weight of the all nodes (initially, it is equal to the number of nodes) in $G$. $G_0$ equals $G$ with the only difference: the weight of the node s becomes equal to $w_s + p_{st} w_t$.

From this, we obtain the following changes in the bounds: $LB_{EDP}$ is increased by $\left(1 - p_{st}\right)$ $w_t[W(G) - w_t]$ and the $UB_{EDP}$ is decreased by $p_{st} w_t [W(G) - w_t]$. Contrary to case of the ATR, the result of removing attached trees highly depends on their structures, so dangling nodes must be removed one at a time. The only known exception with the derived equation is the case of the attached chain [6].

## 7. Using cumulative bounds for network reliability approximation

A simple approximation of the function $\mu(G)$ through its bounds is their average: $\mu(G) = (LB + UB)/2$. However, the bounds tend to the exact solution with different rates. It seems to be reasonable assuming that cross of lines, obtained by the linear approximation of curves for $LB$ and $UB$, may be a better approximation of μ(G). Let us have the bounds $LB_i$ and $uB_i$ at a step $i > 0$ and let $\mu_i(G)$ be the corresponding approximation. The following proportion takes place:

$$(\mu_M - UB_i) : (LB_i - \mu_m) = (UB_i - \widehat{\mu}_i(G)) : (\widehat{\mu}_i(G) - LB_i). \quad (18)$$

Hence, we have

$$\widehat{\mu}_i(G) = \frac{\mu_M \cdot LB_i - \mu_m \cdot UB_i}{\mu_M - UB_i - \mu_m + LB_i}. \quad (19)$$

Let us name this approximation as the approximation based on trends (ABT).

In the case of ATR, $\mu_m = 0$ and $LB = UB = \mu(G)$, so

$$\widehat{R}_i(G) = \frac{LB_i}{1 - UB_i + LB_i}, \quad (20)$$

while in the case of MENC, $\mu_m = 1$ and $\mu_M = C_n^2$, so

$$\widehat{CS}_i(G) = \frac{C_n^2 \cdot LB_i - UB_i}{C_n^2 - UB_i - 1 + LB_i}. \quad (21)$$

At the last step, $LB = UB = \mu(G)$, and from (19) we have that the proposed approximation also equals this value.

In **Figures 8–10**, the behavior of bounds for the EDP of 4x4 lattice ($p = 0.7$), for the MENC of the same lattice with $c$-node 1, and for the probability of the flow transmission (PFT) between two

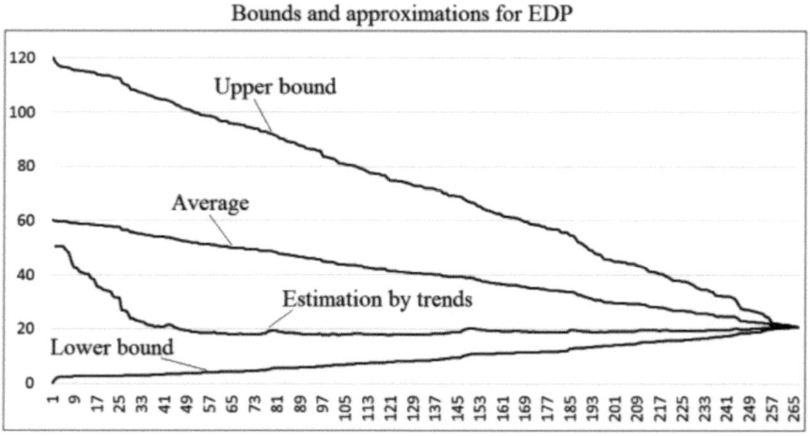


**Figure 8.** The behavior of bounds and approximations of EDP, the exact value is 20.915633.

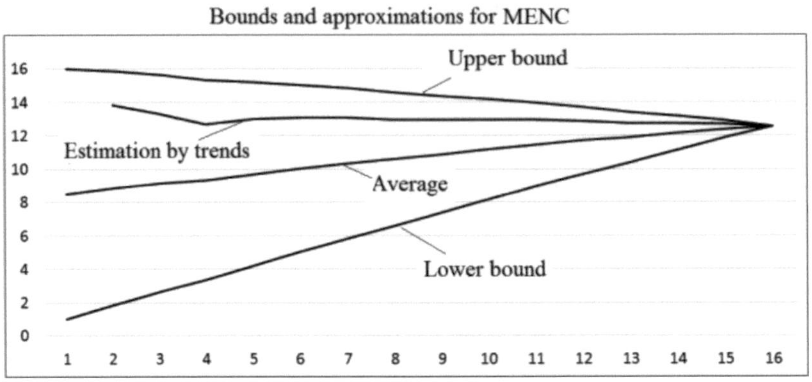


**Figure 9.** The behavior of bounds and approximations of MENC, the exact value is 12.562672.

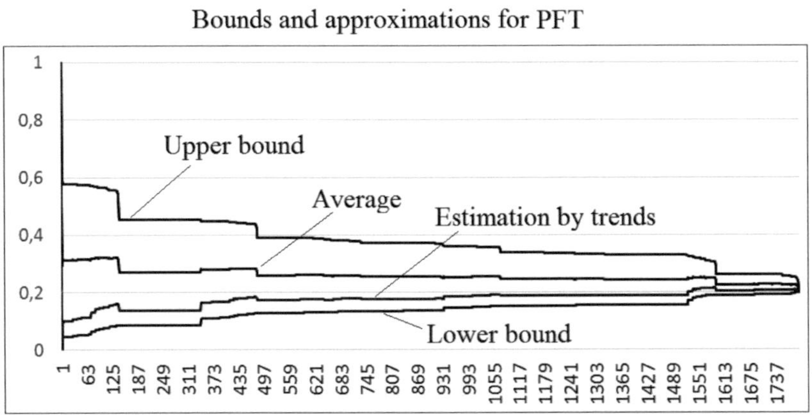


**Figure 10.** The behavior of bounds and approximations of PFT, the exact value is 0.196600.

diagonal corners of this lattice with throughput of all edges 1 and the edges reliability uniformly distributed between 0.5 and 1, are presented, respectively.

As we can see the ABT has become better than the average of *LB* and *UB* from the first step or very fast.

We consider the practical usage of cumulative bounds and ABT in the following section.

## 8. Using cumulative bounds in network topology optimization

When optimizing the network topology by a certain criteria, a fitness function (FF) must be calculated for each alternative. In our case (k-terminal reliability, MENC, and APR) this means using NP-hard algorithms for its obtaining. Note that using approximate algorithms may lead to wrong decisions when structures with close values of FF are compared.

We propose using the cumulative bounds for decision-making about whether a new solution obtained by crossover or mutation deserves including into population.

The ideas are simple (without the loss of generality, we assume that a task is maximizing some function of an unreliable network):

1. A wittingly inappropriate solution may be rejected before carrying out the exhaustive calculation of the corresponding FF. If the upper bound of the FF becomes smaller than the FF value of the worst solution in the current population, then this potential solution is rejected.

2. At initial stages, a new solution may be included into the population based on the LB value. If it exceeds the FF value of the worst solution in the current population, then this new solution substitutes the worst one with assigning ABT as an approximate value of the FF.

Note that inevitable narrowing of bounds of FF in population leads to seldom updating of the population and to a better approximation of FF.

The experiments [30, 31] with network topology optimization by the criteria of the ATR and the DCR show that using the cumulative bounds allow the speeding up of calculations up to two times without loss of precision or with negligible loss (about $10^{-7}$).

## 9. Conclusion

The general approach of obtaining the cumulative bounds of random graph functions is presented. We have shown the possibility of computing the cumulative bounds by partial sums and updating the cumulative bounds in case of applying different methods of reduction and decomposition. For example, various indices of the network reliability were considered: the all-terminal reliability, the diameter constrained reliability, the average pairwise connectivity, and the expected size of a subnetwork that contains a special node. Also, we have described how the cumulative bounds approach can be used for network reliability evaluation and development of evolutionary algorithms for the network topology optimization. Future works can include methods of computing the cumulative bounds for other reliability indices along with methods of reduction and decomposition, further improvement of evolutionary algorithms for the network topology optimization, and parallel algorithms for the network reliability analysis.

## Acknowledgements

This research was supported by the Russian Foundation for Basic Research under grants 17-07-00775, 17-47-540977.

## Author details

Alexey S. Rodionov and Denis A. Migov*

*Address all correspondence to: mdinka@rav.sscc.ru

Institute of Computational Mathematics and Mathematical Geophysics of SB RAS, Novosibirsk, Russia

## References

[1] Colbourn CJ. The Combinatorics of Network Reliability. New York, NY, USA: Oxford University Press; 1987. 160 p

[2] Jereb L. Network reliability: Models, measure and analysis. In: 6th IFIP Workshop on Performance Modeling and Evaluation of ATM Networks; 1998. p. T02/1–T02/10

[3] Moore EF, Shannon CE. Reliable circuits using less reliable relays. Journal of The Franklin Institute. 1956;**262**(4b):191-208

[4] Shooman AM. Algorithms for network reliability and connection availability analysis. In: Electro/International 1995, Boston, MA, USA. 1995, pp. 309-333

[5] Sun F, Shayman MA. On pairwise connectivity of wireless multihop networks. International Journal of Security and Networks. 2007;**2**(1/2):37-49

[6] Rodionov AS, Rodionova OK. Exact bounds for average pairwise network reliability. In: The 7th International Conference on Ubiquitous Information Management and Communication; Malaysia. ACM; 2013. Article No. 45

[7] Cancela H, Petingi L. Diameter constrained network reliability: Exact evaluation by factorization and bounds. In: Int. Conf. on Industrial Logistics; Okinawa, Japan. 2001. p. 359-356

[8] Pandurangan G, Raghavan P, Upfal E. Building low-diameter peer-to-peer networks. IEEE Journal on Selected Areas in Communications. 2003;**21**(6):995-1002

[9] Valiant LG. The complexity of enumeration and reliability problems. SIAM Journal on Computing. 1979;**8**(3):410-421

[10] Bodlaender HL, Wolle T. A note on the complexity of network reliability problems. IEEE Transactions on Information Theory. 2004;**47**:1971-1988

[11] Deuermeyer LA. New approach for network reliability analysis. IEEE Transactions on Reliability. 1982;**31**(4):350-354

[12] Goyal NK, Misra RB, Chaturvedi SK. Snem: A new approach to evaluate terminal pair reliability of communication networks. Journal of Quality in Maintenance Engineering. 2005;**11**(3):239-253

[13] Won JM, Karray F. Cumulative update of all-terminal reliability for faster feasibility decision. IEEE Transactions on Reliability. 2010;**59**(3):551-562

[14] Rodionov AS. Cumulative estimated values of structural Network's reliability indices and their usage. In: Dynamics of Systems, Mechanisms and Machines; Omsk, Russia. IEEE Press; 2016. p. 1-4

[15] Rodionova OK, Rodionov AS, Choo H. Network probabilistic connectivity: Exact calculation with use of chains. In: International Conference on Computational Science and Its Application. Lecture Notes in Computer Science, vol. 3036; Heidelberg: Springer; 2004. p. 565-568

[16] Rodionov AS, Migov DA, Rodionova OK. Improvements in the efficiency of cumulative updating of all-terminal network reliability. IEEE Transactions on Reliability. 2012;**61**(2): 460-465

[17] Wood RK. Triconnected decomposition for computing K–terminal network reliability. Networks. 1989;**19**:203-220

[18] Migov DA. Using node cuts for exact calculation of network probabilistic connectivity. In: Aryn EM, editor. Proc. Int. Conf. on Computational Technologies in Science, Engineering and Education, vol. 2; 20–22–09-2006; Pavlodar, Kazakhstan. 2006. p. 51-58. (in Russian)

[19] Migov DA Random Graph Probabilistic Connectivity Calculation Using Node Cuts [Dissertation] ICM&MG SB RAS, Novosibirsk, Russia: 2008. 97 p. (in Russian) Abstract available from: http://www.sscc.ru/Diss_sov/Migov.pdf

[20] Migov DA, Rodionov AS, Rodionova OK, Choo H. Network probabilistic connectivity: Using node cuts. In: EUC Workshop. LNCS, vol. 4097; Springer; 2006. p. 702-709

[21] Migov DA. Calculation of K-terminal network probabilistic connectivity using 2-node cuts. Bull. Inst. of Comp. Math. Geoph. Ser. Computer Science. Proc. Int. Conf. on Problems of Operation of Information Networks; 7–11-08-2006; Berdsk, Russia. 2006;**6**:134-138. (in Russian)

[22] Migov DA Network probabilistic connectivity computation with use of special kind vertex cuts. In: Proc. Russian Conf. of Young Scientists on Mathematical Modeling and Information Technologies; 1–03–11-2004; Novosibirsk, Russia. ICT SB RAS:2004. (in Russian) Available from: http://www.ict.nsc.ru/ws/YM2004/8566/Paper.htm

[23] Burgos JM, Amoza FR. Factorization of network reliability with perfect nodes I: Introduction and statements. Discrete Applied Mathematics. 2016;**198**:82-90

[24] Burgos JM. Factorization of network reliability with perfect nodes II: Connectivity matrix. Discrete Applied Mathematics. 2016;**198**:91-100

[25] Migov DA. Computing diameter constrained reliability of a network with junction points. Automation and Remote Control. 2011;**72**(7):1415-1419

[26] Migov DA, Rodionov AS. Decomposing graph with 2-node cuts for diameter constrained network reliability calculation. In: 7th Int. Conference on Ubiquitous Information Management and Communication; Malaysia. New York: ACM; 2013. Article No. 39

[27] Migov DA, Nesterov SN. Methods of speeding up of diameter constrained network reliability calculation. In: International Conference on Computational Science and its Application. LNCS, Vol. 9156(2); Springer. 2015. pp. 121-133

[28] Rodionov AS, Rodionova OK. Network probabilistic connectivity: Expectation of a number of disconnected pairs of nodes. In: High Performance Computing and Communications. Lecture Notes in Computer Science, vol. 4208; Berlin: Springer; 2006. p. 101-109

[29] Rodionov AS, Rodionova OK, Choo H. On the expected value of a number of disconnected pairs of nodes in unreliable network. In: Computational Science and its Applications. Lecture Notes in Computer Science, vol. 4707; Berlin: Springer; 2007. p. 534-543

[30] Nechunaeva K, Migov D. Speeding up of genetic algorithm for network topology optimization with use of cumulative updating of network reliability. In: Conference on Ubiquitous Information Management and Communication; Bali, Indonesia. New York: ACM; 2015. p. Article No. 42

[31] Migov DA, Nechunaeva KA, Nesterov SN, Rodionov AS. Cumulative updating of network reliability with diameter constraint and network topology optimization. In: Computational Science and its Applications. Lecture Notes in Computer Science, vol. 9786; Berlin. Springer; 2016. p. 141-152

# Chapter 6

# Spare Parts Forecasting Based on Reliability

Nataša Kontrec and Stefan Panić

Additional information is available at the end of the chapter

http://dx.doi.org/10.5772/intechopen.69608

**Abstract**

Stochastic models for spare parts forecasting have not been widely researched in scientific literature from the aspect of their reliability. In this chapter, the authors present models which analyze standard reliability parameters of technical systems' parts/components. By analyzing system reliability and failure rate, we estimate the required number of spare parts in the moment of expected failure or when reliability falls below the predefined level. Two different approaches based on data availability are presented herewith.

**Keywords:** reliability, spare parts, forecasting methods, Rayleigh model, Weibull model, failure rate, inventory

## 1. Introduction

Technical systems, for example, aircrafts or weapon systems, are typically highly complex and composed of a large number of components and parts. Maintaining those systems adheres to strict rules and procedures for specific part or component. In order for such a system to be successfully maintained, the efficient management of spare part inventory is required, that is, a specific part should be provided in the right place at the right time [1].

Unpredictability of future events, that is, equipment and parts failure, has major impact on this problem. One way to reduce the level of unpredictability is to maintain a sufficient number of spare parts in inventory which results in the increase of warehouse costs and capital trapped in spare parts; another way is the assessment of spare parts inventory by using one of the available models for spare parts forecasting [2].

Forecasting is an essential skill on predicting future events and represents a foundation for every valid assessment. Although forecasts are often deemed expensive and time consuming, a vast number of researchers have been involved in finding novel methods with improved and more accurate assessment in recent years.

Concerning the demand for spare parts, exponential smoothing model and Croston's model are still among the most utilized due to their simplicity. The exponential smoothing model described in Refs. [3, 4] is based on pounder-moving average and can be implemented straightforwardly. In 1972, Croston presented a model based on exponential smoothing but far superior to it [5]. This method was most widely used in industry and is still a part of many software packages for spare parts forecasting. Rao [6] and then Schultz [7] studied Croston's method and proposed certain alterations but with no effect on execution results. Willemain et al. [8], and later Johnston and Boylan [9] proved that for majority of cases Croston's method gives better results than exponential smoothing method. Syntetos and Boyland [10, 11] made a considerable contribution to this scientific field by proposing a modified method based on criticism of Croston's model, calling it biased with reference to spare parts demand per time unit.

Another commonly used method is Bootstrap method—a computer statistical technique which, based on available data sample, creates a large number of new samples of same range as the original sample, by random sampling with replacement from original data set. From the aspect of inventory management, this model was examined in detail in Refs. [12, 13]. In addition, Refs. [14, 15] examine spare parts forecasting using Poisson's model. A conclusion was reached that traditional statistical methods based on analysis of time series can incorrectly assess functional form related to dependent and independent variables.

The nature of the spare parts demand is stochastic, and previously stated models do not always provide most accurate assessments [16]. For that reason, the number of models dealing with assessing the required number of spare parts, based on parameters such as spare part reliability, maintenance possibilities, life span, maintenance costs, and so on, has greatly increased during the past decade.

Due to the increase of system complexity [17], presenting reliability as a quantitative measure is commonly considered. By analyzing system's reliability and failure rate, we can estimate the required number of spare parts in the moment of expected failure or when reliability falls below the predefined level. There are numerous papers on the topic of determining the required number of spare parts, particularly as a part of logistic support [18, 19]. Refs. [20, 21] mostly deal with repairable spare parts or inventory managements with the aim of achieving previously set system reliability. On the other side, quantitative theories based on the theory of reliability were used to estimate the failure rate in order to precisely determine demand for a specific spare part [22–25]. An overview of the abovementioned models in spare parts forecasting with some new approaches proposed has been given in Ref. [26].

In this chapter, we present two models for spare parts forecasting based on the analysis of reliability parameter. Each model can be used depending on data availability.

## 2. Spare parts forecasting using Rayleigh model

Spare parts manufacturers provide only basic information on the part they produce. In this case, we observe the average life span of a part/component ($T_{ut}$) expressed in *hours*, $\mu$, as a stochastic process modeled with Rayleigh's distribution [27]. Probability density function (PDF) of Rayleigh's model is stated in Eq. (1)

$$p(\mu) = \frac{\mu}{\sigma^2} \exp\left(-\frac{\mu^2}{2\sigma^2}\right), \tag{1}$$

where $\sigma$ is the parameter of Rayleigh's distribution determined by relation $E(\mu^2) = 2\sigma^2$. $E(\mu)$ is the mathematical expectation of Rayleigh's random variable $\mu$. Based on that, $T_{ut}$ can be presented as follows:

$$T_{ut} = \int_0^\infty \mu p(\mu) d\mu = \int_0^\infty \frac{\mu^2}{\sigma^2} \exp\left(-\frac{\mu^2}{2\sigma^2}\right) d\mu. \tag{2}$$

With replacement $\mu^2/(2\sigma^2) = x$, Eq. (2) is transformed into

$$T_{ut} = \sqrt{2}\sigma \int_0^\infty x^{1/2} \exp(-x) dx = \sqrt{2}\sigma\Gamma\left(\frac{3}{2}\right). \tag{3}$$

$\Gamma(a)$ is Gamma function [28] and $\Gamma(3/2) = \sqrt{\pi}/2$, so the average life span can be presented as $T_{ut} = \sigma\sqrt{\frac{\pi}{2}}$.

Based on the aforementioned, the PDF of Rayleigh's model can now be presented as follows:

$$p(\mu) = \frac{\mu\pi}{4T_{ut}{}^2} \exp\left(-\frac{\mu^2\pi}{4T_{ut}{}^2}\right). \tag{4}$$

Cumulative density function (CDF) can similarly be determined as

$$F(\mu) = \int_0^\mu p(\mu) d\mu = \int_0^\mu \frac{\mu\pi}{2T_{ut}^2} \exp\left(-\frac{\mu^2\pi}{4T_{ut}^2}\right) d\mu \tag{5}$$

If we substitute $\mu^2\pi/(4T_{ut}^2) = u$ in the previous equation, it can be reduced to

$$F(\mu) = \int_0^{\mu^2\pi/(4T_{ut}^2)} \exp(-u) du = 1 - \exp\left(-\frac{\mu^2\pi}{4T_{ut}^2}\right). \tag{6}$$

The function of part's reliability can further be determined with Eq. (7)

$$R(\mu) = 1 - F(\mu) = \exp\left(-\frac{\mu^2\pi}{4T_{ut}^2}\right). \tag{7}$$

Finally, based on the previous equations, we can define the failure function as a probability that the examined part will cease to perform its function in a specific time interval

$$\lambda(\mu) = \frac{p(\mu)}{R(\mu)} = \frac{\mu\pi}{2T_{ut}^2}. \tag{8}$$

The essence of this model's implementation is in an approach of determining the quantity of spare parts in inventory and costs that occur due to negative level of inventory at the end of usability period of examined part (underage costs) originally presented in Ref. [29]. This approach stresses the stochastic nature of variable $T_{\text{ut}}$. By observing the expected number of variations of this random variable in time interval ($T_{\text{ut}} + dT_{\text{ut}}$) for a given slope $\dot{T}_{\text{ut}}$ in its specific environment d$T_{\text{ut}}$ for $N$ units of stated component, we can then determine the number of components most likely to fail as follows:

$$n = \int_0^{+\infty} p(\mu|\dot{T}_{ut})p(\dot{T}_{ut})d\dot{T}_{ut} = \int_0^{+\infty} p(\mu)\dot{T}_{ut}\frac{1}{\sqrt{2\pi\dot{\sigma}}}\exp\left(-\frac{\dot{T}_{ut}^2}{2\dot{\sigma}^2}\right)d\dot{T}_{ut}. \tag{9}$$

By implementing and substituting $\dot{T}_{\text{ut}}^2/(2\dot{\sigma}^2) = u$, the previous equation is reduced to

$$n = p(\mu)\frac{\dot{\sigma}}{\sqrt{2\pi}}\int_0^{+\infty}\exp(-u)\mathrm{d}u = p(\mu)\frac{\dot{\sigma}}{\sqrt{2\pi}}, \tag{10}$$

where $\dot{T}_{\text{ut}}$ is the Gauss random variable with variance $V(\dot{T}_{\text{ut}}) = \dot{\sigma}$.

Now, given that $\mu$ is a Rayleigh's random variable with mathematical expectation $E(\mu) = T_{\text{ut}}$ and variance $V(\mu) = 2\ T_{\text{ut}}^2/\pi$, then the average number of components $n$, which will be subject to failure in time $T_{\text{ut},}$ can be determined as

$$n = \sqrt{2}T_{\text{ut}}p(\mu) = \frac{\mu\pi}{\sqrt{2}T_{\text{ut}}}\exp\left(-\frac{\mu^2\pi}{4{T_{\text{ut}}}^2}\right). \tag{11}$$

Moreover, the number of spare parts required in inventory can be determined by observing the total time when random variable $\mu$ is below $T_{\text{ut}}$:

$$q = \frac{F(\mu)}{n} = \frac{1-\exp\left(-\frac{\mu^2\pi}{4{T_{ut}}^2}\right)}{\frac{\mu\pi}{\sqrt{2}T_{ut}}\exp\left(-\frac{\mu^2\pi}{4{T_{ut}}^2}\right)} \tag{12}$$

This model can be a part of software for spare parts forecasting as in Ref. [30].

## 3. Spare parts forecasting using Weibull's model

For spare parts forecasting in cases where the total unit time of a product is unavailable but data on previous failure rates are available, Weibull's model can be used [31]. This model is most often used when reliability of a technical system is being determined. The PDF of

Weibull's model with two parameters, shape parameter $\beta$ and characteristic life $\eta$, has the following form:

$$f(t) = \frac{\beta}{\eta}\left(\frac{t}{\eta}\right)^{\beta-1} \exp\left(-\frac{t}{\eta}\right)^{\beta}, \quad t \geq 0, \eta > 0, \beta > 0. \tag{13}$$

The function of reliability can be expressed by the following formula:

$$R(t) = exp\left(-\frac{t}{\eta}\right)^{\beta}, \tag{14}$$

while CDF is

$$F(t) = 1 - \mathrm{R}(t) = 1 - exp\left(-\frac{t}{\eta}\right)^{\beta}. \tag{15}$$

The failure rate function of Weibull's model is

$$\lambda(t) = \left(\frac{\beta}{\eta}\right)\left(\frac{t}{\eta}\right)^{\beta-1}. \tag{16}$$

In relation to Weibull's model, a problem can arise while estimating the distribution parameters. There are multiple ways for parameter estimation but based on Ref. [32] if the available data sample is less than 15, then linear regression for parameter estimation is used, and in the opposite case maximum likelihood estimator (MLE) gives best results.

If linear regression is used, we start from CDF function of Weibull's model expressed as follows:

$$F(t) = 1 - exp\left(\frac{t}{\eta}\right)^{\beta} \Rightarrow 1 - F(t) = \exp\left(\frac{t}{\eta}\right)^{\beta}. \tag{17}$$

Taking the logarithm of previous formula, we get

$$\ln[1 - F(t)] = \ln\left[\exp\left(-\frac{t}{\eta}\right)^{\beta}\right] = \ln\left[\frac{1}{\ln[1 - F(t)]}\right] = \left(\frac{t}{\eta}\right)^{\beta}. \tag{18}$$

If we take the logarithm of the previous expression once again

$$\ln\left[\ln\left[\frac{1}{\ln[1 - F(t)]}\right]\right] = \beta \ln t - \beta \ln \eta \tag{19}$$

and implement

$$y = \ln[-\ln[1 - F(t)]] \tag{20}$$

and

$x = \ln t$, $a = \beta$ and $b = -\beta\ln\eta$, then Eq. (19) can be expressed in the following linear form:

$$y = ax + b \tag{21}$$

If $t_1, t_2, \ldots, t_n$ are observed values of random variable $t$, the procedure for parameter estimation is as follows [33]:

1. Data should be arranged in an ascending order: $t_{(1)} \leq t_{(2)} \leq \cdots \leq t_{(n)}$.
2. Value of empirical function of distribution is then calculated: $\hat{F}(t_{(i)}), \quad i = 1, 2, \ldots, n$.
3. Calculating $y_i = \ln[-\ln[1 - \hat{F}(t_{(i)})]], \quad i = 1, 2, \ldots, n$.
4. Calculating $x_i = \ln t_{(i)}, \ i = 1, 2, \ldots, n$.
5. Drawing points $T_i = (x_i, y_i)$ in the coordinate system and mathematically determine direction $y = \hat{\beta}x + \hat{b}$ which best approximates obtained data set. The most frequently used is the least-square method or linear regression which minimizes the vertical deviation of observed points from the given direction, in the following way:

$$\sum_{i=1}^{N}\left(\hat{a} + \hat{b}x_i - y_i\right)^2 = \min\sum_{i=1}^{N}\left(\hat{a} + \hat{b}x_i - y_i\right)^2 \tag{22}$$

where $\hat{a}$ and $\hat{b}$ are the least-squares estimates of $a$ and $b$, and can be determined with the following formulae:

$$\hat{b} = \frac{\sum_{i=1}^{N} x_i y_i - \frac{\sum_{i=1}^{N} x_i \sum_{i=1}^{N} y_i}{N}}{\sum_{i=1}^{N} x_i^2 - \frac{\left(\sum_{i=1}^{N} x_i\right)^2}{N}} \tag{23}$$

and

$$\hat{a} = \frac{\sum_{i=1}^{N} y_i}{N} - \hat{b}\frac{\sum_{i=1}^{N} x_i}{N} = \overline{y} - \hat{b}\overline{x} \tag{24}$$

6. Based on the equations, it is now simple to determine parameters $\beta$ *i* $\eta$.

In cases where data sample is greater than 15, parameters are determined by maximum likelihood estimator method [33].

By observing $t_1, t_2, t_3, \ldots, t_n$ data sample of size $n$ population with given PDF equation (13), then the joint density of likelihood function can be determined as a product of the densities of each point

$$L(t) = \prod_{i=1}^{n} f(t_i) = \left(\frac{\beta}{\eta}\right)\left(\frac{t_i}{\eta}\right)^{\beta-1} \exp\left(-\left(\frac{t_i}{\eta}\right)^{\beta}\right) = \left(\frac{\beta}{\eta}\right)\left(\frac{t_i}{\eta}\right)^{n\beta-n} \sum_{i=1}^{n} t_i{}^{\beta-1} \exp\left(-\sum_{i=1}^{n}\left(\frac{t_i}{\eta}\right)^{\beta}\right). \quad (25)$$

Taking natural logarithm of both sides gives

$$\ln L = n\ln\left(\frac{\beta}{\eta}\right) + (\beta-1)\sum_{i=1}^{n} t_i - \ln(\eta^{\beta-1}) - \sum_{i=1}^{n}\left(\frac{t_i}{\eta}\right)^{\beta}. \quad (26)$$

By differentiating partially previous equation, we obtain

$$\frac{\partial}{\partial\beta}\ln L = \frac{n}{\beta} + \sum_{i=1}^{n} \ln t_i - \frac{1}{\eta}\sum_{i=1}^{n} t_i{}^{\beta} \ln t_i = 0 \quad (27)$$

and

$$\frac{\partial}{\partial\eta}\ln L = \frac{n}{\eta} + \frac{1}{\eta^2}\sum_{i=1}^{n} t_i{}^{\beta} = 0. \quad (28)$$

From the previous equation, we can estimate $\eta$ as

$$\hat{\eta} = \frac{1}{n}\sum_{i=1}^{n} t_i{}^{\hat{\beta}}. \quad (29)$$

By replacing Eq. (29) into Eq. (27), we get

$$\frac{1}{\beta} + \frac{1}{n}\sum\nolimits_{i=1}^{n} \ln t_i - \frac{\sum\nolimits_{i=1}^{n} t_i{}^{\beta} \ln t}{\sum\nolimits_{i=1}^{n} t_i{}^{\beta}} = 0. \quad (30)$$

Eq. (30) does not have a closed-form solution, so we can estimate shape parameter $\beta$ by using Newton-Raphson's method or any other numeric procedure. After we determined $\hat{\beta}$, by replacing it into Eq. (29) we can calculate $\hat{\eta}$.

Once the parameters of Weibull's model are determined, we can estimate the required number of spare parts that should be in inventory in the given time interval. In order to achieve that, we use the approach presented in Ref. [34]. The PDF of Weibull's distributed failure time is given by Eq. (13), while the PDF of Rayleigh's distribute failure time is given by Eq. (1). Based on those two equations, we can conclude that $\sigma = \frac{\eta}{\sqrt{2}}$, while $\mu = t^{\frac{\beta}{2}}$, wherein $\mu$ is Rayleigh's random and $t$ is Weibull's random variable.

$$p_{t\dot{t}}(t,\dot{t}) = p_{\mu\dot{\mu}}\left(t^{\frac{\beta}{2}}, \dot{t}\frac{\beta}{2}t^{\frac{\beta}{2}-1}\right)|J|. \tag{31}$$

$|J|$ represents Jacobian transformation expressed with Eq. (32)

$$|J| = \begin{vmatrix} \frac{d\mu}{dt} & \frac{d\mu}{d\dot{\mu}} \\ \frac{d\dot{\mu}}{dt} & \frac{d\dot{\mu}}{d\dot{t}} \end{vmatrix} = \frac{\beta^2}{4}t^{\beta-2}. \tag{32}$$

By substituting Eq. (32) in Eq. (31), we obtain Eq. (33)

$$p_{t\dot{t}}(t,\dot{t}) = \frac{\beta^2}{4}t^{\beta-2}p_{\mu\dot{\mu}}(\mu,\dot{\mu}) \tag{33}$$

If we stress the stochastic nature of a specific part's failure rate by observing the expected number of variations of Rayleigh's random variable in interval $(\mu, \mu+d\mu)$ for a given slope $\dot{\mu}$ in specific environment $d\mu$, then the number of spare parts that are most likely to fail can be determined as follows:

$$n = \int_0^{+\infty} \dot{\mu}p_{\mu\dot{\mu}}(\mu,\dot{\mu})d\dot{\mu} = \int_0^{+\infty} \dot{\mu}\frac{\mu}{\sigma^2}e^{-\frac{\mu^2}{2\sigma^2}}\frac{1}{\sqrt{2\pi\sigma^2}}e^{-\frac{\dot{\mu}^2}{2\sigma^2}}d\mu = \int_0^{+\infty} \dot{t}p_{t\dot{t}}(t,\dot{t})d\dot{t}. \tag{34}$$

Based on the previous equations, the number of spare parts that are subject to failure in time period of $[0, t]$ is

$$n = \frac{4\sqrt{2}t^{\frac{\beta}{2}}}{\eta}e^{-\frac{t^\beta}{\eta^2}} \tag{35}$$

As in the previous section, it is now necessary to determine the number of spare parts required to be kept in inventory in time interval $[0, t]$. We will use the approach presented in paper [29] but in this case the quantity will be calculated as a quotient of CDF function of Weibull's random variable and size $n$. More accurately, as parameter $\eta$ marks the time in which 63.2% of units will fail and is approximately equal to MTTF [35], we estimate the time interval in which $t$ is below $\eta$ as

$$q = \frac{F(t)}{n} \tag{36}$$

When we determine the quantity of spare parts that need to be kept in inventory, then in case when we know the price per unit of product based on Newsvendor model [36], we can determine the underage costs, that is, cost per unit of unsatisfied demand as

$$q = \Phi^{-1}\left(\frac{c_u}{c_u + c_o}\right) \tag{37}$$

where $\Phi^{-1}$ presents inverse distribution function (complementary error function), $c_u$ are underage costs, and $c_o$ are the overage costs, which in our case is spare part price.

http://dx.doi.org/10.5772/intechopen.69608

## 4. Case study

Depending on data availability, one of the two recommended approaches will be implemented. If the total unit time ($T_{ut}$) is available for a specific part, we opt to use model for spare parts

| Time | Failure/ censored | Time | Failure/ censored | Time | Failure/ censored | Time | Failure/ censored | Time | Failure/ censored |
|---|---|---|---|---|---|---|---|---|---|
| 40 | F | 2089 | F | 3166 | F | 280 | S | 2137 | S |
| 301 | F | 2097 | F | 3344 | F | 313 | S | 2141 | S |
| 309 | F | 2135 | F | 3376 | F | 389 | S | 2163 | S |
| 943 | F | 2154 | F | 3385 | F | 487 | S | 2183 | S |
| 1070 | F | 2190 | F | 3443 | F | 622 | S | 2240 | S |
| 1124 | F | 2194 | F | 3467 | F | 900 | S | 2314 | S |
| 1248 | F | 2223 | F | 3478 | F | 952 | S | 2435 | S |
| 1281 | F | 2224 | F | 3578 | F | 996 | S | 2464 | S |
| 1303 | F | 2229 | F | 3595 | F | 1003 | S | 2543 | S |
| 1432 | F | 2300 | F | 3699 | F | 1101 | S | 2560 | S |
| 1480 | F | 2324 | F | 3779 | F | 1085 | S | 2592 | S |
| 1505 | F | 2349 | F | 3924 | F | 1092 | S | 2600 | S |
| 1506 | F | 2385 | F | 4035 | F | 1152 | S | 2670 | S |
| 1568 | F | 2481 | F | 4121 | F | 1183 | S | 2717 | S |
| 1615 | F | 2610 | F | 4167 | F | 1244 | S | 2819 | S |
| 1619 | F | 2625 | F | 4240 | F | 1249 | S | 2820 | S |
| 1652 | F | 2632 | F | 4255 | F | 1262 | S | 2878 | S |
| 1652 | F | 2646 | F | 4278 | F | 1360 | S | 2950 | S |
| 1757 | F | 2661 | F | 4305 | F | 1436 | S | 3003 | S |
| 1795 | F | 2688 | F | 4376 | F | 1492 | S | 3102 | S |
| 1866 | F | 2823 | F | 4449 | F | 1580 | S | 3304 | S |
| 1876 | F | 2890 | F | 4485 | F | 1719 | S | 3483 | S |
| 1899 | F | 2902 | F | 4570 | F | 1794 | S | 3500 | S |
| 1911 | F | 2934 | F | 4602 | F | 1915 | S | 3622 | S |
| 1912 | F | 2962 | F | 4663 | F | 1920 | S | 3665 | S |
| 1914 | F | 2964 | F | 4694 | F | 1963 | S | 3695 | S |
| 1918 | F | 3000 | F | 46 | S | 1978 | S | 4015 | S |
| 2010 | F | 3103 | F | 140 | S | 2053 | S | 4628 | S |
| 2038 | F | 3114 | F | 150 | S | 2065 | S | 4806 | S |
| 2085 | F | 3117 | F | 248 | S | 2117 | S | 4881 | S |
| | | | | | | | | 5140 | S |

**Table 1.** Windshield failure time.

forecasting based on Rayleigh's distribution. When historic failure/censored data are available, we use model based on Weibull's distribution.

**Table 1** provides data on specific aircraft model's Windshield, taken from paper [37]. **Table 1** consists of 88 records on the part's failure time and 65 records on censored time expressed in

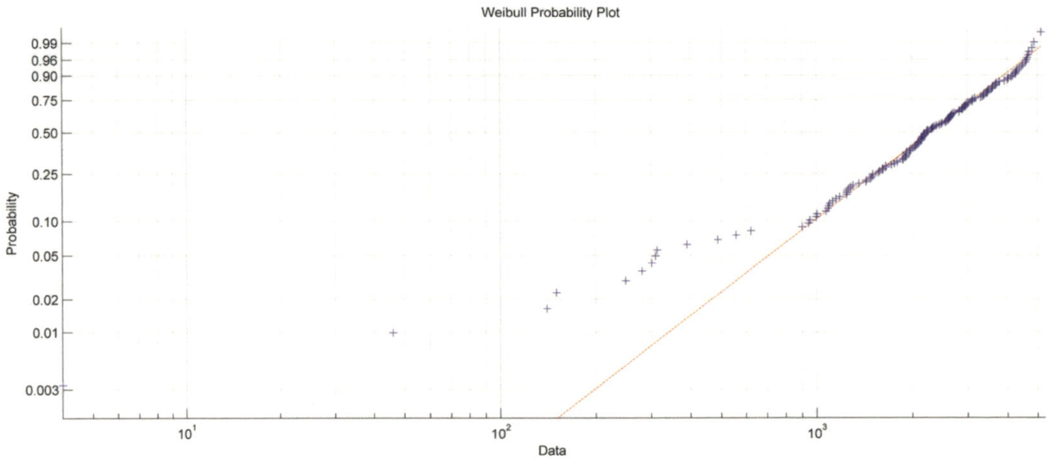


**Figure 1.** Weibull probability plot for **Table 1** data.

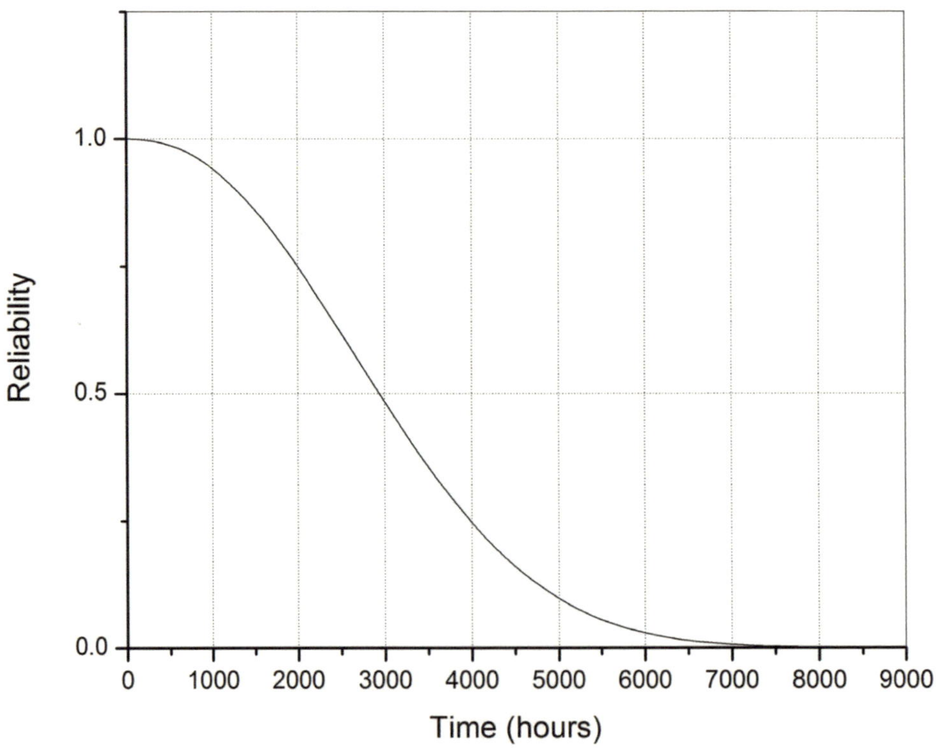


**Figure 2.** Windshield reliability.

flight hours. Censored time (or service time) means that the Windshields have not failed at the time of observation.

Weibull probability plot for data from previous table is presented in **Figure 1**.

As noted in **Figure 1**, records from **Table 1** follow Weibull's distribution, and as their number is greater than 15, based on the previous research, we use maximum likelihood estimation method for the estimation of parameters in Weibull's distribution, described extensively in the previous section.

Implementation of this method results in the value of shape parameter $\beta = 2.28$, and characteristic life parameter $\eta = 3450.54$.

Now, when values of these parameters are known we can determine the reliability of part windshield based on Eq. (15), while its failure function can be determined based on Eq. (16). A graphical representation of these functions is presented in **Figures 2** and **3**.

Finally, we can determine the quantity of spare parts required to be kept in inventory in a given time interval based on Eq. (36). A graphical representation is presented in **Figure 4**.

Based on **Figure 4**, it can be concluded that it is necessary to have one spare part in inventory after 2000 flight hours. The results of this analysis, that is, data on spare part reliability, failure time, and required quantity in inventory can be of great value to decision makers, on questions related to what to keep in inventory and in what quantity, and when to plan maintenance activities in order to prevent the occurrence of failure.

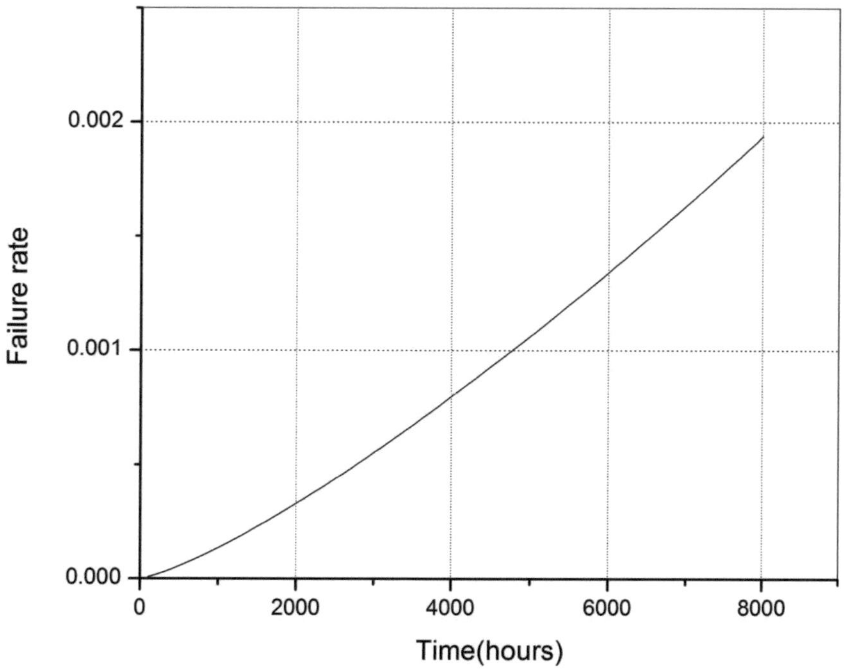


**Figure 3.** Failure rate function of windshield.

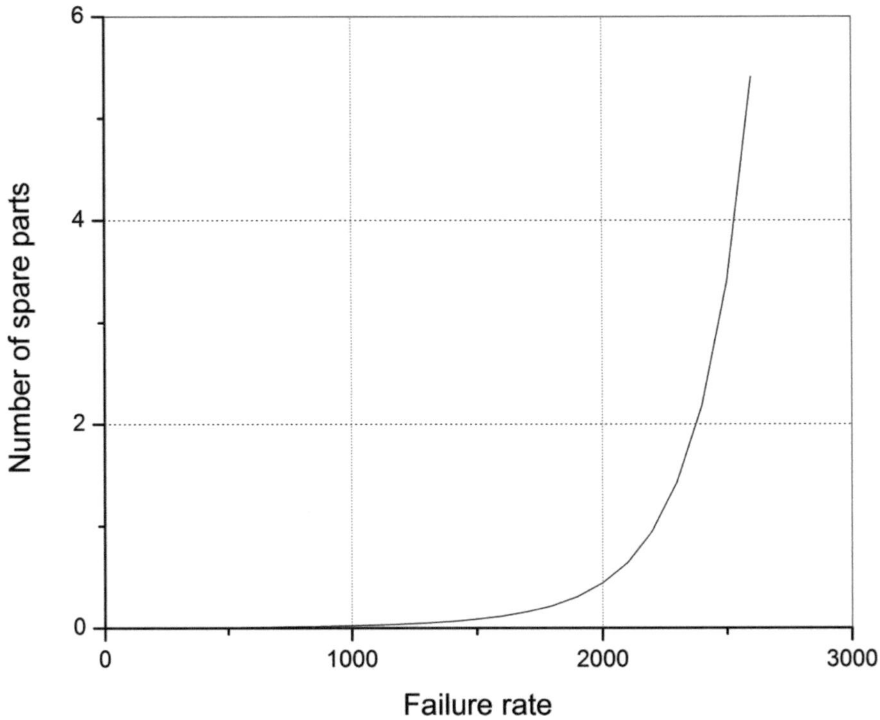


**Figure 4.** Quantity of spare parts in function of time.

The aforementioned models provide possibility for taking into consideration underage costs during decision-making process by using Eq. (37), for those cases when price of the part in issue is available. These costs are difficult to determine objectively. Some consequences of lack of spare parts, for example, damage to company's reputation due to delays, are difficult to express quantitatively.

## 5. Conclusions

This section elaborates on determining spare parts required to be kept in inventory from the aspect of reliability analysis of that part. Depending on data availability, either Rayleigh's or Weibull's method can be used. In case of Weibull's method, two approaches for the assessment of parameters are given, depending on data sample size. Information obtained with this analysis can have a major role in the process of supply management. Based on them, it is possible to reduce costs that occur due to delays, unplanned cancellations, and so on. These models can serve as a solid foundation for the creation of software for spare parts forecasting. Although the emphasis was placed on planning maintenance activities and avoiding delays, all the aforementioned leads to limiting the consequences of suboptimal supply management, that is, minimize the spare parts overstocking. In the case when we are dealing with reparable systems, researches should focus on the determination or minimization of the repair rate of such system.

## Author details

Nataša Kontrec* and Stefan Panić

*Address all correspondence to: natasa.kontrec@pr.ac.rs

Faculty of Science and Mathematics, University of Pristina, Kosovska Mitrovica, Serbia

## References

[1] Gopalakrishnan P, Banerji AK. Maintenance and Spare Parts. PHI Learning Pvt. Ltd; India, 2003. p. 400

[2] Ben-Daya M, Duffuaa SO, Raouf A. Maintenance, Modeling and Optimization. Springer Science & Business Media; New York, US, 2012. p. 474

[3] Gardner Jr, ES. Exponential smoothing: The state of the art. Journal of Forecasting. 1985;**4**:1-28

[4] Gardner Jr, ES. Exponential smoothing: The state of the art—Part II. International Journal of Forecasting. 2006;**22**:637-677

[5] Croston JD. Forecasting and stock control for intermittent demands. Operational Research Quarterly. 1972;**42**(3):289-303

[6] Rao A. A comment on: Forecasting and stock control for intermittent demands. Operational Research Quarterly. 1973;**24**(3):639-640

[7] Schultz CR. Forecasting and inventory control for sporadic demand under periodic review. Journal of the Operational Research Society. 1987;**37**:303-308

[8] Willemain TR, Smart CN, Shockor JH, DeSautels PA. Forecasting intermittent demand in manufacturing: A comparative evaluation of Croston's method. International Journal of Forecasting. 1994;**10**:529-538

[9] Johnston FR, Boylan JE. Forecasting for items with intermittent demand. Journal of the Operational Research Society. 1996;**47**:113-121

[10] Syntetos AA, Boylan JE. On the bias of intermittent demand estimates. International Journal of Production Economics. 2001;**71**:457-466

[11] Syntetos AA, Boylan JE. On the stock control performance of intermittent demand estimators. International Journal of Production Economics. 2006;**103**:36-47

[12] Bookbinder H, Lordahl AE. Estimation of inventory re-order levels using the bootstrap statistical procedure. IIE Transactions. 1989;**21**:302-312

[13] Wang M, Rao SS. Estimating reorder points and other management science applications by bootstrap procedure. European Journal of Operational Research. 1992;**56**:332-342

[14] Ravindran A. Aggregate Capacitated Production Planning in a Stochastic Demand Environment. ProQuest; Purdue University, 2008. p. 153

[15] Hill T, O'Connor M, Remus W. Neural network models for time series forecasts. Management Science. 1996;**42**:1082-1092

[16] Manzini R, Regattieri A, Pham H, Ferrari E. Maintenance for Industrial System. New York, US: Springer Science and Media, 2009

[17] Kumar D. Reliability Analysis and Maintenance Scheduling Considering Operating Conditions [thesis]. Luleå University of Technology; 1996

[18] Chelbi A, Ait-Kadi D. Spare provisioning strategy for preventively replaced systems subjected to random failure. International Journal of Production Economics. 2001;**74**(2): 183-189

[19] Kennedy WJ, Patterson JW, Fredendall LD. An overview of recent literature on spare parts inventories. International Journal of Production Economics. 2002;**76**(2):201-215

[20] Aronis KP, Magou I, Dekker R, Tagaras G. Inventory control of spare parts using a Bayesian approach: A case study. European Journal of Operational Research. 2004;**154**(3): 730-739

[21] Sarker R, Haque A. Optimization of maintenance and spare provisioning policy using simulation. Applied Mathematical Modeling. 2000;**24**(10):751-760

[22] Huiskonen J. Maintenance spare parts logistics: Special characteristics and strategic choice. International Journal of Production Economic. 2001;**71**:125-133

[23] Jardine AKS. Maintenance, replacement and reliability. Ontario, Canada: Preney Print and Litho Inc., 1998

[24] Lewis EE. Introduction to Reliability Engineering, 2nd Edition. John Wiley & Sons, New York, USA: 1996, p. 464.

[25] Xie M, Kong H, Goh TN. Exponential approximation for maintained Weibull distributed component. Journal of Quality in Maintenance Engineering. 2000;**6**(4):260-269

[26] Kontrec N. Primena matematičkih modela kao instrumenta matematičkih tehnologija za procenu zaliha rezervnih delova u avio industriji [dissertation]. Kosovska Mitrovica: Prirodno-matematički fakultet, Univerzitet u Prištini; 2015

[27] Wang H, Pham H. Reliability and Optimal Maintenance. Springer Series in Reliability Engineering; London: Springer, 2006

[28] Gradshteyn IM, Ryzhik I. Table of Integrals, Series and Products. Elsevier, printed in US: Academic Press; 1980

[29] Kontrec N, Milovanovic GV, Panic S, Milosevic H. A reliability-based approach to nonrepairable spare part forecasting in aircraft maintenance system. Mathematical Problems in Engineering. 2015;7 [Article ID731437]. DOI: 10.1155/2015/73143

[30] Kontrec N, Panić S, Milošević H, Djošić D. Software for analyzing reliability and spare parts forecasting in aircraft maintenances system based on Rayleigh model. In: Infoteh; March 2015; Bosnia & Herzegovina: Jahorina. 2015

[31] Walck C. Hand-Book on Statistical Distributions for Experimentalists. Stockholm, Sweden: University of Stockholm; 2007

[32] Nelson WB. Applied Life Data Analysis. Printed in US: John Wiley & Sons; 2004

[33] Smith DJ. Reliability, Maintainability and Risk: Practical Methods for Engineers including Reliability Centred Maintenance and Safety-Related Systems. Oxford, UK: Elsevier; 2011. p. 436

[34] Kontrec N, Petrović M, Vujaković J, Milošević H. Implementation of Weibull's model for determination of aircraft's parts reliability and spare parts forecast. In: Mathematical and Information Technologies, MIT-2016, CEUR Workshop Proceedings; August 28–31, 2016; Vrnjačka Banja, Serbia; 2016

[35] Available from: http://www.barringer1.com/pdf/Chpt1-5th-edition.pdf [Accessed: June 2016]

[36] Hill AV. The Newsvendor Problem. Clamshell Beach Press; US. 2011

[37] Ruhi S, Sarker S, Karim MR. Mixture models for analyzing product reliability data: A case. Springer Plus. 2015;4(634). DOI: 10.1186/s40064-015-1420-x. Article is available online in Springer Plus (open access journal) https://springerplus.springeropen.com/articles/10.1186/s40064-015-1420-x

Chapter 7

# Model Development for Reliability Cannibalization

Bernard Tonderayi Mangara

Additional information is available at the end of the chapter

http://dx.doi.org/10.5772/intechopen.69609

**Abstract**

This chapter looks at cannibalization as a method (procedure) of improving reliability of engineering systems. Cannibalization gives one the opportunity to use resources in the most efficient way. In this chapter, we have explored strategies to reduce the adverse effects of cannibalization on maintenance costs and personnel morale. The strategies developed in this chapter, at least, can be used to determine (1) which types of cannibalizations are appropriate, (2) cannibalization reduction goals and (3) the actions to be taken to meet the cannibalization reduction goals. In this chapter, we also presented a combined analytical and simulation model of a two-line, three-line and *k*-line system when cannibalization is not allowed and when cannibalization is allowed (with and without short interruptions to the system). It is clear from the analytical and simulation results that cannibalization can substantially increase the reliability of the systems where it is allowed. The improvement factor of unreliability obviously exists in systems where cannibalization is allowed as compared to those in which cannibalization is not allowed. Moreover, the improvement factor is larger when we have two-stage cannibalization (short interruptions) than without them.

**Keywords:** cannibalization, system, component, reliability, model

## 1. Introduction

Usually, policy dictates that systems of equipment are dispensed together with spare parts in case of components which are prone to failure. With failed components being replaced and/or repaired, the systems of equipment remain functional subservient to the said policy. Nevertheless, the two exceptions to the preceding policy are [1, 2] as follows: (1) as a result of the high acquisition and holding costs, high-technology manufacturing environments and organizations, which make use of expensive equipment, will not be able to always stock spare parts and (2) when equipment has reached its last life-cycle stage, failure rates have increased, spare

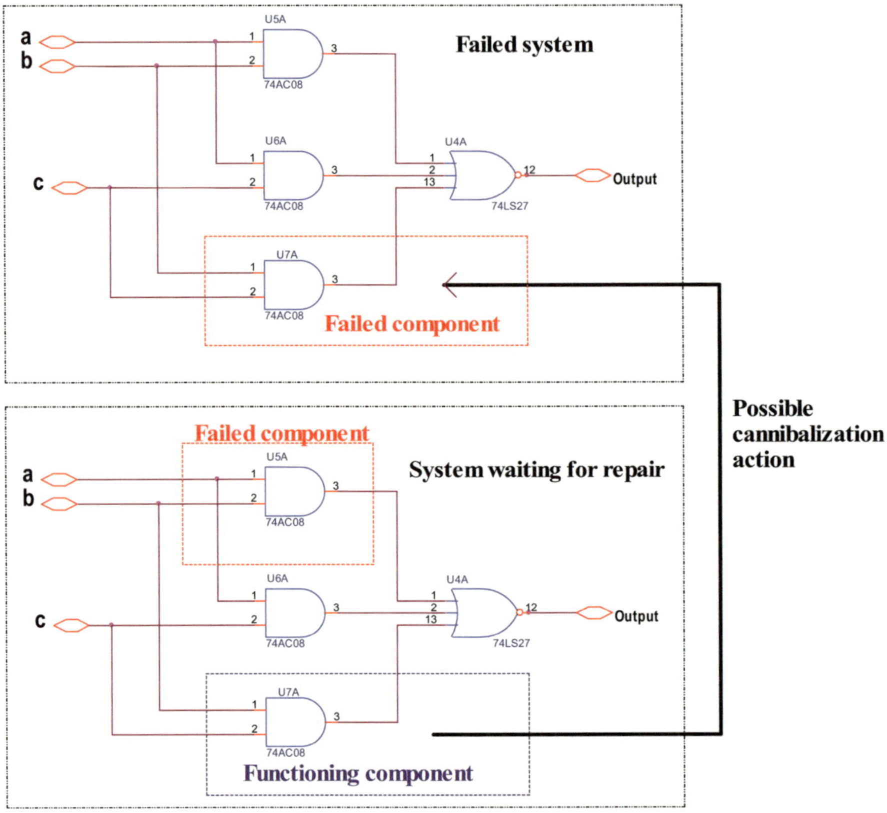


**Figure 1.** The cannibalization concept.

parts become increasingly difficult to acquire and there is a slump in the usage of the said equipment. Therefore, the inclination is not to acquire numerous spare parts as the said equipment will be phased out soon.

One conceivable way of maintaining systems whose spare parts have been depleted or are not available is through cannibalization. Readiness requirements and short maintenance turn-arounds in high-technology environments can be achieved through cannibalization. In general, 'cannibalization' ascribes to the process of removing a failed component from a system and replacing the said component by an operating component (of the same type) extracted from another part of the system. The concept of cannibalization is illustrated in **Figure 1**. The practice of cannibalization has a long history, especially in the military [3].

## 2. Strategies to conduct informed cannibalizations

Cannibalizations, when carried out imprudently, can have unintended negative effects. Cannibalization actions often bear negative connotations because of the following [4, 5]: (1) cannibalizations indicate that there are problems with the spare part supply chain, (2) there is a potential of mechanically damaging systems during cannibalizations and (3) cannibalizations increase the workload of maintenance personnel and if practiced frequently will dampen their morale. Nonetheless, the practice of cannibalization is unlikely to be eliminated completely. Therefore, one has to consider the idea of (1) designated cannibalizations and (2) methodology-informed cannibalizations. These will then cushion against the negative effects of cannibalizations.

The idea of designated cannibalizations means the designation of components in the requirements database as cannibalizable (i.e. easy to cannibalize) or non-cannibalizable (i.e. difficult to cannibalize). It can be noted that the cannibalizable or non-cannibalizable of components is mainly dependent on their type. For example, the cannibalization of a fuse is a trivial task which maintenance personnel can conduct almost invariably rather than wait for long periods of time for the spare part to be delivered. On the other hand, the cannibalization of an aircraft wing spar, for example, is very costly (i.e. in terms of energy, time and money), very dangerous and unfathomed in the world of aircraft maintenance. Cannibalization of components designated as cannibalizable will provide serviceable components when the spare stock is depleted. Cannibalizations of components designated as non-cannibalizable will not be permitted as it may be too costly in labour or time or money, or too risky in that the component (system) may incur mechanical damage during removal.

The methodology-informed cannibalization is an alternative route to estimate optimum rates of cannibalization required to meet the set readiness and operational demand goals, given the system: (1) mean uptime (MUT) (also known as MTTF), (2) MTTR, (3) mean supply response time (MSRT), (4) mean maintenance and supply time (MMST) and (5) mean downtime (MDT). This methodology addresses designated cannibalizations only deemed necessary in the industry. The theoretical model is presented in Section 2.1. This model is an extension of the model developed in Ref. [6]. In the model developed in Ref. [6], it is assumed that the sum of the supply time and maintenance time uniquely classifies the total downtime. Nonetheless, in real-life situations, three categories of downtime, that are exclusive, exist. These are (1) maintenance time, (2) supply time and (3) overlapping MMST. Hence, we extend the model of Ref. [6] to accommodate these different categories of downtime because cannibalizations affect MTTR and MSRT, as well as MMST. In Section 2.2, we provide cannibalizations policy implications based on the results of Section 2.1.

### 2.1. Theoretical model to show the effects of cannibalization on mission time availability of systems

We begin with the following definition of system mission time average availability. The system mission time average availability ($MTAA_{system}$) denotes the mean proportion of mission time the system is functioning. It is assumed that every time a component (system) fails, it is restored to an 'as good as new' condition through repair and the said system mission time average availability is

$$MTAA_{\text{system}} = \frac{MUT}{MUT + MDT} \tag{1}$$

where *MUT* is the mean uptime (also referred to as *MTTF*) and *MDT* is the mean downtime. The *MUT* denotes the mean functioning of the system. The *MDT* is decomposed into three mutually exclusive activities: (1) the *MTTR*;, (2) the *MSRT* and (3) the *MMST*. If we substitute these three variables into Eq. (1), we get

$$MTAA_{\text{system}} = \frac{MUT}{MUT + MTTR + MSRT + MMST} \tag{2}$$

To simplify the calculations, one assumes a zero time for the decision to cannibalize and similarly a zero time for accomplishing the cannibalizations. Therefore, *MSRT* is calculated as follows [6]:

$$MSRT = (1 - GE)(1 - c)\mu \tag{3}$$

where *GE* denotes gross effectiveness—that is, the ratio of the parts required which can be obtained in the supply chain, $c$ denotes the ratio of the parts requests which cannot be obtained in the supply chain and have to be cannibalized and $\mu$ is the mean customer wait time (*CWT*) for spare parts.

The U.S. Navy describes the cannibalization action to be the number of cannibalizations performed per 100 flight hours. This is referred to as the cannibalization rate and is denoted $CANN_{AF}$. In this research, it will suffice to use this definition. The ratio of all the parts requests which are cannibalized, $c$, is determined as follows [6]:

$$c = \frac{CANN_{AF}}{100(1 - GE)\theta} \tag{4}$$

where $\theta$ is the component mean failure rate.

If we substitute Eq. (4) into Eq. (3), we get

$$MSRT = \frac{-\left(\mu\left(CANN_{AF} - (100\theta) + (100GE\theta)\right)\right)}{100\theta} \tag{5}$$

It can be noted that Eq. (5) shows a negative linear relationship between *MSRT* and $CANN_{AF}$. If we hold $\theta$, $\mu$ and *GE* constant, it can be deduced that when the cannibalization rate is higher the MSRT becomes lower.

If we substitute Eq. (5) into Eq. (2) and solve for $CANN_{AF}$, we get

$$CANN_{AF} = \frac{\left(\theta(100GE - 100)\left(\frac{MMST + MTTR + MUT - MUT}{MTAA_{\text{system}}} - \mu(GE - 1)\right)\right)}{\mu(GE - 1)} \tag{6}$$

We impose the following mathematical constraints when calculating with Eq. (6):

$$CANN_{AF} = 0 \ if \ (1 - GE)\mu \leq \frac{MUT}{MTAA_{\text{SYSTEM}} - MUT - MTTR - MMST} \tag{7}$$

$$MTAA_{\text{system}} \leq \frac{MUT}{MUT + MTTR + MMST} \tag{8}$$

It is impossible to achieve an $MTAA_{\text{system}}$ value higher than this. It can only mean that the values set for the parameters in the model are not showing rational thought.

Using Eq. (6), we plot $CANN_{AF}$ as a function of $\mu$ for different values of $MTAA_{\text{system}}$ (with all the other parameters fixed at values given and shown in **Figure 2**. It can be deduced from **Figure 2** that $\mu$ approaches infinity as $CANN_{AF}$ reaches a maximum of 12.31 for an $MTAA_{\text{system}}$ of 0.65; 12.24 for an $MTAA_{\text{system}}$ of 0.60 and 12.17 for an $MTAA_{\text{system}}$ of 0.55.

**Figure 3** shows that a policy to limit cannibalization activities is required. In the example of **Figure 3** (with all the other parameters fixed), it can be seen that a cannibalization rate above 14 does not add any value as the $MTAA_{\text{system}}$ is now 1.

### 2.2. Policy implications of the theoretical model

In this section, we state a number of conclusions based on our simulations using Eqs. (1)–(6). These conclusions have implications on cannibalization policies. The conclusions are as follows:

**a.** The gradient $\frac{\partial CANN_{AF}}{\partial \mu}$ of the graph in **Figure 2** is positive. This shows that the longer $\mu$ is the higher $CANN_{AF}$ is (for different values of $MTAA_{\text{system}}$ all other parameters being held constant). Thus, cannibalization activities may be reduced by decreasing $\mu$;

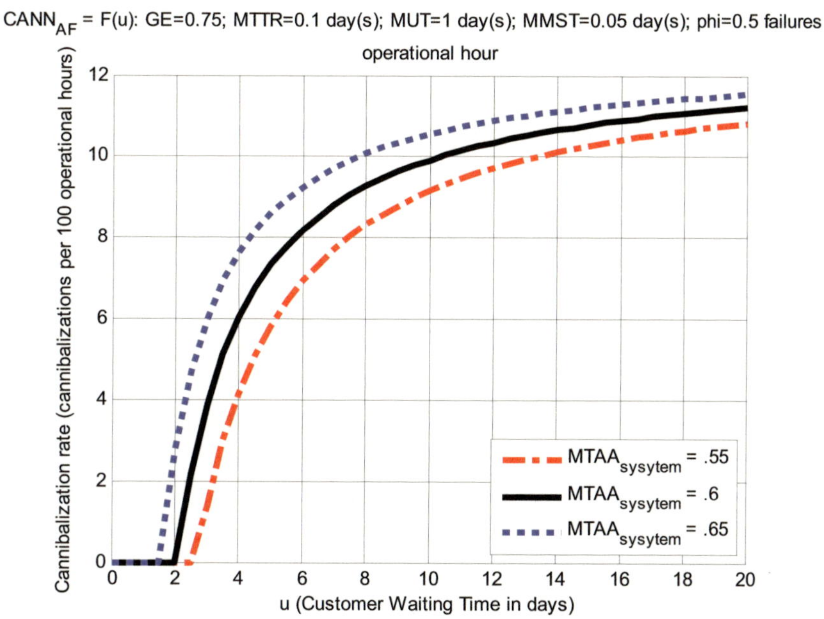


**Figure 2.** $CANN_{AF}$ versus $\mu$ for different $MTAA_{\text{system}}$ values.

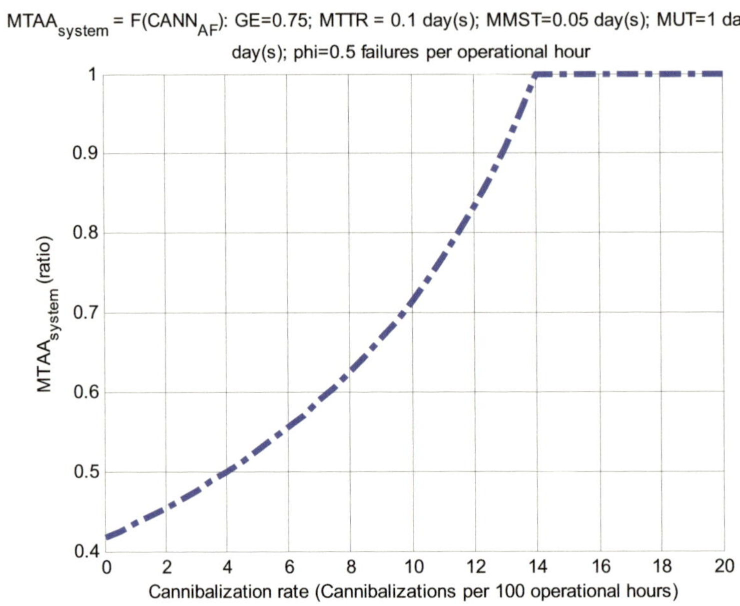


**Figure 3.** $MTAA_{\text{system}}$ versus $CANN_{AF}$.

**b.** The function $MTTR = F(CANN_{AF})$ is positive. Hence, more cannibalization activities imply longer maintenance time;

**c.** The gradient $\frac{\partial CANN_{AF}}{\partial MUT}$ is negative. Thus, it shows that with higher system reliability or with longer mean uptime, cannibalization rates become lower (for different values of $MTAA_{\text{system}}$ with all other parameters being held constant). Thus, cannibalization activities can be reduced when systems are designed taking cognisance of probabilistic design for reliability;

**d.** It can be deduced that the gradient $\frac{\partial CANN_{AF}}{\partial MTTR}$ is positive. It implies that with longer repair time ($MTTR$), the cannibalization rate becomes high (for different values of $MTAA_{\text{system}}$ with all other parameters being held constant). Therefore, cannibalization activities may be reduced with a more efficient maintenance operation system (i.e. better trained and qualified maintenance personnel);

**e.** The gradient $\frac{\partial CANN_{AF}}{\partial GE}$ is negative. It implies that the higher the $GE$, the lower the cannibalization rate (for different values of $MTAA_{\text{system}}$ all other parameters being held constant). Hence, increasing the availability of spare parts in the supply chain reduces cannibalization activities;

**f.** It can be seen from the results that cannibalization activities serve a useful purpose in the maintenance and operation of high performance and complex systems. Cannibalization activities are necessary, viable and cost-effective, only if the optimum cannibalization rate is sought for specific operating parameters; and

g. Lastly, when the actual data on cannibalization rates and other parameters (i.e. the ones related to $MTAA_{\text{system}}$) is available, the models presented in Sections 2 and 3 can be empirically tested.

## 3. Cannibalization revisited: theoretical model and example

This section considers a situation where repair facilities or spare components are not immediately available so that the probability of survival of a system can only be enhanced by extracting needed replacement components from another part of the system. We develop a model of cannibalization for the probability of survival (at time $t$) of a system with $k$ -lines in parallel of $n$ series-connected components when short interruptions to the system are allowed and when short interruptions to the system are not allowed. It is assumed for practical reasons that the lines are identical. Let all components be also identical, with exponentially distributed lifetimes with parameter $\lambda$. We can generalize the approach to the case of non-identical components and lines but the resulting expressions will be extremely cumbersome. We start with two lines as follows:

i. Assume that when only one line is left, the time for replacing the failed component of this remaining line (if one has a spare, e.g. from the failed line) is not allowed. Then, there is no possibility of cannibalization in this system and the survival function can be easily obtained.

ii. The time for replacing the failed component is allowed. Then, when one line fails, all $n-1$ non-failed components of the failed line can be used as spares for the operable line. The corresponding formulas (survival function) are then derived and this is cannibalization.

Then, we consider the case of three lines:

i. No time is allowed for replacing the failed component. However, cannibalization is still performed here. Indeed, when one line fails, we can use $n-1$ spares for the system of two lines (i.e. cannibalization) and when they will be exhausted, then no cannibalization, as in the case with two lines. The formula for probability of survival (at time $t$) is then written for the case with cannibalization and without and compared.

ii. Time for replacing the failed component is allowed. Then, the two-stage cannibalization goes. When the first line fails, $n-1$ non-failed components can be used to maintain the two lines. When this is exhausted and one of the two lines fails, the same process as in the previous case is followed and then the corresponding relationships are obtained.

We generalize the approach to the case when we have more than three lines and obtain the corresponding recurrent equations for survival probabilities in this case that can be solved numerically.

It can be noted that short interruptions to the system give us the possibility to use some components of a system as spares.

### 3.1. Notation

$X$: lifetime of a system.

$S_{kn}^{-}(t)$: probability of survival (at time $t$) of a system with $k$ -lines of $n$ -series-connected components with cannibalization (when short interruptions of the system are not allowed).

$S_{kn}^{+}(t)$: probability of survival (at time $t$) of a system with $k$ -lines of $n$ -series-connected components with cannibalization (when short interruptions of the system are allowed).

$S_{kn}^{nc}(t)$: probability of survival (at time $t$) of a system with $k$ -lines of $n$ -series-connected components when no cannibalization at all is allowed.

$q_{kn}^{-+}(t) = \frac{1 - S_{kn}^{-}(t)}{1 - S_{kn}^{+}(t)}$: improvement factor of unreliability (at time $t$) for $k$ -lines of $n$ -series-connected components with cannibalization due to allowing short interruptions.

$q_{kn}^{nc}(t) = \frac{1 - S_{kn}^{nc}(t)}{1 - S_{kn}^{-(+)}(t)}$: improvement factor due to cannibalization of unreliability (at time $t$) for $k$ -lines of $n$ -series-connected components when initially no cannibalization at all is allowed.

$$k = 1,\ 2,\ \ldots.$$
$$n = 1,\ 2,\ \ldots$$

Note: The improvement factor of unreliability shows how much the unreliability has decreased due to cannibalization being allowed.

### 3.2. One-line system

This section is concerned with the probability of survival (at time $t$) evaluation of standard series network occurring in engineering systems. The series network is the basic building block of the work in this section. In this incident, $n$ number of components form a series network, as illustrated in **Figure 4**. The system fails if any one of the components fails. All components constituting the system must not fail in order to ensure a successful system operation.

The four wheels of a car illustrate a typical example of a series system. The car cannot be driven for practical purposes with any one of the tyres punctured. It therefore follows that these four car tyres form a series system. When one assumes independent and identical components (each component $i(i = 1,2,\ldots,n)$ with a lifetime that is exponentially distributed with failure rate $\lambda$), it then follows that the probability of survival for the series system as shown in **Figure 4** is

$$S_{1n}^{nc}(t) = e^{-n\lambda t} \tag{9}$$

### 3.3. Two-line system

Now consider two identical lines of series-connected components as shown in **Figure 5**. Now, we compute the probability of survival for two cases [7, 8]: (Section 3.3.1) when no short interruptions to the system are allowed (i.e. no possibility of cannibalization: this is indeed

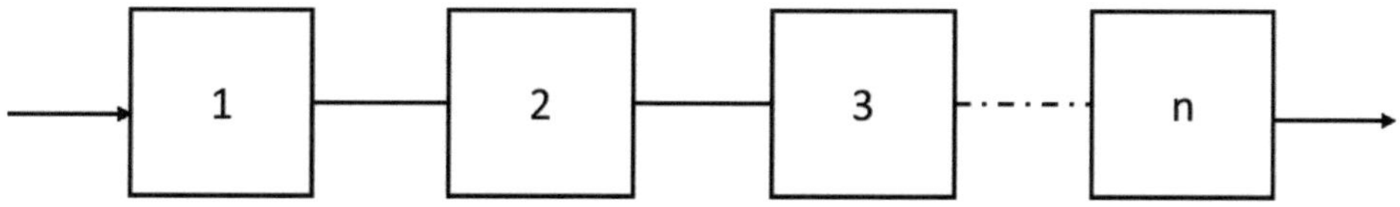


**Figure 4.** One line of series-connected components.

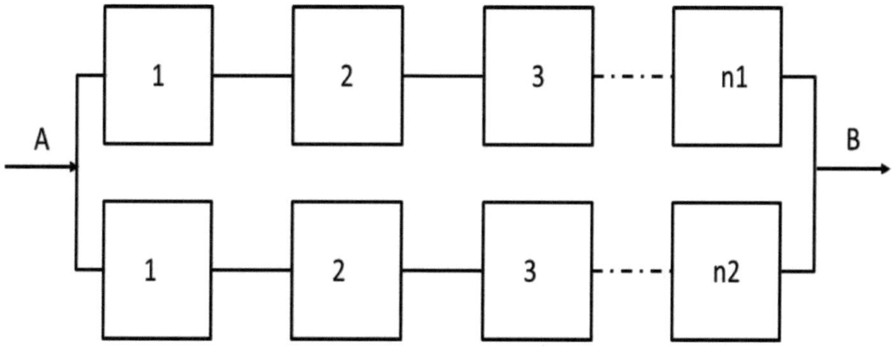


**Figure 5.** Two lines of series-connected components.

the only case without cannibalization, but if we have three or more lines and no interruptions, we already have cannibalization) and (Section 3.3.2) when short interruptions to the system are allowed (i.e. when cannibalization can be executed).

### 3.3.1. No short interruptions to the system are allowed (i.e. no possibility of cannibalization)

The formula for $Pr(X \geq t)$ (i.e. the probability of survival) is written obviously as follows:

$$S_{2n}^{-}(t) = S_{2n}^{nc}(t) = Pr(X \geq t) = 1 - (1 - e^{-n\lambda t})^2 \tag{10}$$

### 3.3.2. Short interruptions to the system are allowed (i.e. cannibalization is allowed)

When one line fails, the time for replacing the failed component of the remaining line is allowed and all $n - 1$ non-failed components of the failed line can be used as spares for the operable line. The cannibalization formula for $Pr\ (X \geq t)$ (i.e. the probability of survival) is written as follows:

$$S_{2n}^{+}(t) = Pr(X \geq t) = e^{-2n\lambda t} + \int_0^t \left( 2n\lambda e^{-2n\lambda x} \sum_{i=0}^{n-1} e^{-n\lambda(t-x)} \frac{\left(n\lambda(t-x)\right)^i}{i!} \right) dx \tag{11}$$

where $e^{-2n\lambda t}$ in the first term in Eq. (11) means that both lines have survived (i.e. by the law of total probability); the integral corresponds to the probability that one line failed and then the remaining line has survived with $n - 1$ spares and $2n\lambda e^{-2n\lambda x}\ dx$ means the density of the first failure of $2n$ components. Then with one line left, there will be no further failures.

One can compare probabilities with and without cannibalization. More appropriately, we compare probabilities of failures. Therefore, we compute the improvement factor of unreliability for the two-line system as $q_{kn}^{nc}(t) = \frac{1 - S_{kn}^{nc}(t)}{1 - S_{kn}^{+}(t)}$ as shown in **Figure 6**.

### 3.4. Three-line system

Now consider three identical lines of series-connected components in a similar manner to the two-line system. Now, we compute $Pr(X \geq t)$ for three cases [7, 8] (Section 3.4.1) when no cannibalization is allowed at all (just three lines of $n$ series parallel-connected components) (Section 3.4.2), when no short interruptions to the system are allowed (cannibalization is made possible here as we are using the operable components of the failed line as spares, as reflected in Eq. (13)) and (Section 3.4.3) when short interruptions to the system are allowed (i.e. when cannibalization is allowed).

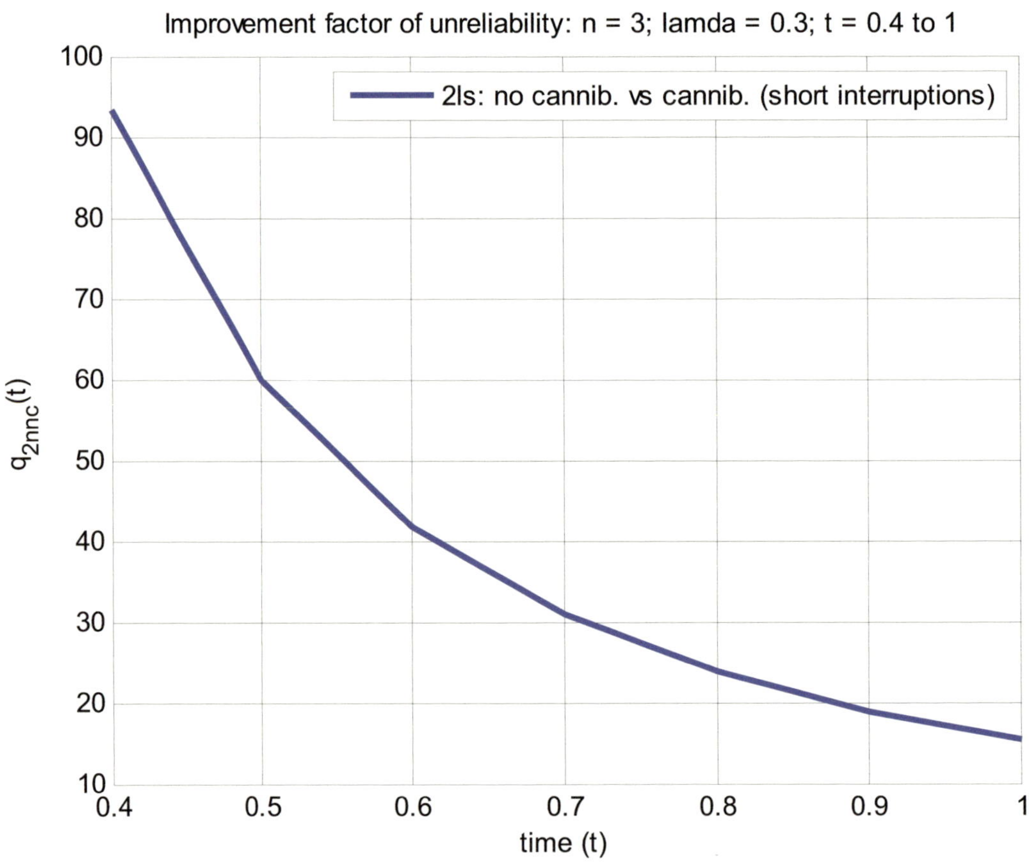


**Figure 6.** Improvement factor of unreliability for a two-line system (comparison of a system with no cannibalization and that with cannibalization when short interruptions to the system are allowed).

*3.4.1. No cannibalization is allowed at all (just three lines of n series parallel-connected components)*

The formula for $Pr(X \geq t)$ (i.e. the probability of survival) is written, obviously, as follows:

$$S_{3n}^{nc}(t) = Pr(X \geq t) = 1 - (1 - e^{-n\lambda t})^3 \tag{12}$$

*3.4.2. No short interruptions to the system are allowed (i.e. cannibalization is made possible here by operable components of the failed line which are used as spares)*

Cannibalization can still be done here. Indeed, when one line fails, we can use $n - 1$ spares for the system of two lines. When the $n - 1$ non-failed components are exhausted, then no cannibalization can be done. The formula is written as follows:

$$S_{3n}^{-}(t) = Pr(X \geq t)$$
$$= e^{-3n\lambda t} + \int_0^t \left(3n\lambda e^{-3n\lambda x}\left(\sum_{i=0}^{n-1} e^{-2n\lambda(t-x)}\frac{\left(2n\lambda(t-x)\right)^i}{i!}\right.\right. \tag{13}$$
$$\left.\left. + \int_0^{t-x}\frac{(2n\lambda)^n}{(n-1)!}y^{n-1}e^{-2n\lambda y}e^{-n\lambda(t-x-y)}dy\right)\right)dx$$

where $e^{-3n\lambda t}$ in the first term in Eq. (13) means that all three lines have survived (i.e. by the law of total probability); the integral corresponds to the probability that one line failed and then the remaining two lines survived with $n - 1$ spares, and $3n\lambda e^{-3n\lambda x}\, dx$ means the density of the first failure of $3n$ components. Then with two lines with $n - 1$ spares and the $n^{\text{th}}$ failure with intensity $2\lambda$ will 'ruin' it and one line will be left and no further failures.

*3.4.3. Short interruptions to the system are allowed (i.e. cannibalization is allowed)*

The two-stage cannibalization goes. When the first line fails, the time for replacing the failed component of the two remaining lines is allowed. $n - 1$ non-failed components of the failed line can be used to maintain the two remaining lines. When these components are exhausted and one of the two lines fails, the $n - 1$ non-failed components of the failed line can be used as spares for the remaining operable line. The corresponding cannibalization formula for $Pr(X \geq 1)$ is written as follows:

$$S_{3n}^{+}(t) = Pr(X \geq t)$$
$$= e^{-3n\lambda t} + \int_0^t \left(3n\lambda e^{-3n\lambda x}\left(\sum_{i=0}^{n-1} e^{-2n\lambda(t-x)}\frac{\left(2n\lambda(t-x)\right)^i}{i!}\right.\right. \tag{14}$$
$$\left.\left. + \int_0^{t-x}\frac{(2n\lambda)^n}{(n-1)!}y^{n-1}e^{-2n\lambda y}\sum_{i=0}^{n-1} e^{-n\lambda(t-x-y)}\frac{\left(n\lambda(t-x-y)\right)^i}{i!}dy\right)\right)dx$$

where $e^{-3n\lambda t}$ in the first term in Eq. (14) means that all three lines have survived; the integral corresponds to the probability that one line failed and then the remaining two lines survived with $n - 1$ spares; $3n\lambda e^{-3n\lambda x}dx$ means the density of the first failure of $3n$ components, and $\frac{(2n\lambda)^n}{(n-1)!}y^{n-1}e^{-2n\lambda y}$ is the density of the $n^{\text{th}}$ event from the Poisson process with rate $2n$. Then with two lines with $n - 1$ spares and the $n^{\text{th}}$ failure with intensity $2\lambda$ will 'ruin' it and one line will be left and no further failures.

Here, we can compare probabilities with no cannibalization at all (just three lines of $n$ -series parallel-connected components) and that with cannibalization (with and without short interruptions). More suitably, we compare probabilities of failures. Hence, we compute the improvement factor of unreliability for the three-line system (i.e. that for no cannibalization and cannibalization with and without short interruptions) as $q_{3n}^{nc}(t) = \frac{1 - S_{3n}^{nc}(t)}{1 - S_{3n}^{(+)}(t)}$ and $q_{3n}^{-+}(t) = \frac{1 - S_{3n}^{-}(t)}{1 - S_{3n}^{+}(t)}$ as depicted in **Figures 7** and **8**, respectively.

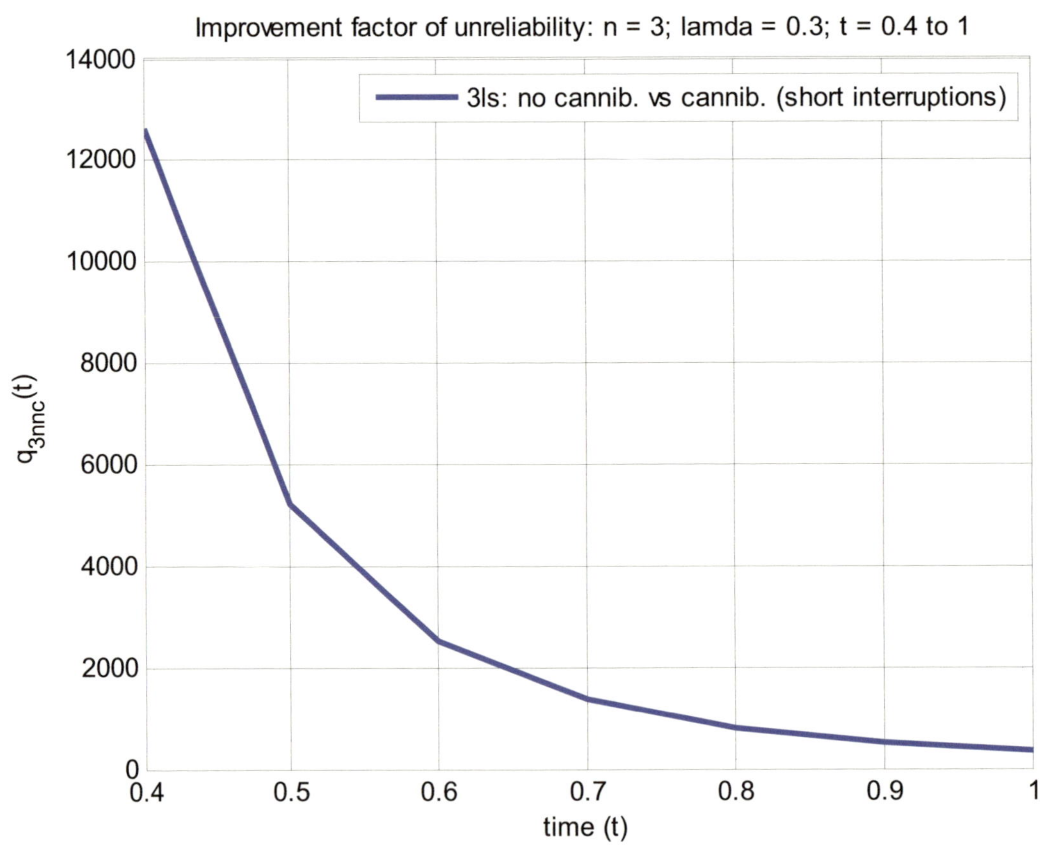


**Figure 7.** Improvement factor of unreliability for a three-line system (comparison of a system with no cannibalization and that with cannibalization when short interruptions to the system are allowed).

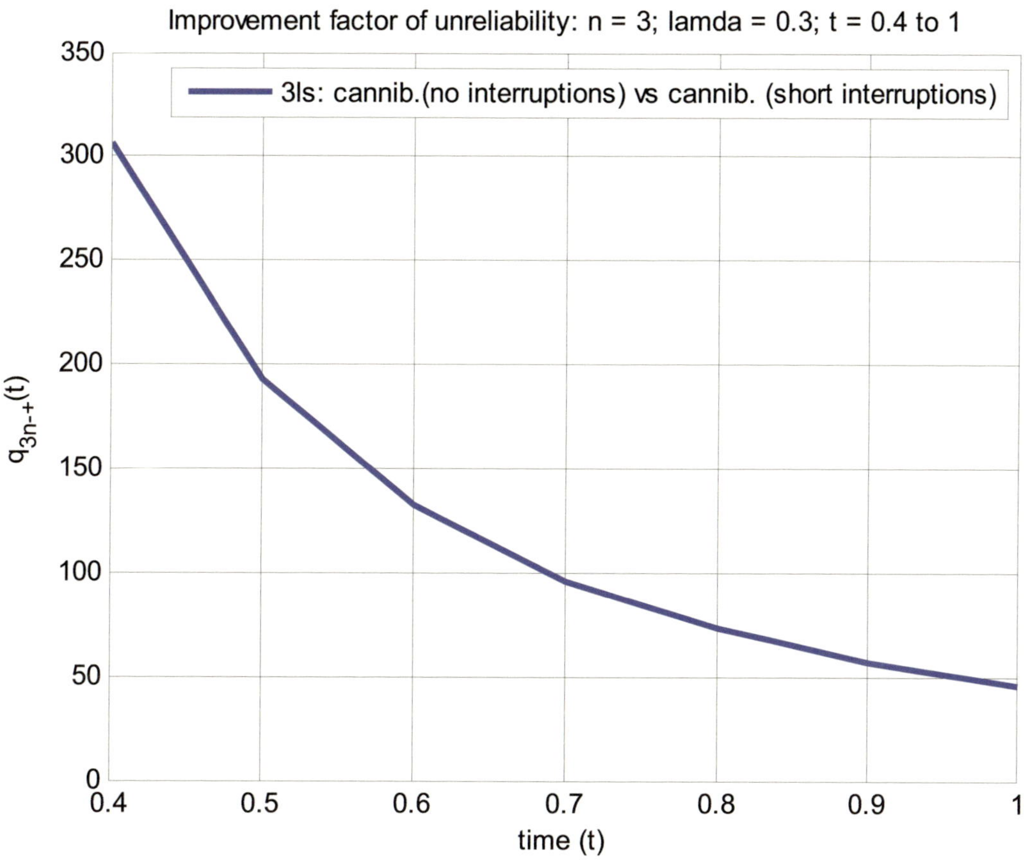


**Figure 8.** Improvement factor of unreliability for a three-line system (comparison of a system with cannibalization when no short interruptions to the system are allowed and that with cannibalization when short interruptions to the system are allowed).

### 3.5. *k*-Line system

Now consider $k$ -identical lines of series-connected components in a similar manner to the two-line system. Now, we compute $Pr(X \geq t)$ for three cases [7, 8]: (Section 3.5.1) when no cannibalization is allowed at all (just $k$ -lines of $n$ -series parallel-connected components) (Section 3.5.2), when no short interruptions to the system are allowed (cannibalization is made possible here as we are using the operable components of the failed line as spares, as reflected in Eq. (16)) and (Section 3.5.3) when short interruptions to the system are allowed (i.e. when cannibalization is allowed).

#### *3.5.1. No cannibalization is allowed at all (just k-lines of n-series parallel-connected components)*

The formula for $Pr(X \geq t)$ (i.e. the probability of survival) is written as follows:

$$S_{kn}^{nc}(t) = Pr(X \geq t) = 1 - (1 - e^{-n\lambda t})^k \tag{15}$$

*3.5.2. No short interruptions to the system are allowed (i.e. cannibalization is made possible here by operable components of the failed lines which are used as spares)*

Cannibalization can still be done here. Indeed, when one line fails, we can use $n - 1$ spares for the system of $k - 1$ lines. When the $n - 1$ non-failed components are exhausted, then no cannibalization can be done. The formula is written as follows:

$$S_{kn}^{-}(t) = Pr(X \geq t)$$

$$= e^{-kn\lambda t} + \int_0^t \left\{ kn\lambda e^{-kn\lambda x} \left( \sum_{i=0}^{n-1} e^{-(k-1)n\lambda(t-x)} \frac{\left((k-1)n\lambda(t-x)\right)^i}{i!} \right.\right. \tag{16}$$

$$\left.\left. + \int_0^{t-x} \frac{\left((k-1)n\lambda\right)^n}{(n-1)!} y^{n-1} e^{-(k-1)n\lambda y} S_{(k-1)n}^{-}(t-x-y)dy \right) \right\} dx$$

where $e^{-kn\lambda t}$ in the first term in Eq. (16) means that all $k$ -lines have survived (i.e. by the law of total probability); the integral corresponds to the probability that one line failed and then the remaining $k - 1$ lines survived with $n - 1$ spares, and $kn\lambda e^{-kn\lambda x}dx$ means the density of the first failure of $kn$ components. Then with $k - 1$ lines with $n - 1$ spares and the $n^{\text{th}}$ failure with intensity $(k - 1)\,\lambda$ will 'ruin' it and one line will be left and no further failures. $S_{(k-1)n}^{-}(t - x)$ is recurrent from the probability of survival (at time $t - x$) of the system with $k - 1$ lines when no short interruptions to the system are allowed (i.e. cannibalization is made possible here as we are using the operable components of the failed line as spares).

*3.5.3. Short interruptions to the system are allowed (i.e. cannibalization is allowed but it was made possible in b) by operable components of the failed lines which are used as spares)*

The $k - 1$ stage cannibalization goes. When the first line fails, the time for replacing the failed component of the $k - 1$ remaining lines is allowed. $n - 1$ non-failed components of the failed line can be used to maintain the $k - 1$ remaining lines. When these components are exhausted and one of the $k - 1$ lines fails, the $n - 1$ non-failed components of the failed line can be used as spares for the remaining operable lines. The process is repeated until one (line) remains, and the time for replacing the failed component of this remaining line (if one has a spare) is not allowed. The corresponding cannibalization formula for $Pr(X \geq t)$ is written as follows:

$$S_{kn}^{+}(t) = Pr(X \geq t)$$

$$= e^{-kn\lambda t} + \int_0^t \left\{ kn\lambda e^{-kn\lambda x} \left( \sum_{i=0}^{n-1} e^{-(k-1)n\lambda(t-x)} \frac{\left((k-1)n\lambda(t-x)\right)^i}{i!} \right.\right. \tag{17}$$

$$\left.\left. + \int_0^{t-x} \frac{\left((k-1)n\lambda\right)^n}{(n-1)!} y^{n-1} e^{-(k-1)n\lambda y} S_{(k-1)n}^{+}(t-x-y)dy \right) \right\} dx$$

where $e^{-kn\lambda t}$ in the first term in Eq. (17) means that all $k$ -lines have survived; the integral corresponds to the probability that one line failed and then the remaining $k-1$ lines survived with $n-1$ spares; $kn\lambda e^{-kn\lambda x}dx$ means the density of the first failure of $kn$ components, and $S^{+}_{(k-1)n}(t-x)$ is the probability of survival (at time $t-x$ of the system with $k-1$ lines when short interruptions to the system are allowed. Then with $k-1$ lines with $n-1$ spares and the $n$th failure with intensity $(k-1)\lambda$ will 'ruin' it and one line will be left and no further failures. Thus, Eq. (17) is a recurrent relationship.

Again, we can compare probabilities with and without cannibalization. More usefully, we compare probabilities of failures. Hence, we compute the improvement factor of unreliability for the $k$-line system (i.e. that for no cannibalization and cannibalization with and without short interruptions) as $q^{nc}_{kn}(t)=\frac{1-S^{nc}_{kn}(t)}{1-S^{(+)}_{kn}(t)}$ and $q_{kn}^{-+}(t)=\frac{1-S^{-}_{kn}(t)}{1-S^{+}_{kn}(t)}$ as depicted in **Figures 9** and **10**, respectively.

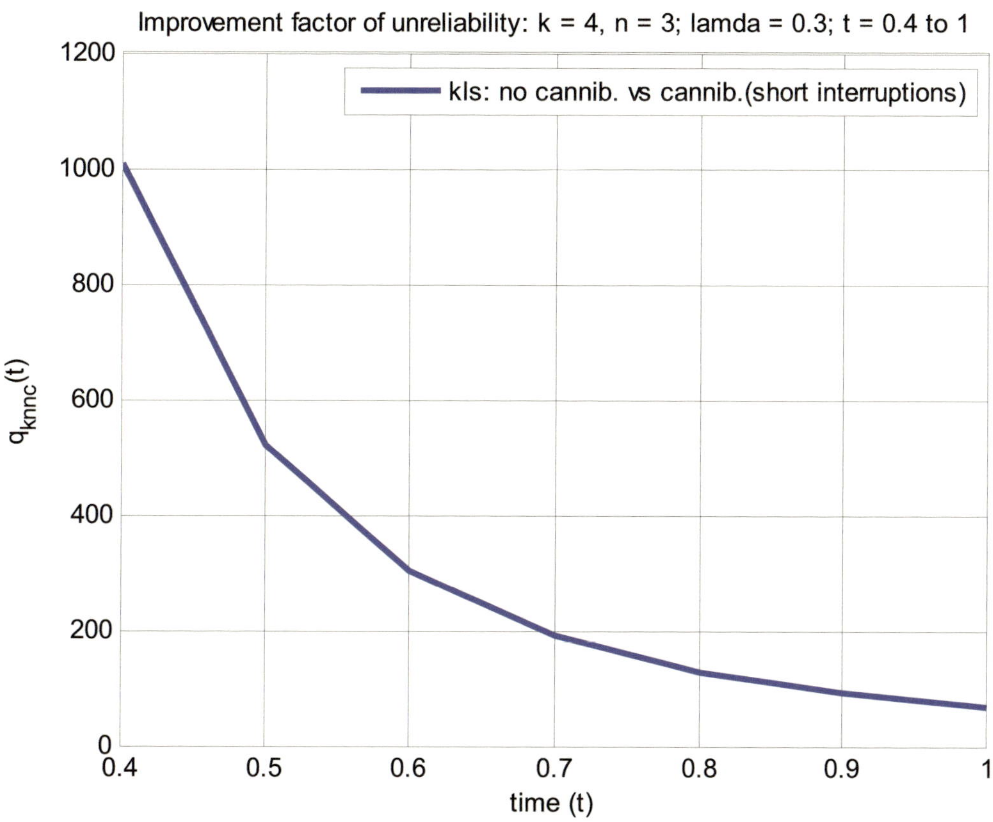


**Figure 9.** Improvement factor of unreliability for a $k$-line system (comparison of a system with no cannibalization and that with cannibalization when short interruptions to the system are allowed).

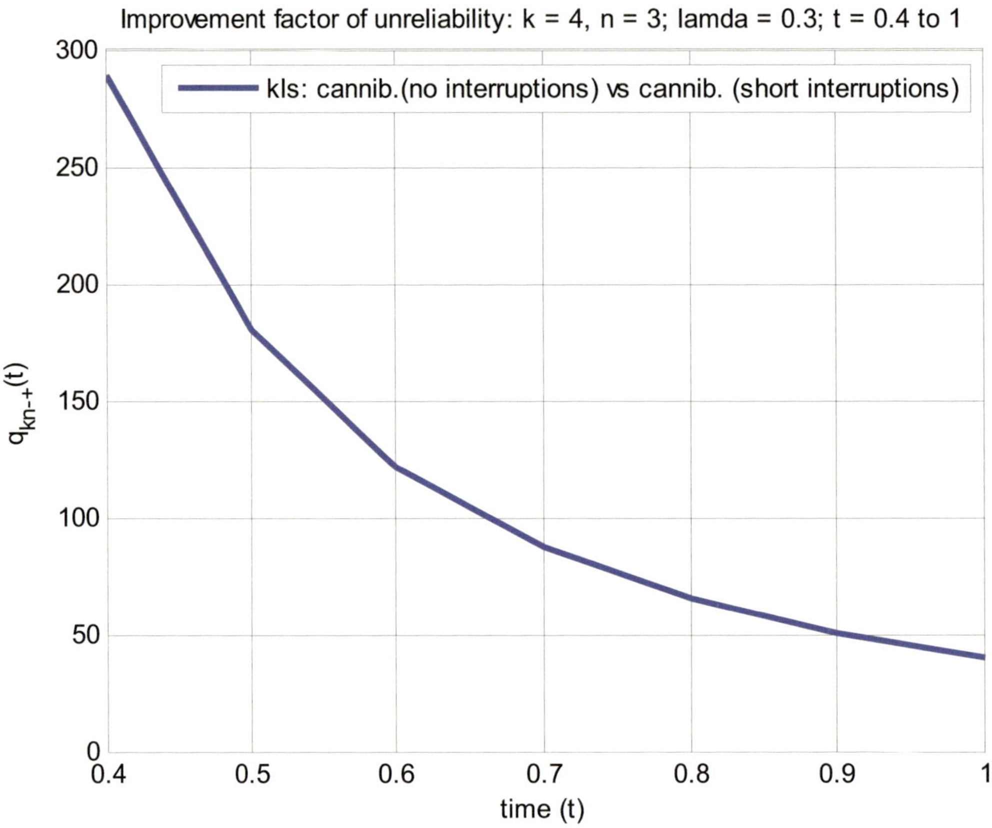


**Figure 10.** Improvement factor of unreliability for a *k*-line system (comparison of a system with cannibalization when no short interruptions to the system are allowed and that with cannibalization when short interruptions to the system are allowed).

### 3.6. Computation results

**Figures 6–10** show the improvement factors of unreliability for the two-line system, three-line system and *k*-line system, respectively. We are looking at reliable systems, where the survival functions should be close to 1. For illustrative purposes, we choose the values of the failure rate and time accordingly. Therefore, it is more effective to compare unreliability of the systems with no cannibalization (i.e. $1 - S_{kn}^{nc}(t)$) and with cannibalization (i.e. with and without short interruptions) (i.e. $1 - S_{kn}^{+}(t)$ and $1 - S_{kn}^{-}(t)$). The improvement factor of unreliability is obtained by dividing the unreliability of the said system without cannibalization (worse quantity) by that of the said system with cannibalization (without short interruptions) (better quantity) and thus is larger than one. This is then compared to that obtained by dividing the unreliability of the said system with cannibalization with short interruptions (worse quantity) by that of the said system with cannibalization (without short interruptions) (better quantity). It can be seen from **Figures 6–10** that the improvement factors of unreliability of the systems in which cannibalization is allowed are better than in those in which it is prohibited.

From simulation results of **Figures 6–10**, it can also be shown that very larger values lead to asymptotically equivalent system performance levels.

## 4. Conclusions

In this chapter, we have explored strategies to mitigate the unfavourable effects that cannibalizations have on the costs of maintaining systems of equipment and also on the morale of maintenance personnel. The methodologies developed in this chapter, at least, can be used to (1) ascertain those cannibalizations that are appropriate, (2) start implementation of goals to reduce cannibalization and (3) enlist actions to be taken in order for cannibalization reduction goals to be met.

We also presented a combined analytic and simulation model of a two-line, three-line and *k*-line system when cannibalization is not allowed and when cannibalization is allowed (with and without short interruptions to the system). It is clear from the analytic and simulation results that cannibalization can substantially increase the reliability of the systems where it is allowed. The improvement factor of unreliability obviously exists in systems where cannibalization is not allowed as compared to those in which cannibalization is allowed. Moreover, the improvement factor is larger when we have two-stage cannibalization (short interruptions) than without them.

## Author details

Bernard Tonderayi Mangara

Address all correspondence to: aragnam@gmail.com

Central University of Technology, Free State (CUT), Bloemfontein, South Africa

## References

[1] Byrkett DL. Units of equipment available using cannibalization for repair-part support. IEEE Transactions on Reliability. 1985;**R34**(1):25–28

[2] Salman S, Cassady CR, Pohl EA, Ormon SW. Evaluating the impact of cannibalization on fleet performance. Quality and Reliability Engineering International. 2007;**23**:445–457

[3] Hirsch WM, Meisner M, Boll C. Cannibalization in multicomponent systems and the theory of reliability. Naval Research Logistics Quarterly. 1968;**15**:331–359

[4] United States General Accounting Office. Report to the Chairman, Subcommittee on National Security, Veterans Affairs, and International Relations, Committee on Government Reform,

House of Representatives [Internet]. November 2001. Available from: https://www.gao.gov/assets/240/233055.pdf [Accessed: March 8, 2017]

[5] de Oliveira N, Ghobbar A A. Cannibalization: How to measure and its effect in the inventory cost. In: Grant I, editor. 27th International Congress of the International Council of the Aeronautical Sciences (ICAS 2010); 19-24 September 2010; Nice, France. Nice, France: International Council of the Aeronautical Sciences; 2010. p. 1-12

[6] Hoover J, Jondrow JM, Trost RS, Ye M. A Model to Study: Cannibalization, FMC, and Customer Waiting Time [Internet]. February 2002. Available from: https://www.cna.org/CNA_files/PDF/D0005957.A2.pdf [Accessed: March 9, 2017]

[7] Finkelstein M. Failure Rate Modelling for Reliability and Risk. UK: Springer-Verlag London; 2008. p. 290. DOI: 10.1007/978-1-84800-986-8

[8] Finkelstein M, Cha JH. Stochastic Modeling for Reliability: Shocks, Burn-in and Heterogeneous Populations. UK: Springer London; 2013. p. 388. DOI: 10.1007/978-1-4471-5028-2

# Chapter 8

# Reliability Evaluation of Drivetrains: Challenges for Off-Highway Machines

Lothar Wöll, Katharina Schick, Georg Jacobs,
Achim Kramer and Stephan Neumann

Additional information is available at the end of the chapter

http://dx.doi.org/10.5772/intechopen.70280

**Abstract**

Downtime of mobile machinery used in fields like construction, earthmoving or mining usually leads to an instant halt of an entire process and can even endanger entire operations. To meet the customer's requirement for high availabilities of their equipment, safeguarding the reliability of overall systems and components is necessary. Since life expectancy of systems and its components strongly depends on the experienced load history, this information needs to be available as accurately as possible to allow reliable lifetime calculation results. Due to the wide range of machines and applications of off-highway machines, determining representative loads is especially challenging. The challenges, in determining both load cycles for the entire system and local component loads, are discussed in this work, along with approaches to face them. Additionally, a method is described, which allows users to quantitatively calculate life expectancy of technical systems in both the concept phase and the later stages of the product life cycle. In the end, two examples are presented in which exemplary challenges are faced.

**Keywords:** reliability, drivetrain, off-highway, load cycles, lifetime

## 1. Introduction

The global sales for off-highway machines are over 100 billion US dollars [1] annually with a wide range of different products, which are often used in logistic processes. Although there are methods available to plan and calculate these processes in the line of supply chain management, over 70% of construction projects cannot be completed within the proposed budget and time [2]. One cause can be the unexpected downtime of construction machinery since in logistic processes the failure of one machine may result in the stop of the entire process chain. These processes include the transport to and from the work site as well as production processes which are performed using mobile machinery. Similar processes can be found in mining operations, both above and below ground and other off-road activities. Even small defects can result in high costs for the operating company and delays due to repair time [3, 4]. Therefore, reliability and availability are important factors for the optimization of projects that rely on off-highway machines.

In order to increase the robustness of processes, the easiest way is to use redundancies on either the system or the component level. However, this option results in high costs and is therefore not preferable. Instead, a detailed knowledge of the systems reliability should be created, which can minimize the total cost of ownership and allow for more robust planning. Methods for the assessment of a system's reliability are often qualitative, such as failure mode and effects analysis (FMEA) and fault tree analysis [5, 6]. These methods do not yield quantitative information about the system's reliability over time, which is necessary for incorporating reliability information in the planning process or work scheduling. For these applications, quantitative reliability evaluation methods are necessary. They can be applied throughout the entire life cycle from the early concept phase to the use of reliability models as a basis for predictive maintenance.

The challenge in determining the quantitative reliability of mobile machinery is in the multitude of influences that determine the loads on the machine, as illustrated in **Figure 1**. The same machine can experience vastly different loads when used at different worksites. The environmental variables, such as temperature, humidity and exposure to sunlight may vary. The ground material and the slope of the path have an influence on the required traction forces. The main influence is the task that is performed by the machine, which can be defined by the loads and typical cycles of movement. The loads on the machine can be transformed into a *stress* on each component. Failures occur when the component stress exceeds the components *strength*, which depends on the components design [7].

This work describes a method to assess the reliability of mobile machinery drivetrains quantitatively, on both a component and a system level. The quantitative approach offers advantages both early in the development process, when different concepts can be compared with each other based on the quantitative reliability, and also in the later stages and in the exploitation phase, when the reliability assessment can be used to enable condition-based and predictive maintenance approaches.

**Figure 1.** Influences on the system reliability of off-highway machines.

The accuracy of the calculated lifetimes depends on the quality of the load assumptions, since the loads determine the damage of the components. The challenges that this crucial part of reliability evaluation creates for off-highway machines, which show a great variety in system concepts and applications, are discussed in this work. Two different systems, to which the method has been applied, will be used as examples for the application of the method and the possible ways to face the challenges in determining representative load cycles and local component loads.

## 2. Lifetime models

The system's reliability depends on the reliability of its machine elements. For many components, the lifetime can be determined through lifetime models, but not all components contribute equally to the overall system reliability. To identify relevant components in an investigated system, a qualitative reliability analysis has to be performed. The simple method *ABC analysis* [7] is well suited for this purpose, see **Table 1**.

Machine elements that are assigned to category A distinguish themselves by having a significant influence on the reliability of the system, to which they contribute. In addition to that, methods for lifetime calculation of these elements are available which provide reliable lifetime predictions. Typical components in common mechanical drivetrains belonging to category A are gears, bearings and shafts. Parts allocated to category B determine the overall life expectancy of the system as well, but their lifetime cannot be calculated as precisely compared to machine elements in category A. B components are, among others, friction clutches and seals. The third category C includes components which neither affect the system lifetime to any extent nor enable a lifetime calculation at all, for example, locking rings. The components in category C are rarely subject to failure, so they usually do not need to be considered.

Failures of components originate from an exceedance of their respective *load capacity* or *strength* by the occurring *loads* or *stresses*. Since both quantities are subject to statistical distribution, the lifetime, which is a result of the proportion of stress and strength, is statistically distributed as well. High volume testing of the investigated system and its components under realistic conditions would yield the most precise information. However, since such tests require an extensive amount of time and resources, this is rarely done during a development process. Therefore, calculation specifications to calculate the load capacity of machine elements have been developed, which are shown in **Table 2** for components in categories A and B.

| | A | B | C |
|---|---|---|---|
| Significance | High | High | Low |
| Lifetime prediction | Reliable | Unreliable | Unreliable |
| Machine elements | Gears, bearings, shafts | Friction clutches, seals, etc. | Locking, rings, etc. |

**Table 1.** Exemplary ABC analysis of mechanical drivetrain components [7, 8].

| | | | | | | | |
|---|---|---|---|---|---|---|---|
| Lifetime | Category A | Gears | DIN 3990 | ISO 6336 | AGMA 2001 | JGMA 6102 | DNV 41.2 |
| | | Shafts | FKM | DIN 743 | | | |
| | | Bearings | DIN ISO 281 basic | | | DIN ISO 281 expanded | |
| | Category B | Friction clutches | Wear | | | | |
| | | Seals | Wear | Temperature degradation | | | |

**Table 2.** Calculation specifications for machine elements.

Cylindrical gears, which are the most common gear type in drivetrains, can be designed and calculated by various available standards and norms [9–14]. These approaches consider three general kinds of failure for gears: pitting, tooth fracture and scuffing. Since scuffing prediction for gears is not advanced enough and furthermore only occurs outside of predefined operating conditions, it is currently not considered in reliability calculations [15, 16]. Flank failure due to pitting can occur on both tooth flanks on each gear tooth. While a tooth fracture ends the lifetime of a gear immediately, pitting does not. In gear-pitting testing, the threshold for pitting for hardened gears is usually 4% of the total flank area [9]. Thus, for each gear, three load capacities per tooth can be calculated and compared with the respective loads: Flanks 1 and 2 (for which the load capacity is usually the same) and the tooth base.

Roller bearings are one of the few machine elements, for which the lifetime calculation has been standardized instead of the load capacity calculation and has been made available in [17]. The basic approach only considers load and a basic load capacity determined by testing, whereas the expanded approach additionally takes lubricant viscosity (temperature adjusted) and particle contamination, as well as revolving speed into account. However, for complex motion behaviour like oscillation, the currently available methods lack accuracy and therefore cannot provide a reliable lifetime prediction [18, 19].

Calculation of load capacity of shafts and axles has been standardized and documented, for example [20, 21]. Shafts in drivetrains are exposed to oscillating stresses due to torque, axial forces and bending moments that depend on the rotation angle. While available methods are designed to prove the strength of shafts, lifetime calculation is possible but not considered to be accurate [22]. The accuracy would improve with finite-element method (FEM) calculations, but the modelling effort and computing time would increase as well.

Friction clutches are allocated in category B since there are many influences that cannot be considered for lifetime calculation yet (e.g. temperature, ageing, topology of disks). However, a rough estimation can be performed, calculating the wear of the clutch disks, proportional to the experienced frictional power and based on static friction coefficients. The clutch is considered as failed, when a predefined state of wear is reached [23].

Since the lifetime calculation of radial shaft seals is currently not possible, these machine elements belong to category B as well. The interactions of the seal with the surrounding elements are very

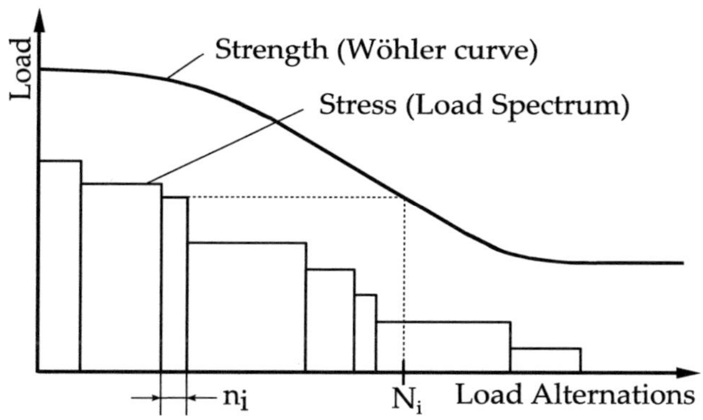


**Figure 2.** Comparison of stress and strength for damage accumulation [8].

complex and therefore hard to quantify. Radial shaft seals can fail not only due to mechanical failure modes (e.g. wear) but chemical failure modes as well (e.g. hardening) [24]. For a rough estimation of seal lifetimes, however, simple models that only consider wear or temperature degradation are available [25].

The previously described models for component strength define a Wöhler curve for each component, which represents the number of load alterations that can be endured at a certain load level. The component stress can be considered in the form of a load spectrum, which consists of the number of load alterations at a certain level the system has to endure, see **Figure 2**.

To assess the component damage caused by the loads in the load spectrum, for each load level the number of load alterations $n_i$ in the load spectrum has to be compared to the number of bearable load alterations $N_i$ from the Wöhler curve which leads to the component damage caused by the experienced loads [26]. To account for temporal distribution of failures, the common procedure is to combine the lifetime with mathematical distributions based on experience, which results in a failure probability function $F(t)$. To describe the failure probability mathematically, the Weibull distribution is commonly used. The shape of the function can be fitted to different distributions by changing the Weibull parameters $b$, $T$ and $t_0$ [7, 8, 25].

To assess the system's reliability, the component's reliabilities have to be combined. This can be done using a suitable system theory. The easiest one is Boole's system theory, which can be applied for non-repairable systems. According to Boole's theory, the system reliability for serial systems without redundancies is simply the product of all component reliabilities [27].

## 3. Method

The previously mentioned processes of component lifetime calculations and system theory are part of a superordinate method, which puts all steps into the context of a system analysis in its entirety [8]. This method is described briefly in the following, and aims to provide a guideline for the calculation of overall reliability of off-highway drive trains, see **Figure 3**.

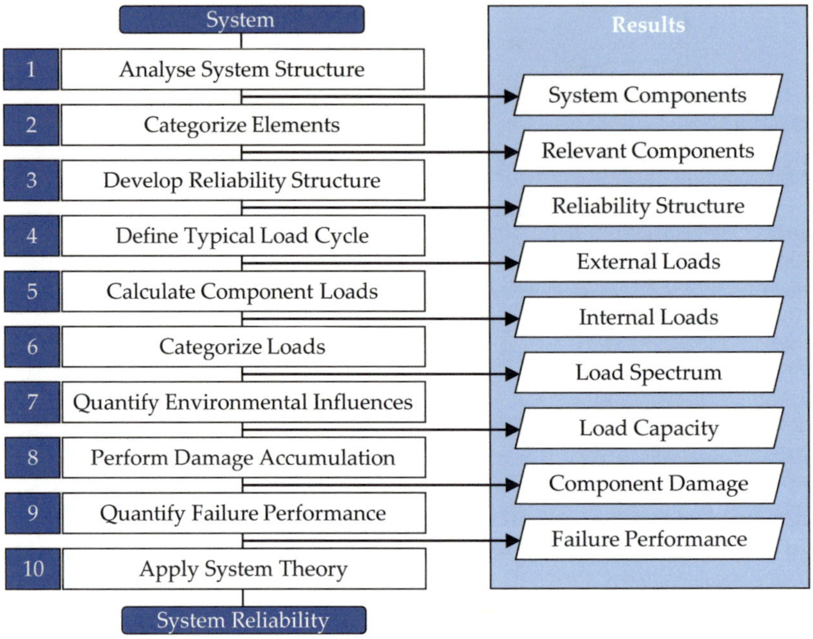


**Figure 3.** Method for quantitative reliability evaluation of drivetrains [8].

When performing a reliability evaluation of an existing system, firstly the system has to be analysed to identify all contained components and to understand how the system works (power flow, moving parts, etc.). The machine elements identified this way need to be categorized (ABC analysis). Since drivetrains usually contain similar components, this only has to be done if components are being used, which are not categorized yet. In preparation of the calculation of the system reliability, the system structure needs to be examined for redundancies. To determine the external loads on a drivetrain, a typical load cycle needs to be defined in order to allow representative calculation results. This topic, as well as the transfer to local component loads including environmental influences and load classification, is discussed in-depth in the next section. After calculating the components' lifetimes by comparing the locally occurring loads with the individuals load capacities, the lifetime and subsequently failure distributions can be obtained. Finally, the failure distributions are combined, taking into account the system reliability structure. The last step is the derivation of the overall system lifetime from the resulting system failure probability. More detailed information can be found in Ref. [8].

## 4. Challenges in load assumptions

In steps four and five of the described method, a representative load cycle and the resulting component loads have to be determined. This is especially challenging during the concept phase, when no measurements can be performed. The particular challenges for off-highway machines will be described in the following.

The term off-highway in drivetrain technology includes a wide range of fields. Typical areas are, for example, construction and mining applications, in which an equally wide range of mobile machines are utilized for various tasks with very different characteristics. Machines in open-pit mining, for example, bucket-wheel excavators or spreaders, are among the largest machines in the world and handle an enormous amount of material and are accordingly subject to immense loads. Other machinery in the general field of earthmoving includes among others excavators, loaders, graders, scrapers and dumpers [28]. Even for these types of vehicles, countless variants exist, for example, excavators, for which crawler undercarriages or wheeled types are common. Besides the technical realization of the drivetrains, other specifics, for example, boom variants (monoblock booms, adjustable booms, embankment booms, etc.), exist [29], which influence the operating behaviour and subsequently the occurring loads on structure and drivetrain. Furthermore, machines, even with same configurations, are utilized for vastly different tasks. Wheel loaders, for example, can be used for material-handling tasks (usually driving continuously in a Y-cycle), transportation of the material of larger distances, grading, material pushing or lifting work. With such a wide bandwidth of operating modes, it becomes clear that dimensioning drivetrains for mobile machines as well as determining life expectancies of such systems remains challenging due to the large variance of occurring loads, even on machines with identical configurations. Uncertain load assumptions are among the main reasons for unplanned downtime [30]. Therefore, representative load cycles as well as precise component loads as a result of the load cycles are necessary for accurate lifetime evaluations of mobile machines.

### 4.1. Load cycles

To build a representative load cycle as the basis for determining the loads on the components, a deployment scenario has to be created, including all possible tasks which the investigated machine can perform. For each task, a representative process pattern has to be defined. The approach supposed to yield most realistic results measures the workflow per task throughout an extended period of time, divided into sections for recurring activities. For the example of a wheel loader, a typical task is material handling in a Y-cycle. In this cycle, the operator fills the bucket by forwarding into a material heap, backs out of the heap, drives forward to an unloading zone, unloads, backs out of the unloading zone and restarts the cycle. Since the driving path differs each time, the path for each Y-cycle should be recorded over a sufficient period of time. Eventually, the sections are superposed and a relevant Y-cycle is created, see **Figure 4**.

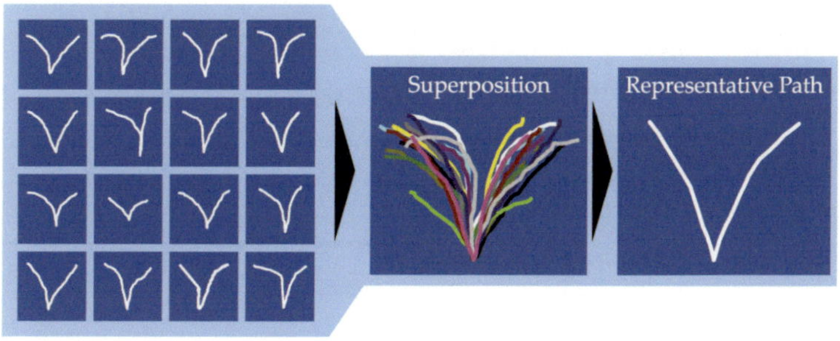


**Figure 4.** Schematic illustration of the determination of a representative driving path.

This procedure has to be performed for each identified task per machine, ideally for not one but multiple job sites, operators, the kind of handled material, ground conditions, etc. If a metrological approach is not possible due to unavailability of machines or if the investigation is performed in the product development process, the representative load cycle can be determined for a similar machine or by simulation without any measurements based on basic knowledge about the future tasks and job site, for example, topology of an open-pit mine.

With either method, a representative workflow is made available for all tasks, including necessary time-dependent load information. With knowledge about the time share of each task in the overall utilization, the representative time functions for each task can subsequently be compiled, as illustrated in **Figure 5**. The compiled load cycle serves as basis for the determination of local component loads in the system, which is discussed in the following section.

### 4.2. Component loads

Since the lifetime of the system depends on the components' lifetimes, defined by the experienced load history, the component loads have to be determined as accurately as possible. The loads can be determined either through measurements, through simulation or through a combination of both methods.

To determine component loads through measurements, either in the field or in a testing environment, extensive measuring equipment has to be applied to the components. For the lifetime calculation, values like forces, torques, rotational speeds, temperatures, etc. have to be known for each component, so that the necessary measuring equipment to capture these many values would be challenging to be incorporated into the system and also very expensive.

As an alternative to measurements, simulations can be performed to determine the component loads. Based on the global load cycle that is investigated, the component loads can be calculated by system models, which can be modelled on different levels of detail and complexity. A basic method is the torque plan, which can be used to determine torques and speeds for gearboxes without considering dynamic loads [31]. For a more detailed simulation, torsional simulations can be performed that consider dynamic effects in the rotational direction, which can be caused by acceleration, gear shift or the system's dynamic behaviour [32]. Through multibody simulations, all degrees of freedom can be considered in the load calculation.

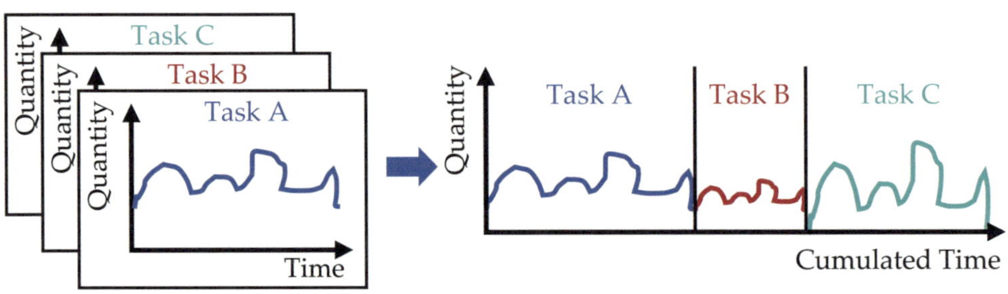


**Figure 5.** Schematic illustration of the compilation into representative load cycle.

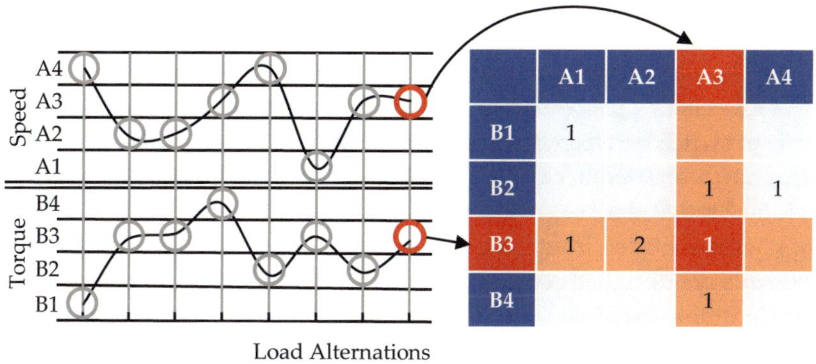


**Figure 6.** Schematic illustration of two parametric time-at-level counting.

The level of complexity that is chosen for the simulation depends on the load cycle that is investigated and on the information that is available about the system. Simulation models that have much detail also require detailed information about the system components. If the system is still in the concept phase, a detailed simulation model may require data that are not yet known. It also has to be considered that simulation models have to be validated to provide accurate and reliable results.

Many systems already have some sensors that provide measurements of some system states. These can be used to improve the quality of simulations by using the available measured data as an input or reference for the validation. This way, the advantages of both methods can be combined.

The measured or calculated loads usually have the form of a load-time function. For the next step, the component models require the data in the form of load per load alternation so that a transformation is necessary. To minimize the calculation effort for the reliability calculation, the load per load alternation can be transformed into a load spectrum using counting methods [33]. For many applications, such as gears and bearings, the combination of speed and torque determines the strain on the component. The time-at-level counting method can be used to define a load spectrum from the continuous data, as illustrated in **Figure 6**. The method defines classes for the values of two parameters, in this case torque and speed. For each load alternation, the combination of torque and speed classes is identified and counted in a matrix. The result of the counting method is the number of load alternations at each load and speed level in the examined time interval.

## 5. Examples of application

To illustrate the discussed challenges for load assumptions for off-highway machinery, two examples have been selected. In the following, the presented method is exemplarily applied to two off-highway drivetrains for different applications. The creation of load cycles and the determination of local component stresses are highlighted.

### 5.1. Power-shift transmission

The following exemplary system is a power-shift transmission designed for mobile machines, such as wheel loaders or dumpers with a power consumption up to 100 kW [8]. The transmission is able to switch between its three forward and three reverse speeds without tractive power interruption with friction clutches. In total, the transmission contains 10 gears, 18 bearings, 7 shafts and 2 radial shaft seals. The initial system analysis reveals gears, shafts and bearings as components allocated in category A and seals and clutches in category B. Since none of the components are designed redundantly, the reliability structure is serial. The representative load cycle in this example is built artificially rather than metrologically. The artificial driving cycle emulates a dumper in a quarry, consisting of five main phases, as illustrated in **Figure 7**. First, after loading material, the dumper drives out of the quarry for which a height profile of the road is being used. In the second section, the loaded dumper is assumed to drive straight ahead. In the third part, the loaded dumper backs up in an unloading zone, unloading and reversing back out. The following fourth segment is similar to the second one, where the empty dumper drives straight ahead, back to the quarry. In the fifth and last phase, the empty dumper drives down to the quarry, following the road profile in reverse. The whole driving cycle lasts 720 s over a distance of approximately 6 km and repeats itself over and over, whereas possible idling times are neglected [8].

The loads on the output of the transmission are calculated by determining the necessary traction force to move the loaded dumper up the quarry and the empty dumper down the quarry. In addition to the weight forces on the slopes, a rolling resistance is assumed, which is 2% of the overall weight force. The empty weight of the dumper is set to 10 t, with a payload capacity of another 10 t. The speed of the vehicle is taken from the maximal acceleration possible from the driving power, subtracting the power to overcome weight forces and rolling forces. The maximal vehicle speed is set to 10 m/s for driving straight ahead and to 7 m/s for uphill/downhill sections [8].

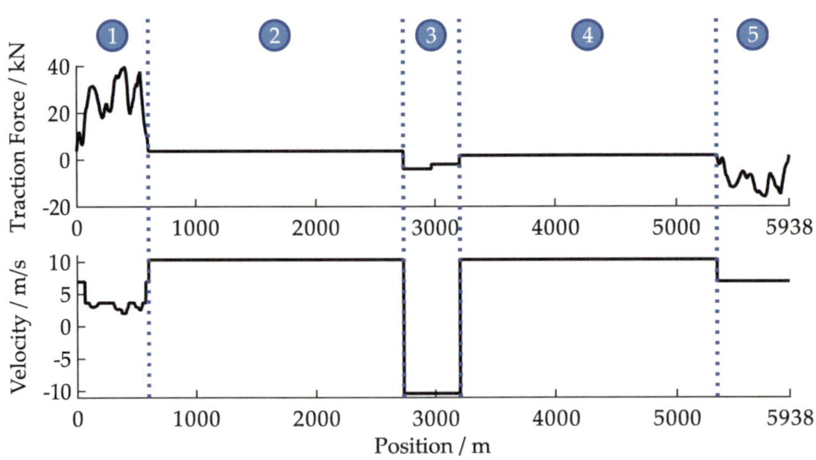


**Figure 7.** Artificial dumper driving cycle [8].

Having torque and speed at the transmission output available, the local component loads and speeds of gears, bearings and shafts can be determined. In this example, a torque plan of the transmission is used, based on the gear ratios of the individual stages, see **Figure 8**.

Loads on the bearings are the result of gear forces and shaft dimensions. For shafts, the critical cross sections and the respective stresses have to be determined according to Ref. [20], which are caused mainly by the gear forces. The friction power in the clutches, which causes the wear considered for its lifetime calculations, is obtained using shifting times, torque and speed differences per shift operation. For the seals, only temperature is considered as an influence on the lifetime. For the environmental influences in this example, a static oil temperature is assumed. The particle contamination factor, which influences the bearing lifetimes according to [17], is assumed to increase over time and drops every 500 operating hours when the oil filter is changed. The load information for the investigated components is categorized using time at level counting which results in load spectra serving as an input for the lifetime calculations. After combination with statistical failure distributions and applying Boole's theory as system theory, it yields a failure distribution plot, illustrated in **Figure 9**. The probability of failure for combined groups of components as well as for the entire transmission is plotted on the ordinate.

The reliability evaluation reveals that the shafts in this example have been designed to be fail-safe and therefore do not appear in the plot. Furthermore, it is highlighted that the gears seem to be designed considerably more durable compared to the other components, according to the load capacity calculation in Ref. [9]. The failure probabilities of the other components are in the same magnitude and define the life expectancy of the transmission. The overall B10 lifetime of the transmission can be derived from the plot with 2961 h which corresponds roughly to a distance of 90,000 km. If the lifetime of the transmission was to be increased, seals, bearings and friction clutches were the components that should be designed more durably [8].

The presented calculation, which yields the transmission lifetime, is repeated for two additional driving cycles, which are basically equal to the initial driving cycle, except the sections in which the dumper drives straight ahead are shortened, so that the total cycle only takes half

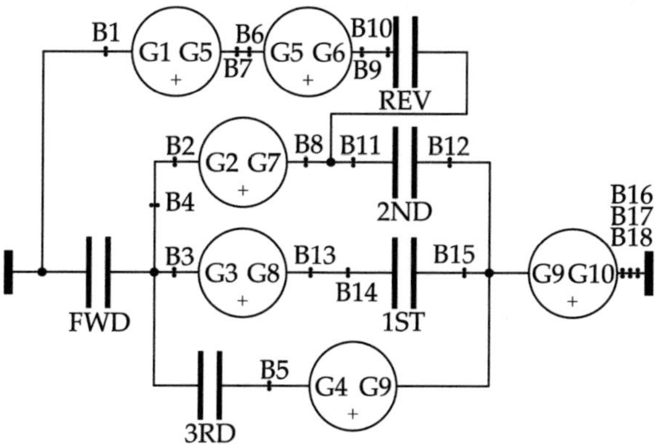


**Figure 8.** Torque plan of investigated power-shift transmission [8].

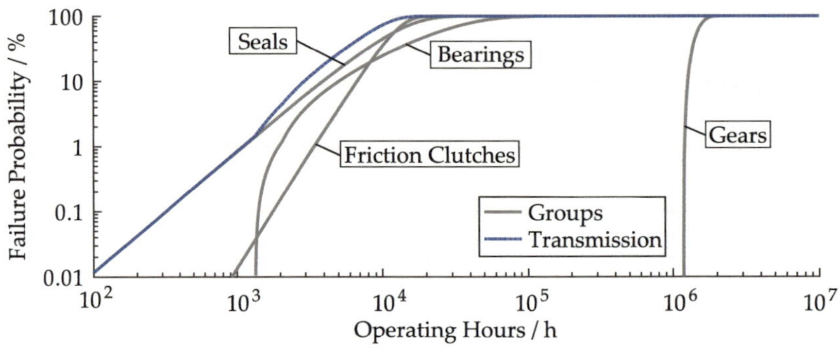


**Figure 9.** Failure probability power-shift transmission [8].

the time (variant 1: 360 s) and extended, so that the driving cycle period is twice as long as before (variant 2: 1440 s). The resulting lifetimes of all three driving cycles for different groups of machine elements as well as the entire system are illustrated in **Figure 10**.

For the short driving cycle, the lifetimes are significantly lower, compared to the initial normal driving cycle. For the long driving cycle, the inverse effect can be observed. Only the lifetime of the seals does not change due to the chosen lifetime model that only considered temperature degradation and is therefore independent of the occurring loads in the transmission. The change in lifetime of the transmission due to the different driving cycles arises from the changed percentages of phases with high loads (uphill/downhill phases) to the section with small loads (straight forward/backward). When executing the short driving cycle, the sections with high loads occur more often, compared to the longer cycles. Therefore, the components in the transmission experience more damage within the same time and subsequently fail sooner [8].

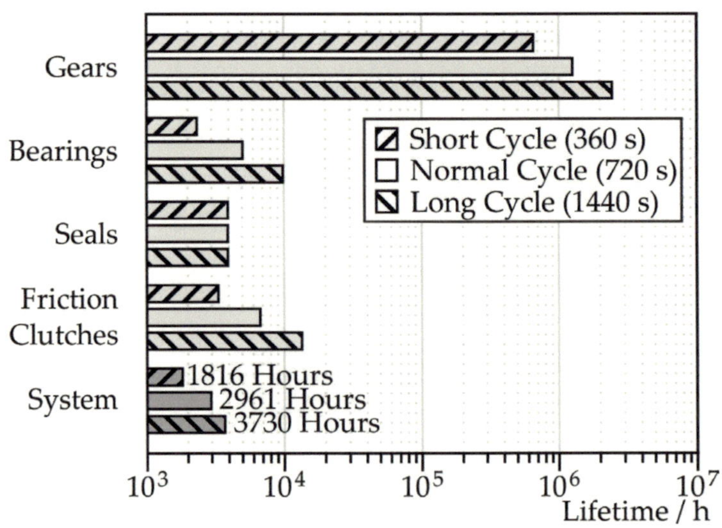


**Figure 10.** Lifetime of the power-shift transmission for different load cycles [8].

This illustrates that the accuracy of the load assumptions has a large influence on the predicted lifetime of the off-highway drivetrain. The more realistic the investigated cycle is for the loads the system experiences during operation, the better the reliability can be calculated. One way to address this problem can be to use sensor data from the machine during operation to define more accurate load spectra. Using simulations that can calculate component loads from few input signals, the accuracy of the reliability evaluation could be increased without installing extensive metrology. Certainly, this would only be possible for existing systems.

### 5.2. Offshore winch

For offshore cranes, a high reliability is important due to the remote locations in which they are operated. In the offshore setting, spare parts and repair equipment are not easily accessible. The reliability of the drivetrain of an offshore winch can be assessed using the proposed method. In cranes for offshore applications, *Active Heave Compensation* (AHC) systems are used to compensate the vertical vessel movement that is induced by the waves. This allows maintaining the payload at a steady position in spite of the wave movement, which is useful when loads are lifted from a ship to a steady platform or the sea floor, as illustrated in **Figure 11**. The compensating motion is generated by winding and unwinding the crane's winch, which creates dynamic loads on the drivetrain.

The drivetrain consists of the main drum with cogwheels on each side that are driven by 10 pinion wheels each. Each pinion is connected to a hydraulic motor by a gearbox, which offers the necessary torque due to two planetary gear stages.

To create a typical load cycle for the AHC mode, all influences on the winch loads have to be identified. The necessary winch movement depends on the wave height over the course of time, so that the vertical wave movement has to be calculated for the chosen operating point. This can be done using the JONSWAP spectrum [34] which yields the wave movement for a chosen wind speed on the Beaufort scale. As a rough estimate, it is assumed that the vessel, and therefore the payload, follows the wave movement directly. This means that the movement to compensate the waves is the opposite of the wave height at each point in time. From this movement, the required acceleration of the winch can be calculated. To calculate the required dynamic torque due to acceleration and deceleration, the winch's moment of inertia has to be known. It is composed of the drum's inertia, the rope's inertia both on the drum and uncoiled and the inertia of the payload on the rope [35].

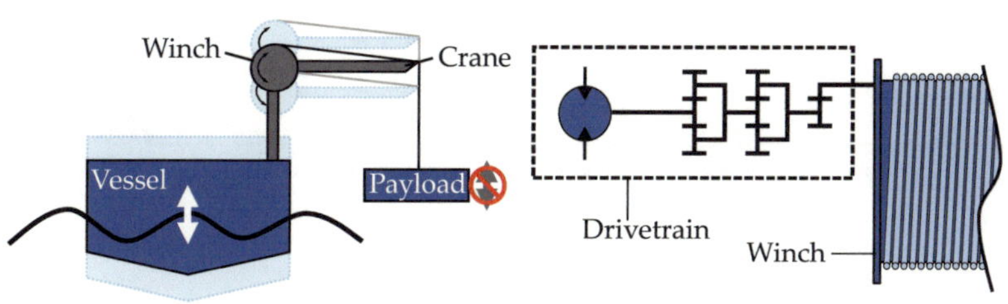


**Figure 11.** Winch movement and offshore winch drivetrain [35].

From the total required torque, the load on all components can be calculated. In this example, only gears and bearings are considered. The gear loads can be calculated using a torque plan, which includes the gear ratios in all gear stages. Since all of the gearboxes are identical and the loads are distributed symmetrically, only one of them is analysed. The torque plan also yields the rotational speeds of all shafts. The bearing loads result from the gear forces and the weight of the shafts. They can be obtained using transfer functions, which can be generated by applying external loads and observing the resulting internal loads that depend on the dimensions of the gears and shafts. The component loads are then categorized and transferred into component stresses using standardized approaches [9, 17].

In the case of the winch, the categorization of the load-time data poses a challenge due to the oscillating motion of the winch. In typical drivetrains, the system performs full revolutions which can then be counted as load alternations. When operating in AHC mode, the drum and the shafts in the drivetrain do not always perform full revolutions so that a modified counting method is used. A new load alternation is counted each time the zero angle position (or multiples of 360°) is crossed, see **Figure 12**. The lifetime models also require the speed and load corresponding to each load alternation. For this application, the mean speed and the mean load between two turning points are considered. This approach offers a conservative estimation of the lifetime.

With the categorized loads, the lifetime of the gears and bearings can be calculated and then used to create a failure probability function. **Figure 13** shows the failure probabilities for the gears and bearings in one gearbox and the combined failure probabilities for the gears and bearings in all gearboxes. The bearings in the system have shorter lifetimes than the gears, which means that they determine the system's reliability. The system's failure probability in this example is equal to the failure probability of the combined bearings. The information about the system can then be used to identify critical components and derive the overall lifetime.

This example shows that due to special circumstances of operation, off-highway drivetrains can pose challenges for the reliability calculation. In the case of the winch, the non-typical oscillation movement creates difficulties for the load classification and the lifetime models of the components. Although the system can be handled by modifying existing methods, it shows that additional research is still required.

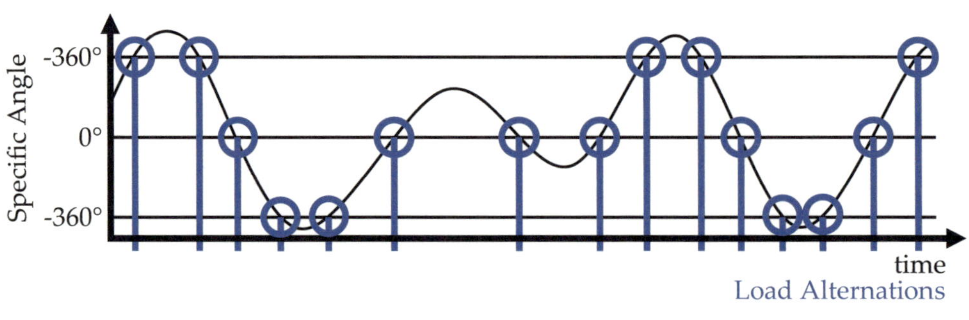


**Figure 12.** Schematic illustration of load alternation counting.

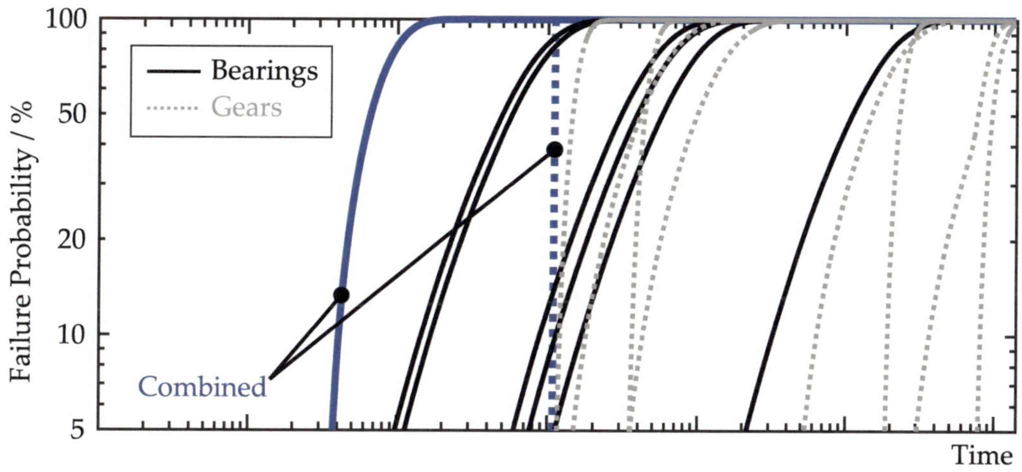


**Figure 13.** Failure distribution of gears and bearings [35].

## 6. Conclusion

Off-highway machines often perform tasks that are embedded into a process chain so that machine downtime causes high costs and delays in the process. Therefore, approaches to assess the reliability of each machine quantitatively offer a possibility for a lower total cost of ownership due to better maintenance planning.

The challenge in the quantitative reliability evaluation of off-highway drivetrains lies in the determination of a representative load cycle, since the accuracy of the lifetime calculation depends on precise load assumptions. Due to the wide range of off-highway machines, their applications with different tasks and varying loads, even for machines with identical configurations, the loads are machine specific. The most precise information could be generated through elaborate measurement campaigns of the machine in the field. Since such approaches are sometimes not possible, especially in early development stages, simulations offer a possible alternative.

The described method for the reliability evaluation is applied exemplarily to two off-highway drivetrains with different fields of applications. The example of a power-shift transmission illustrates that the calculated lifetime depends greatly on the assumed load cycle. Therefore, methods have to be found to address this issue, possibly through an approach that combines measurements with simulations to calculate component loads from few measured signals.

The example of the offshore crane winch shows that the method can be applied to a wide range of systems. The challenge for the reliability calculation for the winch drivetrain lies in handling the oscillating motion during AHC operation. Additional research for better counting methods and more extensive lifetime models is required.

## Author details

Lothar Wöll[1*], Katharina Schick[1], Georg Jacobs[1], Achim Kramer[1] and Stephan Neumann[2]

*Address all correspondence to: lothar.woell@ime.rwth-aachen.de

1 Institute for Machine Elements and Machine Design (IME), RWTH Aachen University, Aachen, Germany

2 IME Aachen GmbH Institut für Maschinenelemente und Maschinengestaltung, Aachen, Germany

## References

[1] Off-Highway Research. The Global Volume and Value Service. August 2012 Update [Internet]. 2012. Available from: http://www.ohr-dev.com/samples-static/The_Global_Volume_and_Value_Service/The_Global_Volume_and_Value_Service_August_2012.pdf [Accessed: 31 May 2017]

[2] Bourn J. Modernizing Construction [Internet]. 2001. Available from: https://www.nao.org.uk/wp-content/uploads/2001/01/000187.pdf [Accessed: 25 May 2017]

[3] Fan Q, Fan H. Reliability analysis and failure prediction of construction equipment with time series models. Journal of Advanced Management Science. 2015;**3**(3):203-210. DOI: 10.12720/joams.3.3.203-210

[4] Peng S, Vayenas N. Maintainability analysis of underground mining equipment using genetic algorithms: Case studies with an LHD vehicle. Journal of Mining. 2014;**2014**: p. 10. DOI: 10.1155/2014/528414

[5] German Institute for Standardisation. DIN EN 60812. Failure Mode and Effects Analysis (FMEA). 2015

[6] German Institute for Standardisation. DIN EN 61025. Fault tree analysis (FTA). 2006

[7] Bertsche B. Reliability in Automotive and Mechanical Engineering. Berlin Heidelberg: Springer-Verlag; 2008. DOI: 10.1007/978-3-540-34282-3

[8] Neumann S, Wöll L, Feldermann A, Strassburger F, Jacobs G. Modular system modeling for quantitative reliability evaluation of technical systems. Modeling Identification and Control. 2016;**37**(1):19-29. DOI: 10.4173/mic.2016.1.2

[9] German Institute for Standardization. DIN 3990. Calculation of load capacity of cylindrical gears. 1987

[10] International Organization for Standardization. ISO 6336. Calculation of load capacity of spur and helical gears. 2006

[11] American Gear Manufacturers Association. AGMA 2001. Fundamental rating factors and calculation methods for involute spur and helical gear teeth. 2004

[12] Japan Gear Manufacturers Association. JGMA 6102. Calculation of surface durability (pitting resistance) for spur and helical gears. 2009

[13] International Organization for Standardization. ISO/TR 13989. Calculation of scuffing load capacity of cylindrical, bevel and hypoid gears—Part 1: Flash temperature method. 2000

[14] Det Norske Veritas. DNV 41.2. Calculation of gear rating for marine transmissions. 2012

[15] Renius T. Last- und Fahrgeschwindigkeitskollektive als Dimensionierungsgrundlage für die Fahrgetriebe von Ackerschleppern. Fortschrittberichte der VDI-Zeitschriften. 1976;**49**(1):1-99

[16] Boog M. Steigerung der Verfügbarkeit mobiler Arbeitsmaschinen durch Betriebslasterfassung und Fehleridentifikation an hydrostatischen Verdrängereinheiten [dissertation]. Karlsruhe: FAST Institut für Fahrzeugsystemtechnik; 2010

[17] German Institute for Standardisation. DIN ISO 281. Rolling bearings—Dynamic load ratings and rating life. 2007

[18] Schwack F, Stammler M, Poll G, Reuter A. Comparison of life calculations for oscillating bearings considering individual pitch control in wind turbines. Journal of Physics: Conference Series. 2016;**753**: p. 10. DOI: 10.1088/1742-6596/753/11/112013

[19] Schmelter R. Über die Lebensdauerberechnung oszillierender Wälzlager. IMW-Institutsmitteilung. 2011;**36**:35-42

[20] German Institute for Standardization. DIN 743. Calculation of load capacity of shafts and axles. 2012

[21] Rennert R, Kullig E, Vormwald M, Esderts A, Siegele D. Rechnerischer Festigkeitsnachweis für Maschinenbauteile. 6th ed. Frankfurt: Forschungskuratorium Maschinenbau; 2012

[22] Nikkel K, Esderts A. Geführte Lebensdauerberechnung für Komponenten der Antriebstechnik in Form eines digitalen Leitfadens. FVA-Heft. 2008;**866**:1-157

[23] Niemann G, Winter H. Maschinenelemente. Band 3: Schraubrad-, Kegelrad-, Schnecken-, Ketten-, Riemen-, Reibradgetriebe, Kupplungen, Bremsen, Freiläufe. Berlin Heidelberg: Springer-Verlag; 1983

[24] Klein B, Kirschmann D, Haas W, Bertsche B. Accelerated testing of shaft seals as components with complex failure modes. In: Reliability and Maintainability Symposium (RAMS); San Jose: IEEE; 2010. DOI: 10.1109/RAMS.2010.5448019

[25] Haas W, Hörl L, Klein B. Betrachtungen zur Zuverlässigkeit und Lebensdauer von Hydraulikdichtungen. Institut für Maschinenelemente, Universität Stuttgart, Stuttgart; 2010

[26] Haibach E. Betriebsfestigkeit—Verfahren und Daten zur Bauteilberechnung. 3rd ed. Berlin Heidelberg: Springer-Verlag; 2006. DOI: 10.1007/3-540-29364-7

[27] Birolini A. Reliability Engineering—Theory and Practice. 8th ed. Berlin: Springer-Verlag; 2017. DOI: 10.1007/978-3-662-54209-5

[28] German Institute for Standardization. DIN EN ISO 6165. Earth-moving machinery—Basic types—Identification and terms and definitions. 2013

[29] Kunze G, Göhring H, Jacob K. Baumaschinen - Erdbau- und Tagebaumaschinen. 1st ed. Wiesbaden: Springer Fachmedien; 2002. DOI: 10.1007/978-3-663-09352-7

[30] Kunze G. Methode zur Bestimmung von Normlastkollektiven für Bau- und Fördermaschinen. Wissensportal baumaschine.de. 2005;**1**: p. 10

[31] Ramm M. Systematische Entwicklung und Analyse stufenlos verstellbarer Getriebe mit innerer Leistungsverzweigung für mobile Arbeitsmaschinen [Dissertation]. Aachen; 2015

[32] Buck G. Probleme bei der Berechnung von Fahrzeuggetrieben mit Lastkollektiven. VDI Berichte. 1973;**159**;37-46

[33] Köhler M, Jenne S, Pötter K, Zenner H. Zählverfahren und Lastannahme in der Betriebsfestigkeit. 1st ed. Berlin: Springer-Verlag; 2012. DOI: 10.1007/978-3-642-13164-6

[34] Hasselmann K et al. Measurements of Wind-Wave Growth and Swell Decay during the Joint North Sea Wave Project. Hamburg: Deutsches Hydrographisches Institut; 1973

[35] Wöll L, Jacobs G, Feldermann A. Sensitivity analysis on the reliability of an offshore winch regarding selected gearbox parameters. Modeling Identification and Control. 2017;**38**(2):51-58. DOI: 10.4173/mic.2017.2.1

Section 2

# Reliability of Electronic Devices

Chapter 9

# Reliability Prediction Considering Multiple Failure Mechanisms

Joseph B. Bernstein

Additional information is available at the end of the chapter

http://dx.doi.org/10.5772/intechopen.69500

**Abstract**

The multiple temperature operational life (MTOL) testing method is used to calculate the failure in time (FIT) by a linear combination of constant-rate failure mechanisms. This chapter demonstrates that, unlike other conventional qualification procedures, the MTOL testing procedure gives a broad description of the reliability from sub-zero to high temperatures. This procedure can replace the more standard single-condition high-temperature operational life (HTOL) by predicting the system failure rate by testing a small number of components over more extreme accelerated conditions for much shorter times than is conventionally used. The result is a much more accurate result for the failure rate, calculating the mean time to failure (MTTF) based on much shorter time-scale testing only a fraction of the number of components. Rather than testing 77 parts for 1000h, a failure rate prediction can be obtained from as few as 15 parts tested for only 200h with reliable results.

**Keywords:** MTTF, MTOL, HTOL, FIT, failure rate, multiple mechanisms

## 1. Introduction to MTOL

Traditional high-temperature operational life (HTOL) test strategy is based on the outdated JEDEC standard that has not been supported or updated for many years. The major drawback of this method is that it is not based on a model that predicts failures in the field. Nonetheless, the electronics industry continues to provide data from tests of fewer than 100 parts, subjected to their maximum allowed voltages and temperatures for as many as 1000h. The result based on zero, or a maximum of one, failure out of the number of parts tested does not actually predict. This null result is then fit into an average acceleration factor (AF), which is the product of a thermal factor and a voltage factor. The result is a reported failure rate as described by the standard failure in time (FIT) model, which is the number of expected failures per billion part hours of operation. FIT is still an important metric for failure rate in today's technology; however, it does not account for the fact that multiple failure mechanisms simply cannot be averaged for either thermal or voltage acceleration factors.

One of the major limitations of advanced electronic systems qualification, including advanced microchips and components,is providing reliability specifications that match the variety of user applications. The standard HTOL qualification that is based on a single high-voltage and high-temperature burn-in does not reflect actual failure mechanisms that would lead to a failure in the field. Rather, the manufacturer is expected to meet the system's reliability criteria without any real knowledge of the possible failure causes or the relative importance of any individual mechanism. More than this, as a consequence of the non-linear nature of individual mechanisms, it is impossible for the dominant mechanism at HTOL test reflect the expected dominant mechanism at operating conditions, essentially sweeping the potential cause of failure under the rug while generating an overly optimistic picture for the actual reliability.

Two problems exist with the current HTOL approach, as recognized by JEDEC in publication JEP122G:

**1.** Multiple failure mechanisms actually compete for dominance in our modern electronic devices and

**2.** Each mechanism has a vastly different voltage and temperature acceleration factors depending on the device operation.

This more recent JEDEC publication recommends explicitly that multiple mechanisms should be addressed in a sum-of-failure-rates approach. We agree that a single point HTOL test with zero failures can, by no means, account for a multiplicity of competing mechanisms.

In order to address this fundamental limitation, we developed a special multiple-mechanism qualification approach that allows companies to tailor specifications to a variety of customer's needs. We call this approach the multiple temperature operational life (MTOL) test at multiple conditions and match the results with the foundry's reliability models to make accurate FIT calculations based on specific customer's environments including voltage, temperature, and speed. The basic strategy is outlined in **Figure 1**. Time to fail models are put into the matrix as failure rates ($\lambda_i$) for each given set of conditions of temperature, voltage, frequency, etc. Then, the left-hand side of the matrix takes measured failure rates as extrapolated from measurements tested on the actual system under investigation. Hence, the relative acceleration factor for each mechanism is calculated based on measurement data rather than from zero-failures, as is traditional done through HTOL.

This new MTOL system allows the FIT value to be calculated with the assumption of not just one but multiple degradation mechanisms that are characterized by multiple acceleration factors. This chapter will describe the advantages of considering multiple failure mechanisms and how they can be linearly combined with a simple matrix solution that accounts for each mechanism proportionally based on data rather than based on a zero-failure result.

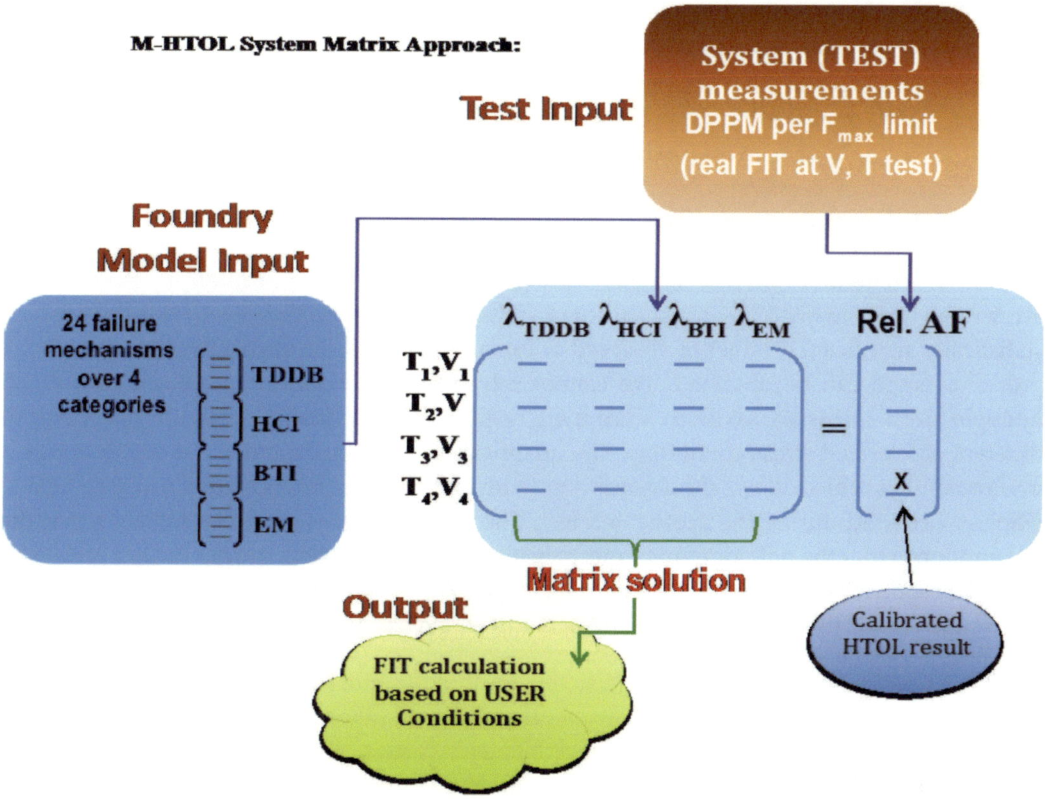


**Figure 1.** Matrix methodology for reliability prediction.

### 1.1. Limitation of traditional HTOL

The semiconductor industry provides an expected FIT for every product that is sold based on operation within the specified conditions of voltage, frequency, heat dissipation, etc. Hence, a system reliability model is a prediction of the expected MTBF's, or as we will use here, mean time to fail (MTTF), for a system that is not replaced as the sum of the FIT rates for every component.

A FIT is defined in terms of an acceleration factor (AF) and MTTF as:

$$FIT = \frac{\#failures}{\#tested * time(hours) * AF} 10^9 = {10^9}/{MTTF(hours) * AF} \tag{1}$$

where #failures and #tested are the numbers of actual failures that occurred as a fraction of the total number of units subjected to an accelerated test per total test time in hours. From a statistical perspective, this calculation would be correct if there is only a single known mechanism that is completely characterized by a single acceleration factor, AF. However, if multiple mechanisms are present, there is no way to average the acceleration factor, and thus, the denominator cannot be characterized as one AF for any set of operating conditions. The true AF must be based on the physics of the actual mechanisms, including different activation energies for different physical processes. Without testing at multiple accelerated conditions, a standard HTOL qualification cannot distinguish effects of more than one thermally activated process, rather only give an approximation for the dominant mechanism at the test condition. The test consists of stressing some number of parts, usually around 77, for an extended time, usually 1000h, at an accelerated voltage and temperature.

In order to excite multiple mechanisms, testing must be performed at multiple conditions of accelerated stress in order to obtain sufficient statistical data. Furthermore, there needs to be a statistically significant number of observed or extrapolated failures during the testing so that a proper average can be obtained. We cannot rely on a "zero failure" pass criterion when multiple mechanisms are involved since there needs to be a distinction between the effects of different accelerated stress conditions. The qualification tests are designed inevitably to result in zero failures, which allows the assumption (with only 60% confidence!) that no more than ½ a failure occurred during the accelerated test. The only fallacy with this approach is that the assumption is that the only dominant mechanism that would be seen during the test is the one with the reported AF. However, if that mechanism is not modelled or observed, there is no way to prove that this mechanism would actually be the cause of a field failure.

We don't need to prove that in most systems, multiple failure mechanisms contribute to the overall reliability of a system. Reliability mathematics assumes that the influences are time-independent, occurring at a constant rate, while each is independent of the others. In reality, most systems experience failures at approximately a constant rate, at least for the first few "random" occurrences. When we consider that the defects responsible for earlier failures are generally distributed in time, the assumption of multiple failure mechanisms makes valid sense as to why the random failures occurring during the useful life of a product will be, in fact, caused by not a single mechanism, but rather by a proportional combination of all the likely failure and wear-out mechanisms. However, due to the physics involved with each cause of failure, each will be accelerated differently depending on the thermal, electrical, or environmental stresses that are responsible for each mechanism. Hence, when an accelerated test is performed at an arbitrary voltage and temperature for acceleration based only on a single failure mechanism, then, only that mechanism will be accelerated. When the failure rate (FIT) is calculated based on the non-occurrence of a failure (i.e., zero failure assumption), then it is naturally over-estimating the reliability by whatever factor was not introduced by the second or third mechanism that was not accounted for in the model.

Unfortunately for the test and qualification industry, the final test procedure and failure rate calculation have not kept pace with the depth of understanding that we have today about the actual failure mechanisms. Also, manufacturing processes are so tightly controlled that each known mechanism is designed to be theoretically non-existent in the field. Thus, naturally, since there is no single mechanism that will cause a known end-of-life, so it is logical that multiple mechanisms will affect the final failure rate. Furthermore, HTOL tests are known to reveal multiple failure mechanisms during final qualification, which would suggest also that no single failure mechanism would dominate FIT in the field. Thus, finally, in order to make a more accurate model for FIT, a preferable approximation should be that all mechanisms contribute and the resulting overall failure distribution resembles combination of *constant failure rate processes* that is consistent with the mil-handbook and JEDEC standards.

### 1.2. MTOL methodology

The key innovation of the multiple temperature operational life (MTOL) testing method is its ability to separate different failure mechanisms so that predictions can be made for any user defined operating conditions. This is opposed to the common approach for assessing device reliability today, using high-temperature operating life (HTOL) testing [1], which is based on the assumption that just one dominant failure mechanism is responsible for a failure of the device in the field [2]. However, it is known that, in reality, multiple failure mechanisms act simultaneously on any system that causes failure based on more than a single mechanism at any time [3].

Our new approach, MTOL, deals with this issue [4]. This method predicts the reliability of electronic components by combining separately measured FITS of multiple failure mechanisms [5]. Our data reveal that different failure mechanisms act on a component in different regimes of operation causing different mechanisms to dominate, depending on the stress and the particular technology. When multiple mechanisms are known to affect the failure a product, then JEDEC standard publication JEP-122G states that "**When multiple failure mechanisms and thus multiple acceleration factors are involved, then a proper summation technique, for example, sum-of-the-failure rates method, is required**." The only question that is not answered by the JEDEC standard is how to "sum" the failure rates. As a practical solution to reaching the desired goal of a calculated failure rate that combines multiple mechanisms, we have proposed the following approach.

Because failure rates are linear and sum linearly only if they are all considered as constant rate processes, they can be combined linearly to calculate the actual reliability as measured in FIT of the system based on the physics of degradation at specific operating conditions. In a more recent publication [6], we present experimental results of the MTOL method tested on both 45 and 28nm FPGA devices from Xilinx that were processed at TSMC (according to the Xilinx data sheets). The FPGAs were tested over a range of voltages, temperature and frequencies. We measured ring frequencies of multiple asynchronous ring oscillators simultaneously during stress in a single FPGA. Hundreds of oscillators and the corresponding frequency counters were burned into a single FPGA to monitor of statistical information in real time. Since the frequency of a ring oscillator itself monitors the device speed and performance, there is no recovery effect, giving a true measure for the effects of all the failure mechanisms in real time. Our results produced an acceleration factor (AF) for each failure mechanism as a function of core voltage, temperature and frequency.

The failure rates of all of the mechanisms were then combined using a matrix to normalize the AF of the mechanisms to find the overall failure in time or FIT of the device. In other words, we found an accurate estimate of the device's mean lifetime and thus the reliability that can be conveniently transposed to other technologies and ASICs and not necessarily only FPGAs, as was the basis of our previous work. In this chapter, we show that the MTOL methodology is general and can apply to any system that is characterized by multiple failure mechanisms, which can individually be treated as approximately occurring at a constant rate, having its own FIT per mechanism.

## 2. Multiple mechanism considerations

The acceleration of the rate of occurrence of a single failure mechanism is a highly non-linear function of temperature and/or voltage as is well known through studies of the physics of failure [3–5]. The temperature acceleration factor ($AF_T$) and voltage acceleration factor ($AF_V$) can be calculated separately for each known mechanism in a combined model. The total acceleration factor of the different stress combinations for each mechanism will be the product of the acceleration factors of temperature and voltage or any other stress-related factor that could include current, frequency, humidity, etc.

$$AF = AF_V * AF_T \tag{2}$$

Calculated acceleration factors (AF) are universally used as the industry standard for device qualification. However, it only approximates a single dielectric breakdown type of failure mechanism and does not correctly predict the acceleration of other mechanisms. Similarly, an acceleration factor can be determined using any other type of stress applied, for example, vibration, radiation, number of cycles, etc. However, when only a single AF is assumed to contribute to the expected time to fail based on the high temperature, high voltage acceleration, there is no way to account for the effect of multiple mechanisms.

The goal here is to improve the approach from standard HTOL to a one where a true "sum of failure rates" model is considered based on a proportional contribution of each mechanism based on its relative influence. Each one mechanism acts on the system in combination with others to cause an eventual failure. When more than one mechanism affects the reliability of a system or component, then the relative acceleration of each one must be defined and calculated at the applied condition. Every potential failure mechanism should be identified, and its unique AF should then be relatively known at a given temperature and voltage so the FIT rate can be approximated separately for each mechanism. Thus, the actual FIT will be the sum of the failure rates per mechanism, as is described by:

$$\mathrm{FIT}_{\mathrm{total}} = \mathrm{FIT}_1 + \mathrm{FIT}_2 + \dots + \mathrm{FIT}_i \tag{3}$$

whereby each mechanism is described by its own FIT, which leads to its own expected failure unit per mechanism, $\mathrm{FIT}_i$. Therefore, it is impossible to accelerate more than one mechanism with a single set of accelerated stress conditions. Thus, requiring that more than a single test is necessary to determine what would be the actual FIT that would be found in any given expected operating conditions.

The qualification of device reliability, as reported by a FIT rate, must be based on an acceleration factor, which represents the failure model for the tested device. Since multiple mechanisms are known to lead to degradation and thus failure in any complex system, it is obvious that a single mechanism model with a single AF assumption will never produce a useful result for reliability prediction. This will be explained by way of example. Suppose there are two identifiable, constant rate competing failure modes (assume an exponential distribution). One failure mode is accelerated only by temperature. We denote its failure rate as $AF_{T1}$ other failure mode is only accelerated by voltage, $AF_{V2}$, and the corresponding failure rate is denoted as

$$FIT = {10^9}/{MTTF_1{*}AF_{T1}} + {10^9}/{MTTF_2{*}AF_{V2}} \tag{4}$$

where the measured Mean Time To Fail (MTTF) (measured in hours) would be different for each mechanism. However, since only one condition of Voltage and Temperature is applied, yet the calculated FIT is based on a combination of two mechanisms, each with its own acceleration factor, then there is now way to determine which mechanism dominates. Because the effective acceleration factor for any given set of test conditions is related to the inverse of the acceleration factor, without separately testing each mechanism, the resulting FIT will have no relation to the actual tested results.

Due to the exponential nature of the acceleration factor as a function of *V* or *T*, or any other stress-inducing parameter, including vibration, humidity, radiation, etc, if only a single parameter is changed, then it is not likely for more than one mechanism to be accelerated significantly compared to the others. As we will see in the next section, at least three mechanisms should be considered, many more perhaps depending on the system and use environment. Also, each voltage and temperature dependencies must be considered separately for each mechanism in order to make a reasonable reliability model for the whole device.

## 3. MTOL test system example

A test system was built in off-the-shelf Xilinx FPGA evaluation boards. The system ran hundreds of internal oscillators at several different frequencies asynchronously, allowing independent measurements across the chip and the separation of current versus voltage induced degradation effects. In order to create a measurable accelerating system, ring oscillators (ROs) consisting of inverter chains were used. The last inverter in the chain is connected to the first, forming a cycle/ring (**Figure 2**). When the number of stages is odd, every sampled cell in the chain will invert its logic level. Additionally, as no clock is fed into the RO, the frequency of the alternating logical states depends just on the internal delay of the cells and the latency of the connections between them, where the frequency of each RO is given by $½NT_p$, where $N$ is the number of inverters and $T_p$ is the propagation delay of each inverter. Each inverter chain was implemented as a complete logical cell using predefined Xilinx primitives, and thus, each ring oscillator was made up of the basic components of the FPGA. When degradation occurred in the FPGA, a decrease in performance and frequency of the RO could be observed and attributed to either increase in resistance or change in threshold voltage for the transistors.

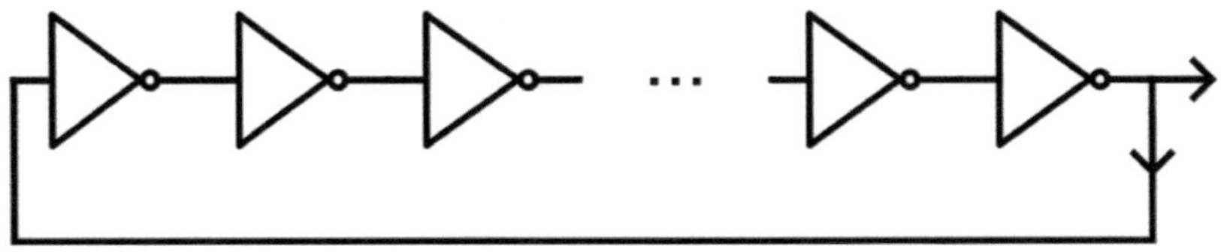

**Figure 2.** Ring oscillator is made of $2N+1$ Inverters connected in a chain.

For optimal testing and chip coverage, different sized ROs were selected, ranging from three inverters, giving the maximum frequency possible in accordance with the intrinsic delays of the FPGA employed (400–700 MHz), and up to 1001-inverter oscillators, giving a much lower frequency (around 800 KHz). The system implemented on the chip starts operating immediately when the FPGA core voltage is connected. Using a wide range of ROs enabled us to measure the frequency and the internal delay of a real, de-facto system on a chip. This allows seeing the frequency dependence of each failure mechanism without any recovery effect. The set of ROs consisted of:

- 150 oscillators of 3 stages
- 50 oscillators of 5 stages
- 20 oscillators of 33 stages
- 3 oscillators of 333 stages
- 1 oscillator of 1001 stages

It is important to note, here, that the size of the ring determines the interdependence of any degradation. The shortest oscillators containing only three stages will have the greatest variability as well as the highest frequency. This is because a shorter critical electrical path will be much more sensitive to minor variations that lead to greater or smaller degradation over time. This means that the lower frequency oscillators containing as many as 1001 stages will average out the effects of individual degradations. Furthermore, the random statistical variability of individual devices will be exaggerated by the statistical distribution in wear-out slopes seen at high frequencies. Thus, we made 150 of the smallest ring size devices, which would need to be averaged to find the average degradation at those frequencies exhibiting more random times to fail. Interestingly, we see that the variability of three-ring oscillators is quite diverse, nearly randomly distributed about an average, whereas the lower frequency rings are much more narrowly distributed, indicating a more predictable time to fail, as compared to circuits having a much shorter critical path.

### 3.1. Testing methods

The testing system was synthesized and downloaded to the FPGA card. The test conditions were predefined for allowing separation and characterization of the relative contributions of the various failure mechanisms by controlling voltage, temperature, and frequency. Extreme core voltages and environmental temperatures, beyond the specifications, were imposed to cause failure acceleration of individual mechanisms to dominate others at each condition, for example, sub-zero temperatures, at very high operating voltages, to exaggerate HCI.

For each test, the FPGA board was placed in a temperature-controlled oven, dedicated to the MTOL testing, with an appropriate voltage set at the FPGA core. The board was connected to a computer via USB and the external clock signal was fed into the chip. The tests performed for 200–500 h, while the device was working in the accelerated conditions. Frequencies of every ring oscillator, of different sizes, were measured. Initial sampling started after one working-hour in the accelerated environment, and then, samples were taken automatically at 5-min intervals. The frequency measurement data were stored in a database from which one could draw statistical information about the degradation in the device performance.

The acceleration conditions for each failure mechanism allowed us to examine the specific effect of voltage and temperature versus frequency on that particular mechanism at the system level and thus define its unique physical characteristics even from a finished product. A close inspection of test results yielded more precise parameters for the acceleration factors (AF) equations and allowed adjusting them to the device under test. Finally, after completing the tests, some of the experiments with different frequency, voltage and temperature conditions were chosen to construct the MTOL Matrix.

### 3.2. Separation of mechanisms

Our tests for various mechanisms included exposing the core of the FPGA to accelerating voltages above nominal. About 45 nm defines the nominal voltage at 1.2 V and for 28 and 20 nm, 1.0 V. Our method of separating mechanisms allowed the evaluation of actual activation energies for the three failure mechanisms, which are hot carrier injection (HCI), bias temperature instability (BTI) and electromigration (EM). We plotted the degradation in frequency and attributed it to one of the three failure mechanisms.

We need to justify our approach for accounting for current in the devices. Both-and HCI have $J^{\gamma}$ factors; however, in a completely packaged CMOS digital circuit, there is no way to directly measure current, *I*, or as current density, *J*. We assume, in our experiments, that the stored gate charge strictly determines current transferred for each switching transition, that is, from a 0 to a 1 and vise verse. Whatever the current is for any state-transition will be same for each transition and the current will be, therefore, directly proportional to the frequency. Hence, the degradation for each transition will be directly proportional to the measured frequency, *f*. The voltage exponent will depend on the frequency, but the exponent, $\gamma$, measured will be the effective voltage acceleration parameter and comes into the equations for EM and HCI at $f^{*}V^{\gamma}$.

The results of our experiments give both $E_A$ and $\gamma$ for the three mechanisms we studied at temperatures ranging from −50 to 150°C. The Eyring model [1] is utilized here to describe the failure in time (FIT) for all of the failure mechanisms. The specific FIT of each failure mechanisms follows these formulae:

$$FIT_{HCI} = fV^{\gamma_{HCI}}e^{-\frac{Ea_{HCI}}{kT}} \tag{5}$$

$$FIT_{BTI} = e^{\gamma_{BTI}V}e^{-\frac{Ea_{BTI}}{kT}} \tag{6}$$

$$FIT_{EM} = fV^{\gamma_{EM}}e^{-\frac{Ea_{EM}}{kT}} \tag{7}$$

The degradation slope, $\alpha$, is measured as the degradation from initial frequency as an exponential decay, approximated by taking the difference in frequency, and divided by initial frequency over the time. In our experiments, we found that when the decay was dominated by BTI, the decay was proportional to the fourth root of time, while HCI and EM, being diffusion-related mechanisms, have decay that is proportional to the square root of time [2], as seen in **Figure 3**.

In the 45nm boards each oscillator, the ring frequency was measured and plotted against the square root of time in 45nm devices. The slope, $\alpha$, was then converted to a FIT for each test as determined by extrapolating the degradation slope to 10% degradation from its initial value. Each set is plotted as an exponential decay dependent on the square root of time as shown by example in **Figure 3**. This slope is then used to find the time to fail as seen in the development of FIT below (Eqs. (8)–(11)). We defined the exponent as $1/n$ so that we can apply the degradation based on the square-root law, as it does the degradation that is dominated by HCI or EM. For 28 and 20nm technology and below, we found that $n=4$ fits much more closely, as seen later with our 1000h evaluation.

$$\alpha_{(slope)} = \frac{\Delta f}{f_0 \times \Delta\sqrt{t}} \tag{8}$$

$$TTF = \left({}^{0.1}/_{\alpha}\right)^{1/n} \tag{9}$$

$$FIT = {}^{10^9}/_{TTF} \tag{10}$$

$$FIT = 10^9(10 * \alpha)^n \tag{11}$$

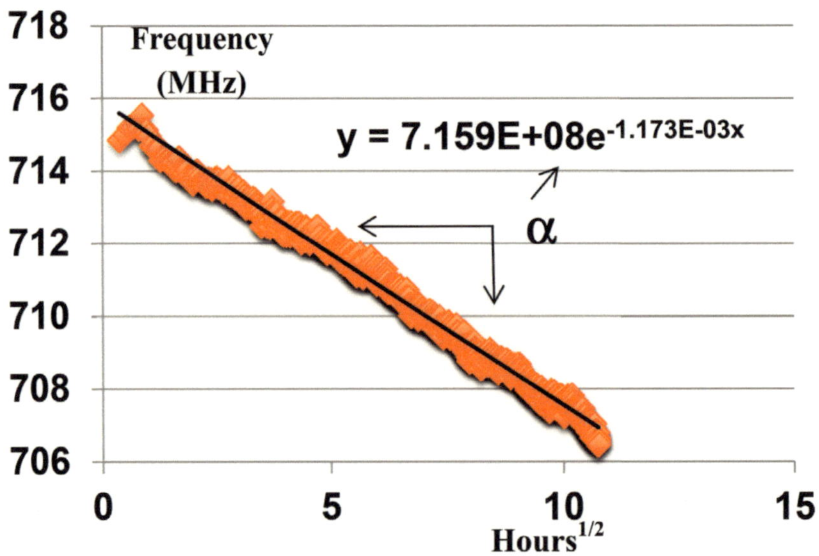


**Figure 3.** Typical frequency versus square root of time showing degradation slope $\alpha$.

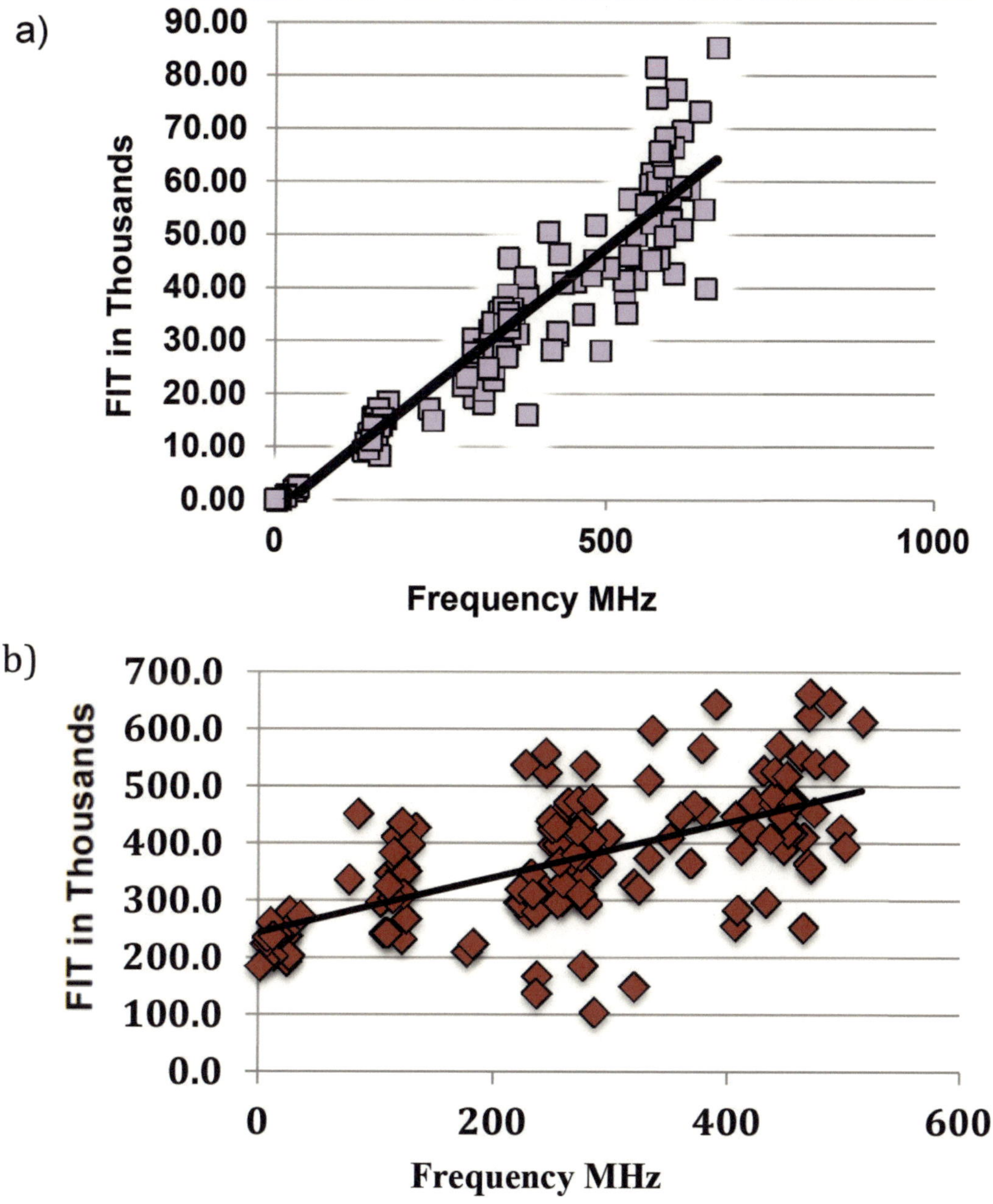


**Figure 4.** Failure rate, FIT/1000, versus frequency in MHz for (a) HCI, stressed at −35°C with 2.0V core voltage and (b) BTI, stressed at 145°C with 2.4V at the core.

The time to fail (TTF) was then calculated for as the square of the inverse slope times the failure criterion, which is 10% degradation in the 45nm technology [1]. Hence, the FIT for each slope is simply determined as the $(10^*\alpha)^2$, where $n = 2$. The average FIT is the metric to determine the reliability since that corresponds to the MTTF in Eq. (9). This FIT value is plotted as a function of the frequency in order to determine the failure mechanisms and to fit the model parameters for each mechanism.

Two typical degradation plots are shown in **Figure 4**(**a** and **b**), the FITs, determined by the slopes, are plotted against frequency in two different experiments. The data demonstrate the clear advantage of RO generated frequencies in a single chip [4]. In the examples of **Figure 4**, we see that FIT is directly proportional to frequency [6], consistent with Eq. (5). **Figure 4(b)** shows a chip that was stressed at high voltage and temperature showing a strong BTI degradation at low frequency and a much shallower slope due to EM in combination with a small HCI effect. Such curves were made for each experiment, incorporating all the oscillators across the chip spanning the range of frequencies, reflecting also the averaging effect of the longer chains. Hence, the variability is much lower than at higher frequencies, demonstrating that the averaging of many variations results in a consistent mean degradation. The slope of FIT versus frequency is then related at low temperatures as occurring only from HCI, while at higher voltages and temperatures, it can be due to BTI [6] and EM. BTI is only responsible for low frequency degradation.

In order to determine the dependence of each mechanism, the activation energy as relating to the temperature factor ($T_F$) and voltage acceleration factors ($V_F$) is determined from Eqs. (2) to (4) and is presented in Ref. [6].

### 3.3. 1000 h extrapolation

We verified that the measurement to 1% degradation over relatively shorter times gives the same slope as longer term measurement that were carried all the way to 1000h. We found that the failure criterion of 10% degradation was reached in these ring oscillators. This is seen in **Figure 5**, where the frequency was recorded at accelerated conditions all the way to 1000h at various voltage and temperatures. The slopes are all very close to $t^{1/4}$, as seen to by typical of the 28nm devices. These are the devices that show only BTI, which is consistent with a ¼

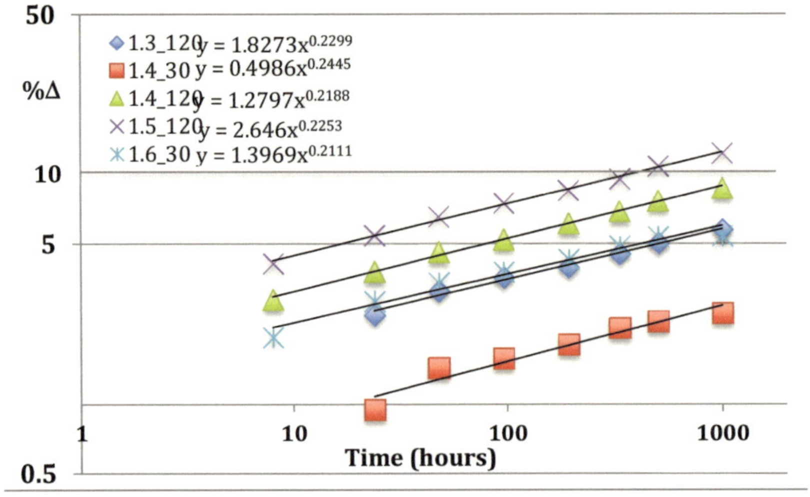


**Figure 5.** 1000h degradation data for 28nm devices over a range of core voltages from 1.3 to 1.6V at 30 and 120°C as indicated.

power law signature. Furthermore, we see that the initial degradation extrapolates to the 10% failure criterion, verifying the approach to measure for only a few hundred hours instead of requiring a complete 1000-h test.

## 4. Linear matrix solution

We assume here that the linear, Poisson, model for constant rate is associated with the probability of failure for each separable mechanism. As we showed in Eq. (3) above, each FIT adds linearly to the other FITs in order to obtain an average system failure rate. By observation of the procedure in **Figure 1**, it is clear that each FIT will have its own value that is uniquely determined by the acceleration factor for each mechanism depending on the voltage (*V*), temperature (*T*), and operational fequency (*F*). For this example, we found that there was no evidence of a time-dependent dielectric breakdown (TDDB), and therefore, we included only HCI, BTI, and EM.

This approach is exactly what JEDEC describes as a sum of failure rates methodology as it sums the expected failure rate of each mechanism distinctly from the other mechanisms. The combination results from actual accelerated life tests where there is an extrapolated mean time to fail based on the known operating conditions of *V*, *T* and *F*. Hence, we are sure to test at a large range of Temperatures, including very high and very low temperatures as well as core voltage as high as will be practical to achieve reliable operation.

Of course, we assume that each component is composed of multiple sub-components, for example, a certain percentage is effectively ring-oscillator, static SRAM, DRAM, etc. Each type of circuit, based on its operation, can be seen to affect the potential steady-state (defect related) failure mechanisms differently based on the accelerated test conditions. However, unlike traditional reliability systems, rather than treat each sub-system or component as an individual source with a known failure rate, we separate the system into distinct mechanisms that is known to have its own acceleration factor with voltage, temperature, frequency, cycles, etc. Hence, the standard system reliability FIT can be modeled using traditional MIL-handbook-217 type of algorithms and adapted to known system reliability tools; however, instead of treating each component as individuals, we propose treating each complex component as a series system of various mechanisms, each with its own reliability.

The matrix is arranged as in **Table 1**. The three most left-hand columns show the temperature, *T*, voltage, *V*, and frequency, *F*, used for the accelerated test. The measured value for FIT is then put in the third column from the right, after the relative (un-normalized) calculations for each mechanism are placed below the column describing the mechanism. Here, they are labeled HCI, BTI, and EM. Any three rows can be used to solve the matrix, and the product of the solution parameters is then put in the FIT column on the right-hand side of the matrix. The three rows that are used to solve the matrix will then have exactly the same solution as the measured FIT values used to calibrate the matrix.

The second from the right-hand column shows the ratio of the extrapolated failure rate and the calculated FIT. These values serve to show the closeness of fit to the model parameters by comparing the other measured FIT values with the calculations. This matrix will have a unique solution that will fit the percentages of each mechanism ($P_i$) with the measured failure rate, FIT.

| $T$ (°C) | $V$ | $F$ (GHz) | HCI | BTI | EM | Measured | Ratio | FIT |
|---|---|---|---|---|---|---|---|---|
| −62.5 | 1.2 | 1 | 99.99% | 0.01% | 0.00% | 30 | 94% | 2.83E+01 |
| 125 | 1.2 | 1 | 0.00% | 86.86% | 13.14% | 997.4 | 102% | 1.01E+03 |
| 153 | 1.2 | 1 | 0.00% | 63.79% | 36.21% | 3672 | 100% | 3.67E+03 |
| −35 | 2.5 | 0.5 | 100.00% | 0.00% | 0.00% | 23,750,000 | 100% | 2.38E+07 |
| 154 | 1.2 | 0 | 0.00% | 100.00% | 0.00% | 2420 | 100% | 2.42E+03 |
| 140 | 2.2 | 0 | 0.00% | 100.00% | 0.00% | 66,200 | 102% | 6.76E+04 |
| −22.5 | 2.8 | 1 | 100.00% | 0.00% | 0.00% | 240,000,000 | 101% | 2.43E+08 |
| 7.3 | 3 | 1 | 100.00% | 0.00% | 0.00% | 156,000,000 | 106% | 1.66E+08 |

**Table 1.** Test results showing proportions of failure mechanisms for given $V$, $T$, and $F$ compared with the calculated as well as the measured failure rate (FIT).

| | Inverse matrix | | $P_i$ |
|---|---|---|---|
| −4.36972E−29 | 4.76285E−18 | −1.10403E−20 | 1.13118E−10 |
| 0 | 0 | 10,946.04333 | 26,489,424.87 |
| 1.19767E+14 | −2040.515932 | −1.15932E+14 | 1.59226E+17 |

**Table 2.** Inverse matrix (left three columns) and respective $P$ values (right-hand column).

Once the parameters for the three mechanisms have been calculated and verified against the other test data, a full set of extrapolated values for FIT can then be calculated using the equations for each mechanism times the same $P$ values used to fit the three exemplary rows. **Table 2** shows the inverse matrix of the values under the three mechanisms with the corresponding $P$ values for HCI, BTI, and EM, respectively.

Since the matrix is linear, as are the calculations for FIT at any given $T$, $V$ and $F$, then the full matrix of actual FIT calculations is simply the sum of each $P$ value times the calculated relative significance of each mechanism. A calculated reliability curve is shown in **Figure 6** across the full range of expected FIT versus temperature for any set of operational conditions shown in **Table 3**.

The unique solution that solves all three equations with the three extrapolated acceleration factors gives a percentage contribution for each of the failure mechanisms. We report the reliability as FIT, which is $10^9$/MTTF for each condition. The percentages for each mechanism are shown, based on the relative contributions that were extrapolated from the physics of failure equations normalized to the measured FIT of each test. Seeing the dispersion of FIT values per test proves that the approximation of a constant rate, meaning a random distribution in time, is the proper statistical model for these results. **Figure 6** shows the resulting FIT as plotted versus temperature (°C) for the measured 45nm technology FPGA.

One advantage of plotting our data as failure rate versus temperature allows one to determine effective activation energy as a function of temperature and stressor parameters, $V$ and $F$. The principle follows the assumption that the failure rate, $\lambda$, is exponentially dependent on the activation energy divided by the absolute temperature, $T$:

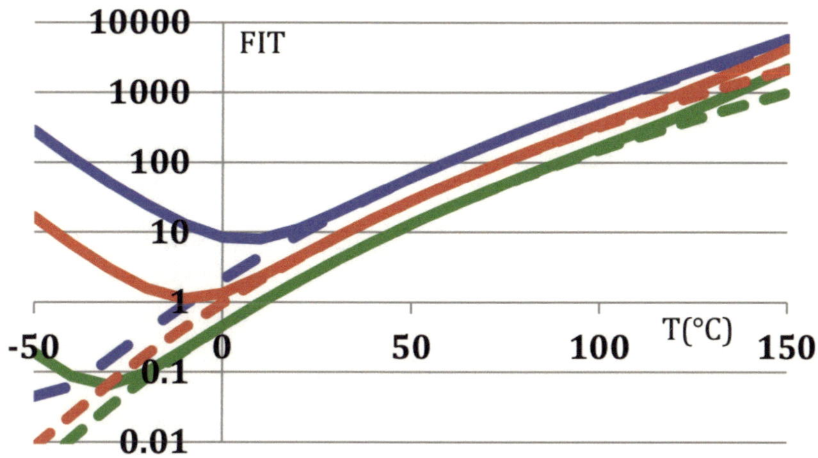


**Figure 6.** Reliability curves for 45nm technology showing FIT versus temperature for voltages. These curves are for 1.0, 1.2, and 1.4V core voltage at 10MHz (dashed) and at 1GHz (solid).

| *T* (°C) | *V* | *F* (GHz) | HCI | BTI | EM | FIT |
|---|---|---|---|---|---|---|
| −50 | 1.2 | 2 | 1.45382E+11 | 2.84438E−10 | 1.99008E−27 | 16.5 |
| −10 | 1.2 | 2 | 6,131,362,305 | 1.61083E−08 | 2.65294E−23 | 1.1 |
| 20 | 1.2 | 2 | 1,006,254,891 | 1.61337E−07 | 6.00169E−21 | 4.4 |
| 30 | 1.2 | 2 | 596,524,778.1 | 3.14239E−07 | 2.8808E−20 | 8.4 |
| 40 | 1.2 | 2 | 365,644,331.1 | 5.86524E−07 | 1.2509E−19 | 15.6 |
| 50 | 1.2 | 2 | 231,020,972.9 | 1.05325E−06 | 4.95957E−19 | 28.0 |
| 80 | 1.2 | 2 | 68,110,854.71 | 4.99845E−06 | 1.9353E−17 | 135.5 |
| 100 | 1.2 | 2 | 33,650,811.61 | 1.22819E−05 | 1.60476E−16 | 350.9 |

**Table 3.** Calculated FIT based on the solved matrix for typical use conditions.

$$FIT = \lambda = \exp\left(-\frac{E_{A,eff}}{kT}\right) \tag{12}$$

If we assume that $E_{A,\text{eff}}$ is a function of $V$ and $F$, we can calculate this from our solutions shown above,

$$-\frac{d\lambda}{d\left(\frac{1}{kT}\right)} = E_{A,eff} * \lambda \tag{13}$$

Hence, if we plot the change in $\lambda$ by (1/kT) divided by $\lambda$ at any temperature, $T$, we get

$$E_{A,eff} = \left(\frac{1}{\lambda}\right) * \frac{-d\lambda}{d\left(\frac{1}{kT}\right)} \tag{14}$$

The advantage of this representation allows a designer to consider the temperature rage as a function of stressor factors that would affect the reliability of a product, especially under extreme conditions. We see very clearly that at low frequencies, the reliability is completely dominated by BTI where the activation energy is around 0.53eV, whereas at very high temperatures and very low temperatures, the effect of frequency becomes dominant. At the very low temperatures, a negative activation energy is seen for higher frequency operation, while at high temperatures, the EM effect becomes more important, both of which are current-related effects; hence, they are frequency dependent.

What is most important to understand about this Matrix solution to linear, constant, failure rate models is that this methodology is not limited to only microelectronics. We must understand that all that is needed are the appropriate physics of failure relations to whatever stresses will be experienced during the expected life of the product. It is also important to know that this method of combining mechanisms is limited to failure mechanisms that have a generally constant rate over time. That is to say that the slope from a Weibull distribution is close to 1. If, however, the failure distribution of a particular mechanism is known to be highly predictable, that is with a wear-out characteristic, having a Weibull slope of 2 or more, then this methodology will not properly work to combine mechanisms. On the other hand, if one mechanism is known to dominate or be the limitation, that one mechanism can be separated from the other more random mechanisms, as shown in **Figure 7** and based on our extrapolation from **Table 3**.

One clear proof from this graph is that it is not possible to choose simply one accelerating temperature and voltage, or any one condition for any accelerated test, expecting that a simple extrapolation can be made based on a single failure mechanism. The mechanisms interact such that any accelerated test will surely give incorrect results, and, thus, the traditional HTOL test is obviously not sufficient for reliability prediction. Furthermore, the MTOL, multiple stressor qualification will give accurate prediction for the failure rate under any given operating conditions from a fraction of the number of samples tested over a much shorter period of time. Hence, this methodology will save a large proportion of the standard qualification procedure and give much more accurate and meaningful results.

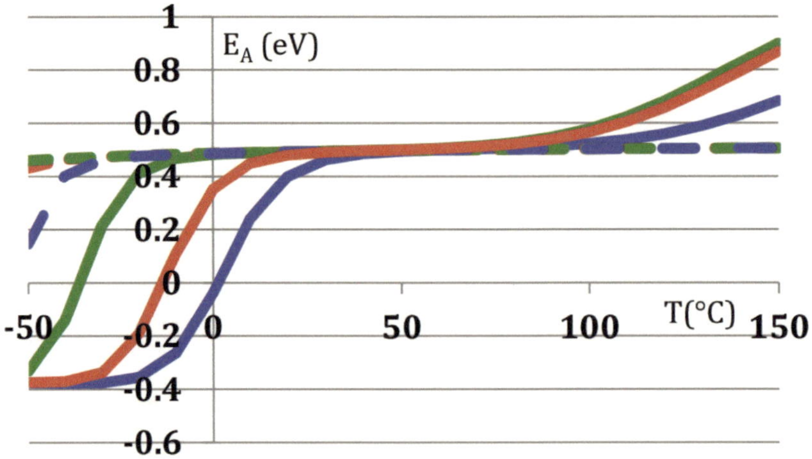


**Figure 7.** Activation energy versus temperature based on the data above in **Figure 6** for the same voltages, 1.0, 1.2, and 1.4V on the 45nm FPGA technology.

## Author details

Joseph B. Bernstein

Address all correspondence to: joeybern@gmail.com

Ariel University, Ariel, Israel

## References

[1] Xilinx, Device Reliability Report, UG116 (v10.3.1), 8 September 2015 (As an example)

[2] Bernstein JB. Reliability prediction from Burn-in Data fit to reliability models. 2014 (ISBN–10: 0–128007–47–8) Academic Press

[3] Bernstein JB. et al. Physics-of-failure based handbook of microelectronic systems, Reliability Information Analysis Center, Utica, NY; 2008 (ISBN-10: 1 933904–29-1)

[4] Bernstein JB, Gabbay M, Delly O. Reliability matrix solution to multiple mechanism prediction. Microelectronics Reliability. 2014;**54**:2951–2955

[5] Bernstein JB, Reliability Prediction for Aerospace Electronics Descriptive Note : Final rept. 15 Jul 014-14 Apr 2015 Accession Number : ADA621707

[6] Joseph B, Bernstein A, Bender BE. Reliability prediction with MTOL. Microelectronics Reliability. 2017;**68**:91–97

Chapter 10

# The Importance of Interconnection Technologies' Reliability of Power Electronic Packages

Sébastien Jacques

Additional information is available at the end of the chapter

http://dx.doi.org/10.5772/intechopen.69611

**Abstract**

This chapter deals with the reliability of die interconnections used in plastic discrete power packages, dedicated to on-board electronic systems used in a wide range of applications such as automotive industry. A complete reliability analysis of two bonding technologies—aluminum wire and ribbon bonding—is proposed. This study is particularly focused on interconnection technologies' aging, when the package is subjected to thermal cycling or power cycling with high-temperature swings. For thermal cycling, the experimental reliability test results highlight that wire bond package aging is about 2.5 faster than the ribbon bond package. For power cycling, this acceleration factor is about 1.5. In both cases and whatever the bonding technique, the failure mechanism of the package is of a fatigue-stress nature. Many failure analysis results show wire bond lift-off. The degradation of the ribbon bond is more difficult to observe. Thermo-mechanical simulations using finite elements show a high stress concentration in the heel area. For the wire-bonding technique, the wire is subjected to repeated flexing and pulling that lead to its lift off. The ribbon-bonding process shows a higher robustness, thanks to a higher contact surface on the die, the low-loop profile and the stiffness of the ribbon.

**Keywords:** discrete power electronic packages, bonding techniques, wire and ribbon, thermal cycling, power cycling, reliability

## 1. Introduction

Over the last few years, due to the global energy crisis and the threat of climate change, transition toward a low carbon electricity system has increasingly become a major issue for all governments around the world [1, 2]. The Intergovernmental Panel on Climate Change (IPCC) has recently reported that greenhouse gas emissions, covered by the Kyoto Protocol, increased by 80% from 1970 to 2010. In particular, the IPCC has specified that the global energy consumption doubled over that period of time. In Europe, with about 20% of global emissions (excluding land use and forestry) in 2013, the transport industry is still a major contributor of greenhouse gases [3, 4].

Despite this worrying background and as pioneer in $CO_2$ emission reduction, European Union (EU) has strengthened its engagement with the development of an efficient and sustainable transport industry. In the automotive sector, electric vehicles (EVs) have gradually become popular over the past decade [5]. In recent years, hybrid EVs (HEVs) have been developed to combine the use of an internal combustion engine with one or more electric motors (EMs) connected to a battery pack. Such hybridization improves the fuel economy, but all of the available energy still comes from the fuel tank. Plug-in HEVs (PHEVs) have then been introduced to displace petroleum energy with multisource electrical energy. Therefore, PHEVs are able to draw power from the electric grid, store it in batteries, and use it for transportation [6]. These batteries play an important role due to their cost-effectiveness, energy and power densities, reliability, and charging time that depend on practical applications. In particular, the lifetime and charging duration strongly depend on the features of the battery charger.

Nowadays, power electronics plays an increasing role in the development of EVs. In particular, it is a key element of the traction inverter, the DC-DC converter used to supply the vehicle's onboard systems, as well as the battery charger [7]. But new important challenges are emerging, including a significant reduction in production costs, greater compactness, and cooling efficiency [8–10]. To achieve these goals, the packaging is of the utmost importance, because it ensures both electric connections of the chip and its power dissipation [11]. With excellent performance of existing power semiconductor devices, driven also by wide bandgap materials, the packaging typically acts as the main limiting factor [12, 13].

Thermal management of power packages is still an important issue to increase the system's lifetime [14]. In particular, it is important to limit thermal gradients and consequently thermomechanical stresses due to multi-physics couplings [15, 16]. The automotive market in Europe has largely adopted qualification according to the AEC-Q101 test flow for discrete components [17]. Such standard highlights that the lifetime has to be equal to 15 years when the power device is subjected to high-temperature variations. One way to address these needs consists in optimizing bonding technologies [18].

In this chapter, a comparative reliability study of two aluminum bonding technologies—wire bonding and ribbon bonding—is presented. At the moment, these bonding technologies are widely used for through-hole and surface-mount power packages. The reliability study focuses on both thermal-cycling and power-cycling tests.

## 2. Interconnection technologies for discrete power packages

### 2.1. Relevance of discrete packages in industrial applications

Packaging is an important step in the assembly process of an electronic chip. A package has the following main functions [19]:

- *Mechanical resistance and die protection*. The aim is to ensure the die disposing (as a function of package dimensions, shape, weight, etc.), and its mechanical and chemical protection against environment (vibrations, temperature variations, dust, moisture, etc.).
- *Consistence with the needs of the application*. A specific interest is granted to electrical properties (electrical insulation), heat transfer capacity, and protection against radiation.
- *Interface between the die and the electrical system (terminals)*.
- *Favorable costs*.

Nowadays, power semiconductor devices can be assembled into two kinds of packages: *discrete packages* and *power modules*. A discrete package contains only one die, while a power module is composed of several dies which are appropriately connected to build one or more basic functions. Although power modules are widely deployed today (e.g., automotive power train, aircraft power distribution, or railway traction inverters), where rated voltage and current can easily reach 6.5 kV and more than 1 kA, respectively, discrete power semiconductor devices still find numerous applications, and especially at power levels up to several kilowatts. D. O. Neacsu has recently highlighted that discrete power packages target the following markets [20]:

- IT (information technology) and consumer electronics for about 34%.
- Consumer appliances for about 30%.
- Industrial equipment for about 24%.
- Automotive for about 12%.

For most of these applications, manufacturing costs must be optimized to reach mass production. The discrete's cost per ampere will always be lower than that of a high-current module because this package is simple to manufacture facilitating series production.

### 2.2. Package manufacturing process and importance of die interconnections

The manufacturing process of a discrete power package (e.g., medium packages such as TO-220, $D^2PAK$, etc.) is composed of the following steps. First of all, a copper lead-frame constitutes the skeleton of the package. It provides both the heat-sink and the terminals (leads). For non-insulated package, the semiconductor die is directly soldered on the lead-frame. Regarding insulated package, a ceramic layer (e.g., alumina) is first soldered on the lead frame to provide insulation, either for safety reasons (if the end user can access the heat-sink) or functional considerations (if the heat-sink is connected to another potential). Then, the die is soldered on the ceramic. Whatever the package (insulated or not), all the layers are stacked up with lead-based solder alloys. These Pb-rich alloys are still used in industry due to the good trade-off between their low cost, and their good thermal and mechanical properties (melting temperature, wettability, thermal conductivity, and coefficient of thermal expansion). The interconnection between the die and the terminals can be done using several technologies: *wire bonding, clip bonding*, or *ribbon bonding*. These techniques will be described in the next sections of this chapter. Encapsulant is the final constituent of the package. Its main role is to protect the die and its interconnections against physical damage (shocks, vibrations, etc.) and external factors (temperature, humidity, etc.). Epoxy is currently the most commonly used organic-resin encapsulant material in use because it offers a beneficial mix of properties and thermal performances at a relatively low cost.

For applications in the low-voltage and high-current ranges, specific attention should be paid to die interconnections. These interconnections necessarily lead to electric resistance, and inductance which is important to minimize to reach performance requirements, especially in high-speed electronics. Their geometric dimensions are also a key factor to optimize.

## 2.3. Wire-bonding technology: scope and reliability limitations

### *2.3.1. Wire-bonding techniques*

Wire bonding is the oldest and the most widespread technology used in industry because this is a straightforward, flexible, and cost-effective solution. Right now, it is estimated that over 90% of the manufactured packages in volume are wire bonded [21]. The wire-bonding technology consists in soldering a wire between two metal parts of the elements that must be interconnected, that is, the leads and the die metallization for a discrete power package. The most established wire materials are aluminum (Al), copper (Cu), and gold (Au) because of their high diffusion rates. For Al wires, some alloys (in the ppm range) can be used, either to harden aluminum (silicon or magnesium alloy) or to reduce corrosion (nickel alloy). Die metallization and wire must be made of the same material to prevent the formation of intermetallic layers.

Two basic techniques are currently used: *wedge wire bonding* and *ball wire bonding*. For each technique, there are three wire-bonding processes: *thermo-compression*, *thermo-sonic*, and *ultrasonic*. Wedge wire bonding typically uses the thermo-sonic and ultrasonic techniques depending on the application requirements, whereas ball wire bonding uses the thermo-compression and thermo-sonic processes.

Thermo-compression bonding requires high temperature (higher than 300°C) and force to deform wire and make bonds. The first wedge wire bonder, which was designed in the mid-twentieth century, used the thermo-compression method. Ultrasonic wedge wire bonding was then introduced in the early 1960s. This process, which combines force and ultrasonic power, is performed at room temperature. In comparison with thermo-compression bonding, the welding time is shorter. Thermo-sonic bonding consists in adjusting heat, force, and ultrasonic power to bond a wire. Nowadays, even if this process was first implemented in a wedge bonder in 1970, thermo-sonic bonding is typically used to bond a gold wire to either a gold or an aluminum surface on a substrate.

Among these three bonding processes, ultrasonic bonding is primarily used for Al wire in power electronics device applications (see **Figure 1**). In that case, the wire diameter range can easily reach 100–500 μm depending on the applications requirements (in particular, current-carrying capacity) and the process compatibility [22].

http://dx.doi.org/10.5772/intechopen.69611

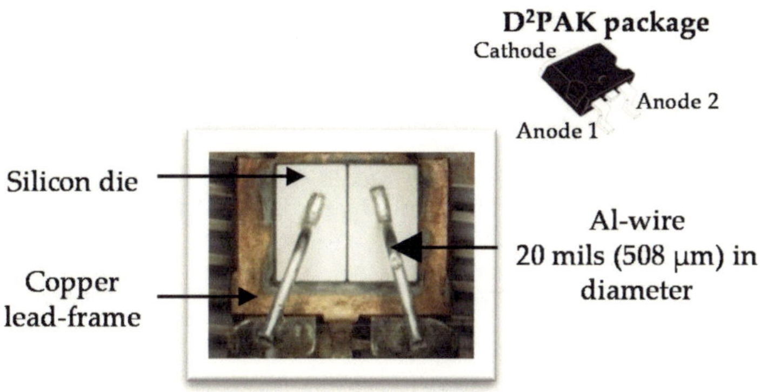


**Figure 1.** 2 × 20 A, 170-V Schottky diode assembled in a D²PAK package using Al-wire bonding.

### *2.3.2. Reliability issues*

Many authors have highlighted that wire-bonding failures of power packages are mainly caused either by shear stresses generated between the die and the wire, or due to repeated flexure of the wire. Two main failure mechanisms can particularly occur: *wire bonding lift-off* and *heel cracking* [23–26].

Wire bond lift-off (see **Figure 2**) occurs due to crack propagation along the interface between the wire and the die. The initiation of a fracture mechanism within the wire tail itself is responsible for the crack development. Its propagation is thermally activated. In particular, during active or passive temperature cycling, the CTE (coefficient of thermal expansion) mismatch between the wire bond material (e.g., aluminum) and die material (e.g., silicon) induces the crack propagation that finally leads to the wire bond lift-off. Many studies have reported the numerical methods to calculate such strength applied onto the wire. Many authors have also reported that it is possible to strengthen wire bond reliability by gluing the wire to the die metallization using a coating layer (e.g., a polyimide cover layer).

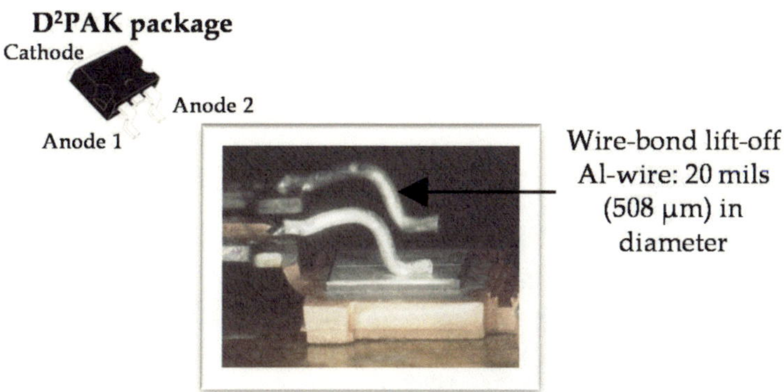


**Figure 2.** Aluminum wire bond lift-off (2 × 20 A, 170-V Schottky diode assembled in a D²PAK package).

Heel cracking is amplified by temperature cycling. In that case, a crack propagation occurs at the wire heel (see **Figure 3**). This phenomenon can lead to partial wire disconnection. As a consequence, the electrical conductivity is not completely achieved. As widely reported in the literature, when a wire bond is subjected to thermal cycles, the wire dilatation induces a flexure. For example, 50°C temperature swing can produce a 10-μm increase of length. However, because the wire is bonded on a metallization layer, it leads to 0.05° additional angle face to the bonded region.

### 2.4. Clip-bonding technology

Clip interconnection may be an alternative to wire bonding because it is a "bond-less" packaging technology. This technique consists in connecting the active area of a die to the package lead frame using a small metal slab which is directly soldered on die's top surface. At the moment, a solid copper bridge, also called "clip," is widely used in discrete power packages (see **Figure 4**) [27].

Copper clip bonding allows both larger possible contact area and lower on-resistance than wires. For example, on a MOSFET (metal-oxide-silicon field effect transistor) device, one copper clip may replace 15 gold wires. In that case, the static drain-source on-resistance ($R_{DS(on)}$) can be reduced by about 30%, while, at the same time, providing an improved current distribution into the device [28].

Clip bonding helps to strengthen the thermal behavior of a power package by providing efficient thermal dissipation from the top of the die to the lead-frame. Therefore, the maximum junction temperature during power device operation can be optimized, which is a key parameter to manage in order to extend its operation life and reliability [29].

Despite these clear advantages, the clip-bonding technology has some drawbacks. A major disadvantage is particularly related to the chemical aspects of the clip soldering. The clip is typically reflow soldered to the die and substrate pads. After clip bonding, it is very important to remove flux residues. This means that the cleaning agent must chemically match the properties of solder paste residues. Thus, special attention should be paid to ensuring a high-reliability bonding process [30].

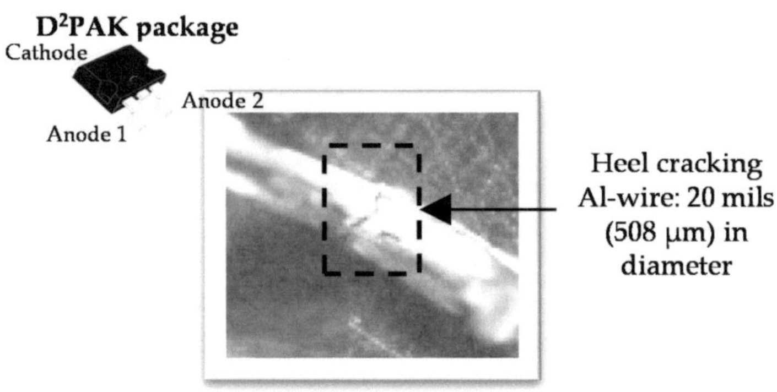


**Figure 3.** Aluminum wire heel cracking (2 × 20 A, 170-V Schottky diode assembled in a D²PAK package).

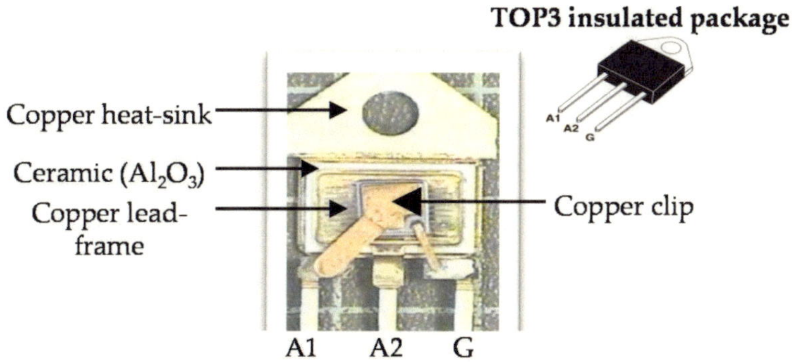


**Figure 4.** 25 A, 1200-V Triac assembled in a TOP3 insulated package using copper clip bonding.

### 2.5. Ribbon bonding as the most efficient and cost-effective interconnection technology

Today's industrial applications require higher current density in electronic power devices. These devices are implemented in more and more compact, lightweight, energy efficient, and cost-effective on-board systems. Die-interconnection technologies have a key role to play in achieving these objectives. They must above all warrant the electrical, thermal, and mechanical stability of the whole package. Their process conditions must also be controlled, stable, reliable, and cost effective.

Ribbon bonds represent currently a very attractive solution for power electronic applications that carry high electrical loads [31–33]. The process consists in interconnecting a semiconductor die to a lead in a power package using a flexible conductive ribbon (see **Figure 5**). Aluminum as interconnect material offers a good compromise between electrical performances (in particular, on-resistance) and cost. Ultrasonic technology is typically used because of its main strengths such as high flexibility and reliability, low cost, and increased productivity.

Compared to aluminum wire bonding, aluminum ribbon bond helps to replace a significant number of parallel wires per device to fulfill the current or necessary on-resistance requirements. For example, one aluminum ribbon (width and thickness equal to 60 mils (1524 μm)

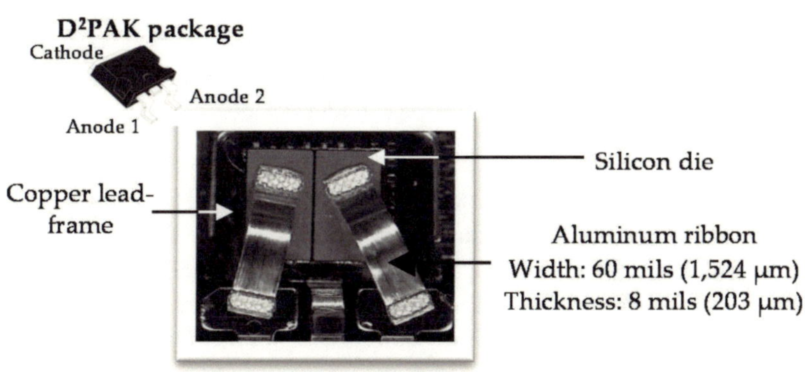


**Figure 5.** 2 × 20 A, 170-V Schottky diode assembled in a D²PAK package using Al-ribbon bonding.

and 8 mils (203 μm), respectively) can replace about two aluminum wires (20 mils (508 μm) in diameter) [34]. In that case, the ribbon limits the risk of non-continuous contact area and inhomogeneous heat dissipation in comparison with multiple wire bonds. Moreover, the high cross-sectional area of aluminum ribbon bond limits parasitic resistances and inductances which can lead to additional losses, especially in high-frequency-switching applications.

# 3. Relevance of aluminum ribbon bonding in temperature-cycling applications

## 3.1. Methodology

### *3.1.1. Experimental reliability test procedure*

A power Schottky diode (2 × 20 A, 170 V, 175°C maximum junction temperature) assembled in a D²PAK package was chosen as a test device for the reliability analysis. Two kinds of experimental tests were performed: *thermal cycling* (passive temperature cycling) and *power cycling* (active temperature cycling). Each test was based on automotive qualification documents such as the AEC-Q101 standard [35]. In particular, this standard describes the cyclic temperature profiles. For thermal cycling, the test conditions are as follows: −65°C/+150°C ($\Delta T_j$ = 215°C), two cycles per hour, 1000 cycles. Regarding power cycling, the devices must be subjected to a junction temperature mismatch at least equal to 100°C during 8572 cycles (one cycle lasts 7 min).

For each test, two sets of samples were used. The first one was composed of 77 units manufactured using the aluminum wire-bonding technique (20 mils (508 μm) in diameter). The second one was composed of 77 devices under test manufactured using the aluminum ribbon-bonding process (width and thickness equal to 60 mils (1524 μm) and 8 mils (203 μm), respectively). It is important to note that all devices under test had the same die size (4.87 mm × 4.24 mm × 280 μm).

Several readouts were carried out during the experimental reliability tests. This means that all units were removed from the test bench at several fixed time intervals. For thermal cycling, the readouts were 100 cycles, 500 cycles, 1000 cycles, 1250 cycles, and 2000 cycles. Regarding power cycling, the devices under test were removed at 4286 cycles and 8572 cycles. For each duration mentioned earlier, the following electrical and thermal parameters were measured for each unit under test: forward voltage drop ($V_F$), reverse leakage current ($I_R$), and junction-to-case thermal resistance ($R_{th(j\text{-}c)}$).

For both sets (the first one using the Al-wire-bonding technique and the second one using the Al-ribbon-bonding process), the evolution of each parameter (in relation to the number of passive or active temperature cycles) can be shown on a normal probability plot (Henry's chart). This chart is typically used to extract the mean and standard deviation of the statistical distribution. Regarding the targeted failure mechanism (bond lift-off), we only focused on the $V_F$-parameter. In particular, its evolution decided when a power device reached its end of life. This parameter was evaluated using the temperature dependence of the forward voltage drop. Regarding the initial $V_F$-values, both sets of 77 units under test were homogeneous before the reliability test launching (average value and standard deviation equal to 0.85 V for 40-A forward current and 5 mV, respectively). The failure criterion we took into consideration was a $V_F$-parameter increase higher than 5% with respect to its initial value. This failure criterion is severe as compared to that of the AEC-Q101 standard, for which allowable shift values within ±20% of the initial readings are tolerated [35]. This aims to establish a reliability analysis as quickly as possible (considering the duration of the passive or active temperature-cycling tests).

### 3.1.2. FEM thermo-mechanical modeling

Thermo-mechanical simulations (for instance, with ANSYS® Workbench software) using the FEM (finite-element method) are commonly used to get a better understanding of devices' failure mechanism.

In this study, a three-dimensional model of the D²PAK package was generated. As mentioned previously, the test device was composed of two Schottky diodes. However, one diode was only taken into consideration to simplify the modeling and optimize the simulation durations. The numerical model was composed of quadrilateral meshed elements with a refined meshing at the wire and especially the heel, as well as the contact area between the silicon die and the wire (see **Figure 6**). The highest element edge length was equal to 200 μm and the lowest one was equal to 20 μm.

The thermo-mechanical properties of the materials are summarized in **Tables 1** and **2** [14]. Regarding the thermal loads, the passive temperature-cycle profile (−65°C/+150°C, rise time, fall time, and cycle duration equal to 15, 15, and 30 min, respectively) was used. For power cycling, a heat flow was adjusted to meet the junction temperature profile described in the previous paragraph. Regarding the mechanical loads, the *X*-, *Y*-, and *Z*-displacements were blocked at the origin of the plan to freely authorize the expansion of materials

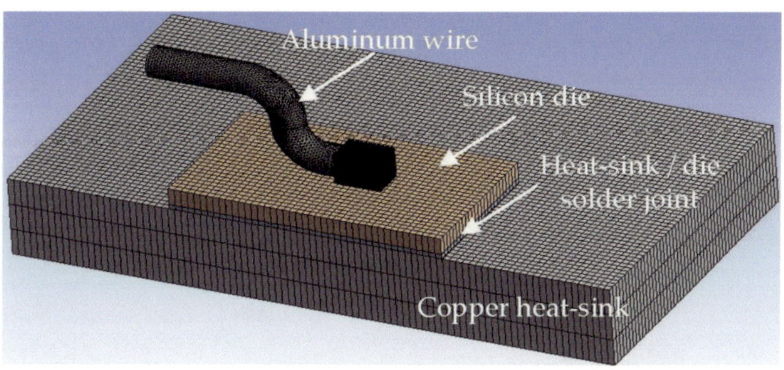


**Figure 6.** FEM model of the D²PAK package (an example of assembly using Al-wire-bonding process).

| | Dimensions | Thermal conductivity ($W.m^{-1}.K^{-1}$) | Specific heat ($J.kg^{-1}.K^{-1}$) | Density ($kg.m^{-3}$) |
|---|---|---|---|---|
| Copper heat-sink | 10.4 × 10.75 × 1.36 mm | 330 | 385 | 8900 |
| Al wire | 508 μm in diameter | 195–0.059 T | 920 | 2680 |
| Al ribbon | Width and thickness equal to 1524 and 203 μm, respectively | | | |
| Silicon die | 4.87 × 4.24 × 280 μm | 156 | 703 | 2330 |
| Epoxy resin | 10.4 × 9.35 × 4.60 mm | 0.75 | 800 | 1820 |
| Solder joint ($PbSn_5Ag_{2.5}$) | 4.87 mm × 4.24 mm × 15 μm | 44 | 130 | 11,070 |

**Table 1.** Typical thermal data of D²PAK materials [14].

## 3.2. Main results and discussion

### *3.2.1. Reliability analysis*

For each reliability test and each set of samples, the distribution of the units' lifetime was fitted with a two-parameter Weibull distribution. At the moment, this law is widely used in reliability engineering due to its versatility and relative simplicity.

From the Weibull cumulative probability density function (*cdf*), in accordance with Eq. (1), it is possible to extract the characteristic lifetime ($\eta$) at which 63.2% of the devices under test in a set failed, and the shape parameter ($\beta$) is also known as the Weibull slope

$$F(t) = cdf = 1 - e^{-\left(\frac{t}{\eta}\right)^{\beta}} \tag{1}$$

- $t$: time to failure.
- $\eta$: characteristic lifetime ($F(t) = 63.2\%$).
- $\beta$: shape parameter (Weibull slope).

For both reliability tests (thermal cycling and power cycling), the two-parameter Weibull analysis is presented in **Figure 7**. **Figure 7** highlights that the D²PAK package failure mode

| | Young's modulus (GPa) | Poisson's ratio | CTE ($10^{-6}$ $K^{-1}$) |
|---|---|---|---|
| Copper heat-sink | 120 | 0.34 | 16.8 |
| Al wire/ribbon | 64 | 0.3 | 23 |
| Silicon die | 130 | 0.28 | 2.6 |
| Epoxy resin | 16.5 | 0.3 | 19 |
| Solder joint ($PbSn_5Ag_{2.5}$) | 20.9–0.04 × T (°C) | 0.3 | 27 |

**Table 2.** Typical mechanical data of D²PAK materials [14].

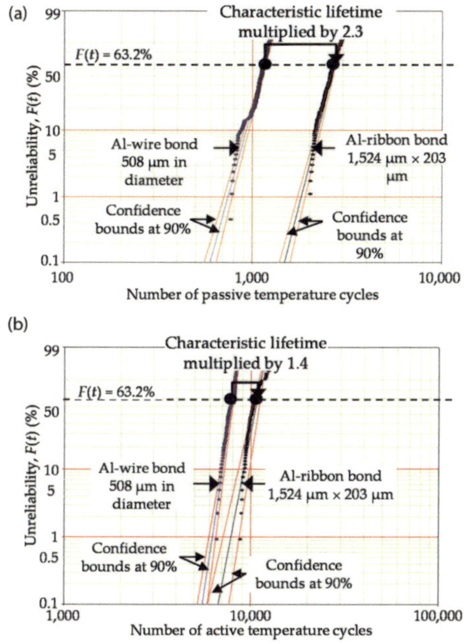


**Figure 7.** Relevance of Al-ribbon-bonding reliability during thermal cycling (a) and power cycling (b).

is of a fatigue-stress nature because the shape parameter ($\beta$) is higher than one, whatever the bonding technique used. For the thermal-cycling tests (see **Figure 7(a)**), the $\beta$-values of the units using Al-wire bonding and Al-ribbon bonding are about 10.5 and 11.5, respectively. Regarding the power-cycling tests (see **Figure 7(b)**), the $\beta$-values of the units using Al-wire bonding and Al-ribbon bonding are about 19.2 and 14.8, respectively. Many failure analysis results show the expected failure mechanism, that is, wire bond lift-off as described in **Figure 2**.

For the thermal-cycling tests, **Figure 7(a)** shows that the characteristic lifetime ($\eta_2$) of the $D^2$PAK assembly using the Al-ribbon-bonding process is about 2.3 times higher than the units using the Al-wire-bonding technique (Al-wire bonding: $\eta_1 \approx$ 1,155 cycles; Al-ribbon bonding: $\eta_2 \approx$ 2,702 cycles). Therefore, the failure acceleration of the wire bond package is more than two times higher than the ribbon bond package.

Regarding the power-cycling tests, **Figure 7(b)** shows that the characteristic lifetime ($\eta_4$) of the $D^2$PAK assembly using the Al-ribbon-bonding process is about 1.4 times higher than that of the units using the Al-wire-bonding technique (Al-wire bonding: $\eta_3 \approx$ 7,899 cycles; Al-ribbon bonding: $\eta_4 \approx$ 10,704 cycles). Thus, the failure acceleration of the wire bond package is about 1.5 times higher than that of the ribbon bond package.

### *3.2.2. Failure mechanism understanding*

This part of the paper focuses on the wire bond lift-off explanation since the degradation of the ribbon bond is more difficult to observe. In the latter case, the number of thermal cycles

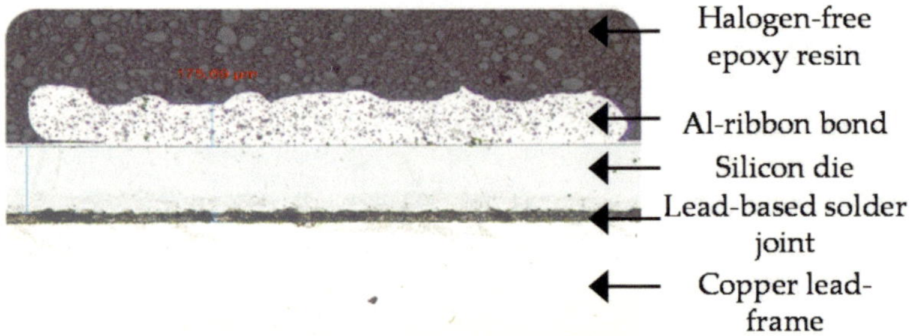


**Figure 8.** Failure analysis result of a D²PAK package using Al-ribbon bonding after 2000 passive temperature cycles (−65℃/+150℃, two cycles/h).

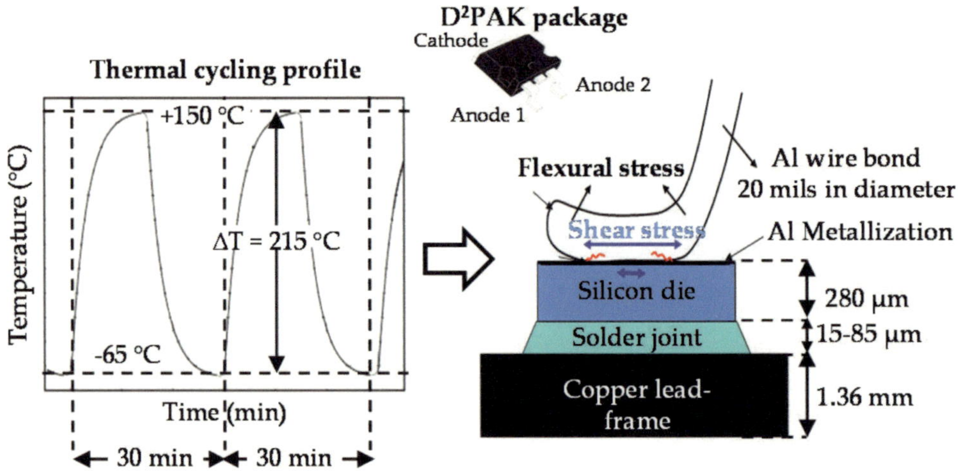


**Figure 9.** Fatigue mechanism explanation of Al-wire bond subjected to high-temperature swings.

must be much higher to have the same failure mechanism. For example, as can be seen in **Figure 8**, for thermal cycling, the ribbon bond lift-off phenomenon cannot be observed, even after 2000 temperature cycles.

The temperature cycles are connected to the stress cycles at the link between the wire and the die. As can be seen in **Figure 9**, the fatigue mechanism is likely due to two phenomena. The first one corresponds to the wire flexure at the bond heel. Indeed, repeated flexing and pulling of the wire occur as the Schottky diode heats and cools during temperature cycling, due to temperature swings. This phenomenon has already been widely described. For example, Meyyappan *et al.* developed a wire fatigue model to predict failure due to flexure in wedge-bonded power modules [36].

The Von Mises stress can typically be used to determine whether an isotropic and ductile material (such as aluminum) will yield when subjected to a complex loading condition. This is accomplished by calculating the Von Mises yield criterion and comparing it to the material's yield stress. **Figure 10** shows the Von Mises stress distribution after one passive temperature cycle (−65℃/+150℃. Rise time, fall time, and cycle duration equal to 15, 15, and 30 min, respectively). The aluminum wire exhibits some stress concentration in the heel.

The wire bond lift-off is also initiated by a thermo-mechanical stress (shear stress) caused by the CTE mismatch between the aluminum bond wire ($CTE_{Al}$ = 23.8 ppm.℃$^{-1}$) and the silicon die ($CTE_{Si}$ = 2.6 ppm.℃$^{-1}$). The number of thermal cycles to failure ($N_f$) is proportional to the maximum strain amplitude ($\Delta\varepsilon$), which depends on the CTE mismatch and the temperature swing ($\Delta T$) in accordance with Eq. (2)

$$N_f \propto \Delta\varepsilon = (CTE_{Al} - CTE_{Si})\Delta T \quad (2)$$

- $N_f$: number of thermal cycles to failure.
- $\Delta c$: maximum strain amplitude.
- $\Delta T$: temperature swing.
- $CTE_{Al}$ = 23.8 ppm.°C$^{-1}$. $CTE_{Si}$ = 2.6 ppm.°C$^{-1}$.

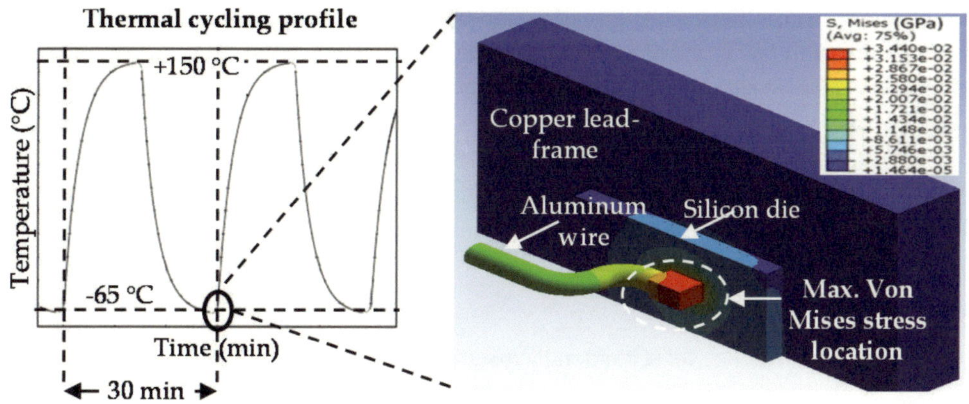


**Figure 10.** Example of Von Mises stress distribution after one passive temperature cycle (−65°C/+150°C, two cycles/h, D²PAK package using Al-wire-bonding process).

The failure analysis results highlight that the bonding rupture starts to progress laterally, finally causing the bond wire to lift off.

The Schottky diodes assembled in a D²PAK package using the Al ribbon-bonding process exhibit a better robustness during thermal cycling. Their characteristic lifetime is about 2.3 times higher than the D²PAK units that use the Al wire-bonding technique. A higher contact surface on the silicon die, the low-loop profile, and the stiffness of the ribbon may allow slowing down crack initiation and propagation between the Al bond and the Al metallization on top of the silicon die.

## 4. Conclusions

This chapter has pointed out the relevance of aluminum (Al) ribbon bonding used in the assembly process of discrete power packages. The reliability analysis has been particularly performed on a surface-mount device ($D^2PAK$ assembly) subjected to thermal cycling and power cycling. For example, at the moment, such package is widely used in automotive applications.

To assess the good performances of Al-ribbon bonding, the reliability analysis has been based on a comparative study between Al-wire bonding and Al-ribbon bonding.

For both thermal cycling and power cycling, the Weibull analysis has highlighted that the failure mode of the $D^2PAK$ package is of a fatigue-stress nature, whatever the bonding technique used. Many failure analysis results have shown wire bond lift-off. The degradation of the ribbon bond is more difficult to observe.

Regarding thermal cycling, the experimental test results have also shown that failure acceleration of the wire bond package is about 2.5 times higher than that of the ribbon bond package. For power cycling, this acceleration factor is about 1.5.

Thermo-mechanical simulations using finite elements have shown that a stress concentration can be observed in the heel area. For the wire-bonding technique, the wire is subjected to repeated flexing and pulling that lead to the wire lift-off. The ribbon-bonding process shows a higher robustness. It can be explained by a higher contact surface on the die, the low-loop profile, and the stiffness of the ribbon.

## Author details

Sébastien Jacques

Address all correspondence to: sebastien.jacques@univ-tours.fr

Research Group on Materials, Microelectronics, Acoustics and Nanotechnology, GREMAN UMR 7347, CNRS INSA, University of Tours, France

## References

[1] Lorubio G, Schlosser P. Euro mix: Current European energy developments and policy alternatives for 2030 and beyond. IEEE Power and Energy Magazine. 2014;**12**(2):65-74. DOI: 10.1109/MPE.2013.2294319

[2] Burmaoglu S, Ozcan S. Evolutionary evaluation of energy and nanotechnology relationship. In: 2016 Portland International Conference on Management of Engineering and Technology; 4-8 September 2016; Portland, OR, USA: IEEE; 2016. pp. 788-794. DOI: 10.1109/PICMET.2016.7806536

[3] European Environment Agency. Annual European Union greenhouse gas inventory 1990-2012 and inventory report 2014, EEA Technical Report. 2014;**9**:19. DOI: 10.2800/18304

[4] Ackermann T, Maria Carlini E, Ernst B, Groome F, Orths A, O'Sullivan J et al. Integrating variable renewables in Europe: Current status and recent extreme events. IEEE Power and Energy Magazine. 2015;**13**(6):67-77. DOI: 10.1109/MPE.2015.2461333

[5] Cheng L, Chang Y, Lin J, Singh C. Power system reliability assessment with electric vehicle integration using battery exchange mode. IEEE Transactions on Sustainable Energy. 2013;**4**(4):1034-1042. DOI: 10.1109/TSTE.2013.2265703

[6] Shojaabadi S, Abapour S, Abapour M, Nahavandi A. Optimal planning of plug-in hybrid electric vehicle charging station in distribution network considering demand response programs and uncertainties. IET Generation, Transmission & Distribution. 2016;**10**(13):3330-3340. DOI: 10.1049/iet-gtd.2016.0312

[7] Pavlovsky M, Guidi G, Kawamura A. Assessment of coupled and independent phase designs of interleaved multiphase buck/boost DC-DC converter for EV power train. IEEE Transactions on Power Electronics. 2014;**29**(6):2693-2704. DOI: 10.1109/TPEL.2013.2273976

[8] Buttay C, Rashid J, Johnson C. M, Udrea F, Amaratunga G, Ireland P et al. Compact inverter designed for high-temperature operation. In: IEEE, editor. Power Electronics Specialists Conference; 17-21 June 2007; Orlando, FL, USA. 2007. DOI: 10.1109/PESC.2007.4342357

[9] Xu NZ, Chung CY. Reliability evaluation of distribution systems including vehicle-to-home and vehicle-to-grid. IEEE Transactions on Power Systems. 2016;**31**(1):759-768. DOI: 10.1109/TPWRS.2015.2396524

[10] Astigarraga D, Ibanez FM, Galarza A, Echeverria JM, Unanue I, Baraldi P et al. Analysis of the results of accelerated aging tests in insulated gate bipolar transistors. IEEE Transactions on Power Electronics. 2016;**31**(11):7953-7962. DOI: 10.1109/TPEL.2015.2512923

[11] Khazaka R, Mendizabal L, Henry D, Hanna R. Survey of high-temperature reliability of power electronics packaging components. IEEE Transactions on Power Electronics. 2015;**30**(5):2456-2464. DOI: 10.1109/TPEL.2014.2357836

[12] Palmer MJ, Johnson RW, Autry T, Aguirre R, Lee V, Scofield JD. Silicon carbide power modules for high-temperature applications. IEEE Transactions on Components, Packaging and Manufacturing Technology. 2012;**2**(2):208-216. DOI: 10.1109/TCPMT.2011.2171343

[13] Mantooth HA, Glover MD, Shepherd P. Wide bandgap technologies and their implications on miniaturizing power electronic systems. IEEE Journal of Emerging and Selected Topics in Power Electronics. 2014;**2**(3):374-385. DOI: 10.1109/JESTPE.2014.2313511

[14] Jacques S, Caldeira A, Batut N, Schellmanns A, Leroy R, Gonthier L. Lifetime prediction modeling of non-insulated TO-220AB packages with lead-based solder joints during power cycling. Microelectronics Reliability. 2012;**52**(1):212-216. DOI: http://dx.doi.org/10.1016/j.microrel.2011.08.017

[15] Medjahed H, Vidal P-E, Nogarede B. Thermo-mechanical stress of bonded wires used in high power modules with alternating and direct current modes. Microelectronics Reliability. 2012;**52**(6):1099-1104. DOI: http://dx.doi.org/10.1016/j.microrel.2012.01.013

[16] Pedersen KB, Pedersen K. Bond wire lift-off in IGBT modules due to thermomechanical induced stress. In: 2012 3rd IEEE International Symposium on Power Electronics for Distributed Generation Systems; 25-28 June 2012; Aalborg, Denmark: IEEE; 2012. pp. 519-526. DOI: 10.1109/PEDG.2012.6254052

[17] Kanert W. Active cycling reliability of power devices: Expectations and limitations. Microelectronics Reliability. 2012;**52**(9-10):2336-2341. DOI: http://dx.doi.org/10.1016/j.microrel.2012.06.031

[18] Jacques S, Leroy R, Lethiecq M. Impact of aluminum wire and ribbon bonding technologies on D2PAK package reliability during thermal cycling applications. Microelectronics Reliability. 2015;**55**(9-10):1821-1825. DOI: http://dx.doi.org/10.1016/j.microrel.2015.06.012

[19] Bajenescu T-M, Bazu MI, editors. Reliability of Electronic Components: A Practical Guide to Electronic Systems Manufacturing. Berlin, Germany: Springer-Verlag Berlin Heidelberg New York ed.; 1999. p. 509. DOI: 10.1007/978-3-642-58505-0

[20] Neacsu DO. editor. Switching Power Converters: Medium and High Power. Boca Raton: CRC Press ed. 2013. p. 589

[21] Zhang GQ, Van Driel WD, Fan XJ, editors. Solid Mechanics and its Applications, Mechanics of Microelectronics. Springer ed. Netherlands: Springer Netherlands; 2006. p. 563. DOI: 10.1007/1-4020-4935-8

[22] Chauhan PS, Choubey A, Zhong Z, Pecht M, editors. Copper Wire Bonding. New York, NY: Springer-Verlag New York ed.; 2014. p. 235. DOI: 10.1007/978-1-4614-5761-9.S

[23] Ciappa M. Selected failure mechanisms of modern power modules. Microelectronics Reliability. 2002;**42**(4-5):653-667. DOI: 10.1016/S0026-2714(02)00042-2

[24] Khatibi G, Lederer M, Weiss B, Licht T, Bernardi J, Danninger H. Accelerated mechanical fatigue testing and lifetime of interconnects in microelectronics. Procedia Engineering. 2010;**2**(1):511-519. DOI: http://dx.doi.org/10.1016/j.proeng.2010.03.055

[25] Ji B, Pickert V, Cao W. In-situ diagnostics and prognostics of wire bonding faults in IGBT modules for electric vehicle drives. IEEE Transactions on Power Electronics. 2013;**28**(12):5568-5577. DOI: 10.1109/TPEL.2013.2251358

[26] Popok VN, Pedersen KB, Kristensen PK, Pedersen K. Comprehensive physical analysis of bond wire interfaces in power modules. Microelectronics Reliability. 2016;**58**:58-64. DOI: http://dx.doi.org/10.1016/j.microrel.2015.11.025

[27] Min Woo DR, Yuan HH, Jie Li JA, Ling HS, Bum LJ, Songbai Z. High power SiC inverter module packaging solutions for junction temperature over 220°C. In: Electronics Packaging Technology Conference; 3-5 December 2014; Marina Bay Sands, Singapore: IEEE; 2014. pp. 31-35. DOI: 10.1109/EPTC.2014.7028383

[28] Wei TB, Kho L, Long LH, Jeng LL, Hang GS. Universal copper clip packaging solution for power management IC. In: Semiconductor Technology International Conference; 15-16 March 2015; Shanghai, China: IEEE; 2015. DOI: 10.1109/CSTIC.2015.7153437

[29] Zhu Y, Chen H, Xue K, Li M, Wu J. Thermal and reliability analysis of clip bonding package using high thermal conductivity adhesive. In: Electronics Packaging Technology Conference; 11-13 December 2013; Singapore: IEEE; 2013. pp. 259-263. DOI: 10.1109/EPTC.2013.6745724

[30] Hwang J, editor. Solder Paste in Electronics Packaging: Technology and Applications in Surface Mount, Hybrid Circuits, and Component Assembly. USA: Springer Science & Business Media ed.; 2012. p. 456. DOI: 10.1007/9781461535287

[31] Ong BYY, Chuah SMC, Luechinger C, Wong G. Heavy Al ribbon interconnect: An alternative solution for hybrid power packaging. In: IMAPS 37th International Symposium on Microelectronics; 14-18 November 2004; Long Beach, CA, USA: IMAPS; 2004. p. 11

[32] Milke E, Mueller T. High temperature behaviour and reliability of Al-Ribbon for automotive. In: Electronics System-Integration Technology Conference; 1-4 September 2008; Greenwich, London, UK: IEEE; 2008. pp. 417-422. DOI: 10.1109/ESTC.2008.4684384

[33] Almagro EIV, Granada HT. Stack bonding technique using heavy aluminum ribbon wires. In: 2008 10th Electronics Packaging Technology Conference; 9-12 December 2008; Singapore: IEEE; 2008. pp. 976-981. DOI: 10.1109/EPTC.2008.4763556

[34] Zhou YN, editor. Microjoining and Nanojoining. Elsevier ed. Cambridge, England: Woodhead Publishing Ltd; 2008. p. 832

[35] Delphi Electronics & Safety, Siemens VDO, and Visteon Corporation. Stress Test Qualification for Automotive Grade Discrete Semiconductors. USA: Automotive Electronics Council; 2005. p. 25

[36] Eleffendi MA, Johnson CM. In-service diagnostics for wire-bond lift-off and solder fatigue of power semiconductor packages. IEEE Transactions on Power Electronics. 2017;**32**(9):7187-7198. DOI: 10.1109/TPEL.2016.2628705

Chapter 11

# Reliability Prediction of Smart Maximum Power Point Converter for PV Applications

Giovanna Adinolfi and Giorgio Graditi

Additional information is available at the end of the chapter

http://dx.doi.org/10.5772/intechopen.72130

**Abstract**

The photovoltaic generation distribution supports new energetic scenario development (Net Zero Energy Cluster and DC microgrids). In this context, smart maximum power point (SMPPT) converter represents innovative systems that are able to monitor operating conditions, communicate energetic production data and signal a fault condition. The Smart Maximum Power Point role becomes crucial, and critical aspects such as efficiency and reliability have to be taken into account from the beginning of the design. In this chapter, the idea is to review different reliability prediction models for electronic components focusing on the military ones with the analysis of a case study related to a Smart Maximum Power Point converter in photovoltaic applications.

**Keywords:** MIL-HDBK-217F photovoltaic, reliability, RIAC 217 Plus

## 1. Introduction

Nowadays, photovoltaic (PV) systems are wide and capillary distributed. With the introduction of digital technologies, a new role emerges for these systems. In new scenarios (microgrids [1] and Net Zero Energy Buildings), PV systems become energetic and information "nodes" of the grid. In this context, the efficiency of PV technologies can be improved by the introduction of a smart maximum power point (SMPPT) converter that is able to maximize each photovoltaic generator power by tracking its Maximum Power Point by real-time impedance matching. These converters are characterized by the presence of a digital microcontroller that is able to "manage" the photovoltaic power but also to assure the execution of ancillary services such as monitoring, diagnostics, communication and so on. Crucial aspects are efficiency and reliability; these issues have to be considered since design starting phase. Many prediction models can be used to estimate reliability indices for Smart Maximum Power Point converters. In this manuscript, the attention is focused on the Military Handbook 217F and its successor RIAC 217 Plus. The case study of a SMPPT Synchronous converter is analyzed.

## 2. SMPPT converter for PV applications

SMPPT converters are DC-DC converters equipped by a suitable control stage. In literature, numerous topological solutions of DC-DC converters [2] have been proposed for PV applications, each of them presenting vantages and critical aspects. Choosing a topology rather than another should be done taking into account the different system parameters in order to determine the optimal solution in terms of cost-benefit analysis.

Topological solution suitable for SMPPT converters can be characterized by synchronous (SR), diode rectification (DR) or interleaved (IL) solutions. In detail, traditional step-up or step-down circuits can be realized with MOSFET for both switching devices, so constituting the SR boost or SR buck. PV application of the most used topology is the boost one reported in the SR, DR and Il versions in **Figure 1**.

IL topology is characterized by a cellular architecture, in which many converters, called cells, are paralleled to create a unique converter. The cells share the same input and output voltages, but each one processes only a fraction of the total system power. One of the primary benefits of a cellular conversion approach is the large degree of input and output ripple cancelation, which can be achieved among cells, leading to reduced ripple in the aggregate input and output waveforms. The active method of interleaving permits to obtain more advantages. In the interleaving method, the cells are operated at the same switching frequency with their switching waveforms displaced

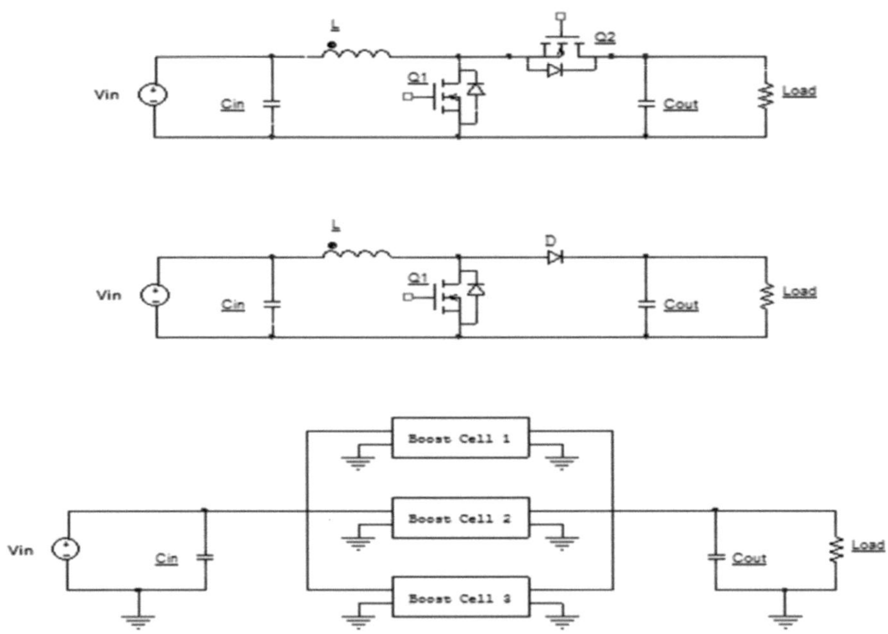


**Figure 1.** SMPPT boost topologies: (a) SR; (b) DR; (c) IL.

in phase over a switching period. The benefits of this technique are due to harmonic cancelation among the cells and include low ripple amplitude and high ripple frequency in the aggregate input and output waveforms. For a broad class of topologies, interleaved operation of N cells yields an N-fold increase in fundamental current ripple frequency, and a reduction in peak ripple magnitude by a factor of N or more compared to synchronous operation. To be effective in cellular converter architecture, however, an interleaving scheme must be able to accommodate a varying number of cells and maintain operation after some cells have failed.

SMPPT converters work in continuous changing conditions as it is evident exploiting historical climatic data series (temperature, irradiance, wind speed) monitored by weather stations [3] and appropriately acquired (**Figure 2**). Since it is difficult to identify the worst operating condition considering both the ambient temperature and the irradiance, the annual frequency of different meteoclimatic conditions can be analyzed in order to identify the most frequent ones.

Both temperature and irradiance effects have to be taken into account to the aim of accurate SMPPT reliability and efficiency estimations. It is possible to identify a joined parameter, the backside temperature $T_{backPV}$, able to represent both temperature and irradiance conditions.

The $T_{backPV}$ formula is reported in Eq. (1).

$$T_{backPV} = T_{PV} + \frac{S}{S_0}\Delta T \tag{1}$$

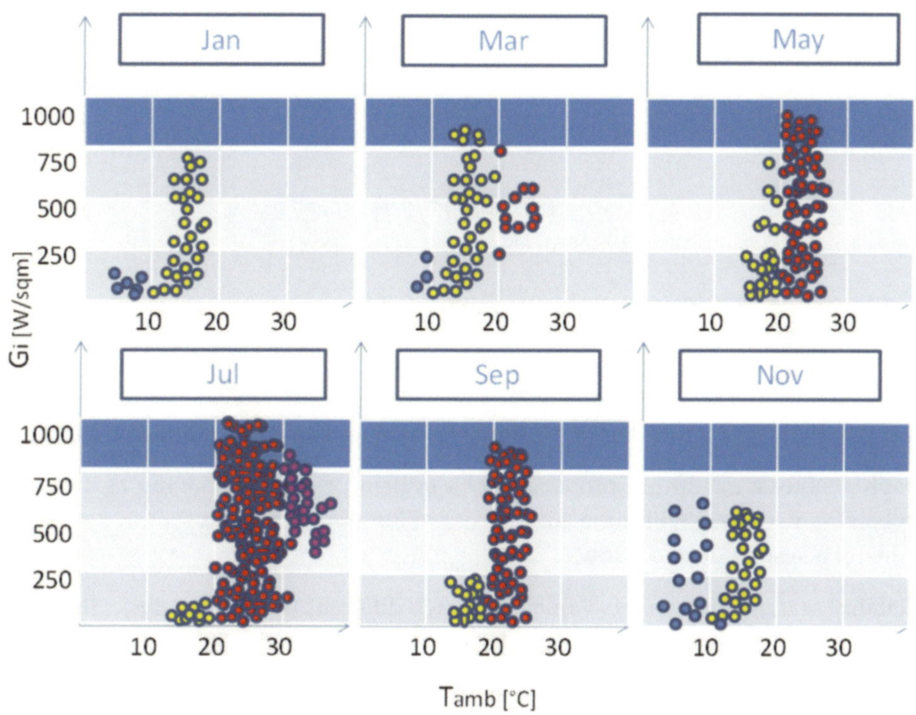


**Figure 2.** Annual distribution of temperature and irradiance.

where

-S is the irradiance at the specific operating condition;

-$S_0$ is 1000 W/$m^2$;

-$T_{PV}$ is the PV generator temperature calculable by means of Eq. (2).

$$T_{PV} = \left(\left(\frac{S}{S_0}\right) * \left(T_1 e^{bw} + T_2\right) + T_a\right) + \frac{S}{S_0}\Delta T \quad (2)$$

where

-$T_a$ is the ambient temperature;

-$T_{1,}$ $T_2$ are empirical coefficients determining upper and lower temperature limits at low wind speeds, in °C;

-b is empirical coefficient determining the rate that module temperature drops as wind speed increases.

## 3. Reliability assessment

The maintenance of correct functioning mode in the time represents an important task for SMPPT converters [4]. This issue involves the concept of reliability evaluation that can be carried out by different reliability prediction model. In detail, the internationally approved definition for the reliability R(t) consists in the probability that an item will perform a required function without failure under stated conditions for a stated period of time. This definition is reported in Eq. (3).

$$R(t) = Pr\{T > t\} \quad (3)$$

where

-t is the mission time;

-T is the unit lifetime.

The reliability value R is a number in the range (0, 1), and its graph is shown in **Figure 3**.

This graph denotes a maximum probability of a system proper functioning at the beginning. This probability decreases with time, and in case of long mission time, the proper functioning probability of a system/device is low.

The reliability of a system can be evaluated by two different types of analysis: the part count analysis (PCA) and the part stress analysis (PSA).

PCA is used in the initial design phase when components and their parameters have not yet been decided. Only the knowledge of the type of components, their level of quality and the environment, in which the system will be used, are necessary.

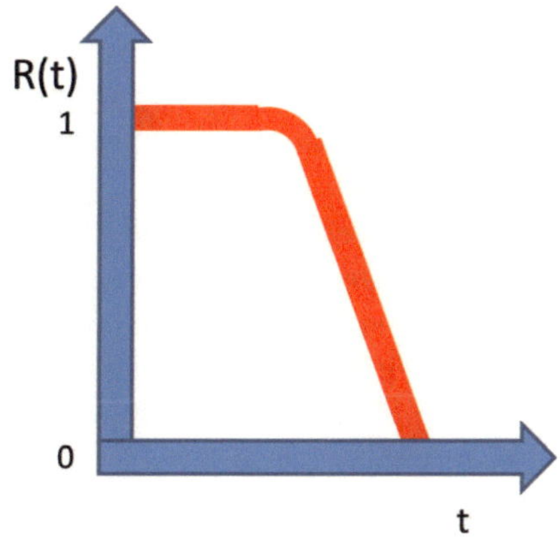


**Figure 3.** Reliability function R(t) graph.

PSA is used in an advanced design phase when the list of system components is known. The failure rate is calculated by considering effects related to the temperature and the electrical stresses to which the devices are subjected.

Various mathematical reliability models [5] are available according to the class of considered components (electronic, electrical, electromechanical devices).

Scientific studies on the reliability of electronic devices have shown that the exponential model represents the suitable one for describing the behavior of these components, as reported in Eq. (4).

$$R(t) = e^{-\lambda t} \tag{4}$$

where

-$\lambda$ is the failure rate;

-t is the mission time.

The exponential distribution is memoryless, and in fact the following expression demonstrates that the probability to have a device lifetime longer than (t + $t_1$) depends only on $t_1$ and it does not depend on t.

$$Pr\{T > t + t_1 \backslash T > t\} = \frac{Pr\{T > t + t_1\}}{Pr\{T > t\}} = \frac{R(t + t_1)}{R(t)} = \frac{e^{-\lambda(t+t_1)}}{e^{-\lambda t}} = e^{-\lambda t_1} \tag{5}$$

In the case of electronic components, this property means that they only break for accidental causes but not for wear.

### 3.1. Reliability prediction model

As said earlier, many reliability prediction models are available.

The most used ones are:

- MIL-HDBK-217: this reliability model [6], published by the United States Navy in 1965, was the only reliability prediction method available at the time; therefore, the reliability communities adopted this tool for their own use. It is probably the most internationally recognized empirical prediction method. As a result, MIL-HDBK-217 became and is still one of the most widely known and used reliability prediction methods. It includes models for a broad range of part types and supports the five most commonly used environments in the telecom industry (Ground Fixed Controlled, Ground Fixed Uncontrolled, Ground Mobile, Airborne Commercial, and Space) plus additional alternatives useful in the military environment. It is based on pessimistic failure rate assumptions. It does not consider other factors that can contribute to failure rate such as burn-in data, laboratory testing data, field test data, designer experience, wear-out and so on.

- The Telcordia prediction model, Reliability Prediction Procedure for Electronic Equipment SR-332 [7], was developed by AT&T Bell Labs in 1997, and it is focused only on electronic equipment. This model (previously known as Bellcore) modified the MIL-HDBK-217 Prediction Model to better represent the equipment of the telecommunication industry by adding the ability to consider burn-in, field and laboratory test data. Although the Telcordia standard was developed specifically for the telecom field, it is used to model products in a number of other industries. A disadvantage is that the predictions are limited to environment that works with temperature between 30 and 65°C.

- The RIAC Handbook 217 Plus model [8], published in 2006, has been developed by the Reliability Information Analysis Center (RIAC) and pointed out by United States Department of Defense as the successor of the MIL-HDBK-217 and the PRISM methodology. The form of this model is quite different from MIL-HDBK-217 and Telcordia SR-332 because 217 Plus considers a different base failure rate for each generic class of failure mechanism. These process factors are determined by a qualitative assessment of process criteria with weighting factors applied.

- FIDES: the reliability methodology FIDES Guide 2004 [9] has been created by FIDES Group, a consortium of French industrialists from the fields of aeronautics and defense (Airbus France, Eurocopter, GIAT Industries, MBDA and THALES). The FIDES methodology is based on the Prediction Model physics of failures supported by the analysis of test data, so it is different from traditional prediction methods, which are exclusively based on the statistical analysis of historical failure data collected in the field, in-house or from manufacturers.

Many of these models allow to calculate reliability in a specific operating condition. These methods are suitable for a large number of application fields characterized by a worst case or a nominal operating case. In photovoltaic applications, neither a nominal operating condition nor a worst case can be identified. PV systems work, in fact, in constantly changing conditions in terms of irradiance, room temperature, wind speed, humidity and so on. At first analysis, a reliability analysis could be carried out on the worst case, but results would not properly characterize a PV plant since the worst-case scenario is certainly not the condition that prevalently occurs. The further difficulty is determined by the operation of PV systems which does not, as in other applications, continue. A PV plant does not work 24 h a day for 365 days a year, but a variable number of hours vary depending on different factors, such as sunshine, season and geographic location.

### 3.1.1. Structure connection types

Another aspect to consider in reliability analysis consists in the structure information and types of connections among the devices of the whole system under investigation. In fact, it is important to consider how the different parts are connected to form the system: In series, in parallel or in series: Parallel combinations. In **Figure 4**, series, parallel and hybrid connections and their respective failure rate formulas are reported.

#### 3.1.1.1. Series connection

A series structure system (**Figure 4(a)**) functions only when all of its parts are correctly functioning. In this case, the event consisting in the "correct operating mode" S of the system is given by the intersection of the "good functioning" events $A_i$ of the i-parts:

$$S = A_1 \cap A_2 \cap A_3 \ldots\ldots \cap A_N \tag{6}$$

In the assumption that $A_i$ events are stochastically independent, the reliability of the "series" system can be calculated by means of Eq. (7).

$$R_s = Pr\{S\} = \prod_{i=1}^{N} Pr\{A_i\} = \prod_{i=1}^{N} R_i \tag{7}$$

where $R_i$ is the ith part reliability.

The series connection failure rate is calculated as reported in the next formula.

$$\lambda_s(t) = \sum_{i=1}^{N} \lambda_i(t) \tag{8}$$

where $\lambda_i$ is the ith part failure rate.

SMPPT converters characterized by buck and boost topologies are examples of series structures.

#### 3.1.1.2. Parallel connection

A parallel structure system (**Figure 4(b)**) fails only when every one of its parts fails. The next formula points out that if $\bar{A}_i$ is the "incorrect operation" of the ith part of the system, the event "incorrect operation" of the system is given by the intersection of the different $\bar{A}_i$.

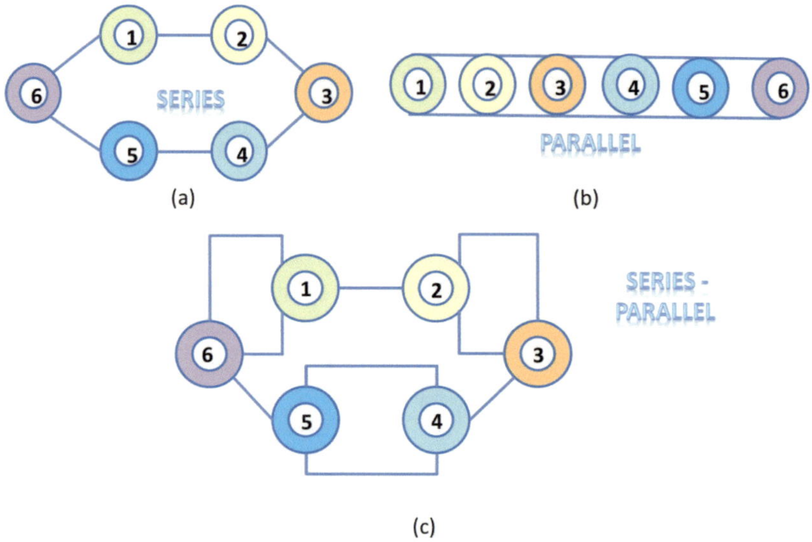


**Figure 4.** Structure connections: (a) series, (b) parallel and (c) series-parallel.

$$\overline{U} = \overline{A_i} \cap \overline{A_2} \cap \overline{A_3} \cap \ldots . \cap \overline{A_N} \tag{9}$$

Under the hypothesis of stochastical independence of the event $\bar{A}_i$, the unreliability of a parallel system ($F_p$) can be calculated by the next equation:

$$F_p = Pr\{\overline{S}\} = \prod_{i=1}^{N} Pr\{\overline{A_i}\} = \prod_{i=1}^{N} F_i \tag{10}$$

The reliability of a parallel structure system is reported in Eq. (11).

$$R_s = Pr\{S\} = 1 - \prod_{i=1}^{N} F_i \tag{11}$$

where $F_i$ is the ith part unreliability.

### *3.1.1.3. Hybrid connection*

In systems with series-parallel connections (**Figure 4(c)**), the reliability and unreliability functions of the different parts have to be calculated in order to achieve the overall reliability. SMPPT IL converters are examples of hybrid structure, so the designer has to calculate the reliability of its series connections and the unreliability of its parallel connections; finally, the whole system reliability can be evaluated.

### *3.1.2. Reliability indices*

Indices used to express devices' reliability performances are as follows:

- the failure rate ($\lambda$) or hazard function that represents the frequency with which a component or a system fails
- the mean time between failure (MTBF) is a measure of how reliable a product is. It is usually given in units of hours. High MTBF values characterize high-reliability products.

The question is:" Which operating condition/s have to be considered to a correct SMPPT reliability evaluation?" The answer is crucial since photovoltaic systems are characterized by operating and nonoperating mode depending on meteorological and climatic conditions and on the day/night alternation. The calculation of reliability indices in a unique working point could not be meaningful. To overcome this drawback, the idea is to propose weighted MTBF formulas obtained by analyzing occurrence of $T_{backPV}$ conditions over an annual time period and identifying the most frequent conditions (**Figure 5**).

In fact, calculating $T_{backPV}$ values and the number of hours they occur, it is possible to obtain $\alpha$ and $\beta$ terms for the weighted MTBF formula:

$$MTBF_{wg} = \alpha_1\, MTBF_{\beta 1} + \alpha_2\, MTBF_{\beta 2} + \alpha_3\, MTBF_{\beta 3} + \alpha_4 MTBF_{\beta 4} + \alpha_5\, MTBF_{\beta 5} \tag{12}$$

In detail, the obtained expression for $MTBF_{wg}$ is reported in Eq. (13).

$$MTBF_{wg} = 0.18\, MTBF_{35\%} + 0.28\, MTBF_{57\%} + 0.19\, MTBF_{71\%} + 0.28 MTBF_{86\%} + 0.07\, MTBF_{100\%} \tag{13}$$

In the followings, the MIL-HDBK-217 and its successor, the RIAC 217 Plus, are analyzed since they represent the most conservative models for reliability prediction.

### 3.2. MIL-HDBK-217F

The Military Handbook 217F can be used for both PCA and PSA analyses.

In case of PCA analysis, the failure rate of the equipment is calculated by the following equation:

$$\lambda_{equip} = \sum_{i=1}^{N} N_i \left(\lambda_g \pi_Q\right)_i \tag{14}$$

where

-$\lambda_g$ is the generic failure rate for the ith component;

-$\pi_Q$ is the quality factor for the ith component;

-$N_i$ is the quantity of the ith component;

-N is the number of different component categories present in the equipment.

Otherwise, in case of PSA analysis, the failure rate of each component is calculated by the following equation:

$$\lambda_p = \lambda_b \ \pi_T \pi_A \pi_Q \pi_E \tag{15}$$

where

-$\lambda_p$ is the failure rate, a specific component;

-$\lambda_b$ is the base failure rate of the component in standard condition;

-$\pi_T$ is the temperature factor;

-$\pi_A$ is the application factor;

-$\pi_E$ is the environment factor.

The procedure and the MIL-HDBK-217F formulas to evaluate failure rates of components constituting SMPPT power stage are reported below. In this manuscript, a case study represented by a SMPPT SR boost converter reliability evaluation is presented. The circuit topology is reported in **Figure 1(a)**. This converter is characterized by series connection among electronic devices. As a consequence, the MTBF index, for a specific operating condition, can be calculated as reported in the following formula:

$$MTBF = \frac{1}{\lambda_{Q1+}\lambda_{Q2} + \lambda_L + \lambda_{Cin} + \lambda_{Cout}} \tag{16}$$

### *3.2.1. MIL_HDBK-217 F MOSFET failure rate*

In detail, MOSFETs are characterized by failure mechanisms such as defects in the substrate, insulation films or metallization, stress at solder connections due to a mismatch of thermal properties of the different materials and excessive electrical stresses and electrostatic discharges.

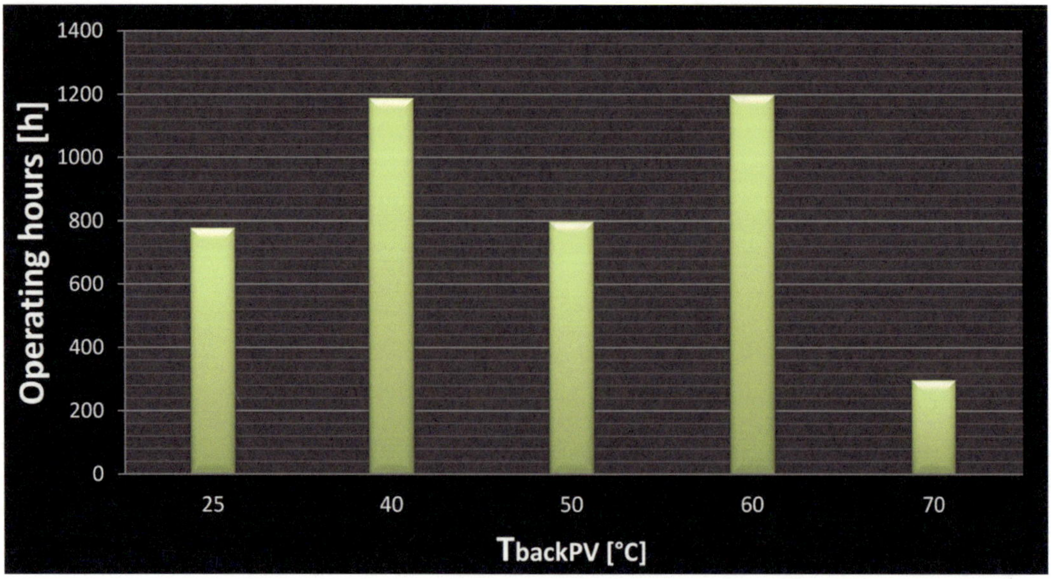


**Figure 5.** $T_{backPV}$ annual distribution.

The MOSFET failure rate $\lambda_p$ must be calculated as shown in paragraph 6.4 of MIL_HDBK-217F. The $\lambda_p$ formula is here reported.

$$\lambda_p = \lambda_b \pi_T \pi_A \pi_Q \pi_E \frac{Failures}{10^6\ hours} \quad (17)$$

The base failure rate $\lambda_b$ depends on the transistor typology as indicated in **Table 1**:

SMPPT converters are constituted by MOSFET transistor, so the relative $\lambda_b$ is 0.012.

The application factor $\pi_A$ that depends on the device nominal power is reported in **Table 2**.

In our case study, the MOSFET nominal power is in the range of 80–100 W, so $\pi_A$ is 8.0.

The quality factor $\pi_Q$ values are reported in **Table 3**.

MOSFETs used in SMPPT application are constituted by plastic cases, so the $\pi_Q$ is 8.0.

| Transistor type | $\lambda_b$ |
|---|---|
| JFET | 0.0045 |
| MOSFET | 0.012 |

**Table 1.** MIL_HDBK-217 F MOSFET base failure rate.

| Application | $\pi_A$ |
|---|---|
| Linear amplification ($P_r < 2W$) | 1.5 |
| Small signal switching | 0.7 |
| Power FETs (non-linear, $P_r \geq 2W$) | |
| $2 \leq P_r < 5\ W$ | 2.0 |
| $5 \leq P_r < 50\ W$ | 4.0 |
| $50 \leq \mathrm{Pr} < 250\ W$ | 8.0 |
| $P_r \geq 250\ W$ | 10 |

**Table 2.** MIL_HDBK-217 F MOSFET application factor values.

| Quality | $\pi_Q$ |
|---|---|
| JANTXV | 0.70 |
| JANTX | 1.0 |
| JAN | 2.4 |
| Lower | 5.5 |
| Plastic | 8.0 |

**Table 3.** MIL_HDBK-217 F MOSFET quality factor values.

The value of the environmental factor $\pi_E$ is indicated in **Table 4** for different types of environment.

The MOSFET of the considered converter is installed on the Earth where it is possible to control both the temperature and the humidity, so it is correct to refer to a "Ground Benign" environment and the relative $\pi_E$ -is 1.0.

The temperature factor $\pi_T$ can be obtained by the following equation:

$$\pi_T = e^{\left(-1925\left(\frac{1}{Tj+273}-\frac{1}{298}\right)\right)} \tag{18}$$

where $T_j$ is the MOSFET junction temperature. For example, considering a $T_j$ value of 50°C, the temperature factor is 1.6.

In the considered case study, the SMPPT converter is a synchronous one. As a consequence, the MOSFET failure rate evaluation reported above has to be carried out for both $Q_1$ and $Q_2$.

In detail, as reported before, the Low Side MOSFET $Q_1$ reaches 50°C and the temperature factor is 1.6, while in case of the High Side MOSFET $Q_2$ reaching 40°C, the temperature factor is 1.4.

So, in conclusion, the MOSFET failure rate can be calculated as follows:

$$\lambda_{Q1} = \lambda_b \pi_T \pi_A \pi_Q \pi_E \frac{Failures}{10^6\ hours} = 0.012 * 1.6 * 8 * 8 * 1 = 1.228 \frac{Failures}{10^6\ hours} \tag{19}$$

$$\lambda_{Q2} = \lambda_b \pi_T \pi_A \pi_Q \pi_E \frac{Failures}{10^6\ hours} = 0.012 * 1.4 * 8 * 8 * 1 = 1.075 \frac{Failures}{10^6\ hours} \tag{20}$$

| **Environment** | $\pi_E$ |
|---|---|
| Ground, Benigh $G_B$ | 1.0 |
| Ground, Fixed $G_F$ | 6.0 |
| Ground, Mobile $G_M$ | 9.0 |
| Naval, Sheltered $N_S$ | 9.0 |
| Naval, Unsheltered $N_U$ | 19 |
| Airborne, inhabited, Cargo $A_{IC}$ | 13 |
| Airborne, Inhabited, Fighter $A_{IF}$ | 29 |
| Airborne, Uninhabited, Cargo $A_{UC}$ | 20 |
| Airborne, Uninhabited, Fighter $A_{UF}$ | 43 |
| Airborne, Rotary Winged $A_{RW}$ | 24 |
| Space, Flight $S_F$ | 0.5 |
| Missile, Flight $M_F$ | 14 |
| Missile, Launch $M_L$ | 32 |
| Cannon, Launch $C_L$ | 320 |

**Table 4.** MIL_HDBK-217 F MOSFET environment factor values.

### 3.2.2. MIL_HDBK-217 F capacitor failure rate

Formulas and tables to calculate capacitors failure rate are reported in the paragraph 10.1 of the MIL-HDBK-217F.

This evaluation is based on Eq. (21).

$$\lambda_p = \lambda_b \pi_T \pi_C \pi_V \pi_{SR} \pi_Q \pi_E \frac{Failures}{10^6\ hours} \tag{21}$$

where

-$\pi_C$ is the capacitance factor;

-$\pi_V$ is capacitor voltage stress factor;

-$\pi_{SR}$ is the series resistance factor.

The numeric values of these parameters are determined as follows: the basic fault rate $\lambda_b$ depends on the type of capacitors in input and output to the SMPPT converter. Considering an aluminum capacitor working at T Celsius degrees, the $\lambda_b$ is 0.00012 and the $\pi_T$ factor is calculated by the following formula:

$$\pi_T = e^{\left(-\frac{0.35}{8.617*10^{-5}}\left(\frac{1}{T+273}-\frac{1}{298}\right)\right)} \tag{22}$$

The $\pi_C$ factor can be calculated by considering the capacitance value in μF and applying Eq. (23); the $\pi_{SR}$ factor depends on the device equivalent series resistance, and its value is 1.0; the quality factor based on the capacitor plastic case is 10.0.

$$\pi_C = C^{0.23} \tag{23}$$

Finally, also in this case, the environment is "Ground Benign" and the $\pi_V$ factor can be obtained by applying the following formula:

$$\pi_V = \left(\frac{S}{0.6}\right)^5 + 1 \tag{24}$$

So considering the SR input and output capacitor of 120 μF, the relative failure rate is reported as follows:

$$\lambda_{Cin} = \lambda_{Cout} = \lambda_b \pi_T \pi_C \pi_V \pi_{SR} \pi_Q \pi_E = 0.00012 * 1 * 3 * 1 * 1 * 10 * 1 = 0.0036 \frac{Failures}{10^6\ hours} \tag{25}$$

### 3.2.3. MIL_HDBK-217F inductor failure rate

Magnetic devices are the most reliable electronic components. The inductor fault rate $\lambda_p$ is calculated by the following equation reported in Section 11.2 of the Military Handbook.

$$\lambda_p = \lambda_b \cdot \pi_T \cdot \pi_Q \cdot \pi_E \tag{26}$$

In detail, the base failure rate $\lambda_b$ is 0.00003 as in **Table 5**.

The temperature factor formula is reported in Eq. (27).

$$\pi_T = e^{\left(-\frac{0,11}{8,617\cdot10^{-5}}\left(\frac{1}{T_{HS}+273}-\frac{1}{298}\right)\right)} \tag{27}$$

where $T_{HS}$ is the hot-spot temperature.

Similar to previous cases, the $\pi_Q$ and the $\pi_E$ factor can be calculated, and their values are 3.0 and 1.0, respectively.

Considering ad inductor of 47 μH and a $T_{HS}$ of 70°C, the device failure rate is as follows:

$$\lambda_L = \lambda_b \cdot \pi_T \cdot \pi_Q \cdot \pi_E = 0.00003 * 1.767 * 3 * 1 = 1.59 * 10^{-4}\,\frac{Failures}{10^6 hours} \tag{28}$$

The obtained failure rate confirms the inductor affordable behavior.

After the calculation of the failure rate of the single components, the evaluation of the whole SMPPT converter failure rate is carried out.

$$MTBF = \frac{1}{\lambda_{Q1+}\lambda_{Q2} + \lambda_L + \lambda_{Cin} + \lambda_{Cout}} = \frac{1}{1.228 + 1.075 + 1.59 * 10^{-4} + 0.0036 + 0.0036} = 0.433 * 10^6 hours \tag{29}$$

The reported procedure can be also applied for different SMPPT topologies. Considering the $T_{backPV}$ distribution over an annual period of time, it is possible to calculate the $MTBF_{wg}$ (Eq. (13)) for the analyzed case study.

$$MTBF_{wg} = 0.18\,MTBF_{35\%} + 0.28\,MTBF_{57\%} + 0.19\,MTBF_{71\%} + 0.28 MTBF_{86\%} + 0.07\,MTBF_{100\%} = 0.2 * 10^6 hours \tag{30}$$

### 3.3. RIAC 217 Plus

The RIAC Handbook 217 Plus reliability prediction model is the updated version of MIL-HDBK-217F. It guarantees compatibility with its predecessor so keeping unchanged industry practice for systems reliability estimations. It also considers components' operating and nonoperating conditions (operating and nonoperating temperatures, duty cycles, cycling rate and so on). Failure rates are calculated as the product of a base failure rate $\lambda_b$ and some $\pi_i$ factors representing the possible i stresses influencing component reliability.

| **Inductor type** | $\lambda_b$ |
|---|---|
| Fixed | 0.00003 |
| Variable | 0.00005 |

**Table 5.** MIL_HDBK-217 F inductor base failure rate.

In **Table 6**, the 217 Plus failure rate formulas are reported.

where

$\lambda_{OB}$ is base failure rate, Operating;

$\lambda_{EB}$ is base failure rate, Environmental;

$\lambda_{TCB}$ is base failure rate, Temperature Cycling;

$\lambda_{SJB}$ is base failure rate, Solder Joint;

$\lambda_{EOS}$ is failure rate, Electrical OverStress;

$\pi_G$ is reliability growth failure rate factor;

$\pi_{DCO}$ is failure rate factor for Duty Cycle, operating;

$\pi_{TO}$ is failure rate factor for Temperature, operating;

$\pi_S$ is failure rate factor for Stress;

$\pi_{DCN}$ is failure rate factor for Duty Cycle, nonoperating;

$\pi_{TE}$ is failure rate factor for Temperature, environmental;

$\pi_{CR}$ is failure rate factor, Cycling Rate;

$\pi_{DT}$ is failure rate factor, Delta Temperature;

$\pi_{SJDT}$ is failure rate factor, Solder Joint Delta Temperature.

Applying the formulas reported in **Table 6** in a similar manner as the MIL-HDBK-217F model, it is possible to evaluate RIAC 217 Plus reliability performance.

Since the thermal stress is one of the most invalidating factors, the attention is here focused on MTBF variations depending on temperature increase. Referring to the reported failure rate formulas, $\pi_{TE}$ and $\pi_{TO}$ factors for different temperature values are shown in **Figure 6**.

In $\pi_{TE}$ factor graph, the device behavior is similar. In case, instead, of $\pi_{TO}$, the temperature factor considered in operating mode, it is evident the temperature strongly influences the switching and magnetic components. It is worth noting that aspect has to be taken into account for an accurate reliability prediction.

| SMPPT electronic component | Failure rate formulas |
|---|---|
| MOSFET | $\lambda_{MOS} = \pi_G(\lambda_{OB}\pi_{DCO}\pi_{TO}\pi_S + \lambda_{EB}\pi_{DCN}\pi_{TE} + \lambda_{TCB}\pi_{CR}\pi_{DT}) + \lambda_{SJB}\pi_{SJDT} + \lambda_{EOS}$ |
| Diode | $\lambda_{diode} = \pi_G(\lambda_{OB}\pi_{DCO}\pi_{TO}\pi_S + \lambda_{EB}\pi_{DCN}\pi_{TE} + \lambda_{TCB}\pi_{CR}\pi_{DT}) + \lambda_{SJB}\pi_{SJDT} + \lambda_{EOS}$ |
| Capacitor | $\lambda_C = \pi_G\pi_C(\lambda_{OB}\pi_{DCO}\pi_{TO}\pi_S + \lambda_{EB}\pi_{DCN}\pi_{TE} + \lambda_{TCB}\pi_{CR}\pi_{DT}) + \lambda_{SJB}\pi_{SJDT} + \lambda_{EOS}$ |
| Inductor | $\lambda_{Inductor} = \pi_G(\lambda_{OB}\pi_{DCO}\pi_{TO} + \lambda_{EB}\pi_{DCN}\pi_{TE} + \lambda_{TCB}\pi_{CR}\pi_{DT}) + \lambda_{EOS}$ |

**Table 6.** RIAC 217 Plus failure rate formulas.

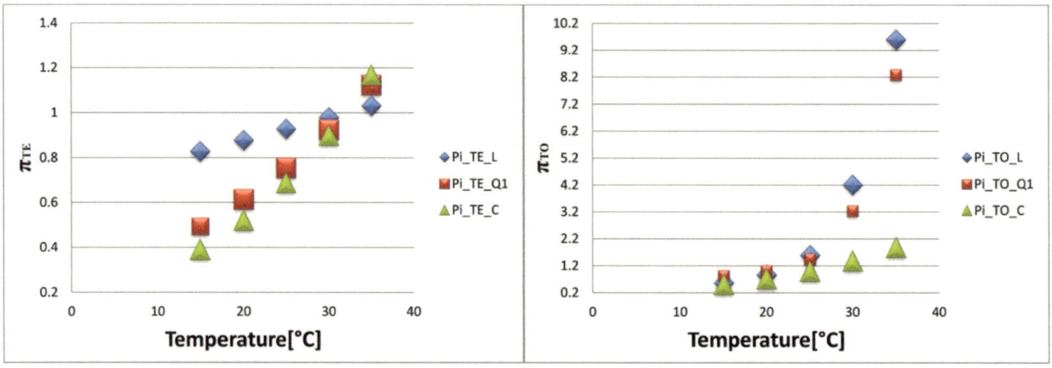


**Figure 6.** RIAC 217+ factors dependence on temperature: (a) $\pi_{TE}$ *vs.* T; (b) $\pi_{TO}$ *vs.* T.

## 4. Reliability comparative analysis among SMPPT topologies

In the last paragraph, the focus is on a comparative analysis among different SMPPT boost converters. In detail, the reliability R(t), considering a period of 25 years (as PV generator one) and an irradiance value of 1000 W/sqm, is calculated for a SR, a DR and an IL boost converter.

Results are reported in Eqs. (31)–(33).

$$R_{SR}(25\ years) = 56\% \tag{31}$$

$$R_{DR}(25\ years) = 68\% \tag{32}$$

$$R_{IL}(25\ years) = 91\% \tag{33}$$

Eqs. (30) and (31) demonstrate that the MOSFET used as High Side switching device in the SR converter appreciably deteriorates the SMPPT reliability performance.

Suitably choosing the IL converter devices, a quasi-redundant or totally redundant structure can be obtained. Such converter is able to assure higher reliability performances than the DR and SR ones as confirmed by Eq. (32).

## 5. Conclusions

In this chapter, reliability prediction models suitable to the evaluation of SMPPT converter performances are considered. The attention is focused on the military models since they provide more conservative predictions with respect to the industrial ones.

A case study about a SMPPT characterized by a SR topology is carried out. In addition, the thermal phenomena influence on the reliability evaluation is showed. Finally, a comparative analysis among SMPPT converters in terms of R (25 years) is obtained, underlining the higher IL boost performances with respect to the SR and DR ones.

## Author details

Giovanna Adinolfi* and Giorgio Graditi

*Address all correspondence to: giovanna.adinolfi@enea.it

ENEA, Italian National Agency for New Technologies, Energy and Sustainable Economic Development, Portici (NA), Italy

## References

[1] Cecati C, Khalid H, Tinari M, Adinolfi G, Graditi G. DC nanogrid for renewable sources with modular DC/DC LLC converter building block. IET power. Electronics. 2017;**9**(5):535-541. DOI: 10.1049/iet-pel.2016.0200

[2] Xiao W, Ozog N, Dunford WG. Topology study of photovoltaic Interface for maximum power point tracking. IEEE Transactions on Industrial Electronics. 2007;**54**(3):1696-1704. DOI: 10.1109/TIE.2007.894732

[3] Ferlito S, Adinolfi G, Graditi G. Comparative analysis of data-driven methods online and offline trained to the forecasting of grid-connected photovoltaic plant production. Applied Energy. 2017;**205**(1). DOI: doi.org/10.1016/j.apenergy.2017.07.124

[4] Calleja H, Chan F, Uribe I. Reliability-oriented assessment of a DC/DC converter for photovoltaic applications. In: Power Electronics Specialists Conference, 2007. PESC 2007. IEEE; 17–21 June 2007; Orlando, USA. IEEE; 2007. DOI: 10.1109/PESC.2007.4342221

[5] Economou M. The merits and limitations of reliability predictions. In: Reliability and Maintainability, 2004 Annual Symposium–RAMS, editors. 26–29 Jan 2004; Los Angeles, CA, USA. USA: IEEE; 2004. DOI: 10.1109/RAMS.2004.1285474

[6] Handbook of Military 217F, 1991, Washington

[7] Telcordia. Reliability prediction procedure for electronic equipment. SR.332, Issue 2, Mar 2016, Sweden

[8] Handbook of 217Plus Reliability Prediction Models. Dec. 2014, Quanterion Solutions, Quanterion Solutions Incorporated, New York

[9] FIDES. Reliability methodology for electronic systems. Guide 2004, Issue A:1-347, Europa

Chapter 12

# Low-Frequency Noise and Resistance as Reliability Indicators of Mechanically and Electrically Strained Thick-Film Resistors

Zdravko Stanimirović

Additional information is available at the end of the chapter

http://dx.doi.org/10.5772/intechopen.69441

**Abstract**

New contemporary applications of thick resistive films are inducing the need to investigate their behaviour under various stressing conditions. On the other hand, there is a growing interest in noise measurements as means of thick-film resistor quality and reliability evaluation and evaluation of degradation under stress. For these reasons, this chapter presents effects of mechanical, electrical and simultaneous mechanical and electrical straining on performances of conventional thick-film resistors that are analysed from micro- and macro-structural, charge transport and low-frequency noise aspects.

**Keywords:** thick-film resistors, mechanical straining, high voltage pulse stressing, resistance, gauge factor, noise index

## 1. Introduction

Present miniaturization trends and ongoing usage of thick-film resistors in sensitive telecommunications equipment have induced the need to investigate their reliability under various straining conditions. The most of the published data dealt with effects of mechanical straining on performances of these complex heterogeneous systems using the piezoresistive effect in thick resistive films for strain gauge realization [1, 2]. On the other hand, performances of standard thick resistive films subjected to unwanted mechanical straining [3–5] have not been sufficiently investigated despite the fact that mechanical straining may take place during all phases of resistor realization, examination and application. In case of high-voltage pulse stressing, the most of the papers investigated effects of trimming of thick resistive films by energy of high-voltage pulses [6–8] and behavioural analysis of surge thick-film resistors [9].

Influence of electrical straining on the reliability of conventional thick resistive films has been seldom investigated. Little attention has particularly been paid to examining effects of the simultaneous impact of these two types of straining with respect to their contrasting effects on resistor performances. In addition, standard low-frequency noise measurements [10–13] are being recognized as useful tools in reliability analysis of thick-film resistors subjected to various straining conditions. For these reasons, this chapter focuses on performance analysis of mechanically, electrically and simultaneously mechanically and electrically strained thick-film resistors based on compositions with three different volume fractions of conducting phase, using standard resistance and low-frequency noise measurements as valuable indicators in reliability evaluation of thick resistive structures under a wide range of extreme working conditions.

## 2. Mechanically strained thick-film resistive structures

Thick resistive films have been known for their piezoresistive properties for more than 40 years. Over the years, strain gauge applications have been topics of the most of the available published data. At first, only the basic piezoresistive characteristics of thick resistive films were examined. Later on, new resistive sensing elements emerged based on novel thick-film inks designed for each specific application [14, 15]. On the other hand, standard thick-film resistors are being continuously used in contemporary electronic equipment that requires high functional capability, improved reliability and environmental stability. These up-to-date applications induced the need to examine performances of standard mechanically strained thick-film resistive structures [5, 16].

Sensitivity of a certain material to mechanical strain is referred as the gauge factor. In case of thick-film resistive structures, gauge factor ($GF$) is defined as the ratio of the relative resistance change ($\Delta R/R$) and the relative change of length of the resistor ($\varepsilon = \Delta l/l$) under influence of mechanical straining:

$$GF = \frac{\Delta R/R}{\varepsilon}. \tag{1}$$

The relation between strain and resistor position on the substrate can be given by the equation [17]:

$$\varepsilon = 12\frac{t\overline{x}}{L^3}d_i, \ d \ll L, \tag{2}$$

where $t$ is substrate thickness, $\overline{x}$ is average position of the thick resistive film with respect to the fixed substrate edge, $d$ is substrate deflection and $L$ is distance between fixed substrate edges. The maximum value for the strain occurs for $\overline{x} = L/2$:

$$\varepsilon_{\max} = \Delta l/l = \frac{6\,d_i\,t}{L^2}. \tag{3}$$

http://dx.doi.org/10.5772/intechopen.69441

Schematic presentation of mechanically strained thick-film resistor is given in **Figure 1**.

Mechanical straining causes a reversible resistance change in thick-film resistors [5, 16]. The reversible resistance change is partially due to change in resistor geometry but mainly due to micro-structure changes. According to 3-D planar random resistor network model [18], transport of electrical charges in thick-film resistive materials takes place via a complex conductive network formed during firing by sintering metal-oxide particles (usually combination of $RuO_2$ and $Bi_2Ru_2O_7$) immerged in the glass matrix. During the sintering process, a number of conducting chains is being formed. These chains consist of clusters of particles (particles that are in contact) and neighbouring particles separated by thin glass barriers (metal-insulator-metal or MIM units). Therefore, the current flow is being determined by metallic conduction and tunnelling through glass barriers. The micro-structure of thick resistive films, determined by the ratio of the conducting and insulating phase, also determines conducting mechanisms present in the film. Performed experiments [16], illustrated by data given in **Table 1**, showed that gauge factor values are greater for resistors realized with compositions with higher sheet resistances, that is, resistor compositions with smaller volume fractions of conducting phase have greater *GF* values. Thick-film resistors based on compositions with high sheet resistances (≥100 kΩ/sq) have lower volume fraction of conducting phase and therefore charge transport is predominantly limited by tunnelling through glass barriers. Resistors based on compositions with low sheet resistances (≤1 kΩ/sq), because of their high volume fraction of conducting phase, are predominantly limited by conducting through clusters of conducting particles and sintered contacts. Resistors realized using compositions with medium values of sheet resistances, such as 10 kΩ/sq compositions, incorporate approximately equally all above-mentioned charge transport mechanisms. Substrate deflection causes resistance increase due to change of charge transport conditions and greatest resistance change leads to greatest *GF* value. Conducting mechanism known as tunnelling through glass barriers is predominantly influenced by mechanical stressing. The bulk modulus of borosilicate glasses is typically between 40 and 80 GPa [19] while $RuO_2$ conducting phase has a bulk modulus of approximately 270 GPa. Since the glass phase is less stiff than the conducting one, tunnelling through glass barriers is more sensitive to the applied straining than conduction through conducting particles and sintered contacts between them. Mechanical straining changes dimensions of the thick resistive film. It cannot alter a number of chains, barriers or contacts in the conducting network. Also, it cannot induce dielectric

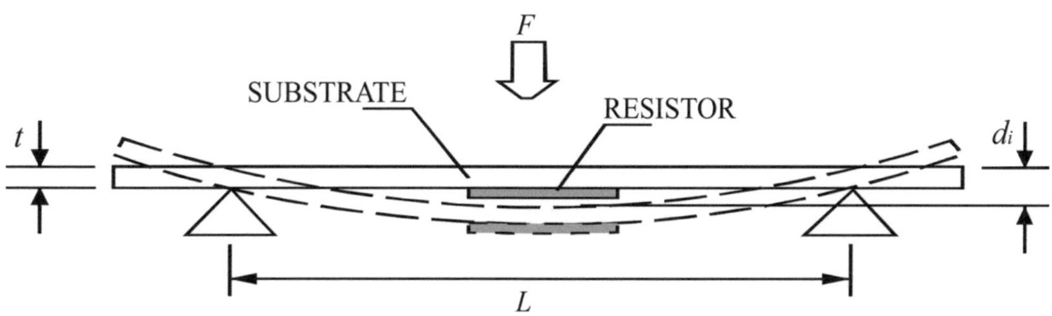


**Figure 1.** Schematic presentation of mechanically strained thick-film resistor.

| $R_{sq}$ (kΩ/sq) | $l$ (mm) | $\overline{R}_i$ (kΩ) | $\overline{\lvert \Delta R/Ri \rvert}$ (%) | $\overline{GF}$ | $\overline{NI}$ (dB) | $\overline{\Delta l}$ (μm) |
|---|---|---|---|---|---|---|
| 1 | 2 | 0.666 | 0.401 | 4.2 | −19.5 | 1.905 |
| 1 | 4 | 1.290 | 0.333 | 3.5 | −26.6 | 3.81 |
| 1 | 6 | 1.49 | 0.303 | 3.2 | −29.6 | 5.715 |
| 10 | 2 | 16.595 | 0.958 | 10.06 | −20 | 1.905 |
| 10 | 4 | 32.81 | 0.945 | 9.92 | −19.5 | 3.81 |
| 10 | 6 | 50.644 | 0.918 | 9.64 | −19.1 | 5.715 |
| 100 | 2 | 276.51 | 1.381 | 14.50 | −1.8 | 1.905 |
| 100 | 4 | 495.83 | 1.266 | 13.30 | −7.3 | 3.81 |
| 100 | 6 | 704.3 | 1.136 | 11.92 | −10.3 | 5.715 |

**Table 1.** Experimental data (nominal sheet resistance $R_{sq}$, resistor length $l$, mean value of initial resistance $R_i$, relative resistance change due to mechanical straining $\Delta R/R_i$, gauge factor GF, noise index $NI$ and resistor length change $\Delta l$) for thick-film resistors with different geometries (width: $w$ = 1 mm, length: $l$ = 2, 4 and 6 mm) subjected to maximal mechanical straining of 400 μm [16].

breakthrough or influence the height of glass barriers that exist between adjacent conducting particles. It can only affect widths of glass barriers present in the film, thus changing the barrier resistance. Alteration of charge transport parameters also reflects on measured noise index values. Noise index values are decreasing with increasing resistor length and increasing for resistors realized with compositions with higher sheet resistances. Thick-film resistors realized using 10 kΩ/sq compositions usually have $GF$ ~10 and stable $NI$ values and therefore are commonly used as strain sensors.

## 3. Electrically strained thick-film resistive structures

Different conditions of thick-film resistor application that induced the need to investigate their behaviour under stress also brought to attention the importance of high-voltage pulse stressing. The most of the available data dealt with trimming of thick resistive films by the energy of high-voltage pulses [6–8], a trimming method based on internal discharges using both thick-film resistor terminations as electrodes for applying the high-voltage energy to the resistor body. Also, several papers explored properties of low-ohm thick-film surge resistors [9] that serve as protection of communication systems. However, little attention has been paid to the influence of high-voltage pulse stressing on structure and noise performances of conventional thick-film resistors [20, 21].

High voltage pulse stressing of thick resistive film causes irreversible resistance change. Experimental data obtained by extensive investigations of performances of thick-film resistors subjected to this type of straining [20, 21] showed that behaviour under strain strongly depends on sheet resistances of resistor compositions used. Resistors based on compositions with low sheet resistances exhibit macro-structural changes that result in irreversible resistance increase. Lack of micro-structural changes is a consequence of the dominant conducting mechanism, conducting through clusters of conducting particles. High-voltage treatment leads to burning and evaporation of the resistive layer. Resistor volume is reduced causing the significant resistance and noise index increase (**Figure 2**).

Resistors realized using compositions with medium values of sheet resistances exhibit initial resistance decrease followed by the significant resistance increase during high voltage pulse stressing (**Figure 3**). Lower pulse amplitudes lead to resistance decrease due to changes in conducting mechanisms, metallic conduction through conducting particles and sintered contacts and tunnelling through glass barriers. High-voltage treatment affects charges captured by traps present in thin glass layers between neighbouring conducting particles or the trap concentration increases [12] due to existence of impurities introduced in insulating layers during firing. In addition, a minor resistance decrease may occur due to the conversion of single chain from non-conductive to conductive state. High-voltage treatment induces electrical field inside MIM unit that is insufficient to provoke dielectric breakthrough and therefore decrease of the resistance due to the increase in a number of contacts between neighbouring particles does not occur. Measured resistances substantially increase when the pulse voltage reaches the critical point when macro-structural changes occur. High-voltage treatment leads to burning and evaporation of the resistive layer thus reducing its volume and causing significant resistance increase similar to the one seen in resistors based on compositions with low sheet resistances. Since the low-frequency noise in thick-film resistors is the consequence of electrical charge transport fluctuations, noise index values are in agreement with resistance behaviour (**Figure 3**). Due to high voltage treatment, conduction is being modulated by electrical charges captured by traps that are not directly involved in conduction, thus altering the height of the potential barriers of MIM units. For these reasons, measured noise index values are more sensitive to changes on micro-structural level than resistance.

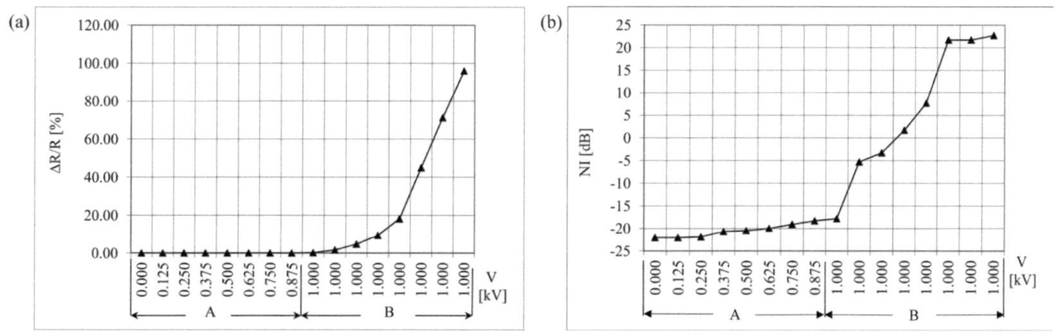


**Figure 2.** Experimental results. (A) Multiple series of 10 pulses with increasing amplitudes, (B) single pulses of the critical amplitude for relative resistance (a) and noise index (b) changes during high-voltage pulse stressing of 1 kΩ/sq thick-film resistor (resistor width: $w = 1$ mm, length: $l = 3$ mm) [21].

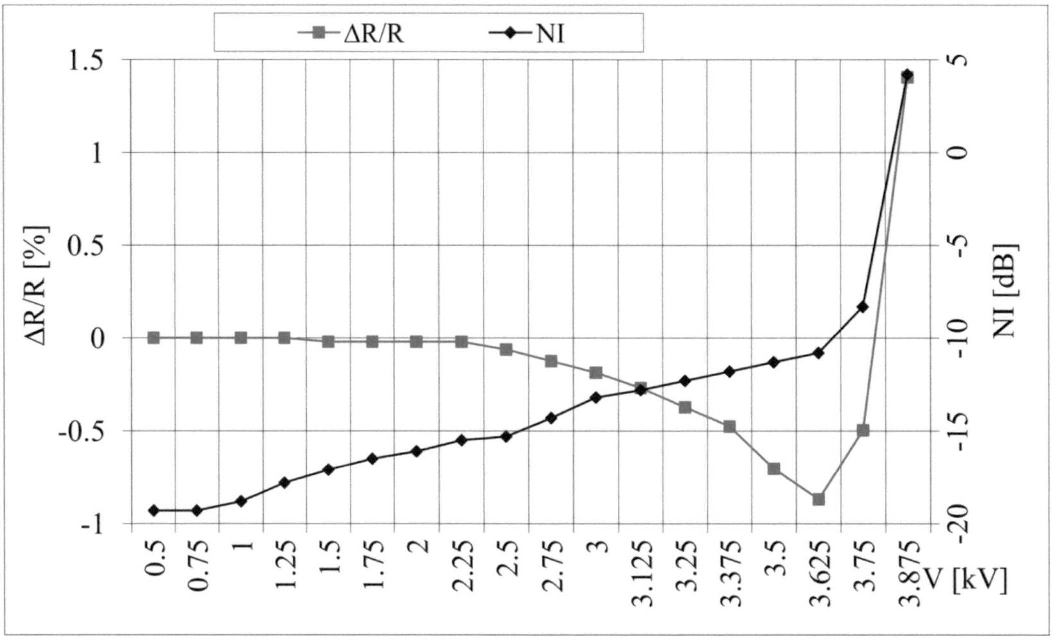


**Figure 3.** Experimental results for relative resistance and noise index changes during high voltage pulse stressing of 10 kΩ/sq thick-film resistor using multiple series of 10 pulses with increasing amplitudes (resistor width: $w$ = 1 mm, length: $l$ = 3 mm) [21].

Established correlation between structural properties and low-frequency noise can also be illustrated using noise spectra measurements (**Figure 4**). The fitting of experimental results for current noise spectra and the following theoretical relation can be performed:

$$S_I(f) = A_0 + \frac{B_0}{f^\gamma} + \sum_i \frac{C_i}{2\pi f_{Ci}(1 + f^2/f_{Ci}^2)}. \tag{4}$$

The first term is the thermal current noise given by:

$$A_0 = \frac{4 k_B T}{R}, \tag{5}$$

where $k_B$ is the Boltzmann's constant, $T$ is the absolute temperature and $R$ is the resistance of thick resistive film. $1/f$ noise is represented by the second term where $B_0$ and $\gamma$ are fitting parameters. The sum of noise spectra of the Lorentzian shape is presented by the third term where $C_i$ and $f_{Ci}$ are characteristic parameters and frequencies, respectively.

$1/f$ noise is the dominant noise component in the total current noise spectrum. Fitting parameters $B_0$ and $\gamma$ can be used in determining its sensitivity to high-voltage treatment. Fitting parameter $\gamma$ has value $\gamma = 1$ both before and after the performed straining. **Figure 5** shows current dependencies of $B_0$. Parameter $B_0$ increases for approximately one order of magnitude with applied

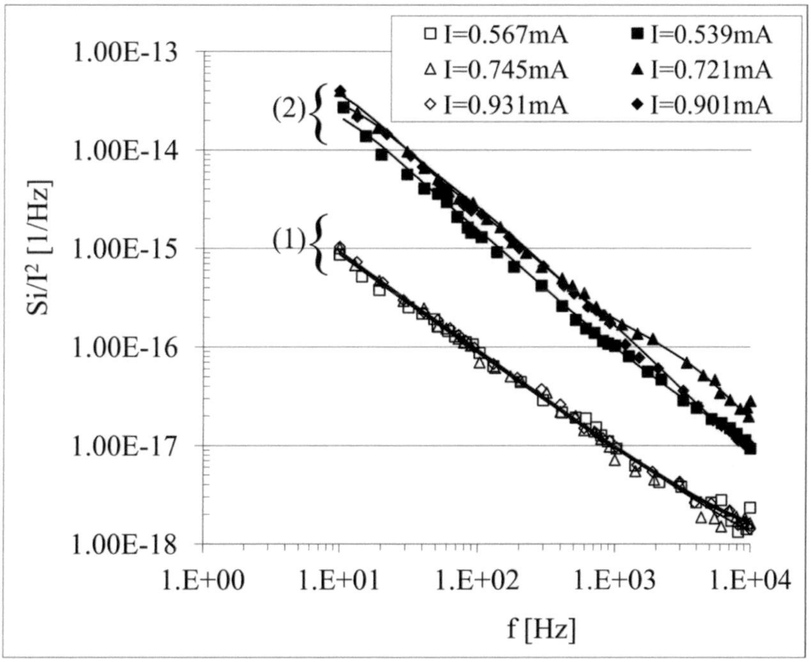


**Figure 4.** Experimental results for normalized current noise spectra before (1) and after (2) high voltage pulse stressing (multiple series of 10 pulses with increasing amplitudes) for 10 kΩ/sq thick-film resistor (full lines, fitting results) [21].

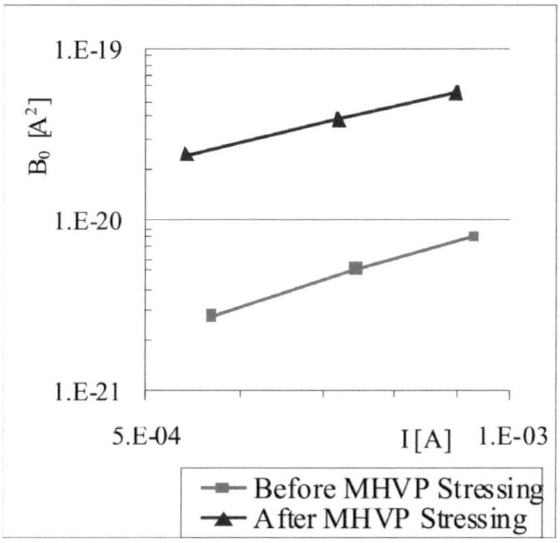


**Figure 5.** 1/*f* noise fitting parameter $B_0$ before and after high voltage pulse stressing of 10 kΩ/sq thick-film resistors [21].

stressing and the current dependence $B_0 \sim I^a$, with exponent $a > 2$ before and $a < 2$ after stressing. Contribution of noise due to mobility fluctuations in clusters and contacts between particles to total 1/*f* noise after high voltage pulse straining are possible causes of the exponent '*a*' change.

Normalized noise amplitude $B_0/I^2$ before and after high-voltage pulse stressing is shown in **Figure 6**. The normalized noise amplitude is dimensionless in opposition to fitting parameter $B_0$. Analogous to data given in **Figure 5**, $1/f$ noise increases for about an order of magnitude with applied high-voltage treatment. After performed fitting procedure, Lorentzian terms caused by fluctuations induced by the presence of traps in insulating layers of MIM units can be observed as slight bends of current noise spectra. Although concealed by $1/f$ noise, Lorentzian terms affect the agreement of measured values and theoretical relation. Current noise spectra analyses, prior to and after high voltage pulse stressing, are in agreement with resistance and noise index behaviour. It is interesting to note that conducting mechanisms seen in resistors with medium sheet resistances combine conducting mechanisms observed in resistors with greater and lower contents of conducting phase. Low-frequency noise measurements that include current noise spectrum and noise index measurements provide results that are far more sensible to macro- and micro-structural changes than measured resistance values. This fact is of special importance in reliability analysis when reversible changes of resistance occur due to straining of thick resistive films.

When thick-film resistors based on high sheet resistance compositions are concerned, due to a low volume fraction of the conducting phase, dominant conduction mechanism is tunnelling through glass barriers. Stressing causes pronounced micro-structural changes, changes barrier resistances and causes significant resistance decrease similar to the ones seen in resistors based medium sheet resistance compositions. Experimental data showed that noise index values are in agreement with resistance behaviour exhibiting an increase of $NI$ values with the resistance progressive decrease and reaching their constant values (saturation) with a stagnation of the resistance decrease (**Figure 7**).

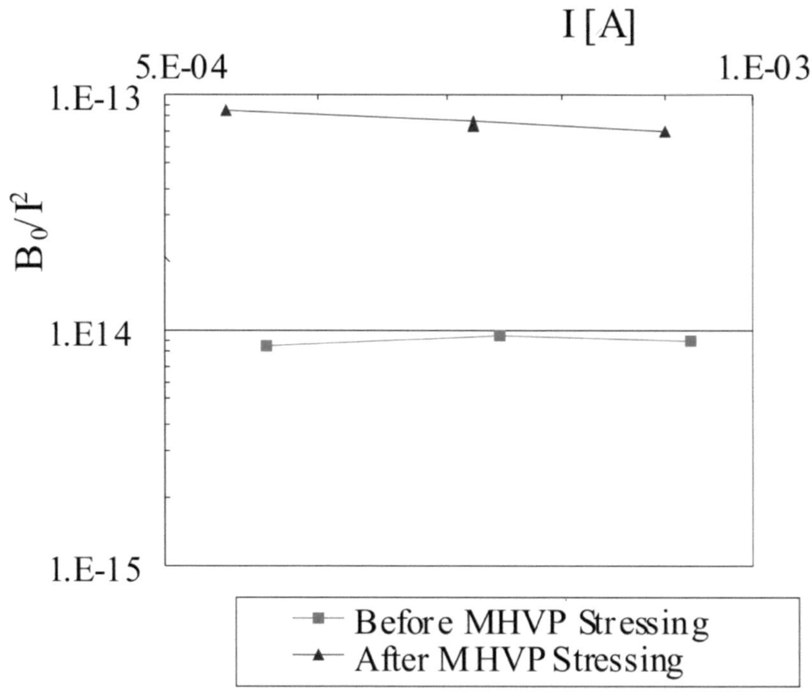


**Figure 6.** Normalized noise amplitude before and after high voltage pulse stressing of 10 kΩ/sq thick-film resistors [21].

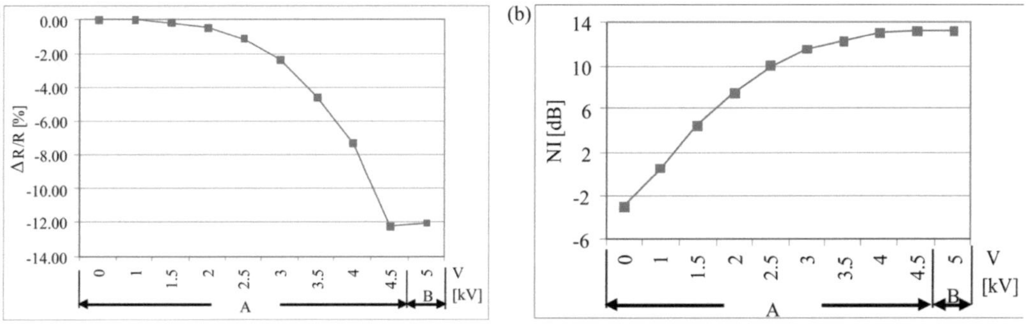


**Figure 7.** Experimental results. (A) Multiple series of 10 pulses with increasing amplitudes, (B) single pulses of the critical amplitude for relative resistance (a) and noise index (b) changes during high-voltage pulse stressing of 100 kΩ/sq thick-film resistors (resistor width: $w = 1$ mm, length: $l = 3$ mm) [21].

## 4. Simultaneous mechanical and electrical straining

Investigations of mechanical and electrical straining of thick-film resistors showed that these two types of straining have opposite effects on behaviour of these complex nanostructures [5, 20, 21]. Examining effects of the simultaneous impact of mechanical and electrical straining on thick-film resistors [16] is of particular interest for sensitive equipment exploitation since simultaneous mechanical and electrical straining may affect resistors capability to withstand high-voltage treatment.

In the case of medium and high sheet resistance compositions, resistance changes of thick resistive films exposed to high-voltage treatment are caused by changes in the micro-structural level [20] that result in decreasing resistance values. In a case of thick-film resistors subjected to mechanical straining, resistance changes are caused by changes in physical dimensions and more dominantly by changes on micro-structural level resulting in increasing resistance values [5, 16]. On the other hand, simultaneous mechanical and electrical straining has opposing effects on performances of thick-film resistors [16]. The ratio of conducting and insulating phase determines sheet resistances of thick resistive films and accordingly micro-structural properties and charge transport conditions. When compositions with a high content of conducting phase are in question, metallic conduction is the dominant conducting mechanism. When the simultaneous impact of these two types of straining are in question, they have opposing effects on tunnelling through insulating layers of MIM unit and accordingly on barrier resistances. In a case of applied mechanical straining, widths of glass barriers are being altered. On the other hand, applied electrical straining affects glass barrier heights. Taking into account the fact that tunnelling is not a dominant conducting mechanism when thick-film resistors with low sheet resistances are concerned, the lack of micro-structure changes is expected. Simultaneous mechanical and electrical straining cause changes in the macro-structure. High-voltage pulse stressing causes visible vaporisation of resistive layers. It decreases volumes of resistors and therefore significantly increases their resistances. Gauge factor changes exhibit increase following the shapes of curves of the resistance changes due to resistor degradation. Noise index values are in agreement with resistance behaviour and show significant increase confirming the fact that noise parameters are very sensitive to structural changes of thick-film resistors, more sensitive than resistance changes (**Figure 8**).

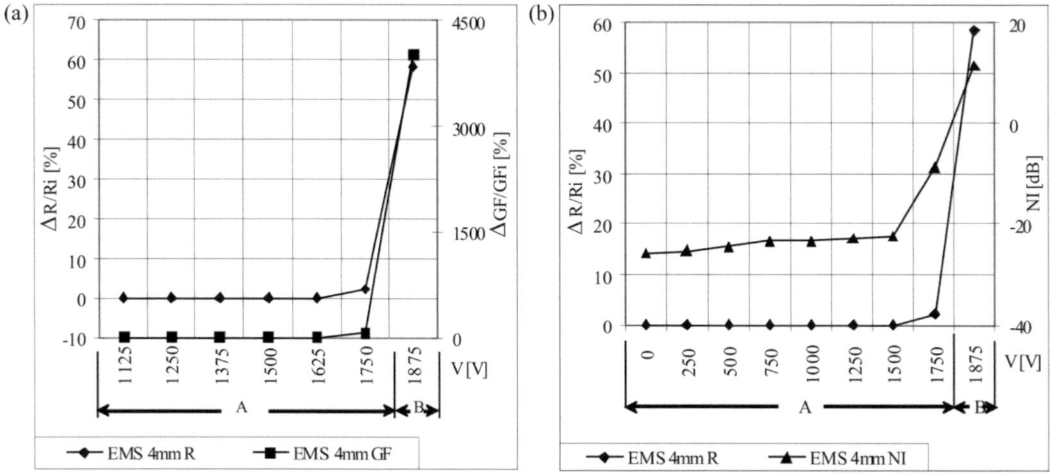


**Figure 8.** Experimental results for relative resistance, gauge factor changes (a) and noise index (b) for simultaneously mechanically and electrically strained 1 kΩ/sq thick-film resistors (resistor width: $w$ = 1 mm, length: $l$ = 4 mm) [16].

In a case of resistor compositions with medium sheet resistances, conduction incorporates both tunnelling through glass barriers and metallic conduction. Change of barrier resistance results in decreasing resistance values that is being succeeded by increasing resistance values caused by alterations on macro-structural level analogous to ones observed in low sheet resistance compositions. *GF* values are stable until degradations on macro-structure level occur and onwards follow the shapes of curves of the resistance changes. Noise index values are very sensitive, registering both micro- and macro-structural degradations of strained thick-film resistors. Experimental results for relative resistance, gauge factor changes and noise index for simultaneously mechanically and electrically strained 10 kΩ/sq thick-film resistors are given in **Figure 9(a)** and **(b)** [16]. In order to illustrate effects of electrical straining alone and simultaneous mechanical and electrical straining, the summary plot for relative resistance change of 10 kΩ/sq resistors is given in **Figure 9(c)** [16]. Figure shows that decreasing resistance values due to changes on micro-structural level are observed in the case of electrically strained and simultaneously electrically and mechanically strained thick-film resistors. Glass barrier height irreversibly changed [20], thus changing barrier resistance value due to applied high-voltage treatment. During simultaneous impact of two different types of straining, glass barriers were affected in two opposing manners: mechanical straining reversibly altered barrier width while electrical straining irreversibly affected barrier height. These opposing effects seem to enhance the ability of thick resistive film to endure high-voltage treatment by extending its lifetime to failure.

In the case of high sheet resistance resistor compositions, small conducting/isolating phase ratio determines dominant conducting mechanism-tunnelling through glass barriers. This small volume fraction of conducting phase leads to pronounced micro-structure changes, changing barrier resistances and causing significant resistance decrease. Gauge factor changes show an increase with the applied straining as well as noise index values. Experimental results for relative resistance, gauge factor changes and noise index for simultaneously mechanically and electrically strained 100 kΩ/sq thick-film resistors are given in **Figure 10**.

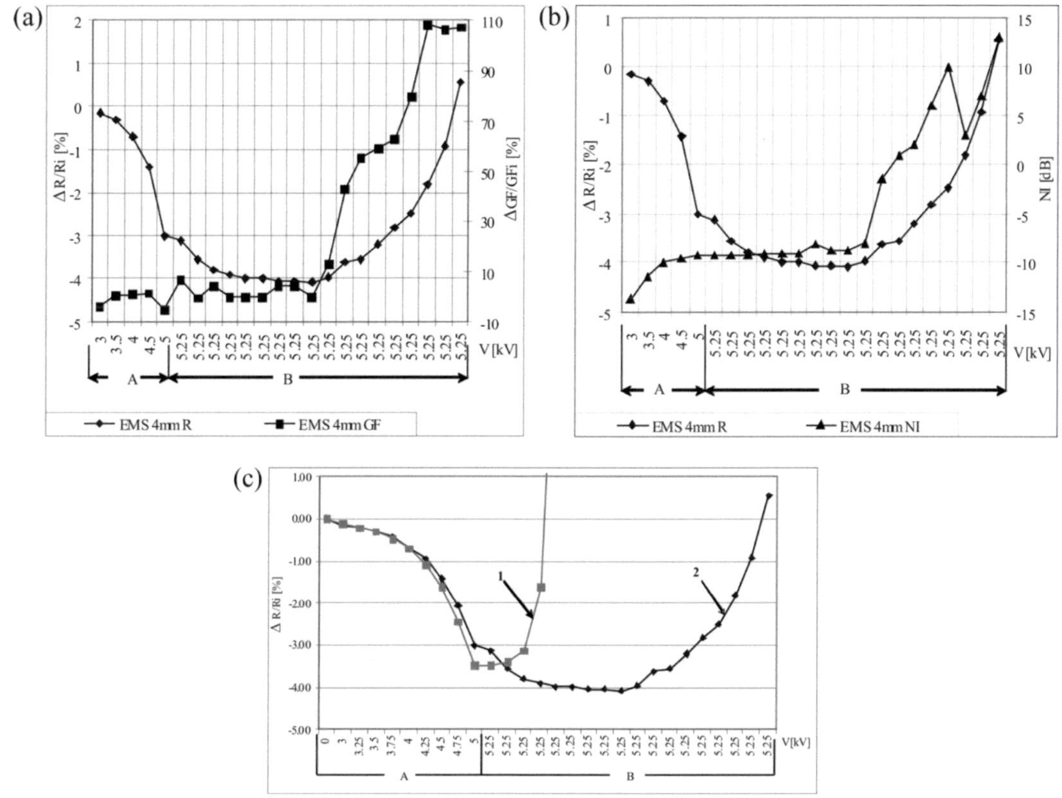


**Figure 9.** Experimentally obtained results for relative resistance, gauge factor changes (a) and noise index (b) for thick resistive films subjected to simultaneous impact of mechanical and electrical straining along with summary plot of relative resistance change (c) for electrically (1) and simultaneously mechanically and electrically strained (2) 10 kΩ/sq thick-film resistors (A—10 pulses per series, B—single pulses, resistor width: $w = 1$ mm, length: $l = 4$ mm) [16].

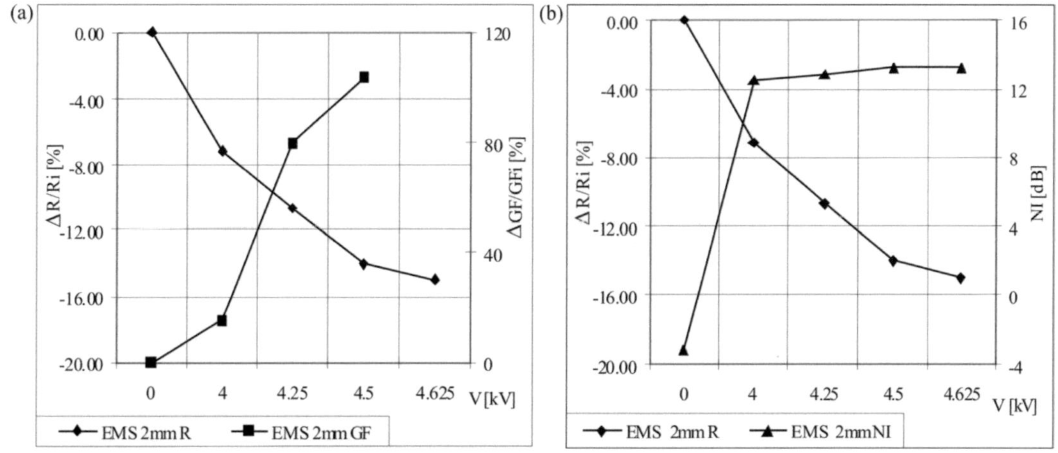


**Figure 10.** Experimental results for relative resistance, gauge factor changes (a) and noise index (b) for simultaneously mechanically and electrically strained (10 pulses per series) 100 kΩ/sq thick-film resistors (resistor width: $w = 1$ mm, length: $l = 2$ mm) [16].

Sources of low-frequency noise in thick resistive films are correlated to charge transport mechanisms [11]; metallic conduction is correlated to resistance fluctuations of contact resistivity and particle resistivity and tunnelling through glass barriers is correlated to noise due to modulation of the Nyquist noise and fluctuations induced by existence of traps in insulating layers of MIM units. **Figure 11** shows experimental results for current noise spectrum before and after simultaneous electrical and mechanical straining of thick resistive films whose experimental results for relative resistance, gauge factor and noise index changes are given in **Figure 9** [16]. Presented data demonstrate that changes on micro-structural level cause initial resistance decrease. Changes on macro-structural level have an opposing effect. Initial resistance decrease is being followed by resistance increase, thus reaching the initial resistance value.

In order to fully comprehend effects correlated to a current noise of simultaneously strained thick resistive films, the fitting procedure was implemented based on experimental data presented in **Figure 11** and theoretical relation (4). As an illustration, the fitting and experimental results for the curve (4) in **Figure 11**, together with contributions of different kinds of noise sources in the total current noise spectrum, are given in **Figure 12**.

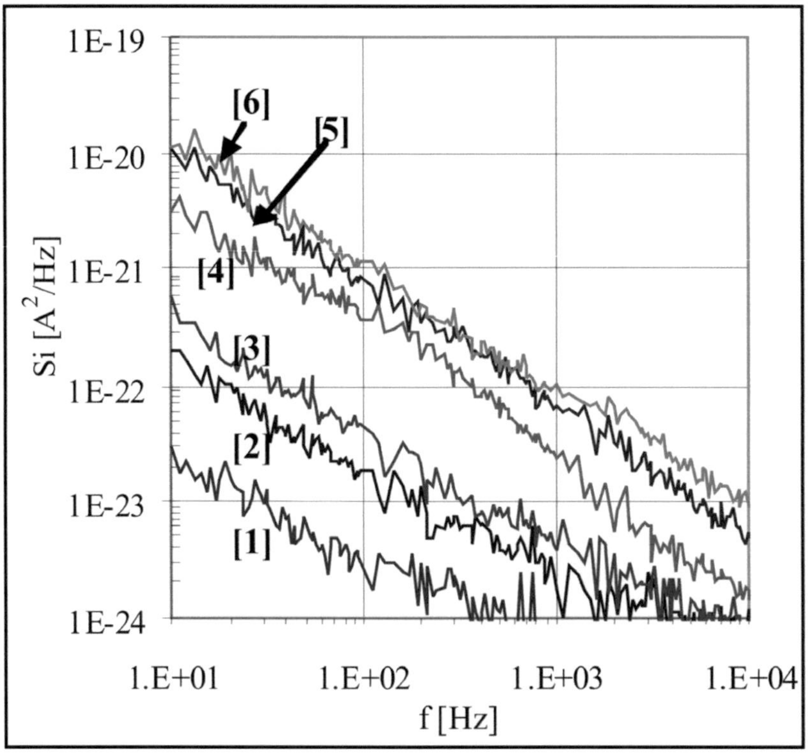


**Figure 11.** Experimental results for current noise spectrum before (1—$I$ = 0.1488 mA, 2—$I$ = 0.2925 mA, 3—$I$ = 0.4452 mA) and after (4—$I$ = 0.1435 mA, 5—$I$ = 0.2917 mA, 6—$I$ = 0.4424 mA) simultaneous mechanical and electrical straining of 10 kΩ/sq thick-film resistors [16].

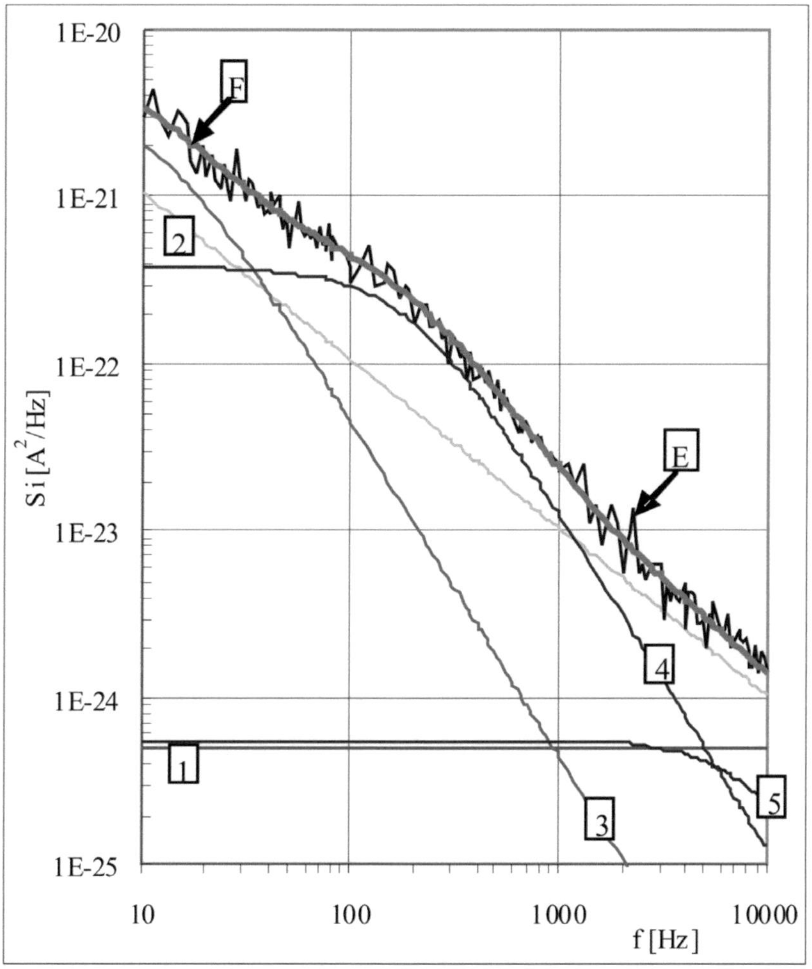


**Figure 12.** Experimental results (*E*) for current noise spectrum and fitting results (*F*) according to Eq. (4) for curve (4) in **Figure 8** with contributions of different kinds of noise sources in the total current noise spectrum (1—thermal current noise, 2—1/*f* noise, 3, 4, 5—noise spectra of the Lorentzian shape) [16].

In the total current noise spectrum, dominant 1/*f* noise includes 1/*f* noise due to particle and contact resistivity fluctuations and fluctuations of potential barrier height caused by Nyquist noise of the insulator [13, 22]. Effects of simultaneous electrical and mechanical straining on 1/*f* noise can be evaluated using 1/*f* noise fitting parameters $\gamma$ and $B_0$. It is found that fitting parameter $\gamma$ has value 1 both before and after simultaneous mechanical and electrical straining while $B_0$ values are presented in **Figure 13**.

It can be seen that parameter $B_0$ increases for approximately one order of magnitude with applied straining. The current dependence $B_0 \sim I^a$, with exponent $a > 2$ before and $a < 2$ after stress is found. This change of the exponent '*a*' could be explained by a greater participation of noise caused by mobility fluctuations in clusters and particles contacts in total 1/*f* noise after stressing. Presence of traps in insulating layers of MIM units is related to Lorentzian

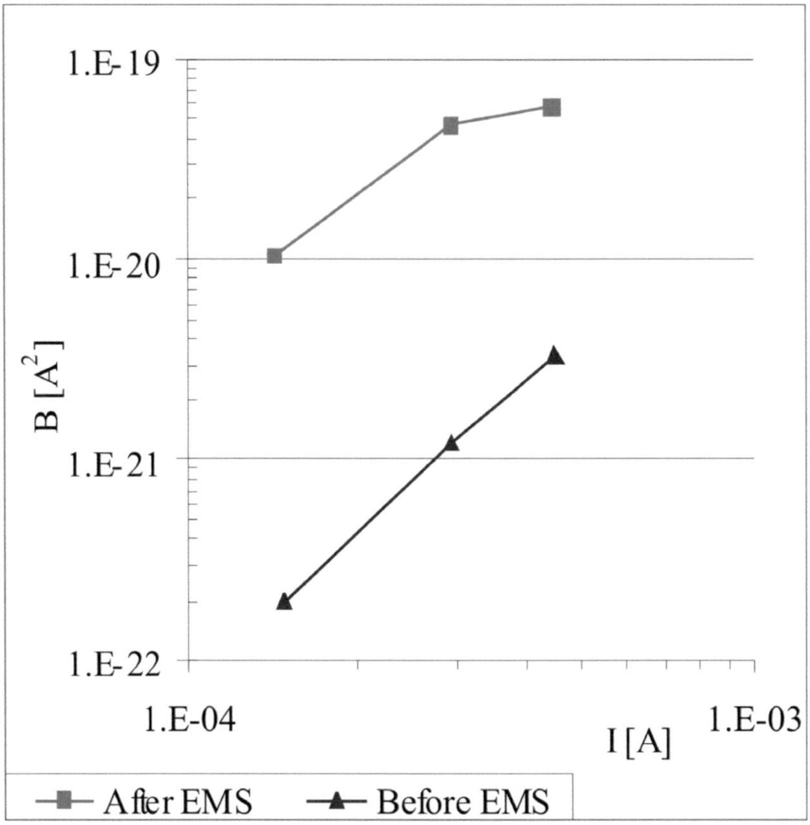


**Figure 13.** $1/f$ noise parameter $B_0$ before and after simultaneous mechanical and electrical straining of 10 kΩ/sq thick-film resistors (resistor width: $w = 1$ mm, length: $l = 4$ mm) [16].

terms present in the total noise spectrum of thick resistive films. Presence of these terms is confirmed by data presented in [23]. Based on current noise spectra measurements of simultaneously electrically and mechanically strained resistors, detailed experimental and numerical analysis proved that straining affects the shape and the level of the noise spectra. Therefore, low-frequency noise parameters are sensitive to degradations induced by applied straining. Obtained results, being in accordance with measured noise index values, are significant for contemporary sensitive applications of thick resistive films. In cases of reversible resistance changes that often disguise degradation processes in thick-film resistors, low-frequency noise measurements can be used as useful tools in detecting the ongoing reliability issues.

## 5. Conclusion

In the fabrication of precise and reliable up-to-date communication systems stability and precise resistance values of widely utilized conventional thick-film resistors are of great importance. Different conditions of their application induced the need to investigate their behaviour under stress, especially under influence of mechanical and electrical straining. Mechanical straining leads to reversible resistance change due to change of charge transport conditions. It predominantly affects tunnelling through glass barriers by changing barrier widths. Electrical straining leads to irreversible resistance change due to barrier height alteration. Simultaneously, mechanically and electrically strained resistors are affected in two opposing manners; mechanical straining reversibly alters barrier widths while electrical straining irreversibly affects barrier heights. Having in mind that tunnelling through glass barriers is primarily affected by simultaneous straining; an impact of the simultaneous straining can be optimally evaluated using resistors with medium sheet resistances that include both metallic conduction and tunnelling through glass barriers. Results presented in this chapter can be viewed as an experimental verification of correlation between resistance, gauge factor and low-frequency noise parameters (noise index and current noise spectra) and changes with resistor degradation due to the impact of these three types of straining. Furthermore, they can be seen as validation of earlier presumptions [24, 25] that standard resistance, noise spectrum and noise index measurements are valuable tools in reliability evaluation of thick resistive films.

## Acknowledgements

The authors would like to thank the Ministry of Education, Science and Technological Development of the Republic of Serbia for supporting this research within projects III44003 and III45007.

## Author details

Zdravko Stanimirović

Address all correspondence to: zdravkos@iritel.com

Institute for Telecommunications and Electronics, IRITEL a.d. Beograd, Belgrade, Republic of Serbia

## References

[1] Prudenziati M, Morten B, Taroni A. Characterization of thick-film resistor strain gauges on enamel steel. Sensors and Actuators. 1982;**2**:17-27. DOI: 10.1016/0250-6874(81)80025-X

[2] Hrovat M, Belavič D, Samardžija Z, Holc J. A characterization of thick film resistors for strain gauge applications. Journal of Materials Science. 2001;**36**(11):2679-2689. DOI: 10.1023/A:1017908728642

[3] Canalli C, Malavasi D, Morten B, Prudenziati M, Taroni A. Piezoresistive effects in thick-film resistors. Journal of Applied Physics. 1980;**51**(6):3282-3288. DOI: 10.1063/1.328035

[4] Shah JS. Strain sensitivity of thick-film resistors. IEEE Transactions on Components, Packaging, and Manufacturing Technology. 1980;**3**(4):554-564. DOI: 10.1109/TCHMT.1980.1135645

[5] Stanimirović Z, Jevtić MM, Stanimirović I. Performances of conventional thick-film resistors subjected to mechanical straining. In: Proceeding of the 24th International Conference on Microelectronics, Niš, Serbia. 2004;**2**:675-678. DOI: 10.1109/ICMEL.2004.1314920

[6] Dziedzic A, Kolek A, Ehrhardt W, Thust H. Advanced electrical and stability characterization of untrimmed and variously trimmed thick-film and LTCC resistors. Microelectronics Reliability. 2006;**46**:352-359. DOI: 10.1016/j.microrel.2004.12.014

[7] Stanimirović Z, Stanimirović I. Effects of high voltage pulse trimming on structural properties of thick-film resistors. Science of Sintering. 2017;**49**:91-98. DOI: 10.2298/SOS1701091S

[8] Stanimirović I, Stanimirović Z. Influence of HVP trimming on primary parameters of thick resistive films. Journal of Materials Science: Materials in Electronics. 2017;**28**:8002-2010. DOI: 10.1007/s10854-017-6504-7

[9] Barker MF. Low ohm resistor series for optimum performance in high voltage surge applications. Microelectronics International. 1997;**43**:22-26. DOI: 10.1108/13565369710800493

[10] Kolek A, Ptak P, Dziedzic A. Noise characteristics of resistors buried in low-temperature co-fired ceramics. Journal of Physics D: Applied Physics. 2003;**36**:1009-1017. DOI: 10.1088/0022-3727/03/081009

[11] Mrak I, Jevtić MM, Stanimirović Z. Low-frequency noise in thick-film structures caused by traps in glass barriers. Microelectronics Reliability. 1998;**38**:1569-1576. DOI: 10.1016/S0026-2714(98)00032-8

[12] Jevtić MM, Stanimirović Z, Mrak I. Low-frequency noise in thick-film resistors due to two-step tunneling process in insulator layer of elemental MIM cell. IEEE Transactions on Components and Packaging Technologies. 1999;**22**(1):120-125. DOI: 10.1109/6144.759361

[13] Kusy A, Szpytma A. On 1/*f* noise in $RuO_2$-based thick resistive films. Solid-State Electron. 1986;**29**:657-665. DOI: 10.1016/0038-1101(86)90148-6

[14] Arshak KI, McDonagh D, Durcan MA. Development of new capacitive strain sensors based on thick film polymer and cermet technologies. Sensors and Actuators A: Physical. 2000;**79**(2):102-114. DOI: 10.1016/S0924-4247(99)00275-7

[15] Tankiewicz S, Morten B, Prudenziati M, Golonka LJ. New thick-film material for piezoresistive sensors. Sensors and Actuators A: Physical. 2001;**95**(1):39-45. DOI: 10.1016/S0924-4247(01)00754-3

[16] Stanimirović Z, Jevtić MM, Stanimirović I. Simultaneous mechanical and electrical straining of conventional thick-film resistors. Microelectronics Reliability. 2008;**48**:59-67. DOI: 10.1016/j.microrel.2006.09.039

[17] Puers B, Sansen W, Paszczynski S: Assessment of thick-film fabrication method for force (pressure) sensors. Sensors and Actuators. 1987;**12**:57-76. DOI: 10.1016/0250-6874(87) 87006-3

[18] Stanimirović Z, Jevtić MM, Stanimirović I. Computer simulation of thick-film resistors based on 3D planar RRN model. In: Proceedings EUROCON Belgrade, Serbia. 2005; **R24**:1687-1690. DOI: 10.1109/EURCON.2005.1630297

[19] Grimaldi C, Ryser P, Strassler S. Gauge factor enhancement driven by heterogeneity in thick-film resistors. Journal of Applied Physics. 2001;**91**(1): 322-327. DOI: 10.1063/ 1.1376672

[20] Stanimirović I, Jevtić MM, Stanimirović Z. High-voltage pulse stressing of thick-film resistors and noise. Microelectronics Reliability. 2003;**43**:905-911. DOI: 10.1016/S0026-2714(03)00094-5

[21] Stanimirović I, Jevtić MM, Stanimirović Z. Multiple high-voltage pulse stressing of conventional thick-film resistors. Microelectronics Reliability. 2007;**47**:2242-2248. DOI: 10.1016/j.microrel.2006.11.017

[22] Kleipenning TGM. On low-frequency noise in tunnel junctions. Solid-State Electron. 1982;**25**:78-79. DOI: 10.1016/0038-1101(82)90100-9

[23] Prudenziati M, Morten B, Masoero A: Excess noise and refiring processes in thick-film resistors. Journal of Physics D: Applied Physics. 1981;**14**:1355-1362. DOI: 10.1088/ 0022-3727/14/7/024

[24] Jevtić MM, Mrak I, Stanimirović Z. Thick-film resistor quality indicator based on noise index measurements. Microelectronics Journal. 1999;**30**:1255-1259. DOI: 10.1016/ S0026-2692(99)00050-6

[25] Jevtić MM, Stanimirović Z, Stanimirović I. Evaluation of thick-film structural parameters based on noise index measurements. Microelectronics Reliability. 2001;**41**:59-66. DOI: 10.1016/S0026-2714(00)00207-9

Section 3

# Power System Reliability and Feasibility

Chapter 13

# Coordination and Selectivity of Protection Devices with Reliability Assessment in Distribution Systems

Marco Antônio Ferreira Boaski, Caio dos Santos,
Mauricio Sperandio, Daniel Pinheiro Bernardon,
Maicon Jaderson Ramos and Daniel Sperb Porto

Additional information is available at the end of the chapter

http://dx.doi.org/10.5772/intechopen.69603

**Abstract**

This chapter provides an overview of the reliability of electricity distribution networks, and its evaluation that is linked with the protection system. In this way, the characteristics of network protection are presented, along with the peculiarities in coordination and device selectivity adjustments. For the assessment of the reliability, we have the methodology of logic-structural matrix (LSM) that integrates the constitution of the network with historical data of faults, so that with this, a model can be elaborated that can evaluate the impact of changes in the system directly on the reliability indicators.

**Keywords:** distribution network reliability, coordination and selectivity, protection system

## 1. Introduction

With the latest technologies and new concepts that have emerged, the electricity distribution system has become more flexible, but all this is reflected in general modifications of operation, planning, study, and analysis [1]. One of the great influencers in this environment is the smart grids, which changed the vision of a strongly static network and inserting versatility in the structure of the networks, using technologies, automation, and methodologies in a coupled manner [2, 3]. The uses of smart grid are wide ranging from load control, self-healing, voltage control, among others [4].

With the possibilities of this development, it inserts new complexities in the network and this can have both positive and negative impacts, due to which there are several factors or functionalities that must be reevaluated. Considering the main function of the power utilities being the transportation of energy to the final customer, we seek to use these new technologies and methods to increase the quality and safety of the service provided. In this way, one of the criteria most considered by the companies is the reliability of the network, both for strategic issues of network operation, and for the existence of financial penalties for not meeting targets [5–7].

The reliability of distribution networks is influenced by several factors like operation, maintenance, and planning [8], but as the reliability of the networks is directly linked to the interruptions in the power supply, one of the systems which has more impact is the protection [9]. Along with new automation, control, and communication technologies, some protection instruments have also advanced [10, 11], enabling reclosing, and different performance curves. In this way, the methodologies for adjusting the coordination and selectivity parameters of the devices must also consider a greater variety of factors, both those that can impact and those that can be impacted by this system.

Thus, it is proposed to evaluate the reliability of networks through their indicators considering the protection system. This will explore the main protection devices and how to obtain the main quality indicators, as well as verify the functionalities and main steps for coordination and selectivity of the device, in addition to modeling the network together with the factors that will directly impact the reliability of the distribution system.

## 2. Distribution network protection devices

The distribution systems of electric energy are composed of long networks, and are mostly aerial and made up of bare cables, due to the extension and exposure can cause failures or faults of various natures. Therefore, means for the protection of these systems become essential, for that, the protection devices are used. Moreover, the most used in distribution networks are the fuses and reclosers [12], which are discussed in the following subsections in detail.

### 2.1. Fuses

The protective devices that are most used in power distribution networks are the fuses. This is due to its low cost when compared to other protection devices and its satisfactory operation for one of the major problems of the distribution networks that refer to the overcurrent from short circuits by contact of the cables between each other, the vegetation or to the ground, and among others.

As a fuse-type protective device, it bases its operation on a metal link with specific characteristics of time versus current; when it reaches the maximum tolerable current, the heat melts the active element and releases the opening of circuit [13].

The construction of a fuse is divided into three elements: *Base*—Consists of an insulating material and serves as an interconnection between the moving parts and the support structure of the device. *Fuse tube*—Consists of an insulating material, serves as a support for the fuse link and is the moving part that promotes the opening between the terminals when the link fuses. *Fuse link*—Consisting of metal alloys with specific characteristics of melting temperature, it is the active element of protection of the equipment, having different types of links, with faster or slower curves depending on the application, the most common are types H and K. One example of fuse is represented in **Figure 1**.

### 2.2. Reclosers

Compared to fuses, reclosers have a relatively high cost, but these devices are more sophisticated and offer wider capabilities in protection, due to the measurement, automation, and the control possibilities. The reclosers have been increasingly used by electric power distributors; this happens due to the possibility of control and telecommunication of this device [14]. Through these characteristics, one can have a real-time control of the network allowing maneuvers for various purposes, thus contributing to the evolution of the smart grid.

The reclosers allow timed automatic reclosers; though this feature associated with fast curves, it is possible to minimize power interruptions caused by transient faults [15] such as contact of tree branches. In most devices, the timing system can be adjusted with fast or slow reclosing operations, or combinations thereof, depending on the need and philosophy of the company. This protection against transient faults also includes the networks protected by the fuses downstream of the reclosers. The construction of reclosers is more complex, and has a chamber for interrupt of electric arc, and there is current sensors and in some cases voltage sensors. One example of recloser is shown in **Figure 2**.

Regardless of the control and communication factors, the operating principle of the recloser is also related to the lack of current and its performance has a behavior described by current curves versus time. However, there are commercial models that are also made up of potential transformers; with this and the ability to communicate, these can protect the distribution system from other possible types of failures.

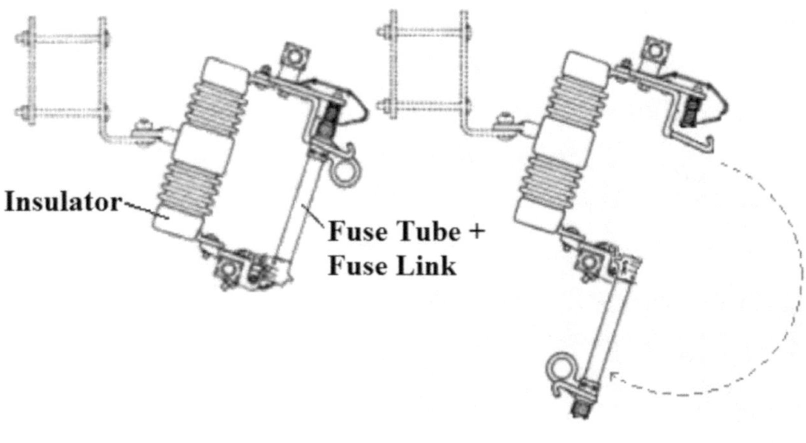


**Figure 1.** Medium voltage fuse.

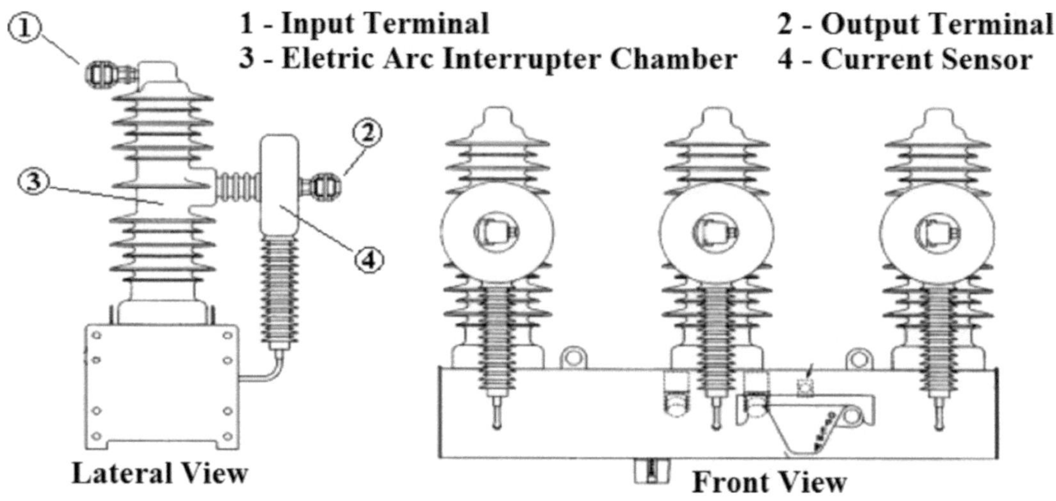


**Figure 2.** Recloser.

### 2.3. Coordination and selectivity

For proper protection of the electrical system, in addition to the use of appropriate devices, must be taken care with the sequence of operation of such equipment. This is necessary due to the fact that along the same network there are several devices and if this sequence of operation does not agree, there will be disconnection of undue loads, thus compromising the reliability of the system and may negatively influence the continuity indicators of the system from the power utility [16, 17].

The selectivity that refers to the sensitivity of the protection devices is evaluated first. This need arises since, that in addition to the interruption capacity of the devices, they need to be sensitized to operate in the minimum short circuit and at the same time allow the normal load current.

The criteria for selectivity of fuses can be summarized by Eq. (1) below:

$$K \cdot I_n \leq I_e \leq \frac{1}{4} \cdot I_{cc\min} \tag{1}$$

where

$I_n$–nominal load current of the circuit(A);

$I_e$–nominal current of the fuse link;

$I_{cc\min}$–minimum value of short-circuit current (A); and

$K$–demand growth rate.

For reclosers, since they have both phase and neutral settings, the settings can be expressed with the following Eqs. (2) and (3)

$$1.5 \cdot I_L \leq I_{pf} \leq \frac{I_{2\Phi F}}{2} \quad (2)$$

$$(0.1 \approx 0.3) \cdot I_c \leq I_p \leq \frac{I_{1\Phi m}}{2} \quad (3)$$

where

$I_L$–nominal load current of the circuit (A);

$I_{pf}$–pick-up current of phase, from recloser;

$I_{2\Phi F}$–biphasic current of short circuit;

$I_p$–pick-up current of neutral, from recloser; and

$I_{1\Phi m}$–single-phase current of short circuit.

In the other step, concerning coordination refers to the evaluation of the sequence of operation of the devices besides evaluating which devices downstream will be sensitized for upstream faults. This process seeks the best sequence of equipment performance, seeking to minimize the area affected by the fault or defect. In order to have the correct sequence, some criteria are used, depending on the type of equipment involved.

For coordination between fuses, we can cite three main criteria:

1. The nominal current of the protected link must always be higher than the rated current of the protective link.
2. Ideally, the protected fuse link (source side) must be coordinated with the protective link (load side) so as not to open first for the maximum short-circuit current at the point of installation of the protective link.
3. The coordination between two serial fuse links is guaranteed if the interruption time of the protective link is at most 75% of the minimum time of fusion of the protected link.

For coordination between fuses and reclosers, we can cite three main criteria that are explained as follows:

1. For all possible fault values within the circuit section protected by the fuse link, the minimum link fusion time must be greater than the recloser opening time multiplied by a factor $K$ characteristic of the recloser, which varies depending on the number of fast operations adjusted and the reclosing time of the circuit.
2. For all possible fault values within the circuit section protected by the fuse link, the total time of the link interruption must be less than the minimum opening time of the recloser in its timed curve (slow curve), by adjusting the recloser for two or more timed operations.
3. In the case of not being able to coordinate between recloser and link for the whole range of short-circuit currents, it will be guaranteed at least the coordination for the condition of

single-phase faults involving contact impedance and maintained the selectivity to 80% maximum circuit.

## 3. Reliability in distribution networks

The term reliability has broad meaning and can refer to different applications in the same system. The authors of Ref. [18] define that the reliability of a system refers to the correct operation with full performance and no failure. Thus, the reliability is influenced by several factors, and may be manageable, such as planning, maintenance, and operation or unforeseen events such as storms or accidents.

For the main current regulations, the concept of reliability in the distribution of electricity is linked to interruptions in energy supply, which may be temporary (momentary) or sustained (permanent) [19]. This can be considered a subgroup of the perturbations that affect the quality of energy, as shown in **Figure 3**. The main disturbances that affect the quality can be frequency variations, noise, transient, harmonic distortions, temporary variation of voltage amplitude and properly the interruptions [20].

For the measurement and evaluation of the reliability of the distribution networks, the reliability indicators are used, these will be explained next. The evaluation of the power utilities is through these indicators, failure to comply with the stipulated values causes a penalty for the company and a discount to the final consumer. Thus in power systems, it is important to keep the reliability indicators at good levels [21].

### 3.1. Reliability indicators

The calculation of the reliability indicators are values that synthesize statistical aggregates, which can be calculated from the historical interruptions that occur in certain regions or groups of consumers of the distribution system [22, 23].

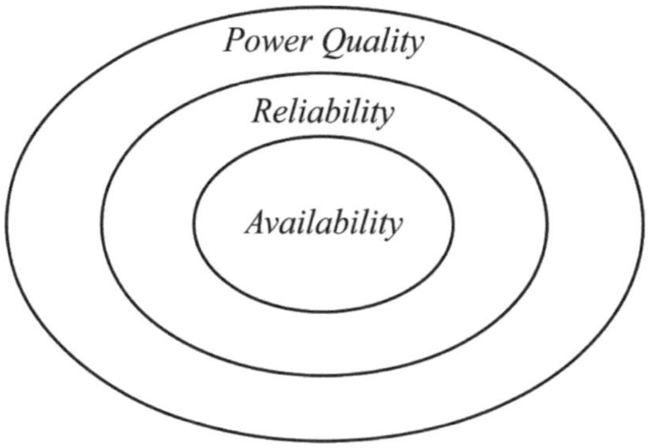


**Figure 3.** Power quality and subgroups.

The main reliability indicators are defined in the guide [24], among which the most used are:

- SAIFI—System average interruption frequency index: This index considers the number of interruptions on a media that consumer or a group or consumers suffer during a period.

$$\text{SAIFI} = \frac{\sum_i N_i}{N_T} \tag{4}$$

where

$N_i$: number of consumer units affected at each interruption; and

$N_T$: total number of consumer units of the group.

- SAIDI—System average interruption duration index: This index considers the duration of the interruptions, thus having an average of hours that a given consumer has his energy supply interrupted during the period.

$$\text{SAIDI} = \frac{\sum_i (r_i \cdot N_i)}{N_T} \tag{5}$$

where

$N_i$: number of consumer units affected at each interruption;

$r_i$: duration of each interruption; and

$N_T$: total number of consumer units of the group.

- ENS—Energy not supplied index: This index relates the total of energy not supplied in all group due interruptions in all the events in a given period.

$$\text{ENS} = \sum_i r_i \cdot L_i \tag{6}$$

where

$r_i$: duration of each interruption; and

$L_i$: energy not supplied in each interruption.

### 3.2. Modeling of reliability in distribution networks

To model the distribution network to calculate its reliability, it is important to be aware that there may be faults that impact different amounts of customers, such as an external fault that affects all consumers or a one fault on a transformer or in a point of the network that only affects a small group [25]. Then there will be faults with different levels of comprehensiveness.

For this modeling, there are different methods such as block classification and analytical simulation [26] and the logic-structural matrix (LSM). The method that best fits with the protection equipment is the logic-structural matrix, due to the presence of the maneuverable equipment installed in the network, on the different groups of consumers. By switching, the equipment can isolate faults with consumers, or reestablish them, based on the maneuvers allowed for fault isolation and network reconfiguration.

The logic-structural matrix is composed of the main data of the following:

- *Annual failure rate* ($\lambda$): This media is obtained through the history of failures of that group.
- *Mean time to restore power supply* (*TR*): This measure is the average time of the restoration of the energy supply, this time is composed of several phases such as the time of displacement of the maintenance team, time of repair, and even the time of waiting between the event of the failure and the authorization of the displacement of the maintenance team
- *Number of customers* (*N*): It refers properly to the number of consumers fed in that region, being able to be fed by transformers or directly connected in the primary network.
- *Load active power* (*L*): Active load, transformers or consumers connected directly to the primary network.

The LSM is composed as follows: the columns are equivalent to the protection or switching equipment of the system and each row is equivalent to the points of the system (these points can be divided according to the needs of the company; they can be transformers, primary consumers or networks extensions). In the cells of the logical-structural matrix, there are initial values of the mean time to power restoration. In order to define these values, it is required to analyze how long it takes to restore the power supply for the corresponding consumers (matrix line), when they are faced with a failure in the distribution network assuming the protective and switching equipments installed on the network (matrix column) [27, 28].

In the presence of switching equipment, one must evaluate the possibilities for switching, isolating defects or transferring loads through these devices. The first possibility is sectionalizing, what corresponds to the isolation of the segment under failure and other associated nodes downstream of a normally closed (NC) from nodes upstream. The mean time to isolate (TI) is computed for consumers on all these upstream nodes. The second option is the transfer of the nodes downstream from the NC switch; when an upstream fault occurs, then the mean time to transfer (TT) is considered for the consumers downstream. The last possibility depends on the existence of a normally open (NO) switch downstream from the NC, and the adjacent feeder must have available technical capacity to receive the loads that will be transferred. For manual switches, the TI and TT also include: mean time of wait (TW) and mean time to travel (TTr). For automatic switches, the TI and TT are much shorter, because there are not TW and TTr. Normally TR > TT > TI.

The protection devices prevent upstream faults or defects from affecting the nodes in downstream of the device. This way, downstream nodes do not have their power supply interrupted, then in the cells of LSM, in these nodes can be placed the number 0.

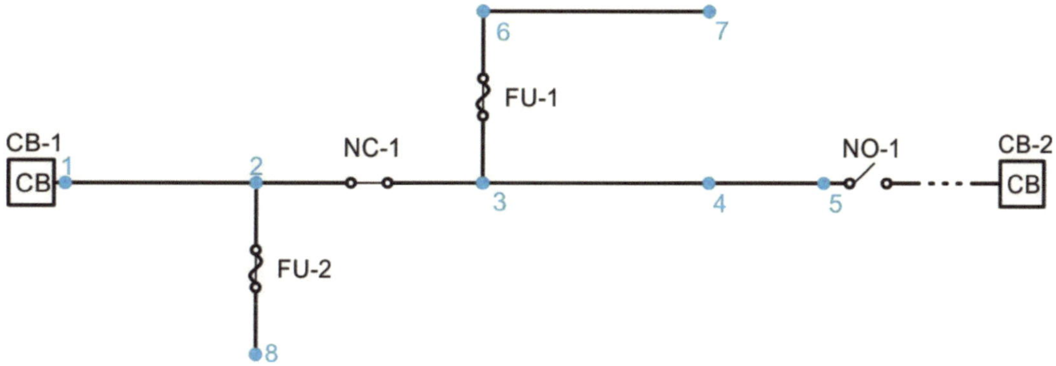


**Figure 4.** Example distribution network.

To illustrate, the logical-structural matrix for the simplified distribution network is shown in **Figure 4**. It is assumed that the NO switch at node 5 is connected to another feeder with the technical capability to receive loads downstream of the NC switch.

**Table 1** shows the construction of logical-structural matrix for the example in **Figure 4**, considering the mean time to power restoration of node *i* (TR*i*), and constants time to isolation (TI), and time to transfer (TT) for each device.

One can note that, for the outage of the circuit breaker (CB-1), the total time to restore power for all consumers is computed, except to those downstream of the NC switch, for which the transfer time to another feeder is considered. For failures downstream of the NC switch, the time to isolate the fault for upstream consumers of the switch and the total time to restore

| Nodes (distribution transformers or primary customers) | Protective and switching equipment | | | |
|---|---|---|---|---|
| | Circuit breaker CB-1 | Switch NC | Fuse FU-1 | Fuse FU-2 |
| 1 | TR1 | TI | 0 | 0 |
| 2 | TR1 | TI | 0 | 0 |
| 3 | TT | TR2 | 0 | 0 |
| 4 | TT | TR2 | 0 | 0 |
| 5 | TT | TR2 | 0 | 0 |
| 6 | TT | TR2 | TR3 | 0 |
| 7 | TT | TR2 | TR3 | 0 |
| 8 | TR1 | TI | 0 | TR4 |

**Table 1.** Logical-structural matrix to the distribution network of **Figure 4**.

power to its downstream customers is computed. Regarding the outage of fuses (FU-1 and FU-2), it only affects its downstream consumers, so the total time to restore power is computed. The upstream nodes are not affected by the fault, not suffering interruption, since the fuse is coordinated to blow before the circuit breaker trips (trip saving scheme).

Then, the matrix values are multiplied by the failure rate of the respective equipment ($\lambda i$), as shown in **Table 2**.

The reliability index is then calculated from the LSM. To calculate the expected value of SAIDI, the terms of each row of **Table 2** are added and then multiplied by the respective amount of consumers in that row, and then the results of all lines are added together and divided by the total number of customers served [28], as follows:

$$\mathrm{ESAIDI} = \frac{\sum_{i=1}^{n}\left(\sum_{j=1}^{m} M_{i,j}\right)\cdot N_i}{N_C} \tag{7}$$

where

ESAIDI = expected value of system average interruption duration index (h/year);

$M_{i,j}$ = element in row $i$ and column $j$ of LSM;

$Ni$ = number of consumers for the row $i$;

$N_C$ = total number of customers served;

$n$ = number of rows; and

$m$ = number of columns.

| Nodes (distribution transformers or primary customers) | Protective and switching equipment | | | |
|---|---|---|---|---|
| | Circuit breaker CB-1 | Switch NC | Fuse FU-1 | Fuse FU-2 |
| 1 | TR1 $\lambda 1$ | TI $\lambda 2$ | 0 | 0 |
| 2 | TR1 $\lambda 1$ | TI $\lambda 2$ | 0 | 0 |
| 3 | TT $\lambda 1$ | TR2 $\lambda 2$ | 0 | 0 |
| 4 | TT $\lambda 1$ | TR2 $\lambda 2$ | 0 | 0 |
| 5 | TT $\lambda 1$ | TR2 $\lambda 2$ | 0 | 0 |
| 6 | TT $\lambda 1$ | TR2 $\lambda 2$ | TR3 $\lambda 3$ | 0 |
| 7 | TT $\lambda 1$ | TR2 $\lambda 2$ | TR3 $\lambda 3$ | 0 |
| 8 | TR1 $\lambda 1$ | TI $\lambda 2$ | 0 | TR4 $\lambda 4$ |

**Table 2.** Logical-structural matrix with times versus failure rate.

The expected value of ENS is straight forward obtained by replacing the number of consumers in Eq. (8) by its respective load, active power of the distribution transformers, ignoring the total number of customers served

$$\text{EENS} = \sum_{i=1}^{n} \left( \sum_{j=1}^{m} M_{i,j} \right) \cdot L_i \tag{8}$$

where

EENS = expected value of energy not supplied (kWh/year);

$M_{i,j}$ = element in row $i$ and column $j$ of LSM;

$L_i$ = average load, maximum demand of active power multiplied by the respective load factor, associated to row $i$ (kW);

$n$ = number of rows; and

$m$ = number of columns.

To obtain the expected value of SAIFI, the process is similar to the SAIDI, requiring only replacement of the logical-structural matrix average times (TR, TI, and TT) by 1, so are considered only the failure rates.

$$\text{ESAIFI} = \frac{\sum_{i=1}^{n} \left( \sum_{j=1}^{m} M_{i,j}^{*} \right) \cdot N_i}{N_C} \tag{9}$$

where

ESAIFI = expected value of system average interruption frequency (failures/year);

$M_{i,j}^{*}$ = element in row $i$ and column $j$ of LSM, without considering the mean times;

$N_i$ = number of customers for the row $i$;

$N_C$ = total number of customers served;

$n$ = number of rows; and

$m$ = number of columns.

## 4. Assessment of protection considering the reliability

In order to evaluate the influence of protection on the reliability of the distribution system, the methodology is shown in the flowchart of **Figure 5** can be used. This generic model seeks that the protection devices can be adjusted for both coordination and selectivity quality. For a better understanding of how the impact of selectivity and coordination on reliability will be, we have an example below, the circuit is shown in **Figure 6**.

Considering a theoretical system, we have the base network shown in **Figure 5** and having the following considerations in the protection:

- *Circuit breaker* (CB-1): It is selective for short circuits in all nodes of the network.
- *Recloser* (R-1): It is selective for short circuits between nodes 4 and 5.
- *Fuse* (FU-1): It is selective for short circuits between nodes 8 and 9.
- *Fuse* (FU-2): It is selective for short circuits between nodes 6 and 7.

Faults may occur with transient (momentary) character as being sustained (permanent) [29]. With the coordination of protection devices, in addition to avoiding a complete shutdown of the system, the effect of transient faults on the network can be reduced through the reclosers. The transient faults are most of the faults that occur in the distribution network, reaching values between 80 and 90% of total faults [30]. To better illustrate this, two LSMs can be established, one for sustained faults (**Table 3**) and the other for momentary faults (**Table 4**).

With the idea of visualizing the impact of device coordination and selectivity on reliability, a simple change that could be implemented would be to selectively the R-1 device for nodes 6 and 7 together with the coordination between the R-1 device and FU-2. With these changes, we would have the following result in the sustained LSM (**Table 5**) and momentary LSM (**Table 6**).

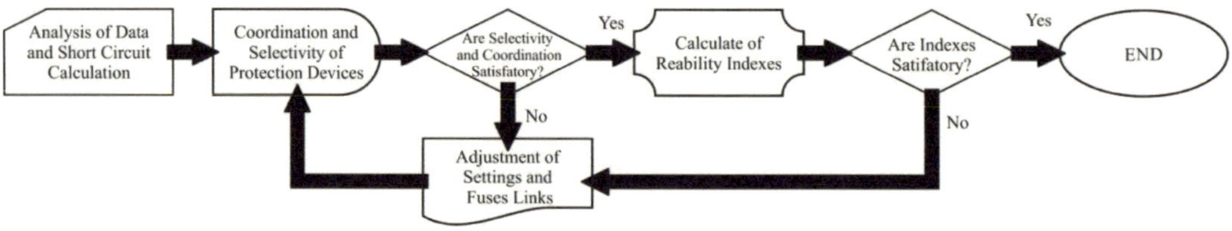


**Figure 5.** Methodology of assessment of protection and reliability.

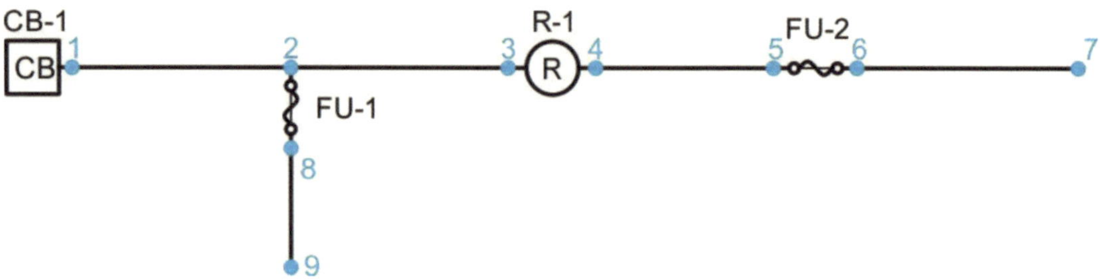


**Figure 6.** Example distribution network.

| Nodes | Protective equipment | | | |
|---|---|---|---|---|
| | Circuit breaker CB-1 | Recloser R-1 | Fuse FU-1 | Fuse FU-2 |
| 1 | TR1 $\lambda_{P1}$ | 0 | 0 | 0 |
| 2 | TR1 $\lambda_{P1}$ | 0 | 0 | 0 |
| 3 | TR1 $\lambda_{P1}$ | 0 | 0 | 0 |
| 4 | TR1 $\lambda_{P1}$ | TR2 $\lambda_{P2}$ | 0 | 0 |
| 5 | TR1 $\lambda_{P1}$ | TR2 $\lambda_{P2}$ | 0 | 0 |
| 6 | TR1 $\lambda_{P1}$ | TR2 $\lambda_{P2}$ | 0 | TR4 $\lambda_{P4}$ |
| 7 | TR1 $\lambda_{P1}$ | TR2 $\lambda_{P2}$ | 0 | TR4 $\lambda_{P4}$ |
| 8 | TR1 $\lambda_{P1}$ | 0 | TR3 $\lambda_{P3}$ | 0 |
| 9 | TR1 $\lambda_{P1}$ | 0 | TR3 $\lambda_{P3}$ | 0 |

**Table 3.** Initial LSM considering sustained faults.

| Nodes | Protective equipment | | | |
|---|---|---|---|---|
| | Circuit breaker CB-1 | Recloser R-1 | Fuse FU-1 | Fuse FU-2 |
| 1 | TR1 $\lambda_{M1}$ | 0 | 0 | 0 |
| 2 | TR1 $\lambda_{M1}$ | 0 | 0 | 0 |
| 3 | TR1 $\lambda_{M1}$ | 0 | 0 | 0 |
| 4 | TR1 $\lambda_{M1}$ | TA2 $\lambda_{M2}$ | 0 | 0 |
| 5 | TR1 $\lambda_{M1}$ | TA2 $\lambda_{M2}$ | 0 | 0 |
| 6 | TR1 $\lambda_{M1}$ | TA2 $\lambda_{M2}$ | 0 | TR4 $\lambda_{M4}$ |
| 7 | TR1 $\lambda_{M1}$ | TA2 $\lambda_{M2}$ | 0 | TR4 $\lambda_{M4}$ |
| 8 | TR1 $\lambda_{M1}$ | 0 | TR3 $\lambda_{M3}$ | 0 |
| 9 | TR1 $\lambda_{M1}$ | 0 | TR3 $\lambda_{M3}$ | 0 |

TA2: Reclosing action time, considering the ability to disconnect and reconnect the circuit with the reclosing device, thus decreasing the shutdown time (TR > TA).

**Table 4.** Initial LSM considering momentary faults.

In short, under the conditions of permanent failure, the system would keep practically the same levels in the reliability indicators. The main impact would be directly on the temporary faults, where for faults in nodes 6 and 7 the re-establishment time would now be only the recloser R-1, no longer the replacement time of the FU-2 fuse. This impact would be strongly noticed in the SAIDI indicator, which considers the duration of faults.

| Nodes | Protective equipment | | | |
|---|---|---|---|---|
| | Circuit breaker CB-1 | Recloser R-1 | Fuse FU-1 | Fuse FU-2 |
| 1 | TR1 $\lambda_{P1}$ | 0 | 0 | 0 |
| 2 | TR1 $\lambda_{P1}$ | 0 | 0 | 0 |
| 3 | TR1 $\lambda_{P1}$ | 0 | 0 | 0 |
| 4 | TR1 $\lambda_{P1}$ | TR2 $\lambda_{P2}$ | 0 | 0 |
| 5 | TR1 $\lambda_{P1}$ | TR2 $\lambda_{P2}$ | 0 | 0 |
| 6 | TR1 $\lambda_{P1}$ | TR2 $\lambda_{P2}$ | 0 | TR4 $\lambda_{P4}$ |
| 7 | TR1 $\lambda_{P1}$ | TR2 $\lambda_{P2}$ | 0 | TR4 $\lambda_{P4}$ |
| 8 | TR1 $\lambda_{P1}$ | 0 | TR3 $\lambda_{P3}$ | 0 |
| 9 | TR1 $\lambda_{P1}$ | 0 | TR3 $\lambda_{P3}$ | 0 |

**Table 5.** Final LSM considering sustained faults.

| Nodes | Protective equipment | | | |
|---|---|---|---|---|
| | Circuit breaker CB-1 | Recloser R-1 | Fuse FU-1 | Fuse FU-2 |
| 1 | TR1 $\lambda_{M1}$ | 0 | 0 | 0 |
| 2 | TR1 $\lambda_{M1}$ | 0 | 0 | 0 |
| 3 | TR1 $\lambda_{M1}$ | 0 | 0 | 0 |
| 4 | TR1 $\lambda_{M1}$ | TA2 $\lambda_{M2}$ | 0 | 0 |
| 5 | TR1 $\lambda_{M1}$ | TA2 $\lambda_{M2}$ | 0 | 0 |
| 6 | TR1 $\lambda_{M1}$ | TA2 $\lambda_{M2}$ | 0 | TA4 $\lambda_{P4}$ |
| 7 | TR1 $\lambda_{M1}$ | TA2 $\lambda_{M2}$ | 0 | TA4 $\lambda_{P4}$ |
| 8 | TR1 $\lambda_{M1}$ | 0 | TR3 $\lambda_{M3}$ | 0 |
| 9 | TR1 $\lambda_{M1}$ | 0 | TR3 $\lambda_{M3}$ | 0 |

**Table 6.** Final LSM considering momentary faults.

## 5. Conclusions

In this chapter, the main issues of reliability in the electric power supply as well as the characteristics and adjustments in coordination and selectivity of protection devices were presented, as well as a brief evaluation of their direct impact on the reliability indicators. In addition, it became clear how important it is to keep reliability levels in the pattern for both operational and financial issues. Finally, it is possible to see a theoretical example where the influence of the protection system on reliability was exposed, where, with the coordination and selectivity of a recloser, a great part of the impact of the temporary faults of a circuit can be

http://dx.doi.org/10.5772/intechopen.69603

reduced. The result, although theoretical, contributed to the validation of the importance of coordination and selectivity of the protection devices of an energy distribution network. This study can contribute directly in the two target areas of the power utilities, since it brings a broad vision of the system, besides the description of a way to model the network and to calculate the main indicators of reliability.

## Acknowledgements

The authors would like to thank the technical and financial support of RGE Sul Power Utility by project "Solução Inovadora para Gerenciamento Ativo de Sistemas de Distribuição" (P&D/ANEEL), Coordination for the Improvement of High Level Personnel (CAPES) and the National Center of Scientific and Technological Development (CNPq).

## Author details

Marco Antônio Ferreira Boaski[1]*, Caio dos Santos[1], Mauricio Sperandio[1], Daniel Pinheiro Bernardon[1], Maicon Jaderson Ramos[2] and Daniel Sperb Porto[2]

*Address all correspondence to: ferreirab.marco@gmail.com

1 Federal University of Santa Maria, Santa Maria, Brazil

2 RGE Sul Power Utility, Brazil

## References

[1] Mello APC, Bernardon DP, Pfitscher LL, Sperandio M, Toller BB, Ramos MJS. Intelligent system for multivariables reconfiguration of distribution networks. In: Presented at the IEEE PES. Conference on Innovative Smart Grid Technologies Latin America (ISGT LA); IEEE; São Paulo, Brazil; 2013

[2] Brown RE. Impact of smart grid on distribution system design. In: Presented at the Power and Energy Society General Meeting—Conversion and Delivery of Electrical Energy in the 21st Century; Pittsburgh, USA 2008 IEEE

[3] Siqueira IP. Estimating the Impact of Wide-Area Protection Systems on Power System Performance and Reliability. In: Presented at the 13th International Conference on Development in Power System Protection 2016 (DPSP); IET 2016

[4] Bernardon DP, Mello APC, Pfitscher LL, Canha LN, Abaide AR, Ferreira AAB. Real-time reconfiguration of distribution network with distributed generation. Electric Power Systems Research. 2014;**107**:59-67

[5] Siirto OK, Safdarian A, Lehtonen M, Fotuhi-Firuzabad M. Optimal distribution network automation considering earth fault events. IEEE Transactions on Smart Grid. 2015;**6**(2): 1010-1018

[6] Elkadeem MR, Alaam MA, Azmy AM. Optimal automation level for reliability improvement and self-healing MV distribution networks. Presented at the 2016 Eighteenth International Middle East Power Systems Conference (MEPCON); 2016; Cairo, Egypt: IEEE; Changsha, China 2017

[7] Xu X, Mitra J, Wang T, Mu L. An evaluation strategy for microgrid reliability considering the effects of protection system. IEEE Transactions on Power Delivery. 2016;**31**(5):1989-1997

[8] Silva LGW, Pereira RAF, Abbad JR, Mantovani JRS. Optimised placement of control and protective devices in electric distribution systems through reactive tabu search algorithm. Electric Power Systems Research. 2008;**78**(3):372-381

[9] Hamman ST, Hopkinson KM, Fadul JE. A model checking approach to testing the reliability of smart grid protection systems. IEEE; Power & Energy Society. 2016

[10] Velásquez MA, Quijano N, Cadena AI. Multi-objective planning of recloser-based protection systems on DG enhanced feeders, In: Presented at the 2015 IEEE PES Innovative Smart Grid Technologies Latin America (ISGT LATAM); Montevideo, Uruguay 2015

[11] Bo Z, Zhao Y, Wang L, Ma X, Ding S, Wei F. The protection cooperation of distribution network based on protection intelligent center and the validation of its reliability. In: Presented at the 2015 5th International Conference on Electric Utility Deregulation and Restructuring and Power Technologies (DRPT); IEEE; Changsha, China 2015

[12] Supannon A, Jirapong P. Recloser-fuse coordination tool for distributed generation installed capacity enhancement. Presented at the Innovative Smart Grid Technologies—Asia (ISGT ASIA); 2015 IEEE; Bangkok, Thailand 2015

[13] Khorshid-Ghazani B, Seyedi H, Mohammadi-ivatloo B, Zare K, Shargh S. Reconfiguration of distribution networks considering coordination of the protective devices. IET Generation, Transmission & Distribution. 2011;**11**(1):82-92

[14] Ramos MJS, Bernardon DP, Comassetto L, Resener M. Analysis of coordination and selectivity of protection systems during reconfigurations of distribution energy systems in real time. In: Presented at the 2013 IEEE PES Conference on Innovative Smart Grid Technologies (ISGT Latin America); São Paulo, BR. IEEE; 2013

[15] Jamali S, Borhani-Bahabadi H. A new recloser time-current-voltage characteristic for fuse saving in distribution networks with DG. IET Generation, Transmission & Distribution. 2017

[16] Comassetto L, Bernardon DP, Canha LN, Abaide AR. Automatic coordination of protection devices in distribution system. Electric Power Systems Research. 2008;**78**(7):1210-1216

[17] IEEE, 242-2001-IEEE Recommended Practice for Protection and Coordination of Industrial and Commercial Power Systems. IEEE; New York, USA 2001

[18] Blischke WR, Murthy DNP. Reliability: Modeling, Prediction, and Optimization. Wiley-Interscience New York, USA 2000

[19] Meneses A, Pinto LC. Quality of supply at the Portuguese Electricity Transmission Grid. In: Presented at the 11th International Conference on Electrical Power Quality and Utilisation (EPQU); IEEE; Lisbon, Portugal 2011

[20] Brown RE. Eletric Power Distribution Reliability. 2nd ed. CRC Press; New York, USA 2009

[21] Wu Y, Li M, Ni M, Zhang Q. Reliability assessment of line protection based on reliability graphs and Monte Carlo simulation. In: Presented at the 2015 5th International Conference on Electric Utility Deregulation and Restructuring and Power Technologies (DRPT); IEEE; Changsha, China 2015

[22] Sailaja CVSS, Prasad PVN. Determination of optimal distributed generation size for losses, protection co-ordination and reliability Evaluation Using ETAP. In: 2016 Biennial International Conference on Presented at the Power and Energy Systems: Towards Sustainable Energy (PESTSE); IEEE; Bangalore, India 2016

[23] Tomsovic FSK. Optimal distribution protection design: quality of solution and computational analysis, International Journal of Electrical Power & Energy Systems.1999;**21**:327-335

[24] IEEE. 1366-2003-IEEE Guide for Electric Power Distribution Reliability Indices. IEEE; 2004

[25] Rocha LF, Borges CLT, Taranto GN. Reliability evaluation of active distribution networks including islanding dynamics. IEEE Transactions on Power Systems. 2016;**32**(2):1545-1552

[26] Zhang J, Kang Q, Huang D, Xiong X, Liu Y, Ma J, et al. Reliability evaluation of the new generation smart substation considering relay protection system. Presented at the 2016 IEEE PES Asia-Pacific Power and Energy Engineering Conference (APPEEC); IEEE Xian, China 2016

[27] Bernardon DP, Sperandio M, Garcia VJ, Canha LN, Abaide AR, Daza EFB. AHP decision making algorithm to allocate remotely controlled switches in distribution networks. IEEE Transactions on Power Delivery. 2011;**26**(3):1884-1892

[28] Neto NK, Abaide AR, Bernardon DP, Canha LN, Oberto M, Pressi R. The application of the logical structural matrix for reliability analysis in a distribution system planning environment. In: Presented at the 23° International Conference on Electricity Distribution-CIRED; Lyon, France 2015

[29] Ferreira GD, Bretas AS, Cardoso G. Optimal distribution protection design considering momentary and sustained reliability indices. In: Presented at the Proceedings of the International Symposium Modern Electric Power Systems (MEPS) 2010; IEEE; Wroclaw, Poland 2010

[30] Blackburn JL, Domin TJ. Protective Relaying: Principles and Applications. CRC-Press New York, USA 2006

Chapter 14

# An Analysis of Software and Hardware Development in the PMU-Based Technology and Suggestions Regarding Its Implementation in the Polish Power Grid

Michał Szewczyk

Additional information is available at the end of the chapter

http://dx.doi.org/10.5772/intechopen.71502

**Abstract**

The ongoing evolution of electric power systems (EPS), especially distribution systems within the EPS structure, is driven by the implementation of the smart grid framework. This requires new approaches and technologies to continue ensuring a reliable and secure supply to end users. Fluctuating output from solar photovoltaic and wind plants can cause voltage and power variations in the feeders. In the power grid framework, phasor measurement units (PMUs) are recognized to be an invaluable aid in ensuring the secure operation and stability of transmission systems. The synchrophasor technique requires a high-accuracy time stamping of all the measurements within the analyzed power system area. It must be emphasized that the harmonic injection from power electronic components such as fluorescent lighting, computers, and power inverters of motors and generators can increase total harmonic distortion (THD) levels on distribution feeders and modify the conventional patterns of voltage and current signals. Therefore, what is vital for the functional reliability of synchronous measurements is the implementation of measurement algorithms, which can realize high-accuracy measurements, both in quasi-static and dynamic EPS operating conditions. This article presents the results of software simulations and hardware tests of measurement algorithms that meet the requirements of the IEEE C37.118™-2011 Standard.

**Keywords:** PMU-based measurements technology, time synchronization, power system operation, reliability, power system adaptive automation

## 1. Introduction

The availability of high-precision timing sources, such as the Global Positioning System (GPS) and the IEEE 1588 compliant network clock sources [1–3] plus the networking capability [4–5] of protective relaying devices and systems, is fundamentally changing the way that many current and future protective relaying applications are or will be implemented. Synchrophasor measurements [6–10], i.e., phasor measurements with high-accuracy time stamping, are under consideration for many future protective relaying applications [11–20]. Synchrophasor measurements are also used in many other power system applications such as wide-area monitoring and situational awareness applications. This chapter focuses primarily on their proper physical realization resulting from the standard requirements. Fulfillment of these requirements can lead to the development of the relaying applications that can either be implemented or are considered for future implementations.

The use of high-accuracy measurements and reliable protective algorithms with adequately high-accuracy and prompt decision-making [21–28] should guarantee an effective operation and protection of the electric power systems (EPS) from the consequences of disturbances also in the power system area characterized by changeable operation frequency over a wide range. The adaptation of measurement and protective algorithms in cases under consideration has the frequency nature, i.e., it concentrates on such a change of the parameters of algorithm operation which guarantees the proper estimation of measurement and criterion values by changeable frequency of the input measurement signals: currents and voltages received from the primary circuit [22–23]. The measurement and protective algorithms used at present in the measurement-protection devices tend to be defined for the constant frequency 50/60 Hz of the input signal, and from the point of view of the synchronous measurement standard requirements, they are characterized by too big inaccuracies when the frequency of the input signals varies over a wide range [21–22]. Another problem for the formulation of the measurement algorithms is the determination of the level of the insensitivity of these algorithms to the existence of interfering components in the input signals ([22, 25, 27]). Several interfering components and factors can occur especially in the input measurement signals during the faults within the EPS ([14–15, 22, 25, 27]). This situation is accompanied by high penetration levels of intermittent dispersed generation (DG) [14–15], and it can significantly affect the EPS operation. The existence of thyristor frequency converter is the source of high harmonics, particularly the odd high 5th, 7th, 9th harmonics with the amplitude reaching up to 10% of the amplitude of the basic component [16]. Moreover, as penetration levels increase, concerns regarding dynamic interactions among DG units are becoming more important. Uncontrolled personal electric vehicle (PEV) charging can lead to the overloading of the distribution equipment and violations of low-voltage limits and their courses. Also, the DC component in the input current signals can reach significant values and has a considerable influence on the accuracy of the measuring algorithms [21–22]. In summary the measuring algorithms implemented in the digital measure-protective systems should be characterized by:

- High accuracy of operation over a wide range of frequency change
- Non-sensitivity or low-level sensitivity to the existence of high harmonics in the input measurement signals
- Low inaccuracy in the case of the existence of the DC component of the high level and longtime decay (in current signals)
- Maximum short time of transient states in the measurements or decision-making

http://dx.doi.org/10.5772/intechopen.71502

## 2. Basic requirements resulting from the series of the standard C37.118™

The IEEE Std 1344™-1995 was the original synchrophasor standard. It was replaced in 2005 by IEEE Std C37.118-2005 [6]. The newest standard has been split into two standards: IEEE Std 37.118.1-2011 [7], covering measurement provisions, and IEEE Std 37.118.2™-2011 [8], covering data communication. In the IEEE standard from 2011, there is a significant change of the requirements for synchronous measurements and additional clarification for the phasor and synchronized phasor definitions. The concepts of total vector error (TVE) and compliance tests are retained and expanded, tests over temperature variation have been added, and dynamic performance tests have been introduced [7–9]. In addition, limits and characteristics of frequency measurement and rate of change of frequency (ROCOF) measurement have been developed.

Phasor representation of sinusoidal signals is commonly used in the AC power system analysis. The sinusoidal waveform defined in Eq. (7):

$$x(t) = X_{\mathrm{m}}(\omega t + \varphi) \tag{1}$$

is commonly represented as the phasor as shown in Eq. (2) [7]:

$$\underline{X} = X_{\mathrm{r}} + \mathrm{j}X_{\mathrm{i}} = \frac{X_{\mathrm{m}}}{\sqrt{2}} e^{\mathrm{j}\varphi} = \frac{X_{\mathrm{m}}}{\sqrt{2}} (\cos\varphi + \mathrm{j}\sin\varphi) \tag{2}$$

where the magnitude is the root-mean-square (RMS) value, $X_m/\sqrt{2}$, of the waveform and the subscripts $r$ and $i$ signify real and imaginary parts of a complex value in rectangular components. This phasor is defined for the angular frequency $\omega$.

TVE is an expression of the difference between a "perfect" sample of a theoretical synchrophasor [7] and an estimate given by the unit under test at the same instant. The value is normalized and expressed as per unit of the theoretical phasor. TVE is defined in the following Eq. (7):

$$TVE(n) = \sqrt{\frac{\left(\hat{X}_{\mathrm{r}}(n) - X_{\mathrm{r}}(n)\right)^2 + \left(\hat{X}_i(n) - X_{\mathrm{i}}(n)\right)^2}{(X_r(n) + X_i(n))^2}} \tag{3}$$

where $\hat{X}_{\mathrm{r}}$ and $\hat{X}_{\mathrm{i}}$ are the sequences of estimates given by the unit under test and $X_{\mathrm{r}}(n)$ and $X_{\mathrm{i}}(n)$ are the sequences of theoretical values of the input signal at the times (n) assigned by the unit to those values. Synchrophasor measurements shall be evaluated using the TVE criterion of Eq. (3).

A phasor measurement unit (PMU) shall also calculate and be capable of reporting frequency and ROCOF [7]. For this measurement, the following standard definitions are used. If there is given a sinusoidal signal, as in the Eq. (7):

$$x(t) = X_{\mathrm{m}} \cos\left[\Psi(t)\right] \tag{4}$$

the frequency is defined as follows:

$$f(t) = \frac{1}{2\pi}\frac{\Psi(t)}{\mathrm{d}t} \tag{5}$$

Then, the ROCOF (rate of change of frequency) is defined:

$$\mathrm{ROCOF}(t) = \frac{df(t)}{\mathrm{d}t} \tag{6}$$

As mentioned above, the series of the Standard C37.118™ was developed for synchronized phasor measurement systems in power systems. They define a synchronized phasor (synchrophasor), frequency, and rate of change of frequency (ROCOF) measurements. They also describe time tag and synchronization requirements for measurements of all three of these quantities. Next, C37.118™ specifies methods for evaluating these measurements and requirements for compliance with the standard under both static and dynamic conditions. The "old" 2005 version of the standard, commonly followed by equipment manufacturers and system integrators, specifies the performance of phasor measurements only under steady-state conditions. The latest revision of the standard (the 2011 update) extends the synchrophasor definition. It also specifies measurement requirements and test conditions. Steady-state requirements of TVE and ROCOF (resulting from the IEEE C37.118.1–2011 Standard) are included in [7, 9].

The standard C37.118.1–2011 defines the measurement response time and measurement delay time [7]. The measurement response time is the time to transition between two steady-state measurements before and after a step change is applied to the input. The PMU shall support data reporting (by recording of output). This reporting shall be done at submultiples ($F_s$) of the nominal power-line (system) frequency (**Table 1**) [7].

According to the standard, the support for other reporting rates is optional and includes higher rates like 100/s or 120/s or rates lower than 10/s such as 1/s. The rates lower than 10/s are not subject to dynamic requirements. In this case no filtering is required, and lower-rate data (<10/s) can be provided directly by selecting every nth sample from a higher-rate stream. The reporting rate and performance class are often the largest factors influencing the accuracy of the measurements. These determine the measurement window to be used, filtering, and the length of the interval over which an event will be reported.

To comply with the standard, a PMU shall provide synchrophasor, frequency, and ROCOF measurements that meet the requirements in a given class [7]. These requirements shall be met at all times and under all configurations irrespective of whether the PMU function is a stand-alone physical unit or included as part of a multifunction unit. So, all compliance tests are to be performed with all parameters set to standard reference conditions, except those being varied

| **System frequency** | **50 Hz** | | | **60 Hz** | | | | | |
|---|---|---|---|---|---|---|---|---|---|
| Reporting rates ($F_s$, frames per second) | 10 | 25 | 50 | 10 | 12 | 15 | 20 | 30 | 60 |

**Table 1.** Required PMU reporting rates [7].

http://dx.doi.org/10.5772/intechopen.71502

as specified for the test. The reference condition specified for each test is the value of the quantity being tested when others are unvarying. Only the parameters specified for each requirement shall be varied as the effects shall be considered independently. Reference conditions for all tests are defined in [7].

## 3. General characteristics of filtration

The main objective of filtration in the measuring and automation devices within the EPS is to pass signal components lying in a selected frequency band and to suppress signal components of the other frequencies. It is because of the existence of the interfering components in the input signals which can occur especially during faults within the EPS, high penetration levels of intermittent DG, and today's widely used thyristor frequency converters.

Generally, a linear filter can be described by the equation:

$$y(t) = \int_{-\infty}^{\infty} x(t-\tau)\ h(\tau)d\tau \tag{7}$$

where $y$ and $x$ are the output and input signals and $h$ is the response function of the filter.

After the Fourier transform of the Eq. (7) is obtained:

$$Y(j\omega) = X(j\omega)\ H(j\omega) \tag{8}$$

Eqs. (7) and (8), respectively, characterize the filter completely in the time and frequency domain. The function $h(\tau)$ is important for the characteristics of the filter like the function $H(j\omega)$ which represents its frequency spectrum.

In principle, the shorter the interval of the transient state of the filter, the worse the filtration properties. Thus, better filtration properties lead to the extension of time to the stable output response. In practice, it is a trade-off between the short time of the transient state and high quality of filtration, depending on the application requirements. In particular the short transient states of the filtration are required in the power system protection devices. Short stabilization time should be kept in order to take fast and proper decisions during the disturbance, the change of network configuration, the discharge of power, etc. Therefore, it is recommended that filters should be used with fast-decay $w$-functions or $w$-functions described in the period $T_w$ and outside this range: taking a value 0. Concluding, filtering of input measurement signals should provide:

- Fast stabilization of output signals after a step change of an input signal (ensuring short time of the measurement)
- Effective elimination or dumping of the interfering (unusable) components
- Possible low computational effort (quasi real-time measurement)

Digital filtration is based on the discrete values of an input signal. Using this technique, a discrete output signal can be obtained and is dependent on the input signal and filter characteristics. The characteristics of the filter can be described using the frequency and time characteristics.

Considering the frequency characteristics of filters, both analog and digital, the following are groups of filters:

- Low-pass filters
- Band-pass filters
- High-pass filters
- Stopband filters

From this group, low-pass and band-pass filters are especially widely used in the power system protection automation.

Analyzing the time characteristics, two groups of digital filters can be distinguished:

- Non-recursive filters of finite impulse response (FIR)
- Recursive filters of infinite impulse response (IIR)

A digital non-recursive filter can be described by Eq. (9):

$$y(n) = \sum_{k=0}^{p-1} a_{(k)} x(n-k) \tag{9}$$

where $x$ and $y$ are the input and output samples, $a_{(k)}$ is the filter coefficients, and $p$ is the number of samples in the window.

Eq. (9) of a non-recursive filter is a discrete convolution of filter coefficients and an input signal, limited to $p$ samples of the input signal.

A digital non-recursive filter can be described by Eq. (10):

$$y(n) = \sum_{k=0}^{p-1} a_{(k)} x(n-k) + \sum_{k=1}^{w} b_{(k)} y(n-k) \tag{10}$$

where $a_{(k)}$ and $b_{(k)}$ are the filter coefficients and $w$ is number of coefficients $b_{(k)}$, $w < (p - 1)$.

Eq. (10) shows that actual response of that filter is not only a function of an input signal and filter coefficients but also a function of $w$-numbers of the previous values of the output signal. After using the Z-transform, we can obtain:

- For a non-recursive filter:

$$H(z) = \frac{Y(z)}{X(z)} = \sum_{k=0}^{p-1} a_{(k)} z^{-k} \tag{11}$$

- For a recursive filter:

$$H(z) = \frac{Y(z)}{X(z)} = \frac{\sum_{k=0}^{p-1} a_{(k)} z^{-k}}{1 - \sum_{k=1}^{w} b_{(k)} z^{-k}} \quad (12)$$

To obtain the discrete spectra of the filter, the operator $z$ shall be substituted by the equivalent operator $e^{-j\omega \ T_i}$:

- For a non-recursive filter:

$$H^*(j\omega) = \sum_{k=0}^{p-1} a_{(k)} e^{-jk\omega \ T_i} \quad \text{for } \omega < \frac{\pi}{T_i} \quad (13)$$

- For a recursive filter:

$$H(j\omega) = \frac{\sum_{k=0}^{p-1} a_{(k)} e^{-jk\omega \ T_i}}{1 - \sum_{k=1}^{w} b_{(k)} e^{-jk\omega \ T_i}} \quad \text{for } \omega < \frac{\pi}{T_i} \quad (14)$$

where $T_i$ is the sampling period.

Recursive filters are rarely used in power system automation. The cause is a long stabilization time of an output signal after a step change of an input signal and a relatively high sensitivity to the slight changes of values of the coefficients $a_{(k)}$ and $b_{(k)}$.

The possibilities of optimizing non-recursive filters are limited. The optimization due to the required frequency response, the required high dynamics, and the coefficients of the filter and quick transition from band-pass to stopband frequently leads to high-order filters that are difficult to implement in real time. Therefore, optimization of non-recursive filtering should be pursued by selecting the permissible length of the measurement window and possibly a simple form of the coefficients function (coefficients $a_{(k)}$) that define the required spectral characteristics.

## 4. Experimental setup

In this study, MATLAB with Simulink and Signal Processing Toolbox was used for software simulations. MATLAB is a popular programming language developed by MathWorks [29]. MATLAB can analyze data, develop algorithms, and create models and applications. The Simulink toolbox is a block diagram environment for multidomain simulation and model-based design. It supports simulation, automatic code generation, and continuous test and verification of embedded systems. Simulink provides a graphical editor, customizable block libraries, and solvers for modeling and simulating dynamic systems. It is integrated with MATLAB. It makes it possible to incorporate MATLAB algorithms into models and export simulation results to MATLAB for further analysis. The Signal Processing Toolbox provides industry-standard algorithms and applications for analog and digital signal processing (DSP). Among other things, it makes it possible to visualize signals in time and frequency domains, compute FFTs for spectral analysis, and design FIR and IIR filters. Algorithms in the toolbox can be used as a basis for developing custom algorithms.

The second stage of this study was performed using microprocessor-based Automatic Relay Test System ARTES 440 II manufactured by KoCoS Company (**Figure 1a**), and a digital fault recorder RZ-40 (**Figure 2**), manufactured by Energotest, with the implemented functionality of a PMU unit (both algorithms and other special functions). The ARTES 440 II [30] is used for carrying out general operating tests and for testing the configured excitation and tripping characteristics of various protection devices, such as distance protection relays, overcurrent

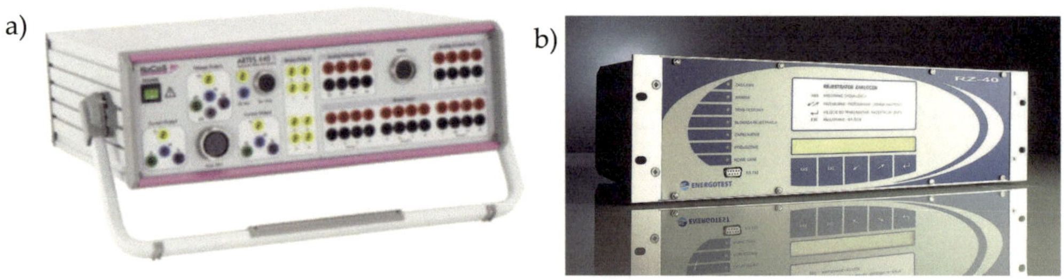


**Figure 1.** Real photos of (a) ARTES 440 II [30] and (b) RZ-40 fault recorder [31].

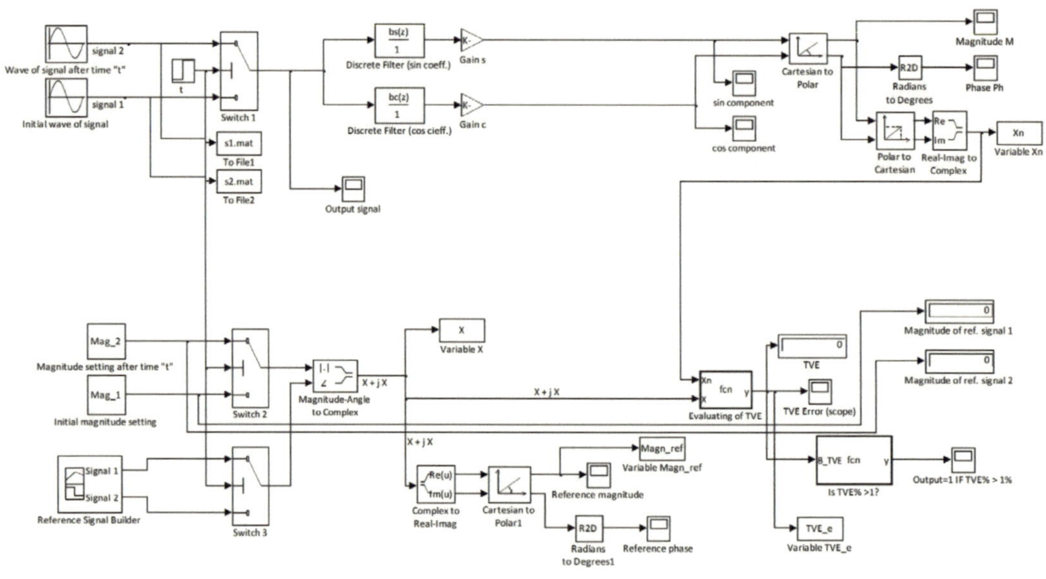


**Figure 2.** MATLAB/Simulink model used in the software simulations.

relays, and voltage and frequency relays. In addition, the test instrument can also be used as a three-phase function generator which is freely configurable with regard to amplitude, frequency, and phase relation. As mentioned earlier, ARTES 440 II has four voltage and six current amplifiers whose output signals can be set independently from one another as regards amplitude, frequency, and phase angle. The test quantities are calculated from the parameters entered via the software and are supplied to the device under test by means of digital-analog converters and amplifiers. Because the test quantities are generated synthetically, they are unaffected by disturbances in the incoming supply.

The currents generated by the test system are available via the sockets of the current output group. The six current channels are freely configurable as regards phase, amplitude, and frequency. The current amplifiers of the ARTES 440 II provide a maximum test current of 25 A per channel. If higher test currents are required, the amplifier outputs can be operated in parallel. The output values of the current and voltage amplifiers are monitored by the system during tests (internal feedback measurement). If the output values do not agree with the setpoint values, the software issues a warning to this effect. Connections to the device under test are made via safety sockets or a multipole generator output socket. Detailed parameters of the test system are presented in **Table 2**.

Disturbance (fault) recorders have been in use for a number of years and have evolved from analog recording devices to units using digital signal processing and recording techniques. Digital records can be easily collected, transmitted, stored, printed, and analyzed [31]. Also, RZ-40 (**Figure 1b**), a fault recorder selected for this study, typically contains directly measured analog channels, as well as event or binary channels. This allows the recorder to capture the time sequence of analog power system quantities, along with breaker contacts, logic-state changes, event contacts, etc. State-of-the-art recorders typically include calculated analog quantities and logic functions to ensure that pertinent power system information is captured during an event. RZ-40 fault recorder is designed for logging of instantaneous values of voltages and currents as well as binary signals in electric power structures during failures or disturbances. It has many additional functions, e.g., event recorder and electrical value meter. RZ-40 has many functions and wide-open hardware and software structure. Each of the functions can be used independently without lowering parameters of other functions. That is why it was possible to implement additional software and functions of PMUs to fulfill the requirements of the IEEE the C37.118™ series of the Standard.

| **General** | **Frequency range**<br>**Transient signals**<br>**Frequency resolution**<br>**Frequency accuracy** | **DC to 3 kHz**<br>**DC to 4 kHz**<br>**0.001 Hz**<br>**Error < ±0.01%** | **Phase angle**<br>**Phase resolution**<br>**Phase accuracy** | **0 to 360°**<br>**0.001°**<br>**Error < ±0.1°[1]** |
|---|---|---|---|---|
| Voltage amplifiers | Resolution THD | 13 mV < 0.05%[1] | Accuracy | Error < ±0.05%[1] |
| Current amplifiers | Resolution THD | 1 mA < 0.05%[1] | Accuracy | Error < ±0.05%[1] |

[1]For the frequency range of 10 to 200 Hz.

**Table 2.** Detailed accuracy and resolution parameters of the analog outputs of ARTES II [30].

The other benefit of using this digital fault recorder to realize the functions of PMUs is the fact that many of them are already working (especially in the Polish Power Grid infrastructure). Thus, the implementation of synchronous measurements using these devices will have low hardware costs of the initial wide-area measurements infrastructure. The only thing which needs to be done is to update the software of the functioning devices to the newest one which has the optimized and modified algorithms realizing the PMU functionality. Additionally, existing and future infrastructure facilitates making long-term pilot programs. These programs can lead to the development of new and long-awaited functions to ensure high-power grid availability, stability, and reliability, especially in the power system automation scope. Therefore, the main objective of this study is to initially check the PMU properties of this device according to the IEEE the C37.118™ series of the Standard.

### 4.1. Software simulations in MATLAB

The first stage of the study was to work out a model in MATLAB/Simulink environment (**Figure 3**). This model was used to simulate various tests according to the requirements of C37.118™ series of the Standard, especially C37.118.1–2011 [7–9]. It has been used exclusively for research and teaching purposes to demonstrate the properties of orthogonal filters.

On the basis of several other considerations and analyses, e.g., [21–22], full-cycle filters with sine and cosine widows were implemented. Because of the necessity to evaluate TVE (Eq. (3)), there is a reference signal builder in the simulator. The magnitude, phase, and frequency of input signals are freely configurable. The step output "t" is used to switch from signal 1 to signal 2 after a configured time *t*. Signals 1 and 2 are taken out to Files 1 and 2, a full signal sequence is visible on the scope "Output signal." A generated signal sequence is delivered to the full-cycle filters block with sine and cosine windows. The required coefficients are loaded from MATLAB workspace according to the frequency of the time window. Then, the orthogonal components are used to compute magnitude (M) and phase (Ph). Calculated values of the magnitude and phase (in complex variable $X_n$) are delivered to the function block "evaluating of TVE." The slope of TVE can be viewed on the scope and is exported to variable TVE_e. Additionally, the B_TVE function block is used for evaluation purposes if TVE is below its reference value of 1%. The model presented in **Figure 2** does not have the ROCOF

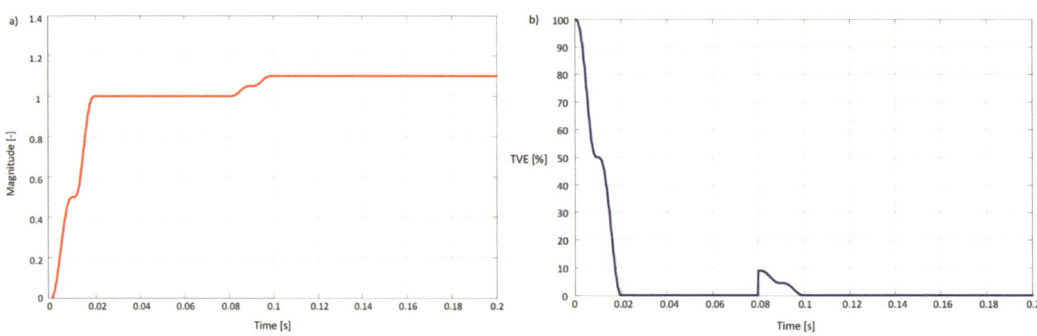


**Figure 3.** Magnitude and TVE course after a step change of the magnitude in t = 80 ms (a) magnitude course and (b) TVE course.

measurement module. Although the implementation of ROCOF measurements poses no problems (and indeed this was done), the main tests in the dynamic conditions were carried out on a real hardware device, and only a simplified model is presented in **Figure 2**.

Several testing conditions have been simulated using the developed model. Sample results of these tests are presented in **Figures 3** to **6**. All simulations were realized for the sampling frequency $f_s = 4$ kHz. **Figure 3** presents magnitude and the evaluated TVE course after a step change of input signal magnitude of +10% (in relative values), for the nominal frequency of input signal ($f_1 = 50$ Hz) and the window frequency of filter ($f_w$) equal to the nominal frequency. This step change was performed in 80 ms from the beginning of simulation.

**Figure 4** shows the courses of TVE for the same $f_1$ and $f_w$ parameters as before: there is no change of the magnitude, but there is a step change of phase of $\pi/6$ (**Figure 4a**) and $5\pi/6$ (**Figure 4b**) radians. Next, again $f_i$ and $f_w$ are equal to 50 Hz: there is no change of the magnitude and phase, but there is a step change of frequency of $-0.5$ Hz (**Figure 4c**) and $-1$ Hz (**Figure 4d**). Finally, **Figure 5** shows the situation, when the frequency of the filter window ($f_w$ = 48.78 H) is almost fully correlated with the frequency of the input signal ($f_1 = f_w = 48.78$ Hz (**Figure 5a**)) and closely correlated ($f_1 = 48.5$ Hz, $f_w = 48.78$ (**Figure 5b**)). Of course for the last test, the coefficients of the filter are different from the first three tests and evaluated and loaded from the MATLAB workspace. In the example, for the selected frequency $f_w = 48.78$ Hz, there are 82 samples in the filter window, and the filter window is almost

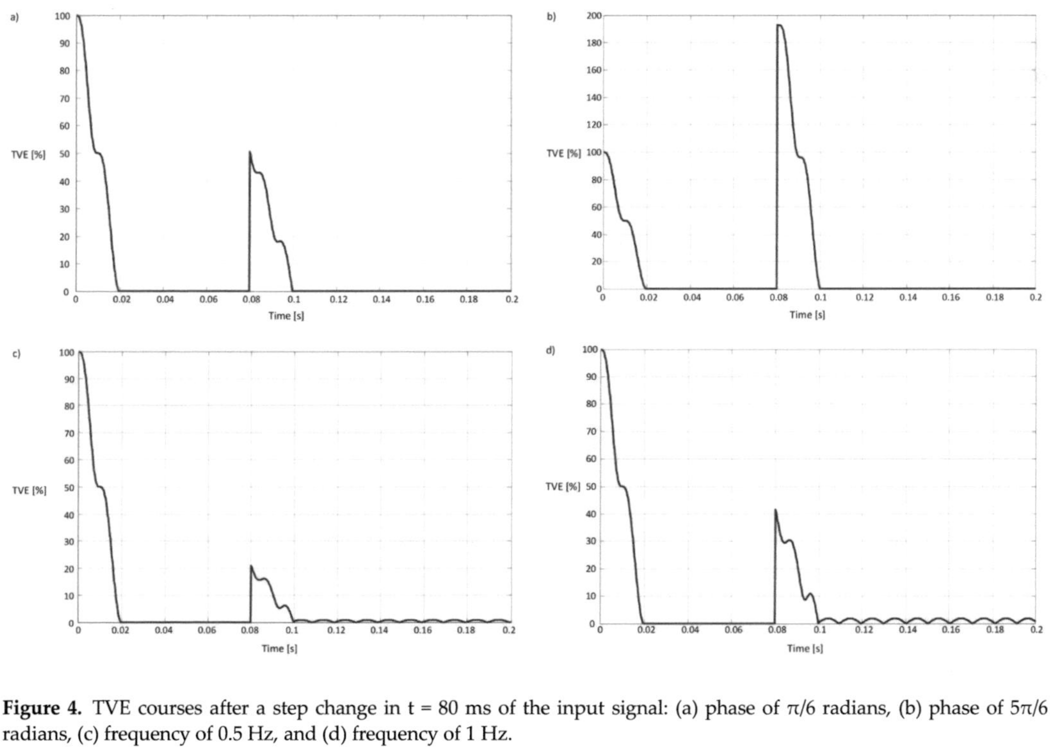


**Figure 4.** TVE courses after a step change in t = 80 ms of the input signal: (a) phase of $\pi/6$ radians, (b) phase of $5\pi/6$ radians, (c) frequency of 0.5 Hz, and (d) frequency of 1 Hz.

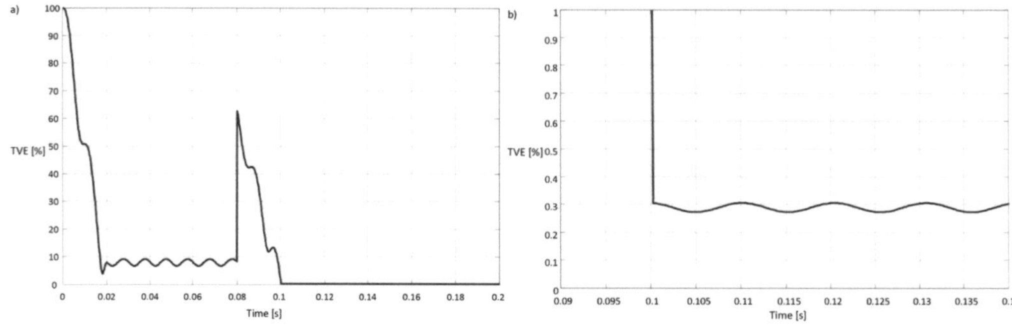


**Figure 5.** TVE courses for filter window frequency $f_w = 48.78$ Hz and the input signal of frequency: (a) $f_1 = 48.78$ and (b) $f_1 = 48.5$ (zoomed).

fully matched to the frequency of the input signal also equal to 48.78 Hz. The number of samples is a natural number.

Therefore, taking the sampling frequency of 4 kHz, the other two nearest frequencies of the fundamental component of an input signal, for which the window filter can be almost fully correlated with the input signal, are 49.28 (81 samples in the filter window) and 48.19 (83 samples in the filter window). It should be emphasized that this phenomenon is dependent on the sampling frequency. When the sampling frequency is 2 kHz, there are 40 samples in the window (for the frequency of the fundamental component equal to 50 Hz). The nearest two natural numbers of samples are 39 and 41. Evaluating 2000/39 gives the frequency of the window equal to 51.28 Hz and 2000/41 gives $f_w = 48.78$ Hz. Comparing these frequencies with the ones for $f_s = 4$ kHz (relatively 50.63 Hz and 49.28 Hz), it can be noticed that lowering the sampling frequency leads to wider frequency ranges in which the algorithm works with the nonmatched frequencies of windows and generates inaccuracies exceeding the limits defined by the standard. On the other hand, raising the sampling frequency leads to the increase in the computational effort and time.

Analyzing software simulations, it can be seen that for the assumed sampling frequency, TVE exceeds the limit of 1% for the frequencies differing from the nominal more than about 0.5 Hz. Therefore, the adaptive technique using switching between filter windows frequencies can be considered to comply with the standard limitations as well as with other solutions described in [21–22].

### 4.2. Testing and validation of the RZ-40 device by ARTES II

**Figures 6–8** show sample hardware tests carried out with the ARTES II and RZ-40 devices. These tests were realized for various conditions and cases described in the IEEE C37.118.1–2011 Standard.

Each of the tests is realized in 1 second. In the half of this period (500 ms from the beginning of test), there is a step change of the given quantity. **Figure 6** shows the courses of the magnitude

http://dx.doi.org/10.5772/intechopen.71502

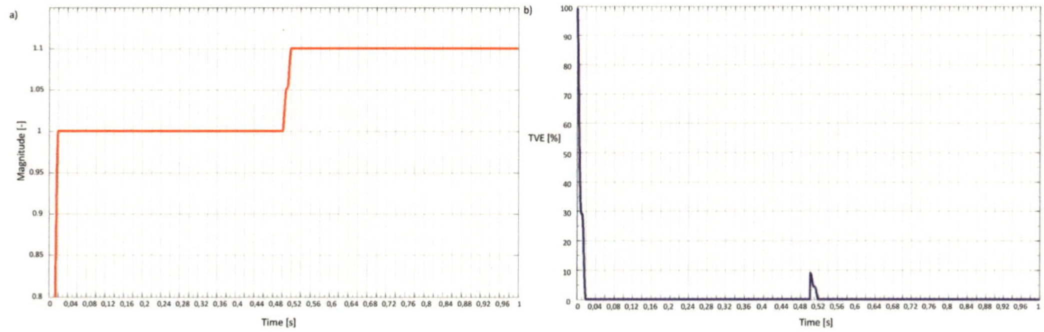


**Figure 6.** Magnitude and TVE course after a step change of the magnitude in t = 500 ms: (a) magnitude recorded in RZ-40 device and (b) TVE course (evaluated).

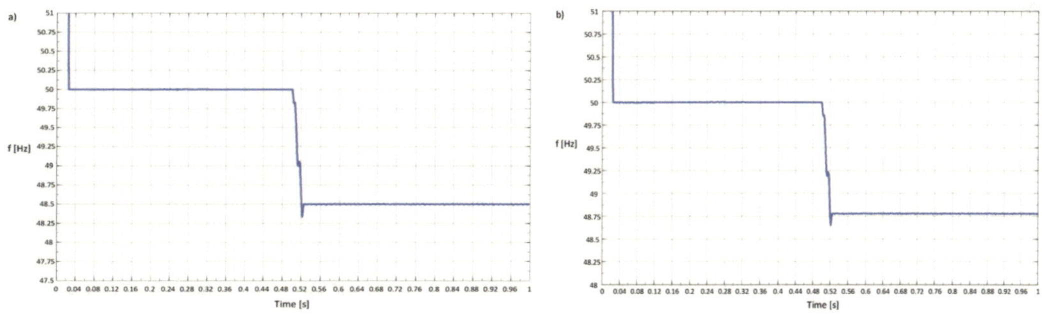


**Figure 7.** Frequency courses evaluated in RZ-40 device after a step change of the input signal frequency in t = 500 ms: (a) $f_1$ = 45.5 Hz and (b) $f_1$ = 45.78 Hz.

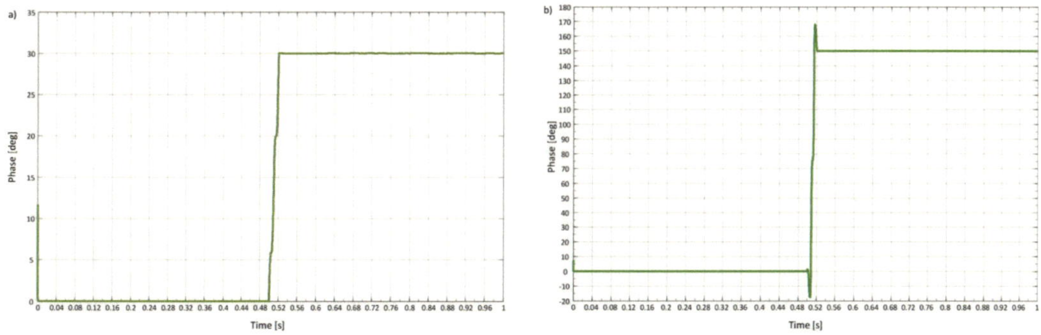


**Figure 8.** Phase courses recorded in RZ-40 device after a step change of the input signal phase in t = 500 ms: (a) from 0 to 30° and (b) from 0 to 150°.

(**Figure 6a**) and TVE (**Figure 6b**) during the step change of that first on the level + 10% (relative values). The first course is recorded using the disturbance recorder ability implemented in the ET Manager software supplied by the manufacturer. The TVE is evaluated using the recorded values. **Figure 7** presents the evaluated frequency for 48.5 Hz (**Figure 7a**) and 48.78 (**Figure 7b**). Lastly, **Figure 8** shows the phase courses for the step change of phase from 0 to 30° (**Figure 8a**) and from 0 to 150° (**Figure 8b**).

As it can be observed, there are full-cycle window filters implemented in the RZ-40 devices. It can be concluded from the time period between the occurrence of a step change (0.50 s) and the moment of the stable response to this step (0.52 s). Any fluctuations in the initial period of tests should not be considered as they are the results of transient states of starting simulations (MATLAB/Simulink) or transients after injecting voltages on the RZ-40 inputs when the previous input magnitude was 0.

As mentioned previously, the RZ-40 has been tested in accordance with the C37.118.2011™ Standard and the "2014" amendment. In particular, the implemented measurement algorithms focused on providing high accuracy and stability of frequency determination across a wide range of frequency changes in measurement signals. **Tables 3** and **4** show the results of frequency estimation for a monoharmonic signal with set fundamental frequencies (45 to 55 Hz in 1 Hz steps) for successive 50 samples. Similar measurements were made for further 100 and 200 samples. The results of the measurements are shown in **Figures 9** and **10**. Based on the obtained results, it can be stated that the algorithm is characterized by high-frequency estimation accuracy (average error of measurement below 2 mHz) and high stability (maximal standard deviation at 1 mHz for the extreme frequency of the tested range). The dynamic tests

| | 45 Hz | 46 Hz | 47 Hz | 48 Hz | 49 Hz | 50 Hz |
|---|---|---|---|---|---|---|
| Frequency (mean) [Hz] | 44,998,237 | 45,998,779 | 46,999,410 | 47,998,901 | 48,999,969 | 49,999,111 |
| Standard deviation [mHz] | 0,894 | <0,001 | 0,840 | <0,001 | <0,001 | 0,415 |
| Mean freq. Error [mHz] | −1763 | −1221 | −0,590 | −1099 | −0,031 | −0,889 |
| Average percentage error [−] | −0,004 | −0,003 | −0,001 | −0,002 | <0,001 | −0,002 |

**Table 3.** Results of frequency estimation in the RZ-40 for the 50 successive samples (1).

| | 51 Hz | 52 Hz | 53 Hz | 54 Hz | 55 Hz |
|---|---|---|---|---|---|
| Frequency (mean) [Hz] | 50,998,157 | 51,999,145 | 52,998,279 | 53,999,208 | 54,998,621 |
| Standard deviation [mHz] | 0,395 | 0,000 | 0,395 | 0,344 | 0,720 |
| Mean error [mHz] | −1843 | −0,855 | −1721 | −0,792 | −1379 |
| Average percentage error[−] | −0,004 | −0,002 | −0,003 | −0,001 | −0,003 |

**Table 4.** Results of frequency estimation in the RZ-40 for the 50 successive samples (2).

http://dx.doi.org/10.5772/intechopen.71502

performed indicated excellent properties of the algorithm for potential applications in the power system. The results obtained have also been confirmed by independent tests carried out at the Fraunhofer Institute in Magdeburg using a reference device[1].

Synchrophasor measurements should be synchronized with the UTC time with the accuracy sufficient to meet the accuracy requirements of the C37.118 Standard [3]. Time-stamp error should be of just a few (single) μs. In order to fulfill this requirement, the accuracy of the time source should be about ten times higher than the level of the expected time error. To synchronize the RZ-40 unit, the GPS receiver EM-406A and then EM-506 was taken as the source of the time signal. The EM-406A is a 20-channel GPS receiver, and the EM-506 one is a 48-channel receiver. High accuracy of measurements and high precision of time stamps make it possible to develop many new or improved applications. One of them is the SmartLoad system [13, 31].

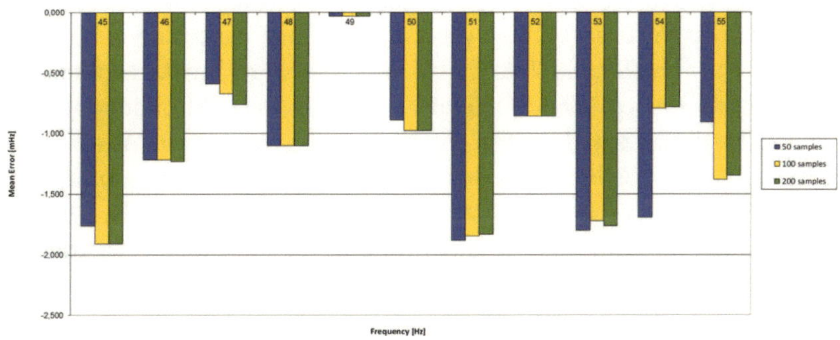


**Figure 9.** Mean error of frequency estimation.

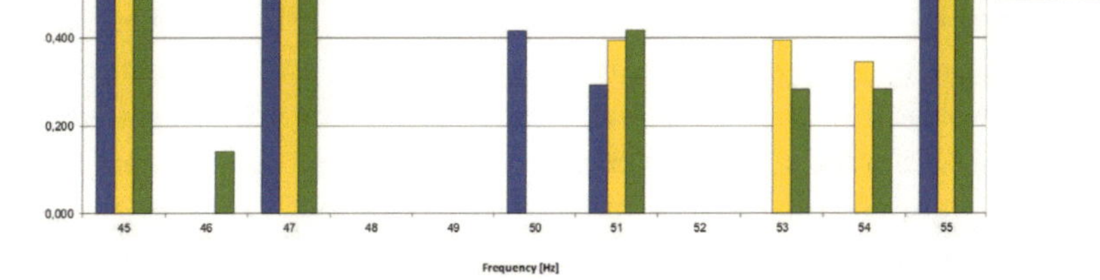


**Figure 10.** Standard deviation of frequency estimation.

[1]Tests commissioned by Energotest Sp. z o.o.

This system is an advanced system designed to make a quasi real-time balance of power and adaptive load-off in the case of a possible active power deficit in the supervised area of a power network. This system has already been implemented. Some new implementations are under development and will be the subject of subsequent publications. It seems that they can significantly improve the reliability of the power system's operation with the benefits of a well-implemented PMU-based measurements technology.

## 5. Discussion and conclusions

The IEEE C37.118.1.2011 Standard defines a phasor measurement unit (PMU), which can be a stand-alone physical unit or a functional module within another physical unit. The Standard does not specify any hardware, software, or a method for computing phasors, frequency, or ROCOF. In all likelihood the best solution, especially in the initial period of implementation, is to incorporate the PMU functionality into the existing hardware devices commonly used in the power grid. This minimizes costs and facilitates realizing potential pilot programs. The pilot programs seem to be the essential part of the PMU and synchronous measurement studies. Considering the RZ-40 unit is by no means accidental. This unit has a modular and scalar hardware architecture. It is very easy to change the input analog cards, DSP module, and communication module or to update the firmware of the unit including the measurement algorithms and other functionalities [28]. Last but not least, it is also a very popular unit in the Polish Power Grid.

Analyzing the requirements for the synchronous measurements resulting from the C37.118™ series of the Standard, significant changes can be noticed between the 2005 and 2011 versions. The requirements in the version of the Standard dated 2005 were defined only for static conditions and were very imprecise. This leads to their free interpretation by manufactures because several of them were not directly defined or defined on the levels which give many different solutions allowing compliance with the Standard. Although many algorithms which are used may comply with the 2005 Standard, they may have different properties in the dynamic stages, especially during the fast transient states. That is why the Standard from 2011 emerges. It is much more precise, and it defines the requirements both in the static and dynamic conditions. It also defines many important time limits for the measurement algorithms realizing synchronous measurements. It was concluded that the compliance problems with the Standard in the case of wide-area measurements are dependent on their target applications. It was almost impossible to build the wide-area protection systems based on the Standard of 2005. As a matter of fact, it defined some practical requirements in steady conditions, but it is the requirements in dynamic states that are most vital for the protections. In addition these "dynamic" requirements should be comparable in different devices realizing PMU functions. All this makes synchronous measurements very hard to realize, mainly as regards factors related to the PMU-expected higher functionalities. Simulations realized in this study indicate the possibility of developing fast and reliable adaptive measurement algorithms to comply with the C37.118.1–2011 requirements. Several tests confirm that for the RZ-40 device. Detailed analyses of the 2011 Standard point that some of the requirements are very hard or even impossible to meet. Some of the difficulties in meeting the requirements probably

http://dx.doi.org/10.5772/intechopen.71502

result from simple typo-errors they contain. Other requirements are feasible but hard to implement. They cause computational problems resulting from increasing the response time and error on the output. This seems to be confirmed by Amendment 1 to the Standard released at the end of March 2014 [9] entitled "Modification of Selected Performance Requirements," in which many of the required parameters, mainly applying to the dynamic requirements, are revised. It should be emphasized that most of these parameters were met by the RZ-40 during the tests discussed in this study for limits defined in 2011. Summing up, on the basis of the tests defined in the Standard and conducted in this study, it is possible to meet the Standard requirements by the devices like RZ-40. It should be observed that these are "synthetic tests" and probably many other devices can comply with the Standard. On the other hand, most of these "others" were only tested and confirmed by the manufacturers to comply with the standard of 2005. Therefore, long-term pilot tests are essential to compare the properties of different devices with the PMU functionality. These tests should be conducted using different PMU devices working in the same locations of the power grid. The initial forecasting time of such tests is from about 6 months to a year. Further research is needed to analyze the results. Some of the most noticeable changes to the power distribution landscape are being driven by the proliferation of intermittent DG, PEVs, microgrids, and power electronic components. The application of PMUs in distribution systems is still under-explored or is at the implementation stage. This is mostly due to the recognized technical problems (explained and revised in this study) and economic factors. However, since the drivers of change and needs for implementation are rapidly increasing, there is a growing interest in applying PMUs in the EPS, which will consequently engender a new body of research.

## Acknowledgements

The author would like to thank Energotest Sp. z o.o. for providing opportunity to use the RZ-40 device in his study and contributing in the software and hardware development, both Energotest Sp. z o.o. and KoCoS Company for the invaluable hardware and software support as well as hardware and software components.

## Author details

Michał Szewczyk

Address all correspondence to: michal.szewczyk@polsl.pl

Silesian University of Technology, Gliwice, Poland

## References

[1] IEEE Std 1588™-2008. IEEE Standard for a Precision Clock Synchronization Protocol for Networked Measurement and Control Systems

[2] PC37.238. IEEE Draft Standard Profile for Use of IEEE 1588 Precision Time Protocol in Power System Applications

[3] Szewczyk M. Time synchronization for synchronous measurements in electric power systems with reference to the IEEE C37.118TM standard - selected tests and recommendations. Przegląd Elektrotechniczny. 2015;**4**:144-148. DOI: 10.15199/48.2015.04.32

[4] Szewczyk M. Selected analyses of teletransmission and teleinformatic structures in electrical power. Przegląd Elektrotechniczny. 2014;**3**:1-5. DOI: 10.12915/pe.2014.03.1

[5] Szewczyk M. Analysis of the requirements of reliability and quality for systems and data transmission equipment in modern power systems. Przegląd Elektrotechniczny. 2014;**3**:84-89. DOI: 10.12915/pe.2014.03.1

[6] C37.118 rev. IEEE Standard for Synchrophasor Measurements for Power Systems. 2005

[7] C37.118.1 rev. IEEE Standard for Synchrophasor Measurements for Power Systems. 2011

[8] C37.118.2 rev. IEEE Standard for Synchrophasor Measurements for Power Systems. 2011

[9] IEEE Standard for Synchrophasor Measurements for Power Systems, Amendment 1. Modification of Selected Performance Requirements. 27 March, 2014

[10] Szewczyk M. Standard requirements for systems realizing synchronous measurements in the power system infrastructure. Przegląd Elektrotechniczny. 2014;**3**:80-83. DOI: 10.12915/pe.2014.03.16

[11] Wache M, Murray DC. Application of Synchrophasor measurements for distribution networks. IEEE Power and Energy Society General Meeting. Jul. 2011:1972-1975

[12] Sanchez-Ayala G, Aguerc JR, Elizondo D, Lelic M. Current trends on applications of PMUs in distribution systems, innovative smart grid technologies (ISGT). 2013. IEEE. Feb. 2013;**24-27**:1-6

[13] Szewczyk M. Conditions for the improvement and proper functioning of power system automation equipment in the present and the expected future structure of the electric power sector. Przegląd Elektrotechniczny. 2015;**5**:179-186. DOI: 10.15199/48.2015.05.40

[14] Halinka A, Szewczyk M. Distance protections in the power system lines with connected Wind Farms. Gesche Krause, editor. From Turbine to Wind Farms - Technical Requirements and Spin-Off Products. InTech; April 2011:135-160. DOI: 10.5772/15955

[15] Halinka A, Rzepka P, Szablicki M, et al. Impact of proper functioning of power system automation for the safety of power system in the consideration of the new technical and economical solutions in polish power grid. Przegląd Elektrotechniczny. 2011;**2**:140-143

[16] Borghetti A, Nucci CA, Paolone M, Ciappi G, Solari A. Synchronized phasors monitoring during the islanding maneuver of an active distribution network. IEEE Transactions on Smart Grid. March 2011;**2**(1). Article number 5680630:70-79

[17] Meng W, Wang X, Wang Z, Kamwa I. Impact of causality on performance of Phasor measurement unit algorithms. IEEE Transactions on Power Systems. 2017:1-11. DOI: 10.1109/TPWRS.2017.2734662

[18] Roscoe AJ, Burt GM, McDonald JR. Frequency and fundamental signal measurement algorithms for distributed control and protection applications. IET Generation, Transmission and Distribution. 2009;**3**(5):485-495

[19] Kundu P, Pradhan AK. Wide area measurement based protection support during power swing. International Journal of Electrical Power & Energy Systems. December 2014;**63**:546-554. DOI: 10.1016/j.ijepes.2014.06.009

[20] Shima KS, Kimb ST, Leec JH, et al. Detection of low-frequency oscillation using synchrophasor in wide-area rolling blackouts. International Journal of Electrical Power & Energy Systems. December 2014;**63**:1015-1022. DOI: 10.1016/j.ijepes.2014.06.069

[21] Rebizant W, Szafran J, Wiszniewski A. Digital Signal Processing in Power System Protection and Control. Springer-Verlag London; 2013. DOI: 10.1007/978-0-85729-802-7

[22] Halinka A., Sowa P, Szewczyk M. Measurement algorithms of selected electric parameters in wide range of frequency change. Proc. of SIP2001. Honolulu. USA: 155–159

[23] Kamwa I, Samantaray SR, Joos SR. Wide frequency range adaptive phasor and frequency PMU algorithms. IEEE Transactions on Smart Grid. March 2014;**5**(2). Article number 6575186:569-579

[24] Chakir M, Kamwa I, Le H. Extended C37.118.1 PMU algorithms for joint tracking of fundamental and harmonic phasors in stressed power systems and microgrids. IEEE Transactions on Power Delivery. June 2014;**29**(3). Article number 6810880:1465-1480

[25] Phadke AG, Kasztenny B. Synchronized phasor and frequency measurement under transient conditions. IEEE Transactions on Power Delivery. 2009;**24**(1):89-95

[26] Roscoe AJ, Burt GM, Rietveld G. Improving frequency and ROCOF accuracy during faults, for P class Phasor measurement units. 2013 IEEE International Workshop on Applied Measurements for Power Systems, AMPS 2013 – Proceedings 2013. Article number 6656233. 97–102

[27] Premerlani W, Kasztenny B, Adamiak M. Development and implementation of a synchrophasor estimator capable of measurements under dynamic conditions. IEEE Transactions on Power Delivery. January 2008;**23**(1):109-123

[28] A Study of the Measurement-Path Accuracy of Digital Protection Devices in a Wide Range of the Frequency Change. Unpublished report of the research project sponsored by the Polish National Research Committee. Research manager of the project: Michał Szewczyk

[29] www.mathworks.com

[30] www.kocos.de

[31] www.energotest.com.pl

Chapter 15

# Power System Reliability: Mathematical Models and Applications

Rabah Medjoudj, Hassiba Bediaf and Djamil Aissani

Additional information is available at the end of the chapter

http://dx.doi.org/10.5772/intechopen.71926

**Abstract**

This chapter deals with power systems reliability including technical, economical, and decisional aspects. Knowing that almost 90% of failures occur in the distribution systems, great interest was dedicated to this part of the system, and the first work was oriented to reliability indices defined as objectives to attempt and as performance measures in the electricity market. Some works deal with the managers' behavior, and the customers reactions are modeled using economic criteria in uncertain future and inspired from game theory. When studying components, degradation models were introduced and combined with the effects of socks to study the reliability changing during system operation. In some works, the correlation between maintenance policies and reliability aspects was highlighted. In a recent work, considering the importance of new technologies integration and renewable energy insertion to power systems, it was revealed that reliability aspects and energy sustainability are two fundamental issues of progress in a given society.

**Keywords:** power systems reliability, distribution functions, degradation modeling, maintenance, decision-making

## 1. Introduction

In general way, power system reliability addresses the issues of service interruption and power supply loss. In several cases, it is defined as an objective to attempt in terms of indices directly related to the customer. Typical reliability index values for US utilities are SAIFI, SAIDI, and CAIDI. Over time, they become standard values for evaluating the reliability of electrical systems and used in several publications. Medjoudj et al. [1], in their recent publication, defined other indices as reliability subcriteria in their decision-making attributes, giving the best model of a smart energy grid. These indices are discussed in Section 2 with an application to a real case study. After this classical definition, some works have integrated data analysis and processing taking into account the calculation of distribution parameters such as those used for Weibull and Weibull-Markov processes. These works were vulgarized with applications in power system reliability by Van Casteren et al. [2] and Medjoudj et al. [3]. The most techniques used in power system reliability optimization and processing is the reliability centered preventive maintenance (RCM). Several publications have highlighted that in most cases of multicomponent systems, the maintenance actions arrive very early without any effects on the system or very late, that is, the need of curative maintenance with its negative consequences. This issue is also treated in the case of a differentiated service of reliability in the case of customers with different requirements of the reliability level. These two concepts are discussed in Section 3. Recent publications have highlighted the interest of combining reliability attributes for maintenance actions in the case of degrading systems and components [4]; however, a novel work developed recently at the LaMOS unit research dealing with multiple degradation processes is applied to power switchgear and is discussed in Section 4. The discussions and the conclusion highlighting the place of reliability in a power energy smart grid are given in Section 5.

## 2. Power system reliability indices

In the beginning, the methods used were classical to evaluate reliability indices of distribution systems such as failure frequency, mean failure times, mean time between failure and energy not supplied. These indices help decision makers to define technical and management measures to perform systems. After that was introduced the notion of loss of load probability (LOLP), which has a lot of applications in load modeling and electrical parameters dimensioning. It is significant for any power enterprise to analyze customer satisfaction. A variety of indices have been developed to measure reliability and its cost in power systems area such as loss of load probability (LOLP), loss of load expectation (LOLE), expected frequency of load curtailment (EFLC), expected duration of load curtailment (EDLC), expected duration of a curtailment (EDC), and expected energy not supplied (EENS) [1].

### 2.1. Loss of load probability

LOLP is an expected value, sometimes calculated on the basis of the peak hourly load of each day and sometimes on each hour's load (24 in a day). Moreover, in the beginning, LOLP is used to characterize the adequacy of generation to serve the load on the bulk power system; it does not directly model the reliability of the transmission and distribution system where the majority of outages actually occur [5]. Nourelfeth and Ait Kadi [6] have recently made that the LOLP is usually used to estimate the reliability index. Considering $S$ and $D$ as the supply and the load demand respectively, they compute the reliability of a multistate system (MSS) as:

$$R = \Pr(S \geq D) Or\ R = 1 - LOLP \tag{1}$$

Using the well-known formulation of the LOLP given in several publications and discussed in the following section, they generalized the MSS reliability index $R$

$$R = \frac{1}{\sum_{j=1}^{M} T_j} \sum_{j=1}^{M} \Pr(S \geq D_j) T_j \tag{2}$$

where the operation period T is divided into M intervals, and each interval has duration $T_j$ and a required demand level $D_j$. In the same context, as advancement in reliability applications, Taboada et al. [7] have generalized the use of LOLP as a reliability index. Using a series-parallel system, they calculate the availability of each part of the power including transmission and distribution system using LOLP model.

#### 2.1.1. *Loss of load probability formulation*

The generation system reliability calculations are based mainly on two analytical methods, which differ by the load model used. The first method is the calculation of the LOLP; there the load is given by the load duration curve. The second method is the frequency and duration approach, by which, besides the probability, the frequency and duration of load levels higher than the generation capacity can be determined. The loss of load probability method associates each value of MW outage with a certain cumulative probability, thus producing a capacity outage table. The expected load loss of the system is obtained from the capacity outage table, and a daily load peak variation curve is derived from the daily load curves. The LOLP, number of days on which capacity is insufficient, is obtained by adding the probability that the amount of capacity on forced outage, on day $i$ is greater than or equal to the reserve on day $i$, for all days of the period being studied.

For a system state where the remaining generating capacity is $C_j$, the percentage of time $t_j$ during which the load demand exceeds $C_j$ can be determined from the load curve $L$. The overall probability that the load demand will not be met is called the loss of load probability and is given by the Eq. [4]:

$$LOLP = \sum_j P[C = C_j]\ P[L > C_j] = \sum_j \frac{p_j\ t_j}{100} \tag{3}$$

where $p_j$ is the probability associated to the number of the failed generating units at time $t_j$, and it is formulated through the following development.

In addition to complete failures, generating units may experience partial failure, when they continue to operate but at reduced capacity levels. They are also taken out of service from time to time for preventive maintenance. Using a simple two-state model for the operation of a unit, its failure probability is given depending on its failure outage rate (FOR), which can be assumed as the unit steady-state unavailability denoted $\overline{A}$. If at time $t_j$, $r$ units have failed from a total of $n$ identical and independent installed units in the generating system, the probability $p_j$is given by:

$$p_j = \binom{n}{r}\overline{A}^r\left(1-\overline{A}\right)^{n-r} \tag{4}$$

A case of unequal size of the units can appear.

*2.1.2. Purposes*

*2.1.2.1. First purpose*

It is well known that availability is a measure of success used primarily for repairable systems. For nonrepairable systems, availability $A(t)$ equals reliability $R(t)$. In repairable systems, $A(t)$ will be equal to or greater than $R(t)$. In the optimistic case, the availability is greater than the reliability. Following the Levitin and Lisnianski development for a multistate generating system (MSGS), the availability expectation is the function of demand $D$ and may be defined as [8]:

$$E_A\left(D_j, T_j\right) = \frac{1}{\sum_{j=1}^{M} T_j}\sum_{j=1}^{M} A\left(D_j\right)T_j \tag{5}$$

The index $(1 - E_A)$ is often used and treated as loss of load probability and can be written as:

$$LOLP = (1 - E_A)\ Or\ E_A = 1 - LOLP \tag{6}$$

*2.1.2.2. Second purpose*

This purpose highlights the correlation between the system reliability, the energy availability, and the loss of load probability. To understand this correlation, we consider a multistate generating repairable system MSGS connected to a load L, and on a given period of time, we draw two curves representing the evolutions of the system available capacity (SAC) and the hourly system load (HSL), respectively, as shown in **Figure 1**. Depending on the states of generating units (up or down) that involve partial or total failure of a simple unit or of several units, the appearance of dips in the same curve reflects units' breakdowns, and the resumption to the initial level of capacity indicates that repairs were made. One of the most reliability indices that concerns more the utility than the customer is the energy not supplied (ENS) given by the dashed lines under the curve. Their corresponding time intervals denote durations, where the consumption exceeded the production, and therefore, we have loss of load. The decreasing level of system reliability is highlighted by degrading state, corresponding to each decreasing in the SAC curve behavior.

The generating system failures can occur in two ways: either through unit failures or through load increases. There is a loss of load when the demand is greater than the supply. However, there is a loss of supply when a failure occurs in the upstream of the load point. It is important

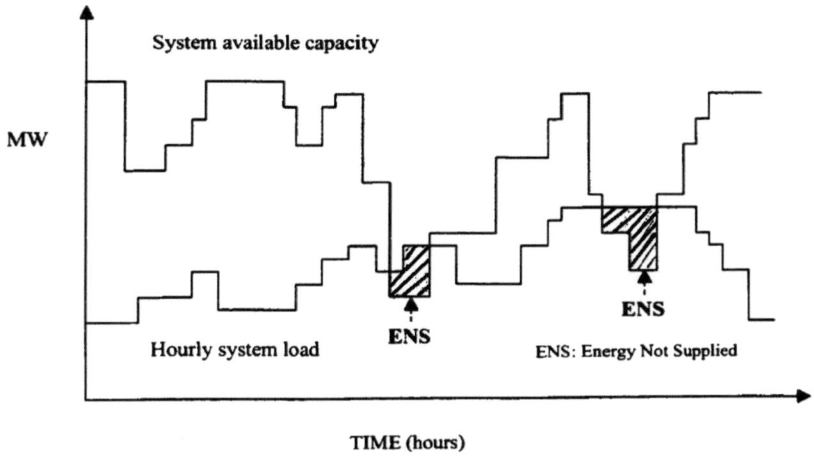


**Figure 1.** Superimposition of the system available capacity and the load model.

to retain this difference. In this context, two relevant questions are to be asked and the answers were given in [9, 10] with a case study application, such as: What happens when load increases? What is the consequence of generating system failure?

### 2.2. Frequency and duration indices

Almost every electricity utility computes reliability indices on an annual basis. The most important reliability indices involving decision-making criteria are given as follows [1]:

The Expected Frequency of Load Curtailment in (fault/yr):

$$EFLC = \sum_{k=1}^{n} \lambda_k \tag{7}$$

The Expected Duration of Load Curtailment in (hrs/yr):

$$EDLC = \sum_{k=1}^{n} \lambda_k T_k \tag{8}$$

The Expected Energy Not Supplied in (kWh/yr):

$$EENS = L.(EDLC) \tag{9}$$

where $\lambda_k$, $T_k$ are failure rate and failure duration of an item $k$ and $L$ is the load curtailed at a considered load point, respectively. Application is done for a part of the distribution system of Algiers city (Algeria). Considering the electrical characteristics (network topology, section length, power value at load points and the fault search method) and reliability parameters mentioned earlier, the overall system reliability indices are computed.

### 2.3. Reliability indices improvement

To improve the reliability level, technical and organizational measures are considered during system planning and operation. The actions currently carried out are as follows: intensifying the operations of maintenance; networks reorganization, looping and meshing systems, and automation of networks. In [11], some options are added such as load transfer between feeders, undergrounding circuits, and replacement of aging equipment. From a practical standpoint, this application allows to highlight the goodness of each measure to the system performances by a simple comparison of reliability indices. The results of reliability indices improvement are published in [1].

## 3. Interruptions modeling and reliability service differentiation

In several studies dealing with electrical distribution system reliability, the objective often sought by the energy distributor is the balance between the required reliability level and its cost. In the following, we develop two important points of view in the reliability of the electrical systems: that relating to modeling and that of the differentiation of electricity prices according to the level of reliability required with a minimal guaranteed reliability level for customers without any prior requirement [12].

### 3.1. Interruption modeling using the Weibull-Markov process

In the last decade, a novel vision of interruption modeling in power systems was developed and consists of the Weibull-Markov process. The purpose is to model the failure and operating data according to Weibull distribution proprieties, while retaining those assigned to the Markov model where the system occupies discrete states. This process was initially developed by Van Castaren [2] and was applied successfully by Pivatolo [13] and Medjoudj et al. [3]. Applications were made to highlight maintenance policies gathered on three types of actions: namely nondestructive action which does not improve reliability level but slows the system degradation. This action, denoted $(1a)$ as a minor maintenance, is characterized by an improvement factor $m_1$. A second action is considered and denoted $(2b)$ and can touch some of the components of a system up to their replacement. To this action is associated an improvement factor $m_2$, and the maintenance is a major one. The third and final proposed maintenance action is on the renewal of equipment, and it is assumed to be perfect, and after its implementation, the system is assumed as good as new, and it is denoted $(2p)$. From a practical standpoint, this action is highlighted by taking $m_1$ and $m_2$ equal to the unity. This notion is introduced by Tsai et al. [14] for a mechatronic system and applied for power systems by Medjoudj et al. [3]. In this part of the section, we introduce the concept of the differentiated reliability with an application to the case of an electrical MV/LV substation. Starting from the expression of reliability function expressed by a desired threshold, the need of performing preventive maintenance action at time is decided regarding the behavior of this function at the coming stage of maintenance. Then, the choice of the type of action to perform is dictated by the value of the maximum benefit brought by this action. Threshold reliability is allocated to the opposite risk of system failure occurrence.

### 3.1.1. Reliability data analysis

Considering the formulations of mean up time and mean down time of an item is given respectively by:

$$MUT = t_m - t_b \int_0^{t_m} h(t)dt \tag{10}$$

and

$$MDT = t_a - t_b \int_0^{t_m} h(t)dt, \tag{11}$$

where, $t_m$, $t_a$ and $t_b$ are respectively, the preventive maintenance($PM$) interval, the $PM$ and corrective maintenance ($CM$) times on replacement; the operational availability is defined as:

$$A = \frac{t_m - t_b \int_0^{t_m} h(t)dt}{t_m + t_a} \tag{12}$$

Subsequently, the $PM$ interval for maximizing the availability can be derived by differentiating Eq. 9 to time $t_m$, such as $\frac{dA}{dt} = 0$, and the differential result is:

$$(t_m + t_a)h(t_m) - \int_0^{t_m} h(t)dt = \frac{t_a}{t_b} \tag{13}$$

For data treatment and statistical processing, forced and planned outages are collected over 17 years of system operation continuously at the national company of electricity and gas center (SONELGAZ) of Bejaia city, Algeria. For an MV/LV transformer which is a critical item of an electrical substation, the estimated parameters and the adequate probability distribution functions are listed in **Table 1**.

The obtained results show that, based on the Kolmogorov Smirnov ($KS$) test [15], $d_{ks}$ is lower than $d_{(n,0.05)}$, the Weibull distribution is not rejected; however, with the exponential law, $d_{ks}$ is greater than $d_{(n,0.05)}$, the hypothesis is not accepted. In **Table 2** are gathered reliability indices, where both the feeder failure frequency ($F_i$) and the transformer failure rate ($h$) are added. The nonacceptation of the exponential distribution is comforted by the review results of reference [16], where the authors state that the exponential law, usually used to describe failures, is not always 100% suitable for electricity distribution systems.

In this study, it is assumed that the substation failures are due either to the transformer or to the internal cable connector failures. A three states diagram (working, failure and maintenance) is dressed for the life cycle modeling of the substation as shown in **Figure 2**.

Let $X_{12}$, $X_{13}$, $X_1$, $X_2$, $X_3$ be the random variables representing the duration of the operation until failure, the duration of the operation until maintenance, the duration of the operation (state $S_1$), the duration of the interruption (state $S_2$) and the duration of the maintenance (state $S_3$), respectively. The estimated parameters of the random variables following the Weibull distributions are listed in **Table 3**.

| Component or Subsystem | $n$ | Distribution | Parameters | $d_{ks}$ | $d_{(n,0.05)}$ | Decision |
|---|---|---|---|---|---|---|
| MT/LV Transformer | 17 | Weibull | $\beta = 2.459579$; $\eta = 5.909754 \times 10^6$ | 0.2264 | 0.308 | Not rejected |
| (Tr) | | Exponential | $\lambda = 0.007785467$ | 0.3136 | | Rejected |

**Table 1.** Distribution functions parameters estimation.

| Component or Subsystem | *MUT* (hours) | *MDT* (hours) | *MTBF* (hours) | *A* | $F_i$ (1/year) | *h* (1/year) |
|---|---|---|---|---|---|---|
| MT/LV Transformer | 87358.33 | 9.97 | 87368.32 | 0.9998 | 0.10024 | 0.00778 |

**Table 2.** Reliability indices of the power transformer.

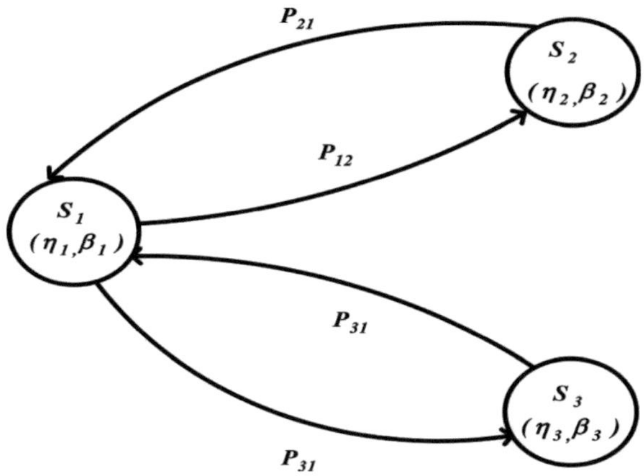


**Figure 2.** Three states diagram.

| Variable | *N* | Distribution | Parameters | $d_{ks}$ | $d_{(n,0.05)}$ | Decision |
|---|---|---|---|---|---|---|
| $X_{12}$ | 27 | Weibull | $\beta = 1.0644$; $\eta = 3.2827 \times 10^6$ | 0.1356 | 0.25438 | Not rejected |
| $X_{13}$ | 27 | Weibull | $\beta = 1.6689$; $\eta = 7.2589 \times 10^5$ | 0.2173 | 0.25438 | Not rejected |
| $X_2$ | 27 | Weibull | $\beta = 0.6764$; $\eta = 196.00159$ | 0.1219 | 0.25438 | Not rejected |
| $X_3$ | 27 | Weibull | $\beta = 1.03894$; $\eta = 123.6005$ | 0.1219 | 0.25438 | Not rejected |

**Table 3.** Parameters estimation of the random variables following Weibull distributions.

### *3.1.2. Reliability under preventive maintenance*

The improvement of maintenance to reliability is developed using two factors, and the selection of the action to do for the components on every *PM* stage is decided by maximizing system benefit in maintenance. Depending on the percent of the survival parts of system when it is maintained, the reliability function is:

$$R_j(t) = R_{0,j}.R_{V,j}(t) \tag{14}$$

where $R_{0,j}$ is the initial reliability of the $j^{th}$ stage and $R_{V,j}(t)$ is the reliability degradation of surviving parts on this stage. Considering periodical $PM$ which interval is $t_m$, the reliability of surviving parts is defined as:

$$R_{V,j}(t) = R\left(\frac{1}{m_1}(t-(j-1)t_m\right) \tag{15}$$

With: $(j-1)t_m \le t \le jt_m$ and $m_1$, $(0 < m_1 \le 1)$ is the improvement factor of action $(1a)$.

To model the reliability of systems following$PM$, the effects of various actions on $R_{0,j}$ and $R_{V,j}$ must be evaluated.

$$R_{0,j} = R_{f,j-1} = R_{0,j-1}.R(t_m) \tag{16}$$

where $R_{0,j-1}$, $R_{f,j-1}$ indicate the initial and final reliability values of the system on the $(j-1)^{th}$ stage.

Action $(1b)$ can improve the surviving parts of the system and also recover the failed parts. Generally, the impact of this action on the failed parts can be measured by an improvement factor $m_2$, which is also set between 0 and 1 representing the restored level except the surviving parts. According to the definition, the initial reliability on the action $(1b)$ can be expressed as:

$$R_{0,j} = R_{f,j-1} + m_2\left(R_0 - R_{f,j-1}\right) \tag{17}$$

where $R_0$ denotes the initial reliability of the new system.

The system reliability is expressed as:

$$R_j(t) = R_{0,j}\, exp\left[(t-(j-1)t\_m)/(m\_1\ \eta)\right]^{\beta} \tag{18}$$

where $R_{0,j}$ is the initial reliability of the $j^{th}$ stage and $m_1$is the improvement factor of action $(1a)$. The benefit of component maintenance on the $j^{th}$ stage is defined as [14]:

$$B_{i,k} = \frac{\int_{t_j}^{\infty} R_{i,j+1}(t)dt - \int_{t_j}^{\infty} R_{i,j}(t)dt}{C_{i,k}} \tag{19}$$

where $i, k$ denote, respectively, the $i^{th}$ subsystem or component and the maintenance action considered and $C_{i,k}$, the action cost. The advantageous one will correspond to the maximum of the benefit, that is, $B_i^* = \text{Max}(B_{i,k})$. Once the action of maintenance is defined and retained, the availability of the system at any stage is processed as:

$$A_{s,j} = \frac{T - t_{b,m} \sum_{i=1}^{n} \int_{t_{j-1}}^{t_j} h_{i,j}(t)dt}{T + \sum_{i}^{n} t_{i,k,a}} \tag{20}$$

where $n$ is the number of components or subsystems and $t_{i,k,a}$ is the time of the $PM$ actions $(1a)$, $(1b)$ and $(2p)$ and $T$, the cycle time. In the following are described the different types of $PM$ actions in the case of the power transformer.

- Action $(1a)$: cleaning, lubricating, tightening and oil-level verification,
- Action $(1b)$: oil and internal cable connectors' $(I.C.C)$ replacement,
- Action $(2p)$: transformer replacement.

Parameters are needed to compute the benefits such as the distribution function parameters $(\beta, \eta)$ $PM$ and $CM$ times ($t_a$ and $t_b$), maintenance actions costs ($C_{1a}$, $C_{1b}$, and $C_{2P}$) listed in **Table 4**, and the threshold value of reliability ($R_{th}$ = 0.8).

The obtained results are: ($t_m$ $(I.C.C)$ = 715 days, $t_m$ $(Tr)$ = 2415 days); however, the maintenance interval for the system is $T_m = \min\{715, 2415\} = 715$ days. The maintenance action to retain is based on the maximum benefit value, and the results at different maintenance stages are listed in **Table 5** using the following notations:

*: no maintenance is needed,

0: nothing to do (after an inspection),

1: action $(1a)$ is carried out,

2: action $(1b)$ is carried out,

3: action $(2p)$ is carried out,

$R((j+1)T_m)$: The instantaneous reliability at $(j+1) \times T_m$.

The results listed in **Table 5** can be interpreted as follows: at every maintenance stage, verify for each component if its reliability for the coming stage is greater or equal to the $R_{th}$.

• If the condition is realized, the decision is doing nothing. For the example of the transformer, at $j = 1$, the reliability is $R((j+1) \times T_m) = R(2 \times T_m) = 0.9272 > R_{th} = 0.80$; however, no maintenance is needed for the stage $j = 1$.

• If no, compute the benefit for each action proposed and choose the maximum value. For example, for the threshold value $R_{th} = 0.95$, at the first stage, $j = 1$, the reliability at the coming

| $\beta$ | $\eta$ | $t_a(days)$ | $t_b(days)$ | $m_1$ | $m_2$ | $C_{1a}$ $ | $C_{1b}$ $ | $C_{2p}$ $ |
|---|---|---|---|---|---|---|---|---|
| 2.45 | 4103.99 | 3.5 | 28 | 0.80 | 0.90 | 600 | 1500 | 8600 |

**Table 4.** The useful parameters for benefit evaluation.

| Stage | Action proposed | $R((j+1)T_m)$ | | | Benefit $ | | | Action Retained | | |
|---|---|---|---|---|---|---|---|---|---|---|
| | | $R_{th}=0.8$ | $R_{th}=0.9$ | $R_{th}=0.95$ | $R_{th}=0.8$ | $R_{th}=0.8$ | $R_{th}=0.95$ | $R_{th}=0.8$ | $R_{th}=0.8$ | $R_{th}=0.95$ |
| 1 | $(1a)$ | 0.9272 | 0.9272 | 0.9272 | ■ | ■ | 0.4378 | 0 | 0 | 2 |
| | $(1b)$ | | | | ■ | ■ | 0.8285 | | | |
| | $(2p)$ | | | | ■ | ■ | 0.1601 | | | |
| 2 | $(1a)$ | 0.8155 | 0.8155 | 0.9234 | ■ | 2.0944 | 0.6371 | 0 | 1 | 2 |
| | $(1b)$ | | | | ■ | 1.7618 | 0.9393 | | | |
| | $(2p)$ | | | | ■ | 0.3254 | 0.1827 | | | |
| 3 | $(1a)$ | 0.6618 | 0.8286 | 0.9230 | 3.0284 | 1.2526 | 0.6601 | 1 | 2 | 2 |
| | $(1b)$ | | | | 2.5859 | 1.8010 | 0.9394 | | | |
| | $(2p)$ | | | | 0.4710 | 0.3326 | 0.1844 | | | |
| 4 | $(1a)$ | 0.7287 | 0.9220 | 0.9222 | 1.7524 | ■ | 0.6830 | 2 | 0 | 2 |
| | $(1b)$ | | | | 2.1457 | ■ | 0.9393 | | | |
| | $(2p)$ | | | | 0.3945 | ■ | 0.1861 | | | |
| 5 | $(1a)$ | 0.9200 | 0.8056 | 0.9222 | ■ | 1.5305 | 0.7058 | 0 | 1 | 2 |
| | $(1b)$ | | | | ■ | 1.0851 | 0.9393 | | | |
| | $(2p)$ | | | | ■ | 0.3429 | 0.1877 | | | |
| 6 | $(1a)$ | 0.8038 | 0.8032 | 0.9219 | 3.1444 | 1.5599 | 0.7514 | 1 | 1 | 2 |
| | $(1b)$ | | | | 2.6712 | 0.3430 | 0.9392 | | | |
| | $(2p)$ | | | | 0.4881 | 0.3188 | 0.1910 | | | |
| 7 | $(1a)$ | 0.6455 | 0.8721 | 0.9215 | 1.7271 | 1.3359 | 0.7741 | 2 | 1 | 2 |
| | $(1b)$ | | | | 2.1735 | 0.6047 | 0.9392 | | | |
| | $(2p)$ | | | | 0.4016 | 0.3687 | 0.1927 | | | |
| 8 | $(1a)$ | 0.7181 | 0.8376 | 0.9211 | 1.7271 | 1.3359 | 0.7741 | 2 | 1 | 2 |
| | $(1b)$ | | | | 2.1735 | 0.6047 | 0.9392 | | | |
| | $(2p)$ | | | | 0.4016 | 0.3687 | 0.1927 | | | |
| 9 | $(1a)$ | 0.9190 | 0.7916 | 0.9207 | ■ | 1.3303 | 0.7967 | 0 | 1 | 2 |
| | $(1b)$ | | | | ■ | 0.9271 | 0.9392 | | | |
| | $(2p)$ | | | | ■ | 0.4300 | 0.1944 | | | |
| 10 | $(1a)$ | 0.8024 | 0.7385 | 0.9203 | ■ | 1.3084 | 0.8193 | 0 | 1 | 2 |
| | $(1b)$ | | | | ■ | 1.2177 | 0.9391 | | | |
| | $(2p)$ | | | | ■ | 0.4867 | 0.1960 | | | |

**Table 5.** Maintenance plan depending on reliability thresholds.

stage of maintenance $R((j+1)\times T_m) = R(2\times T_m) = 0.9272 < R_{crit} = 0.95$ and actions $((1a), (1b))$ benefits are $(0.4378, 0.8285)$, respectively. The action $(1b)$ is retained looking at the maximum value of the benefit.

The risk management is highlighted by thresholds of reliability. Depending on the reliability level reached, or fixed a priori, maintenance operations can be decided. The objectives are the determination of maintenance frequencies on an item and consequently their costs. It will be remarked that a high level of reliability is required (i.e., the risk of failure is minimized), the maintenance frequency increases, and subsequently, the cost increases.

## 4. Competing failure processes of oil circuit breaker

The components constituting a high voltage oil circuit breaker (HVOCB) are subject to various degradations, namely the aging of the insulating oil in the arc's extinguishing chamber, the contacts wear out and the sharp breakdown of bus bars supports. In this section, we have modeled the behavior of this item as subject to a competing three degradation processes by using the Markov state diagram as given in **Figure 3**. The states were defined using thresholds of degradation parameters. To the degradation processes was associated a shock process highlighting the effects of short circuit arrivals on the HVOCB when defaults occur at the downstream feeder. The novelty in this work is outlined by the use of three-dimensional matrix to show the possible states, where the HVOCB can sojourn.

### 4.1. Case of three degradation processes modeling

We consider that the processes of degradation are modeled using continuous probability functions, and the operating condition of the system is characterized by a number of states which space is noted by $\Omega_\mu$.

Following the Li and Pham theory [17], we consider the three state spaces $\Omega_1$, $\Omega_2$,and $\Omega_3$ corresponding to the degradation processes $Y_1(t), Y_2(t),$ and $Y_3(t)$, respectively. After obtaining the state spaces $\Omega_1$, $\Omega_2$, and $\Omega_3$, we develop a methodology to establish a relationship between the states of the system $\Omega_\mu$, the set of degradation states and catastrophic state due to shocks arrivals $\{\Omega_1, \Omega_2, \Omega_3, F\}$.

The study deals with the three processes of degradation $Y_1(t)$, $Y_2(t)$, and $Y_3(t)$ combined with the shock process denoted $D(t)$ as given in **Figure 3**. The sets of states are represented by $\Omega_1 = M_1, \ldots, 1_1, 0_1$, which corresponds to the degradation 1 with$(M_1 + 1)$ states, $\Omega_2 = M_2, \ldots,$

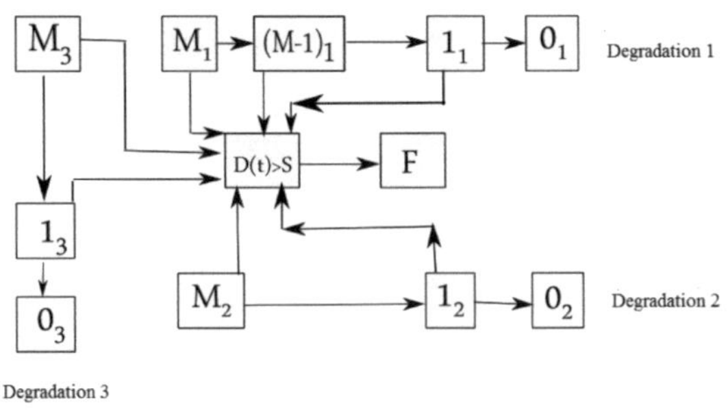


**Figure 3.** Diagram of transition states of a system subject to four failure processes.

http://dx.doi.org/10.5772/intechopen.71926

$1_2, 0_2$, corresponds to the degradation 2 with $(M_2 + 1)$, and $\Omega_3 = M_3, \ldots, 1_3, 0_3$ corresponds to the degradation 3 with $(M_3 + 1)$ states. **Figure 3** shows the transition between states of a system submitted to four failure processes.

The equivalence relations between degradation states $\Omega_1 = M_1, \ldots, 1_1, 0_1$; $\Omega_2 = M_2, \ldots, 1_2, 0_2$ and $\Omega_3 = M_3, \ldots, 1_3, 0_3$, and their corresponding intervals are given as follows:

Degradation Process 1:

$0 < Y_1(t) \leq W_M \Longrightarrow$State $M_1$.

$W_M < Y_1(t) \leq W_{M-1} \Longrightarrow$State $(M-1)_1$.

.

.

.

$W_2 < Y_1(t) \leq W_1 \Longrightarrow$State $1_1$.

$G_1 = W_1 < Y_1(t) \Longrightarrow$State $0_1$.

Degradation Process 2:

$0 < Y_2(t) \leq A_M \Longrightarrow$State $M_2$.

$A_M < Y_2(t) \leq A_{M-1} \Longrightarrow$State $(M-1)_2$.

.

.

.

$A_2 < Y_2(t) \leq A_1 \Longrightarrow$State $1_2$.

$G_2 = A_1 < Y_2(t) \Longrightarrow$State $0_2$.

Degradation Process 3:

$0 < Y_3(t) \leq Z_M \Longrightarrow$State $M_3$.

$Z_M < Y_3(t) \leq Z_{M-1} \Longrightarrow$State $(M-1)_3$.

.

.

.

$Z_2 < Y_3(t) \leq Z_1 \Longrightarrow$State $1_3$.

$G_3 = Z_1 < Y_3(t) \Longrightarrow$State $0_3$.

The states' space of the system is defined by $\Omega_\mu = \{M, \ldots, 1, 0, F\}$ with $(M+2)$ states. In this part, we develop a function that generates the relation between the states' space of the system

$\Omega_\mu$ and degradation states' space $\{\Omega_1, \Omega_2, \Omega_3, F\}$ . For example, at a given time t, it is assumed that the degradation process 1 is at the state $i_1 \in \Omega_1$ and the degradation process 2 is at state $j_2 \in \Omega_2$ and the process of degradation 3 is in the state $k_3 \in \Omega_3$. It is assumed that at the present time, the system is not at fault condition (catastrophic state $F$). Thus, state $F$ can be ignored for the moment, we must therefore seek a function relationship between $\Omega$ and$\{\Omega_1, \Omega_2, \Omega_3\}$. Instead of $\Omega_\mu$ and $\{\Omega_1, \Omega_2, \Omega_3, F\}$. The operation can be described by the mathematical function formulated as follows:

$$f : R = \Omega_1 \times \Omega_2 \times \Omega_3 \rightarrow \Omega = \{M, ...., 1, 0\} \tag{21}$$

where $R = \Omega_1 \times \Omega_2 \times \Omega_3 = \{(i_1, j_2, k_3) \mid i_1 \in \Omega_1, j_2 \in \Omega_2, k_3 \in \Omega_3\}$

The function $f$ is defined by: $\Omega_1 \times \Omega_2 \times \Omega_3 \ \Rightarrow f \Rightarrow H$.

The matrix $H$ represented in **Figure 4** gives information about the resulting states space and component of $(M + 1)$ elements corresponding to each space leaving by the function $f$. The line at the top of the matrix $H$ represents the states of the degradation process 1, the right column of the matrix represents the states of the degradation process 2, and the top page of the matrix $H$ represents the degradation process 3. The elements of the matrix $H$ represent the states $f(i_1, j_2, k_3) = L$.

We note that in the matrix H, some elements are zeros and it can be assumed that when degradation 1 is in a certain advanced state $I_1(0_1 < I_1 < M_1)$, the degradation 2 is also in a certain weak state $I_2(0_2 < I_2 < M_2)$, and the degradation 3 is also in a certain weak state $I_3(0_3 < I_3 < M_3)$, it is considered as a failure condition. We also notice that $f(M_1; M_2; M_3) = M$, and initially, the system is in a perfect state.

We define time until failure by: $T = \inf\{ t : Y_1(t) > G_1, Y_2(t) > G_2, Y_3(t) > G_3$ where $D > S\}$.

It is important to know that the life of the system depends on a single process among the three degradations and of that of the shock. However, the system failure is caused by the process

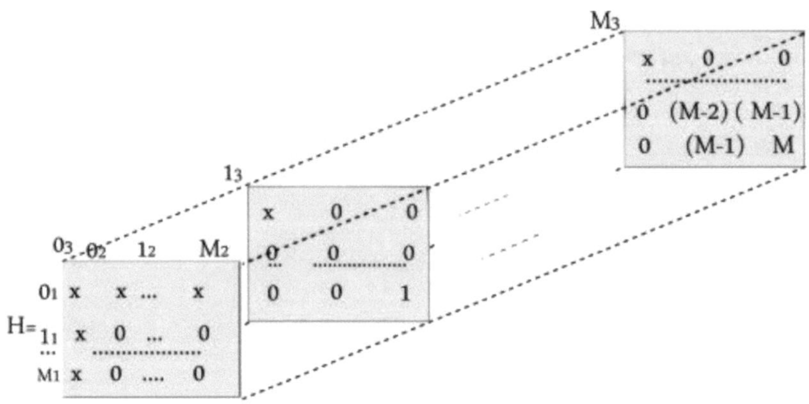


**Figure 4.** Three-dimensional matrix of system' states, where $I_1 \in \Omega_1$, $J_2 \in \Omega_2$, $K_3 \in \Omega_3$, and $L \in \Omega$.

that occurs first, exceeding its critical value corresponding to the level which can bring the system back to failure.

### 4.2. System case study modeling and application

Initially, the system is considered in good states of operation$(M_1, M_2, \text{and } M_3)$. It can pass first, to the degradation states $\left((M-1)_1, (M-1)_2, (M-1)_3\right)$ or to the state of catastrophic failure (state $F$), due to random shock. When the system reaches the first state of degradation, it can either remain in this state or go to the second degradation state $\left((M-2)_1, (M-2)_2, (M-2)_3\right)$, or it passes to state $F$. The same process is repeated at each degradation stage with the exception of the states $(0_1, 0_2, 0_3)$

Assumptions:

1. The system occupies $(M+2)$ states, where 0 and $F$ are the states of failure, state $i$ is a degradation state, $1 < i < M$;
2. No repair or maintenance is carried out on the system;
3. $Y_i(t);\ i = 1, 2, 3;$ is a not decreasing and not negative function. With respect to time $t$, it corresponds to an irreversible accumulation of damage;
4. $Y_i(t);\ i = 1, 2, 3;$ and $D(t)$ is statistically independent implying that the state of one process will have no effect on the other state;
5. At time $t = 0$, the system is at state $M$;
6. The system may fail due to;

- Degradation process if: $Y_i(t) > G_i$ for $i = 1, 2, 3$;
- Random shock process (the system passes to the condition of the catastrophic failure state $F$), if: $D(t) = \sum_{t=1}^{N(t)} Xt > S$; with $G_i$ and $S$ are critical level of degradation and shocks, respectively.

The reliability function is defined as follows:

$$
\begin{aligned}
R_M(t) &= P(state \geq 1) \\
&= \sum_{i=1}^{M} P_i(t) \\
&= P\{Y_1 \leq G_1, Y_2 \leq G_2, Y_3 \leq G_3, D(t) \leq S\}
\end{aligned}
\tag{22}
$$

The system will fail if any of the degradation rates exceeds the critical level $G_i$; $i = 1, 2, 3$ or the process of shock also exceeds the critical level $S$.

The system subject to three displacement processes is defined by:

1. The process of increasing degradation representing the wear of the contacts of the circuit breaker is denoted by $Y_1(t)$;

2. The process of increasing degradation representing the aging of oil insulating circuit of the circuit breaker is denoted by $Y_2(t)$;

3. The degradation process of bus bars supports is denoted by $Y_3(t)$;

4. A random process of cumulative shock damages is given by $D(t) = \sum_{t=1}^{N(t)} Xt.$

We obtain a system with four competing degradation processes. For the wear of the contacts: $Y_1\ (t)$, $M_1 = 2$; for the aging of the oils $Y_2\ (t)$: $M_2 = 2$; and for the degradation of the supports: $Y_3\ (t)$, $M_3 = 3$; the system fails if the process of degradation $Y_i(t)$ exceeds a level $G_i$; $i = 1, 2, 3$ or the process of damages cumulates shocks, and D (t) exceeds the level S. It is assumed that the state spaces associated to $Y_1$, $Y_2$ and $Y_3$ are $\Omega_1 = \{2_1, 1_1, 0_1\}$, $\Omega_2 = \{2_2, 1_2, 0_2\}$, $\Omega_3 = \{3_3, 2_3, 1_3, 0_3\}$, respectively. Consequently, the space of the system is defined as follows: $\Omega = \{3; 2; 1; 0; F\}$. Thus, the function f is defined as being the Cartesian product of three sets following $f : R = \Omega_1 \times \Omega_2 \times \Omega_3 \rightarrow \Omega = \{3, 2, 1\}$, and the result is illustrated by the three-dimensional matrix given in **Figure 5**.

The implementation of the abovementioned models under the MATLAB software which has given the results of the probabilities of the sojourns in different states and the system reliability changing is shown in **Figure 6**. The system (oil circuit breaker, bus bars) is in good condition for 3 years (1100 days) with a probability greater than 0.9. After this period, the latter decreases exponentially to reach the zero value after 4 years (1490 days) without any maintenance actions.

The evolution of the probability of the system in degradation state 2 is complementary to that of the probability of the degradation state 3. Indeed, during 3 years of operation, the probability of being in state 2 is zero. Then, it increases exponentially to reach its maximum value of 0.41 up to 4 years (1280 days). As this state is transient, its probability function decreases to 0 after 5 years. For the system reliability, we note that during the first 8 years, it is expected to decrease by 20% due to the random shock process that governs the system during this period.

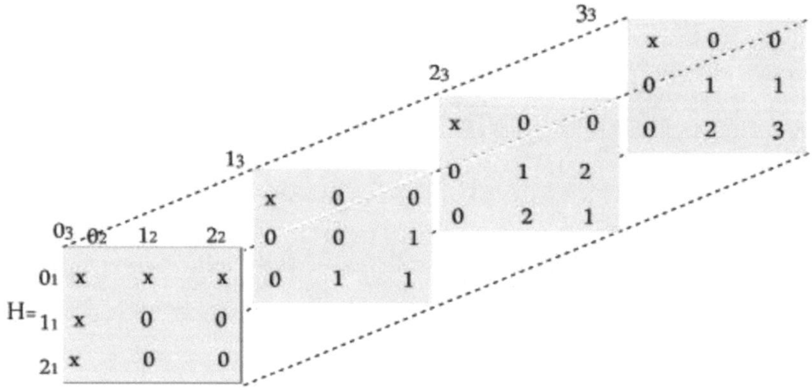


**Figure 5.** The three-dimensional matrix of the studied system.

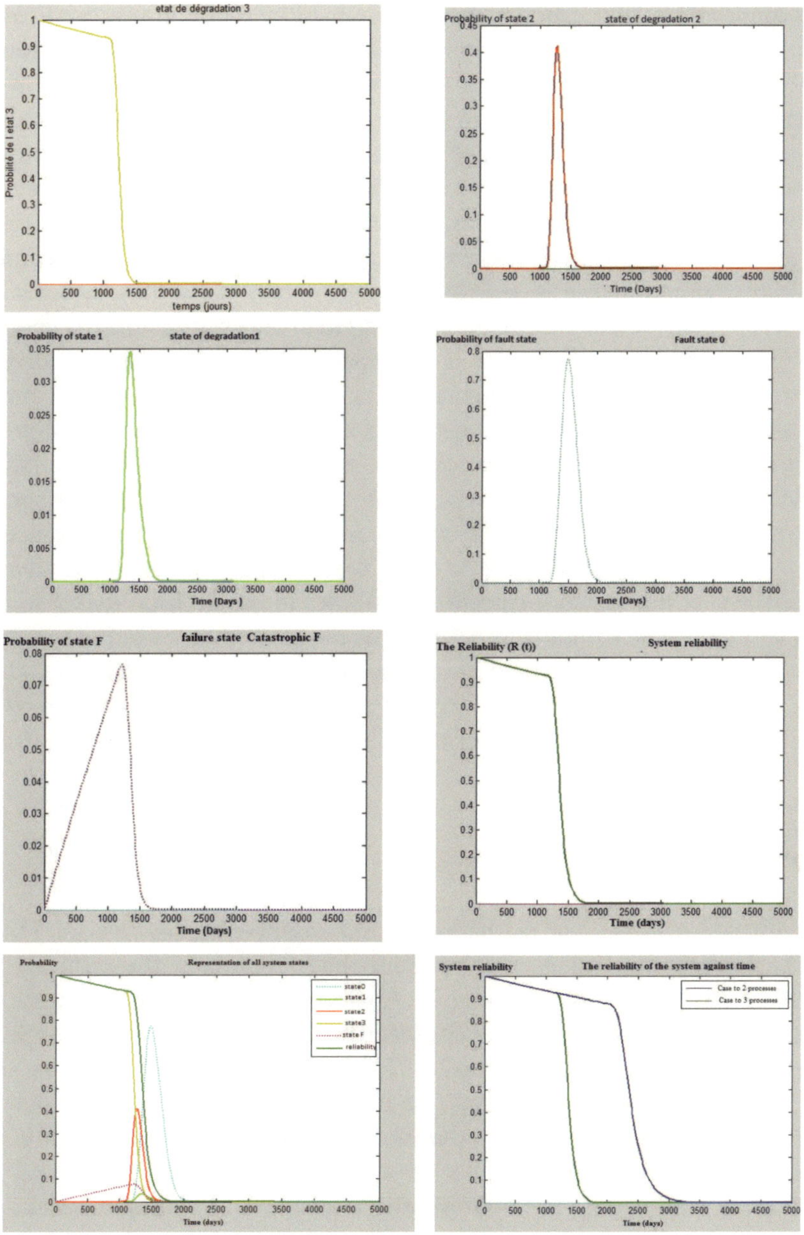


**Figure 6.** Probabilities of the states and the system reliability changing.

## 5. Conclusion and discussions

Recently researchers in electrical systems have proposed differentiated electricity service based on reliability and have shown some inconveniences to apply it into a real case. In the same location are connected both consumers with high reliability requirements, with an agreement to pay more and others who are not concerned. Because the technical measure proposed is to add a reliable feeder, the differentiation is not quite possible. Our proposition consists of organizational measure and is oriented to maintenance actions on MV/LV substations, which are directly connected to the end users of the network. The differentiated service is directly related to the reliability of the substation where the improvement is a function of maintenance actions and the frequency of interventions. For statistical considerations and for interruption (forced or scheduled outages) modeling, we have applied the Weibull-Markov approach rather than the Markov method, which is usually used for the case of electrical systems. It has been proven that it is possible to maintain in another way than the classical one based on systematic preventive maintenance. In this chapter, it is shown that the maintenance is decided on the reliability level and benefit bases. Another critical component of electrical substation is studied using competing failure processes and consists a circuit breaker. The reliability aspects are formulated in the bases of oil aging, contacts wear and bus bars support degradation. Investigations conducted by Pham in a theoretical framework have been applied successfully to complement system such as electrical system. The models applied on simple numerical examples have been validated by application to a real case engineering. During system operation, the results analysis of the network current state allows to the decision maker to reach better information and target the equipment that reduces the performances of the system and practicing suitable maintenance actions. Recent studies in energy sustainability and smart energy grid have revealed that reliability is the main criterion taken into account by decision makers in electricity market behavior and a performance index for electric utilities classifications.

## Acknowledgements and tributes

In 1988, a group of professors in mathematics and engineering (A. Aissani, D. Aissani, K. D. Haim, A. Boubakeur, and A. Ouabdeslam) had organized the national conference named MFSI at the University of Bejaia, in Algeria, where during 2 days, the notion of reliability was vulgarized in the field of engineering. From that manifestation was born the group of work in power system reliability FSE2. Over than 1500 various works were conducted in engineer, master and doctorate theses dealing with all the aspects of power systems reliability and with a large cooperation with other universities and various manufactures and services. A great number of applications were done around the power systems including production, transportation and distribution parts. In recent years, a lot of novelties were developed compared to what is done over the word, such as the Weibull-Markov modeling in data analysis, nonparametric distributions in switching components behavior, Box and Jenkins models in blackouts forecasting and reliability aspects in smart grids development and multicriteria optimization. The results were valorized in a great number of international conference proceedings and in valuable international journals. This chapter dealing with power system reliability constitutes an interesting opportunity to express our acknowledgments and tributes to these pioneers of reliability in Algeria for what they have given for research.

## Author details

Rabah Medjoudj[1]*, Hassiba Bediaf[2] and Djamil Aissani[3]

*Address all correspondence to: r.medjoudj66@gmail.com

1 Lamos Research Unit, Faculty of Technology, Electrical Engineering Department, University of Bejaia, Bejaia, Algeria

2 SOPERIE, Electrical Engineering Society, Bejaia, Algeria

3 Lamos Research Unit, Faculty of Exact Sciences, Operational Research Department, University of Bejaia, Bejaia, Algeria

## References

[1] Medjoudj R, Aissani D, Haim KD. Power customer satisfaction and profitability analysis using multi-criteria decision making methods. International Journal of Electrical Power and Energy Systems (Elsevier). 2013;**45**(1):331-339. DOI: 10.1016/j.ijepes.2012.08.062

[2] Van Castaren J. Reliability assessment in electrical power systems: The Weibull-Markov stochastic model. IEEE Transactions on Industry Applications. 2000;**36**(6)

[3] Medjoudj R, Aissani D, Boubakeur A, Haim KD. Interruption modeling in electrical power distribution systems using Weibull-Markov model. Proceedings of the Institution of Mechanical Engineers (IMEchE), Part O: Journal of Risk and Reliability. 01-06-2009;**223**:145-157. DOI: 10.1243/1748006XJRR215

[4] Iberraken F, Medjoudj R, Medjoudj R, Aissani D. Combining reliability attributes to maintenance policies to improve high-voltage oil circuit breaker performances in the case of competing risks. Proceedings of the Institution of Mechanical Engineers, Part O: Journal of Risk and Reliability 1748006X15578572. First published on 02-04-2015. DOI: 10.1177/1748006X15578572

[5] Kuech JD, Kirby BJ, Overholt PN and Markel LC. Measurement practices for reliability and power quality. U.S Department of energy. ORNL/TM-2004/91

[6] Nourelfeth M, Ait-Kadi D. Optimization of series-parallel multi-state systems under maintenance policies. Reliability Engineering and System Safety. 2007;**92**:1620-1626

[7] Taboada HA, Espiritu JF, Coit DW. Design allocation of multistate series-parallel systems for power systems planning: A multiple objective evolutionary approach. Proceedings IMEchE, Part O: Journal of Risk and Reliability. 2008:381-391

[8] Levitin G, Lisnianski A. Multi-state system reliability analysis and optimization (UGF and GAA). In: Pham H, editor. Chapter 4, Handbook of Reliability Engineering. Springer; 2003

[9] Medjoudj R. Reliability Aspects of Electrical Systems: Semi-Markovian Modeling and Maintenance Optimization. Doctorate thesis: University of Bejaia, Algeria; 2009

[10] Iberraken F, Medjoudj R, Aissani D. Mathematical models to support the issue of electrical blackouts in the context of smart grids. Lecture Notes on Information Theory, Engineering and Technology Publishing. December 2014;**2**(4). DOI: 10.12720/lnit.2.4.295-301 (LNIT,ISSN: 2301-3788, www.lnit.org ), 2014

[11] Endrenyi J. Reliability Modeling in Electric Power Systems. Toranto: Wiley & Interscience; 1978

[12] Brown ER, Marshall MW. The Cost of Reliability. ABB consulting report. 2001

[13] Pievatolo A. The downtime distribution after a failure of a system with multistate independent components. IMATI-CNR Technical report, Presented at the Mathematical Methods on Reliability conference, MMR, Glasgow, 2007

[14] Tsai YT, Wang KS, Tsai LC. A study of availability centred preventive maintenance for multi-component systems. Reliability Engineering and System Safety. 2004;**84**:261-269

[15] Dhilon BS. Design Reliability, Fundamentals and Applications. USA: CRC press LLC; 1999

[16] Billinton R, Goel R. An analytical approach to evaluate probability distributions associated with the reliability indices of electric distribution systems. IEEE Transactions on Power Delivery. 1986;**1**(3):245-251

[17] Li W, Pham H. Reliability modeling of multi-state degraded systems with multi-competing failures and random shocks. IEEE Transactions on Reliability. 2005;**54**(2):297-230

Chapter 16

# Techno-Economic Feasibility Study of Autonomous Hybrid AC/DC Microgrid System

Atanda K. Raji

Additional information is available at the end of the chapter

http://dx.doi.org/10.5772/intechopen.69582

**Abstract**

Distributed generation technology based on diesel generators often has been considered as a viable solution to providing power to remote areas, but the sky-rocketing of diesel fuel price and the increasing cost of delivery to such remote sites have called for providing a sustainable solution that is environmentally friendly, economical, affordable, and easily accessible. To this end, the use of locally available energy resources is accepted as a sustainable solution in providing electricity for rural and remote settlements. The system cost of wind and solar energy systems is continuously decreasing because of the increase in the acceptance and deployment of the energy systems based on these renewable energy resources. A standalone hybrid AC/DC electric power system is designed, modeled, simulated, and optimized in HOMER Pro. HOMER is a Hybrid Optimization Model of Electric Renewable that enables the comparison of electric and thermal power production technologies across an extensive variety of applications. Both cycle-charging and load-following dispatched strategies are investigated. Plausible selected system components ratings are chosen for the simulation to ensure that there is enough search space for HOMER Pro to obtain an optimal system configuration. Net present cost (NPC) is used as an economic metric to assess the optimal configuration that is technically feasible.

**Keywords:** HOMER, wind, solar PV, renewable energy system, electrochemical devices, net present cost, economics analysis, Sub-Sahara Africa

## 1. Introduction

Energy access and affordability is a very big concern for the majority of the Africa populace. With over 620 million people without access to modern electricity in the Sub-Saharan Africa alone [1], different innovative solutions are being considered. Grid extension is definitely not an option because of the sparse population distribution, which makes the economic viability of such action infeasible. Distributed generation technology based on diesel generators has often been considered as a viable solution but the sky-rocketing of diesel fuel price and the increasing cost of delivery to remote sites have called for providing a sustainable solution that is environmentally friendly, economical, affordable, and easily accessible. To this end, the use of locally available energy resource has been accepted as a sustainable solution in providing electricity for rural and remote settlements. Matching the consumer demand profile and renewable energy production always has been perceived as enormous problem but can be mitigated by incorporating adequate sized energy storage systems [2, 3].

An autonomous electric power system is one that is not connected to the utility for one or all of the reasons given in the last paragraph. Wind and solar resources are local energy resources that can be efficiently and economically harnessed to provide electric power to remote site electricity consumers. While the cost of such renewable resources are cost-free, but the system required for harnessing and processing the energy plus the balance of system (BOS) is expensive. However, the system cost of wind and solar energy systems have been continuously decreasing because of the increase in the acceptance and deployment of the energy systems based on renewable resources. For the many Africa people who do not have access to modern energy system, it has been proven that high-quality electric power can be delivered to remote areas using local energy resources such as wind, solar, biomass, and so on [4–6].

Many of the historical technological advantages that resulted in AC power system to dominate over DC power systems are not valid any longer. For example, in the electric power generation sector, many distributed energy resources, such as electrochemical storage systems, fuel cells system, and photovoltaic systems, produce efficient and economical DC power. Many of the modern electrical and electronic loads, as well as energy storage systems, are either internally DC power operated or worked equally well with DC power sources and connect to the AC systems through power electronic converters [7]. Overall improved system efficiency, reduced system cost, increased reliability, and reduced system footprint are attained when multiple conversion stages are eliminated in externally operated AC power system for internally operated DC power devices. Also, many power quality issues, such as harmonics and unbalances, are not present in DC power systems [8, 9]. Advancements in power electronic technology continue unabated, leading to more applications in modern systems such as VSC-HVDC, electric vehicles, wireless power transfer, fuel cells vehicles, and so on [10, 11].

The design of hybrid renewable energy systems has been extensively studied, and numerous optimization techniques, such as genetic algorithm (GA), linear programming, etc., have been implemented for optimum economic and technical feasibility [12–16]. Many software packages, such as HOMER Pro [9], RETScreen [17], and Hybrid2 [18], have been developed for the proper selection of appropriate generation technologies and capacity sizing. This software makes a multi-objective optimization process easy to carry out to arrive at a robust and reliable decision. HOMER Pro is user-friendly software produced by the national renewable energy laboratory by the department of energy in the USA. It uses hourly data for the assessment of hybrid renewable energy systems and performs optimization based on net present cost (NPC). HOMER Pro is used in this study because of its proven ability to design, model, simulate, and perform sensitivity analysis of complex mix of different energy technologies, thermal and electrical load, and storage devices. Both cycle-charging and load-following dispatch strategies are available to choose from or use simultaneously as utilized in this study [19, 20].

The rest of the chapter is organized as follows. A brief discussion of two different dispatch strategies used in hybrid energy systems is provided in Section 2 of this chapter. All the energy system components of the studied autonomous hybrid AC/DC microgrid system are briefly discussed in Section 3. Section 4 explains the capability and functionalities of Hybrid Optimization Model of Electric Renewable (HOMER) as powerful software that is designed and optimized for modeling, simulating, optimizing, and performing sensitivity analysis of micropower systems. The system model and simulation process of the studied hybrid system is comprehensively detailed in Section 5 of this chapter. The system simulation results and discussion and conclusion are given in the penultimate and Section 7 of this chapter, respectively.

## 2. Cycle-charging and load-following dispatch strategy

Two system dispatch strategies are employed in HOMER Pro namely: cycle-charging and load-following dispatch strategy. Under the load-following dispatch strategy, a power generator produces only enough to serve the load and does not charge the battery bank. For the cycle-charging dispatch strategy, the generator run at full power to serve the load and any excess is used to charge the battery [7].

Tuglie and Torelli [21] developed a new load following dispatch algorithm that correlates system imbalances with unbalanced transaction in a bilateral transaction-based market. The developed algorithm is implemented on the IEEE 30-bus test system under a variety of operating conditions. In Bizon [22], a hybrid power source comprising wind energy conversion system, solar Photovoltaic system, and fuel cell system with energy storage system is designed and optimized for load-following dispatch strategy. The battery/supercapacitor hybrid energy storage system operates as an auxiliary source for supplying the power deficit based on dynamic balance strategy. An optimal energy management system for a stand-alone wind turbine/photovoltaic/hydrogen/battery hybrid system with supervisory control based on fuzzy logic is presented in Pablo et al [23]. The energy management system in Ref. [23] uses a fuzzy logic control to satisfy the energy demanded by the load and maintain the state-of-charge of the battery and the hydrogen tank level between desired target margins, while trying to optimize the utilization cost and lifetime of the energy storage system. A new method to model and control the aggregated power demand from a population of thermostatically controlled load, with the goal of delivering ancillary services such as frequency regulation and load following, is developed by Callaway [24]. It is shown in the paper that identified models perform only marginally better than the theoretical model. Subho and Sharma [25] developed a hybrid system with cycle charging dispatch strategy using swarm optimization algorithm for a remote area in India. A comparative study of particle swarm optimization (PSO), genetic algorithm, and HOMER is performed in Ref. [25]. PSO is judged to be better than GA and HOMER in terms of the minimum cost of electricity generation. In another paper by Subho and Sharma [26], different optimization algorithms and software packages are used to evaluate the economics of a hybrid renewable energy system. It is concluded that cycle charging strategy is most cost-effective compared to the load following and peak shaving dispatch strategies.

Both dispatch strategies are employed in this study to determine which is economically viable and technically feasible.

## 3. Energy system components

For the considered autonomous electric power system model in this study, all the components and energy resources involved are briefly discussed. The components related to this work are Solar PV, wind energy conversion system (WECS), electrochemical battery storage device, and bidirectional power electronics converter.

### 3.1. Solar PV energy system

Solar PV technology uses semiconductor material, such as silicon, to convert solar radiation energy to direct current (DC) electrical power. The generated DC power is regulated by charge regulator to charge electrochemical battery storage device. The electrochemical storage device supplies power directly to the electrical load if it is a DC load. If not, an inverter is used to convert the DC power supplied by the electrochemical storage device into an AC power for the AC load.

PV cell is usually modeled using a single diode or double diodes, but a single-diode model is shown in **Figure 1**. Series and shunt resistances represented in the figure are used to represent the leakage current and the inherent loss due to internal electrical connection, respectively [27–29].

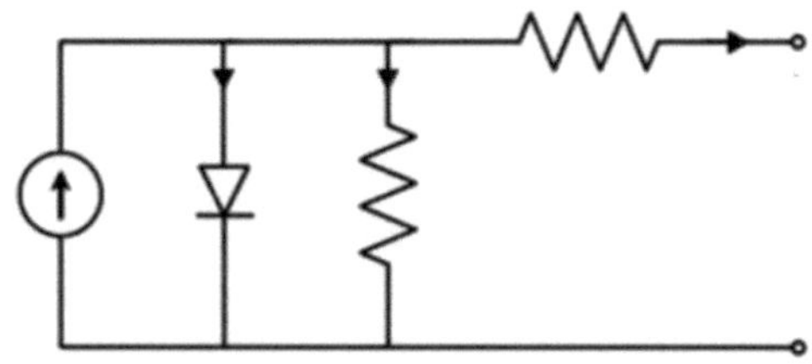

**Figure 1.** Single-diode model of solar cell.

The insolation intensity is strongly related to the solar cell current, and the cell operating temperature is inversely proportional to the output voltage across the solar cell. These factors basically limit the maximum power that can be generated by the solar cell [30].

Homer Pro calculates the power output of the PV array using equation 1 below.

$$P_{PV} = f_{PV} Y_{PV} \frac{I_T}{I_S} \tag{1}$$

Where $f_{PV}$ is the PV derating factor, $Y_{PV}$ is the rated capacity of the PV array (kW), $I_T$ is the global solar radiation (beam plus diffuse) incident on the surface of the PV array (kW/m$^2$), and $I_S$ is the standard amount of radiation used to rate the capacity of the PV array given as 1 kW/m$^2$.

A generic flat plate PV system is selected for the modeled hybrid microgrid from the HOMER Pro components library with a rated capacity of 40 kW. The capital cost of the PV is \$3000/kW$_p$, replacement cost of \$2000, and operation and maintenance cost of \$10.00/year.

### 3.2. Wind energy conversion system

A blowing wind consists of kinetic energy carried by the atmospheric air. The moving air is converted to rotational kinetic energy by the rotor blade of the wind turbine system. The rotating rotor is coupled via a solid shaft to an electrical generator to generate AC electric power. The extraction of the power contained in the moving is limited by the aerodynamic efficiency of the rotor blade as illustrated in **Figure 2**.

The theoretically generated turbine power limit of which is 59% is known as the Betz limit. Most of the modern wind turbines are practically in the 30–40% range. Equation 1 below gives the relationship between the extracted power and other parameters and variable of a wind

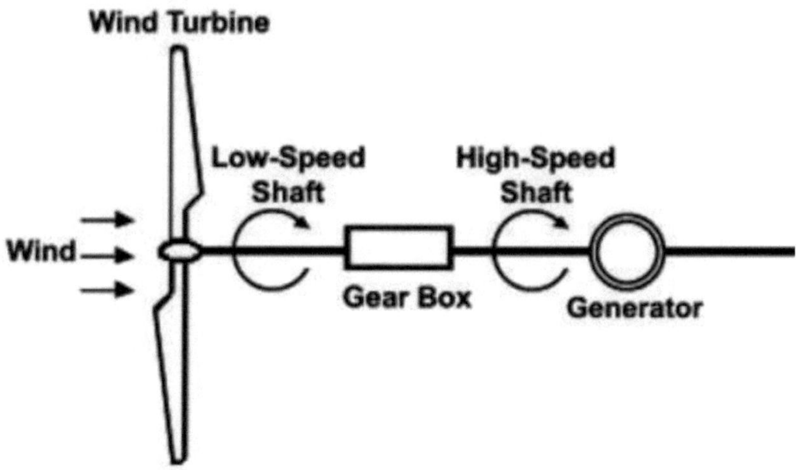


**Figure 2.** Wind energy conversion system (WECS) adopted from Ref. [31].

energy conversion system. The wind speed is strongly coupled to the available power, whereas the power coefficient depends on the tip speed ration (TSR) and the blade pitch angle [10, 16].

$$P = \frac{1}{2} A \rho C_p V^3 \tag{2}$$

$A$ is the rotor swept area, $\rho$ is the air density, $C_p$ is the power coefficient that is dependent on the rotor blade pitch angle and tip speed ratio, and $V$ is the average wind speed.

HOMER Pro determines the output power of the wind turbine in a four-step process. In the first step, it determines the average wind speed for the hour at the anemometer height by referring to the wind resource data. In the second step of the process, it determines the corresponding wind speed at the turbine's height using either the logarithmic law or the power law (the logarithmic law is used in this study). Thirdly, it refers to the turbine's power curve to calculate its output power value at that hourly wind speed assuming standard air density. Finally, it multiplies that output power value by the air density ratio, which is the ratio of the actual air density to the standard air density. HOMER Pro also assumes that air density ratio is constant throughout the year.

A generic wind turbine model of 10 kW rated power is selected in HOMER Pro for the study. The capital cost is $5000/kW, replacement cost of $5000/kW, operation, and maintenance cost of $500/year, operational lifetime of 20 years. The hub height of the generic wind turbine model is 24 m above sea level. The power curve of the selected turbine is shown in **Figure 3**.

### 3.3. Electrochemical battery energy storage system

A common electrochemical energy storage used in autonomous and emergency standby is Lead-Acid battery that has proven to be robust, cost-efficient, and cost-effective. However, many new battery technologies are being developed and utilized for practical applications with improved operational characteristics but are currently more expensive than the traditionally used battery technology [7]. With the interest in electric vehicle, lithium-ion battery technology is in the pole position to be the mainstream electrochemical technology

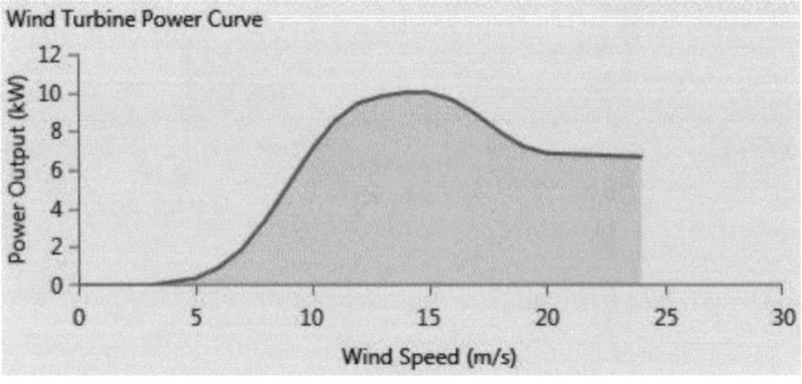


**Figure 3.** 10 kW generic wind turbine power curve (HOMER Pro).

for various applications. Among the operational parameters of Lead-Acid battery, maximum state of discharge (DOD), minimum state of charge (SOC), cycle times, round trip efficiency, maximum charge rate, maximum discharge rate are the most important. This battery technology also needs regular maintenance and should be well ventilated especially if it is not the sealed type [7].

A generic 1 kWh lead-acid battery is selected from the Homer Pro library for the hybrid microgrid presented in this study. The nominal voltage of the battery is 12 V configured in a string of 20 batteries (240 Vdc) to conform to the bidirectional rectifier rated input voltage. The maximum capacity and round trip efficiency of the battery are 83.4 Ah and 80% respectively. The capital cost is $300/kWh, replacement cost of $200/kWh, operation and maintenance cost of $20/unit/year.

### 3.4. Static power converters

Power electronics converters are needed for various power conversions and conditioning to optimally match electrical power source with electrical and electronic loads. Four main types of such converters are available: AC-AC, AC-DC, DC-DC, and DC-AC power converter [7]. A bidirectional power electronic converter is considered in this study because the wind energy conversion system is modeled as an AC power source, whereas the energy storage and solar PV are modeled as DC power source. Such power electronic converter allows power to flow bidirectionally between the AC and DC bus. Such functionality is possible by adequately

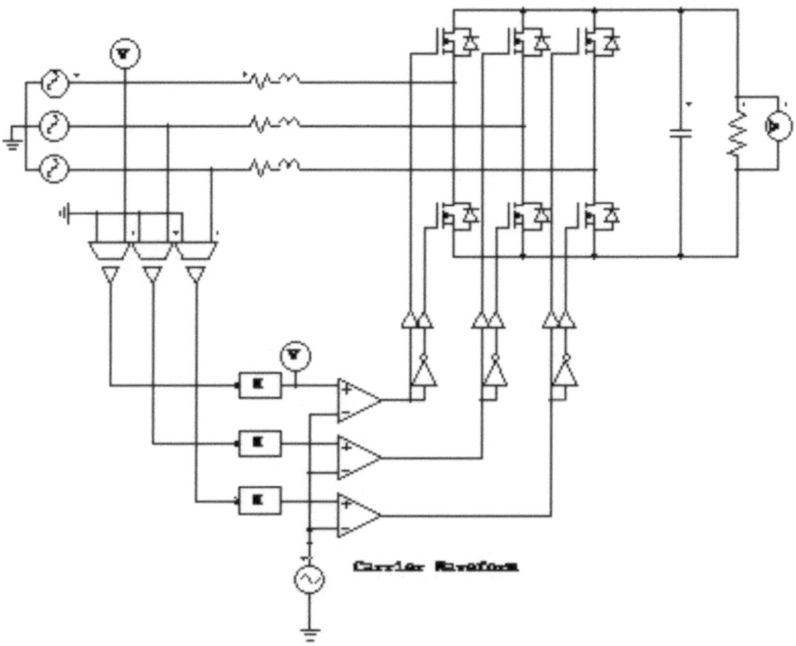


**Figure 4.** Three-phase pulse-width modulation (PWM) rectifier.

modulating the semiconductor switches that are embedded in the power converter topology [7, 32]. Such configuration that operates as an active end rectifier is shown in **Figure 4**.

HOMER Pro assumes that the inverter and rectifier capacities are not surge capacities that the device can tolerate for only short periods of time, but rather, constant capacities that the device can withstand for as long as needed [7].

A Leonics MTP41FP 25 kW 240V DC bidirectional power converter is selected for the studied hybrid AC/DC microgrid to interconnect the DC and AC bus. The capital cost of the bidirectional converter is $600/kW, replacement cost of $600, operation and maintenance cost of $0/year. The lifetime of the selected converter is 10 years and its inverter efficiency is 96% while the rectifier relative capacity is chosen to be 80% and its efficiency is 94%.

**Table 1** shows the associated costs of all the system model components. These costs are used in Section 4 to perform the system simulation in HOMER Pro software.

| System component | Capital cost per unit of rated value | Replacement cost per unit | Operating and maintenance cost per unit per year |
|---|---|---|---|
| PV panels | $3000 | $2000 | $10 |
| Wind turbine | $5000 | $5000 | $500 |
| Bidirectional power converter | $600 | $600 | $0 |
| Electrochemical battery storage system | $300 | $200 | $20 |

**Table 1.** Costs of system model components [33].

## 4. HOMER pro simulation environment

HOMER Pro software is able to evaluate off-grid or grid connected power system design, choose the best system based on cost, technical requirements, or environmental considerations. It can also simulate many design configurations under market uncertainty and evaluate risk plus the ability of the modeler to choose the best addition or retrofit for an existing system [7].

HOMER Pro software is used in this research investigation. Wind energy conversion system, solar PV system, bidirectional converter, electrochemical batteries, AC primary load, and renewable resources are modeled and analyzed. Energy resources data, system components data, typical community load profile, and system components costs are some of the important parameters that are used as inputs in the HOMER Pro model. AC and DC bus are created in the HOMER Pro model to facilitate the connection of different types of locally available resources. HOMER Pro is able to optimize feasible electric power system configurations that are technically possible [7]. Sensitivity analysis of feasible configurations is performed to give an indication of the system robustness which is reported in Section 4 that follows.

## 5. System modeling and simulation

The investigated hybrid AC/DC microgrid system model architecture is shown in **Figure 5**. The system consists of two electrical buses: AC and DC bus. To ensure improved overall system efficiency, 25% of total electrical load is the DC load that is connected to the DC bus. The wind turbine and solar energy system are connected to the AC and DC bus, respectively. In modeling the battery, HOMER Pro includes in the battery model, a charge regulator to ensure that the specified operating condition is not violated via the SOC and DOD characteristic curves. The interlinking power converter connects the two buses together by acting as a rectifier or inverter depending on the power state of the hybrid system. In the rectifier mode, it transfers power from the AC bus to the DC bus whenever there is excess power in the AC bus and there is instant demand of power from the DC bus. Alternatively, in the inverter mode of operation, power is transferred from the DC bus to the AC bus whenever there is excess power on the DC and there is instant demand of power from the AC bus. In both modes of operation, the SOC of the energy storage system is to be taken into consideration. The dispatch strategy can be set in HOMER Pro. Load-following and cycle-charging dispatch strategies are available for selection in HOMER Pro. Under the load-following strategy, generators produce only enough power to serve the load, and do not charge the energy storage device. For the cycle-charging, the excess energy available after serving the load is used to charge the battery [2, 7, 34].

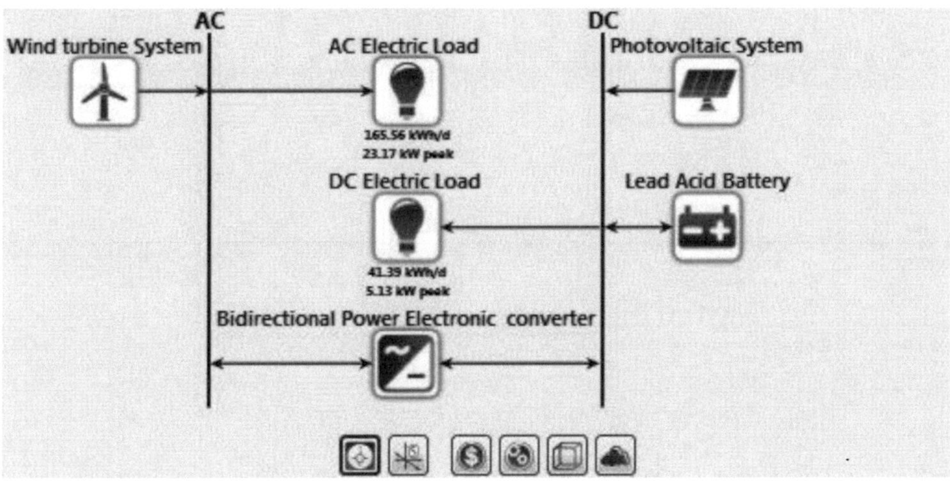


**Figure 5.** Autonomous AC/DC hybrid microgrid system.

The simulation process models the micropower system under investigation, and the optimization process determines the optimal system configuration that satisfies the modeler-specific constraints at the lowest total net present cost. The sensitivity analysis functionality of HOMER Pro allows energy planner, which energy technology combinations are optimal under different conditions [7, 9].

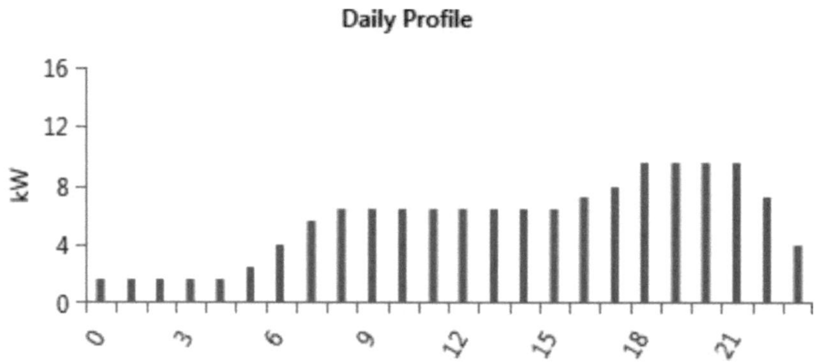


**Figure 6.** Community daily AC load profile.

**Figure 7.** Community daily DC load profile.

| Electrical loads | | |
|---|---|---|
| | **AC load** | **DC load** |
| Average consumed energy | 165.56 | 14.39 |
| Average power (kW) | 6.9 | 0.6 |
| Peak power (kW) | 23.17 | 2.67 |
| Load factor | 0.3 | 0.22 |

**Table 2.** Community load profile data.

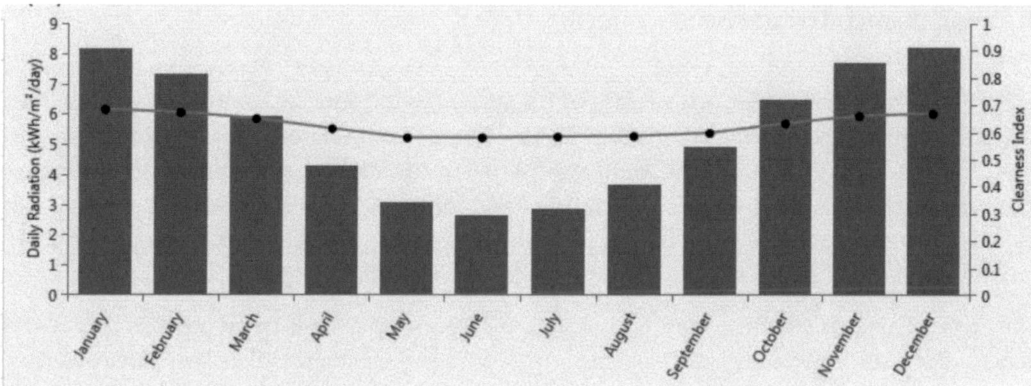


**Figure 8.** Monthly average solar global horizontal irradiance of Cape Town South Africa.

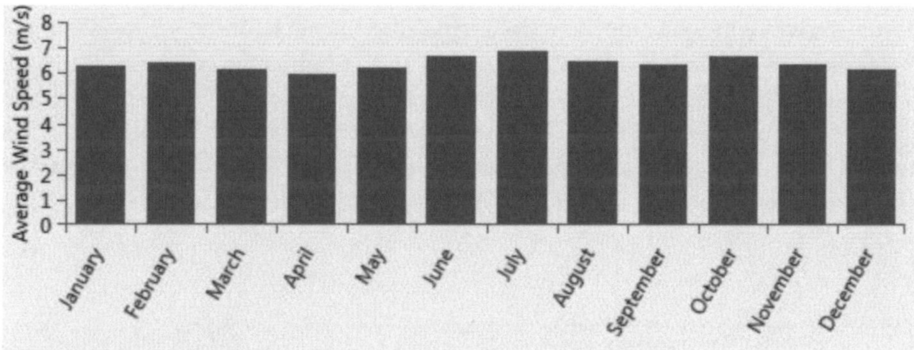


**Figure 9.** Average wind speed of Cape Town South Africa.

The AC load daily and DC load profiles are shown in **Figures 6** and **7**, respectively. The data used for this plot are shown in **Table 2**. The community load profile used in this study is the reference load date available in the HOMER Pro library [9].

The wind and solar resource data are obtained from the NASA Surface Metrological and Solar Energy database website. **Figure 8** depicts the monthly average solar global horizontal irradiance data. These monthly average data are averaged over 22-year period (July 1983–June 2005) [35]. The solar irradiance of Cape Town is low during the winter months of April, May, June, and July as shown in **Figure 8**. The wind speed data are also obtained from NASA Surface Metrological and Solar Energy database website. Wind speed at 50 m above the surface of the earth for terrain similar to airport, and monthly averaged values over 10 years period (July 1983–June 1993) are used in this work. These data are displayed in **Figure 9** (NASA Surface Meteorology and Solar Energy).

The Cape Town average wind speed as reported by the NASA database is nearly constant throughout the year, with average speed hovering around 6 m/s.

## 6. Results and discussions

The hybrid microgrid system consisting of PV system, wind turbine, bidirectional converter, lead-acid battery, and DC and AC load shown in **Figure 4** is simulated in HOMER Pro software environment. Both cycle-charging and load-following dispatched strategies are investigated. Plausible selected system component ratings are chosen for the simulation to ensure that there is enough search space for HOMER Pro to obtain an optimal system configuration. Net present cost is used as an economic metric to assess the optimal feasible configuration.

The optimal system configuration with load-following dispatch strategy consists of a 50-kW generic flat plate PV system, 20 kW wind turbine, 15 kW bidirectional power electronic converter, and 160 kWh lead-acid battery consisting of eight string. The initial capital cost of this optimal configuration is $307 000. The total present cost is $558,717, with levelized cost of energy (COE) being $0.3652, and the operational cost is of $11 363 throughout the project lifetime. The number of autonomy is 11 days. For more simulation results, see **Figure 10**.

The optimal system configuration with cycle-charging dispatch strategy consists of a 30-kW generic flat plate PV system, 20 kW wind turbine, 15 kW bidirectional power electronic converter and 140 kWh lead-acid battery consisting of seven strings. The initial capital cost of this optimal configuration is $241,000. The total present cost is $453,517, with levelized cost of energy (COE) being $0.2973, and the operation cost is of $9 593 throughout the project lifetime. The number of autonomy is 10 days. For more simulation results, see **Figure 11**.

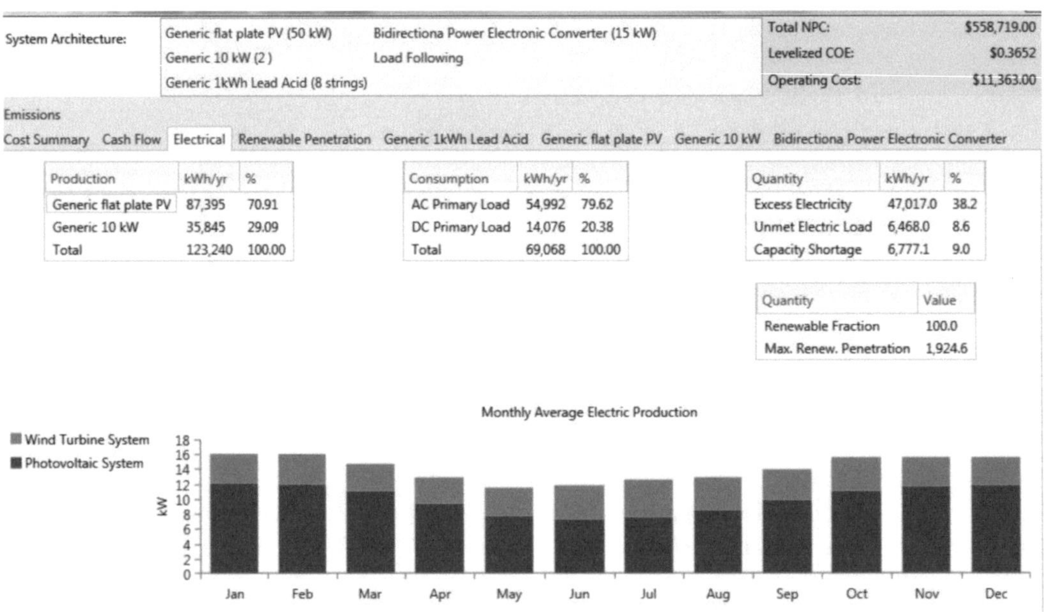


**Figure 10.** Simulation results of the optimal architecture with load-following dispatch strategy.

http://dx.doi.org/10.5772/intechopen.69582

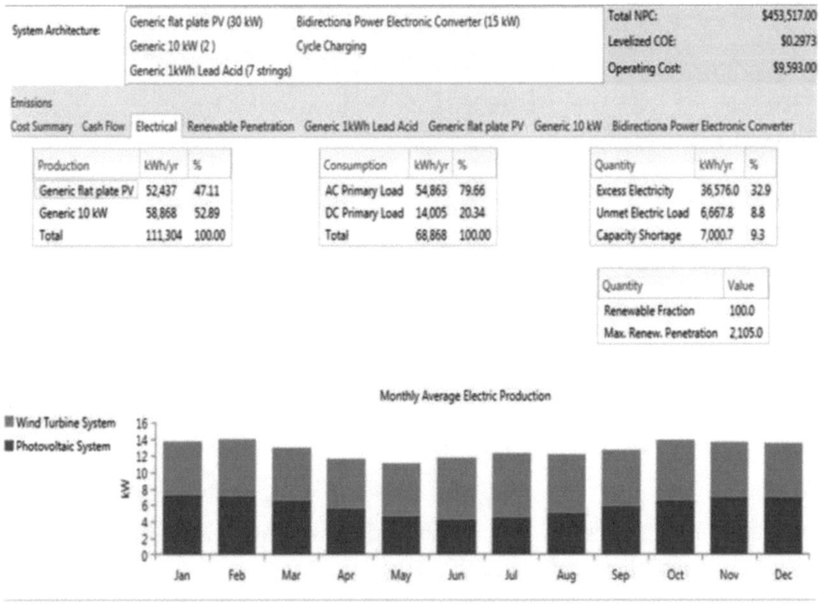


**Figure 11.** Simulation results of the optimal architecture with cycle-charging dispatch strategy.

For the same technically feasible configuration, the total net present cost and the initial capital cost of the load-following dispatch strategy are more expensive than the cycle-charging dispatch strategy. The levelized cost of energy of optimal configuration with load-following strategy is higher compared to the one with cycle-charging dispatch strategy. In the periods of extended cloudy season and low solar insolation, the configuration with load-following dispatch strategy offers a marginally extended period of autonomy.

## 7. Conclusion

The use of renewable energy as local energy resources enhances the accessibility and affordability of energy to unconnected areas around the world. Renewable energies, such as wind, solar, and biomass, have the potential to reduce the energy poverty of rural areas and sites very far from the grid. Being local resources, they also provide higher energy security to the populace.

The technical and economic analysis of the designed hybrid AC/DC micropower system studied in this work reveals the optimal electric power system configuration that is technically and economically viable using the net present cost as the econometric index. It is concluded that it is marginally better to dispatch load-following strategy for higher net present cost and autonomy compared to cycle-charging strategy dispatch. For lower levelized cost of energy (COE) and initial capital cost, it is marginally better to consider a cycle-charging dispatch strategy compared to a load-following dispatch strategy.

## Author details

Atanda K. Raji

Address all correspondence to: rajia@cput.ac.za

Centre for Distributed Power and Electronic Systems, Cape Peninsula University of Technology, Bellville Campus, Cape Town, South Africa

## References

[1] Cha J. Electrifying Sub-Saharan Africa Sustainably with Renewable Energy. SAIS Perspectives; April 2011. http://www.saisperspectives.com/2017issue/renewable-energy-africa (Accessed 8-06-2017)

[2] Raji AK, Kahn MTE. Analysis of distributed energy resources for domestic electricity users. Journal of Southern Africa. 2012;**2**(1):50-55

[3] Shaahid SM, Elhadidy MA. Opportunities for utilization of stand-alone hybrid (photovoltaic + diesel + battery) power systems in hot climates. Renewable Energy. 2003;**28**(11): 1741-1753

[4] Bokanga GM, Raji A, Kahn MTE. Design of a low voltage DC microgrid system for rural electrification in South Africa. Journal of Energy in Southern Africa. 2014;**25**(4):9-14

[5] Schmidt K, Patterson DJ. Benefits of load management applied to an optimally dimensioned wind/photovoltaic/diesel/battery hybrid power system. Proceedings of solar, Australian and New Zealand Solar Energy Society. 1997; Paper 139. 1-6

[6] Kellogg WD, Nehrir MH, Venkataramanan G, Gerez V. Generation unit sizing and cost analysis for stand-alone wind, photovoltaic and hybrid wind/PV systems. IEEE Transactions on Energy Conversion. 1998;**13**(1):70-75

[7] Farret FA, Godoy SM. Integration of Alternative Sources of Energy. New Jersey: John Wiley & Sons; 2006. p. 471

[8] Venter C, Raji A, Adonis M. The DC house for low power households-DC-DC converter analysis. In: Proceedings of Domestic Use of Energy International Conference (DUE); 30 March–1 April 2015; Cape Town, South Africa. 143-146

[9] HOMER (The Hybrid Optimization Model for Electric Renewables). http://www.nrel/gov/HOMER (Accessed 10-02-2015)

[10] Raji AK, Kahn MTE. Fuel cell systems as an efficient domestic distributed generation unit. In: Proceedings of Domestic Use of Energy International Conference (DUE); 2-4 April, 2013; Cape Town, South Africa. 91-94

[11] Borowy B, Salameh Z. Methodology for optimally sizing the combination of a battery bank and PV Array in a Wind/PV Hybrid system. IEEE Transactions on Energy Conversion. 1995;**11**(2):367-375

[12] Kusakana K, Vermaak HJ, Numbi BP. Optimal sizing of a hybrid renewable energy plant using linear programming. In: Proceedings of IEEE PES Power Africa 2012 Conference and Exposition; 09-13 July 2012; Johannesburg, South Africa. http://ieeexplore.ieee.org/stamp/stamp.jsp?arnumber=6498608 (Accessed 23-04-2016)

[13] Connoly D, Lund H, Mathiesen BV, Leahy M. A review of computer tools for analysing the integration of renewable energy into various energy systems. Applied Energy. 2010;**87**:1059-1082

[14] Bagul AD, Salameh ZM, Borowy B. Sizing of a stand-alone hybrid wind photovoltaic system using a three-event probability density approximation. Solar Energy. 1996;**56**(4)

[15] Yang HX, Lu L, Zhou W. A novel optimization sizing model for hybrid solar– wind power generation system. Solar Energy. 2007;**81**(1):76-84

[16] Mellit A, Kalogirou SA, Hontoria L, Shaari S. Artificial intelligence techniques for sizing photovoltaic systems: a review. Renewable and Sustainable Energy Reviews. 2009; **13**(2):406-419

[17] Retscreen Expert. https://www.nrcan.gc.ca/energy/software-tools/7465 (Accessed 11-02-2015)

[18] Hybrid2. http://www.umass.edu/windenergy/research/topics/tools/software/hybrid2 (Accessed 10-02-2015)

[19] Choi B-Y, Noh Y-S, HyokJi Y, Lee B-K, Won C-Y. Battery-integrated power optimizer for PV-battery. In: Proceedings of IEEE Vehicle Power and Propulsion Conference; 9-12 October 2012; Seoul, Korea. http://ieeexplore.ieee.org/stamp.jsp?arnumber=6422686 (Accessed 20-01-2017)

[20] Li C-H, Zhu X-J, Cao G-Y, Sui S, Hu M-R. Dynamic modeling and sizing optimization of stand-alone photovoltaic power systems using hybrid energy storage technology. Renewable Energy. 2009;**32**(3):815-826

[21] Tuglie ED, Torelli F. Load following control schemes for deregulated energy markets. IEEE Transaction on Power Systems. 2006;**21**(4):1691-1697

[22] Bizon N. Load-following mode control of a standalone renewable/fuel cell hybrid power source. Energy Conversion and Management. 2014;**77**:763-772

[23] Garcia P, Torreglosa JP, Fernandez LM, Jurado F. Optimal energy management system for stand-alone wind turbine/photovoltaic/hydrogen/battery hybrid system with supervisory control based on fuzzy logic. International Journal of Hydrogen Energy. 2013;**38**:14146-14158

[24] Callaway DS. Tapping the energy storage potential in electric loads to deliver load following and regulation, with application to wind energy. Energy Conversion and Management. 2009;**50**:1389-1400

[25] Subho U, Sharma MP. Development of hybrid energy system with cycle charging strategy using particle swarm optimization for a remote area in India. Renewable Energy. 2015;**77**:586-598

[26] Subho U, Sharma MP. Selection of a suitable energy management strategy for a hybrid energy system in a remote rural area of India. Energy. 2016;**94**:352-366

[27] McGowan JG, Manwell JF. Hybrid wind/PV/diesel system experiences. Renewable Energy 1999;**16**(1-4):928-933

[28] Chedid R, Saliba Y. Optimization and control of autonomous renewable energy systems. International Journal of Energy Research. 1996;**20**(7):609-624

[29] Protogeropoulos C, Brinkworth BJ, Marshall RH. Sizing and techno-economical optimization for hybrid solar photovoltaic/wind power systems with battery storage. International Journal of Energy Research. 1998;**21**(6):465-479

[30] Sissine F. Renewable energy: Background and issues. In: the 110th Congress Resources 2008. http://nationalaglawcenter.org/wp-content/uploads/assets/crs/RL343162.pdf (Accessed 15-05-2017)

[31] Javashri R, Kumudini Devi RP. Effect of tuned unified power flow controller to mitigate the rotor speed instability of fixed-speed wind turbines. Renewable Energy. 2009; **34**(3):591-596

[32] Kellogg WD, Nehrir MH. Venkataramanan G, Gerez V. Generation unit sizing and cost analysis for stand-alone wind photovoltaic and hybrid wind/PV systems. IEEE Transactions on Energy Conversion 1998;**13**(1):70-75

[33] Online Eco Store Online. Available online http://www.sustainable.co.za

[34] Raji AK. Impacts of electricity rates on the economics of rooftop PV system. International Journal of Advanced Research in Engineering. 2016;**2**(2):1-6

[35] NASA Surface Meteorology and Solar Energy. http://eosweb.larc.nasa.gov/sse/ (Accessed 10-02-2017)

Section 4

# Maintenance

Chapter 17

# Resource Planning to Service Restoration in Power Distribution Systems

Magdiel Schmitz,
Maria Clara Ferreira Almeida da Silva,
Vinícius Jacques Garcia, Daniel Bernardon,
Lynceo Favigna Braghirolli and Júlio Fonini

Additional information is available at the end of the chapter

http://dx.doi.org/10.5772/intechopen.69573

**Abstract**

Whenever there are extreme weather events, electric power distribution systems are generally affected largely because they are highly subject by their constructive nature: overhead networks. In this context, the management of maintenance actions is generally referred to as emergency service order, usually associated with a lack of supply and requiring human intervention. The key issue for the resource planning refers to an estimation of service time that allows for more assertive planning possible. This chapter proposes a predictive modelling of emergency services for resource planning when considering the geographic dispersion of such services and also the time windows that comprise the amount of service time demanded. After presenting the methodological procedures, a case study depicts the application of the proposed method in order to support proactive service routing.

**Keywords:** power distribution systems, decision-making, predictive modelling, emergency services, service restoration

## 1. Introduction

Certain regions are subject to frequent adverse weather events of sufficient intensity to cause damage to the electrical networks, requiring human intervention to verify and eliminate the causes of such disruption and thus leading to the creation of emergency service orders. These orders, with a high degree of uncertainty and urgency, along with commercial orders, with a previously defined route, become a task of great relevance for the network operation center of the electric distribution systems and qualifying even more the highly dynamic environment involved. The key issue for the resource planning, with regard to the service crews, refers to an estimation of service time that allows for more assertive planning possible, improving the crew throughput.

Considering the need to define a typical behaviour to those orders, this chapter proposes a predictive modelling of emergency services for resource planning when considering the geographic dispersion of such services and also the time windows that comprise the amount of service time demanded. The methodology to resource planning presented in this chapter is based on the predictive modelling of emergency service by defining random variables to denote how much service time is demanded in each portion of the whole geographic area considered and also how important such demands may be in the sense of reliability indexes, as number of customers and the amount of power not supplied.

Following this definition of random variables, one may obtain the service time to attend emergency services stratified by each region and also restricted to a certain time window, that is, 1-h interval. With this information, a decision-making process may be conducted to define the number of hours demanded by maintenance crews that will be needed to attend all the services in the 1-h interval. The whole picture when assuming 1 day demand will be possible by collecting all these 1-h intervals, furnishing the geographic location and the on-site service time to a possible proactive-routing approach.

In order to show how the methodology may be applied, a case study is developed considering a horizon of emergency service occurrences, from a given Brazilian Power Utility. The purpose of this case study is not only about the past but also about how to update the case study of actual occurrences in such a way that they may be considered for future estimations.

The process of conducting statistical analyses to construct the time series for each given 1-h interval is depicted in a graphic user interface also allowing the timeline description of the estimation of service time dispersed over the geographic area assumed, followed by the decision-making process of selecting the most relevant random variables to denote the service time demanded.

## 2. Service restoration by repair crews

Although the advent of smart grids has conducted to increasing the level of automation [1], the need for human intervention is particularly necessary in cases of extreme weather events or even in the occurrence of collisions in overhead networks, which become exposed by itself and may cause the lack of power supply.

This interruption refers to the switching from normal to the emergency operating condition [2], in all of the networks or even in a restricted part of it, conducting customers to the lack of power supply and affecting continuity indexes. In addition, companies still have the billing process affected by not addressing the load demand for a certain period.

The use of repair crews becomes imperative precisely to overcome the limitations that interfere with the remote control equipment, since the insulation of the defect can even be performed automatically but its correction requires human interference.

From this crew management comes the well-known emergency dispatching problem to repair crews already scheduled with pre-programmed services, which leads to a change of course and describes the problem of dynamic routing of vehicles [3]. Exactly in the context of service operations in electric power distribution utilities, the problem of this work is defined in [4].

The nature of the optimization problem has its roots on minimizing the waiting time between the detection of the defect and the arrival of crews at the service location, because the less this time the less the time without supply, thus contributing to the reliability of the electric power distribution system.

Another aspect worth mentioning is the random nature of events: the occurrence of emergency orders. In case of a reactive system, each new pending emergency order, or a set of them, gives rise to the problem of dynamic routing [5]. In addition, the routing problem associated with the context of this work also presents an important particularity: the service time in each service location, which is exactly the estimated service time for resolution. It follows that the dispatch problem is closed to the problem of minimum latency [6].

From the conclusions of [7], it can be observed that a decision that is completely reactive, responding to the disturbances generated by the creation of emergency orders, can lead to a sequence of mistaken decisions due to the loss of the 'global vision' related to the whole problem depicted after analysing all the past events over the day.

This chapter aims to anticipate the occurrence of emergency services in order to provide a proactive approach to dispatch, in which emergency occurrences can be assumed to be probable and therefore likely to be included in routing solutions of programmed orders.

## 3. Demand forecasting

The routing problem is usually addressed by assuming that attributes (customer demands, travel and service times, orders criticality) are previously known. However, most of the real cases present some level of uncertainty [8]. Anticipating these uncertainties already in planning phase is important because it allows a more precisely routing plan and favouring the crew management, operational cost reduction and also avoiding penalties from regulatory agencies through improvement of reliability indicators and the revenue from the energy sale, by reducing supply disruptions.

According to [9], the data of vehicle-routing problems can be 'Static and deterministic', 'Static and stochastic', 'Dynamic and deterministic' and 'Dynamic and stochastic'. Exactly, this last class is that related to this work, referred many times as partially dynamic, this approach assumes that some of the unknown data are in stochastic format. A proactive planning can be realized by transforming the stochastic knowledge in dummy consumers with expected service time and temporal and spatial position.

With such definitions, the next section describes how to address the uncertainty by performing demand-forecasting techniques.

### 3.1. Demand forecasting

Considering the external service orders of electric power utilities, we treat two different types of situations, the commercial and the emergency orders. The commercial ones have a typical input behaviour, known deadline and, as a consequence of these characteristics, can be planned in advance, without technical or legal problems involving power failures. By contrast, the emergency orders are randomly generated, with the corresponding dynamic character [10].

As to the attendance capacity, all the available crews have commercial and emergency demand during the workday, overlapping one each other and respecting the attendance priority. This behaviour can be seen in **Figure 1**, where there is a considerable difference between the planned route and the route performed mainly due to the occurrence of emergency orders: they have higher priority with respect to commercial orders and cause modification on the route every time they are considered.

These constant changings on planned route by knowing customers and their attributes progressively classify the problem addressed as dynamic and stochastic [7]. One possible alternative that is adopted in this work is associated with an attempt to predict the emergency occurrences and answering when and where they may occur [11].

Indispensable in the planning and in the strategies for decision-making, demand forecasting oriented towards a future period should rely on precise techniques [11]. Of the commonly used techniques to perform the forecasting, the exponential-smoothing approach can be applied to predict continuous and discrete variables.

The exponential smoothing provides the service time to the analysis horizon, using Eq. (1), according to [12]

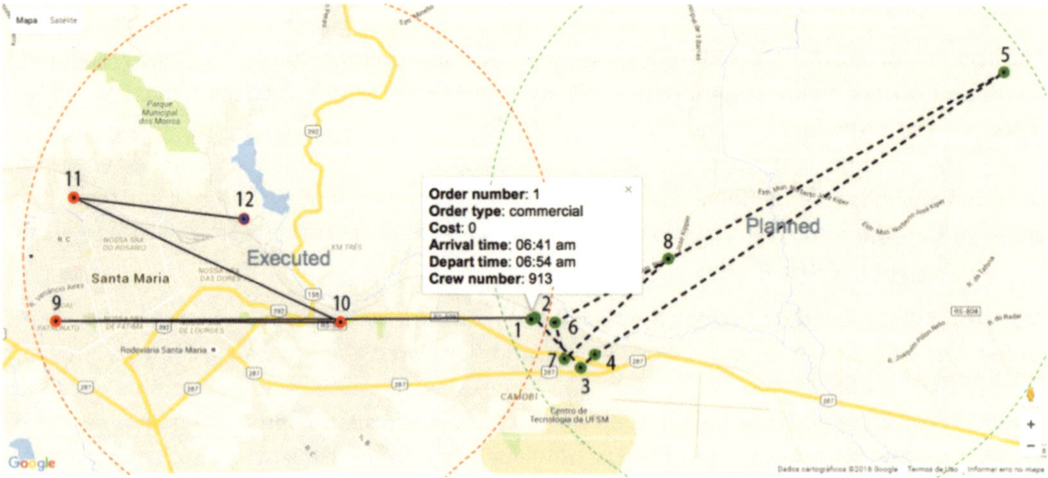


Figure 1. Routing of crews with emergency-order fulfilment.

$$F_{t+1} = \alpha x_t + (\alpha)F_t \tag{1}$$

where $t$ = actual period; $F_{t+1}$ = future forecasting; $x_t$ = demand in the period $t$; $F_t$ = forecasting to $t$; $\alpha$ $(0 < \alpha < 1)$ is a smoothing constant.

With these definitions, what one has is the theoretical basis to realize the forecasts of time of service from the available time series.

However, it is necessary to consider that only the service time forecasting is not enough to answer the questions pointed out by this work, which are the service time related to a certain part of the geographic area considered and also related to a time interval defined [13, 14]. Exactly these challenges will be illustrated in the next section, which describes the methodological procedures that are employed.

## 4. Proposed methodology

Following the contribution by Ferrucci et al. [7], inspired by the paper of Ichoua et al. [15], the approach proposed in this work also considers the classification of orders information that has already occurred, based at service time of each request, as pointed out in **Figures 2** and **3**.

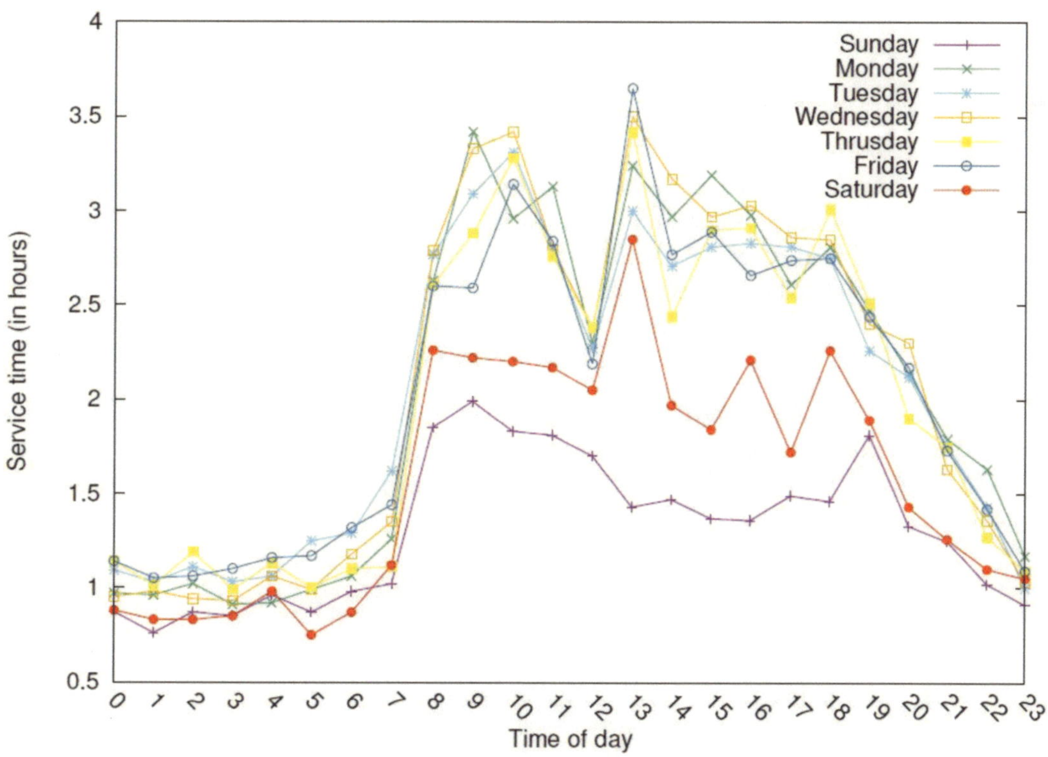


**Figure 2.** Variation analysis from the historical information. Day of the week and time of the day.

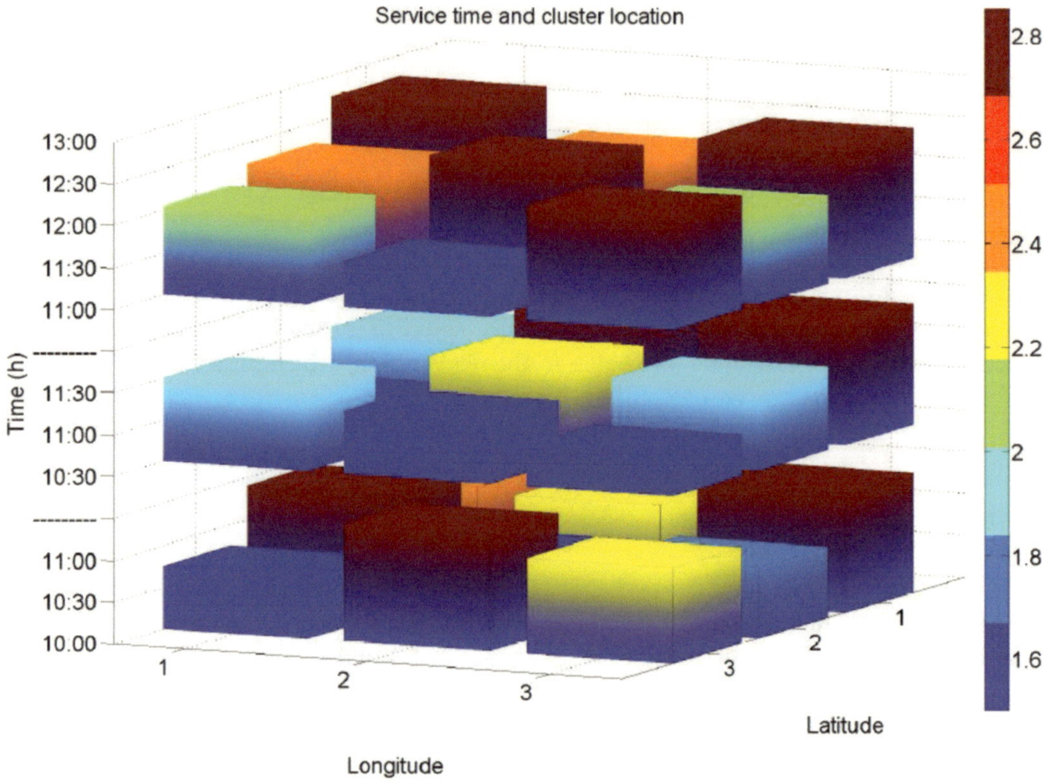


**Figure 3.** Variation analysis from the historical information. Geographical and time interval.

There is a well-defined pattern from the requested service time for each day of the week, very similar to business days unlike non-business days. In addition, **Figure 2** also shows that request demand is higher from 8 am to 8 pm. The last feature with regard to the influence on future requests is related to the geographical location: **Figure 3** depicts several layers, each one referring to one time interval, with their corresponding surface composed by cuboids; each cuboid's height refers to a level of service time required in the respective area.

The division of the geographical area and the identification of time windows to define the required service time are performed by stratifying and separating the historical data in order to identify a particular time series for each part of the geographical area and also particular for each interval of 1 day of the week.

### 4.1. Demand-forecasting methodology

A natural treatment to deal with emergency-order uncertainties would be to optimize the expected values. However, the proposed methodology approaches a restrictive model of chances [16], where each viable solution must be at an acceptable and predetermined level of failure. In the presented model (see Algorithm 1), this constraint is represented by a coefficient, corresponding to the maximum value of variation allowed. This artifice prevents exceptional disturbances from leading to distortions in the statistic analysis process.

The relationship between the scale of the geographical area considered and the coefficient of variation is performed both temporally and spatially. In order to increase the forecasting accuracy, historical data are stratified to each quadrant of the area considered for each day of the week and for each hour of the day. The methodology to proceed with this stratification is detailed in the algorithm of **Table 1**.

Beta (β) corresponds to the maximum value of variation allowed, normally 0.5, without considering those regions without emergency-order occurrences. Therefore, the referred coefficient of variation has the following limits:

$$0 < \text{coefficient of variation} \leq \beta \quad (2)$$

Instead of manually testing multiple values for the coefficient of variation, the algorithm minimizes a nonlinear function obtained from an interpolation in order to determine the best value for that coefficient. This procedure aims to minimize the size of the quadrants obtained taking into account the constraint related to the coefficient of variation.

The main contribution of this work is the stratification of the data in order to have a more accurate forecast, since the coefficient of variation is calculated for each defined quadrant.

## 5. Case study

This section describes the computational study, performed to evaluate the proposed model. The model presented in algorithm of **Table 1** was implemented in the computational tool *GNU Octave*, chosen for its efficiency features in numerical processing and matrix computation, as well as being open-source software.

### 5.1. Instance

The model was tested using an actual historical data set, provided by a power distribution utility from southern Brazil. Constituted of 9408 emergency orders in a period of 12 months, the sample is endowed with several attributes, as latitude, longitude, year, month, day, weekday, hour of day and service time, used for the analysis performed.

### 5.2. Parameters setting

*Size:* Vector representing the values referring to the side of the area analysed in the calculation. This measure is in kilometres. The higher the number of analysed values, the greater is the accuracy of the function obtained by the interpolation.

Another point to be considered is that the lower the value reported on size vector: the minimum value for the area side size may vary, depending on the study region, that is, urban or

**Algorithm 1—Geographic area optimization**

```
Input: Database reading containing the following data for each order: latitude, longitude, year, month, day, day of week, time and service time.
Function: Temporal clustering of orders per hour and day of the week.
Input: size vector = [0.4:0.2:1]; #Vector with area side size values
for size = 0.4 to 1 do
    Construction of the grid with the definition of the geographical positions for the size considered.
    Order assignment to the respective areas based on the geographical position.
    for area = 1 to n do #Where n is the total of areas
        Select a specific area of the grid.
        for day of the week = 1 to 7 do
            for hour = 1 to 24 do
                Calculation of the coefficient Of Variation for the selected area.
            end-for
        end-for
    end-for
end-for
#Thus, we have a matrix with the coefficients of variation associated with the respective area side size, for each area of the grid, day of the week and time of day.
#Elimination of data with high randomness
for day of the week = 1 até 7 do
    for hour = 1 to 24 do
        for size = 0.4 to 1 do
            for 0 < coefficient Of Variation <= beta do
                Calculation of the mean Coefficient Of Variation for all areas
            end-for
        end-for
        Determine the characteristic function by the Lagrange interpolation procedure.
        Find the minimum of the function size (mean Coefficient Of Variation) in the range: 0 <= mean Coefficient Of Variation<=beta
    end-for
end-for
```

**Table 1.** Geographic area optimization algorithm.

rural area. Therefore, the following vector sizes were chosen, measured in kilometres: sizeVector = [0.4 0.6 0.8 1].

*Beta* ($\beta$): As previously informed, it corresponds to the maximum value to the coefficient of variation allowed to ensure the lowest acceptable variance. This value was set to 0.5.

*Interpolation*: In order to obtain a polynomial function that best represents the analysed data set, the Lagrange interpolation is used. It is important to highlight that the interpolation can be made with any other function which best describes the data set. The Lagrange interpolation method can construct a polynomial of $n$ degree, which coincides with a given function in $n + 1$ points, according to the description of the method next presented by [17].

Lagrange interpolation theorem: Given $n$+1 distinct points, $z_0, z_1, \ldots, z_n$ and $n$+1 values, $w_0, w_1, \ldots, w_n$, there exists a unique polynomial $p_n(z) \in P_n$ for which

$$p_n(z_i) = w_i, i = 0, 1, \ldots n \tag{3}$$

Let us introduce the following polynomials of degree $n$:

$$l_n(z) = \frac{(z - z_0)(z - z_1)\ldots(z - z_{k-1})(z - z_{k+1})\ldots(z - z_n)}{(z_k - z_0)(z_k - z_1)\ldots(z_k - z_{k-1})(z_k - z_{k+1})\ldots(z_k - z_n)} \tag{4}$$

where $k = 0,1,2,\ldots,n$.

Now let

$$G_n = (z - z_0)(z - z_1)\ldots(z - z_n) \tag{5}$$

Then

$$G_k' = (z_k - z_0)(z_k - z_1)\ldots(z_k - z_{k-1})(z_k - z_{k+1})\ldots(z_k - z_n) \tag{6}$$

and hence it follows from Eq. (4) that

$$l_k(z) = \frac{G_n(z)}{(z - z_k)G'_n(z_k)} \tag{7}$$

Eq. (7) becomes

$$p_n(z) = \sum_{k=0}^{n} w_k \frac{G_n(z)}{(z - Z_k)G_n'(zk)} \tag{8}$$

Eq. (8) is known as the *Lagrange Interpolation Formula,* and the polynomials $l_k(z)$ are called the *interpolating* or *sampling* of the Lagrange interpolation. In this problem, the $z_n$ points are the *size* sample, and the $w_n$ values the *coefficients of variation.*

### 5.3. Results

**Table 2** presents the optimization result, with the cell values corresponding to the coefficient of variation and the area side size.

| Hour of the day | Day of the week | | | | | | |
|---|---|---|---|---|---|---|---|
| | 1 | 2 | 3 | 4 | 5 | 6 | 7 |
| 1 | (NaN,NaN) | (NaN,NaN) | (NaN,NaN) | (NaN,NaN) | (NaN,NaN) | (NaN,NaN) | (NaN,NaN) |
| 2 | (0, 0.6) | (NaN,NaN) | (NaN,NaN) | (NaN,NaN) | (0, 0.6) | (NaN,NaN) | (NaN,NaN) |
| 3 | (NaN,NaN) | (NaN,NaN) | (NaN,NaN) | (0, 0.6) | (NaN,NaN) | (NaN,NaN) | (NaN,NaN) |
| 4 | (NaN,NaN) | (NaN,NaN) | (NaN,NaN) | (NaN,NaN) | (NaN,NaN) | (NaN,NaN) | (NaN,NaN) |
| 5 | (NaN,NaN) | (NaN,NaN) | (NaN,NaN) | (NaN,NaN) | (NaN,NaN) | (NaN,NaN) | (NaN,NaN) |
| 6 | (NaN,NaN) | (NaN,NaN) | (NaN,NaN) | (NaN,NaN) | (0.5, 0.44207) | (NaN,NaN) | (NaN,NaN) |
| 7 | (NaN,NaN) | (0.5, 0.4) | (0, 0.4) | (0, 0.5) | (0, 0.66667) | (0, 0.6) | (NaN,NaN) |
| 8 | (0, 0.5) | (0, 0.6) | (0, 0.4) | (0, 0.4) | (0, 0.4) | (0.5, 0.4) | (0, 0.4) |
| 9 | (0.5, 0.4) | (0.5, 0.4) | (0, 0.4) | (0,5, 0.4) | (0, 0.4) | (0.5, 0.4) | (0, 0.4) |
| 10 | (0.5, 0.4) | (0.5, 0.4) | (0,24203, 0.4) | (0.28101, 0.4) | (0.5, 0.4) | (0, 0.4) | (0.27432, 0.4) |
| 11 | (0.5, 0.4) | (0.5, 0.4) | (0,5, 0.4) | (0, 0.4) | (0.5, 0.4) | (0.5, 0.4) | (0, 0.4) |
| 12 | (0, 0.4) | (0, 0.4) | (0.5, 0.4) | (0, 0.4) | (0, 0.4) | (0.5, 0.4) | (0, 0.4) |
| 13 | (0, 0.4) | (0.42894, 0.4) | (0.5, 0.4) | (0.42321, 0.4) | (0, 0.4) | (0, 0.4) | (0, 0.4) |
| 14 | (0.5, 0.4) | (0.5, 0.4) | (0, 0.4) | (0, 0.4) | (0, 0.4) | (0.5, 0.4) | (0.5, 0.4) |
| 15 | (0, 0.4) | (0.5, 0.4) | (0, 0.4) | (0, 0.4) | (0, 0.4) | (0, 0.4) | (0.5, 0.4) |
| 16 | (0.22726, 0.47263) | (0, 0.4) | (0.13927, 0.4) | (0, 0.4) | (0, 0.4) | (0.4, 0.4) | (0.5, 0.4) |
| 17 | (0, 0.4) | (0.10581, 0.48655) | (0, 0.4) | (0.5, 0.4) | (0.5, 0.4) | (0.5, 0.4) | (0, 0.4) |
| 18 | (0.13379, 0.4) | (0, 0.4) | (0, 0.4) | (0, 0.4) | (0.5, 0.4) | (0, 0.4) | (0.11458, 0.4) |
| 19 | (0.5, 0.4) | (0, 0.4) | (0,33325, 0.4) | (0.5, 0.4) | (0.5, 0.4) | (0, 0.4) | (0.5, 0.4) |
| 20 | (0.5, 0.4) | (0.5, 0.4) | (0,5, 0.4) | (0.5, 0.4) | (0. 0.4) | (0, 0.4) | (0, 0.4) |
| 21 | (0, 0.4) | (0, 0.4) | (0.30439, 0.4) | (0,5, 0.4) | (0.28037, 0.4) | (0.5, 0.4) | (0, 0.4) |
| 22 | (0, 0.4) | (0.5, 0.4) | (0.5, 0.4) | (0.23448, 0.4) | (0.5, 0.4) | (0, 0.4) | (0.5, 0.4) |
| 23 | (0.5, 0.4) | (0, 0.66667) | (0.41580, 0.4) | (0, 0.4) | (0, 0.4) | (0.5, 0.4) | (0, 0.4) |
| 24 | (0, 0.54451) | (0, 0.66667) | (0, 0.4) | (0, 0.4) | (0.5, 0.4) | (0.5, 0.4) | (0, 0.54451) |

Note that the NaN indicates that there was no coefficient of variation within the established limits.

**Table 2.** Optimization result (coefficient of variation, area side size).

The results presented in **Table 2** show that not all hours of the day respect the coefficient of variation constraint: one can note that periods with more incidences of emergency orders (e.g. from 19 to 23 h) presented more precise data, analytically we can see that in the period with less emergency-orders occurrence (e.g. from 1 to 6 h), the data did not present the necessary reliability.

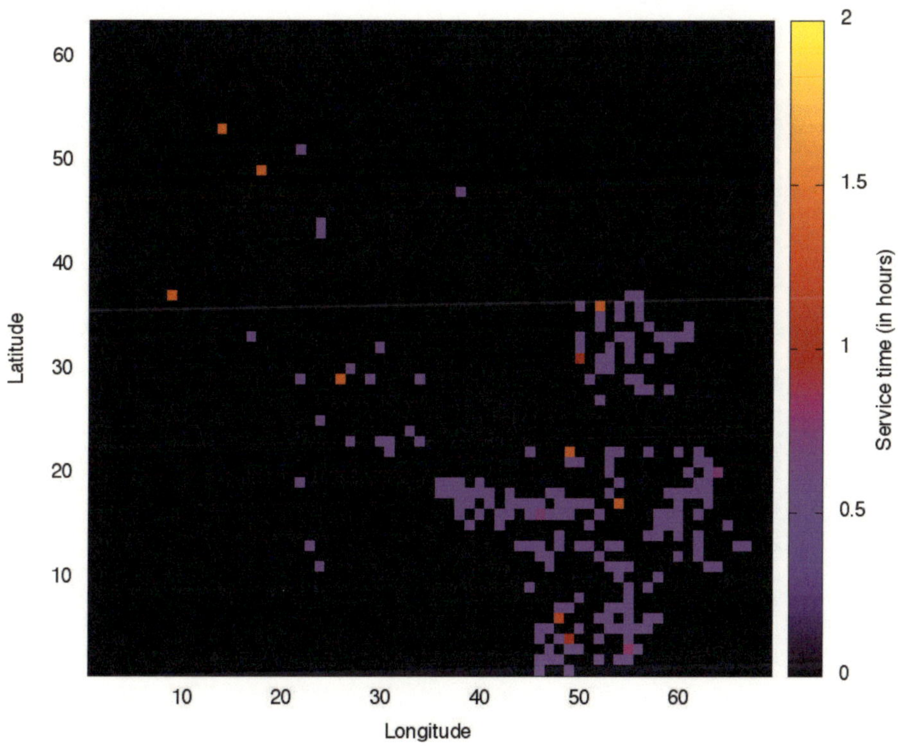


**Figure 4.** Service time history at Tuesday/19 h.

**Figure 4** presents one solution, with the service time required for emergency-order attendance in the corresponding geographical area and restrict to a given hour of the day, in this case, it is a Tuesday evening, 19 h. From **Table 2**, one notes that the side area size of each quadrant is 0.4 km, with a coefficient of variation equal to 0.33325. It is worth noting that the dispersion depicted in **Figure 4** is for only those quadrants that respect the coefficient of variation constraint.

From the results of service times of **Figures 4–8**, one may note that the more precise the stratification, and the consequent reduction in the size of the side of the area (0.4 km), the greater the dispersion of the areas with the greatest demand. Such behaviour makes it more precise to define the area with the highest demand.

**Figures 9–12** depict the number of selected areas to forecast service time, considering a Tuesday and varying the area side size from 0.4 to 1 km. An important relationship between area side size and the number of selected areas: as area side size increases, the number of areas decreases and also the accuracy to define the dummy nodes to be further used in a routing solution approach.

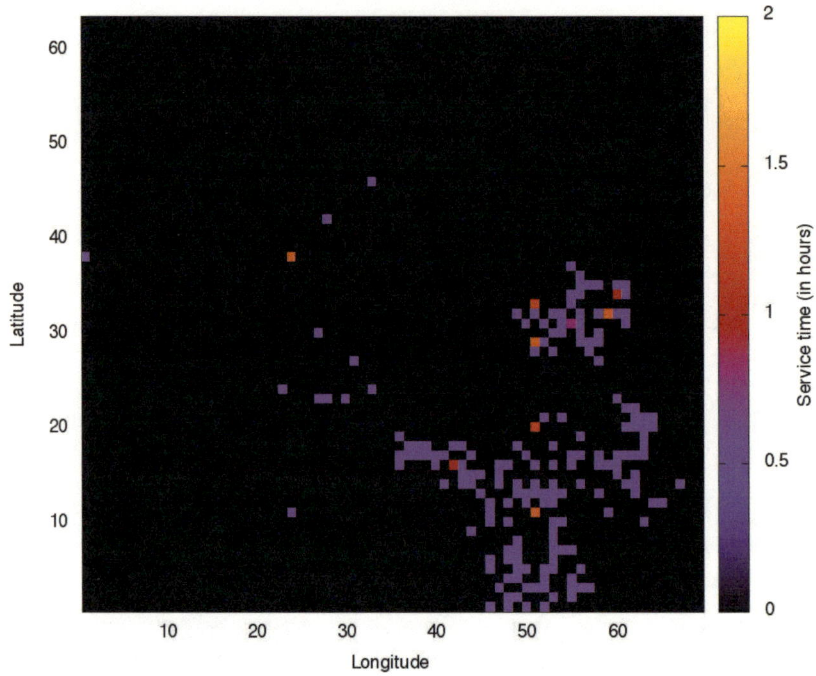


**Figure 5.** Service time history at Tuesday/20 h.

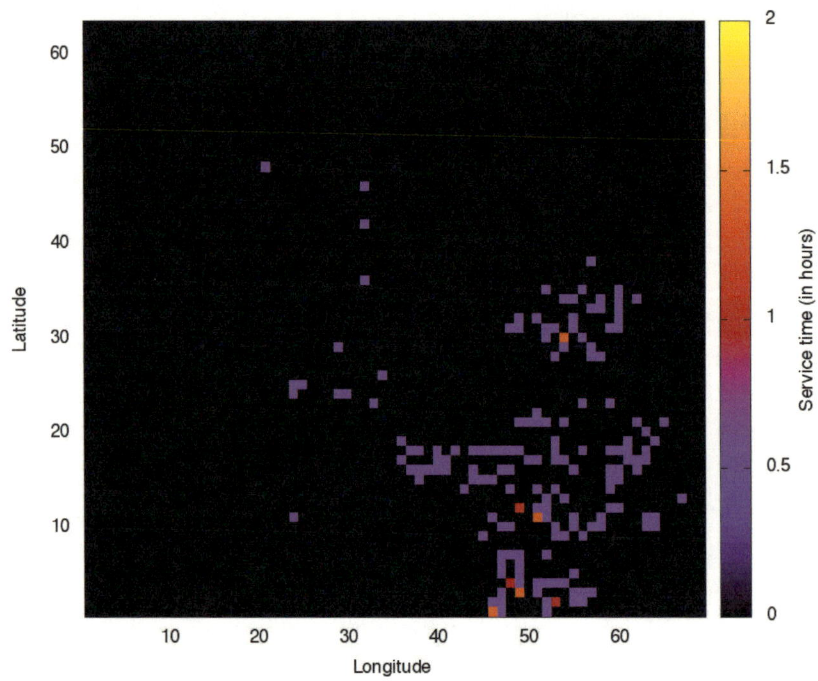


**Figure 6.** Service time history at Tuesday/21 h.

http://dx.doi.org/10.5772/intechopen.69573

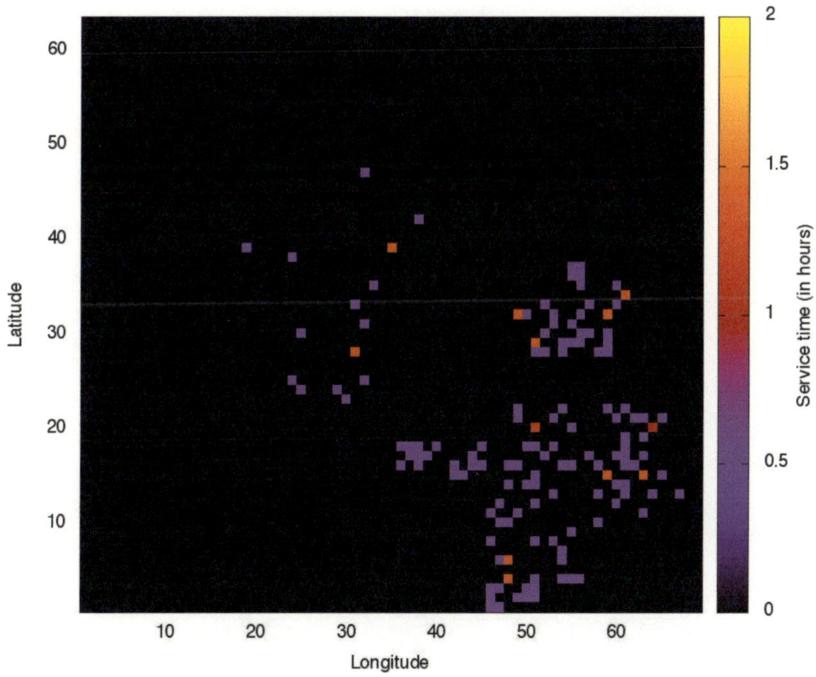


**Figure 7.** Service time history at Tuesday/22 h.

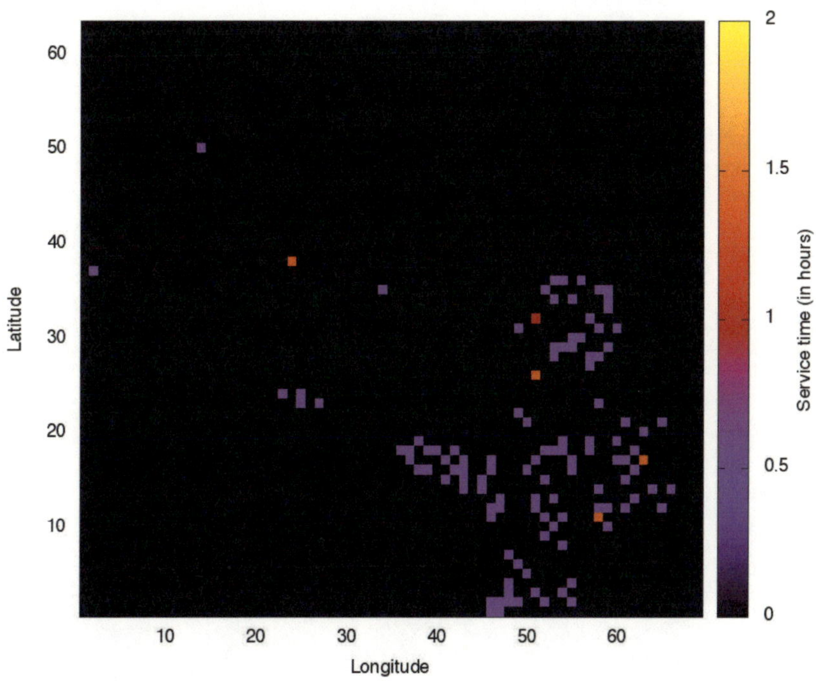


**Figure 8.** Service time history at Tuesday/23 h.

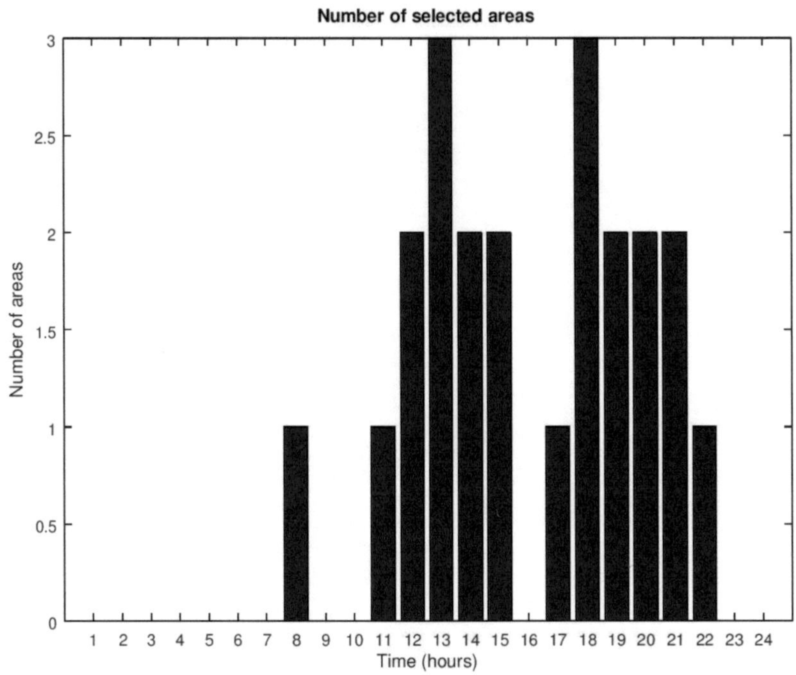


**Figure 9.** Number of selected areas at Tuesday for area side size of 0.4 km.

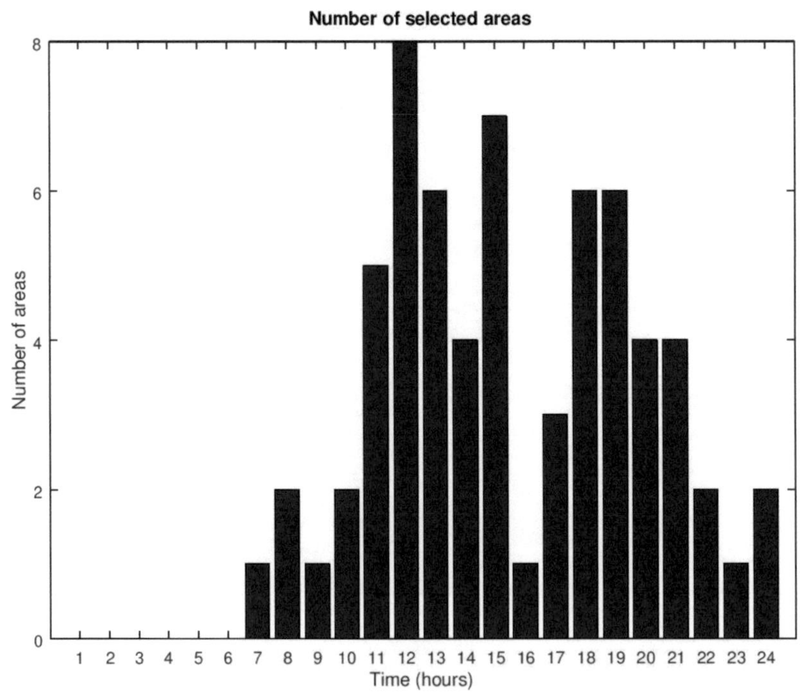


**Figure 10.** Number of selected areas at Tuesday for area side size of 0.6 km.

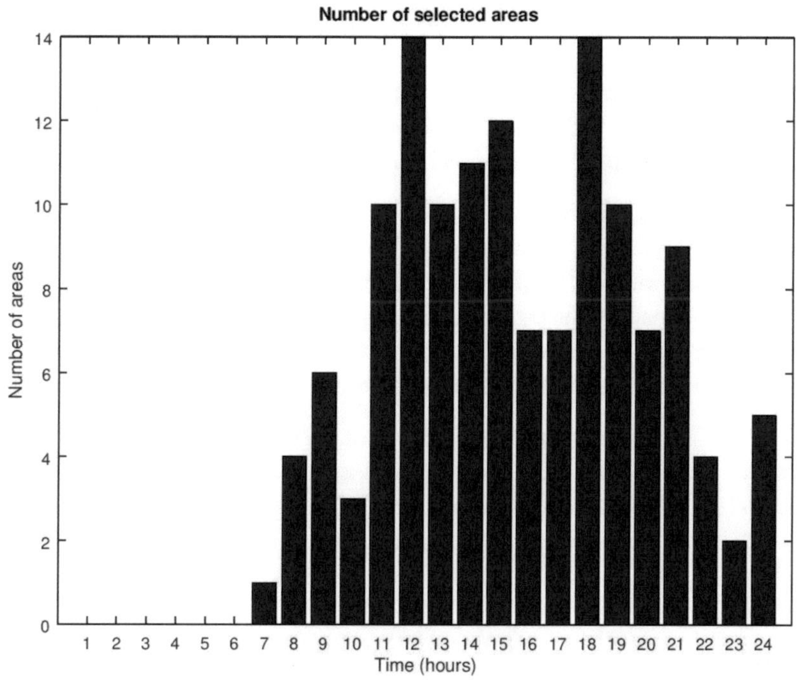


**Figure 11.** Number of selected areas at Tuesday for area side size of 0.8 km.

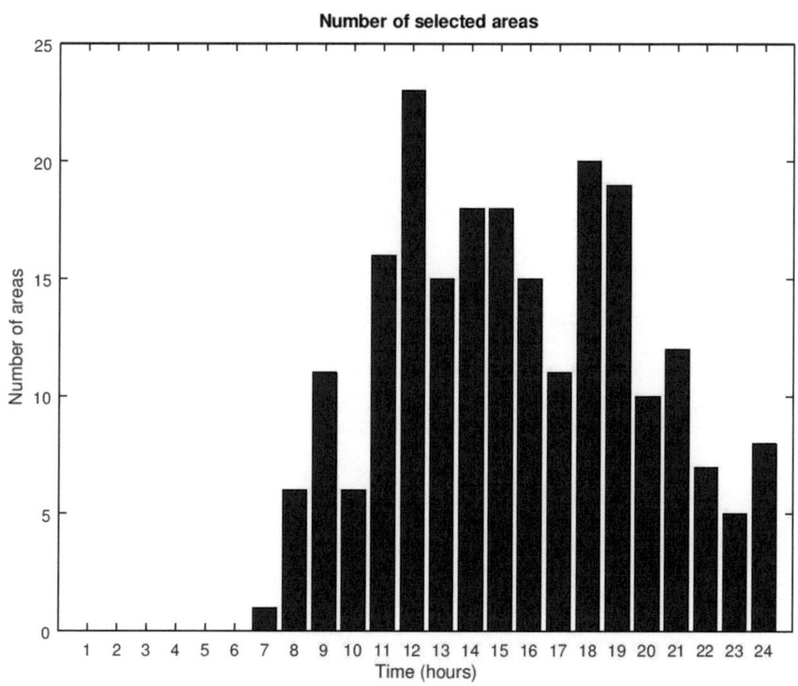


**Figure 12.** Number of selected areas at Tuesday for area side size of 1 km.

## 6. Conclusion

An inherent issue of the vehicle-routing problem is due to uncertainty of real-time planning attached to the emergency orders, where the latency of an order is minimized generally conflicting with the goal of minimizing travel time, since the pre-established route is changed.

This work has presented an approach to support proactive real-time vehicle routing, by using an algorithm to estimate service time demand related to emergency orders and considering geographical attributes and time windows. From the results obtained by this approach, one may successfully use dummy nodes obtained by the forecast method proposed to be included in a further real-time routing solution. The main purpose is to support this solution in order to minimize the waiting time and the latency, by decreasing the displacements over the route directions over the day.

Computational results indicate that the forecast of service demand can be performed with great precision when optimizing the geographical area considered, showing that the integration of derived stochastic knowledge results in a more accurate dimensioning of the service teams, and significant reductions in the discrepancy between the planned and executed routes may be obtained.

## Acknowledgements

The authors of this chapter would like to thank the Power Distribution Company RGE Sul by the technical and financial support on the Research and Development project called 'Dynamic operations planning', linked to the National Electric Energy Agency(ANEEL).

## Author details

Magdiel Schmitz[1], Maria Clara Ferreira Almeida da Silva[1], Vinícius Jacques Garcia[1]*, Daniel Bernardon[1], Lynceo Favigna Braghirolli[1] and Júlio Fonini[2]

*Address all correspondence to: viniciusjg@gmail.com

1 Federal University of Santa Maria, Rio Grande do Sul, Brazil

2 RGE Sul Power Utility, Rio Grande do Sul, Brazil

## References

[1] Arif A, Wang Z, Wang J, Chen C. Power distribution system outage management with co-optimization of repairs, reconfiguration, and DG dispatch. In: Transaction on Smart Grid, editor. IEEE; 2017. p. 1

[2] Murphy L, Wu F. Comprehensive analysis of distribution automation systems. Technical Report M90/72. Berkeley: University of California; 1990

[3] Psaraftis HN. In: Golden BL and Assad AA. (Eds). Vehicle routing: Methods and studies. Dynamic Vehicle Routing Problems. North Holland, Amsterdam, The Netherlands: Dalctraf. 1988. pp. 223–248

[4] Garcia VJ, Bernardon DP, Abaide AR, Bassi OA, Dhein G. Reliability assessment by coordinating maintenance vehicles in electric power distribution systems. Procedia-Social and Behavioral Sciences. 2014;**111**:1045–1053

[5] Psaraftis HN. Dynamic vehicle routing: Status and prospects. Annals of Operations Research. 1995;**61**:143–164

[6] Angel-Bello F, Alvarez A, García I. Two improved formulations for the minimum latency problem. Applied Mathematical Modelling. 2013;**37**(4):2257–2266

[7] Ferrucci F, Bock S, Gendreau M. A pro-active real-time control approach for dynamic vehicle routing problems dealing with the delivery of urgent goods. European Journal of Operational Research. 2013;**225**(1):130–141

[8] Gounaris C, et al. An adaptive memory programming framework for the robust capacitated vehicle routing problem. Transportation Science. 2014

[9] Toth P, Vigo D. Vehicle Routing Problems, Methods, and Applications. DEI, University of Bologna. Bologna Italy. 2nd ed. SIAM; 2014

[10] Taha HA. Operations Research: An Introduction. Pearson/Prentice Hall; New Jersey. 2007. p. 557

[11] Makridakis S. METAFORECASTING: Ways of improving forecasting accuracy and usefulness. International Journal of Forecasting. 1988;**4**:467–491

[12] Hillier FS, Lieberman GJ. Introduction to Operations Research: Cases Developed. New York, NY: McGrawHill; 2001. p. 6

[13] Guimarães I, Garcia VJ, Bernardon DP, Foninni J. Emergency prediction in electric utilities: A case study from South Brazil. In: European Conference on Modelling and Simulation; Regensburg, Germany; 2016. p. 30

[14] Garcia VJ, et al. Emergency work orders in electric distribution utilities: From business process definition to quantitative improvements. In: 47th International Universities' Power Engineering Conference; 2012

[15] Ichoua S, Gendreau M, Potvin J. Exploiting knowledge about future demands for real-time vehicle dispatching. Transportation Science. 2006;**40**(2):211–225

[16] Braaten S, et al. Heuristics for the robust vehicle routing problem with time windows. Expert Systems with Applications. 2017;**77**:136–147

[17] Zayed AI, Butzer PL. Chapter 3—Lagrange interpolation and sampling theorems. In: Marvasti FA, editor. Nonuniform Sampling: Theory and Practice. Springer Science + Business Media, LCC; Kluwer Academic/Plenum Publishers, New York. 2001. pp. 123–127

# Chapter 18

# Imperfect Maintenance Models, from Theory to Practice

Filippo De Carlo and Maria Antonietta Arleo

Additional information is available at the end of the chapter

http://dx.doi.org/10.5772/intechopen.69286

**Abstract**

The role of maintenance in the industrial environment changed a lot in recent years, and today, it is a key function for long-term profitability in an organization. Many contributions were recently written by researchers on this topic. A lot of models were proposed to optimize maintenance activities while ensuring availability and high-quality requirements. In addition to the well-known classification of maintenance activities—preventive and corrective—in the last decades, a new classification emerged in the literature regarding the degree of system restoration after maintenance actions. Among them, the imperfect maintenance is one of the most studied maintenance types: it is defined as an action after which the system lies in a state somewhere between an "as good as new" state and its pre-maintenance condition "as bad as old." Most of the industrial companies usually operate with imperfect maintenance actions, even if the awareness in actual industrial context is limited. On the practical definition side, in particular, there are some real situations of imperfect maintenance: three main specific cases were identified, both from literature analysis and from experience. Considering these three implementations of imperfect maintenance actions and the main models proposed in the literature, we illustrate how to identify the most suitable model for each real case.

**Keywords:** imperfect maintenance, optimization model, reliability

## 1. Introduction

Maintenance is defined as "the combination of all technical, administrative and managerial actions during the lifecycle of an item intended to retain it in or restore it to, a state in which it can perform the required function" [1]. Maintenance is everywhere, when there are systems, machines, elements that we use every day, requiring specific actions for functioning correctly, since degradations and failures reduce the effectiveness in their use. The industrial sector is perhaps one of the most interested in maintenance actions since companies need to guarantee their required productivity targets. Despite this, maintenance has not always had the attention it deserved and for many years it was a "Cinderella function" [2], but recently its role was revalued, and since the 1960s, it was considered as a specific organizational unit [3]. According to Sherwin [2], the reasons why maintenance was a "Cinderella function" for so many years are mostly historical and can be overcome by new information technology (IT). IT, in fact, simplifies data acquisition and analysis of systems requiring maintenance activities, whereas integrated IT allows mathematical optimization of many aspects related to maintenance, such as costs and availability [2].

One of the strategies to adopt in order to increase awareness of maintenance importance in industrial and productive contexts is to prove its effectiveness [4]. It is essential to measure maintenance performance, for the justification of investments in this function [5] and for strategic thinking for asset managers. However, in the existing literature, only the internal efficiency is measured even if the maintenance contribution toward a total business goal should be measured (both external effectiveness and internal efficiency). Life-cycle profit (LCP) could be a fair measure of overall effectiveness for highlighting both value and cost of maintenance [2]. Maintenance, in fact, consists of so many activities that it is difficult to quantify its benefits at an individual activity level, whereas at a macro level, it is difficult to find the best trade-off between costs and benefits for the company profit [6].

Despite all that, today it is accepted that maintenance is a key function for long-term profitability in an organization [7], and for this reason, organizations are treating maintenance as an important part of their business [8], in the same way as other functions like production, marketing, sales and so on. Effective maintenance management, in fact, requires a multidisciplinary approach where maintenance is viewed strategically from the overall business perspective [9]. Companies invest in early reliability estimation also in order to setup their on-field service [10].

The maintenance role evolution in industrial context could be summarized through **Figure 1**, proposed by Furlanetto and Mastriforti [11].

Corrective maintenance (CM) is certainly the approach with the lowest prevention and engineering contributions since it simply reacts to an occurred failure. Moving from CM to prevention maintenance, it goes from a defeatist and passive approach to a more "aggressive" one, where engineering principles and the aim of preventing future failures are the most important aspects in maintenance management.

Therefore, maintenance engineering is nowadays a very important function in industrial context. It is usually defined as "a staff function whose prime responsibility is to ensure that maintenance techniques are effective, equipment is designed and modified to improve maintainability, on-going maintenance technical problems are investigated, and appropriate corrective and improvement actions are taken" [12]. It is a very strategic resource essential for ensuring production capacity, product quality and best lifecycle cost. It is related to Business Continuity and to High-Reliability Organization (HRO) too: the choice of a specific maintenance policy, in fact, defines a more or less compliant maintenance approach to the HRO paradigm [13].

http://dx.doi.org/10.5772/intechopen.69286

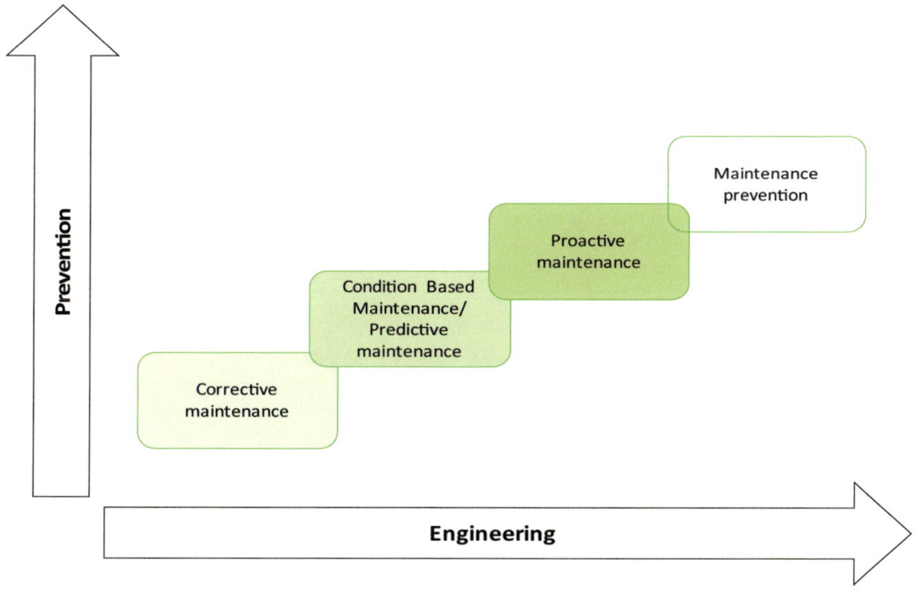


**Figure 1.** From failure to prevention [11].

Maintenance engineering is one of the three main maintenance organizational processes, as shown in **Figure 2**, adapted from Furlanetto et al. [16].

Maintenance has a unique role also in the latest 4th Generation Industrial Revolution (Industry 4.0), which focuses on intelligent products and production processes [14]. The new digital industrial technology defines a transformation where sensors, machines, workpieces and Information Systems are connected along the value chain: internet allows communication between humans and machines in cyber-physical systems (CPS). The interaction with surrounding systems is able to create self-aware and self-learning machines condition, with consequent improvement of overall performance and maintenance management [15]. It is clear that maintenance processes must be integrated in an effective way in the Industry 4.0 framework. Digitalization and networking, in particular, shall provide an increasing amount of actor and sensor data which can support the continuous monitoring of the process and of the machines conditions. Data, in fact, can be recorded and transmitted in real time to the cloud for predictive maintenance analysis.

Maintenance engineering is strictly connected to maintenance management and maintenance implementation. In fact, the maintenance implementation process gets orders from maintenance management, to which refer on their progress. Maintenance management sends information to maintenance engineering function for reliability and maintenance analysis. Then, maintenance engineering, through tools and specific software, defines maintenance plans to be transferred to maintenance management function and, finally, to maintenance implementation [16].

It is well recognized, however, that in many asset-intensive industries, maintenance costs are an important part of the operational cost [5], both for preventive and for corrective maintenance activities. Even if corrective and preventive maintenance is the most known maintenance

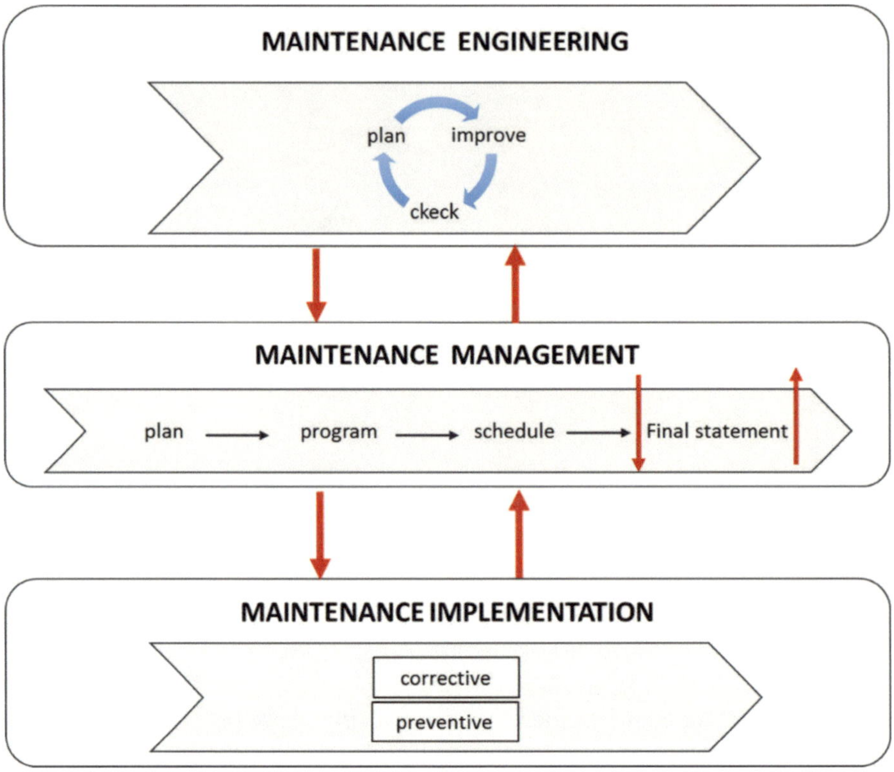


**Figure 2.** Role of maintenance engineering in maintenance organization process [16].

types, in the last decades, new kinds of maintenance were identified in the literature, as better explained in the following sections.

### 1.1. Maintenance and reliability

Reliability and maintenance are strictly connected: reliability models, in particular, define the main reliability properties of a system, which are essential for maintenance management.

Here, we show a brief summary of the most important reliability considerations useful in maintenance theory.

First of all, reliability is defined as the probability that a component (or an entire system) will perform its function for a specified period of time, when operating in its design environment [17]: reliability definition, therefore, requires an unambiguous criterion for distinguishing operation from non-functioning states and the exact definition of environmental conditions. In this way, reliability will depend only on time.

The most significant reliability parameters are:

- $f(t)$: probability density function (pdf)
- $F(t)$: cumulative distribution function (cdf), also known as failure function or unreliability function

- $R(t)$: reliability function
- $\lambda(t)$: hazard function, defined as the trend of the instantaneous failure rate at time $t$ of an element that has survived up to that time t.

Hazard function is usually presented as a time depending curve, called the "bathtub" curve, which is useful for determining three main failure types corresponding to three different life stages of a component. The life of a device has a first part with a failure rate relatively high and a descending shape (decreasing failure rate DFR) because of the so-called early failures depending on potential manufacturing defects (design or installation defects, for example); the central part of its life corresponds to the useful life, in which it is usually assumed that the failure rate does not change over time (constant failure rate CFR); finally, the end of life corresponds to the wear-out phase with an increasing failure rate (IFR).

The reliability properties of a component might be represented with a specific probability distribution model in each one of the main phases (DFR, CFR, IFR): the early stage is modelled through the Weibull distribution, the random failure stage is defined by a negative exponential distribution and finally, for the wear-out failures, the normal distribution is the most suitable one, since its failure rate is always a monotonically increasing function of time (so the normal distribution is an IFR distribution) [18]. Obviously, the most suitable distribution has to be selected according to the real failure rate dependency on time.

The knowledge of the reliability features of a system is the starting point for the definition of the maintenance and replacement models to apply. Many models are shown in the literature [19]: all of them can fall into some categories, such as the age replacement policy, the block replacement policy, the periodic preventive maintenance policy, the failure limit policy, the opportunistic maintenance policy, the inspection policy [20], etc.

Reliability parameters are used in maintenance optimization models, too. For a long time, in fact, the cost was the only parameter considered for their optimization, in line with the old maintenance concept (maintenance as a necessary evil and the only maintenance is corrective). Today, however, both costs and reliability measures (such as availability) are present in maintenance optimization models. In literature, there are many contributions on maintenance optimization models, and the review papers on this topic are really interesting [21, 22]. Some models, for example, try to optimize maintenance activities according to risk management [23]: the main aim is the minimization of the effects of failures on the organization's main objectives.

Maintenance models could be quite difficult to apply because of lack of data on maintenance actions and faults [24]. At the same time, cost data could be harder to have, especially with respect to indirect costs since, for example, it could be very difficult to quantify intangible aspects like the benefits of maintenance.

Furthermore, models are usually complex to apply since several constraints and several objectives usually affect the optimization of maintenance policies [22]: simulation approach could be a viable alternative to the analytical one.

In conclusion, we can say that maintenance optimization models could be very useful in a maintenance management process because they allow to consider all the main maintenance objectives (ensuring system function, ensuring system life, ensuring safety, ensuring human wee-being) [6]. At the same time, the input data must be corrected in accordance with the facts. The best suitable model could be identified among the several models proposed by literature.

## 2. The "new" maintenance classification and imperfect maintenance

Imperfect maintenance (I.M.) is a new kind of maintenance approach, which has spread in the last decades, as an alternative to the common classification of maintenance, proposed for example by the EN 13306:2010 [1]. This last one considers two main classes of maintenance approaches: corrective maintenance (CM) and preventive maintenance (PM). The distinguishing element in this classification is the time in which maintenance activity is performed and in particular the relationship between the occurrence of a fault and the maintenance activity. PM is an action done before the occurrence of the fault, while each maintenance action carried out after the fault is called CM. A CM action is performed at unpredictable time points (it is not known the failure time of a component), and its main function is to put the item into a state in which it can perform the required function. If CM is carried out immediately after the occurrence of the fault, we speak of *immediate* maintenance, while if it is postponed (scheduled with other maintenance actions or with down production period), we speak of *deferred* maintenance. Since CM costs are three or four times higher than the PM costs [25], it could make sense to delay CM activities waiting for a situation in which the time of repair has a negligible impact on the unavailability. On the other side, PM activities could be carried out at predetermined intervals of time or number of units of use (*predetermined maintenance*) or according to information on system degradation supplied by condition monitoring, inspection and test activities (*Condition Based Maintenance*).

In addition to this well-known classification, in the last decades, a new categorization emerged in the literature. It dates back to 70s and is based on the item restoration degree after maintenance actions. In Wang and Pham [34], an example of this classification is proposed. Both the "classic" and "new" classifications are summarized in **Figure 3**: it shows that both corrective and preventive maintenance could be:

- Perfect maintenance: Maintenance action that restores a system to an "As Good As New" (AGAN) condition. Considering the main parameters defining the reliability features of the system, we could say that after a perfect maintenance action, a system has the same lifetime distribution and the same failure rate function of a new one. For this reason, generally, the replacement of a system by a new one is a perfect repair.
- Imperfect maintenance: Maintenance action which makes a system not AGAN but younger: upon an imperfect maintenance action, the system lies in a state somewhere between AGAN and its pre-maintenance condition.
- Minimal maintenance: It restores a system to an "As Bad As Old" (ABAO) condition and, therefore, to the same failure rate as before the maintenance action. First of all, Barlow and Hunter [26] studied minimal repair proposing two preventive maintenance policies.

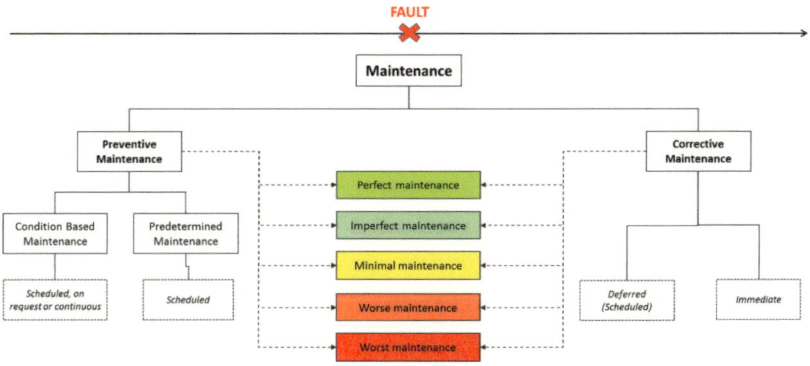


**Figure 3.** Classification of maintenance activities according to the time in which maintenance activity is performed (preventive maintenance vs. corrective maintenance) and to the system restoration degree (perfect, imperfect, minimal, worse, worst maintenance).

- Worse maintenance: It defines actions accidentally causing a worsening operating condition of the system. The system failure rate or the actual age increases after a worse maintenance action.
- Worst maintenance: It is similar to worse maintenance but, in addition, it causes fault or breaks of the system too.

This classification suggests that each "new" kind of maintenance action could be considered worse than the previous one. For example, imperfect maintenance causes a restoration degree lower than perfect maintenance but higher than minimal maintenance and so on.

The new maintenance activities classification is an accurate reflection of the reality since the maintained components are not always restored to an as good as new state [27]. At the same time, it has a huge impact on the reliability and maintenance theory, since maintenance has always been perfect or minimal, and therefore, reliability and optimization maintenance models were defined for these two extreme situations. The introduction of new maintenance ideas required new models, proposed in the literature in the last years.

In literature, there are many contributions on imperfect maintenance, and, at the same time, worse and worst maintenance were studied, for example, by Kay [28], Nakagawa and Yasui [29] and Chan and Downs [30], which considered a kind of preventive maintenance causing the failure of the maintained equipment.

Among the maintenance types considering the item restoration degree after maintenance actions, this chapter focuses on imperfect maintenance (IM).

### 2.1. Imperfect maintenance literature review

The first interest in imperfect maintenance dates back to the second part of 1970s thanks to researchers like Kay [28], Ingle and Siewiorek [31], Chaudhuri and Sahu [32], Chan and Downs [30] and Nakagawa [33]. In particular:

- Kay [28] dealt with the effectiveness of preventive maintenance concept [28].
- Ingle and Siewiorek [31] presented the imperfect recovery concept for multiprocessor systems: they suggested to use a factor C called coverage, that is, the condition probability that a system recovers successfully after a failure [31].
- Chaudhuri and Sahu [32] suggested a model to define the optimum preventive maintenance intervals both for perfect and imperfect PM [32].
- Chan and Downs [30] presented two criteria for preventive maintenance analysis (the maximization of the steady-state availability and the minimization of the expected maintenance cost per unit time under specific assumptions). The authors suggested a state transition diagram to represent preventive maintenance, considering a probability $p$ of not restoring the component in an AGAN condition but in a failed state (worst preventive maintenance) [30].
- Nakagawa is one of the most interested researchers for this topic. He supposes that preventive maintenance is imperfect; then, he considers this kind of maintenance for defining optimum preventive maintenance policies [33].

These are only some examples, and many other researchers showed interest in imperfect maintenance topic.

Even if many works on imperfect maintenance regard the one-unit system [34], this concept could be also applied the multi-components system, which is the most common configuration in real problems.

Researchers show interest in imperfect maintenance even now, as highlighted by the number of publications of the more recent years (a systematic literature review was conducted on the web search engines Scopus, IEEE Xplore, Google Scholar and Web of Science). Analysing the last contributions, it is possible to underline a greater interest in IMpractical implications. Some more recently published papers were analysed in order to understand which are the more recent trends and interest in IM topic. Some of them are here shown:

- Sanchez et al. [35] consider the problem of testing and maintenance activities optimization with uncertainty in the imperfect maintenance modelling. The proposed methodology is applied to a stand by safety-related system of a nuclear power plant. It shows the importance of considering uncertainties in the modelling of imperfect maintenance since it impacts on system unavailability and cost [35].
- Mabrouk et al. [36] propose a model to determine the optimum PM scheduling strategy for a leased equipment, considering that both PM and CM are imperfect. Since a leasing agreement is considered, some penalty costs are in the model (when the total expected equipment downtime due to maintenance activities in the lease period are greater than a pre-specified value) [36].
- Pandey et al. [37] propose a mathematical model for decision-making on selective maintenance actions under imperfect repair, for binary systems (i.e. they are either working or failed). Since it is usually difficult to do all the required maintenance actions during the maintenance break, the focus is both on the optimal use of the available resources (budget, repairman and time) and on the maximization of the next mission reliability [37].

- Le and Tan [38] studied the optimal maintenance strategy of systems subject to a degradation condition, subject to imperfect maintenance. They suggest a combined approach including both inspection and continuous monitoring activities to improve system reliability [38].

IM is increasingly considered in maintenance optimization problem: researchers are more and more aware that this "new" kind of maintenance must be considered both in theoretical and in practical problems.

### 2.2. Imperfect maintenance models proposed in literature

As shown in Section 1.1, the definition of an optimal maintenance policy for a system requires that many factors have to be considered, like the maintenance policies, the system architectures (single-unit, series, parallel, k-out-of-n, complex), the shut-off rules for series systems (which is the dependence of a failed component on the others), the maintenance degree (perfect, imperfect, minimal, worse, worst) and the maintenance costs. It is clear that both reliability parameters and costs are essential to optimize maintenance activities management.

The first step, however, is the definition of the reliability parameters of the system after an imperfect maintenance action, since the system itself will be not AGAN.

In literature, many different models were proposed: most of them are summarized in Ref. [39]. In the following paragraphs, the main categories will be explained.

#### 2.2.1. *(p, q) rule method and its variants*

These methods use a probability parameter to model imperfect maintenance. According to the $(p, q)$ rule method, a component, after a maintenance activity, could return to the AGAN state with probability $p$ and to the ABAO state with probability $q = 1 - p$. It is clear that if $p = 1$ maintenance is perfect, whereas if $p = 0$ maintenance is minimal. This method was proposed by Nakagawa at the end of 1970s [33] and then resumed by other authors: Brown and Proschan [40], for example, under specific assumptions, obtained important results. They considered an item repaired each time it fails (with negligible repair time), in a perfect way with probability $p$ or in a minimal way with probability $1 - p$. It is assumed that no perfect repairs occur in $[0, t)$ so that the item, at $t$, behaves as an item of age $t$, with failure rate $\lambda(t)$. Let $F$ be the life time distribution of the item and $\lambda$ its failure rate. Under these assumptions, they proved that the time distribution function of the time between successive perfect maintenance actions is

$$F_p(t) = 1 - (1 - F(t))^p \tag{1}$$

and the corresponding failure rate is $\lambda_p(t) = p\lambda(t)$.

Furthermore, they proved that, since the original failure rate function is simply multiplied by $p$, then also $F_p$ has this property for $0 < p < 1$ [40].

This model was subsequently changed, and other methods were obtained:

- $[p(t),q(t)]$ rule method: It was proposed by Block et al. [41] and differs from the previous method only by time dependence: considering a one-unit system subjected to corrective maintenance activities with negligible repair time, the item is restored to a AGAN state with probability $p(t)$ while to an ABAO state with probability $q(t) = 1-p(t)$, where $t$ is the age of the item (the time since last perfect maintenance) [41].

- $[p(n, t), q(n, t), s(n, t)]$ rule method: This model considers another parameter affecting on the effectiveness of maintenance activities, that is $n$, the number of failures since replacement. Therefore, according to this model, after a repair, a system will be in an AGAN state with probability $p(n, t)$, or in an ABAO state with probability $q(n, t) = 1-p(n, t)$. A third possibility is considered: the repair could be unsuccessful with probability $s(n, t) = 1 - p(n, t) - q(n, t)$ [42].

### 2.2.2. *Improvement factor method*

The improvement factor method considers that an imperfect maintenance PM activity can reduce the age of a system from $t$ to $t/\beta$, with a new reliability of $R(t/\beta)$.

This method was proposed by Malik [43] in order to consider that the failure rate curve changes after preventive maintenance activities (in particular, the failure rate after PM lies between AGAN and ABAO) according to a parameter called improvement factor $\beta$ [43]. In **Figure 4**, there is a representation of failure rate when minimal, perfect and imperfect repair are performed.

Over time, many improvement factors were proposed, since the restoration effect depends on several factors like the operating time of the system, the PM interval, the related costs, the number of PM carried out and so on. Lie and Chun [44], for example, considered an improvement factor to measure the restoration effect depending on PM cost and age of the system [44] considering the improvement factor as a variable of the model. Another example is represented by Chan and Shaw [45], which considered two types of failure rate reduction: the first one is fixed (so that we have always the same reduction of the failure rate), whereas the second one is proportional (all the reductions of failure rate are proportional) [45].

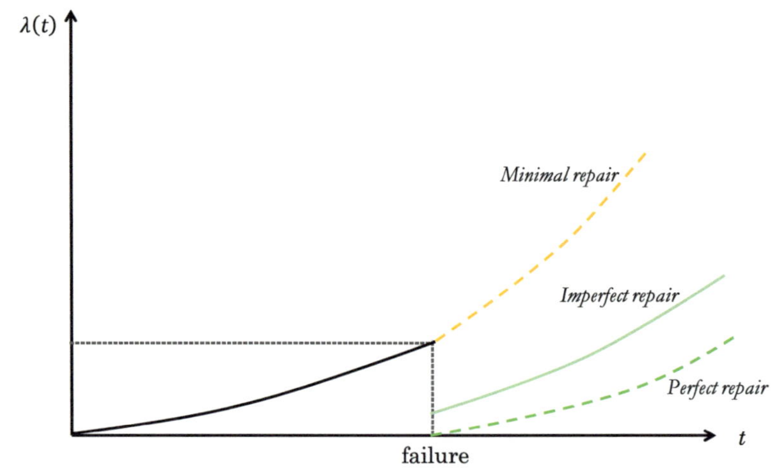


**Figure 4.** Minimal, perfect and imperfect repair according to the improvement factor method.

### 2.2.3. *Virtual age method*

Virtual age defines the restoration level achieved after a repair of a system. It depends on the operation time and on the number of maintenance activities performed. This method was proposed by Kijima et al. [46], which suggested that a system with virtual age $V \geq 0$ behaves as if it was a new system which reached the age $v$ without failure [46]. The main idea of the virtual age models, in fact, is to evaluate failure intensity considering virtual age instead of the real time [47].

According to Kijima, two main virtual age types could be identified: the first one assumes that the $n$th repair is able to remove the damage related to the time between the $(n{-}1)$th and the $n$th failures. The virtual age after the $n$th repair is:

$$v_n = v_{n-1} + q\, x_n \tag{2}$$

$v_n$ is the virtual age immediately after the $n$th repair, $q$ is a parameter representing the effect of the $n$th repair (the quality of an intervention involving the component) and $x_n$ is the time between the $(n{-}1)$th and the $n$th failures. Obviously, if $q = 0 = 0$ the model explains a perfect maintenance, while if $q = 1=1$, we have a minimal maintenance.

With the second virtual age model, it is assumed that the $n$th repair is able to remove the cumulative damage of both current and previous failures, so that the virtual age after the $n$th repair is:

$$v_n = q(v_{n-1} + x_n) \tag{3}$$

**Figure 5** shows a representation of the relationship between actual age and virtual age [48].

### 2.2.4. *Shock model method*

This model considers a unit that is subject to shocks randomly in time. At $t = 0$ its damage level is equal to zero, while the damage level will increase over time upon occurrence of shock events. Obviously, each damage increases the current damage level of the unit: the unit fails when its cumulative damage is greater than a specified threshold value [49].

Kijima and Nakagawa [49] used this approach to model imperfect preventive maintenance, suggesting that each PM reduces the damage level by $100(1 - b)\%$, $0 \leq b \leq 1$, of total damage.

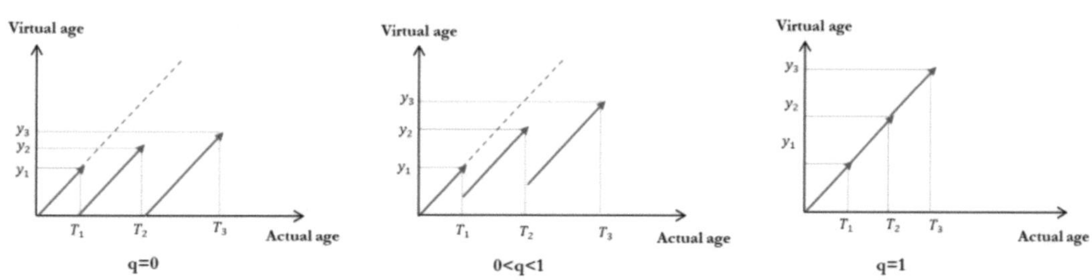


**Figure 5.** Virtual age vs. Actual age for $q = 00$, $0 < q < 1$, $q$=1.

Even then, it is possible to trace back this situation to perfect and minimal repair: if $b = 1$, the PM is minimal, whereas if $b = 0$, the PM is perfect.

### 2.2.5. (α, β) rule method

This IM treatment method was proposed by Wang and Pham [50]. According to this model, upon each repair, the lifetime of a system will be reduced to a fraction $0 < \alpha < 1$ of its immediately previous one, and all lifetimes are independent (the lifetime decreases with the number of repairs) [50]. For the repair time, which is non-negligible and upon repair, the model supposes that the next restoration time becomes a multiple $\beta > 1$ of its current one. Finally, all repair times are independent.

**Figure 6** shows how, according to this model, when the number of repairs increases, the time to repair increases too, while the lifetimes decreases.

### 2.2.6. Hybrid model method

The hybrid model proposed by Lin et al. considers, in particular, the effects of PM activities in terms of how the hazard rate function and the effective age of the equipment are changed by the PMs [51]. The model is "hybrid" because it derives from two PM models proposed by Nakagawa [52]: the Hazard Rate PM model and the Age Reduction PM model.

The first one considers that the hazard rate function of a system is an increasing function of time when there are no PM activities, while each PM resets the $\lambda(t)$ to zero and causes a faster growing of $\lambda(t)$ itself after each additional PM action. The failure rate after the $i$th PM action becomes $a_i\lambda(t)$, given that $\lambda(t)$ is the hazard rate in the previous period, $a_i \geq 1$.

On the other side, the Age Reduction PM model evaluates the effective age of a system after the $i$th PM action as a fraction of its effective age just prior to this PM. The effective age after the $i$th PM action is $b_iE_i$: $0 \leq b_i < 1$ is the improvement factor in the effective age due to the $i$th PM action, while $E_i$ is the effective age of the system just prior to the $i$th PM action [52].

The hybrid model merges these two aspects, as better explained by **Figure 7**.

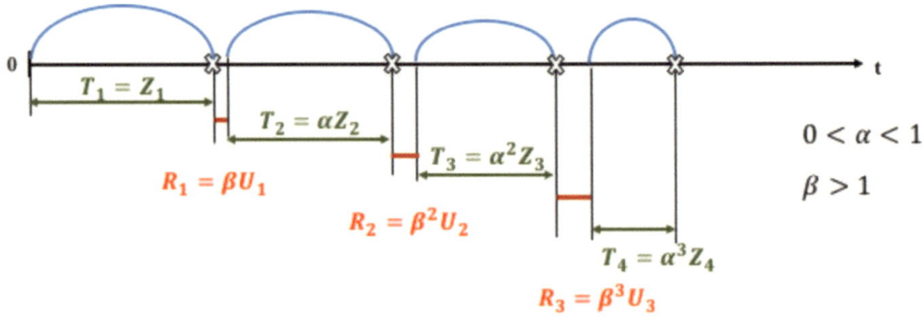


**Figure 6.** (α, β) rule or quasi-renewal process method.

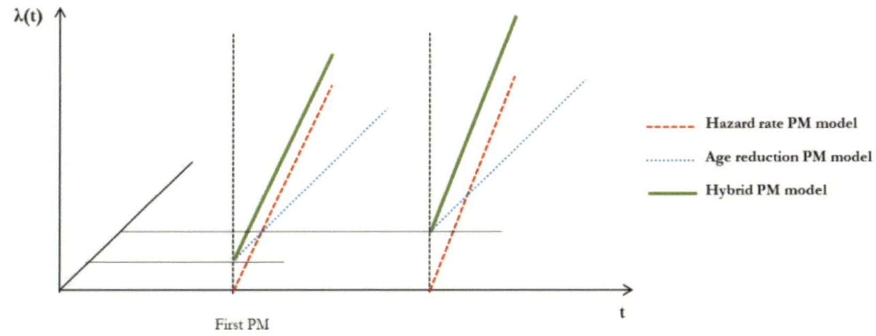


**Figure 7.** Comparison between the Hazard Rate PM model, the Age Reduction PM model [52] and the Hybrid PM model [51].

## 3. Imperfect maintenance in real cases

The theoretical definition of imperfect maintenance needs to be identified in real applications in an industrial setting. Literature analysis was the starting point, but the experience shared by reliability technicians was essential in order to identify some typical actual situations in which imperfect maintenance is indeed verified. Three main specific situations are proposed and summarized in **Figure 8**.

- "Conscious" imperfect maintenance: This is a case of an aware imperfect maintenance action. The reasons are various, and among them, it is possible to consider the use of unsuitable spare parts. When a fault occurs, for example, the required new component could not be available because of cost or time reasons. In this case, in order to reactivate the normal operation situation, it could be required to adopt alternative solutions like reconditioned part or "non-original" ones. Inventory management of spare parts, in fact, plays a key role in the maintenance management: their specific demand is usually unknown and random, even if it should be filled almost instantly or as soon as possible [53]. At the same time, for specific expensive components with a high risk of obsolescence, storage is not the best solution. Considering the issues related to the inventory management of spare parts, product recovery could be considered as a better option if compared to the acquisition of a new one. Since several remanufacturing options are possible, then several spare parts qualities could

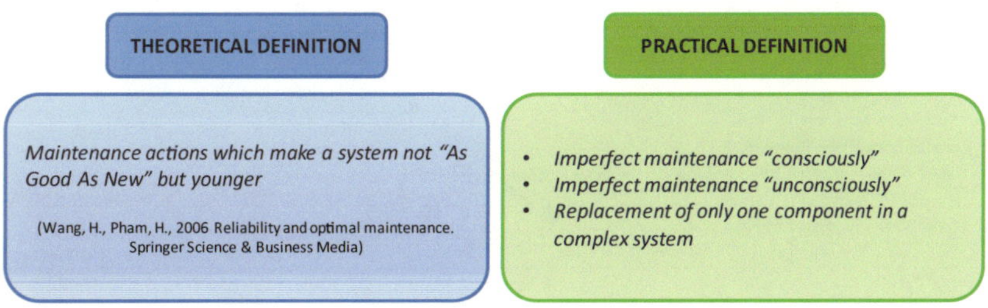


**Figure 8.** Imperfect maintenance theoretical vs. practical definition.

derive [54]. This quality level has a huge impact on the reliability of the system subject to maintenance, since the reconditioned part is not in the AGAN condition. The spare parts management just shown is very common in real contexts, and in literature, there are many papers on the topic. Just as examples, we can cite two contributions: Boudhar et al. [55] considered the problem of choosing the spare parts' quality to be used at each replacement minimizing the total cost; Boudhar et al. [54], an extension of the previous paper, proposed an optimal maintenance policy considering both the reliability and maintenance aspects and the quality of spare parts used for maintenance. Hence, it is possible to say that reliability and maintenance are strictly connected to reverse logistic and remanufacturing [54]. Another example of Imperfect maintenance done consciously, always related to spare parts, regards the use of "non-original" parts such as adapted part of the similar system (different brand, different technical features and so on). Both this last example and the others presented in this section could be related to maintenance expenditure problems like suggested by Helvic [56].

- "Unconscious" imperfect maintenance: This example is strictly connected to the maintenance operator skills. Contrary to the previous case, this one refers to an unknown "imperfect" action caused by incompetence, lack of attention and errors. According to some researches, in fact, some causes for imperfect, worse and worst maintenance could be ascribed to the skills of the maintenance technicians [29, 56], such as repairing the wrong part, only partially repairing the faulty part, repairing (partially or completely) the faulty part but damaging adjacent parts, incorrectly assessing the condition of the unit inspected and deferring maintenance actions.
- Replacement of only one component in a complex system: This situation is very common in real context; it refers to the case in which, after a practicality period, where all the components have a specific degradation state, if a component breaks, it is usually the only one replaced with a new one. The global system reliability must be defined with imperfect maintenance models since the whole system in not in an As Good As New condition. This is explained in **Figure 9**.

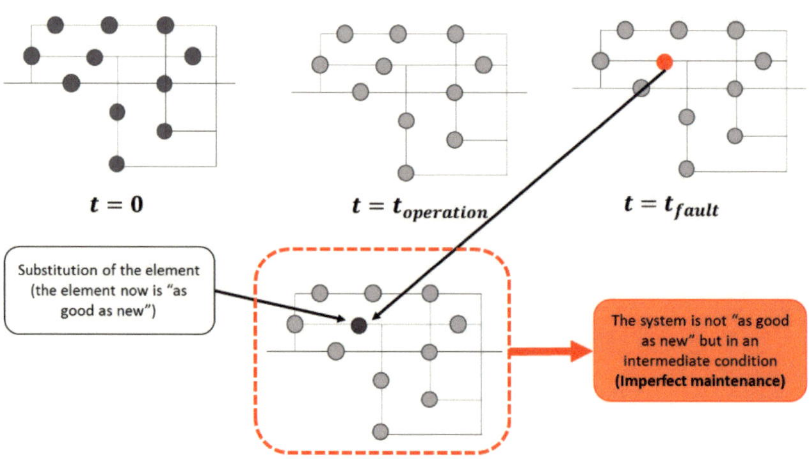


**Figure 9.** Representation of one of the real imperfect maintenance case (replacement of only one component in a complex system).

http://dx.doi.org/10.5772/intechopen.69286

## 4. Analysis of real imperfect maintenance cases through the most appropriate optimization model

The imperfect maintenance models proposed in the literature and shown in Section 2.2 are useful if used to describe the reliability features of real systems subject to imperfect maintenance actions. There are many models and many kinds of imperfect maintenance, so a question could be: "Which model should I apply for a correct evaluation of the reliability of a system subjected to imperfect maintenance?

This section proposes the most suitable models for the three real cases of imperfect maintenance. These are summarized in **Table 1**.

We can see how the $(p, q)$ rule method is suitable to model all the cases in which there is uncertainty on the maintenance action effectiveness. For this reason, it could be used in all the real situations of IM. This is also true for the Shock Model method.

The improvement factor method seems to be highly suitable for the "Conscious" IM, since this method considers some parameters that could be linked to a conscious decision of IM (the cost of maintenance activities, from which the improvement factor could be derived, is linked, for example, to the use of unsuitable spare parts).

The virtual age method fits better to the "Conscious" IM. This method, in fact, can model how much of the damage related to the time between the $(n-1)$th and the $n$th failures, the maintenance

| Real cases of IM | IM model |
|---|---|
| "Conscious" imperfect maintenance | • $(p, q)$ rule method<br>• Improvement factor method<br>• Virtual age method<br>• Shock model method<br>• $(\alpha, \beta)$ rule method<br>• Hybrid model method |
| "Unconscious" imperfect maintenance | • $(p, q)$ rule method<br>• Improvement factor method<br>• Virtual age method<br>• Shock model method<br>• $(\alpha, \beta)$ rule method<br>• Hybrid model method |
| Replacement of only one component in a complex system | • $(p, q)$rule method<br>• Shock model method |

**Table 1.** Proposal of the most suitable methods to treat each real case of IM.

action is able to remove: using not a new spare part (but, for example, an already used component, with its own wear level) only a part of the damage of the system will be removed.

The ($\alpha$, $\beta$) rule method is highly appropriate both for the "Unconscious" and the "Conscious" IM: the reduction of lifetime of a system and the growth of time to repair with the number of repairs might refer both to human error in maintenance activities and to cumulative damages because of non-suitable spare parts. The same reasoning applies to the hybrid model method.

## 5. Conclusions

In the last years, a new maintenance classification was introduced: it suggests various maintenance types that are: perfect, minimal, imperfect, worse and worst [34]. Imperfect maintenance (IM), in particular, is defined as maintenance action that makes a system not "as good as new" but younger: upon an imperfect maintenance action, the system lies in a state somewhere between AGAN and its pre-maintenance condition.

In addition to the theoretical definition, some real cases for IM were identified, considering both literature analysis and maintenance technicians' experience. Each one of them could be studied and optimized through appropriate maintenance models. Among the IM models in the literature, there is a proposal to find the most suitable one for describing each real situation. From this approach, a correct reliability and availability estimation can be developed in real cases.

## Author details

Filippo De Carlo* and Maria Antonietta Arleo

*Address all correspondence to: filippo.decarlo@unifi.it

University of Florence, Florence, Italy

## References

[1] UNI EN 13306, Maintenance—Terminology. 2010

[2] Sherwin D. A review of overall models for maintenance management. Journal of Quality in Maintenance Engineering. 2000;**6**:138-164

[3] Furlanetto, Luciano. Manuale di manutenzione degli impianti industriali e servizi. Franco Angeli, Milano (1998). ISBN 9788846408662

[4] Aditya Parida, Uday Kumar, Diego Galar, et al. Performance measurement and management for maintenance: A literature review. Journal of Quality in Maintenance Engineering. 2015;**21**:2-33

[5] Parida A, Chattopadhyay G. Development of a multi-criteria hierarchical framework for maintenance performance measurement (MPM). Journal of Quality in Maintenance Engineering. 2007;**13**:241-258

[6] Dekker R. Applications of maintenance optimization models: A review and analysis. Reliability Engineering & System Safety. 1996;**51**:229-240

[7] Al-Sultan KS, Duffuaa SO. Maintenance control via mathematical programming. Journal of Quality in Maintenance Engineering. 1995;**1**:36-46

[8] Liyanage JP, Kumar U. Towards a value-based view on operations and maintenance performance management. Journal of Quality in Maintenance Engineering. 2003;**9**:333-350

[9] Murthy DNP, Atrens A, Eccleston JA. Strategic maintenance management. Journal of Quality in Maintenance Engineering. 2002;**8**:287-305

[10] De Carlo F, Borgia O, Tucci M. Accelerated degradation tests for reliability estimation of a new product: A case study for washing machines. Proceedings of the Institution of Mechanical Engineers, Part O: Journal of Risk and Reliability. Epub ahead of print 9 September 2013. DOI: 10.1177/1748006X13500650

[11] Furlanetto L, Mastriforti C. Outsourcing e global service: Nuova frontiera della manutenzione. FrancoAngeli. 2000

[12] Mobley K, Higgins L, Wikoff D. Maintenance Engineering Handbook. McGraw Hill Professional. 2008. ISBN-10: 0071826610

[13] Andriulo S, Arleo MA, de Carlo F, et al. Effectiveness of maintenance approaches for High Reliability Organizations. 15th IFAC Symposium on Information Control in Manufacturing. 2015;**48**:466-471

[14] Brettel M, Friederichsen N, Keller M, et al. How virtualization, decentralization and network building change the manufacturing landscape: An Industry 4.0 Perspective. International Journal of Mechanical Engineering & Technology. 2014;**8**:37-44

[15] Lee J, Kao H-A, Yang S. Service innovation and smart analytics for industry 4.0 and Big data environment. Procedia CIRP. 2014;**16**:3-8

[16] Furlanetto L, Garetti M, Macchi M. Ingegneria della manutenzione. Strategie e metodi: Strategie e metodi. FrancoAngeli. 2015

[17] De Carlo, F. (2013). Reliability and maintainability in operations management. INTECH Open Access Publisher

[18] Modarres M, Kaminskiy MP, Krivtsov V. Reliability Engineering and Risk Analysis: A Practical Guide, Third Edition. CRC Press. 2016

[19] Wang H. A survey of maintenance policies of deteriorating systems. European Journal of Operational Research. 2002;**139**:469-489

[20] De Carlo F, Borgia O, Tucci M. Risk-based inspections enhanced with Bayesian network. Proceedings of the Institution of Mechanical Engineers, Part O Journal of Risk and Reliability. 2011;**225**(3):375-386

[21] Sharma A, Yadava GS, Deshmukh SG. A literature review and future perspectives on maintenance optimization. Journal of Quality in Maintenance Engineering. 2011;**17**:5-25

[22] Vasili M, Hong TS, Ismail N, et al. Maintenance optimization models: A review and analysis. Optimization. 2011. http://ieomsociety.org/ieom2011/pdfs/IEOM173.pdf (Accessed 19 October 2016)

[23] Apeland S, Aven T. Risk based maintenance optimization: Foundational issues. Reliability Engineering & System Safety. 2000;**67**:285-292

[24] Dekker R, Scarf PA. On the impact of optimisation models in maintenance decision making: The state of the art. Reliability Engineering & System Safety. 1998;**60**:111-119

[25] Chitra T. Life based maintenance policy for minimum cost. In: Reliability and Maintainability Symposium, 2003. Annual. 2003, pp. 470-474

[26] Barlow R, Hunter L. Optimum preventive maintenance policies. Operations Research. 1960;**8**:90-100

[27] Horenbeek AV, Pintelon L, Muchiri P. Maintenance optimization models and criteria. International Journal of Systems Assurance Engineering. 2011;**1**:189-200

[28] Kay E. The effectiveness of preventive maintenance. International Journal of Production Research. 1976;**14**:329-344

[29] Nakagawa T, Yasui K. Optimum policies for a system with imperfect maintenance. IEEE Transactions on Reliability. 1987;**5**:631-633

[30] Chan PKW, Downs T. Two criteria for preventive maintenance. IEEE Transactions on Reliability. 1978;R-27:272-273

[31] Ingle AD, Siewiorek DP. Reliability models for multiprocessor systems with and without periodic maintenance. DTIC Document. 1976. http://oai.dtic.mil/oai/oai?verb=getRecord&metadataPrefix=html&identifier=ADA034854 (Accessed 31 March 2015)

[32] Chaudhuri D, Sahu KC. Preventive maintenance interval for optimal reliability of deteriorating system. IEEE Transactions on Reliability. 1977;R-26:371-372

[33] Nakagawa T. Imperfect preventive-maintenance. IEEE Transactions on Reliability. 1979; R-28:402-402

[34] Wang H, Pham H. Reliability and optimal maintenance. Springer Science & Business Media. 2006. https://books.google.it/books?hl=it&lr=&id=KeJAAAAAQBAJ&oi=fnd&pg=PA2&dq=reliability+and+optimal+maintenance+pdf&ots=hEBlskdDEp&sig=057lbpKOvZiZQLKJr9p1Xp0058U (Accessed 31 March 2015)

[35] Sanchez A, Carlos S, Martorell S, et al. Addressing imperfect maintenance modelling uncertainty in unavailability and cost based optimization. Reliability Engineering & System Safety. 2009;**94**:22-32

[36] Mabrouk AB, Chelbi A, Radhoui M. Optimal imperfect maintenance strategy for leased equipment. International Journal of Production Economics. 2016;**178**:57-64

[37] Pandey M, Zuo MJ, Moghaddass R. Selective maintenance scheduling over a finite planning horizon. Proceedings of the Institution of Mechanical Engineers Part O Journal of Risk and Reliability 2016;**230**:162-177

[38] Le MD, Tan CM. Optimal maintenance strategy of deteriorating system under imperfect maintenance and inspection using mixed inspection scheduling. Reliability Engineering & System Safety. 2013;**113**:21-29

[39] Pham H, Wang H. Imperfect maintenance. European Journal of Operational Research. 1996;**94**:425-438

[40] Brown M, Proschan F. Imperfect repair. Imperfect Maintenance Models, from Theory to Practice. 1983;**20**:851-859

[41] Block HW, Borges WS, Savits TH. Age-dependent minimal repair. Journal of Applied Probability. 1985;**22**:370-385

[42] Makis V, Jardine AK. Optimal replacement policy for a general model with imperfect repair. Journal of the Operational Research Society. 1992;**43**:111-120

[43] Malik MAK. Reliable preventive maintenance scheduling. AIIE Transactions. 1979; **11**:221-228

[44] Lie CH, Chun YH. An algorithm for preventive maintenance policy. IEEE Transactions on Reliability. 1986;**35**:71-75

[45] Chan J-K, Shaw L. Modeling repairable systems with failure rates that depend on age and maintenance. IEEE Transactions on Reliability. 1993;**42**:566-571

[46] Kijima M, Morimura H, Suzuki Y. Periodical replacement problem without assuming minimal repair. European Journal of Operational Research.1988;**37**:194-203

[47] Guo H, Liao H, Pulido J. Failure process modeling for systems with general repairs. In: The 7th International Conference on Mathematical Methods in Reliability: Theory, Methods and Applications (MMR2011) . 2001. http://www.reliasoft.com/pubs/2011_MMR_failure_process_model.pdf (Accessed 28 November 2016)

[48] Jacopino AG. Generalisation and Bayesian Solution of the General Renewal Process for Modelling the Reliability Effects of Imperfect Inspection and Maintenance based on Imprecise Data. 2005. http://drum.lib.umd.edu/handle/1903/3168 (accessed 28 November 2016)

[49] Kijima M, Nakagawa T. A cumulative damage shock model with imperfect preventive maintenance. Naval Research Logistics. 1991;**38**:145-156

[50] Wang H, Pham H. A quasi renewal process and its applications in imperfect maintenance. International Journal of Systems Science. 1996;**27**:1055-1062

[51] Lin D, Zuo MJ, Yam R. Sequential imperfect preventive maintenance models with two categories of failure modes. Naval Research Logistics (NRL). 2001;**48**:172-183

[52] Nakagawa T. Sequential imperfect preventive maintenance policies. IEEE Transactions on Reliability. 1988;**37**:295-298

[53] Inderfurth K, Kleber R. An advanced heuristic for multiple-option spare parts procurement after end-of-production. Production and Operations Management. 2013;**22**:54-70

[54] Boudhar H, Dahane M, Rezg N. New dynamic heuristic for the optimization of opportunities to use new and remanufactured spare part in stochastic degradation context. Journal of Intelligent Manufacturing. 2017;**28**:437-454. DOI: 10.1007/s10845-014-0989-1

[55] Boudhar H, Dahane M, Rezg N. Joint optimisation of spare parts demand and remanufacturing policy under condition-based maintenance for stochastic deteriorating manufacturing system. IFAC Proceedings. 2013;**46**:414-419

[56] Helvic BE. Periodic maintenance on the effect of imperfectness. In: 10thInternational Symposium on Fault-Tolerant Computing. 1980. pp. 204-206

Chapter 19

# A Decision Support System for Planning and Operation of Maintenance and Customer Services in Electric Power Distribution Systems

Carlos Henrique Barriquello, Vinícius Jacques Garcia, Magdiel Schmitz, Daniel Pinheiro Bernardon and Júlio Schenato Fonini

Additional information is available at the end of the chapter

http://dx.doi.org/10.5772/intechopen.69721

**Abstract**

This chapter aims to present the design and development of a decision support system (DSS) for the analysis, simulation, planning, and operation of maintenance and customer services in electric power distribution system (EPDS). The main objective of the DSS is to improve the decision-making processes through visualization tools and simulation of real cases in the EPDS, in order to allow better planning in the short, medium, and long term. Therefore, the DSS helps managers and decision-makers to reduce maintenance and operational costs, to improve system reliability, and to analyze new scenarios and conditions for system expansion planning. First, we introduce the key challenges faced by the decision-makers in the planning and operation of maintenance and customer services in EPDS. Next, we discuss the benefits and the requirements for the DSS design and development, including use cases modeling and the software architecture. Afterwards, we present the capabilities of the DSS and discuss important decisions made during the implementation phases. We conclude the chapter with a discussion about the obtained results, pointing out the possible enhancements of the DSS, future extensions, and new use cases that may be addressed.

**Keywords:** decision support system, system reliability, visualization tools, maintenance, customer services, electric power distribution

## 1. Introduction

A reliable and efficient power grid plays a central role in the infrastructure of a country, with a huge impact to its economic and social progress. An uninterrupted, secure, and efficient power flow since generators to consumers is not only desirable but an important requirement for a reliable power grid, where electricity must be produced and delivered in the right amount at the right time.

Although electrical energy can be stored on energy storage systems (EESs), like batteries, to be available at the consumer side in case of a network fault, this is yet not an economically viable solution for bulky quantities [1]. Albeit rapid progress has been observed in the development of EES in the last years, mainly driven by the requirements imposed by the introduction of renewable energy resources into the electric power system, several drawbacks are still present. Most EES technologies are not yet mature, have very limited lifetime, low power/energy density and/or low storage capacity, and always have a round-trip efficiency penalty due to the required charge/discharge cycles [2].

Even in case of higher EES availability at the consumer side due to the current trend of technology evolution and price reduction, there is also the increasing penetration of renewable generation sources near to the consumer, known as distributed generation (DG). As such, DG sources are being adopted at a growing rate by the consumers; they also bring more challenges to the system operators and to the distribution infrastructure, imposing a higher demand for network reliability, mainly at distribution level [3].

This situation is even more challenging when considering the aging infrastructure of distribution systems, uncertainties and regulatory changes, the entry of new players into the energy market (e.g. prosumer), the constant pressure for cost reduction, and the more stringent requirements of reliability. On the contrary, the modernization of the distribution system with the deployment of information and communication technologies (ICTs) and the integration with operational technologies (OTs) is enabling a smarter grid and helping the distribution system operators (DSOs) to face those challenges in a better condition. Nevertheless, the deployment of those smart grid technologies, such as advanced metering infrastructure (AMI) and supervisory control and data acquisition systems (SCADA), also brings new decision variables to the DSOs and, consequently, the need of decision support systems (DSSs) to help them in their decision-making processes [4].

In the literature, several works have been published with proposals of DSS targeted to support the decision-making of the different players in the electric power systems, such as electric utilities, producers, consumers, and market regulators. In Refs. [5] and [6], DSSs were proposed for assisting the competitors in their trading strategies for the electricity market. Also, for the utility companies, there are proposals of DSS for the planning and expansion of the distribution system network based on geographic information system (GIS) [7, 8], for power grid operation and planning considering meteorological conditions [9], for analyzing disturbances of electrical power distribution transformers to diagnosing failures [10], and for assisting utilities to comply with the limits of sulfur dioxide ($SO_2$) emissions [11]. Additionally, some DSSs have been proposed for the producers and consumers, including support for the calculation of the solar energy potential of surfaces in urban landscapes [12], for managing energy-efficient buildings [13], and also for assisting consumers to participate in demand response programs [14]. There are still proposals of DSS for assisting regulators in deciding about the installation of DG facilities [15] and about the deregulation of the electricity market [1].

Different from prior works, the main contribution of the DSS proposed here is in supporting the distribution system operators in the planning and the execution of customer and maintenance services, which are introduced in the next section. In the sequel, the design and the development of the DSS are reported in the third section. The evaluation of the DSS is left for the fourth section, followed by our concluding remarks at the end of the chapter.

## 2. Customer and maintenance services in the electric power distribution system

The electric utilities are responsible for maintenance services required to maintain the distribution network operating satisfactorily. Accordingly, the utility shall allocate its material and human resources (e.g. maintenance crews) in order to attend its customers and to fix the network as soon as possible in case of emergency (e.g. equipment failures and blackouts). However, due to the resources constraints, there is a need to prioritize emergency orders over normal maintenance orders, while taking the best decision possible regarding the available maintenance teams and the required response time for each order in the service queue.

The decision problem that must be solved by system operators working at the utility company is called the emergency dispatching and routing problem (EDRP) and can be stated as follows. Given pending emergency work orders and normal work orders in a service queue, and a set of available maintenance teams, to allocate the emergency orders to the crews in order to minimize the waiting time, which is defined as the sum of the travel time and the execution times of normal orders under execution or scheduled to be executed before the pending emergency orders.

It is important to note that the ERDP is dynamic by nature, given that new emergency orders may arrive in an unpredictable manner, changing the inputs of the problem and, thus, requiring a real time and continuous decision-making process. Therefore, a new dispatching solution must be found during the execution of the previously defined routes, possibly requiring changes to the routes assigned to the teams and the insertion of new emergency orders into their service queue. On the other hand, the decision process must take into account several conflicting criteria, such as the minimization of the required time to complete the emergency orders and the reduction of the cost incurred by the changes in the assigned routes. Note that the routing cost includes the delays of the previously assigned services and the possible decrease of the work efficiency, which is the ratio of effective service time to the total spent time (i.e. service times plus travel times). The relationship among the decision variables and decision outputs is represented in **Figure 1**, while the overall decision process is illustrated in **Figure 2**.

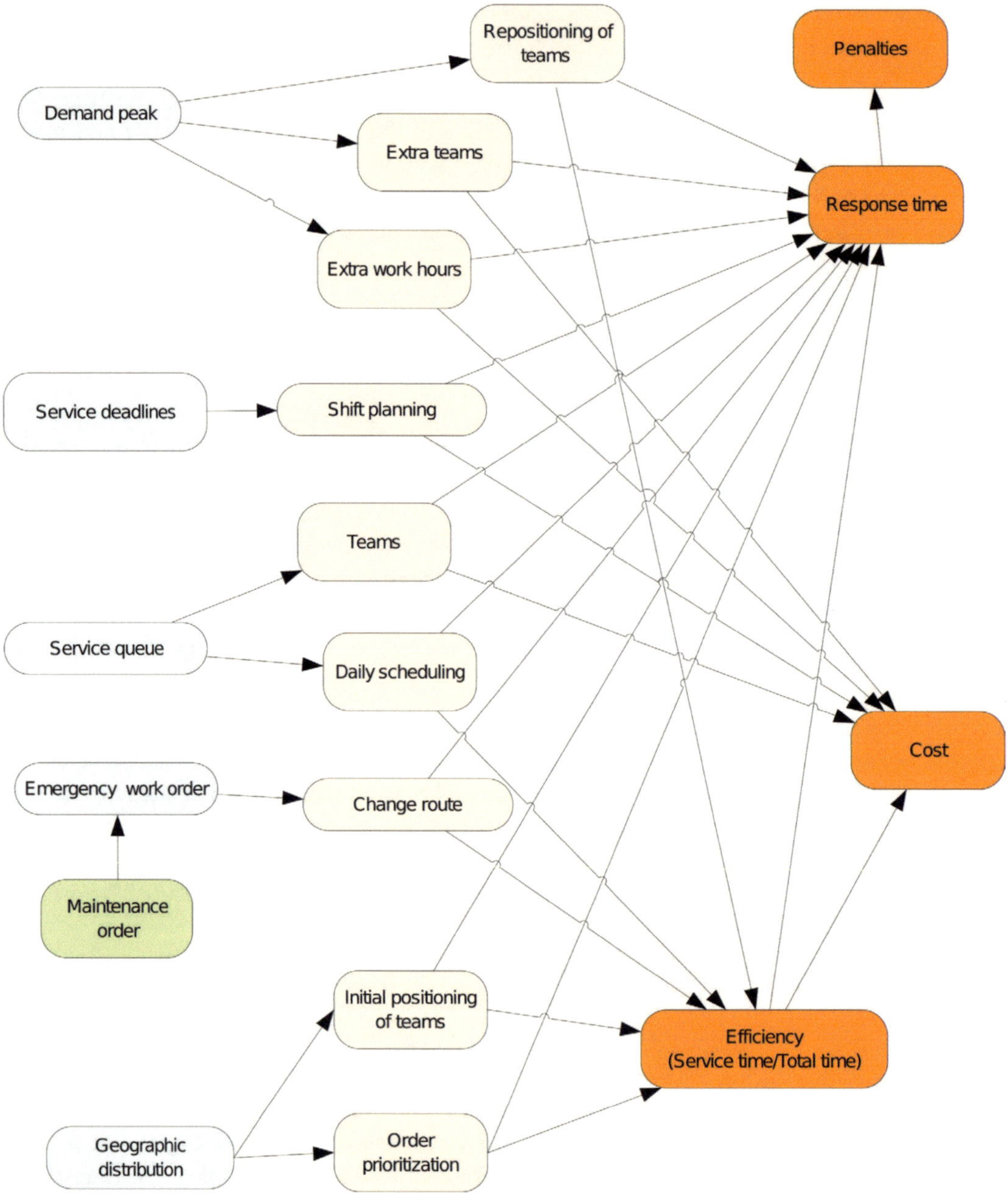


**Figure 1.** Diagram of the relationship among decision variables.

As shown in **Figure 1**, the main goals of the decision-maker are the increasing of the work efficiency and the reducing of the operational cost, the response time, and, consequently, the penalties incurred. Yet, the response time influences the calculation of the electric power distribution system (EPDS) reliability indices, such that the reduction of the response time has

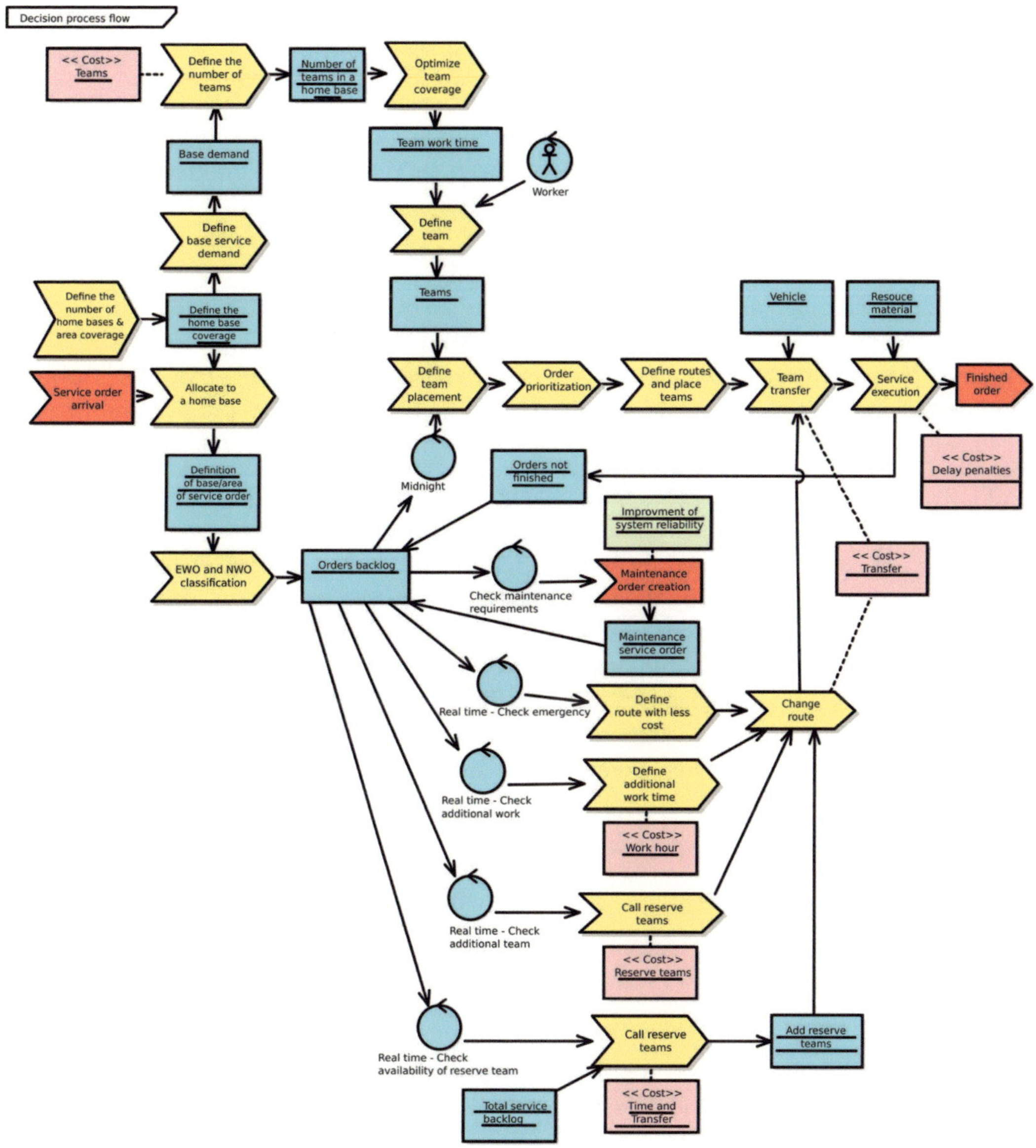


**Figure 2.** Decision process flowchart.

a positive impact in the improvement of the indices. Therefore, in fact, the decision-making process affects ultimately the EPDS reliability.

The assessment of the EPDS reliability is done by a set of reliability indices, whose definitions and forms of calculation can be found in Ref. [16]. In general, the reliability indices are measurements of the system performance and represent the frequency, duration, and magnitude of the interruptions of the electric energy supply. To this end, their calculations may take into account the number of customers, the connected load, the duration of the interruption, the amount of power interrupted, and the frequency of interruptions.

In summary, the following reliability indices can be used: System Average Interruption Frequency Index (SAIFI), System Average Interruption Duration Index (SAIDI), Customer Average Interruption Duration Index (CAIDI), Customer Total Average Interruption Duration Index (CTAIDI), Customer Average Interruption Frequency Index (CAIFI), Average Service Availability Index (ASAI), Customers Experiencing Long Interruption Durations (CELID), Average System Interruption Frequency Index (ASIFI), Average System Interruption Duration Index (ASIDI), Momentary Average Interruption Frequency Index (MAIFI), Momentary Average Interruption Event Frequency Index (MAIFIE), Customers Experiencing Multiple Sustained Interruption, and Momentary Interruption Events (CEMSMIn). However, the most commonly used indices are SAIFI, SAIDI, CAIDI, and ASAI, which provide information about the average EPDS performance [16].

SAIFI and SAIDI give information, respectively, about the average frequency and duration of sustained interruptions experienced by the customers over a predefined period of time, usually 1 year, and a predefined area. Mathematically, SAIFI is the ratio between the total number of customers affected by the interruption and the total number of customers served in that area, while SAIFI is the ratio between the sum of the number of customers affected by each interruption multiplied by the restoration time of each interruption and the total number of customers served in that area. SAIFI and SAIDI are, respectively, given in Eqs. (1) and (2), where $Ni$ is the number of customers interrupted for each sustained interruption event, $Nt$ is the total number of customers served for the area, and $ti$ is the restoration time for each interruption

$$SAIFI = \frac{\sum Ni}{Nt} \tag{1}$$

$$SAII = \frac{\sum ti \times Ni}{Nt} \tag{2}$$

CAIDI, by its turn, gives information about the average time needed to restore service. Mathematically, it is the ratio between the sum of the durations of customer interruptions and the total number of customers served. Equivalently, it is the ratio between SAIDI and SAIFI, as given in Eq. (3)

$$CAIDI = \frac{SAIDI}{SAIFI} \tag{3}$$

Finally, ASAI represents a fraction of time of the defined reporting period, usually 8760 h (equivalent to 1 year), which the average customer has received power. It is given in Eq. (4), where $Nhy$ is the number of hours in 1 year and is equal to 8760 h in a non-leap year and 8784 h in a leap year

$$ASAI = \frac{Nt \times Nhy - \sum ti \times Ni}{Nt \times Nhy} = 1 - \frac{SAIDI}{Nhy} \tag{4}$$

## 3. Decision support system design and development

The proposed DSS, named Pladin (a contraction of "dynamic planner"), was designed using a three-tier architecture based on a client/server model, as shown in **Figure 3**. The client/server model was chosen to allow the DSS to support its deployment in a cloud infrastructure. According to Ref. [17], cloud-based systems are attractive for utility companies due to the business model (i.e. memory/computation elasticity, scalability, economies of scale, pay-as-you-go model, etc.), beyond offering more efficiency, productivity, and anywhere access. Moreover, the deployment of a DSS in cloud is also interesting due to the possibly large variability of the required computing resources (e.g. memory and computing power) and the complete integration with web-based GIS portals when required [12].

In order to implement the proposed architecture, modern web technologies have been selected and employed according to the requirements of each layer of the DSS, as shown in **Figure 4**. On the server side, the decision models, the database access, and the middleware were implemented in Java language, where the middleware follows a Spring MVC [18] design pattern and the database access is managed by the Hibernate ORM tool [19]. On the client side, the user interface was implemented in Javascript language, following the React JS [20] and the Redux JS [21] frameworks. Additionally, the client/server data exchange is implemented as a web service Application Programming Interface (API) based on the Representational State Transfer (REST) architecture [22] using JavaScript Object Notation (JSON) data format [23].

The next step in the development process of the DSS was the design of the user interface, where special attention has been given to provide to the user visualization tools, such as maps and graphs, which are valuable for assisting her/him in the decision-making process given the spatial and temporal complexity of the problem at hand. Thus, the user interface has been prototyped and developed, according to the modeled use cases, which are shown in **Figure 5**.

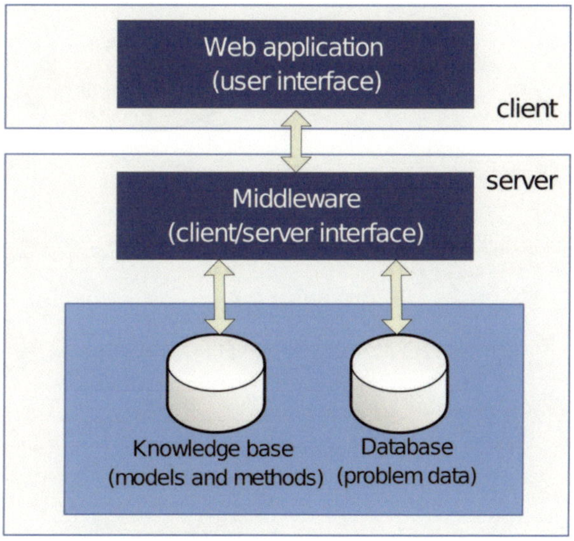


**Figure 3.** DSS three-layer architecture based on a client/server model.

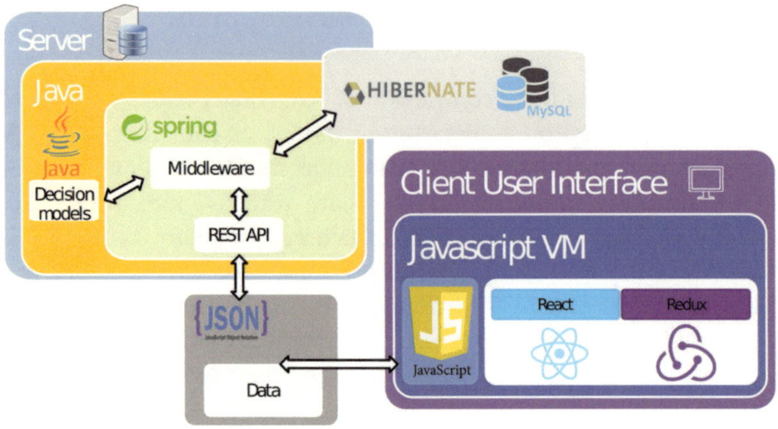


**Figure 4.** Web technologies used for client and server developments.

Due to the foreseen complexity of the user interface and the evolving requirements, it was decided to develop it as single-page application (SPA). Single-page applications are web applications written in Javascript that almost fully execute on the web browser at the client side, thus similar to desktop applications (i.e. applications that run natively). Some of the advantages of SPAs are flexibility, easy maintenance and deployment, rich visual experience, and a clear separation of concerns.

When developing an SPA, a choice of some architecture can ease the development and the maintenance of the application. There are basically four main architectures used in SPAs: model-view-controller (MVC), model-view-presenter (MVP), model-view-view-model (MVVM), and Flux. Flux is the most recent one and brings several advantages over the other architectures, mainly compared to the well-known MVC.

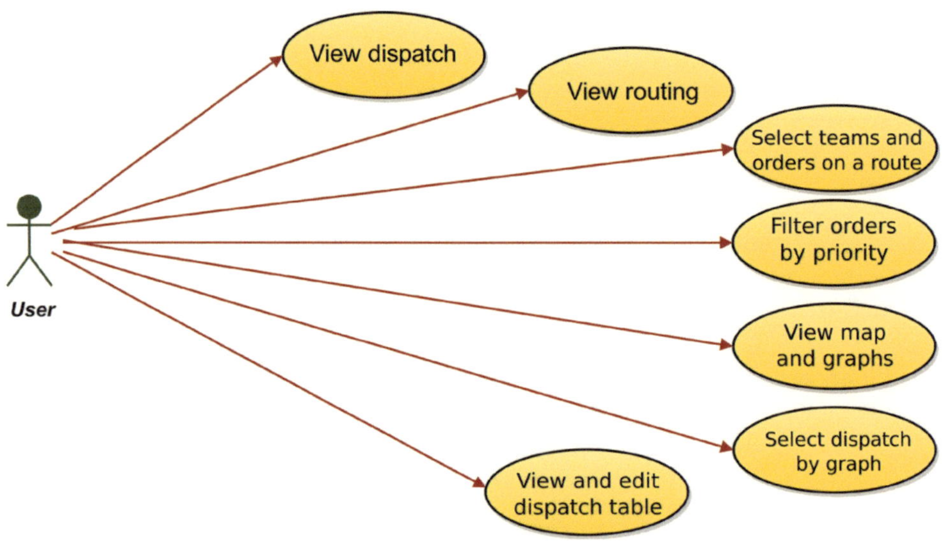


**Figure 5.** DSS use cases modeling.

In the MVC architecture, as shown in **Figure 6**, the model is responsible for maintaining and updating the application data, the view displays the data to the user (in some visual form) and receives his/her inputs, and the controller is in charge of taking the user inputs to update the model and to notify the view of the updates of the mode. The main principle of the MVC is the separation between the three components, which allows the implementation of different views for the same model and improves testability of the components. However, the MVC architecture is not easily scalable due to the need of adding more instances of the components (models, views, and controllers) as the application grows and the increasing difficulty of reasoning about the logic of the application due to the bidirectional data flows among the components.

The lack of the scalability of the MVC motivated the introduction of the Flux architecture, which adheres to the philosophy of unidirectional data flow. In Flux, as shown in **Figure 7**, there are four components: the action creators, the dispatcher, the stores, and the views. The dispatcher, the stores, and the views are components with independent inputs and outputs, while the action creators are components that create, as the name suggests, special objects known as actions, which carry the data of the application (payload) and a unique property that identifies the action type.

The dispatcher is a central hub that manages all the data flow in the application by maintaining a registry of callbacks into the stores. These callbacks are used as a mechanism for distributing the actions to the stores, thus inverting the control of the application. Actions are generated in the application usually in response to the user interaction with the views or may come from the server. Then, each generated action goes to the stores via the dispatcher and is processed in the stores. The stores contain the application state (data) and logic, and are somewhat similar to the models in MVC. However, the stores can manage the state of many objects and not a single one as models in MVC.

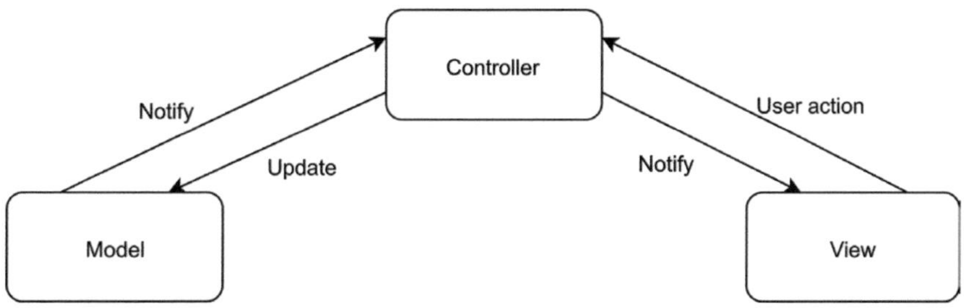


**Figure 6.** MVC architecture.

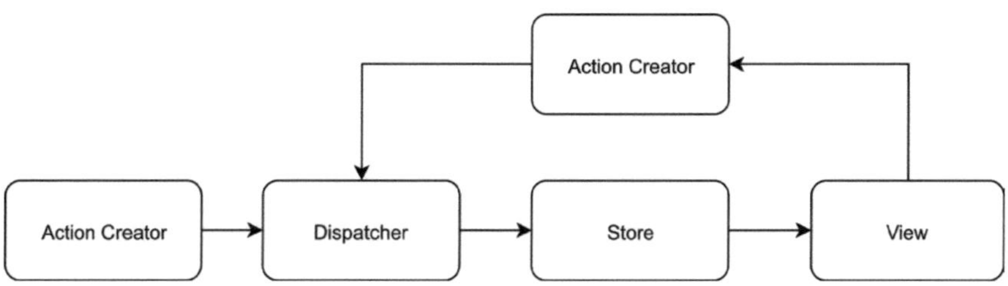


**Figure 7.** Flux architecture.

After receiving an action, a store interprets it according to the action's type using a switch statement. Then, the store updates its state and broadcasts an event declaring that its state has changed. In turn, special kinds of view, which are known as controller views, listen for events broadcast by the stores that they depend on. These controller views sit on top of a hierarchy of views and are able to obtain the data from the store and pass it down to their descendant views. Usually, the entire state of store is passed down the chain of views, so that views can take the part of the state they need and update themselves accordingly. Finally, views can create new actions based on the user interaction, closing the flow loop.

Basically, the Flux architecture avoids the complexities of the MVC architecture, mainly due to the restriction of unidirectional data flow in the application, which make easy for reasoning about the application workings. However, it is possible to simplify the Flux architecture even further in order for using a single store in the application. This concept is behind of Redux, which is an implementation of the Flux architecture, with added constraints in order to make it a predictable state container for the application, as shown in **Figure 8**.

Redux follows three fundamental principles: single source of truth, read-only state, and changes made with pure functions [21]. The first principle means that state of the whole application is stored in a single store as an object tree, which allows for easier debugging of the application. The second principle implies that it is only possible to change the state emitting actions, which are processed one by one in a strict ordering, avoiding race conditions and simplifying logging and debugging. And the third principle means that a new state of the application is created by a reducer, which is a function that takes as inputs the previous state and the action, and returns the next state without mutating the previous state. Therefore, because the state transitioning in Redux is done by functions (i.e. reducers) instead of event emitters as in the original Flux architecture, the Redux architecture does not have the concept of a dispatcher and allows easier composing of reducers. In fact, due to the several qualities of Redux compared to other alternatives architectures, such as simplicity, scalability, and predictability, it has been chosen as the basis architecture of the proposed DSS.

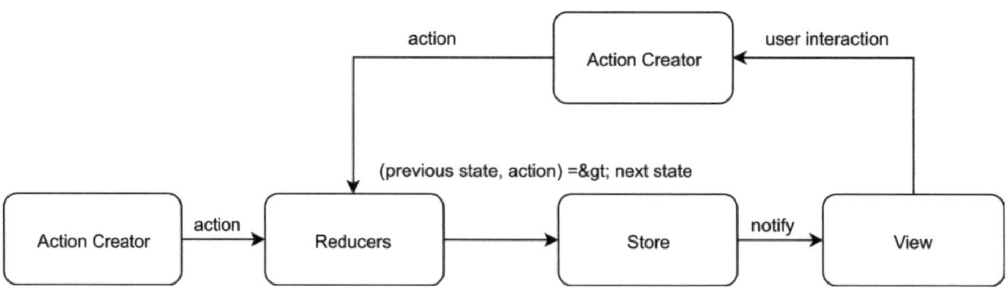


**Figure 8.** Redux architecture.

## 4. Evaluation of the developed decision support system

In this section, implemented use cases are presented and discussed. The first one is the "View dispatch" case, where the user can have an overview of a dispatch solution for a given instance of the dispatching problem. The instance is based on the user selection of the position of an operating base and the target date. Then, the user can select from a list one of the available dispatching solutions for the chosen day in order to have a graphical overview of its characteristics, such as the number of crews and orders, the classification and the priority assignment of the orders and the distribution of the execution and travel times by orders and by crews. A sample screen of this use case is shown in **Figure 9**.

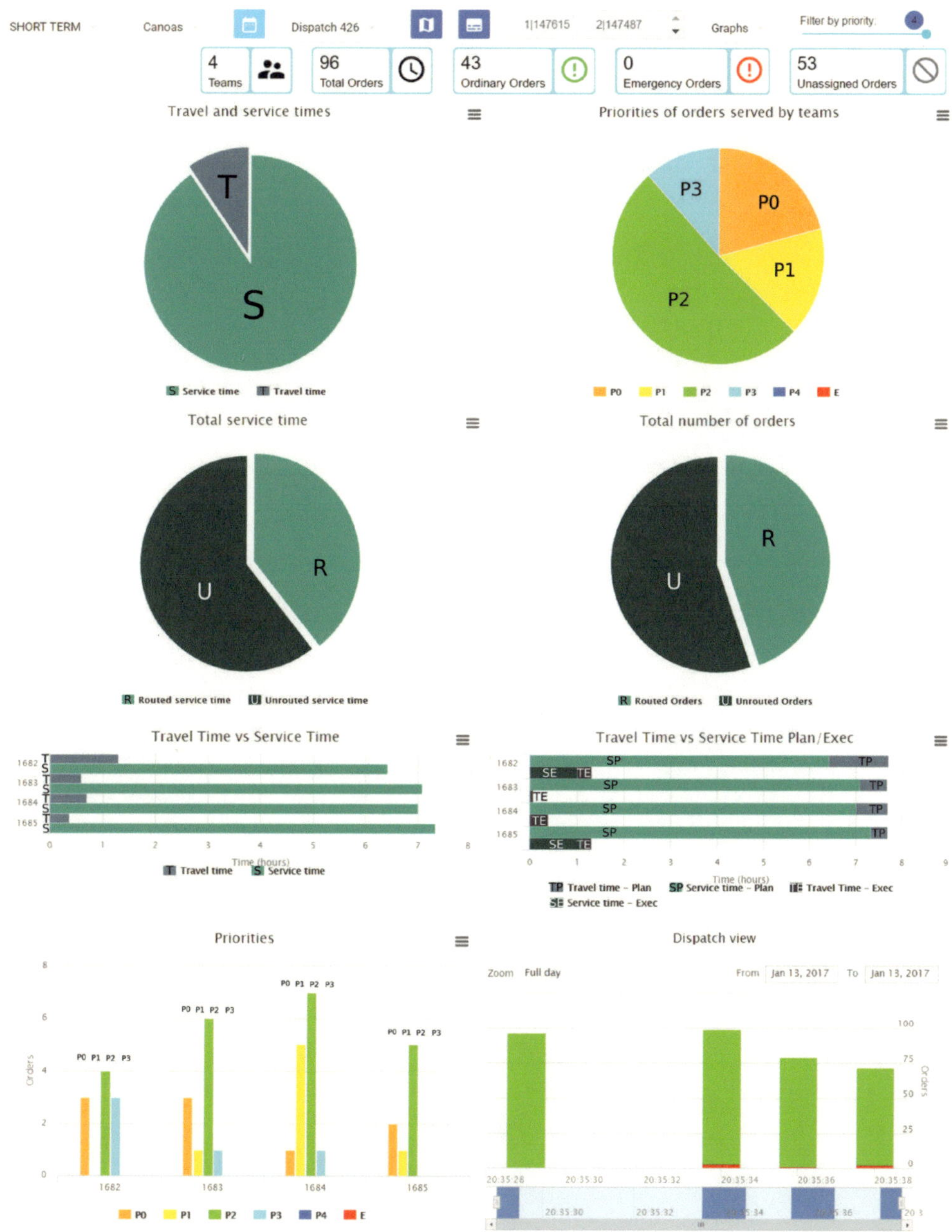


**Figure 9.** Sample screen of "View dispatch" use case of the Pladin DSS.

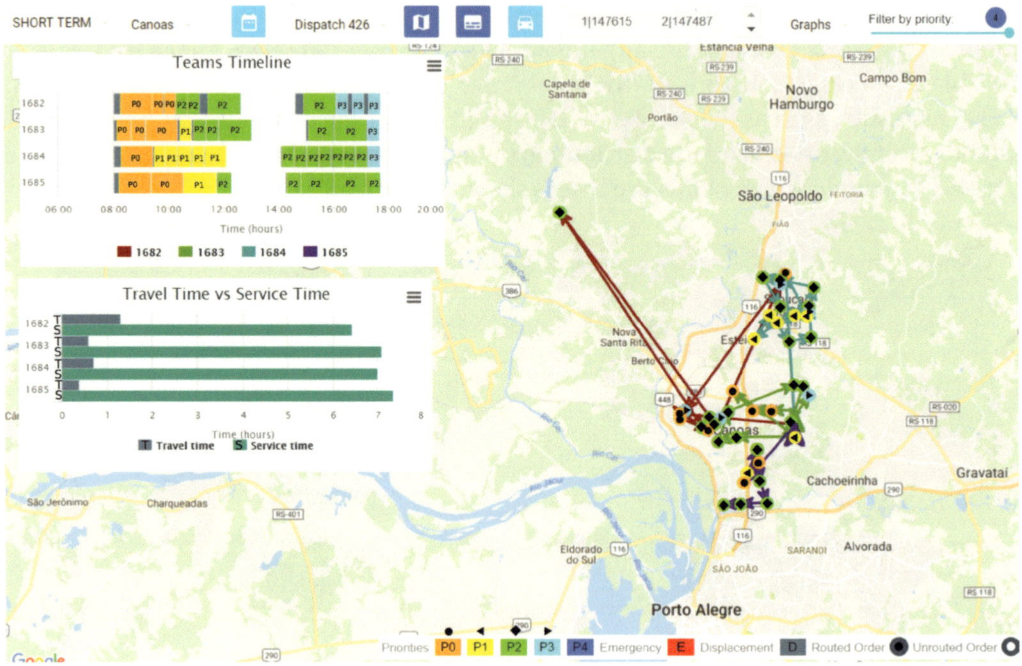


**Figure 10.** Sample screen of the "View routing" and "Show map and graphs" use cases.

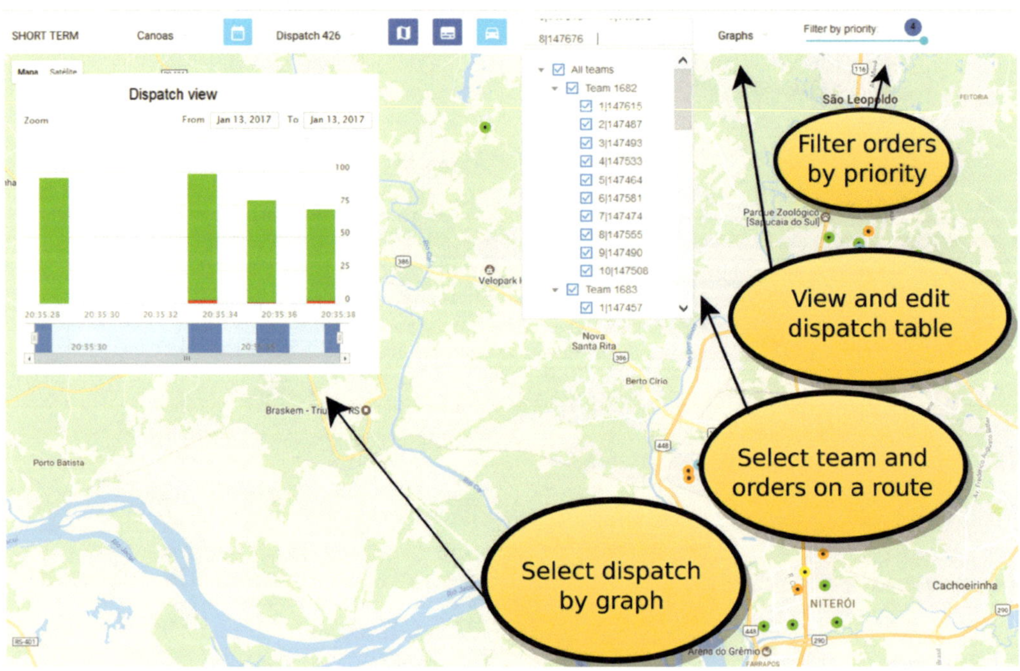


**Figure 11.** Sample screen highlighting the use cases "Filter orders by priority" and "Select teams and orders on a route," "Select a dispatch by graph," and "View and edit dispatch table".

After the user has selected a dispatching solution, he/she also can view the routes taken for each crew, according to the "View routing" use case. This use case allows the user to analyze the quality of the proposed dispatching solution with assistance of graphical tools, which are provided according to the use case "View map and graphs." A sample screen including both use cases is shown in **Figure 10**.

From the screen shown in **Figure 10**, the user also can access four other functionalities, which are implementations for the use cases "Filter orders by priority," "Select teams and orders on a route," "Select dispatch by graph," and "View and edit dispatch table." These use cases are highlighted in **Figure 11**.

Using the Pladin DSS, the user can better take his/her decisions for the planning and the execution of customer and maintenance services in the electric power distribution system. Therefore, he/she can appropriately allocate the available resources (e.g. teams) for improving the efficiency of the teams by reducing travel and response times. Thus, leading ultimately to reduce operational costs and improving the distribution system reliability.

## 5. Conclusions

In this chapter, we have introduced a web-based decision support system aimed to aid distribution system operators for planning and executing customer and maintenance services in the electric power distribution system. The DSS provides visualization tools for supporting the user decision to better allocating the available resources in order to solve the emergency dispatching and routing problem in real time.

In addition, the user interface of the DSS also allows for tweaking the input parameters, in order to change the solution for improving the work efficiency and reducing operational costs. Moreover, the user has access to facilities for analyzing what-if scenarios and studying the influence of the decision variables in the solution provided by the DSS.

Finally, due to the simplicity and the scalability of the chosen application architecture, it is expected that future extensions and enhancements will be added for improving it. Currently, the DSS is being used by the system operators working at a utility company located at the South Brazil, and thus the enhancements will be added after receiving the feedback of the current users. Possible enhancements and new use cases include the addition of more visualization tools and analytics services, artificial intelligence techniques, and the integration of more information sources.

## Acknowledgements

The authors would like to thank the technical and financial support of RGE Sul Power Utility by project "Planejamento Dinâmico de Operações" (P&D/ANEEL), Coordination for the Improvement of High Level Personnel (CAPES), and the National Center of Scientific and Technological Development (CNPq).

## Author details

Carlos Henrique Barriquello[1]*, Vinícius Jacques Garcia[1], Magdiel Schmitz[1], Daniel Pinheiro Bernardon[1] and Júlio Schenato Fonini[2]

*Address all correspondence to: barriquello@gmail.com

1 Federal University of Santa Maria, Santa Maria, Brazil

2 RGE Sul, Brazil

## References

[1] Bergey PK, Ragsdale CT, Hoskote M. A decision support system for the electrical power districting problem. Decision Support Systems. 2003;**36**(1):1-17. DOI: http://dx.doi.org/10.1016/S1344-6223(02)00033-0

[2] Aneke M, Wang M. Energy storage technologies and real life applications – A state of the art review. Applied Energy. 1 October 2016;**179**:350-377. ISSN 0306-2619. Available from: http://dx.doi.org/10.1016/j.apenergy.2016.06.097

[3] Ipakchi A, Albuyeh F. Grid of the future. IEEE Power and Energy Magazine, March–April 2009;**7**(2):52-62. DOI: 10.1109/MPE.2008.931384

[4] Doğdu E et al. Ontology-centric data modelling and decision support in smart grid applications a distribution service operator perspective. In: Proceedings of the IEEE International Conference on Intelligent Energy and Power Systems (IEPS), 2-6 June 2014; Kiev. New York: IEEE; 2014. pp. 198-204. DOI: 10.1109/IEPS.2014.6874179

[5] Sancho J, Sánchez-Soriano J, Chazarra JA, Aparicio J. Design and implementation of a decision support system for competitive electricity markets. Decision Support Systems. March 2008;**44**(4):765-784. ISSN 0167-9236, http://dx.doi.org/10.1016/j.dss.2007.09.008

[6] Sueyoshi T, Tadiparthi GR. An agent-based decision support system for wholesale electricity market. Decision Support Systems. January 2008;**44**(2):425-446. ISSN 0167-9236, http://dx.doi.org/10.1016/j.dss.2007.05.007

[7] Yao FS, Zhang XQ, Zhang Y and Wang TH. Computer decision-making support system for power distribution network planning based on geographical information system. In: Proceedings of the International Conference on Electricity Distribution, 10-13 December 2008; Guangzhou. New York: IEEE; 2009. pp. 1-6. doi: 10.1109/CICED.2008.5211676

[8] Luo F et al. A practical GIS-based decision-making support system for urban distribution network expansion planning. In: Proceedings of the International Conference on Sustainable Power Generation and Supply; 6-7 April 2009; Nanjing. New York: IEEE; 2009. pp. 1-6. DOI: 10.1109/SUPERGEN.2009.5348306

[9] Li L, Yao-qiang X, Xing-zhi W, Kai W. Study on analysis and decision support system of power grid operation considering meteorological environment based on big data and GIS. In: Proceedings of the International Conference on Electricity Distribution (CICED); 10-13 Aug. 2016; Xi'an. New York: IEEE; 2016, pp. 1-6. DOI: 10.1109/CICED.2016.7576045

[10] Nina DLF, Neto JVdF, Ferreira EFM, Santos AMd. Hybrid support system for decision making based on MLP-ANN, IED and SCADA for disturbances analysis of electrical power distribution transformers. In: 2013 UKSim 15th International Conference on Computer Modelling and Simulation; Cambridge; 2013. pp. 12-20. DOI: 10.1109/UKSim.2013.147

[11] Ghandforoush P, Sen TK, Wander M. A decision support system for electric utilities: Compliance with clean air act. Decision Support Systems. October 1999;**26**(4):261-273. ISSN 0167-9236. Available from: http://dx.doi.org/10.1016/S0167-9236(99)00056-1

[12] Boulmier A, White J, Abdennadher N. Towards a Cloud Based Decision Support System for Solar Map Generation. In: Proceedings of the IEEE International Conference on Cloud Computing Technology and Science (CloudCom); 12-15 Dec. 2016; Luxembourg City,.New York: IEEE; 2017, pp. 230-236. DOI: 10.1109/CloudCom.2016.0047

[13] Perea E et al. Decision support system for distributed energy resources and efficient utilisation of energy in buildings. In: Proceedings of the 22nd International Conference and Exhibition on Electricity Distribution (CIRED 2013); 10-13 June 213; Stockholm. New York: IEEE; 2013, pp. 1-4. DOI: 10.1049/cp.2013.0557

[14] Sianaki OA, Hussain O, Dillon T, Tabesh AR. Intelligent Decision Support System for Including Consumers' Preferences in Residential Energy Consumption in Smart Grid. In: Proceedings of the Second International Conference on Computational Intelligence, Modelling and Simulation; 28-30 Sept. 2010; Bali. New York: IEEE; 2011, pp. 154-159. DOI: 10.1109/CIMSiM.2010.84

[15] Monteiro C et al. Spatial decision support system for site permitting of distributed generation facilities. In: Proceedings of the IEEE Porto Power Tech Proceedings (Cat. No.01EX502); 10-13 Sept. 2001; Porto. New York: IEEE; 2002. pp. 6 pp. vol.3. DOI: 10.1109/PTC.2001.964892

[16] IEEE Guide for Electric Power Distribution Reliability Indices. In: IEEE Std 1366-2012 (Revision of IEEE Std 1366-2003), May 31 2012. New York: IEEE; 2012. vol., no., pp.1-43. DOI: 10.1109/IEEESTD.2012.6209381

[17] Qin YB, Housell J, Rodero I. Cloud-Based Data Analytics Framework for Autonomic Smart Grid Management. In: Proceedings of the International Conference on Cloud and Autonomic Computing; 8-12 Sept. 2014; London. New York: IEEE; 2015, pp. 97-100. DOI: 10.1109/ICCAC.2014.39

[18] Spring MVC official website [Internet]. 2017. Available from: spring.io [Accessed: 2017-02-19]

[19] Hibernate ORM official website [Internet]. 2017. Available from: hibernate.org/orm [Accessed: 2017-02-19]

[20] React JS official website [Internet]. 2017. Available from: facebook.github.io/react [Accessed: 2017-02-19]

[21] Redux JS official website [Internet]. 2015. Available from: redux.js.org [Accessed: 2017-03-10]

[22] Fielding RT. Chapter 5: Representational State Transfer (REST). Architectural styles and the design of network-based software architectures [Ph.D.]. Irvine: University of California; 2000

[23] Bray T. RFC7159—The JavaScript Object Notation (JSON) Data Interchange Format [Internet]. March 2014. Available from: tools.ietf.org/html/rfc7159

Chapter 20

# Optimum Maintenance Policy for Equipment over Changing of the Operation Environment

Ibrahima dit Bouran Sidibe and Imene Djelloul

Additional information is available at the end of the chapter

http://dx.doi.org/10.5772/intechopen.72334

**Abstract**

This chapter investigates optimization of maintenance policy of a repairable equipment whose lifetime distribution depends on the operating environment severity. The considered equipment is undergone to a maintenance policy which consists of repairing minimally at failure and maintaining after operating periods. The periodic maintenance is preventive maintenance (PM) and allows reducing consequently the equipment age but with higher cost than minimal repair. In addition, the equipment has to operate at least in two operating environments with different severity. Therefore, in this analysis, the equipment lifetime distribution function depends on the operating severity. Under these hypotheses, a mathematical modeling of the maintenance cost per unit of time is proposed and discussed. This cost is mathematically analyzed in order to derive optimal periods between preventive maintenance (PM) and the optimal condition under which these exist.

**Keywords:** minimal repair, preventive repair, repairable equipment, several operating environments

## 1. Introduction

To reduce the failure risk of production equipments, preventive maintenance or replacement activities should be performed in appropriate schedules. The search of these appropriate schedules has led to the development and implementation of maintenance optimization policies for stochastic degrading production equipments. Indeed, the literature on this matter is already extensive, growing rapidly and also very heterogeneous. Accordingly, this chapter focuses only to some relevant and fundamental works on the maintenance theory. Early in [1, 2], several models appeared on the optimization of replacement or maintenance policies on infinite time horizon. In these works, the authors mainly discussed about the optimality conditions of theses maintenance models. Subsequently to these works, many extensions of the previous models were proposed on finite time span [3, 4] and also on infinite time horizon in the literature. For survey, the reader may refer, for example, to [5–8] and the references therein. We note that in most cited works, the authors assumed that the equipment lifetime distribution is parametrically characterized and well known. However, Coolen and his coauthors [9, 10] showed that this assumption impacts clearly the optimal replacement age and its cost per unit of time when the equipments undergo an age replacement policy (ARP). Recently in [11], de Jonge et al. pointed out also the weakness of the assumption on knowing of the equipment lifetime distribution and proposed a parametric modeling of ARP for new equipment with an uncertainty on the parameters of the equipment lifetime distribution. In this work, de Jonge and his coauthors used Bayesian approach to model the uncertainty on the parameters and figured out that this uncertainty has effects on the optimal policy (age and cost) under ARP.

Another way, most existing models merely rely only on a classical assumption which states that the operating environment is steady and has not any effect on the equipment characteristic and its lifetime distribution. Roughly speaking, they assume that the degradation process is the same during the equipment's life cycle. This is a restrictive assumption in many industrial areas where production equipment may have experiences under different operating environments with their own severity degree that impacts the equipment performance. For example, the degradation process of the mining machinery is impacted by the severity level of the environment where the machinery is being exploited. Another example may be the engines used for oil extraction. The degradation process of such equipment depends on whether they are operated onshore or offshore. In some other industries, production equipments are first operated in a given environment and then moved to another location where this latter might be more or less severe than the first. In the same way, many companies operate their equipments at home for several years before shipping them to their subsidiaries in other countries where they would be subjected to more severe operating conditions. Therefore, suitable maintenance strategies, integrating the heterogeneous operating conditions, should be developed to assess the degradation of such equipments.

In this chapter, a preventive maintenance is investigated for such equipment subject to random failures. The equipments are assumed to have an experience under two operating environments. In fact, each operating environment is characterized by its own degree of severity, which impacts the equipment lifetime distribution. Therefore, the equipment lifetime distributions follow then a different distribution depending on the operating environments. To reduce the failure occurrence risk during operating under both operating environments, the equipment undergoes to an periodic preventive maintenance (PM). However, the equipment is subjected to minimal repair at failure. The objective consists then on evaluating the optimal age to perform periodic preventive repair in order to minimize the expected maintenance cost per unit of time. This expected cost is induced by the costs of minimal and preventive repairs. This policy was already discussed by Nakagawa in [12], in which Nakawaga considered that the equipment lifetime remains the same during the operation. Nakagawa analyzed mathematically the periodic and sequential maintenance policies. Therefore, our chapter can be considered as an extension of Nakawaga work.

The remainder parts of the chapter are organized as follows. The analyzed problem is briefly introduced in Section 2. This section proposes a mathematical formulation of the total maintenance cost. Section 3 focuses on the maintenance cost analysis in order to derive the optimal conditions which ensure minimal total cost per unit of time. In this same section, an heuristic is proposed to find the optimal number and period between preventive actions on both environments. Numerical experiments are conducted to illustrate the proposed approach on the one hand, and on the other hand, the accuracy and robustness of model are demonstrated through the simulation in Section 4. At the end, a conclusion and future works are drawn in the last section.

## 2. Mathematical formulation of the maintenance cost

In this section, modeling of the maintenance policy is going to be proposed. This modeling takes into account different hypotheses of our analysis. In fact, our equipment has to be used under two operating environments with different severities denoted by $j = 1$ and $j = 2$ which stand, respectively, for the first and second environments. Therefore, the equipment spends T1 and T2 respectively in operating environment 1 and 2. Therefore, the operation duration is the combination of both durations $T_1 + T_{2.}$ The equipment operates successively on both environments in order to perform its missions. During this operation, the equipment undergoes by two types of maintenance actions. Roughly speaking, the equipment is going to be repaired minimally at failure and preventively after some $x_j$ operating periods. The minimal repair costs $c_{mj}$ and allows that the equipment reaches the same reliability just before its failure. However, the preventive repair costs $C_{pj}$ such as $C_p >> c_{mj}$. Therefore, this preventive repair impacts the equipment according to its age and its hazard function. First, the preventive action reduces the equipment age to zero. Second, the preventive action modifies the hazard function such as the hazard function after repair becomes higher than its hazard before. That involves that the wear-out process of the equipment degrades more after the preventive action than before **Figure 1**.

### 2.1. Preventive maintenance cost

During operation, the equipment undergoes by preventive action after each $x_1$ and $x_2$ unit of time, respectively, on the first and second environments. Each of these preventives actions costs $C_{p1}$ on the first and $C_{p2}$ on the second environment. In addition, the number of preventive actions is $n_1$ and $n_{2,}$ respectively, on the first and second environments. Therefore, the total preventive repair costs

$$C_{TP} = n_1{}^* c_{p1} + n_2{}^* c_{p2}, \tag{1}$$

during the length of operation $(T_1 + T_2) = (n_1{}^* x_1 + n_2{}^* x_2)$.

### 2.2. Minimal repair cost

The minimal repair is performed regardless of the preventive actions. The minimal repair is performed at failures in order that equipment reaches the same reliability just before failing. Each minimal repair costs $c_{m1}$ and $c_{m2,}$ respectively, on the first and second environments.

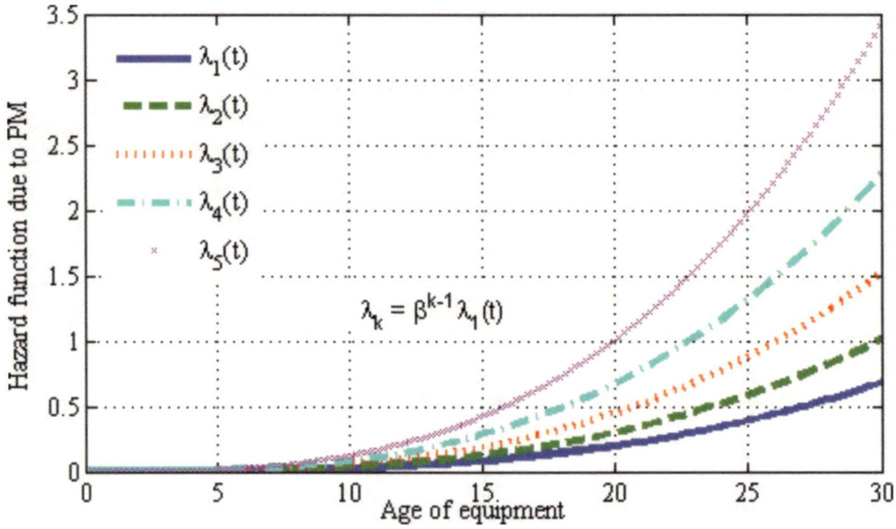


**Figure 1.** Evolution of hazard function due to preventive maintenance.

Therefore, the cost of minimal repair, on the $k^{th}$ interval with a duration $x_j$, is product of expected number of failure by the cost of a minimal repair $c_{mj}$. From Thompson analysis [13], the expected number of renewal on the interval $[0; x_j]$ coincides with the integration of hazard function on $[0; x_j]$. Then, the minimal repair costs during the $k^{th}$ interval are given by

$$C_{mj} = \int_0^{x_j} \lambda_{j,k}(t)dt, \tag{2}$$

$$= -c_{mj} \log R_{j,k}(x_j). \tag{3}$$

where $\lambda_{j,k}(t)$, and $R_{j,k}(t)$ stand for the hazard and the reliability functions of the equipment on the $k^{th}$ and during the $j^{th}$ environment. Therefore, the total minimal cost on the first environment is

$$Ct_{m1} = \sum_{k=1}^{n_1} \int_0^{x_1} \lambda_{1,k}(t)dt, \tag{4}$$

$$= -c_{m1} \sum_{k=1}^{n_1} \log R_{1,k}(x_1). \tag{5}$$

We also deduce the total minimal cost on the second environment as follows

$$Ct_{m2} = \sum_{k=n_1+2}^{n_1+n_2+1} \int_0^{x_2} c_{m2}\lambda_{2,k}(t)dt, \tag{6}$$

$$= -c_{m2} \sum_{k=n_1+2}^{n_1+n_2+1} \log R_{2,k}(x_2). \tag{7}$$

In addition, the operation on the $(n_1 + 1)^{th}$ period also implies a minimal cost. In fact, on this period, the equipment operates on both environments. On the first environment, the equipment operates on $y$ units of time before moving on to the second environment such as $y < x_1$. The minimal cost during this operation is

$$C_{tmy} = c_{m1} \int_0^y \lambda_{1,n_1+1}(t)dt \tag{8}$$

$$= -c_{m1} \log R_{1,n_1+1}(y). \tag{9}$$

After, the equipment moves on to the second environment to operate between $[y; y + x_2]$. In addition, we point out that the second operating environment can be more or less severe than the first. Therefore, to ensure the continuity of reliability function between both operating environments, a transfer function $\phi(t)$ is introduced and defined such as:

$$\begin{aligned} R_{1,n_1+1}(t) &= R_{2,n_1+1}(\phi(t)), \\ \phi(0) &= 0. \end{aligned} \tag{10}$$

That involves a minimal cost on this period

$$C_\phi = c_{m2} \int_{\phi(y)}^{\phi(y)+x_2} \lambda_{2,n_1+1}(t)dt \tag{11}$$

$$= -c_{m2}\left(\log R_{2,n_1+1}(x_2 + \phi(y)) - \log R_{2,n_1+1}(\phi(y))\right). \tag{12}$$

To reduce the complexity during computing, we assume that the duration $y = 0$. That involves a total minimal which clearly depends on the Eqs. (5), (7), (12). The total minimal cost on all operating duration is defined by addition

$$C_{Tm} = Ct_{m1} + Ct_{m2} + C_\phi. \tag{13}$$

Indeed, the hypothesis $y = 0$ also impacts the number of preventive actions. In fact, under this latter hypothesis, the number of preventive actions becomes $n_1 + n_2 - 1$ instead of $n_1 + n_2$ as we indicated in Eq. (1). The total preventive is going to cost

$$C_{TPh} = \begin{cases} n_1 C_{p1} + (n_2 - 1)C_{p2}, \\ (n_1 - 1)C_{p1} + n_2 C_{p2}. \end{cases} \tag{14}$$

Eq. (14) is equivalent to

$$C_{TPh,\gamma} = (n_1 - 1 + \gamma)C_{p1} + (n_2 - \gamma)C_{p2}, \tag{15}$$

where $\gamma = 1$ stands for the fact that at the end of $n_1^{th}$ period the equipment is repaired before moving to the second environment, while $\gamma = 0$ corresponds to the reverse.

### 2.3. Total maintenance cost

From previous Eqs. (13) and (14), we deduce a mathematical formulation of the total maintenance cost according to the set of parameters $(n_1, n_2, x_1, x_2)$ as follows:

$$C(n_1, n_2, x_1, x_2) = \frac{C_{Tm} + C_{TPh,\gamma}}{nx_1 + n_2x_2}. \quad (16)$$

Based on the equation, the next section is going to analyze the optimality according to the different parameters such as the number and the duration between the preventive repairs.

## 3. Optimality analysis

Herein, the maintenance cost is rewritten in order to integrated the impacts of preventive maintenance (PM) on the equipment lifetime distribution. We assume that a preventive action allows to reduce the age of equipment to zero and increase the hazard function. **Figures 1** and **2** point out the impact of PM on the equipment hazard and reliability functions. The hazard function is defined after PM as follows

$$\lambda_{j,k}(t) = \beta_j \lambda_{j,k-1}(t), \quad (17)$$

where $j = 1, 2,$ and $\beta_j > 1$. Under these hypotheses, Eq. (5), which represents the total minimal cost on the first environment, is rewritten as

$$C_{tm1} = -c_{m1} \frac{1 - \beta_1^{n_1}}{1 - \beta_1} \log R_1(x_1), \quad (18)$$

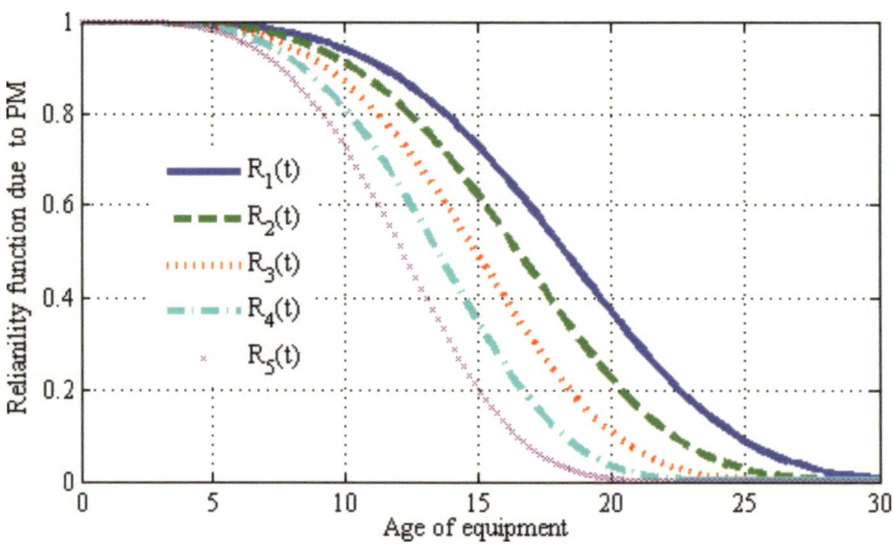


**Figure 2.** Evolution of reliability function due to preventive maintenance.

with

$$R_1(x_1) = R_{1,1}(x_1). \tag{19}$$

In the second operating environment, the hazard function at $(n+1)^{th}$ is a consequence of $n_1 - 1 + \gamma$ PM in first and $1 - \gamma$ in the second environment.

$$\begin{aligned}
\lambda_{2,n_1+1}(t) &= \beta_1^{n_1-1+\gamma}\beta_2^{1-\gamma}\lambda_2(t)\\
\lambda_{2,n_1+2}(t) &= \beta_1^{n_1-1+\gamma}\beta_2^{2-\gamma}\lambda_2(t)\\
\ldots &= \ldots\\
\lambda_{2,n_1+n_2}(t) &= \beta_1^{n_1-1+\gamma}\beta_2^{n_2-\gamma}\lambda_2(t)
\end{aligned} \tag{20}$$

with

$$\lambda_2(t) = \lambda_{2,1}(t). \tag{21}$$

The total cost due to the minimal repair in the second environment becomes

$$C_{mt2} = -c_{m2}\beta_1^{n_1-1+\gamma}\beta_2^{1-\gamma}\frac{1-\beta_2^{n_2}}{1-\beta_2}\log R_2(x_2). \tag{22}$$

By considering Eqs. (15), (18), and (19), the total cost per unit of time is rewritten as follows

$$\begin{aligned}
C(n_1,n_2,x_1,x_2) &= \frac{1}{n_1x_1+n_2x_2}\left((n_1-1+\gamma)C_{p1} + (n_2-\gamma)C_{p2}\right)\\
&-\frac{1}{n_1x_1+n_2x_2}\left(c_{m1}\frac{1-\beta_1^{n_1}}{1-\beta_1}\log R_1(x_1)\right)\\
&-\frac{1}{n_1x_1+n_2x_2}\left(c_{m2}\beta_1^{n_1-1+\gamma}\beta_2^{1-\gamma}\frac{1-\beta_2^{n_2}}{1-\beta_2}\log R_2(x_2)\right).
\end{aligned} \tag{23}$$

### 3.1. Optimality according to $n_1$ and $n_2$

Let us assume that there is a pair $(n_1, n_2)$ that provides the minimal cost per unit according to the Eq. (20) for given periods $(x_1, x_2)$ between preventive repairs. Then, corresponding cost has to remain the unique lowest bound relative to other pairs of integer. This implies that cost at $(n_1, n_2)$ must be better than the costs from the successive pairs $\{(n_1+1, n_2), (n_1-1, n_2)\}$; $\{(n_1, n_2+1), (n_1, n_2-1)\}$ and $\{(n_1+1, n_2+1), (n_1-1, n_2-1)\}$. The existence and uniqueness of the pairs are analyzed through some propositions.

#### *3.1.1. Local optimality*

The local optimality concerns the direct neighbors of the optimal pair such as $\{(n_1+1, n_2), (n_1-1, n_2)\}$ and $\{(n_1, n_2-1), (n_1, n_2+1)\}$. Let pose that

$$L_1(n_1|n_2) = C(n_1, n_2, x_1, x_2) - C(n_1 + 1, n_2, x_1, x_2), \tag{24}$$

$$L_2(n_2|n_1) = C(n_1, n_2, x_1, x_2) - C(n_1, n_2 + 1, x_1, x_2), \tag{25}$$

**Proposition 1** *If the lifetime distribution functions are increasing failure rate (IFR) and* $L_1(1|n_2) > 0$, *then there exists a unique optimal number of PM* $n_1$ *in the first environments in which this* $n_1$ *ensures the minimal cost per unit time for a fixed pair* $(x_1, x_2)$ *and* $n_2$.

**Proof**. As the maintenance cost per unit of time is minimal for $(n_1, n_2)$, then we have

$$\begin{cases} C(n_1, n_2, x_1, x_2) \le C(n_1 - 1, n_2, x_1, x_2), \\ C(n_1, n_2, x_1, x_2) < C(n_1 + 1, n_2, x_1, x_2). \end{cases} \tag{26}$$

This system is equivalent to

$$\begin{cases} L_1(n_1 - 1|n_2) \ge 0, \\ L_1(n_1|n_2) < 0. \end{cases} \tag{27}$$

with

$$\begin{aligned} L_1(n_1 - 1|n_2) &= -n_2 x_2 C_{p1} + x_1\left((\gamma - 1)C_{p1} + (n_2 - \gamma)C_{p2}\right) \\ &-x_1\left(c_{m_1}\frac{1-\beta^{n_1}}{1-\beta_1}\log R_1(x_1) + c_{m2}\beta_1^{n_1-1+\gamma}\beta_2^{1-\gamma}\frac{1-\beta_2^{n_2}}{1-\beta_2}\log R_2(x_2)\right) \\ &+(n_1x_1 + n_2x_2)\left(c_{m1}\beta_1^{n_1-1}\log R_1(x_1) + c_{m2}\beta_1^{n_1-2+\gamma}\beta_2^{1-\gamma}\frac{\beta_1 - 1}{1-\beta_2}(1-\beta_2^{n_2})\log R_2(x_2)\right). \end{aligned}$$

In fact

$$\lim_{n_1 \leftarrow +\infty} L_1(n_1|n_2) = -\infty,$$

and

$$\begin{aligned} L_1(n_1|n_2) - L_1(n_1 - 1|n_2) &= (n_1x_1 + n_2x_2)\left(c_{m1}\beta_1^{n_1}\log R_1(x_1)\right) \\ &+(n_1x_1 + n_2x_2)\left(c_{m2}\beta_1^{n_1-1+\gamma}\beta_2^{1-\gamma}\frac{\beta_1 - 1}{1-\beta_2}(1-\beta_2^{n_2})\log R_2(x_2)\right), \end{aligned}$$

The right-hand side of the previous equation shows that $L_1(n_1|n_2) - L_1(n_1 - 1|n_2) < 0$. This implies that $L_1(n_1|n_2)$ decreases with $n_1$. If $L_1(1|n_2) > 0$, then there exists a unique $n_1$ which verifies condition (23) and ensures the minimal cost per unit time for given $n_2$.

**Proposition 2** *If the lifetime distribution function of equipment on both environments is IFR and* $L_2(1|n_1) > 0$, *then there exists a unique optimal number of PM* $n_2$ *in the second environment in which this number ensures the minimal cost per unit time for corresponding fixed pair* $(x_1, x_2)$ *and* $n_2$.

**Proof**. As the cost maintenance per unit time is minimal for $(n_1, n_2)$, then we have

$$\begin{cases} C(n_1, n_2, x_1, x_2) \leq C(n_1, n_2 - 1, x_1, x_2), \\ C(n_1, n_2, x_1, x_2) < C(n_1, n_2 + 1, x_1, x_2). \end{cases} \tag{28}$$

This is equivalent to

$$\begin{cases} L_2(n_2 - 1|n_1) \geq 0, \\ L_2(n_2|n_1) < 0. \end{cases} \tag{29}$$

with

$$\begin{aligned} L_2(n_2 - 1|n_1) &= -n_1 x_1 C_{p2} + x_2((n_1 - 1 + \gamma)C_{p1} - \gamma C_{p2}) \\ &-x_2\left(c_{m1}\frac{1-\beta_1^{n_1}}{1-\beta_1}\log R_1(x_1) + c_{m2}\beta_1^{n_1-1+\gamma}\beta_2^{1-\gamma}\frac{1-\beta_2^{n_2}}{1-\beta_2}\log R_2(x_2)\right) \\ &+(n_1x_1 + n_2x_2)\left(c_{m2}\beta_1^{n_1-1+\gamma}\beta_2^{n_2-\gamma}\log R_2(x_2)\right). \end{aligned}$$

This equation implies

$$\lim_{n_2 \to \infty} L_2(n_2|n_1) = -\infty,$$

and

$$L_2(n_2|n_1) - L_2(n_2 - 1|n_1) = (n_1x_1 + n_2x_2)\left(c_{m2}\beta_1^{n_1-1+\gamma}\beta_2^{n_2-\gamma+1}\log R_2(x_2)\right) < 0.$$

Therefore, $L_2(n_2|n_1)$ decreases with and for $L_2(1|n_1) > 0$, we have a unique $n_2$ in which the total per unit of time is minimal for fixed $n_1$.

### 3.1.2. Global optimality

The global optimality compares the optimal pair to $\{(n_1 + 1, n_2 + 1), (n_1 - 1, n_2 - 1)\}$. Let us pose that

$$L_3(n_1, n_2) = C(n_1, n_2, x_1, x_2) - C(n_1 + 1, n_2 + 1, x_1, x_2). \tag{30}$$

**Proposition 3** *If the lifetime distribution functions are IFR and $L_3(1,1) > 0$, then there exists a unique optimal number of PM $(n_1, n_2)$ in which this ensures the minimal cost per unit time for a fixed pair $(x_1, x_2)$.*

**Proof**. As the cost is minimal for $(n_1, n_2)$, then

$$\begin{cases} C(n_1, n_2, x_1, x_2) \leq C(n_1 - 1, n_2 - 1, x_1, x_2), \\ C(n_1, n_2, x_1, x_2) < C(n_1 + 1, n_2 + 1, x_1, x_2). \end{cases} \tag{31}$$

This is equivalent to

$$\begin{cases} L_3(n_1-1,n_2-1)\geq 0, \\ L_3(n_1,n_2)<0. \end{cases} \tag{32}$$

with

$$L_3(n_1-1,n_2-1)=(x_1+x_2)((\gamma-1)C_{p1}-\gamma C_{p2})$$
$$-(x_1+x_2)\left(c_{m1}\frac{1-\beta_1^{n_1}}{1-\beta_1}\log R_1(x_1)+c_{m2}\beta_1^{n_1-1}\beta_2^{1-\gamma}\frac{1-\beta_2^{n_2}}{1-\beta_2}\right)$$
$$+(n_1x_1+n_2x_2)\left(c_{m1}\beta^{n_1-1}\log R_1(x_1)-c_{m2}\frac{1-\beta_1-\beta_2^{n_2-1}(1-\beta_2)}{1-\beta_2}\beta_1^{n_1-2}\beta_2^{1-\gamma}\right).$$

With

$$L_3(+\infty,+\infty)=-\infty,$$

and

$$L_3(n_1,n_2)-L_3(n_1-1,n_2-1)<0.$$

Therefore, $L_3(n_1,n_2)$ decreases with $(n_1,n_2)$ and for $L_3(1,1)>0,$ we have a unique pair $(n_1,n_2)$ in which the total per unit of time is minimal.

### 3.2. Optimality according to $x_1$ and $x_2$

For given number of preventive actions $(n_1,n_2)$, the optimal durations $(x_1,x_2)$ between preventive actions in both environments have to verify

$$\begin{cases} \frac{\partial}{\partial x_1}C(n_1,n_2,x_1,x_2)=0 \\ \frac{\partial}{\partial x_2}C(n_1,n_2,x_1,x_2)=0; \end{cases} \tag{33}$$

This implies

$$\begin{cases} c_{m1}\frac{1-\beta_1^{n_1}}{1-\beta_1}\lambda_1(x_1) & =n_1C(n_1,n_2,x_1,x_2), \\ c_{m2}\beta_1^{n_1-1+\gamma}\beta_2^{1-\gamma}\frac{1-\beta_2^{n_2}}{1-\beta_2}\lambda_2(x_2) & =n_2C(n_1,n_2,x_1,x_2). \end{cases} \tag{34}$$

By dividing, we obtain

$$\frac{\lambda_1(x_1)}{\lambda_2(x_2)}=\left(\beta_1^{n_1-1+\gamma}\beta_2^{1-\gamma}\right)\frac{n_1}{n_2}\frac{c_{m2}}{c_{m1}}\frac{1-\beta_2^{n_2}}{1-\beta_1^{n_1}}\frac{1-\beta_1}{1-\beta_2} \tag{35}$$

**Proposition 4** *If the lifetime functions of the equipment are Weibull-distributed in both environments with the same shape parameter b, then the optimal interval between PM is defined as*

$$\frac{x_1}{x_2} = Cste^* \left(\frac{n_1}{n_2}\right)^{1/(b-1)}. \tag{36}$$

**Proof**. As lifetime functions are Weibull-distributed with the same parameter $b$, then the hazard functions are defined as follows

$$\lambda_1(x_1) = \frac{b}{\eta_1}\left(\frac{x_1}{\eta_1}\right)^{b-1}, \tag{37}$$

$$\lambda_2(x_2) = \frac{b}{\eta_2}\left(\frac{x_2}{\eta_2}\right)^{b-1}, \tag{38}$$

and from Eq. (32), we deduce

$$\frac{x_1}{x_2} = \left(\frac{c_{m2}}{c_{m1}}\left(\beta_1^{n_1-1+\gamma}\beta_2^{1-\gamma}\right)\frac{1-\beta_2^{n_2}}{1-\beta_1^{n_1}}\frac{1-\beta_1}{1-\beta_2}\right)^{1/(b-1)}\frac{\eta_1}{\eta_2}\left(\frac{n_1}{n_2}\right)^{1/(b-1)}. \tag{39}$$

The uniqueness is tough to establish due to the number of parameters and the complexity of the proposed cost model here. To make the research of optimal solution easy, we propose a handy heuristic based on the optimal derived conditions in this chapter. The next section describes step by step the proposed heuristic which leads to a suitable solution for our optimization problem.

### 3.3. Numerical resolution of problem

Herein, an algorithm is drawn in order to find the optimal pairs for $(n_1, n_2)$ and $(x_1, x_2)$. The optimal pairs ensure the minimal cost per unit time defined by Eq. (20). Moreover, the existence of these optimal pairs is discussed in the previous sections. The proposed heuristic makes switching between the research of pairs ($(n_1, n_2)$ and $(x_1, x_2)$). This algorithm converges surely toward the pair that ensures the minimal cost according to the conditions deduce from the Eq. (20). The next section presents an application of our approach. The algorithm is on the previous propositions and defined as follows.

---

**Algorithm 1** Compute the optimal pairs of number $(n_1, n_2)$ and periods $(x_1, x_2)$ of PM.

---

Initialize the pair $(n_1, n_2)_0 = (1, 1)$.

Put $(n_1, n_2) = (n_1, n_2)_0$

STEP (A) Research optimal $(x_1, x_2)$ for given $(n_1, n_2)$.

Compute $L_1(1|n_2)$, $L_2(1|n_1)$ and $L_3(1, 1)$.

**if** $L_1(1|n_2) > 0$ **then**

Research $n_1(a)$ which verifies condition (24) is verified

$n_1(1) = n_1(a)$ and $n_2(1) = n_2$.

$C_1 = C(n_1(1), n_2(1))$

**else**

$\{L_1(1|n_2) < 0\}$

$C_1 = \infty$

**if** $L_2(1|n_1) > 0$ **then**

Research $n_2(b)$ which verifies conditions (26).

$n_1(2) = n_1; n_2(2) = n_2(b)$

$C_2 = C(n_1(2), n_2(2))$

**else** $\{L_2(1|n_1) < 0\}$

$C_2 = \infty$

**if** $L_3(1,1) > 0$ **then**

Research $n_2(c)$ which verifies $L_3(n_1(c), n_2(c))$ (29).

$n_1(2) = n_1(c); n_2(2) = n_2(c)$

$C_3 = C(n_1(2), n_2(2))$

**else**

$\{L_3(1,1) < 0\}$

$C_3 = \infty$

$C_{min} = \text{Min}\{C_1, C_2, C_3\}$

$m = m + 1$

$(n_1, n_2)_m = \{(n_1(i), n_2(i)|C_i = C_{min}\}$

**if** $(n_1, n_2)_m = (n_1, n_2)_{m-1}$ **then**

$(n_1, n_2) = (n_1, n_2)_m$

Keep corresponding $(x_1, x_2)$ **else**

$\{(n_1, n_2)_m \neq (n_1, n_2)_{m-1}\}$

$(n_1, n_2) = (n_1, n_2)_m$

Go to step (A)

**end if**

End.

## 4. Numerical application

We consider an equipment whose lifetime distribution function is Weibull with the same shape parameter $b = 2.0$. The equipment has to be used on two environments with different severity. Their severity depends on the scale parameter, such as in first the scale is $\eta_1 = 20$, while $\eta_2 = 10$ stands for the scale parameter in the second environment. This implies that the second environment is twice more severe than first. To reduce the risk of equipment failure of the failure, the equipment undergoes periodic, preventive maintenance. The preventive maintenance costs $C_{p1} = 100$ and $C_{p2} = 150$, respectively, on the first and second. The preventive actions impact the lifetime distribution of equipment. The impact factors due to PM are equal to $\beta_1 = 1.85$ in first and $\beta_2 = 2.5$ in the second environment. In addition, the equipment is minimally repaired at failure. The costs of minimal repair are in both environments $c_{m1} = 80$ and $c_{m2} = 70$. Based on this information, we are going to solve the optimization problem in order to find the number and duration period between PM on each environment which ensure a minimal cost per unit of time. With these parameters, the minimal cost reaches 10.37. This minimal cost involves $n_1 = 1$ and $n_2 = 1$ preventive maintenance (PM) respectively in the first and second environments. The durations between each PM are $x_1 = 26.06$ and $x_2 = 3.03$.

## 5. Conclusion

This chapter shows how to solve Nakagawa maintenance policy problem for an equipment which operates simultaneously on two environments. Each environment impacts the lifetime distribution function of our equipment. Nakagawa's maintenance problem is modeled under lifetime distribution changing in operation. The proposed model is deeply analyzed in order to derive the conditions under which optimal pairs exist and are reachable. To reach these pairs, algorithm was proposed to find the optimal solution for the periodic preventive maintenance on infinite horizon. The model is handy and suitable for production equipments which have to experience under different operating environments with their own severity degree that impacts the equipment performance such as onshore or offshore.

For future work, we plan to propose a statistical modeling by ignoring the hypothesis on the knowledge of the equipment lifetime distribution and perform an extension of the analysis by considering an finite-time horizon/span.

## Author details

Ibrahima dit Bouran Sidibe[1]* and Imene Djelloul[2,3]

*Address all correspondence to: bouransidibe@gmail.com

1 Centre de Formation et de Perfectionnement en Statistique (CFP-STAT) Bamako, Mali

2 Higher School of Applied Sciences of Algiers, Place des Martyres, Alger, Algerie

3 Manufacturing Engineering Laboratory of Tlemcen (MELT), Abou Bekr Belkaid University of Tlemcen, Tlemcen, Algeria

## References

[1] Barlow R, Hunter L. Optimum preventive maintenance policies. Operations Research. 1960;**8**(1):90-100

[2] Barlow R, Proschan F. Mathematical Theory of Reliability. New York: John Wiley & Sons; 1965

[3] Nakagawa T, Mizutani S. A summary of maintenance policies for a finite interval. Reliability Engineering & System Safety. 2009;**94**(1):89-96

[4] Nakagawa T. Advanced Reliability Models and Maintenance Policies. Springer series in reliability engineering, London: Springer; 2008

[5] Cho D, Parlar M. A survey of maintenance models for multi-unit systems. European Journal of Operational Research. 1991;**51**(1):1-23

[6] Dekker R. Applications of maintenance optimization models: A review and analysis. Reliability Engineering & Systems Safety. 1996;**51**(3):229-240

[7] Jardine A, Tsang A. Maintenance, Replacement, and Reliability: Theory and Applications. Boca Raton: CRC/Taylor & Francis; 2006

[8] Lugtigheid D, Jardine A, Jiang X. Optimizing the performance of a repairable system under a maintenance and repair contract. Quality and Reliability Engineering International. 2007;**23**(8):943-960

[9] Coolen F, Coolen-Schrijner P, Yan K. Non-parametric predictive inference in reliability. Reliability Engineering & System Safety. 2002;**78**(2):185-193

[10] Coolen-Schrijner P, Coolen F. Non-parametric predictive inference for age replacement with a renewal argument. Quality and Reliability Engineering International. 2004;**20**(3): 203-215

[11] de Jonge B, Klingenberg W, Teunter R, Tinga T. Optimum maintenance strategy under uncertainty in the lifetime distribution. Reliability Engineering & System Safety. 2015;**133**: 59–67.

[12] Nakagawa T. Periodic and sequential preventive maintenance policies. Journal of Applied Probability. 1986;**23**(2):536-542

[13] Thompson WA. On the foundations of reliability. Technometrics. 1981;**23**(1):1-13

For EU product safety concerns:
IN TECH d.o.o., Prolaz Marije Krucifikse Kozulić 3, 51000 Rijeka, Croatia,
info@intechopen.com or visit our website at intechopen.com.

# Meet the editor

Dr Parinya Sanguansat received the B.Eng., M.Eng and Ph.D. degrees in Electrical Engineering from Chulalongkorn University, Thailand, in 2001, 2004 and 2007, respectively. From 2007 to 2011, he was a member of the Faculty of Information Technology, Rangsit University, Thailand. He is currently an Assistant Professor at the Faculty of Engineering and Technology, Panyapiwat Institute of Management, Thailand. His research interests include machine learning, pattern recognition and image processing.

# Contents

# Preface

It is more than a century since Karl Pearson invented the concept of Principal Component Analysis (PCA). Nowadays, it is a very useful tool in data analysis in many fields. PCA is the technique of dimensionality reduction, which transforms data in the high-dimensional space to space of lower dimensions. The advantages of this subspace are numerous. First of all, the reduced dimension has the effect of retaining the most of the useful information while reducing noise and other undesirable artifacts. Secondly, the time and memory that used in data processing are smaller. Thirdly, it provides a way to understand and visualize the structure of complex data sets. Furthermore, it helps us identify new meaningful underlying variables.

Indeed, PCA itself does not reduce the dimension of the data set. It only rotates the axes of data space along lines of maximum variance. The axis of the greatest variance is called the first principal component. Another axis, which is orthogonal to the previous one and positioned to represent the next greatest variance, is called the second principal component, and so on. The dimension reduction is done by using only the first few principal components as a basis set for the new space. Therefore, this subspace tends to be small and may be dropped with minimal loss of information.

Originally, PCA is the orthogonal transformation which can deal with linear data. However, the real-world data is usually nonlinear and some of it, especially multimedia data, is multilinear. Recently, PCA is not limited to only linear transformation. There are many extension methods to make possible nonlinear and multilinear transformations via manifolds based, kernel-based and tensor-based techniques. This generalization makes PCA more useful for a wider range of applications.

In this book the reader will find the applications of PCA in many fields such as taxonomy, biology, pharmacy, finance, agriculture, ecology, health, architecture. It also includes the core concepts and the state-of-the-art methods in data analysis and feature extraction.

Finally, I would like to thank all recruited authors for their scholarly contributions and also to InTech staff for publishing this book, and especially to Mr.Oliver Kurelic, for his kind assistance throughout the editing process. Without them this book could not be possible. On behalf of all the authors, we hope that readers will benefit in many ways from reading this book.

**Parinya Sanguansat**
Faculty of Engineering and Technology, Panyapiwat Institute of Management
Thailand

# 1

# Kernel Methods for Dimensionality Reduction Applied to the «Omics» Data

Ferran Reverter, Esteban Vegas and Josep M. Oller
*Department of Statistics, University of Barcelona*
*Spain*

## 1. Introduction

Microarray technology has been advanced to the point at which the simultaneous monitoring of gene expression on a genome scale is now possible. Microarray experiments often aim to identify individual genes that are differentially expressed under distinct conditions, such as between two or more phenotypes, cell lines, under different treatment types or diseased and healthy subjects. Such experiments may be the first step towards inferring gene function and constructing gene networks in systems biology.

The term "gene expression profile" refers to the gene expression values on all arrays for a given gene in different groups of arrays. Frequently, a summary statistic of the gene expression values, such as the mean or the median, is also reported. Dot plots of the gene expression measurements in subsets of arrays, and line plots of the summaries of gene expression measurements are the most common plots used to display gene expression data (See for example Chambers (1983) and references therein).

An ever increasing number of techniques are being applied to detect genes which have similar expression profiles from microarray experiments. Techniques such clustering (Eisen et al. (1998)), self organization map (Tamayo et al. (1999)) have been applied to the analysis of gene expression data. Also we can found several applications on microarray analysis based on distinct machine learning methods such as Gaussian processes (Chu et al. (2005); Zhao & Cheung (2007)), Boosting (Dettling (2004)) and Random Forest (Diaz (2006)). It is useful to find gene/sample clusters with similar gene expression patterns for interpreting the microarray data.

However, due to the large number of genes involved it might be more effective to display these data on a low dimensional plot. Recently, several authors have explored dimension reduction techniques. Alter et al. (2000) analyzed microarray data using singular value decomposition (SVD), Fellenberg et al. (2001) used correspondence analysis to visualize genes and tissues, Pittelkow & Wilson (2003) and Park et al. (2008) used several variants of biplot methods as a visualization tool for the analysis of microarray data. Visualizing gene expression may facilitate the identification of genes with similar expression patterns.

Principal component analysis has a very long history and is known to very powerful for the linear case. However, the sample space that many research problems are facing, especially the sample space of microarray data, are considered nonlinear in nature. One reason might be that the interaction of the genes are not completely understood. Many biological pathways are still beyond human comprehension. It is then quite naive to assume that the genes should be connected in a linear fashion. Following this line of thought, research on nonlinear dimensionality reduction for microarray gene expression data has increased (Zhenqiu et al. (2005), Xuehua & Lan (2009) and references therein). Finding methods that can handle such data is of great importance if as much information as possible is to be gleaned.

Kernel representation offers an alternative to nonlinear functions by projecting the data into a high-dimensional feature space, which increases the computational power of linear learning machines, (see for instance Shawe-Taylor & Cristianini (2004); Scholkopf & Smola (2002)).

Kernel methods enable us to construct different nonlinear versions of any algorithm which can be expressed solely in terms of dot products, known as the kernel trick. Thus, kernel algorithms avoid the explicit usage of the input variables in the statistical learning task. Kernel machines can be used to implement several learning algorithms but they usually act as a black-box with respect to the input variables. This could be a drawback in biplot displays in which we pursue the simultaneous representation of samples and input variables.

In this work we develop a procedure for enrich the interpretability of Kernel PCA by adding in the plot the representation of input variables. We used the radial basis kernel (Gaussian kernel) in our implementation however, the procedure we have introduced is also applicable in cases that may be more appropriated to use any other smooth kernel, for example the Linear kernel which supplies standard PCA analysis. In particular, for each input variable (gene) we can represent locally the direction of maximum variation of the gene expression. As we describe below, our implementation enables us to extract the nonlinear features without discarding the simultaneous display of input variables (genes) and samples (microarrays).

## 2. Kernel PCA methodology

KPCA is a nonlinear equivalent of classical PCA that uses methods inspired by statistical learning theory. We describe shortly the KPCA method from Scholkopf et al. (1998).

Given a set of observations $\mathbf{x}_i \in \mathbb{R}^n, i = 1, \ldots, m$. Let us consider a dot product space $F$ related to the input space by a map $\phi : \mathbb{R}^n \to F$ which is possibly nonlinear. The feature space $F$ could have an arbitrarily large, and possibly infinite, dimension. Hereafter upper case characters are used for elements of $F$, while lower case characters denote elements of $\mathbb{R}^n$. We assume that we are dealing with centered data $\sum_{i=1}^{m} \phi(\mathbf{x}_i) = 0$.

In $F$ the covariance matrix takes the form

$$\mathbf{C} = \frac{1}{m} \sum_{j=1}^{m} \phi(\mathbf{x}_j)\phi(\mathbf{x}_j)^\top.$$

We have to find eigenvalues $\lambda \geq 0$ and nonzero eigenvectors $\mathbf{V} \in F \backslash \{0\}$ satisfying

$$\mathbf{CV} = \lambda \mathbf{V}.$$

As is well known all solutions $\mathbf{V}$ with $\lambda \neq 0$ lie in the span of $\{\phi(\mathbf{x}_i)\}_{i=1}^m$. This has two consequences: first we may instead consider the set of equations

$$\langle \phi(\mathbf{x}_k), \mathbf{CV} \rangle = \lambda \langle \phi(\mathbf{x}_k), \mathbf{V} \rangle, \tag{1}$$

for all $k = 1, \ldots, m$, and second there exist coefficients $\alpha_i$, $i = 1, \ldots, m$ such that

$$\mathbf{V} = \sum_{i=1}^{m} \alpha_i \phi(\mathbf{x}_i). \tag{2}$$

Combining (1) and (2) we get the dual representation of the eigenvalue problem

$$\frac{1}{m} \sum_{i=1}^{m} \alpha_i \left\langle \phi(\mathbf{x}_k), \sum_{j=1}^{m} \phi(\mathbf{x}_j) \left\langle \phi(\mathbf{x}_j), \phi(\mathbf{x}_i) \right\rangle \right\rangle = \lambda \sum_{i=1}^{m} \alpha_i \langle \phi(\mathbf{x}_k), \phi(\mathbf{x}_i) \rangle,$$

for all $k = 1, \ldots m$. Defining a $m \times m$ matrix $K$ by $K_{ij} := \left\langle \phi(\mathbf{x}_i), \phi(\mathbf{x}_j) \right\rangle$, this reads

$$K^2 \boldsymbol{\alpha} = m\lambda K \boldsymbol{\alpha}, \tag{3}$$

where $\boldsymbol{\alpha}$ denotes the column vector with entries $\alpha_1, ..., \alpha_m$. To find the solutions of (3), we solve the dual eigenvalue problem

$$K\boldsymbol{\alpha} = m\lambda \boldsymbol{\alpha}, \tag{4}$$

for nonzero eigenvalues. It can be shown that this yields all solutions of (3) that are of interest for us. Let $\lambda_1 \geq \lambda_2 \geq \cdots \geq \lambda_m$ the eigenvalues of $K$ and $\boldsymbol{\alpha}^1, ..., \boldsymbol{\alpha}^m$ the corresponding set of eigenvectors, with $\lambda_r$ being the last nonzero eigenvalue. We normalize $\boldsymbol{\alpha}^1, ..., \boldsymbol{\alpha}^r$ by requiring that the corresponding vectors in $F$ be normalized $\left\langle \mathbf{V}^k, \mathbf{V}^k \right\rangle = 1$, for all $k = 1, ..., r$. Taking into account (2) and (4), we may rewrite the normalization condition for $\boldsymbol{\alpha}^1, ..., \boldsymbol{\alpha}^r$ in this way

$$1 = \sum_{i,j}^{m} \alpha_i^k \alpha_j^k \left\langle \phi(\mathbf{x}_i), \phi(\mathbf{x}_j) \right\rangle = \sum_{i,j}^{m} \alpha_i^k \alpha_j^k K_{ij} = \left\langle \boldsymbol{\alpha}^k, K\boldsymbol{\alpha}^k \right\rangle = \lambda_k \left\langle \boldsymbol{\alpha}^k, \boldsymbol{\alpha}^k \right\rangle. \tag{5}$$

For the purpose of principal component extraction, we need to compute the projections onto the eigenvectors $\mathbf{V}^k$ in $F$, $k = 1, ..., r$. Let $\mathbf{y}$ be a test point, with an image $\phi(\mathbf{y})$ in $F$. Then

$$\left\langle \mathbf{V}^k, \phi(\mathbf{y}) \right\rangle = \sum_{i=1}^{m} \alpha_i^k \langle \phi(\mathbf{x}_i), \phi(\mathbf{y}) \rangle, \tag{6}$$

are the nonlinear principal component corresponding to $\phi$.

### 2.1 Centering in feature space

For the sake of simplicity, we have made the assumption that the observations are centered. This is easy to achieve in input space but harder in $F$, because we cannot explicitly compute the mean of the mapped observations in $F$. There is, however, a way to do it.

Given any $\phi$ and any set of observations $\mathbf{x}_1, ..., \mathbf{x}_m$, let us define

$$\bar{\phi} := \frac{1}{m}\sum_{i=1}^{m}\phi(\mathbf{x}_i)$$

then, the points

$$\tilde{\phi}(\mathbf{x}_i) = \phi(\mathbf{x}_i) - \bar{\phi} \tag{7}$$

will be centered. Thus the assumption made above now hold, and we go on to define covariance matrix and dot product matrix $\tilde{K}_{ij} = \left\langle \tilde{\phi}(\mathbf{x}_i), \tilde{\phi}(\mathbf{x}_j) \right\rangle$ in $F$. We arrive at our already familiar eigenvalue problem

$$m\tilde{\lambda}\tilde{\boldsymbol{\alpha}} = \tilde{K}\tilde{\boldsymbol{\alpha}}, \tag{8}$$

with $\tilde{\boldsymbol{\alpha}}$ being the expansion coefficients of an eigenvector (in $F$) in terms of the centered points (7)

$$\tilde{\mathbf{V}} = \sum_{i=1}^{m}\tilde{\alpha}_i\tilde{\phi}(\mathbf{x}_i). \tag{9}$$

Because we do not have the centered data (7), we cannot compute $\tilde{K}$ explicitly, however we can express it in terms of its noncentered counterpart $K$. In the following, we shall use $K_{ij} = \left\langle \phi(\mathbf{x}_i), \phi(\mathbf{x}_j) \right\rangle$. To compute $\tilde{K}_{ij} = \left\langle \tilde{\phi}(\mathbf{x}_i), \tilde{\phi}(\mathbf{x}_j) \right\rangle$, we have:

$$\begin{aligned}\tilde{K}_{ij} &= \left\langle \phi(\mathbf{x}_i) - \bar{\phi}, \phi(\mathbf{x}_j) - \bar{\phi} \right\rangle \\ &= K_{ij} - \frac{1}{m}\sum_{t=1}^{m}K_{it} - \frac{1}{m}\sum_{s=1}^{m}K_{sj} + \frac{1}{m^2}\sum_{s,t=1}^{m}K_{st}.\end{aligned}$$

Using the vector $\mathbf{1}_m = (1, ..., 1)^\intercal$, we get the more compact expression

$$\tilde{K} = K - \frac{1}{m}K\mathbf{1}_m\mathbf{1}_m^\intercal - \frac{1}{m}\mathbf{1}_m\mathbf{1}_m^T K + \frac{1}{m^2}(\mathbf{1}_m^\intercal K\mathbf{1}_m)\mathbf{1}_m\mathbf{1}_m^\intercal.$$

We thus can compute $\tilde{K}$ from $K$ and solve the eigenvalue problem (8). As in equation (5), the solution $\tilde{\alpha}^k$, $k = 1, ..., r$, are normalized by normalizing the corresponding vector $\tilde{\mathbf{V}}^k$ in $F$, which translates into $\tilde{\lambda}_k \left\langle \tilde{\alpha}^k, \tilde{\alpha}^k \right\rangle = 1$.

Consider a test point $\mathbf{y}$. To find its coordinates we compute projections of centered $\phi$-images of $\mathbf{y}$ onto the eigenvectors of the covariance matrix of the centered points,

$$\begin{aligned}\left\langle \tilde{\phi}(\mathbf{y}), \tilde{\mathbf{V}}^k \right\rangle &= \left\langle \phi(\mathbf{y}) - \bar{\phi}, \tilde{\mathbf{V}}^k \right\rangle = \sum_{i=1}^{m}\tilde{\alpha}_i^k \left\langle \phi(\mathbf{y}) - \bar{\phi}, \phi(\mathbf{x}_i) - \bar{\phi} \right\rangle \\ &= \sum_{i=1}^{m}\tilde{\alpha}_i^k\big( \langle \phi(\mathbf{y}), \phi(\mathbf{x}_i) \rangle - \langle \bar{\phi}, \phi(\mathbf{x}_i) \rangle - \langle \phi(\mathbf{y}), \bar{\phi} \rangle + \langle \bar{\phi}, \bar{\phi} \rangle \big) \\ &= \sum_{i=1}^{m}\tilde{\alpha}_i^k \Big\{ K(\mathbf{y}, \mathbf{x}_i)) - \frac{1}{m}\sum_{s=1}^{m}K(\mathbf{x}_s, \mathbf{x}_i)) - \frac{1}{m}\sum_{s=1}^{m}K(\mathbf{y}, \mathbf{x}_s) + \frac{1}{m^2}\sum_{s,t=1}^{m}K(\mathbf{x}_s, \mathbf{x}_t) \Big\}.\end{aligned}$$

Introducing the vector

$$\mathbf{Z} = \Big(K(\mathbf{y}, \mathbf{x}_i)\Big)_{m\times 1}. \tag{10}$$

Then,

$$\begin{aligned}\Big(\Big\langle\tilde{\phi}(\mathbf{y}), \tilde{\mathbf{V}}^k\Big\rangle\Big)_{1\times r} &= \mathbf{Z}^\intercal\tilde{\mathbf{V}} - \frac{1}{m}\mathbf{1}_m^\intercal K\tilde{\mathbf{V}} - \frac{1}{m}(\mathbf{Z}^\intercal\mathbf{1}_m)\mathbf{1}_m^\intercal\tilde{\mathbf{V}} + \frac{1}{m^2}(\mathbf{1}_m^\intercal K\mathbf{1}_m)\mathbf{1}_m^\intercal\tilde{\mathbf{V}}\\ &= \mathbf{Z}^\intercal\Big(\mathbf{I}_m - \frac{1}{m}\mathbf{1}_m\mathbf{1}_m^\intercal\Big)\tilde{\mathbf{V}} - \frac{1}{m}\mathbf{1}_m^\intercal K\Big(\mathbf{I}_m - \frac{1}{m}\mathbf{1}_m\mathbf{1}_m^\intercal\Big)\tilde{\mathbf{V}}\\ &= \Big(\mathbf{Z}^\intercal - \frac{1}{m}\mathbf{1}_m^\intercal K\Big)\Big(\mathbf{I}_m - \frac{1}{m}\mathbf{1}_m\mathbf{1}_m^\intercal\Big)\tilde{\mathbf{V}},\end{aligned} \tag{11}$$

where $\tilde{\mathbf{V}}$ is a $m \times r$ matrix whose columns are the eigenvectors $\tilde{\mathbf{V}}^1, ..., \tilde{\mathbf{V}}^r$.

Notice that the KPCA uses only implicitly the input variables since the algorithm formulates the reduction of the dimension in the feature space through the kernel function evaluation. Thus KPCA is usefulness for nonlinear feature extraction by reducing the dimension but not to explain the selected features by means the input variables.

## 3. Adding input variable information into Kernel PCA

In order to get interpretability we add supplementary information into KPCA representation. We have developed a procedure to project any given input variable onto the subspace spanned by the eigenvectors (9).

We can consider that our observations are realizations of the random vector $X = (X_1, ..., X_n)$. Then to represent the prominence of the input variable $X_k$ in the KPCA. We take a set of points of the form $\mathbf{y} = \mathbf{a} + s\mathbf{e}_k \in \mathbb{R}^n$ where $\mathbf{e}_k = (0, ..., 1, ..., 0) \in \mathbb{R}^n$, $s \in \mathbb{R}$, where $k$-th component is equal 1 and otherwise are 0. Then, we can compute the projections of the image of these points $\phi(\mathbf{y})$ onto the subspace spanned by the eigenvectors (9).

Taking into account equation (11) the induced curve in the eigenspace expressed in matrix form is given by the row vector:

$$\sigma(s)_{1\times r} = \Big(\mathbf{Z}_s^\intercal - \frac{1}{m}\mathbf{1}_m^\intercal K\Big)\Big(\mathbf{I}_m - \frac{1}{m}\mathbf{1}_m\mathbf{1}_m^\intercal\Big)\tilde{\mathbf{V}}, \tag{12}$$

where $\mathbf{Z}_s$ is of the form (10).

In addition we can represent directions of maximum variation of $\sigma(s)$ associated with the variable $X_k$ by projecting the tangent vector at $s = 0$. In matrix form, we have

$$\frac{d\sigma}{ds}\Big|_{s=0} = \frac{d\mathbf{Z}_s^\intercal}{ds}\Big|_{s=0}\Big(\mathbf{I}_m - \frac{1}{m}\mathbf{1}_m\mathbf{1}_m^\intercal\Big)\tilde{\mathbf{V}} \tag{13}$$

with

$$\frac{d\mathbf{Z}_s^\intercal}{ds}\Big|_{s=0} = \Big(\frac{d\mathbf{Z}_s^1}{ds}\Big|_{s=0}, ..., \frac{d\mathbf{Z}_s^m}{ds}\Big|_{s=0}\Big)^\intercal$$

and, with

$$\begin{aligned}
\left.\frac{d\mathbf{Z}_s^i}{ds}\right|_{s=0} &= \left.\frac{dK(\mathbf{y},\mathbf{x}_i)}{ds}\right|_{s=0} \\
&= \left(\sum_{t=1}^{m} \frac{\partial K(\mathbf{y},\mathbf{x}_i)}{\partial y_t}\frac{dy_t}{ds}\right)\Big|_{s=0} \\
&= \sum_{t=1}^{m} \left.\frac{\partial K(\mathbf{y},\mathbf{x}_i)}{\partial y_t}\right|_{\mathbf{y}=\mathbf{a}} \delta_t^k = \left.\frac{\partial K(\mathbf{y},\mathbf{x}_i)}{\partial y_k}\right|_{\mathbf{y}=\mathbf{a}}
\end{aligned}$$

where $\delta_t^k$ denotes the delta of Kronecker. In particular, let us consider the radial basis kernel: $k(\mathbf{x},\mathbf{z}) = \exp(-c\,\|\mathbf{x}-\mathbf{z}\|^2)$ with $c > 0$ a free parameter. Using above notation, we have

$$K(\mathbf{y},\mathbf{x}_i) = \exp(-c\,\|\mathbf{y}-\mathbf{x}_i\|^2) = \exp(-c\sum_{t=1}^{n}(y_i - x_{it})^2)$$

When we consider the set of points of the form $\mathbf{y} = \mathbf{a} + s\mathbf{e}_k \in \mathbb{R}^n$,

$$\begin{aligned}
\left.\frac{d\mathbf{Z}_s^i}{ds}\right|_{s=0} &= \left.\frac{\partial K(\mathbf{y},\mathbf{x}_i)}{\partial y_k}\right|_{\mathbf{y}=\mathbf{a}} \\
&= -2cK(\mathbf{a},\mathbf{x}_i)(a_k - x_{ik})
\end{aligned}$$

In addition, if $\mathbf{a} = \mathbf{x}_\beta$ (a training point) then

$$\left.\frac{d\mathbf{Z}_s^i}{ds}\right|_{s=0} = -2cK(\mathbf{x}_\beta,\mathbf{x}_i)(x_{\beta k} - x_{ik})$$

Thus, by applying equation (12) we can locally represent any given input variable in the KPCA plot. Moreover, by using equation (13) we can represent the tangent vector associated with any given input variable at each sample point. Therefore, we can plot a vector field over the KPCA that points to the growing directions of the given variable.

We used the radial basis kernel in our implementation however the procedure we have introduced is also applicable to any other smooth kernel, for instance the Linear kernel which supplies standard PCA analysis.

## 4. Validation

In this section we illustrate our procedure with data from the leukemia data set of Golub et al. (1999) and the lymphoma data set Alizadeh et al. (2000).

In these examples our aim is to validate our procedure for adding input variables information into KPCA representation. We follow the following steps. First, in each data set, we build a list of genes that are differentially expressed. This selection is based in accordance with previous studies such as (Golub et al. (1999), Pittelkow & Wilson (2003), Reverter et al. (2010)). In addition we compute the expression profile of each gene selected, this profile confirm the evidence of differential expression.

Second, we compute the curves through each sample point associated with each gene in the list. These curves are given by the $\phi$-image of points of the form:

$$\mathbf{y}(s) = \mathbf{x}_i + s\mathbf{e}_k$$

where $\mathbf{x}_i$ is the $1 \times n$ expression vector of the $i$-th sample, $i = 1, ..., m$, $k$ denotes the index in the expression matrix of the gene selected to be represented, $\mathbf{e}_k = (0, ..., 1, ..., 0)$ is a $1 \times n$ vector with zeros except in the $k$-th. These curves describe locally the change of the sample $x_i$ induced by the change of the gene expression.

Third, we project the tangent vector of each curve at $s = 0$, that is, at the sample points $\mathbf{x}_i$, $i = 1, ..., m$, onto the KPCA subspace spanned by the eigenvectors (9). This representation capture the direction of maximum variation induced in the samples when the expression of gene increases.

By simultaneously displaying both the samples and the gene information on the same plot it is possible both to visually detect genes which have similar profiles and to interpret this pattern by reference to the sample groups.

### 4.1 Leukemia data sets

The leukemia data set is composed of 3051 gene expressions in three classes of leukemia: 19 cases of B-cell acute lymphoblastic leukemia (ALL), 8 cases of T-cell ALL and 11 cases of acute myeloid leukemia (AML). Gene expression levels were measured using Affymetrix high-density oligonucleotide arrays.

The data were preprocessed according to the protocol described in Dudoit et al. (2002). In addition, we complete the preprocessing of the gene expression data with a microarray standardization and gene centring.

In this example we perform the KPCA , as detailed in the previous section, we compute the kernel matrix with using the radial basis kernel with $c = 0.01$, this value is set heuristically. The resulting plot is given in Figure 1. It shows the projection onto the two leading kernel principal components of microarrays. In this figure we can see that KPCA detect the group structure in reduced dimension. AML, T-cell ALL and B-cell ALL are fully separated by KPCA.

To validate our procedure we select a list of genes differentially expressed proposed by (Golub et al. (1999), Pittelkow & Wilson (2003), Reverter et al. (2010)) and a list of genes that are not differentially expressed. In particular, in Figures 2, 3, 4 and 5 we show the results in the case of genes: `X76223_s_at`, `X82240_rna1_at`, `Y00787_s_at` and `D50857_at`, respectively. The three first genes belong to the list of genes differentially expressed and the last gene is not differentially expressed.

Figure 2 (top) shows the tangent vectors associated with `X76223_s_at` gene, attached at each sample point. This vector field reveals upper expression towards T-cell cluster as is expected from references above mentioned. This gene is well represented by the second principal component. The length of the arrows indicate the strength of the gene on the sample position despite the dimension reduction. Figure 2 (bottom) shows the expression profile of

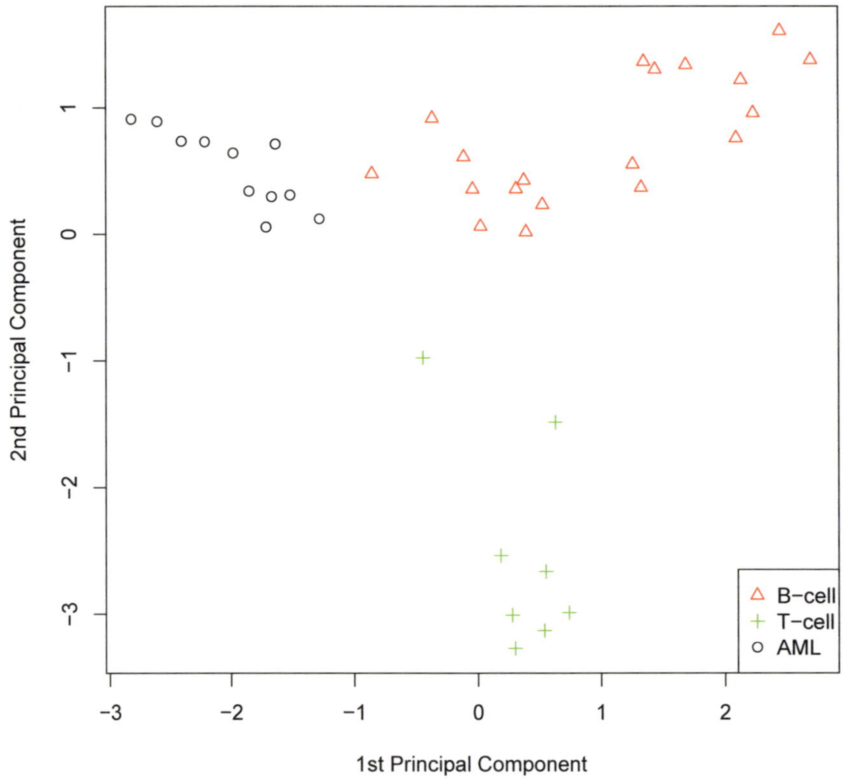


Fig. 1. Kernel PCA of Leukemia dataset.

X76223_s_at gene. We can observe that X76223_s_at gene is up regulated in T-cell class. This profile is agree with our procedure because the direction in which the expression of the X76223_s_at gene increases points to the T-cell cluster.

Figure 3 (top) shows the tangent vectors associated with X82240_rna1_at gene attached at each sample point. This vector field reveals upper expression towards B-cell cluster as is expected from references above mentioned. Figure 3 (bottom) shows the expression profile of X82240_rna1_at gene. We can observe that X82240_rna1_at gene is up regulated in B-cell class. This profile is agree with our procedure because the direction in which the expression of the X82240_rna1_at gene increases points to the B-cell cluster.

Figure 4 (top) shows the tangent vectors associated with Y00787_s_at gene attached at each sample point. This vector field reveals upper expression towards AML cluster as is expected from references above mentioned. Figure 4 (bottom) shows the expression profile of Y00787_s_at gene. We can observe that Y00787_s_at gene is up regulated in AML class. This profile is agree with our procedure because the direction in which the expression of the Y00787_s_at gene increases points to the AML cluster.

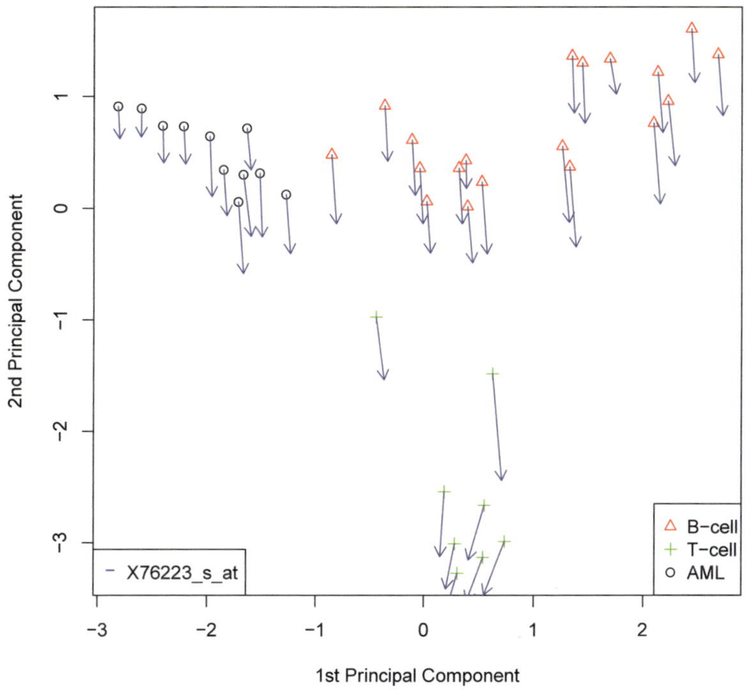


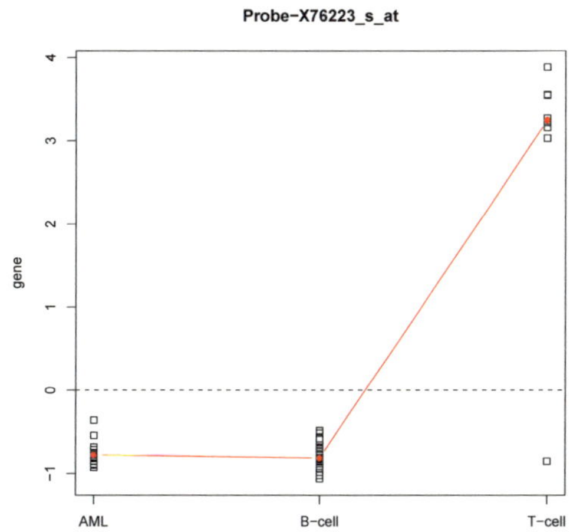


Fig. 2. (Top) Kernel PCA of Leukemia dataset and tangent vectors associated with X76223-s-at gene at each sample point. Vector field reveals upper expression towards T-cell cluster. (Bottom) Expression profile of X76223-s-at gene confirms KPCA plot enriched with tangent vectors representation.

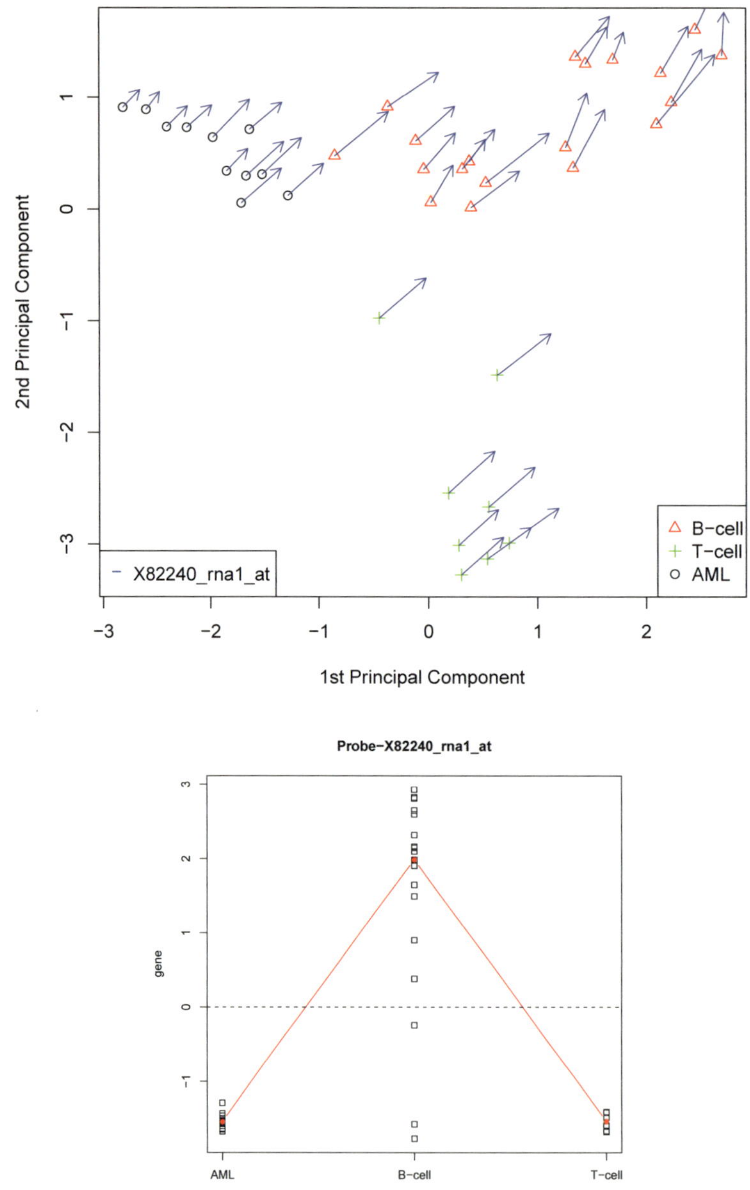


Fig. 3. (Top) Kernel PCA of Leukemia dataset and tangent vectors associated with X82240-rna1-at gene at each sample point. Vector field reveals upper expression towards B-cell cluster. (Bottom) Expression profile of X82240-rna1-at gene confirms KPCA plot enriched with tangent vectors representation.

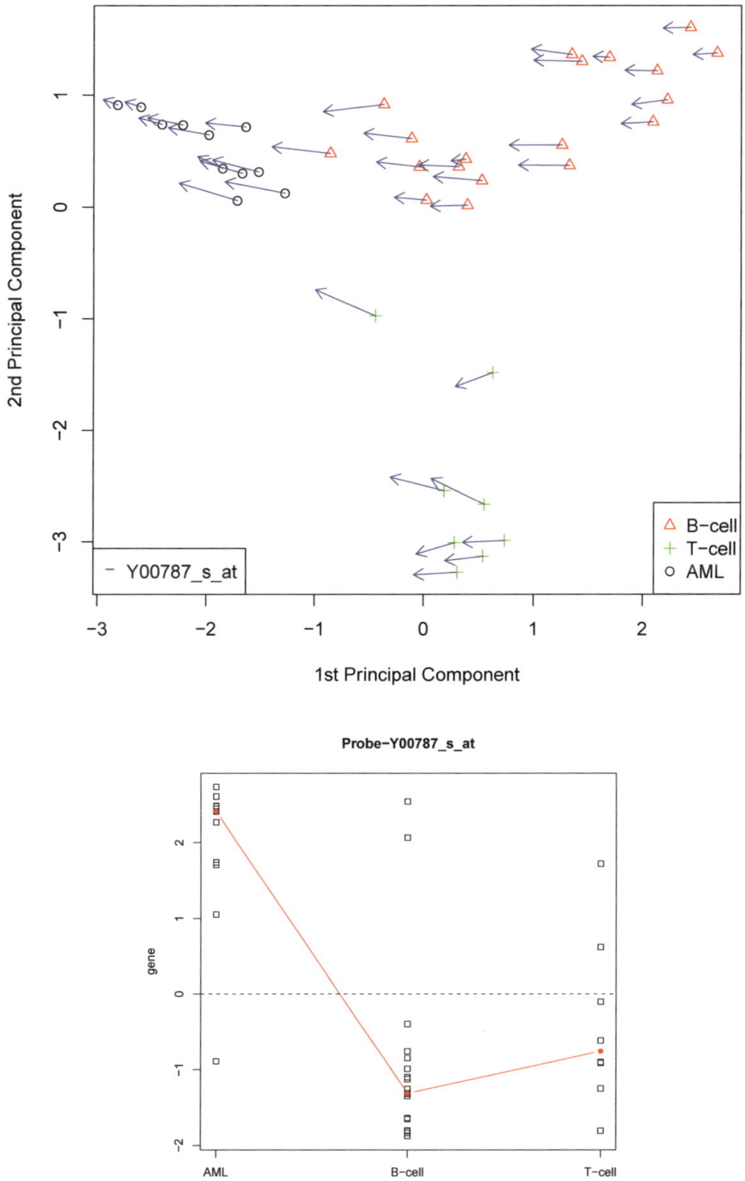


Fig. 4. (Top) Kernel PCA of Leukemia dataset and tangent vectors associated with Y00787-sat gene at each sample point. Vector field reveals upper expression towards AML cluster. (Bottom) Expression profile of Y00787-sat gene confirms KPCA plot enriched with tangent vectors representation..

Figure 5 (top) shows the tangent vectors associated with `D50857_at` gene attached at each sample point. This vector field shows no preferred direction to any of the three cell groups. The arrows are of short length and variable direction in comparison with other genes showed in previous Figures. Figure 5 (bottom) shows a flat expression profile of `D50857_at` gene. This profile is agree with our procedure because any direction of expression of the `D50857_at` gene is highlighted.

### 4.2 Lymphoma data sets

The lymphoma data set comes from a study of gene expression of three prevalent lymphoid malignancies: B-cell chronic lymphocytic leukemia (B-CLL), follicular lymphoma (FL) and diffuse large B-cell lymphoma (DLCL). Among 96 samples we took 62 samples 4026 genes in three classes: 11 cases of B-CLL, 9 cases of FL and 42 cases of DLCL. Gene expression levels were measured using 2-channel cDNA microarrays.

After preprocessing, all gene expression profiles were base 10 log-transformed and, in order to prevent single arrays from dominating the analysis, standardized to zero mean and unit variance. Finally, we complete the preprocessing of the gene expression data with gene centring.

In this example we perform the KPCA , as detailed in the previous section, we compute the kernel matrix with using the radial basis kernel with $c = 0.01$, this value is set heuristically. The resulting plot is given in Figure 6. It shows the projection onto the two leading kernel principal components of microarrays. In this figure we can see that KPCA detect the group structure in reduced dimension. DLCL, FL and B-CLL are fully separated by KPCA.

To validate our procedure we select a list of genes differentially expressed proposed by (Reverter et al. (2010)) and a list of genes that are not differentially expressed. In particular, in Figures 7, 8, 9 and 10 we show the results in the case of genes: `139009`, `1319066`, `1352822` and `1338456`, respectively. The three first genes belong to the list of genes differentially expressed and the last gene is not differentially expressed.

Figure 7 (top) shows the tangent vectors associated with `139009` gene attached at each sample point. This vector field reveals upper expression towards DLCL cluster as is expected from references above mentioned. This gene is mainly represented by the first principal component. The length of the arrows indicate the influence strength of the gene on the sample position despite the dimension reduction. Figure 7 (bottom) shows the expression profile of `139009` gene. We can observe that `139009` gene is up regulated in DLCL cluster. This profile is agree with our procedure because the direction in which the expression of the `139009` gene increases points to the DLCL cluster.

Figure 8 (top) shows the tangent vectors associated with `1319066` gene attached at each sample point. This vector field reveals upper expression towards FL cluster as is expected from references above mentioned. This gene is mainly represented by the second principal component. Figure 8 (bottom) shows the expression profile of `1319066` gene. We can observe that `1319066` gene is up regulated in FL class. This profile is agree with our procedure because the direction in which the expression of the `1319066` gene points to the FL cluster.

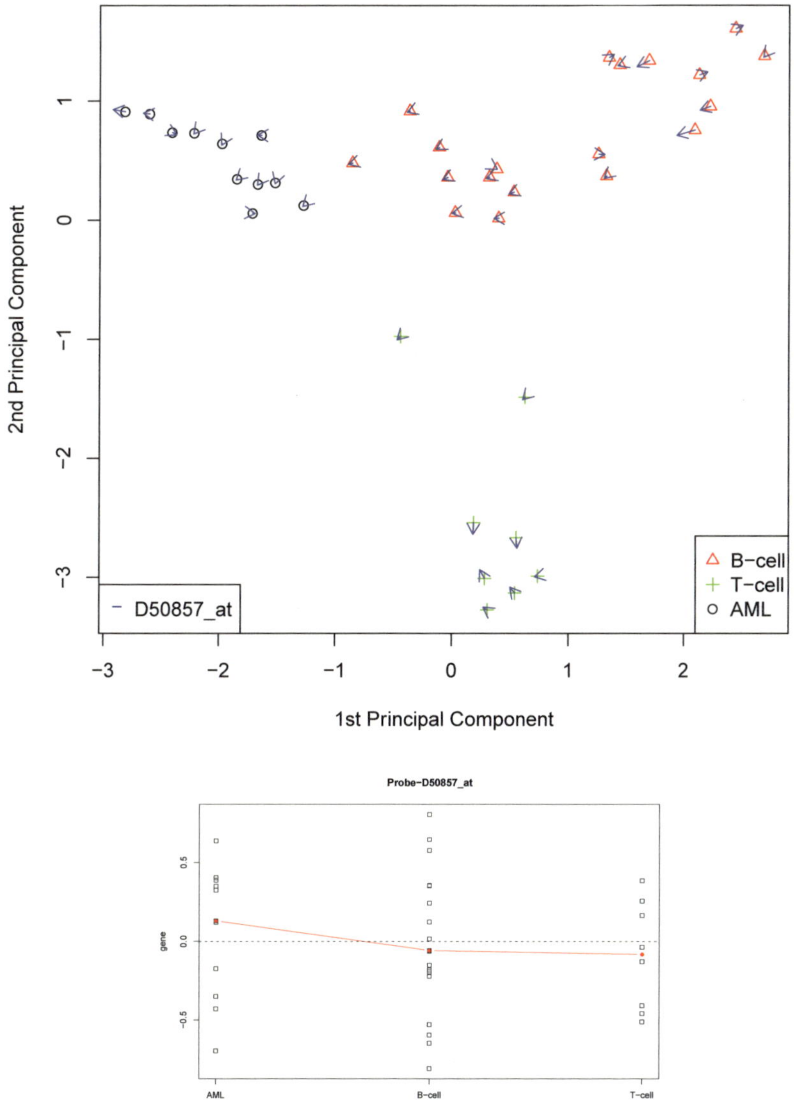


Fig. 5. (Top) Kernel PCA of Leukemia dataset and tangent vectors associated with D50857-at gene at each sample point. Vector field shows no preferred direction. (Bottom) Flat Expression profile of D50857-at gene confirms KPCA plot enriched with tangent vectors representation.

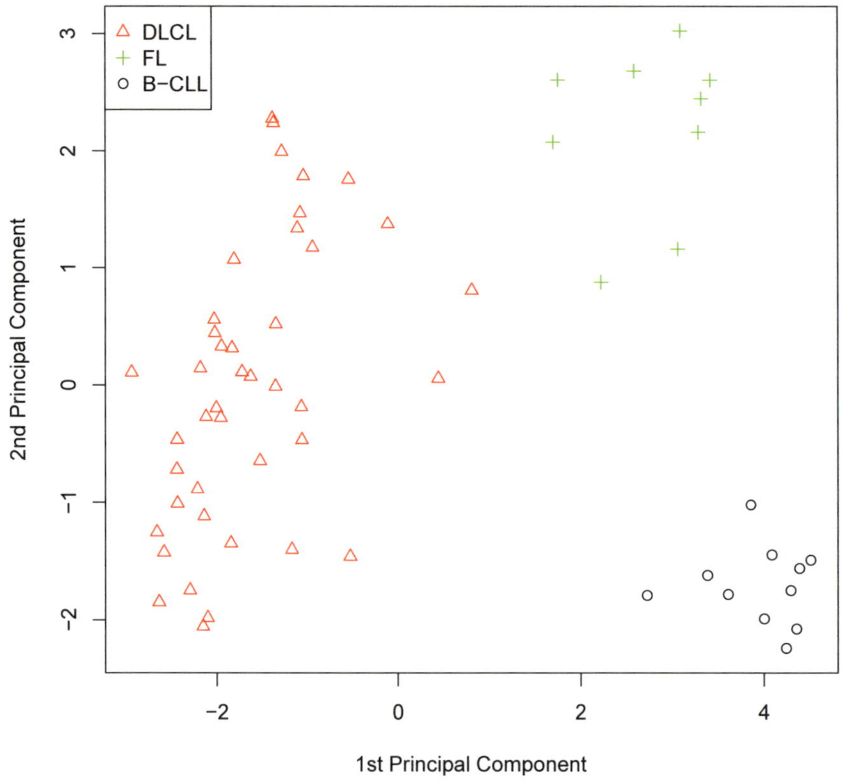


Fig. 6. Kernel PCA of Lymphoma dataset.

Figure 9 (top) shows the tangent vectors associated with `1352822` gene attached at each sample point. This vector field reveals upper expression towards B-CLL as is expected from references above mentioned. Figure 9 (bottom) shows the expression profile of `1352822` gene. We can observe that `1352822` gene is up regulated in B-CLL class. This profile is agree with our procedure because the direction in which the expression of the `1352822` gene increases points to the B-CLL cluster.

Figure 10 (top) shows the tangent vectors associated with `1338456` gene attached at each sample point. This vector field shows no preferred direction to any of the three cell groups. The arrows are of short length and variable direction in comparison with other genes showed in previous Figures. Figure 10 (bottom) shows a flat expression profile of `1338456` gene. This profile is agree with our procedure because any direction of expression of the `1338456` gene is highlighted.

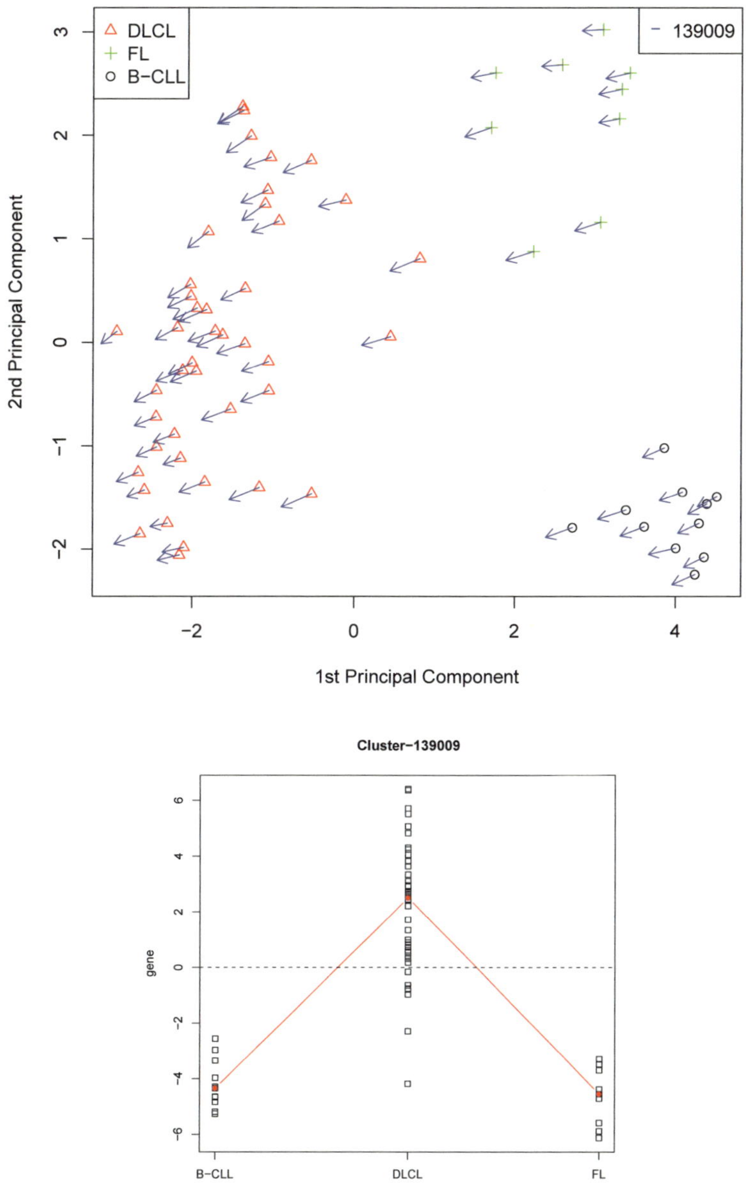


Fig. 7. (Top) Kernel PCA of Leukemia dataset and tangent vectors associated with 139009 gene at each sample point. Vector field reveals upper expression towards DLCL cluster. (Bottom) Expression profile of 139009 gene confirms KPCA plot enriched with tangent vectors representation.

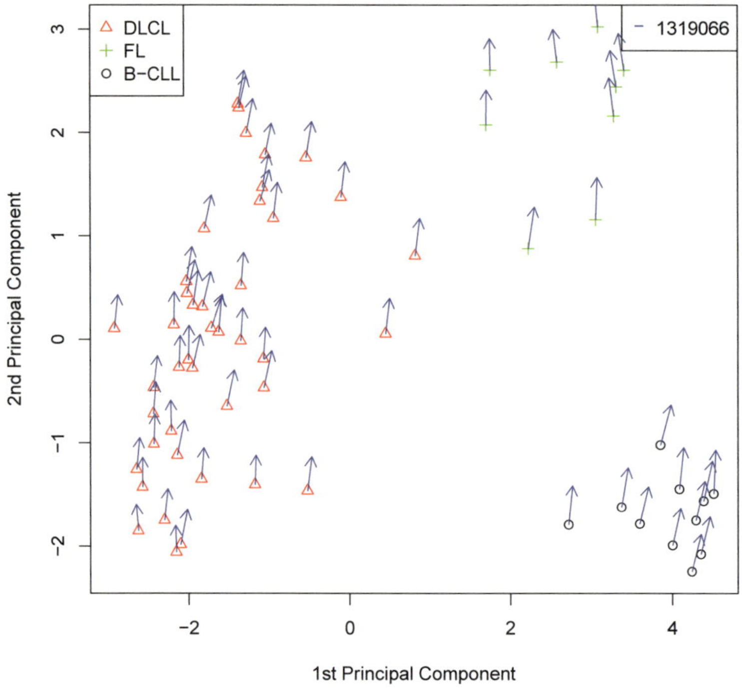


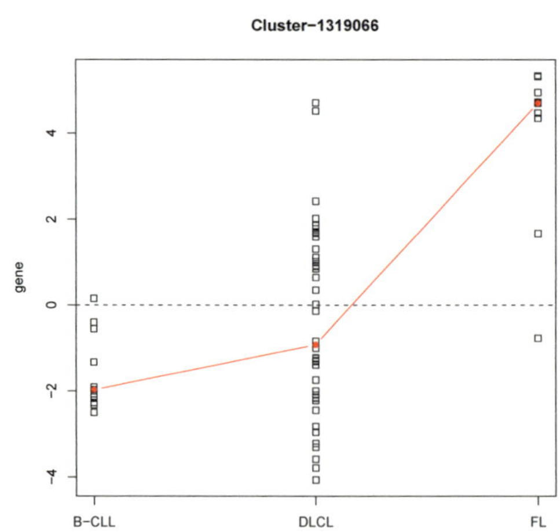


Fig. 8. (Top) Kernel PCA of Leukemia dataset and tangent vectors associated with 1319066 gene at each sample point. Vector field reveals upper expression towards FL cluster. (Bottom) Expression profile of 1319066 gene confirms KPCA plot enriched with tangent vectors representation.

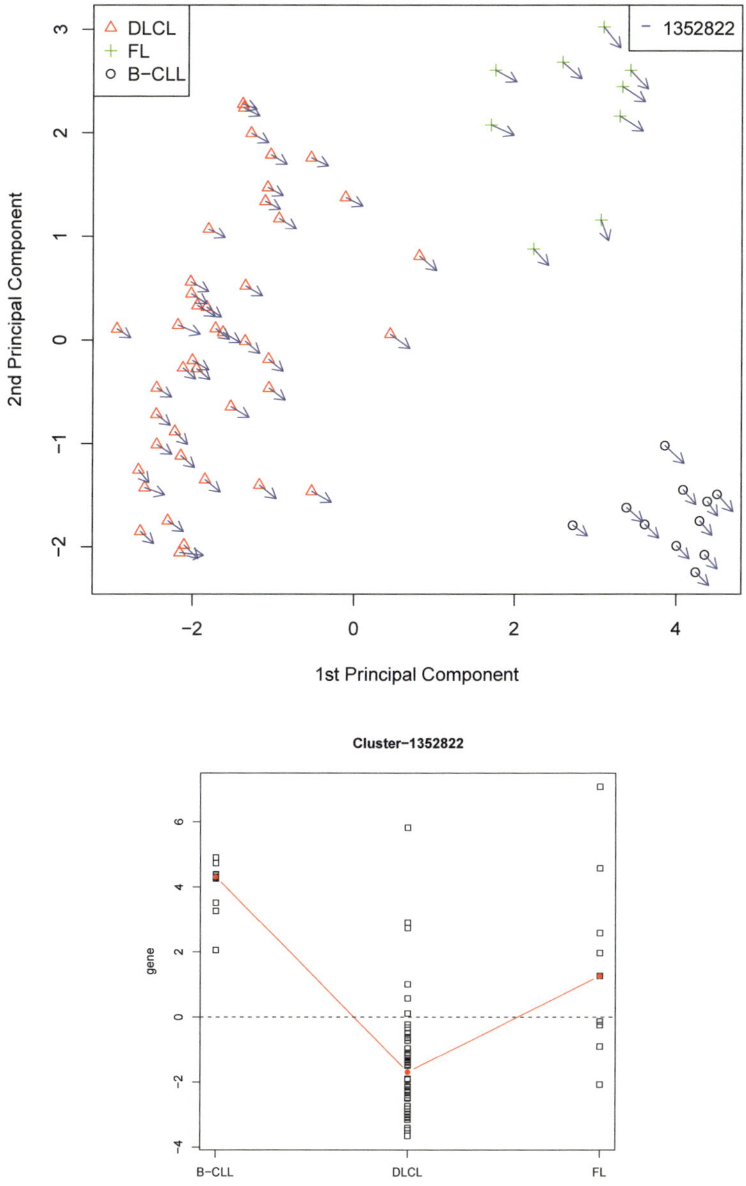


Fig. 9. (Top) Kernel PCA of Leukemia dataset and tangent vectors associated with 1352822 gene at each sample point. Vector field reveals upper expression towards B-CLL cluster. (Bottom) Expression profile of 1352822 gene confirms KPCA plot enriched with tangent vectors representation.

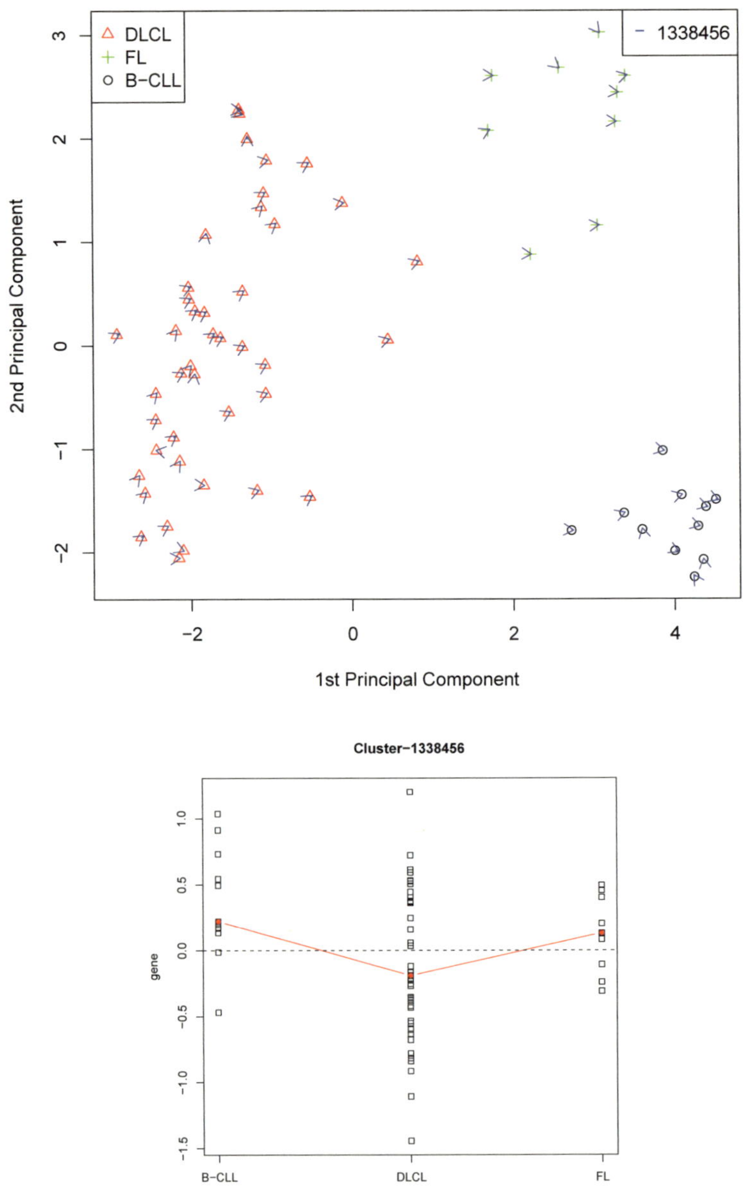


Fig. 10. (Top) Kernel PCA of Leukemia dataset and tangent vectors associated with 1338456 gene at each sample point. Vector field shows no preferred direction. (Bottom) Flat expression profile of 1338456 gene confirms KPCA plot enriched with tangent vectors representation.

## 5. Conclusion

In this paper we propose an exploratory method based on Kernel PCA for elucidating relationships between samples (microarrays) and variables (genes). Our approach show two main properties: extraction of nonlinear features together with the preservation of the input variables (genes) in the output display. The method described here is easy to implement and facilitates the identification of genes which have a similar or reversed profiles. Our results indicate that enrich the KPCA with supplementary input variable information is complementary to other tools currently used for finding gene expression profiles, with the advantage that it can capture the usual nonlinear nature of microarray data.

## 6. References

Alizadeh, A.A.; Eisen, M.B.; Davis, R,E.; Ma, C.; Lossos, I.S.; Rosenwald, A.; Boldrick, J.C.; Sabet, H.; Tran, T.; Yu, X.; Powell, J.I.; Yang, L.; Marti, G.E.; Moore, T.; Hudson, J.J.; Lu, L.; Lewis, D.B.; Tibshirani, R.; Sherlock, G.; Chan, WC.; Greiner, T.C.; Weisenburger, D.D.; Armitage, J.O.; Warnke, R.; Levy, R.; Wilson, W.; Grever, M.R.; Byrd, J.C; Bostein, D.; Brown, P,O. & Staudt, LM. (2000). Different type of diffuse large b-cell lymphoma identified by gene expression profiling. *Nature*, 403:503–511.

Alter, O.; Brown, P,O. & Botstein, D. (2000). Singular value decomposition for genome-wide expression data processing and modeling. *Proc. Natl. Acad. Sci.* USA. 97(18), 10101-10106.

Chambers, J.M.; Cleveland, W.S.; Kleiner, B. & Tuckey, P.A. (1983) *Graphical Methods for Data Analysis statistics/probability*. Wadsworth.

Chu, W.; Ghahramani, Z.; Falciani. F. & Wild, D: (2005) Biomarker discovery in microarrays gene expression data with Gaussian processes. *Bioinformatcis*. 21(16), 3385–3393.

Dettling, M. (2004). BagBoosting for tumor classificcation with gene expression data. *Bioinformatcis*. 20(18), 3583–3593.

Diaz-Uriarte, R. & Andres, S.A. (2006) Gene selection and classification of microarray data using random forest. *BMC Bioinformatcis*. 7:3, 1–13.

Dudoit, S.; Fridlyand, J. & Speed, T,P. (2002). Comparison of discrimination methods for the classification of tumours using gene expression data. *J. Am. Statist. Soc.* 97:77–87.

Eisen, M.B.; Spellman, P,T.; Brown, P.O. & Botstein, D. (1998). Cluster analysis and display of genome-wide expression patterns. *Proc. Natl. Acad. Sci.* USA. 95(25):14863–14868.

Fellenberg, K.; Hauser, N.C.; Brors, B.; Neutzner, A. & Hoheisel, J. (2001): Correspondence analysis applied to microarray data. *Proc. Natl. Acad. Sci.* USA. 98(19) 10781–10786.

Golub, T.R.; Slonim, D.K.; Tamayo, P.; Huard, C.; Gaasenbeek, M.; Mesirov, J.P.; Coller, H.; Loh, M.L.; Downing, J.R.; Caligiuri, M.A.; Bloomfield, C.D. & Lander, E.S. (1999). Molecular classification of cancer: Class discovery and class prediction by gene expression profiling. *Science*. 286(5439):531–537.

Park, M.; Lee, J.W.; Lee, J.B. & Song, S.H. (2008) Several biplot methods applied to gene expression data. *Journal of Statistical Planning and Inference*. 138:500–515.

Pittelkow, Y.E. & Wilson, S.R. (2003). Visualisation of Gene Expression Data - the GE-biplot, the Chip-plot and the Gene-plot. *Statistical Applications in Genetics and Molecular Biology*. Vol. 2. Issue 1. Article 6.

Reverter, F.; Vegas, E. & Sanchez, P. (2010) Mining Gene Expressions Profiles: An integrated implementation of Kernel Principal Components Analysis and Singular Value Decomposition. *Genomics, Proteomics and Bioinformatics*. 8(3):200–210.

Shawe-Taylor, J. & Cristianini, N. (2004). *Kernel Methods for Pattern Analysis*. Cambridge University Press.

Scholkopf, B.; Smola, A.J. & Muller, K.R. (1998) Nonlinear Component Analysis as a Kernel Eigenvalue Problem. *Neural Computation*. 10:1299-1319.

Scholkopf, B.; Smola, A.J. (2002). *Learning with Kernels - Support Vector Machines, Regularization, Optimization and Beyond*. Cambridge, MA. MIT Press.

Tamayo, P.; Solni, D.; Mesirov, J.; Zhu, Q.; Kitareewan, S.; Dmitrovsky, E.; Lander, E.S. & Golub, T.R. (1999). Interpreting patterns of gene expression with self-organizing maps: Methods and application to hematopoietic differentiation. *Proc. Natl. Acad. Sci.* USA. 96(6):2907–2912.

Xuehua, L. & Lan, S. (2009). Kernel based nonlinear dimensionality reduction for microarray gene expression data analysis. *Expert Systems with Applications* 36:7644-7650.

Zhao, X. & Cheung, L.W.K. (2007) Kernel-Imbedded Gaussian processes for disease classification using microarrays gene expression data. *BMC Bioinformatcis*. 8:67:1–26.

Zhenqiu, L., Dechang, C. & Halima B. (2005). Clustering gene expression data with kernel principal components. *Journal of Bioinformatics and Computational Biology*. 3(2):303–316.

# 2

# Principal Component Analysis in the Era of «Omics» Data

Louis Noel Gastinel
*University of Limoges, University Hospital of Limoges, Limoges France*

## 1. Introduction

### 1.1 Definitions of major «omics» in molecular biology and their goals

The «omics» era, also called classically the post-genomic era, is described as the period of time which extends the first publication of the human genome sequence draft in 2001 (International Human Genome Sequencing Consortium, 2001; Venter et al., 2001). Ten years after that milestone, extensive use of high-throughput analytical technologies, high performance computing power and large advances in bioinformatics have been applied to solve fundamental molecular biology questions as well as to find clues concerning human diseases (cancers) and aging. Principal «omics», such as Gen-*omics*, Transcript-*omics*, Prote-*omics* and Metabol-*omics*, are biology disciplines whose main and extremely ambitious objective is to describe as extensively as possible the complete class-specific molecular components of the cell. In the «omics» sciences, the catalog of major cell molecular components, respectively, genes, messenger RNAs and small interfering and regulatory RNAs, proteins, and metabolites of living organisms, is recorded qualitatively as well as quantitatively in response to environmental changes or pathological situations. Various research communities, organized in institutions both at the academic and private levels and working in the «omics» fields, have spent large amounts of effort and money to reach. standardization in the different experimental and data processing steps. Some of these «omics» specific steps basically include the following: the optimal experimental workflow design, the technology-dependent data acquisition and storage, the pre-processing methods and the post-processing strategies in order to extract some level of relevant biological knowledge from usually large data sets. Just like Perl (Practical Extraction and Report Language) has been recognized to have saved the Human Genome project initiative (Stein, 1996), by using accurate rules to parse genomic sequence data, other web-driven. programming languages and file formats such as XML have also facilitated «omics» data dissemination among scientists and helped rationalize and integrate molecular biology data.

Data resulting from different «omics» have several characteristics in common, which are summarized in Figure 1: (a) the number of measured variables n (SNP, gene expression, proteins, peptides, metabolites) is quite large in size (from 100 to 10000), (b) the number of samples or experiments p where these variables are measured associated with factors such as the pathological status, environmental conditions, drug exposure or kinetic points (temporal experiments) is rather large (10 to 1000) and (c) the measured variables are organized in a matrix of n x p dimensions. The cell contents of such a matrix usually record a metric (or numerical code) related to the abundance of the measured variables. The observed data are acquired keeping the lowest amount of possible technical and analytical variability. Exploring these «omics» data requires fast computers and state-of-the-art data visualization and statistical multivariate tools to extract relevant knowledge, and among these tools PCA is a tool of choice in order to perform initial exploratory data analysis (EDA).

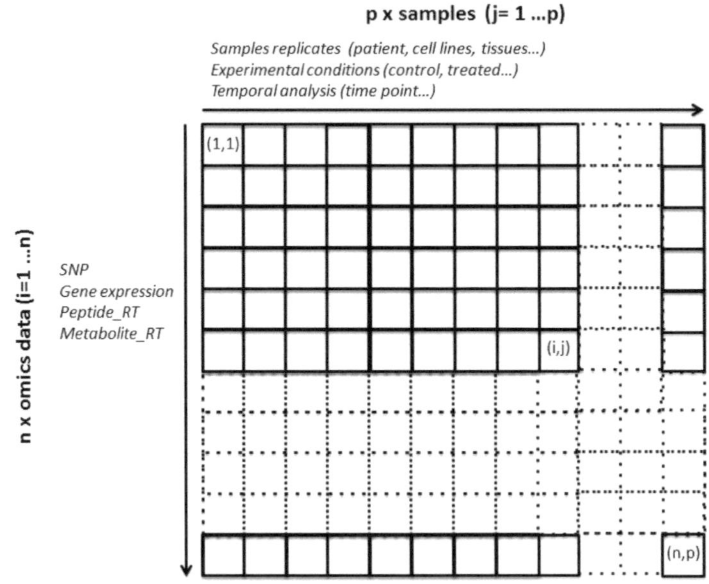


Fig. 1. General organization of raw «omics» data represented in a n x p matrix.

Rows contain the measured quantitative variables (n) and columns contain the samples or experimental conditions tested (p) from which variables n are measured and for which grouping information or factors is generally present. Each cell (i,j) of this matrix contains a measured quantitative information which is usually the abundance of the molecule under study.

### 1.1.1 Genomics and genetics data are different

Genomics and genetics data are of different types. Genomics data are related mainly to the collection of DNA sequences modeled as linear strings composed of the four nucleotides symbolized by the letters A, C, G and T (bases). These strings are usually obtained following large sequencing efforts under the supervision of academic and private consortia. NextGen sequencing technologies are used to acquire the data and, specialized softwares are used to assemble sequences in one piece in order to complete an entire genome of thousands of megabases long. The final result of these extensive and costly efforts is the establishment of the genome sequence of all living organisms and particularly the human genome. Genomics has been particularly successful these last few years in determining micro-organism genomes such as bacteria and viruses. Genomics is regularly used in academic research and even proposes on-demand service for the medical community to obtain emerging pathological genomes (SRAS, toxic strains of *Escherichia coli* …) that allow a fast medical response. Genomics aims to attain the technical challenge of obtaining 99.99% accuracy at the sequenced nucleotide level, and completeness and redundancy in the genome sequence of interest. However, the understanding or interpretation of the genome sequence, which means finding genes and their regulatory signals as well as finding their properties collected under the name "annotations", are still the most challenging and expensive tasks.

Genetics data, or genotype data, are related to the sequencing efforts on the human genome, particularly at the individual level. Genetics data record that the status of some nucleotides found at a certain position in the genome are different from one person to another. These base- and position-specific person-to-person variations are known as SNP or Single Nucleotide Polymorphism. When the frequency of the variation in a population is greater than 1%, this variation is considered as a true polymorphism possibly associated with traits (phenotypes) and genetic diseases (mutations). Moreover this information is useful as a genetic biomarker for susceptibilities to multigenic diseases or ancestrality and migration studies.

### 1.1.2 Transcriptomics data

Transcriptomics data consist in the recording of the relative abundance of transcripts or mature messenger RNAs representing the level of gene expression in cells when submitted to a particular condition. Messenger RNAs are the gene blueprints or recipes for making the proteins which are the working force (enzymes, framework, hormones…) in a cell and allow the cell's adaptation to its fast changing environment. Transcriptomics give a snapshot of the activity of gene expression in response to a certain situation. Generally mRNA abundances are not measured on an absolute scale but on a relative quantitative scale by comparing the level of abundance to a particular reference situation or control. Raw transcriptomics data associated with a certain gene g consist in recording the ratio of the abundances of its specific gene transcript in two biological situations, the test and the control. This ratio reveals if a particular gene is over- or under- expressed in a certain condition relative to the control condition. Moreover, if a set of genes respond together to the environmental stress under study, this is a signature of a possible common regulation control (Figure 2). Furthermore, transcriptomics data are usually organized as for other «omics» data as large tables of n x p. cells with p samples in columns and n genes in rows (Figure 1). A data pre-processing step is necessary before analyzing transcriptomics data. It consists in $\log_2$ intensity ratios transformation, scaling the ratios across different experiments, eliminate outliers. Multivariate analysis tools, particularly PCA, are then used to find a few genes among the thousands that are significantly perturbed by the treatment. The signification level of the perturbation of a particular gene has purely statistical value and means that the level of measured variation in the ratio is not due to pure chance. It is up to the experimentalist to confirm that it is truly the biological factor under study, and not the unavoidable variation coming from technical or analytical origin inherent to the acquisition method, that is responsible for the observations. To estimate this significance level it is absolutely necessary to measure ratios on a certain replicative level, at least three replicates per gene and per situation. ANOVA and multiple testing False Discovering Rate (FDR) estimates are generally used. Further experimental studies are mandatory to confirm transcriptomics observations. Moreover, Pearson correlation coefficient and different linkage clustering methods are used for each gene in order to perform their hierarchical clustering and to group genes with similar behavior or belonging to the same regulation network.

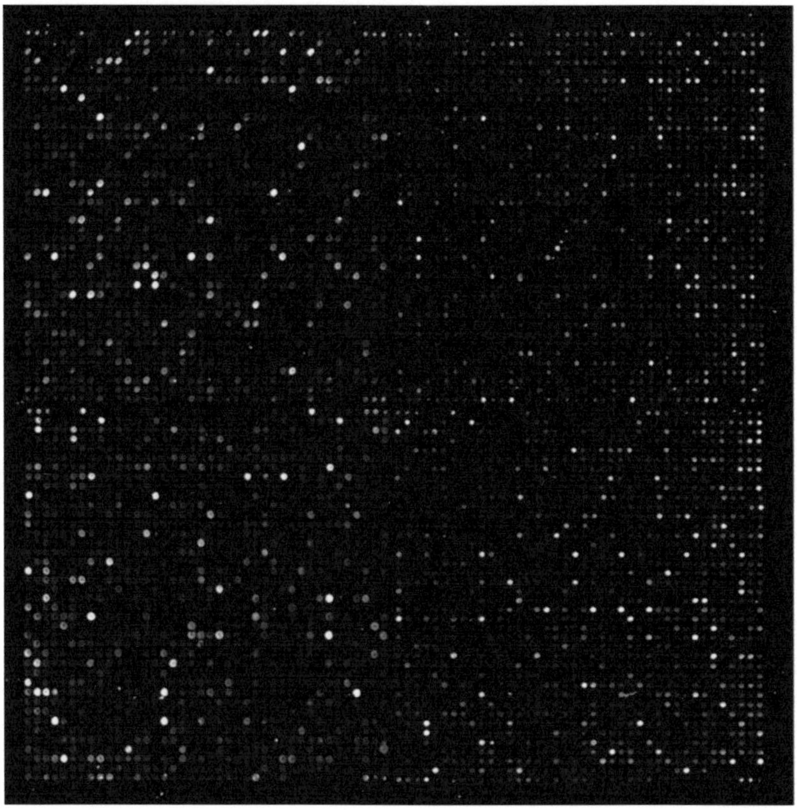

Fig. 2. A picture of a DNA microarray used in high-throughput transcriptomics.

DNA chip of 18 x 18 mm In size containing 6400 yeast gene Specific sequences organized as a matrix in which gene coordinates (x,y) are known. After hybridization with transcripts labeled respectively with green and red fluorochromes from two situations (treated versus untreated), 2 images in red and green fluorescence are recorded and superposed. Spot intensity seen on this image is then mathematically converted to a ratio of relative abundance of gene expression in the two situations under study (DeRisi et al., 1997).

### 1.1.3 Proteomics and metabolomics data

Proteomics and metabolomics data consist in measuring absolute or relative abundances of proteins and metabolites in the organism, tissue or cells after their proper biochemical extraction. These two fundamental and different classes of molecules are important for preserving cell integrity and reactivity to environment changes. These molecules are generally recognized and their abundances measured by mass spectrometry technologies after a liquid (HPLC) or gas (GC) chromatographic separation is performed to lower the high complexity level of analytes in the sample under study. Proteins have the large size of a few thousands of atoms and weigh a few thousands of Daltons (1 Dalton is the mass of a hydrogen atom) in mass, contrary to metabolites that are smaller molecules in size and mass (less than 1000 Daltons). Mass spectrometers are the perfect analytical tool to separate physically ionized analytes by their mass-to-charge ratio (m/z) and are able to record their abundance (peak intensity). Mass spectrometry data are represented graphically by a spectrum containing abundances versus m/z ratios or by a table or a peak list with two columns containing m/z and abundances after performing a de-isotopic reduction step and a noise filtration step.

Because of the large size of protein molecules, entire proteins should be cut in small pieces, called peptides, of 10-15 amino acids by using a protease enzyme trypsin. These peptides then have the right size to be analyzed directly by mass spectrometers. Peptide abundances are recorded. and their sequences even identified by collision-induced fragmentation (CID) breaking their peptide bonds, which some mass spectrometers instruments can perform (Triple Quadrupole mass spectrometrer in tandem, MALDI TOF TOF, Ion traps).

Raw data from metabolomics and proteomics studies originating from mass spectrometry techniques have the same basic contents. However, contrary to previous «omics», analytes are first separated by a chromatographic step and one analyte is characterized by its unique retention time (rt) on the separation device, its mass-to-charge ratio (m/z) and its abundance (a). This triad (rt - m/z - a) is a characteristic of the analyte that is measured accurately and found in the final «omics» data matrix n x p. Because of the separation step, multiple chromatography experiments should be normalized on both the scale of abundance and the scale of retention time to be further compared. A relevant multiple alignment of the chromatographic separations of different p samples is necessary and is performed by using sophisticated methods and models (Listgarten & Emili, 2005). This alignment step consists in recognizing which analyte is recorded in a given retention time bin and in a given m/z bin. Analytes found in common in the chosen bin are by definition merged in intensity and considered to be the same analyte. The same m/z analyte is recorded across multiple chromatographic steps and should be eluted at the same rt with some degree of tolerance both on rt (a few minutes) and on m/z (a few 0.1 m/z). The rows in the prote- and metabol- «omics» final matrix n x p contain the. proxy "m/z_rt," or "feature" and on the columns are the samples where the analytes come from. The cell content of this matrix record the abundance. "m/z_rt" is a set of analytes which have the same m/z with the same retention time rt, hopefully only one. Data can also be visualized as a 3D matrix with 3 dimensions: rt, m/z and abundances (Figure 3). For convenience it is the "m/z_rt" versus the 2D sample matrix which is further used in EDA for sample comparisons. The absolute value of intensity of the m/z analyte with retention rt corresponds to the mass spectrometry response given by its detector (cps).

## 1.2 Technologies generating «omics» data, their sizes and their formats

### 1.2.1 Genetics data format

Genome-wide studies using genetics data consist in recording the status of a particular DNA position or genotype in the genome called SNP or Single Nucleotide Polymorphism among few thousand of genes for a certain number of samples. The SNP status is obtained by accurately sequencing genomic DNA and recording its sequence in databases such as Genbank (www.ncbi.nlm.nih.gov/genbank). The SNP status is then coded by a simple number, 0, 1, 2, according to the nature of the nucleotide found at the genome's particular

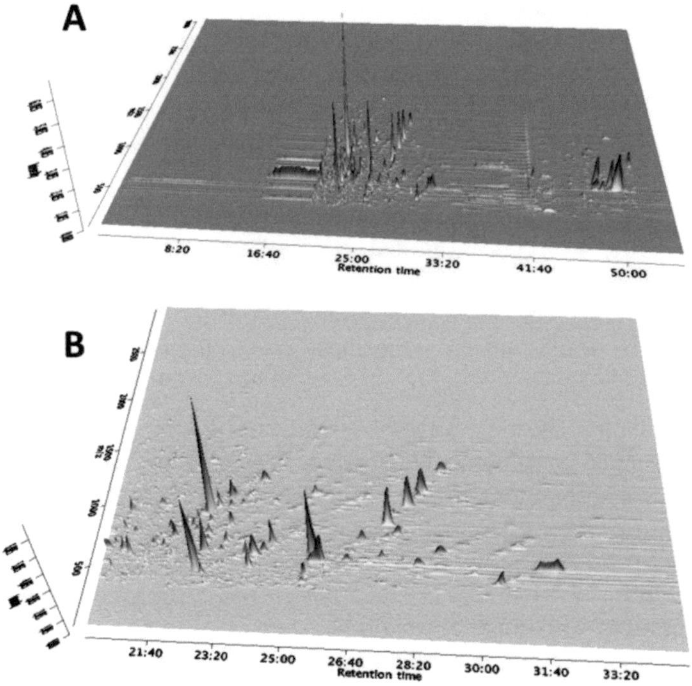


Fig. 3. A 3D representation of a mass spectrum of a liquid chromatographic separation in LC-MS typical analysis of proteomics and metabolomics data.

(A) Urinary native peptides without noise filtration and (B) with noise filtration are shown on a smaller time scale (20 to 34 minutes). These spectra were obtained using MZmine 2.1 with raw data converted first to the mzxml format (Pluskal et al., 2010).

position. It is not rare for the n x p matrix used in genetics data to have for dimension n=500000 SNP positions recorded for p=1000 individuals grouped according to ethnical, geographical or disease status. SNP positions, sequence, type and frequencies are maintained and accessible on different websites such as dbSNP (www.ncbi .nlm.nih.gov/projects/SNP), the International HapMap project (hapmap.ncbi.nlm.nih.gov), the SNP consortium (snp.cshl.org), the Human Gene Mutation Database or HGMD (www.hgmd.org), the 1000 Genomes project (www.1000genomes.org), the Pharmacogenomics database PharmGKB (www.pharmgkb.org) and the genotype-phenotype association database. GWAS Central (www.gwascentral.org). This information is particularly relevant in order to attempt SNP associations to disease status or health conditions. In recent human genetic studies, genotype data have been harvested, consisting in collecting for a few thousand human samples of different classes (ethnic groups, disease status groups, and so on) all the SNP profiles for particular genes (or even better all the genome). Algorithms such as EIGENSOFT suite is used to find statistically acceptable genotype-phenotype associations (Novembre & Stephens, 2008; Reich et al, 2008). The suite contains the EIGENSTRAT tool which is able to detect and correct for population bias of allele frequency, also called stratification, and suggests where the maximum variability resides among the population. PCA was demonstrated as a. valuable tool for detecting population substructure and correcting for stratification representing allele frequency differences originating from ancestry between the considered population before associating SNPs profile and disease status (Price et al., 2006). These studies were recently published for making qualified inferences about human migration and history.

### 1.2.2 Transcriptomics data formats

In order to analyze in parallel the large population of mRNAs or transcriptomes that a cell is expressing, a high-throughput screening method called DNA microarrays is used today. These DNA chips, some of which are commercially available (ex: Affymetrix), contain imprinted on their glass surface, as individualized spots,. thousands of short nucleic acid sequences specific of genes and organized in matrices to facilitate their location. (Figure 2) Pangenomic DNA chips contain sequences representing ALL the genes known today for a certain species (a few tens of thousands). These chips are hybridized with equal quantity of mRNA or complementary DNA copies of the mRNA prepared from control and treated samples, including fluorescent red (treated) and green (control) nucleotide analogs, in order to keep track of the sample origin. After subsequent washing steps, green and red fluorescence signals present on the chip are measured and the red-to-green ratio is calculated for each gene. The colors of the spots are from red (treated) to green (control) indicating over- and under- abundance of gene expression in the treated condition. A yellow color indicates an equal abundance of gene expression (no effect of condition) and a black spot indicates absence of gene expression in both conditions. Major free-access transcriptomics databases are the Stanford microarray database (smd.stanford.edu) and the NCBI GEO omnibus (www.ncbi.nlm.nih.gov/geo). The size of these arrays depends on the gene population under study. It is not rare to study transcriptomics on n= 7000 genes (yeast) or more on pangenomic arrays n = 20000 - 30000 (Arabidopsis, humans, mice ...). The number of DNA microarrays p is generally of the order of a few tens to a few hundreds, taking into account experimental replicates.

Alternative techniques exist to study gene expression, but they are not applied on a large- or genomic-wide scale as DNA microarrays,. and they are used in order to confirm hypotheses given by these later experiments. Among them, the technique using qRT-PCR or quantitative Reverse Transcription Polymerise Chain Reaction or its semi-high throughput variant called microfluidic cards (AppliedBiosystems) allow to quantify gene expression focused on 384 selected genes in one sample.

### 1.2.3 Proteomics and metabolomics data formats

Numerous mass spectrometry technologies are available today to perform proteomics and metabolomics analyses in specialized laboratories. These «omics» have not yet attained the mature status and the standardization level that transcriptomics has now attained, particularly at the level of data acquisition, data storage and sharing, as well as data analysis. However, some consortia, such as the human proteomic organization HUPO (www.hupo.org) and PeptidesAtlas (www.peptidesatlas.org), are spending a considerable amount of efforts and money to find standardization rules. One of the main difficulties in working with these «omics» data resides in maintaining intra- and inter-laboratory reproducibility. The second difficulty is that few mass spectrometers associated with the chromatographic separation devices are able to record a quantitative signal that is directly proportional to analyte abundance. Semi quantitative data are generally obtained with matrix-assisted laser desorption ionization (MALDI) and quantitative signal is better obtained with electrospray ionization (ESI) methods. The use of relevant working strategies is necessary to lower technical and analytical variabilities, and this is also accomplished through the use of numerous replicates and internal standards with known or predictive mass spectrometry behaviors. The third difficulty is inherent to the commercially available instruments for which data acquisition and processing use computational tools and proprietary data formats. There are, however, a few format converters that are accessible, among them OpenMS (open-ms.sourceforge.net) and TransProteomicPipeline (tools.proteomecenter.org). These techniques are used extensively with the aim of detecting and quantifying biomarkers or molecular signals. specific to drug toxicity, disease status and progression, sample classification, and metabolite pathways analysis. The size of proteomics and metabolomics matrices depends on the accuracy level measured on the analytes mass and the range of mass under study. n varies from a. few hundreds to a few tens of thousands of m/z analytes and the p dimension of experiments or samples is largely dependent on the biological question. (a few tens).

Alternative techniques to confirm proteomic expression use protein chips and immunoprecipitation techniques that are antibody dependent. Another mass spectrometry technique, which is Liquid Chromatography coupled to Selected Reaction Monitoring (LC-SRM), is also used to confirm the level of expression of a particular analyte focusing on its physico-chemical properties as well as its chemical structure. In this case a high specificity and sensibility are generally obtained because the mass spectrometer records the signal related to both occurrences of the presence of one analyte mass (precursor ion) and the presence of one of its specific fragments obtained by collision-induced dissociation (CID).

## 2. Exploratory data analysis with PCA of «omics» data

### 2.1 The principles of PCA in «omics»

The original. n x p matrix (Figure 1) or its transposed p x n (Figure 4) contains raw data with n generally much larger than p. The data should be preprocessed carefully. according to the nature and accuracy of the data. In the «omics», particularly proteomics and metabolomics, autoscaling and Pareto normalizations are the most used (Van der Berg et al, 2006). Autoscaling is the process of rendering each variable of the data (the «omics» item) on the same scale with a mean of 0 and a standard deviation of 1. PCA consists in reducing the normalized n x p matrix to two smaller matrices, an S score matrix and an L loading matrix. The product of scores S and the transposed loadings L′ matrix plus a residual matrix R gives the original n x p matrix X according to the formula $X = S*L' + R$. PCs are the dimension (d) kept for S and L matrices and numbered $PC_1$, $PC_2$, $PC_3$.... according to the largest variance they capture. $PC_1$ captures most of the variability in the data followed by $PC_2$ and $PC_3$ (Figure 4). PCA helped originally to detect outliers. PCA axis capture the largest amount of variability in the data that scientists in the «omics» fields want to interpret and to relate to biological, environmental, demographic and technological factors (variability in replicates). Therefore variance in the higher PCs is often due to experimental noise, so plotting data on the first two to three PCs not only simplifies interpretation of the data but also reduces the noise. The scoring plot displays the relationship existing between samples, meaning that two similar samples will be distantly close in the PC space. Furthermore, the loading plot displays the relationship between the «omics» items (genes, peptides, proteins, metabolites, SNPs). A strong correlation between items will be expressed by a linear arrangement of these points in the loading matrix. Moreover PCA biplots representing score and loading scatter plots superposed together are useful to detect the importance of particular loadings («omics» measured items) responsible for separating these sample clusters.

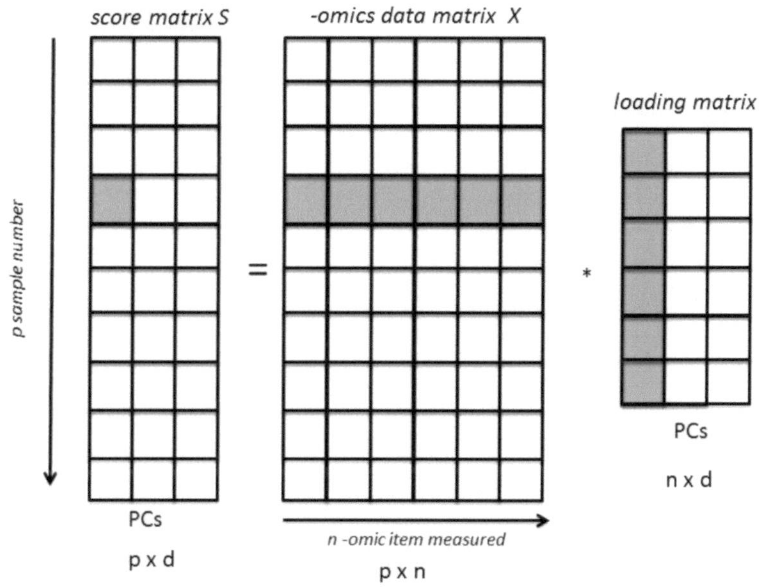


$s_{4,1} = x_{4,1} \cdot l_{1,1} + x_{4,2} \cdot l_{2,1} + x_{4,3} \cdot l_{3,1} + x_{4,4} \cdot l_{4,1} + x_{4,5} \cdot l_{5,1} + x_{4,6} \cdot l_{6,1}$

Fig. 4. Relationship between the X «omics» data matrix X, the S score matrix and the L loading matrix in principal component analysis.

Here the original «omics» matrix has been transposed for convenience, with n=6 being the number of omic items experimentally measured and p = 10 being the number of samples considered. d=3 is the number of PCs retained in the model, facilitating its graphical exploration. The highlighted row in X and column in L show what is required to generate a $PC_1$ score for sample 4.

PCA could reveal main patterns in the data but can detect some systematic non-biologically related or unwanted biologically related bias defined as batch effects. The existence of batch effects in «omics» data is being more and more recognized to frequently misguide biological interpretations. A large number of softwares can calculate these S and L matrices for large data sets. A PLS toolbox from Eigenvector research (www.eigenvector.com) running under MATLAB (www.matworks.com) contains representative 2D or 3D graphics of PCs space. Moreover, statistical indexes such as Q residues allow to estimate the disagreement behavior of some variables (samples) in the model, and Hotelling's T2 indexes measures the multivariate distance of each observation from the center of the dataset. R (cran.r-project.org) contains, in the statistical package included in the basic R installation, the prompt() function, which performs PCA on the command line. Particular «omics» -specific softwares containing sophisticated normalization, statistics and graphic options for proteomics and metabolomics data are available, such as DAnteR (Poplitiya et al., 2009) and MarkerView (www.absciex.com).

### 2.2 Interactive graphic exploratory data analysis with Ggobi and rggobi

Scatter plots are still the simplest and most effective forms of exploratory analyses of data but are limited to a pairwise comparison with just two samples in any one scatterplot diagram. Ggobi (www.ggobi.org) and rggobi, an alternative of ggobi with R GUI interface (www.ggobi.org/rggobi), are free tools that allow doing a scatter plot matrix with some limitation, as they graphically display a small number of explicative variables (less than 10). Ggobi and rggobi have an effective way of reducing a large multivariate data matrix into a simpler matrix with a much smaller number of variables called principal component or PCs without losing important information within the data. Moreover, this PC space is graphically displayed dynamically as a Grand Tour or 2D tour. Moreover, samples can be specifically colored or glyphed by using a "brushing" tool according to their belonging to some factors or categorical explicative variables (patient status, sample group, and so on…). Moreover, unavailable measurements (NA) are managed by Ggobi by using simple value replacements (mean, median, random) as well as sophisticated multivariate distribution modeling (Cook & Swayne, 2007).

### 2.3 PCA for «omics» data

#### 2.3.1 PCA for genetics data

PCA was used almost 30 years ago by Cavalli-Sforza L.L in population genetics studies to produce maps summarizing human genetic variation across geographic regions (Menozzi et al., 1978). PCA is used also in genotype-phenotype association studies in order to reveal language, ethnic or disease status patterns (Figure 5). Recently it has been shown that these studies are difficult to model with PCA alone because of the existence of numerous unmeasured variables having strong effects on the observed patterns (Reich et al., 2008; Novembre & Stephens, 2008). When analyzing spatial data in particular, PCA produces highly structured results relating to sinusoidal functions of increasing frequency with PC numbers and are sensitive to population structure, including distribution of sampling locations . This observation has also been seen in climatology. However PCA can reveal

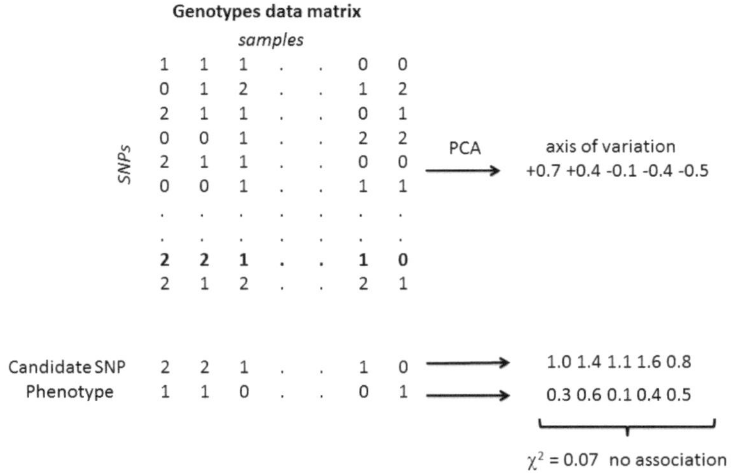


Fig. 5. The EIGENSTRAT algorithm from the EIGENSOFT suite.

some patterns on the data, and reliable predictions require further genetic analysis and integration with other sources of information from archeology, anthropology, epidemiology, linguistic and geography (François et al., 2010).

Genotype data consists of a n x p matrix where p individuals are recorded for their n SNPs. in their genomes. PCA is applied to infer continuous axes of genetic variation. A single axis of variation is indicated here. A genotype at a candidate SNP and phenotype are continuously adjusted by amounts attributable to ancestry along each axis. A $\chi^2$ show here no significant association for this particular SNP (Price et al., 2006).

### 2.3.2 PCA for transcriptomics data

Gene expression array technology has reached the stage of being routinely used to study clinical samples in search of diagnostic and prognostic biomarkers. Due to the nature of array experiments, the number of "null-hypotheses" to test, one for each gene, can be huge (a few tens of thousands). Multiple testing corrections are often necessary in order to. screen non-informative genes and reduce the number of null-hypotheses. One of the commonly used methods for multiple testing control is to calculate the false discovery rate (FDR) which is the ratio of the number of false rejections among the total number of rejections. FDR adjustment on raw p-values is effective in controlling false positives but is known to reduce the ability to detect true differentially expressed genes.

In transcriptomics studies, PCA is often used for the location of genes relative to each other in a reduced experiment space. Genes are plotted with respect to the two orthogonal linear combinations of experiments that contain the most variance (Lu et al., 2011). Transcriptomics also use other multivariate tools for classification and clustering (Tibshirani et al., 2002). A very fast and effective classification strategy is linear discriminant analysis. In classification problems there are positive training examples that are known members of the class under. study and negative training examples that are examples known not to be members of the class. The test examples are compared to both sets of training examples, and the determination of which set is most similar to the test case is established. In this process the test example is "classified" based on training examples. Clustering is a commonly used categorizing technique in many scientific areas using K-means grouping technique. Using this approach the user can cluster data based on some specified metric into a given number of clusters. Users can cluster arrays or genes as desired into a pre-specified number of clusters. The algorithm has a randomized starting point so results may vary from run to run.

### 2.3.3 PCA for proteomic and peptidomic data

#### 2.3.3.1 Urinary peptides and biomarker discovery study

PCA was used in order to distinguish urine samples containing or not pseudo or artificial spiked-in analytes or pseudo biomarkers (Benkali et al., 2008). The objectives were to analyze variations in the data and distinguish their sources. These variations could arise from (a) experimental variations due to changes in the instrument or experimental conditions, (b) real variations but of no interest in the primary objective, such as male versus female subjects, drug treatments, metabolites of a therapeutic agent... and (c) relevant differences that reflect changes in the system under study (spiked-in or not spiked-in). The experiment consisted in using human urines from 20 healthy volunteers splitted in two groups of ten, one which was spiked-in with few synthetic peptides at a certain variable concentration and the other without. Urines were processed using the same peptide extraction solid phase extraction (SPE) protocol, by the same experimentalist, and peptide compositions were recorded by off-line nanoLC-MS MALDI TOF/TOF. Data were processed with MarkerView software version 1.2 (www.absciex.com). PCA preprocessing consisted in using Pareto scaling without weighing and no autoscaling because Pareto scaling is known to reduce but not completely eliminate the significance of intensity, which is appropriate for MS because larger peaks are generally more reliable and all variables are equivalent. Different scaling methods are worth trying because they can reveal different features in the data with peak finding options and Pareto normalization (Van der Berg et al., 2006).

More than 5000 features (or m/z analytes) were retained from which respective abundances were observed. The n x p matrix contains n= 5000 and p = 20 samples. Scores and loading on PCs were calculated with 3 PCs capturing 80.4% of total data variability. Figure 6 shows $PC_1$ (70.6%) versus $PC_2$ (7.4%) (Figure 6A), as well as. $PC_1$ (70.6%) versus $PC_3$ (2.4%) (not shown). Sample points in the scoring scatterplot were colored according to. their group assignment before analysis (unsupervised). PCs scores on the $PC_1$-$PC_2$ projection axis allowed us to define the A9 sample as an outlier behaving as an unspiked B group sample (labeling tube error perhaps). We had to discard this sample for the rest of the analysis. This analysis was carried out on samples blinded to categorical label (spiked and unspiked) and the coloring of samples on the graphic was only carried out after the PCA. Spiked samples (A samples) are in red color and unspiked samples in blue color (B samples). The high positive value of loadings (green points) in the $PC_1$ and $PC_2$ axes are associated with features (or m/z analytes) most responsible to discriminate the two sample groups. The relative abundance of the spiked analyte of m/z = 1296.69 and its two $^{13}C$. isotopically stable labeled variants, 1297.69 and 1298.69, is shown in the spiked group (A samples, red points) and in the unspiked group (B samples, blue points). Moreover they tend to lie close to straight lines that pass through the origin in the loading plots (Figure 6A). These points (green points) are correlated because they are all the isotopic forms of the same spiked compound. The same is observed for other spiked analytes (904.46, 1570.67, 2098.09 and 2465.21). Finally superposed mass spectra from 20 samples of both groups show the relative abundance of analytes (panel B and insert focused on m/z = 1296.69 analyte and its natural $^{13}C$ isotopes).

Variables with large positive $PC_1$ loadings are mainly in group A (spiked samples) and absent or at lower intensities in group B. $PC_1$ separates both groups but $PC_2$ seems to separate both groups A and B in two half-groups (Figure 6A). What is the nature of the variation captured by $PC_2$ where some loadings (900.40, 1083.54 and 1299.65) give high positive $PC_2$ values and negative $PC_1$ values ?. The examination of Figure 7 shows that these analytes show a progessive increase in their intensity with a gradient following the order of their analysis in the instrument. The data were acquired in the order they are displayed (from left to right) and group A members were acquired before members of group B, which introduces a bias or a batch effect in these data. To avoid this effect, the samples should be acquired in a random order,. with group members mixed.

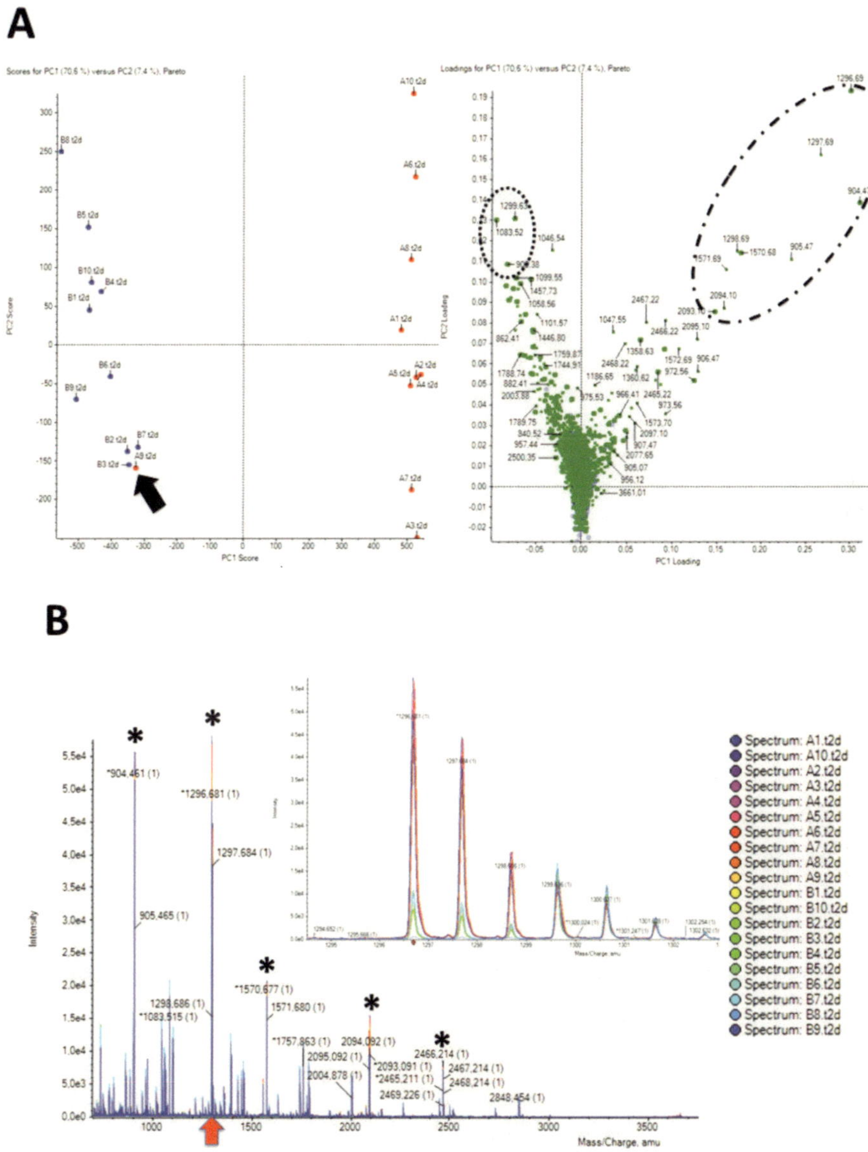


Fig. 6. PCA of 20 urine samples spiked or not spiked with synthetic peptides (pseudo biomarkers).

(A) Scores and loadings plots of $PC_1$ and $PC_2$ axes show a good separation of group A (spiked,. colored in red) from group B (not spiked, colored in blue). The A9 sample (black arrow) is an outlier and behaves as a B member group. It should be removed from the analysis. Loading plots show the 5000 analytes (green points) from which the majority are not contributing to the variability (0,0). Some analytes contribute to the large positive variation in the $PC_1$ axis (spiked peptides) and to the positive $PC_2$ (bias effect). (B) Superposition of the 20 spectra of urine samples after their alignment. with. a symbol (*) indicating the m/z of spiked peptides. The insert corresponds to the enlargement of the spectra located at the red arrow in the spectra, showing the abundance of. the 1296.69. and their $^{13}C$ isotopes among the 20 samples, particularly in the A group.

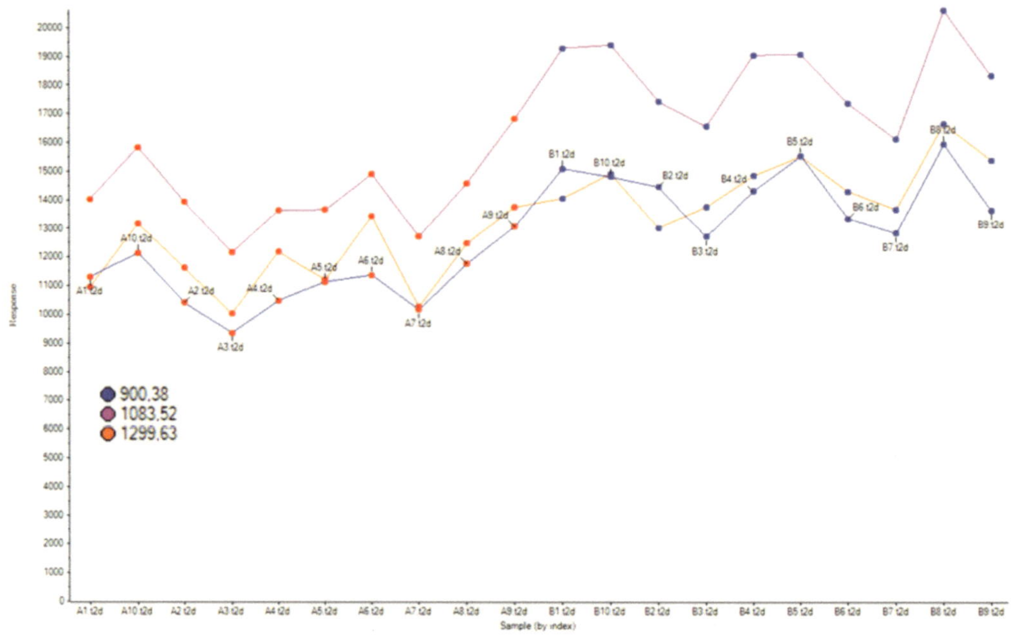


Fig. 7. A variation. recognized in the $PC_2$ axis probably due to a batch effect in the sample processing. Analytes. 900.38, 1083.52 and 1299.63. responsible for the positive value of scores in the $PC_2$ axis (Figure 6, panel A) see their intensity increase slightly during their acquiring time (from left to right), signaling a probable batch effect.

### 2.3.3.2 Microbial identification and classification by MALDI-TOF MS and PCA

MALDI-TOF MS mass spectrometry has recently been used as a technique to record. abundance of proteins extracted from different phyla of bacteria with the aim of finding phylum-specific patterns and use them to classify or recognize these bacteria in a minimum culture time. Sauer, S. has pioneered the technique of rapid extraction of proteins from alcohol, strong acid treatment or direct transfer from single colonies of bacteria, with or without the need to cultivate them (Freiwald & Sauer, 2009). Ethanol/Formic acid extraction of proteins of two clones of each. 6 bacteria strains, *Klebsiela pneumonia* (KP), *Acinetobacter baumanii*, (AB), *Lactobacillus plantarum* (LP), *Pseudomonas aeruginosa* (PA), *Escherichia coli* MG (MG), *Bacillus subtillis* (BS) were prepared. Mass spectra of proteins were recorded in five analytical replicates in the range 4000 to 12000 Daltons (Figure 8A). Major extracted proteins come from abundant ribosomal proteins. The natural variants in their amino acid sequence are responsible for the differences of masses in the peaks observed in the spectra, and their abundance is characteristic of the bacteria. Moreover, 6 bacteria clones (X1 to X6) were blindly analyzed in triplicate. PCA was used in order to distinguish the axes of greater variability in the data.

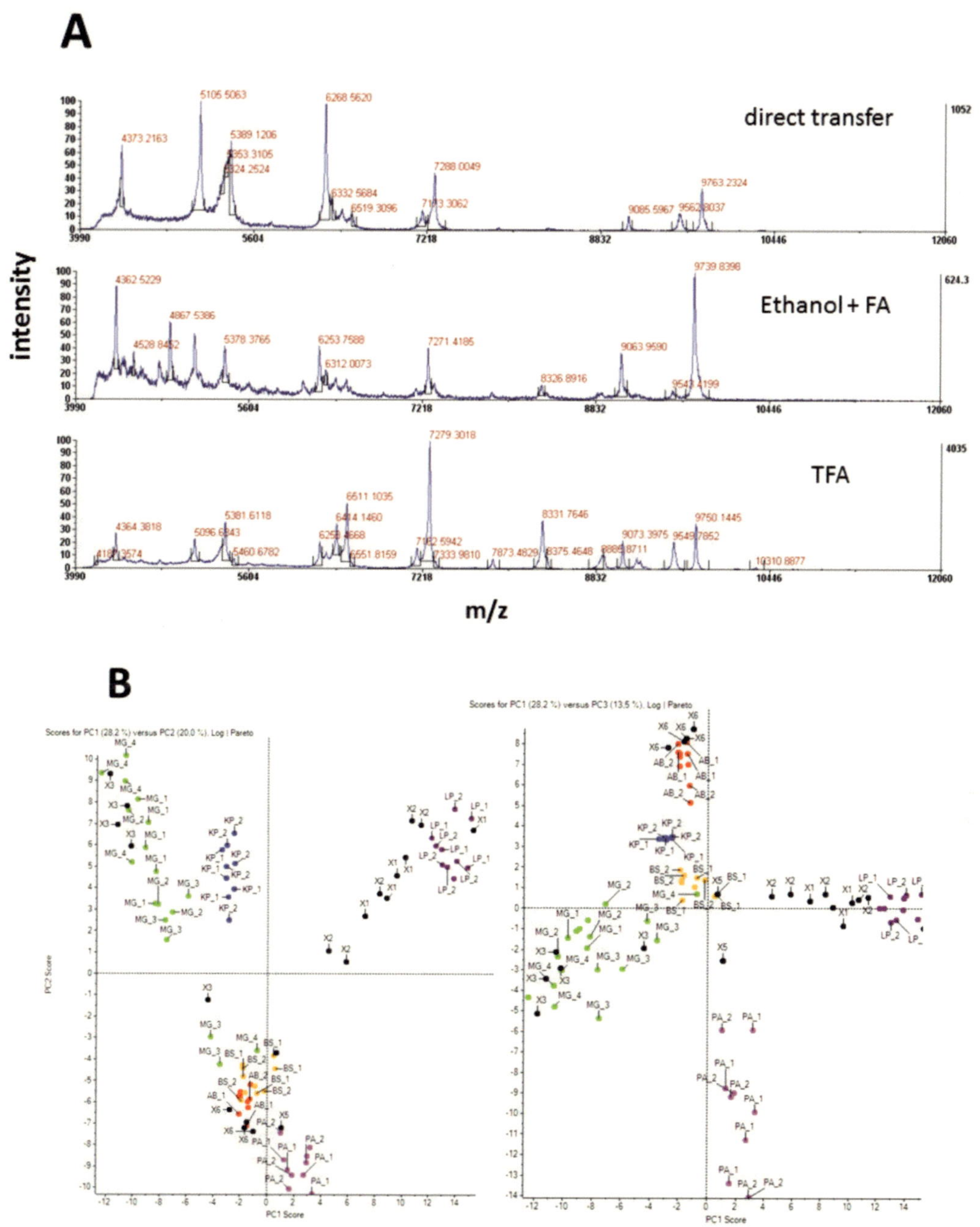


Fig. 8. MALDI TOF MS spectra of protein extracts of 6 bacteria strains and PCA.

(A) MS spectra of *Escherichia coli* MG clone proteins extracted with 3 different methods, direct transfer, ethanol/Formic acid (FA). and Trifluoroacetic acid(TFA). Peak abundances. may vary according to the extraction methods, from bacteria strains and from analytical variability. (B) Scores of $PC_1$ versus $PC_2$ and $PC_1$ versus $PC_3$ of the full 6 bacteria dataset,. including unknown X1 to X6 samples. (black points). (unpublished Maynard & Gastinel, 2009).

For this analysis, $PC_1$ takes 28.2% of variability, $PC_2$ 20% and $PC_3$ 15%, for a total of 63.2% captured in the model (Figure 8B). The $PC_1$ axis separates LP and PA bacteria strain. from. KP, MG, BS and AB strains. The $PC_2$ and $PC_3$ axes separate. KP and LP from BS , AB and PA. but not so well for the MG strain. Moreover, X samples are 100% separated in their correct respective clusters. The broad distribution of samples in the score plot is probably due to the relatively poor analytical reproducibility in the ionization of. MALDI TOF MS analysis. From the loading score plots (not shown) few proteins of particular m/z responsible for this bacteria strain separation have been recognized in protein databases by their annotation. Among them, the 5169 Daltons protein is attributed to the 50S ribosomal protein annotated as L34 of *Acinetobater baumanii.*

### 2.3.4 PCA for metabolomics data

Humic acids are one of the major chemical components of humic substances, which are the major organic constituents of soil (humus), peat, coal, many upland streams, dystrophic lakes, and ocean water. They are produced by biodegradation of dead organic matter. They are not a single acid; rather they are a complex mixture of many different acids containing carboxyl and phenolate groups so that the mixture behaves functionally as a dibasic acid or, occasionally, as a tribasic acid. Humic acids can form complexes with ions that are commonly found in the environment, creating humic colloids. Humic and fulvic acids (fulvic acids are humic acids of lower molecular weight and higher oxygen content than other humic acids) are commonly used as a soil supplement in agriculture, and less commonly as a human nutritional supplement. Humic and fulvic acids are considered as soil bioindicators and reflect an equilibrium between living organic and non-organic matters.

Mass spectrometry has been used to estimate signature analytes and patterns specific to some soils (Mugo & Bottaro, 2004). Fulvic acids were prepared from a soil using different extraction protocols resulting in 5 samples,. H1, H1H2, EVM1, EVM2 and EAA. Are these extraction protocols similar and which analytes are they extracting more efficiently? MALDI MS spectra from 150 to 1500 m/z range were recorded in the presence of the MALDI matrix alpha-cyano-4-hydroxycinammic acid (CHCA). Normalization of intensities were done with the 379 m/z analyte in common to these samples, and Pareto scaling was chosen during the alignment process performed by MarkerView. Figure 9A shows that the PCA analysis reveals poor separation of samples with $PC_1$ explaining 25.8%, $PC_2$ 18.7% and $PC_3$ 14.8%. of variability (a total of 59.3% captured). Samples are not so well separated by the first PC axis, demonstrating the large influence of factors other than soil extraction differences (chemical precipitation, physical precipitation, filtration). Discriminant Analysis associated with PCA (supervised PCA-DA) was attempted to further separate these known 5 groups (Figure 9B). This supervised technique means that it uses class information based on the assigned sample group to improve their separation. Figure 9B shows a dramatic improved separation but this may be based on noise. Peaks which are randomly more intense in one group as compared to another can possibly influence the results, and careful examination of loading plots as well as analyte profiles across the samples is necessary to avoid batch effects. This analysis is also affected by samples incorrectly assigned to wrong group and outliers.

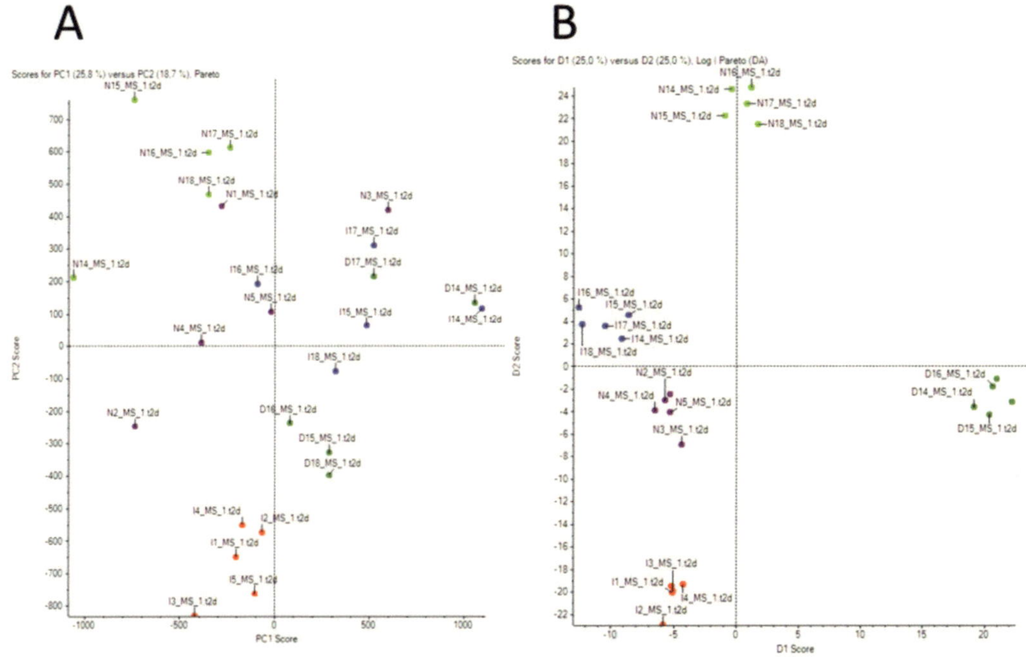


Fig. 9. Mass spectrometric study of fulvic acids profiles in the range 150 to 1000 m/z present in 5 sample preparations and analyzed in five analytical replicates by PCA.

(A) Unsupervised PCA and (B) PCA-DA or supervised PCA with group information included before reducing data. Only the first and second PC axes are shown for the score plots. (unpublished Basly & Gastinel, 2009).

In Figure 9, PCA (A) and PCA-DA (B) show a different pattern of the variability in the dataset. $PC_1$. axis does not discriminate H1H2 (green) , EVM1 (blue) and EAA (violet) as for. the $PC_2$ axis. The $PC_2$ axis discriminates however EVM2 (pale green). and H1 (red). Keeping these 5 different protocols as 5 different classes, PCA-DA (Figure 9B) however discriminates quite well these these 5 preparation analyses in quintuplicates. EVM1 and EAA are still the closest group. Loading score plots reveal which analytes are the most favored in a particular extraction method relative to another.

## 3. How can PCA help to reveal batch effects in «omics» data?

### 3.1 What are batch effects?

Batch effect is one overlooked complication with «omics» studies and occurs because high-throughput measurements are affected by multiple factors other than the primary tested biological conditions (Leek et al, 2010; Leek & Storey, 2008). These factors are included in a comprehensive list among which are laboratory conditions, reagents batches, highly trained personnel differences, and hardware maintenance. Batch effect becomes a problem when these conditions vary during the course of an experiment, and it becomes a major problem when the various batch effects are possibly correlated with an outcome of interest and lead to incorrect conclusions (Ransohoff, 2005; Baggerly, et al., 2004). Batch effects are defined as a sub-group of measurements that have qualitatively different behaviors across conditions and are primarily unrelated to the biological or scientific variables under study. Typical batch effect is seen when all samples of a certain group are measured first, and when all samples of a second group are measured next. Batch effect occurs too when a particular batch of reagent (ex: Taq polymerase enzyme for PCR experiments) is used with all samples of the first group, and another reagent batch is used with all samples of the second group. Typical batch effects are also seen when an experimentalist/technician acquires all samples from the first group and a different experimentalist/technician works with the other group or when the instrument's characteristics (example for MALDI mass spectrometry: laser or detector replacements) used to acquire the data have been deeply modified. Data normalization generally does not remove batch effect unless normalization takes into account the study design or takes into account the existence of a batch problem.

### 3.2 How to find evidence of batch effects

The first step in addressing batch and other technical effects is to develop a thorough and meticulous study plan. Studies with experiments that run over long periods of time, and large-scale, inter-laboratory experiments, are highly susceptible to batch effects. Intra-laboratory experiments spanning several days and several personnel changes are also susceptible to batch effects. Steps necessary to analyze batch effects require different. levels of analysis, according to the recent review of Leek J.T (Leek et al., 2010). What follows are some of the recommended actions:. Performing a hierarchical clustering of samples that assigns a label to the biological variables and to the batch surrogates estimates, such as laboratory and processing time; plotting individual features (gene expression, peptides or metabolites abundances). versus biological variables and batch surrogates using ggobi for example; calculating principal components of the high-throughput data and identifying components that correlate with batch surrogates. If some batch effects are present in the data, artifacts must be estimated directly, using surrogate variable analysis (SVA) (Leek et al., 2007). Recently, the EigenMS algorithm has been developed and. implemented within a pipeline of bioinformatic tools of DanteR in order to correct for technical batch effects in MS proteomics data analysis (Polpitya et al., 2008; Karpievitch et al., 2009). The algorithm uses an SVA approach to estimate systematic residual errors using singular value decomposition taking account primary biological factors and substracting those estimates from raw data in the pre-processing data analysis. The estimated/surrogate variables should be treated as standard covariates in subsequent analyses or adjusted for use with tools such as Combat (Johnson & Li, 2007). After adjustments that. include surrogate variables (at least processing time and date), the data must be reclustered to ensure that the clusters are not still driven by batch effects.

### 3.3 How to avoid batch effects

Measures and steps must be taken to minimize the probability of confusion between biological and batch effects. High-throughput experiments should be designed to distribute batches and other potential sources of experimental variation across biological groups. PCA of the high-throughput data allows the identification of components that correlate with batch surrogate variables.

Another approach to avoid and prevent batch effects is to record all parameters that are important for the acquisition of the measures and the relevant information related to demographic and grouping factors. The structure of a database under MySQL with attractive web graphic user interface (GUI) should be conceived at the same time as the study design is defined. Such a database was constructed for a mass spectrometry based biomarker discovery project in kidney transplantation in a French national multicenter project. The BiomarkerMSdb database structure contains 6 linked tables: Demographic data, Peptide Extraction Step, Liquid Chromatography Separation, Probot Fractionation, Spectrometry Acquisition and Data Processing. Figure 10 shows the details of the form that the user must complete to record demographic data of patients enrolled in this project. This approach, with an internet interface used to facilitate data exchange between laboratories enrolled in the project, allows to keep track of essential parameters that could interfere with future interpretations of the results. At minimum, analyses should report the processing group, the analysis time of all samples in the study, the personnel involved, along with the biological variables of interest, so that the results can be verified independently. This is called data traceability.

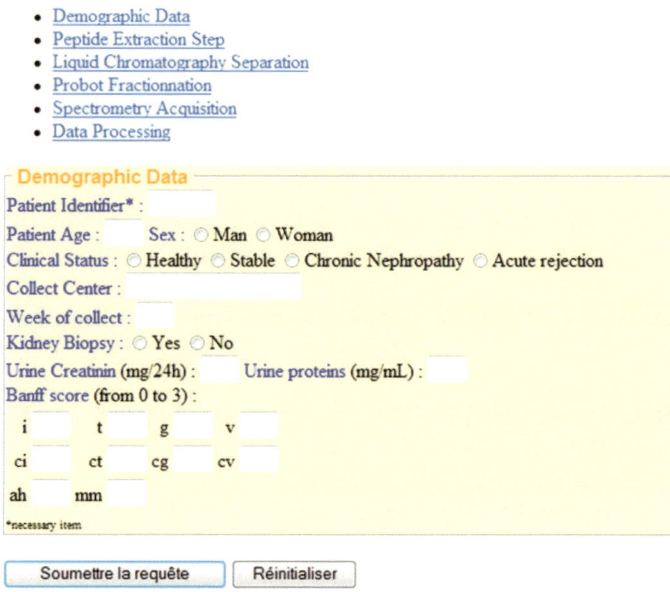
- Demographic Data
- Peptide Extraction Step
- Liquid Chromatography Separation
- Probot Fractionnation
- Spectrometry Acquisition
- Data Processing

Demographic Data
Patient Identifier* :
Patient Age : Sex : Man Woman
Clinical Status : Healthy Stable Chronic Nephropathy Acute rejection
Collect Center :
Week of collect :
Kidney Biopsy : Yes No
Urine Creatinin (mg/24h) : Urine proteins (mg/mL) :
Banff score (from 0 to 3) :
i t g v
ci ct cg cv
ah mm
*necessary item

Soumettre la requête Réinitialiser

Fig. 10. Extract of the internet form used to interact with BiomarkerMSdb, a relational database under MySQL constructed to record essential parameters involved in a biomarker discovery project using LC-MS strategies. This form is used to fill one of the 6 tables of the database called. "Demographic Data"(unpublished Moulinas & Gastinel, 2011).

## 4. Conclusion and perspectives

Observational and experimental biology is confronted today with a huge stream of data or "omics" data acquired by sophisticated and largely automated machines. This data is recorded under a digital format that has to be stored safely and shared within the scientific community. One of the challenges of modern biologists is to extract relevant knowledge from this large data. For that purpose, biologists not only should be acquainted with the use of existing multivariate and statistical data mining tools. generally used by meteorologists, economists, publicists and physicists but also should conceive their own specific tools. Among the tools available for multivariate analysis of «omics» data, Principal Component Analysis (PCA) as well as PLS and PCA-DA derivatives have demonstrated their utility in visualizing patterns in the data. These patterns consist sometimes in detecting outliers that spoiled data and that could be removed from them. PCA quantify major sources of variability in the data and allows to show which variables are most responsible for the relationship detected between the samples under study. Moreover, unwanted sources of variability can be revealed as batch effects and partially corrected by surrogate variable analysis (SVA) and EigenMS approaches. However, there are some limitations to using PCA in "omics" data. These limitations result from the large choice of methods of the data pre- and post-processing and the technical difficulty in displaying graphically all the data. Ggobi and rggobi allow to display quite large data using Grand tour and 2D tour, showing dynamic projections with the ability to color and glyph points according to factors (brushing). PCA is an invaluable tool in the preliminary exploration of the data and in filtering or screening them according to noise, outliers and batch effects before using other multivariate tools such as classification and clustering. An appropriate educational program should be pursued in universities in order to expose the theory and praticability of these tools to future biologists.

## 5. References

Baggerly, K.A., Edmonson, S.R., Morris, J.S. & Coombes, K.R.(2004) High-resolution serum proteomic patterns for ovarian cancer detection. *Endocrine-Related Cancer*, 11, 583-584.

Basly, J.C & Gastinel, L. (2009) Influences of extraction protocols in fulvic acids preparations from soils. *Master 2 Training Report*, GREASE Department, University of Limoges, France.

Benkali, K., Marquet,P.,. Rerolle, J., LeMeur,Y. & Gastinel, L. (2008) A new strategy for faster urine biomarker identification by nano-LC-MALDI-TOF/TOF mass spectrometry. *BMC Genomics*, 14, 9, 541-549.

Cavalli-Sforza,L.L., Menozzi,P. & Piazza,A. (1994) The History of Geography of Human Genes, Princeton University Press, ISBN: 9780691087504, NJ, USA.

Cook, D & Swayne, DF (2007) Interactive and dynamic graphics for data analysis with R and ggobi. Springer, ISBN: 978-0-387-71761-6, NJ, USA.

DeRisi,J.L., Vishwanath, R.I & Brown, P.O. (1997) Exploring the Metabolic and Genetic Control of the Gene Expression on a Genomic Scale. *Science*, 278, 681 - 686.

François, O., Currat,M., Ray, N., Han, E., Excoffier, L.& Novembre, J. (2010) Principal component analysis under population genetic models of range expansion and admixture. *Molecular Biological Evolution*, 27, 6, 1257-1268.

Freiwald, A. & Sauer, S (2009) Phylogenetic classification and identification of bacteria bymass spectrometry", *Nature protocols*, 4, 5, 732-742.

International Human Genome Sequencing Consortium (2001) Initial sequencing and analysis of the human genome, *Nature*, 409,860-921.

Johnson, W.E. & Li, C (2007) Adjusting batch effects in microarray expression data using empirical Bayes methods. *Biostatistics*, 8,1,118-127.

Karpievitch, Y.V., Taverner, T., Adkins, J.N., Callister, S.J., Anderson, G.A., Smith, R.D. & Dabney, A.R. (2009) Normalization of peak intensities in bottom-up MS-based proteomics using singular value decomposition. *Bioinformatics*, 25, 9, 2573-2580.

Leek,J.T., Scharpf, R.B., Corrada -Bravo, H., Simcha, D., Langmead, B., Johnson, W.E., Geman, D., Baggerly, K. & Irizarry,R.A (2010) Tackling the widespread and critical impact of batch effects in high-throughput data. *Nature Genetics*, 11, 733-739.

Leek,J.T. & Storey, D.J (2008) A general framework for multiple testing dependence. *Proceeding of the National Academy of Sciences*, 105, 48, 18718-18723.

Listgarten, J. & Emili, A. (2005) Statistical and Computational methods for comparative proteomic profiling using liquid chromatography-tandem mass spectrometry. *Molecular and Cellular Proteomics*, 4.4, 419-434.

Lu, J., Kerns, R.T., Peddada, S.D. & Bushel, P.R. (2011) Principal component analysis-based filtering improves detection for Affymetrix gene expression. *Nucleic Acids Research*, 2011, 1-8.

Maynard,C. & Gastinel, L. (2010) Phylogenetic relationship between bacteria revealed by MALDI TOF mass spectrometry. *Life Science Undergraduate Training Report*, Microbiology Department, University of Limoges, France.

Menozzi, P., Piazza, A & Cavalli-Sforza,L (1978) Synthetic maps of human gene frequencies in Europeans. *Science*, 201,4358, 786-792.

Mugo, S.M. & Bottaro, C.S. (2004) Characterization of humic substances by matrix-assisted laser desorption/ionization time-of-flight mass spectrometry. *Rapid Communication in. Mass Spectrometry*. 18,20, 2375-2382.

Moulinas, R & Gastinel,L (2011) How to detect batch effect in mass spectrometry analysis - Constitution of BiomarkerMSdb. *Master 1 SVT Report* , University of Limoges, France.

Novembre, J. and Stephens, M. (2008) Interpreting principal component analyses of spatial population genetic variation. *Nature Genetics*, 40, 5, 646-649.

Pluskal, T., Castillo, S., Villar-Briones, A., & Oresic, M. (2010) MZmine 2: Modular framework for processing, visualizing, and analyzing mass spectrometry-based molecular profile data. BMC *Bioinformatics* 11:395-405.

Poplitiya,A.D., Qian, W-J., Jaity, N., Petyuk, V.A., Adkins, J.N., Camp II, D.G., Anderson, G.A. and Smith, R.D (2008) DAnTE: A statistical tool for quantitative analysis of - "omics" data. *Bioinformatics*, 24,13, 1556-1558.

Price, A.L., Patterson, N.J., Plenge, R.M., Weinblatt, M.E., Shadick, N.A. & Reich, D. (2006). Principal components analysis corrects for stratification in genome-wide association studies. *Nature Genetics*, 38,8,904-908.

Ransohoff, D.F. (2005) Lessons from controversy: ovarian cancer screening and serum proteomics. *Journal of The National Cancer Institute*,97, 4, 315-317.

Rao,P.K. & Li, Q (2009) Principal component analysis of proteome dynamics in iron-starved mycobacterium tuberculosis. *J. Proteomics Bioinformatics*, 2,1, 19-31.

Reich, D., Price, A.L. and Patterson (2008) Principal component anaysis of genetic data. *Nature Genetics*, 40, 5, 491-492.

Sah, H.N. & Gharbia, S.E. (2010) *Mass Spectrometry for Microbial Proteomics*, Wiley,ISBN: 978-0-470-68199-2,UK.

Stein, L. (1996) How Perl saved the Human Genome Project, in *The Perl Journal,* 1,2, available from: www.foo.be/docs/tpj/issues/vol1_2/tpj0102-0001.html.

Tibshirani, R., Hastie, T., Narasimhan, B & Chu, G. (2002) Diagnosis of multiple cancer types by shrunken centroids of gene expression, *Proceedings of the National Academy of Sciences,* USA. 99, 10, 6567-6572.

Van der Berg, R., Hoefsloot, H.C.J., Westerhuis, J.A., Smilde, A.K. & Ven der Werf, M.J(2006) Centering, scaling and transformations: improving the biological information content of metabolomics data. *BMC Genomics,* 7, 142-156.

Venter, J.G. et al, (2001) The sequence of the Human Genome, *Science,* 291, 1305-1351.

# 3

# Chemometrics (PCA) in Pharmaceutics: Tablet Development, Manufacturing and Quality Assurance

Ingunn Tho[1] and Annette Bauer-Brandl[2]
*[1]University of Tromsø*
*[2]University of Southern Denmark*
*[1]Norway*
*[2]Denmark*

## 1. Introduction

Pharmaceutical tablets are subject to special regulations regarding their quality. Obvious examples are requirements for tablet performance in terms of pharmacological effect, which is closely connected to the uniformity of drug substance content in each tablet, to the disintegration properties of the tablet into smaller particles after intake, and to the rate of dissolution of the drug substance from the tablet or particles (drug release). For each product on the market, a set of quality criteria and their specific limits are defined by the relevant regulatory health authorities, e.g. Food and Drug Administration (FDA) in the USA, or European Medicines Agency (EMA) for Europe.

Following the steps of the development of a new tablet, we distinguish six main phases:

a. design and synthesis of the new pharmaceutically active drug substance (active pharmaceutical ingredient, API),
b. characterisation and quality assurance of the bulk raw materials, i.e. the API and additives (excipients),
c. optimisation of the composition of the formulation with respect to API and excipients,
d. scale-up to larger throughput (industrial scale),
e. optimisation and validation of the processing steps for the manufacturing; monitoring processing steps during manufacture,
f. quality assurance of the finished product, the tablet.

Tablets have been produced for more than 100 years in large quantities, and their quality has always been an issue; therefore the technical factors that influence tablet properties have been studied extensively. It has immediately become obvious that the number of factors is large and that there is a lot of interaction between them. Therefore, pharmaceutical development in general, and tablet manufacture in particular, have been one of the first areas where factor analysis and Design of Experiments (DoE) have been introduced.

The current international regulations for pharmaceutical products, namely the ICH (International Conference of Harmonisation) guidelines (ICH Q8) including Quality by

Design (QbD) approaches together with Process Analytical Technology (PAT) initiated by the FDA in 2004 have drawn even more attention to the use of state-of-the-art science and technology (PAT-Initiative, 2004; Guidance for industry, 2004). Improvement of the manufacturing of drug preparations is thereby initiated with the aim to replace end-product control of random samples by real-time quality control of every single item.

It is nowadays a prerequisite for the registration of a new tablet preparation to show how the use of systematic studies during the development phase has led to a good understanding of the processes, how rational decisions have been made during the development process with regard to an optimum product, how the manufacturing processes can be controlled within set limits, and to estimate the robustness of the processes, i.e. what a certain variability of any ingoing parameters means for the properties of the final product. Therefore, it is very common to integrate factor analysis and multivariate projection methods such as PCA in all stages of pharmaceutical development and manufacture in general. However, tablets are a particularly prominent example thereof, because all stages from the preparation of the bulk powder, granulation, compression and coating widely influence the properties of the final marketed product. Systematic studies within a reasonable design spaces and a good scientific understanding of the processes reduce uncertainty and create the basis of flexible and risk-based regulatory decisions. Therefore, chemometric methods, such as DoE and multivariate analysis (MVA) are - if nothing else of necessity - frequently and extensively used in pharmaceutical product and process development.

This chapter focuses exclusively on fundamental studies for the rational development of tablet formulations of general interest. Therefore, it will restrict to examples of commonly used excipients and to simple manufacturing methods. The interested reader is addressed to specialized textbooks for further reading about production technologies and quality assurance (e.g. Bauer-Brandl & Ritschel, 2012; Gad, 2008).

## 2. Definitions and background information

### 2.1 Pharmaceutical tablets

Tablets contain both the active pharmaceutical ingredient, API, for obvious reasons, and different types of excipients to generate both manufacturability and product performance. In a standard procedure, both types of powders (API and excipients) are homogeneously mixed and granulated, the granules finally mixed with more excipients (glidants, lubricants, etc), upon which the final blend is compressed into tablets. The tablets will, in contact with water or fluids of the gastrointestinal tract, disintegrate into the granule/powder again, before the drug substance is released. Figure 1 depicts the basic steps of production of a standard tablet and processes upon contact with water or gastrointestinal fluids.

### 2.2 Quality of tablets

For pharmaceutical tablets, quality has many aspects: We have to distinguish between the tablets as such, and the drug product, which includes the packaging and the product description. Table 1 lists some of the most important quality criteria for a tablet product. Any of these may be used as a quality attribute for optimisation of a tablet product.

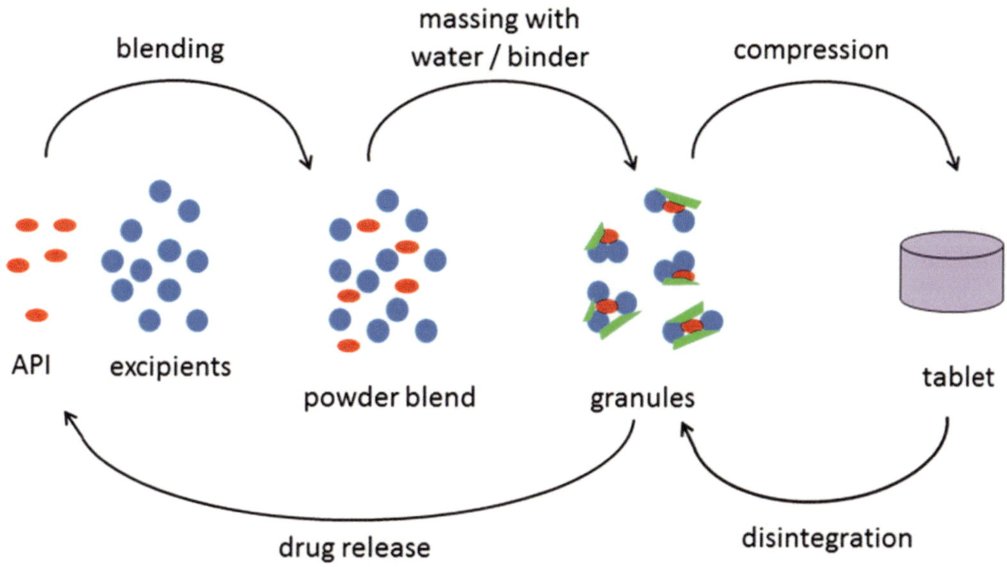


Fig. 1. Schematic drawing of the basic production steps of a tablet (upper arrows), and the disintegration and release of drug from the tablet (lower arrows)

| *Appearance of the tablets* | *Composition* | *Properties of the tablets* | *Properties of the drug product* |
|---|---|---|---|
| Shape | Identity of drug substance(s) | Mechanical strength | Outer package identity |
| Dimensions: Height, Diameter | Additives (excipients) | Ease to break | Label identity |
| Engravings (Symbols, numbers, etc) | Quantitative drug content per tablet | Disintegration behaviour in water | Leaflet identity |
| Colour | Total mass per tablet | Drug release rates in different media | Tamper-proof inner package |
| Surface appearance (shiny, smooth) | Coating composition | (mouth feeling) | Safe storage conditions |
| No specks, no spots | Coating thickness | No capping, no layering | No sign of tampering |
| No cracks, no chipping | | | No defect |

Table 1. Some important quality criteria of tablets as a drug product

For the marketing of a drug product, all the specifications need to be authorized by the respective local health authorities. Thereby, the actual values of quality requirements are defined for each single product individually, depending on the significance of the different criteria for this particular product and on special rules in the respective country.

### 2.3 Tablet development

The development of a new tablet formulation and production process takes its start in the properties of the active ingredient(s). For formulation (composition) development, it is of importance to choose the most suitable set of excipients to mix with the APIs to obtain a tablet product of certain quality attributes. In process (handling) development, including the choice of the processing method, all the processing parameters (such as temperatures, sequence of addition of excipients, mixing conditions, duration of the processing steps etc.) contribute to rather complex interrelated relationships between critical process parameters and quality criteria. The entire production process is typically divided into several steps, each of which is individually optimised.

Figure 2 shows general principles of how to control step-wise production. Choice of production method, optimisation of production steps, and control schemes have traditionally been handled by practical experience and by empirical development methods.

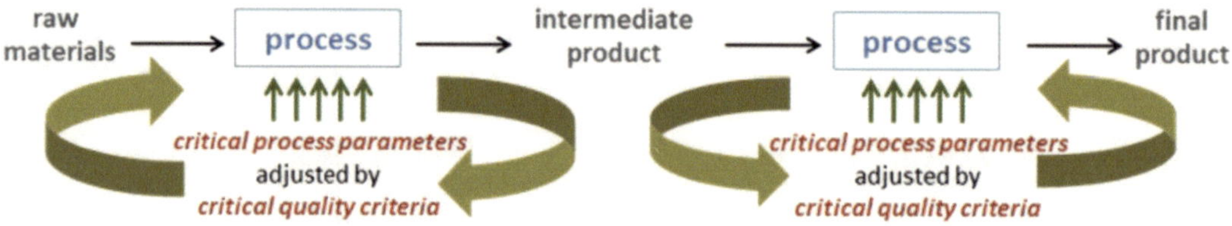


Fig. 2. Schematic drawing of production step control

### 2.4 A Brief overview of processes and parameters involved

In the following the elaboration will be restricted to basic tablets produced by a standard technique. Table 2 shows the impact of the main production steps on the quality of standard tablets. A schematic depiction of the sequential arrangement of some commonly used processing steps for standard tablets is given in Figure 3 together with a selection of related important processing parameters. Furthermore, monitoring of some machine parameters is also included.

### 2.5 The combination of DoE and MVA

Statistical experimental design, also called design of experiments (DoE), is a well-established concept for planning and execution of informative experiments. DoE can be used in many applications connected to tablet development and manufacture. One may, for instance, use designs from the *fractional factorial* or *central composite* design families, when addressing problems in which the influence of process parameters (temperature, pressure, time, etc.) on product properties are monitored and optimized. Another main type of DoE application concerns the preparation and modification of mixtures or formulations, where the challenge is that the parameters are not independent but components add up to 100%. This involves the use of *mixture designs* for changing composition and exploring how such changes will affect the properties of the formulation (Eriksson et al., 1998). Also designs in which both process and mixture factors are varied simultaneously are frequently seen. For tablet development and manufacture, one or several of the process parameters and/or critical quality attributes listed above, are likely to be used as optimization goals.

| *Tablet Property* | *Raw materials* | *Granulation* | *Drying* | *Final blending* | *Compression* |
|---|---|---|---|---|---|
| Dimensions | ++ | + | + | + | ++ |
| Appearance | + | + | ++ | + | ++ |
| Total mass per tablet | + | + | ++ | ++ | - |
| Drug content (assay) | + | + | + | ++ | - |
| Mechanical strength | ++ | ++ | ++ | ++ | ++ |
| Disintegration | ++ | ++ | + | ++ | ++ |
| Drug release rate | ++ | ++ | + | ++ | ++ |
| Stability; degradation | + | - | ++ | - | - |
| Microbiological quality | ++ | - | ++ | - | - |

+: important; ++: very important; -: important in special cases only

Table 2. Impact of Processing Steps on Tablet Quality

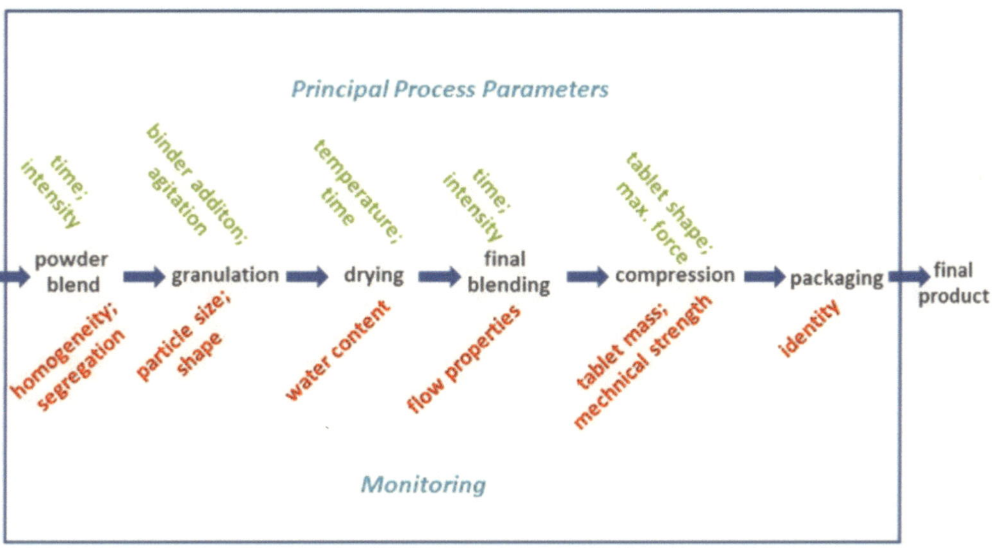


Fig. 3. A schematic view of the course of main processing steps, processing parameters and monitoring features of tablet production

Multivariate projection methods, such as principal component analysis (PCA) and partial least square (PLS) regression, are combined with DoE in formulation or process development for screening of a large number of influence parameters (fractionated factorial design) and further optimization of the important parameters, to study their interactions and possible non-linear behaviour (central composite design, Box Behnken, D-optimal, mixture designs).

## 3. PCA in exploring new molecular entities

QSAR and QSPR respectively (quantitative structure activity/property relationship) interlink biological effects or physical properties to the molecular structure of the respective substance. The best known example is the design and synthesis of new drug molecular entities in medicinal chemistry where the aim is that the drug molecule (API) interacts with the molecular structures at the site of action (e.g. receptors, enzymes). Another important step for a drug substance to act as a useful remedy is the uptake of such molecules into the body, which is a prerequisite to reach the receptor - has led to systematic PCA based studies of structure permeability relationships (e.g. intestinal barriers for the uptaker of oral drugs, or transport though the blood brain barrier; BBB partitioning).

Similar PCA methods have also been applied in formulation development by means of calculated quantum chemical descriptors for excipients (see Section 4.1.1.).

## 4. PCA in formulation development and processing

For the development of the formulation (composition) and the process (handling), all the experimental parameters during the entire manufacturing (such as chemical composition, processing steps, and processing conditions) contribute to the properties of the final product. In the following, examples of recent studies are briefly discussed, organized according to separate steps, although the entire flow of processing steps affects the product properties.

### 4.1 Excipient choice

As a rule, excipients have multiple functionality depending on the types of (other) materials present in the formulation, composition of the blends, and processes used. Well-known examples are filler-binders and lubricant-glidants.

A set of basic studies including four groups of excipients (filler, binder, disintegrant, lubricant) with several examples for each, in total almost one hundred excipients, were screened with respect to tablet properties (good mechanical strength and short disintegration time) (Gabrielsson et al., 2000; 2004). DoE combined with PCA and PLS were used in screening and optimisation: PCA provides a general overview over relationships and PLS models quantify the relationships between excipient properties and responses (tablet properties). Such mapping of excipients can be useful for selection of the appropriate material for a new formulation.

#### 4.1.1 Examples of functional polymers as excipients

Microcrystalline cellulose (MCC) is a frequently used excipient with filler/binder function. It is derived from native cellulose through hydrolysation and purification. Using DoE and stepwise multiple regression analysis, particle properties and the functionality of different MCC products can be quantitatively related to the hydrolysis conditions (Wu et al., 2001).

As another example the formulation of pectin pelletisation masses using different granulation liquids is shown in the form of a PCA score plot mapping the experimental area. Correlation between properties of the masses and the respective additives was found using quantum chemical molecular descriptors (QSAR) of the additives. Partial least square regression methods, mainly PLS-2 models, were used to quantify the effects (Tho et al., 2002). Based on the models, a rational choice of the composition of the formulation can be made.

### 4.2 Mixing / blending

Homogeneous powder blends are a prerequisite of the predefined dose of API per tablet. In optimized blending processes, homogeneity of the bulk is achieved.

However, segregation tendencies - becoming significant even for initially homogeneous bulk powder blends when these materials are transported - are even more important to consider. The effect of powder properties on segregation was studied, and PCA models developed to connect material properties to segregation tendency. A comparison of the multivariate approach and univariate approaches reveals that the latter leads to incomplete models (Xie et al., (2008)).

### 4.3 Particle size and powder flow

Multivariate latent variable methods have also been applied to mixture modelling of pharmaceutical powders. Particle size distribution was shown to predict powder flow (Mullarney and Leyva, 2009). Particle size distribution along with particle shape information has also been used to predict bulk packing and flow behavior (Sandler & Wilson, 2010).

### 4.4 Granulation

Wet granulation is classically done in high shear mixers or in fluidized bed equipment, or in a recent approach as a continuous process.

A basic investigation of general interest into the fluid bed granulation process is described by Dias and Pinto (2002), who used DoE and cluster analysis to find the relevant processing parameters and relations to product properties.

Another basic study of the fluid bed granulation process is described by Otsaka et al. (2011). PCA was first used to describe the relationships between a set of granule properties and to identify the properties that produce optimum tablet properties. In a second step, regression modelling was used to optimize the granulation conditions.

For wet granulation in high shear mixers, there are two different possibilities: The first one is a separate wet massing step and subsequent drying in another machine. The alternative is a single-pot set-up, where the drying step can be conducted in the same piece of equipment. Giry et al. compare these two processes (Giry et al., 2009) with respect to product properties using a set of formulations. This is a typical DoE and PCA application. Both processes led to products with only slight differences in properties. However, the robustness of the processes was different.

An alternative to wet granulation methods is dry granulation, basically by compression of dry powder blends to make granules. A dry granulation process conditions was studied within a PLS framework to predict selected granulation properties (Soh et al. 2008).

#### 4.5 Drying

In-line monitoring of the water content is essential in order to find the end point of the drying step for optimum water content of granules. NIR (near infrared) spectroscopy is particularly sensitive to water content, and using PCA/PLS models enable prediction of the actual water content. In addition, the NIR spectra are also sensitive to other granule properties such as particle size. However, in bulk material of a wide particle size, the actual water content of the granules varies significantly with size. Nieuwmeyer et al. (2007) developed a method for characterization of both water content and granule size during fluid bed drying based on NIR spectroscopy.

#### 4.6 Final blend; lubrication

The final blend is typically produced by blending solid lubricant into the granules. The problem with too long mixing times ("overmixing") is that lubricant particles scale off and cover too much of the granule surface, which leads to weak tablets. As an example to monitor the blending profile (the degree of lubricant coverage) over time, and correlate this to tablet hardness, NIR was used (Otsuka & Yamane, 2006). Principal component regression was employed to model and predict the hardness of the tablets.

#### 4.7 Tablet compression

Compression of the materials into tablets is a complex processing step, which is not yet fully understood. Amongst others, influences of particle size, shape, and structure on the deformation mechanism are difficult to separate and quantify. It has been shown that it is possible to derive basic material deformation characteristics during the compression step from in-line processing data using model materials (Klevan et al., 2010). Due to the complexity of the deformation of powder beds and the large number of parameters that may be regarded, screening of such in-line derived parameters for the most useful set of parameters is necessary in order to avoid over-determined functions (Andersen at al., 2006). Examples are discussed using a set of commonly used excipients of most diverse properties within the useful materials (Haware et al., 2009a; Roopwani & Buckner, 2011). The effect of particle engineering with regard to the tablet properties, such as tablet tensile strength, studied by PLS models (Otsuka & Yamane, 2006; Haware et al., 2010) using cross validation (jack-knifing). Prediction of tablet mechanical properties using PLS model can be made from NIR spectra of the blends (Otsuka & Yamane, 2009) or from in-line compression parameters derived on examples of very similar properties (Haware et al., 2009b).

Figure 4 shows a PCA biplot as an example how particle properties of different commercially available lactose qualities spread out in a design space. The data together with in-line compression parameters were used to predict the mechanical strength of tablets made with different blending ratios (formulations) and under different experimental conditions of compression (Haware et al., 2009b).

#### 4.8 Combination of processes

Polizzi and García-Munoz (2011) proposes a quantitative approach to simultaneously predict mechanical properties of particles, powders, and compacts of a pharmaceutical blend, based on the raw materials. They used a two-step, multivariate modeling approach created using historical physical data of APIs, excipients, and multi-component blends. The physical properties for each individual component were first transformed using a PCA technique to place them in a multivariate design space and capture property correlations. The scores from these PCA models were then weighted by the blending ratios prior to PLS regression versus actual measured blend properties. This method produced a complete prediction of all the material properties simultaneously which was shown to be superior to the prediction performance observed when applying linear ideal-mixing.

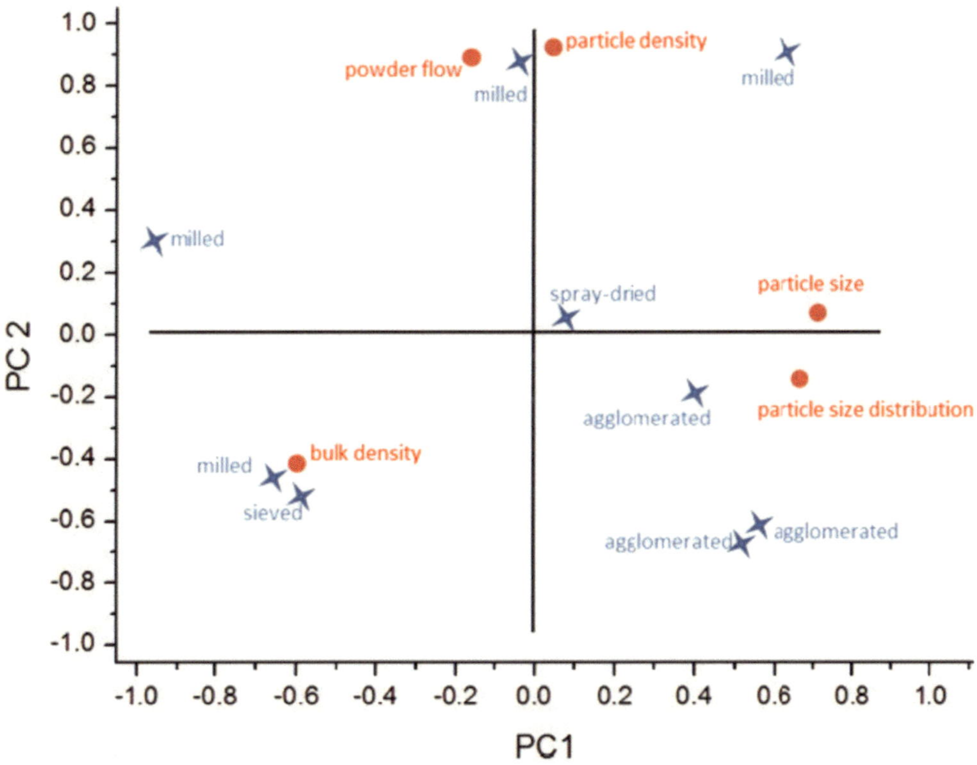


Fig. 4. Example of a PCA biplot showing powder characteristics of commercially available lactose grades spread out in the design space (modified after Haware et al. 2009b).

## 5. PCA in quality assurance of tablets

### 5.1 Spectroscopic methods

Molecular vibrational spectroscopy techniques, such as infrared (IR), near infrared (NIR) and Raman, are characterisation methods that have been applied to monitor both physical and chemical phenomena occurring during the processing as well as for off-line characterisation of raw materials and end-products. These techniques produce data of high dimensionality, since each sample is described with hundreds or even thousands of variables (wavelengths). Multivariate projection techniques, such as PCA are frequently combined with spectroscopic methods to enable detection of multivariate relationships between different variables such as raw materials, process conditions, and end products. Also well-known spectroscopic methods that are easier to interpret, e.g. UV (ultraviolet) spectroscopy and X-Ray diffraction, have been suggested to benefit from the use of PCA.

Both NIR and Raman spectroscopic measurements can be done in a very fast (within seconds) and non-destructive way, making them suitable tools for real-time process monitoring. They are frequently used for in-line monitoring of processes, e.g. particle size growing during granulation or the water content in the granule (Räsänen & Sandler, 2007; Rantanen 2007). NIR is also used for monitoring the final quality of the single tablets, because it is possible to quantify the active ingredient in a none-destructive manner (Tabasi et al., 2008), and can be applied to indentify counterfeit drugs (Rodionova et al., 2005). Another example is the prediction of tablet hardness based on NIR spectra of powders with lubricants (Otsuka & Yamane, 2006, 2009). For further applications of vibrational spectroscopy and chemometric methods in pharmaceutical processes, the reader is referred to the reviews by Gendrin et al. (2008), De Beer et al. (2011) and Rajalahti & Kvalheim (2011).

### 5.2 PCA in drug release

Sande and Dyrstad (2002) used PCA/PCR for parameterisation of kinetic drug release profiles, and showed that the combination of PCA and SIMCA was useful for classification of formulation variables based on the entire profiles. Also multi-way principal component analysis (MPCA) has been used to study entire tablet dissolution profiles and detection of shifts upon accelerated stability (Huang et al., 2011). Korhonen et al. (2005) calculated physicochemical descriptors (using VolSurf software) of different model drugs and correlated those to the respective dissolution profiles from matrix tablets. Again, the entire dissolution profiles without any parametric curve fitting were used in the models. This is a great advantageous, since the contribution of different variables tends to change in the course of the drug release event due to the dynamic nature of drug release from the matrix tablets.

Another example of the non-parametric curve description of drug release from a sustained release matrix is the approach suggested by Kikuchi and Takayama (2010). They use non-parametric descriptors and non-parametric differences between release curves for optimization (24 h linear release) and prediction. Surface plots are used for prediction of composition of the matrix tablets.

## 6. Monitoring and sensors

Combination of instrumentation and multivariate analysis provides powerful tools for effective process monitoring and control enabling detection of multivariate relationships between different variables such as raw materials, process conditions, and end products. Multivariate methods play an important role in process understanding, multivariate statistical process control, default detection and diagnosis, process control and process scale-up.

Continuous monitoring of process parameters allows continuous regulation and adjustment. Given rational development of the parameters as a background, a statistical process control is possible in order control the variation in a process within specification limits.

Furthermore, non-statistical variation caused by a specific event can be identified and the special causes thereof studied, with a possibility to eliminate these for the future.

In the case of tableting, continuous monitoring of the maximum compression force for each single tablet has been a standard in industrial production for more than 30 years. The force values are used as a measure for the individual tablet mass (and given homogeneity of the tableting material, also the dosing of each tablet). It is possible to sort out single bad tablets from the production line. A feedback loop needs to readjust the correlation between force and dosing based on sample analysis. By this, each tablet is held within the specification limits. Further development on tablet machine instrumentation includes measuring forces and displacements at different machine parts, and acoustic sensors in order to catch effects connected to lubrication problems.

Statistical Process Control (SPS) is the best way to take advantage of real-time monitoring. Based on the experience from process development, upper and lower warning and control limits can be set, where predefined action is taken: further trend analysis, actual changing of parameters, searching for the cause of an outlier. Figure 5 illustrates how SPS may be implemented in process monitoring. Trending analysis and feedback loops are necessary features to readjust these limits if necessary. In cases where secondary measures are used, reassuring the calibration with other measures is frequently conducted on a regular basis. A typical example is the relationship between compression force and mass of tablets. This is directly connected to the bulk density of the tableting material, which may change during the course of the actual tableting process. In many cases increased density is observed over time due to bulk material transport in the machine combined with machine vibrations. The mass of the tablets therefore needs to actually be checked on a balance.

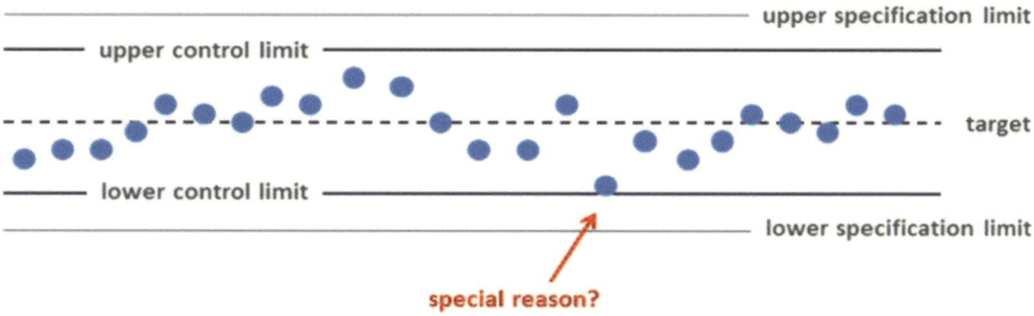


Fig. 5. Principles of Statistical Process Control (SPS) based on process parameter monitoring

Frequently used sensors based on NIR or Raman spectroscopy identify the chemical or physical variation between intermediates and products, particularly in the case of large numbers of samples:

- particle size distribution of powders and granules,
- granulation process monitoring,
- powder mixing optimization,
- scale-up of powder mixing,
- tablet properties (e.g. drug content).

These methods can be used to collect all batch characteristics, model all batches, detect outliers, and check future batches against historic batch models (in-line monitoring). These are prerequisites to diagnose out-of-limit-batches.

## 7. Future of tablet manufacture development

Traditionally, pharmaceutical manufactures are batch-based production methods with sample-based end-product control. This includes a sampling procedure with all its drawbacks. The future will be in the real-time in-line monitoring during processing, in order to be able to control the processing steps rather than end-point controls on the product. This will consequently lead to continuous production processes, which can be conducted in equilibrium. Continuous granulation has been developed already for a number of processes, including fluid-bed granulation, extrusion, and roller-compaction. These days, complete continuous production lines from the powder to the tablet are being introduced to the market.

The vision is to have such continuous production all the way from API synthesis to the final drug product ready for the market. The advantage will be a controlled and stable process with less product variation compared to traditional batch-based production methods. Furthermore, sample-based post-production quality control will be unnecessary and discarding or reworking of bad batches will not happen again. The real-time-release of the product will also become possible.

## 8. Conclusion

Although pharmaceutical tablets have been produced for more than 100 years on an industrial scale, there are up to day a number of unresolved challenges in tablet development, manufacturing and quality assurance. The basic factors that influence the product properties are widely known. However, depending on the individual composition of the tablet, numerous factors may have different impact on product quality. Furthermore, there is commonly a large interaction between many of these factors. Depending on composition, processing conditions for the different steps and environmental conditions the properties of the final bulk material (tableting blend) can differ widely. The actual tablet compression step, again committed to numerous processing factors, in addition would have large impact on the product quality. These complex relationships open a multivariate perspective and require suitable statistical methods.

For the tableting procedure, the steps that have been studied include powder blending, granulation, particle egalisation, lubrication, compression, coating, and drug release studies. Such step-wise studies have brought light into the impact of the parameters and their interactions and increased the understanding of the respective processes.

Pharmaceutical products underlie special requirements for their quality set by the health authorities. Therefore, the pharmaceutical development has been one of the first areas where factorial analysis and PCA/PLS have even become compulsory.

Moreover, the requirements for pharmaceutical drug products are under ongoing revision. It is a general trend that the requirements set higher standards with each new revision in order to assure safety and efficacy of the products. Some of the most important recent developments are the introduction of "Quality by Design" and PAT. It is a prerequisite to understand the processes, factors and interactions for the entire production line. Figure 6 depicts how the statistical tools are related to each other, and used interconnected in tablet development, manufacture and quality assurance.

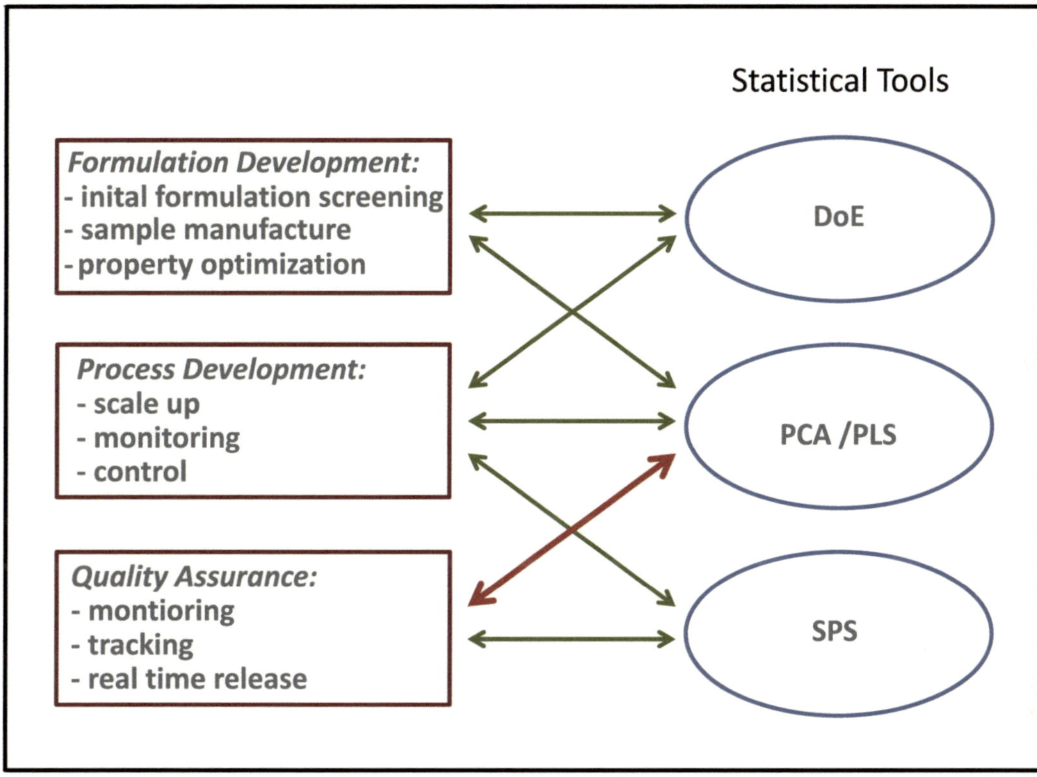


Fig. 6. A schematic overview of which statistical tools mainly are used in tablet development, manufacture and quality assurance

The development in the regulatory guidelines combined with new technology has found its expression in the efforts towards continuous production of pharmaceutical products, and in the forefront thereof the tablet manufacture due to both its relevance and complexity.

## 9. References

Andersen, E., Dyrstad, K., Westad, F. & Martens, H. (2006). Reducing over-optimism in variable selection by cross-model validation, *Chemometrics and Intelligent Laboratory Systems* Vol. 84 (No. 1-2): 69-74.

Bauer-Brandl A & Ritschel W.A. (2012). *Die Tablette: Handbuch der Entwicklung, Herstellung und Qualitatssicherung*, 3rd edition, Editio Cantor Verlag, Aulendorf, Germany.

De Beer, T., Burggraeve, A., Fonteyne, M., Saerens, L., Remon, J.P. & Vervaet, C. (2011). Near infrared and Raman spectroscopy for the in-process monitoring of pharmaceutical production processes, *International Journal of Pharmaceutics* Vol. 417 (No. 1-2): 32-47.

Dias, V. & Pinto, J.F. (2002). Identification of the most relevant factors that affect and reflect the Quality of Granules by application of canonical and cluster analysis, *Journal of Pharmaceutical Sciences* Vol. 91 (No. 1): 273-281.

Eriksson, L., Johansson, E. & Wikström, C. (1998). Mixture design- design generation, PLS analysis, and model usage, *Chemometrics and Intelligent Laboratory Systems* Vol. 43 (No. 1-2): 1-24.

Gabrielsson, J., Nyström, Å. & Lundstedt, T. (2000). Multivariate methods in developing an evolutionary strategy for tablet formulation, *Drug Development and Industrial Pharmacy* Vol. 26 (No. 3): 275-296.

Gabrielsson, J., Lindberg N-O., Pålsson, M., Nicklasson, F., Sjöström, M. & Lundtsedt, T. (2004). Multivariate methods in the development of a new tablet formulation; optimization and validation, *Drug Development and Industrial Pharmacy* Vol. 30 (No. 10): 1037-1049.

Gad, S.C. (ed.) (2008). *Pharmaceutical Manufacturing Handbook: Production and Processes, Section 6: Tablet Production*, Wiley-Interscience, Hoboken, New Jersey, USA, pp 879-1222.

Gendrin, C., Roggo, Y. & Collet, C. (2008). Pharmaceutical applications of vibrational chemical imaging and chemometrics: A review, *Journal of Pharmaceutical and Biomedical Analysis* Vol. 48 (No. 3-4): 533–553.

Giry, K., Viana, M., Genty, M., Louvet, F., Désiré, A., Wüthrich, P. & Chulia, D. (2009). Comparison of single pot and multiphase high shear wet granulation processes related to excipient composition, *Journal of Pharmaceutical Sciences* Vol. 98 (No. 10): 3761-3775.

Guidance for Industry (2004). PAT - A Framework for innovative pharmaceutical manufacturing and quality assurance. Available from: www.fda.gov/downloads/ Drugs/GuidanceComplianceRegulatoryInformation/Guidances/UCM070305.pdf [Last accessed 30. Oct. 2011]

Haware, R., Bauer-Brandl, A. & Tho, I. (2010). Comparative Evaluation of the Powder and Compression Properties of Various Grades and Brands of Microcrystalline Cellulose by Multivariate Methods, *Pharmaceutical Development and Technology* Vol. 15 (No. 5): 394-404.

Haware, R., Tho, I. & Bauer-Brandl, A. (2009a). Application of multivariate methods to compression behavior evaluation of directly compressible materials, *European Journal of Pharmaceutics and Biopharmaceutics* Vol. 72 (No. 1): 148-155.

Haware, R., Tho, I. & Bauer-Brandl, A. (2009b). Multivariate analysis of relationships between material properties, process parameters and tablet tensile strength for α-lactose monohydrates, *European Journal of Pharmaceutics and Biopharmaceutics* Vol. 73 (No. 3): 424-431.

Huang, J., Goolcharran, C. & Ghosh, K. (2011). A Quality by Design approach to investigate tablet dissolution shift upon accelerated stability by multivariate methods, *European Journal of Pharmaceutics and Biopharmaceutics* Vol. 78 (No. 1): 141-150.

ICH Q8 - Pharmaceutical Development, Available from: http://www.ich.org/fileadmin/ Public_Web_Site/ICH_Products/Guidelines/Quality/Q8_R1/Step4/Q8_R2_Guideline.pdf [Last accessed 30. Oct. 2011

Kikuchi, S. & Takayama, K. (2010). Multivariate statistical approach to optimizing sustained release tablet formulations containing diltiazem HCl as a model highly water-soluble drug, *International Journal of Pharmaceutics* Vol. 386 (No. 1): 149-155.

Klevan, I., Nordstroem, J., Tho, I. & Alderborn, G., (2010). A statistical approach to evaluate the potential use of compression parameters for classification of pharmaceutical powder materials, *European Journal of Pharmaceutics and Biopharmaceutics* Vol. 75 (No. 3): 425-435.

Korhonen, O., Matero, S., Poso, A. & Ketolainen, J. (2005). Partial Least Squares Projections to latent structure analysis (PLS) in evaluation and prediciting drug release from starch acetate matrix tablets, *Journal of Pharmaceutical Sciences* Vol. 94 (No. 12): 2716-2730.

Mullarney, M.P. & Leyva, N. (2009). Modeling pharmaceutical powder-flow performance using particle-size distribution data. *Pharmaceutical Technology* Vol. 33 (No. 3): 126–134.

Nieuwmeyer, F.J.S., Damen, M., Gerich, A., Rusmini, F., van der Voort Maarschalk, K. & Vromans, H. (2007). Granule characterization during fluid bed drying by development of a near infrared method to determine water content and median granule size, *Pharmaceutical Research* Vol. 24 (No. 10): 1854-1861.

Otsuka, T., Iwao, Y., Miyagishima, A. & Itai, S. (2011). Application of PCA enables to effectively find important physical variables for optimization of fluid bed granulation conditions, *International Journal of Pharmaceutics* Vol. 409 (No. 1-2): 81-88.

Otsuka, M. & Yamane, I. (2006). Prediction of tablet hardness based on near infrared spectra of raw mixed powders by chemometrics, *Journal of Pharmaceutical Sciences* Vol. 98 (No. 7): 1425-1433.

Otsuka, M. & Yamane, I. (2009). Prediction of tablet hardness based on near infrared spectra of raw mixed powders by chemometrics: Scale-up factor of blending and tableting process, *Journal of Pharmaceutical Sciences* Vol. 98 (No. 11): 4296-4305.

Polizzi, M.A. & García-Munoz, S. (2011). A framework for in-silico formulation design using multivariate latent variable regression methods, *International Journal of Pharmaceutics* Vol. 418 (No. 2): 235-242.

Roopwani, R. & Buckner, I. (2011). Understanding deformation mechanisms during powder compaction using principal component analysis of compression data, *International Journal of Pharmaceutics* Vol. 418 (No. 2): 227-243.

PAT - Initiative (2004). Available from: http://www.fda.gov/AboutFDA/CentersOffices/CDER/ucm088828.htm [Last accessed 30. Oct. 2011]

Rajalahti, T. & Kvalheim O.M. (2011). Multivariate data analysis in pharmaceutics: A tutorial review, *International Journal of Pharmaceutics* Vol. 417 (No. 1-2): 280-290.

Rantanen, J. (2007). Process analytical applications of Raman spectroscopy, *Journal of Pharmacy and Pharmacology* Vol. 59 (No. 2): 171–177.

Rodionova, O.Y., Houmøller, L.P., Pomerantsev, A.L., Geladi, P., Burger, J., Dorofeyev, V.L. & Arzamastsev, A.P. (2005). NIR spectrometry for counterfeit drug detection. A feasibility study, *Analytica Chimica Acta* Vol. 549 (No. 1-2): 151–158.

Räsänen, E. & Sandler, N. (2007). Near infrared spectroscopy in the development of solid dosage forms, *Journal of Pharmacy and Pharmacology* Vol. 59 (No. 2): 147–159.

Sande, S.A. & Dyrstad, K. (2002). A formulation development strategy for multivariate kinetic responses, *Drug Development and Industrial Pharmacy* Vol. 28 (No. 5): 583-591.

Sandler, N. & Wilson, D. (2010). Prediction of granule packing and flow behavior based on particle size and shape analysis. *Journal of Pharmaceutical Sciences* Vol. 99 (No.): 958–968.

Soh, J.L.P., Wang, F., Boersen, N., Pinal, R., Peck, G.E., Carvajal, M.T., Cheney, J., Valthorsson, H. & Pazdan, J. (2008). Utility of multivariate analysis in modeling the effects of raw material properties and operating parameters on granule and ribbon properties prepared in roller compaction, *Drug Development and Industrial Pharmacy* Vol. 34 (No. 10): 1022–1035.

Tabasi, S.H., Fahmy, R., Bensley, D., O'Brien, C. & Hoag, S.W. (2008). Quality by design, part I: application of NIR spectroscopy to monitor tablet manufacturing process. *Journal of Pharmaceutical Sciences* Vol. 97 (No. 9): 4040–4051.

Tho, I., Anderssen, E., Dyrstad, K., Kleinebudde, P. & Sande, S.A. (2002). Quantum chemical descriptors in the formulation of pectin pellets produced by extrusion/ spheronisation, *European Journal of Pharmaceutical Sciences* Vol. 16 (No. 3): 143-149.

Wu, J.-S., Ho, H.-O. & Sheu, M.-T. (2001). A statistical design to evaluate the influence of manufacturing factors on the material properties and functionalities of microcrystalline cellulose, *European Journal of Pharmaceutical Sciences* Vol. 12 (No. 4): 417-425.

Xie, L., Shen, M., Augsburger, L.L., Lyon, R.C., Khan, M.A., Hussain, A.S. & Hoag, S.W. (2008). Quality by Design: Effects of testing parameters and formulation variables on the segregation tendency of pharmaceutical powder, *Journal of Pharmaceutical Sciences* Vol. 97 (No. 10): 4485-4497.

# 4

# Pharmacophoric Profile: Design of New Potential Drugs with PCA Analysis

Érica C. M. Nascimento and João B. L. Martins
*Universidade de Brasília, LQC, Instituto de Química*
*Brazil*

## 1. Introduction

Searching for the pharmacophoric profile based on the concepts of chemical and structural contribution of the receptor active sites as well as receptor ligand interactions are fundamental for the development of new potential drugs for several different diseases and body dysfunctions such as degenerative brains disorders, Alzheimer disease, Parkinson's, diabetes Mielittus, cancer and many others known sickness conditions.

Basically, a pharmacophore describes the main molecular features regarding the recognition of a ligand by a biological macromolecule. In accord to the IUPAC (International Union of Pure and Applied Chemistry) Pharmacophore can be considered as "the largest common denominator shared by a set of active molecules to specific biological target"(Wermuth et al., 1998).

A pharmacophore is defined by the pharmacophoric descriptors that are a set of electronic, structural, electrostatic and topological properties. Those descriptors can be obtained by experimental and theoretical studies. Experimental studies like X-ray crystallography, spectroscopic measurements are used to define some atomic properties of a molecule, but need a previous job - a synthesis of such molecule. Thus, in this way theoretical studies eliminate this previous job the *in silico* experiments or the theoretical studies are important tools to know a molecule before this one gets synthesized.

In modern computational chemistry, pharmacophores are used to define the essential features of one or more molecules with the same biological activity. In the past few years, it has been common that the drug discovery has a contribution from pharmacophore modeling to identify and develop new potential molecules with desired biological effect (Steindl et al., 2006).

Therefore, in order to enhance the insights to identify potential new medicinal drugs many computational strategies are widely used (Steindl et al., 2006; Tasso, 2005). Mathematics and chemical molecular modeling applied on bioinformatics are important examples of these strategies. Multivariate analysis (principal component analysis - PCA) and quantum chemistry calculations (density functional theory - DFT) are some of them, and can led to the identification of the main information required to describe the essential pharmacophore profile.

PCA has been used to find new potential molecules in a series of biological systems (Carotti et al., 2003; de Paula et al., 2009; Nascimento, et al., 2011). It is a method abundantly used in a lot of multivariate analysis (Jolliffe, 2002). This method has two main aims: to decrease variable sets in the problems of multivariate data and to select the best properties (linearly independents) that describe the system (principal components). According to Jolliffe (Jollife, 2002) the central idea of PCA analysis "is to reduce the dimensionality of a data set in which there are a large number of interrelated variables, while retaining as much as possible of the variation present in the data set. This reduction is achieved by transforming to a new set of variables, the principal components, which are uncorrelated, and which are ordered so that the first *few* retain most of the variation present in *all* of the original variables. Computation of the principal components reduces to the solution of an eigenvalue-eigenvector problem for a positive-semidefinite symmetric matrix".

Although it may seem a simple technique, PCA can be applied in a lot of problems with several variables, such as the problems involving the determination of the pharmacophoric profile of a particular class of molecules.

There are different pretreatment methods for the data analysis, e.g., mean-centering and autoscaling. This work covers the second method. The autoscaling is very similar to the mean-centering with additional paths.

First, it is calculated the mean of the data matrix **X**, where the columns correspond to the variables and the lines are the samples (or objects). The mean is the j$^{th}$ column vector of a data matrix:

$$\bar{x}_n = \frac{1}{M}\sum_{m=1}^{M} x_n \tag{1}$$

After mean-centering the standard deviation (Equation 2) for each column is calculated and the autoscaling is obtained by the division of standard deviation (Equation 3).

$$s_n^2 = \frac{1}{M-1}\sum_{m=1}^{M}(x_n - \bar{x}_n)^2 \tag{2}$$

$$\bar{x}_{ij} = \frac{x_{ij} - \bar{x}_i}{s_i} \tag{3}$$

Within the autoscaling all variables are adjusted to the same range, such that each variable has a zero mean, while the standard deviation is one. Therefore, autoscaling is important for data set where the variables have different units.

DFT is a method to investigate the electronic structure of many-body systems, like atoms and molecules (Cramer, 2004). This method can be used to calculate electronic and structural properties of molecules with biological activity like new drugs to treat a disease. Applying functional that are spatially dependent on the electronic density of multi-electronics systems.

The application of human and economic resources to improve life quality has promoted benefits and has led to the increasing on the number of diseases known as age-dependent.

Among these diseases the most frequently are the dementias, diabetes and cardiovascular disease. Among the several types of dementia existing and the most common incidents in the population over 60 years of age are Alzheimer's disease (AD), vascular dementia and Pick's disease. It is estimated that 45 million people worldwide have some kind of dementia, among these, 18 million have symptoms characteristic of Alzheimer's disease. AD is a disease that has no determined cause, and therefore their treatment is based on drugs that only treat the disease in order to remediate the cognitive functions and slow down the degenerative advance. (Sugimoto, 2002)

AD is a kind of dementia that affects neural cognitive function. Thus, one of the most widely used strategies for the treatment of this disease consists in blocking off the hydrolysis (Figure 1) of the neurotransmitter acetylcholine in cholinergic synapses of neurons, this strategy is called cholinergic hypothesis, nowadays a major target in the development of drugs that benefits and improves the chemical balance of the concentration of acetylcholine in patients with AD (Tezer, 2005; Camps, 2005; Costantino, 2008).

Some drugs act as inhibitors of AChE, among them, tacrine first drug approved by FDA for the treatment of AD (Sugimoto, 2002; Ul-haq, 2010), followed by donepezil (Sippl, 2001), rivastigmine (Sugimoto, 2002; Ul-haq, 2010) and galantamine (Sippl, 2001). Other drugs have been clinically studied and tested for use in the treatment of AD, such as physostigmine (Sippl, 2001; Ul-haq, 2010; Bartolucci, 2006). Some others are being tested and are promising candidates for the approval, including, Huperzine A (Patrick, 2005; Barak, 2005), metrifonate (Sippl, 2001) and phenserine (Sippl, 2001; Ul-haq, 2010). These drugs are indicated to treat mild to moderate stages of AD, when the patient still has independent cognitive activity.

Fig. 1. Hydrolysis of acetylcholine

Chemically, the inhibitors mentioned above have in common the inhibitory action of AChE, but have different structure and chemical nature. The geometries of some AChEIs are shown in Figure 2. Organophosphates, alkaloids and acridines, are some of the classes of drugs that inhibit acetylcholinesterase and are used in the treatment of AD. This raises the question: what are the electronic, structural, topological and chemical properties that correlate these different classes of drugs as inhibitors of the same biological target?

The brain is the most complex system of the human body and main focus of biochemical and neuromolecular studies of the twenty -first century. The research and development of drugs that act more effectively in the treatment and the study of brain's diseases mechanisms are subject to various areas of knowledge.

PCA have been used to find new potential molecules in a series of biological systems (Carotti, 2003, de Paula, 2009; de Paula, 2007; Li, 2009; Nascimento, 2011; Nascimento, 2008; Rocha, 2011; Steindl, 2005). A methodology based on PCA for the pharmacophore identification of the acetylcholinesterase inhibitors (AChEI) was recently used (de Paula, 2009; de Paula, 2007; Fang, 2011; Nascimento, 2011; Nascimento, 2008). In this chapter the electronic, structural, topological and chemical properties of studied molecules, were taken from the theoretical procedures and related to the activity by means of the multivariate analysis PCs that provides the pharmacophoric profile of these AChEIs.

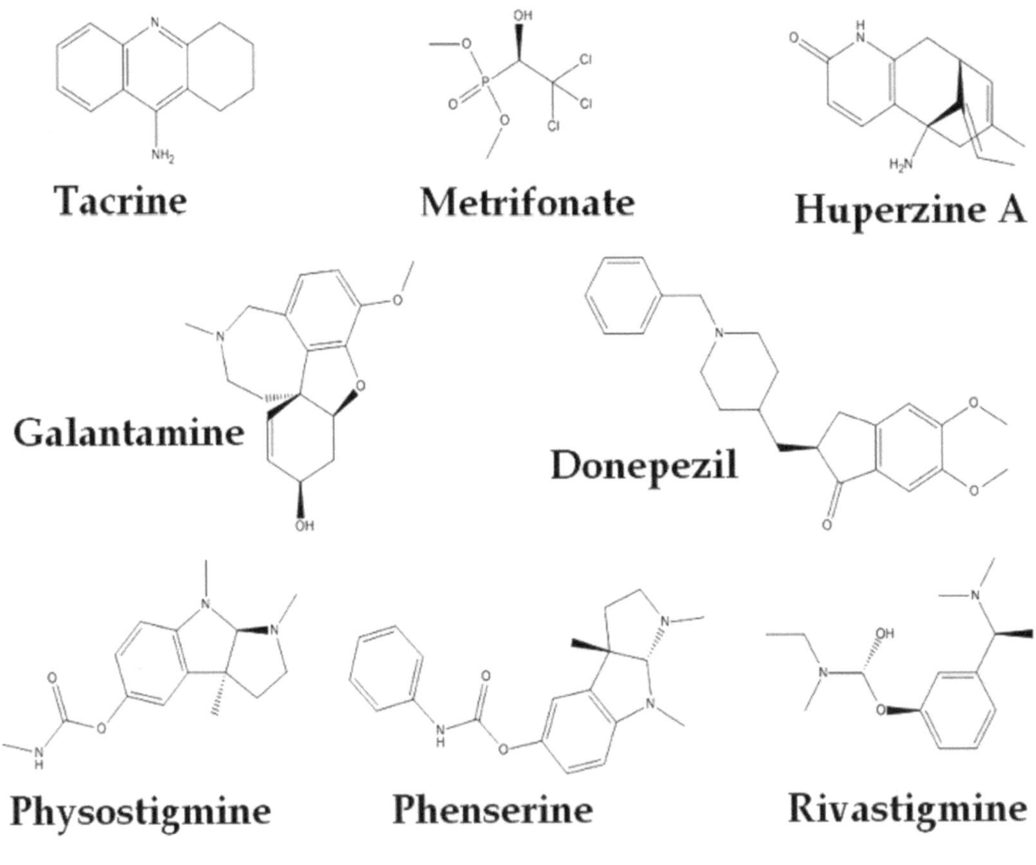


Fig. 2. Structures of some acetylcholinesterase inhibitors.

## 2. Mapping the pharmacophoric profile of acetylcholinesterase inhibitors

### 2.1 Acetylcholinesterase inhibitors, the AChEIs molecules

The cholinesterase inhibitors can be used for various purposes, as chemical warfare weapons such as sarin gas, in diseases treatment, e.g., the Alzheimer's disease treated with rivastigmine. There were used two different concepts to the rational design of inhibitors of this enzyme:

- drugs that have the molecular recognition in the catalytic site of AChE, doing preferably covalent bonds with the hydroxyl of the serine residue 200, this bond may be reversible or irreversible;
- drugs that are recognized by the to active site of the enzyme, which act by blocking the entry of the natural substrate, the neurotransmitter acetylcholine at the active site of AChE.

Some drugs that inhibit the function of the enzyme have also neuroprotective function, acting more effectively in the AD treatment. These drugs act in a negative way in the process of hyperphosphorylation of β-amyloid protein, inhibiting the breakdown of this protein and consequently the formation of amyloid plaques.

The inhibitors tacrine (THA), donepezil (E2020), rivastigmine (RIVA), galantamine (GALA), metrifonate (METRI), dichlorvos (DDVP), phenserine (PHEN), physostigmine (PHYSO), huperzine A (HUPE) and the tacrine dimer (DIMTHA) been studied theoretically by means of molecular modeling. From these studies, the mapping of pharmacophoric profile of these AChEIs was important to elucidate the correlation of such molecules as drugs to treating Alzheimer's disease.

## 2.2 Molecular modeling

Molecular modeling is one of the fields of theoretical chemistry that has been a great development in recent decades. The rational drug design is an area of application of molecular modeling of high relevance. Predict computationally a molecule with pharmacological activity and a drug candidate brings the expectation of reducing the search time and the cost in the molecular synthesis process and the clinical tests *in vitro* and *in vivo*.

The molecular modeling is characterized by applications of classical mechanics, quantum mechanics and stochastic methods like Monte Carlo in chemical and biochemical systems. One possibility is the prediction of spatial behavior and chemical interaction between a receptor and its biological substrate (ligand). Another is the calculation of molecular properties relevant to determining how this interaction takes place. Modeling the system receptor-ligand (RL) is possible to obtain important information about the interactions. These RL interactions establish the functionality of a given molecule. However, only the studies of interactions do not provide enough parameters to the development of new ligand.

The properties of a molecule can be obtained by theoretical calculations using classical mechanics and quantum mechanics. In the molecular mechanics approach, the classical mechanics deals with the problem without taking into account the electrons. Thus, electronic properties can only be treated by methods of electronic structure by means of quantum mechanical calculations. The main properties that describe a molecule are classified according to how it is defined spatially.

With the development of computational tools (hardware and software) it is possible today to treat complex molecular systems using approximate methods of quantum theory. Quantum mechanics considers that the energy of a molecule can be described as a sum of terms of energy resulting from electrostatic interactions between core and electrons (kinetic and potential energy). The fundamental postulate of quantum mechanics describes a wave function for each chemical system where a mathematical operator can be applied to this function resulting in observable properties of such system. Equation 4 is the focus of the study of quantum methods for the description of chemical systems and is known as the Schrödinger equation, independent of time and without relativistic corrections;

$$H\Psi = E\Psi \tag{4}$$

where **H** is a mathematical operator that returns the energy eigenvalue of the system as an operator called the Hamiltonian. **Ψ** is the wave function that describes the system state from the coordinates of the positions of nuclei and electrons of the atoms in the molecule. **E** is the eigenvalue, and represents the total energy of the system.

### 2.2.1 Computational modeling of AChEIs

The theoretical studies of AChEIs were performed using the Gaussian03 program package (Frisch, 2004) in order to determine the best electronic and geometrical parameters of AChEIs molecules. Was have used DFT at the B3LYP hybrid functional level with 6-31+G(d,p) basis set. The geometries of the target drugs were first optimized using internal coordinates in order to minimize all geometric data.

AChEI structures were taken from Protein Data Bank (PDB). Specifically, the structure of AChEI complexed with AChE included the inhibitors tacrine (PDB code 1ACJ), galantamine (PDB code 1DX), donepezil (PDB code 1EVE), tacrine dimer (PDB code 2CKM), huperzine A (PDB code 1VOT) and rivastigmine (PDB code 1GQR). However, the three-dimensional structures of physostigmine, phenserine, metrifonate and dichlorvos were not found in a complex with AChE, and so the structures were modeled using the GaussView 4.1 program.

In order to analyze the activity of such AChEIs, we have computed the electronic and structural properties of those molecules: dipole, frontier molecular orbital energies Highest Occupied Molecular Orbital and Lowest Unoccupied Molecular Orbital (HOMO, HOMO-1, LUMO and LUMO+1), charge of heteroatoms, charge of most acid hydrogens, volume, distance between the most acid hydrogens - H-H, molecule size, LogP (partition coefficient/lipophilicity), LogS (solubility), number of H-bond acceptors, number of H-bond donors, number of aromatic rings, gap (HOMO-LUMO), and map of electrostatic potential (MEP). The charge used was from the ChelpG (i.e., the charge from electrostatic potentials using a grid-based method) population analysis.

## 2.3 Electronic structure of AChEI molecules

Table 1 shows the values for some calculated properties for the studied AChEIs. The values of the HOMO orbital energies are different when comparing the organophosphate DDVP and METRI (higher values) and other AChEIS. Thus, the values of the orbital HOMO-1 energy for these organophosphates are almost similar while the values of this energy for other AChEIS, structurally different, are between -6.04 and -6.75, indicating that this orbital may be important in process of interaction of these drugs with acetylcholinesterase.

Properties such as volume, size of the molecule, logP and dipole have different values for each AChEI. Although some of these molecules are of the same class, the calculated values for the same property changes sharply, as the volume between DDVP and METRI.

| | DDVP | METRI | DIMTHA | THA | HUPE | GALA | E2020 | PHYSO | RIVA | PHEN |
|---|---|---|---|---|---|---|---|---|---|---|
| Dipole (D) | 3.04 | 2.93 | 2.09 | 3.51 | 6.08 | 1.67 | 2.48 | 1.08 | 2.53 | 1.85 |
| Charge $H^+$ (ua) | 0.070 | 0.438 | 0.260 | 0.339 | 0.365 | 0.371 | 0.136 | 0.333 | 0.103 | 0.281 |
| Charge N (ua) | 0 | 0 | -0.320 | -0.766 | -1.013 | -0.632 | -0.484 | -0.619 | -0.378 | -0.753 |
| Charge O( ua) | -0.638 | -0.639 | 0 | 0 | -0.670 | -0.685 | -0.554 | -0.659 | -0.594 | -0.505 |
| HOMO (eV) | -6.86 | -8.11 | -5.90 | -5.76 | -5.90 | -5.53 | -5.95 | -5.46 | -5.79 | -5.23 |
| HOMO-1 eV) | -8.57 | -8.30 | -6.71 | -6.56 | -6.75 | -6.04 | -6.05 | -6.08 | -6.53 | -6.43 |
| LUMO (eV) | -0.54 | -1.59 | -1.47 | -1.28 | -1.30 | -0.48 | -1.53 | -0.33 | -0.43 | 0.52 |
| LUMO+1(eV) | 0.42 | 0.70 | -0.58 | -0.46 | 0.36 | -0.14 | -0.48 | -0.07 | 0.21 | 0.51 |
| GAP (eV) | 6.32 | 6.52 | 4.47 | 4.49 | 4.60 | 5.53 | 4.41 | 5.12 | 5.36 | 5.75 |
| H-H (Å) | 1.796 | 2.307 | 1.998 | 1.683 | 1.625 | 2.373 | 2.342 | 2.324 | 2.460 | 2.250 |
| Volume ($Å^3$) | 185 | 207 | 606 | 236 | 286 | 329 | 454 | 321 | 312 | 332 |
| Size (Å) | 7.849 | 7.068 | 19.386 | 9.516 | 9.051 | 10.290 | 12.281 | 12.927 | 11.242 | 14.808 |
| logP | 1.66 | 0.80 | 3.88 | 3.13 | 2.60 | 1.39 | 4.14 | 1.94 | 2.86 | 2.99 |
| logS | -1.44 | -1.77 | -4.98 | -3.16 | -2.47 | -2.23 | -4.93 | -2.44 | -1.89 | -4.2 |
| PSA($Å^2$) | 44.8 | 55.8 | 58.5 | 38.9 | 58.9 | 41.9 | 38.8 | 44.8 | 32.8 | 44.9 |
| H-donnor | 0 | 1 | 4 | 2 | 2 | 1 | 0 | 1 | 0 | 1 |
| H-acceptor | 4 | 4 | 4 | 2 | 4 | 4 | 4 | 5 | 4 | 5 |
| Aromatic ring | 4 | 0 | 4 | 2 | 1 | 2 | 2 | 2 | 1 | 1 |

$H^+$: most acid hydrogen of the molecule. H-H: distance between the most acid hydrogens.

Table 1. Electronic and structural data of AChEIs from B3LYP/6-31+G(d,p) level of calculation.

The distribution of frontier orbitals in the molecule is important to determine which atoms are most relevant for the interaction between the ligand and the receptor, during the activity of such drug with the biological target. Figure 3 shows the regions of the frontier orbitals of RIVA. The LUMO of RIVA is located in an unusual region for orbitals poor in electrons, since the benzyl ring is by definition the region of high electronic density. RIVA have the HOMO and LUMO in the opposite region of carbamate moiety.

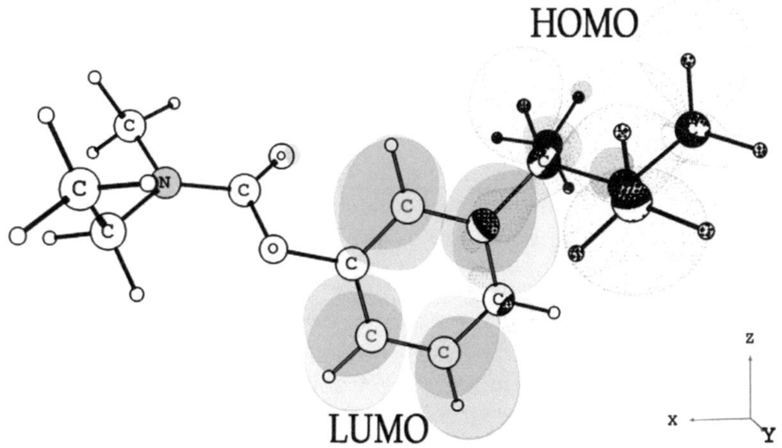


Fig. 3. HOMO and LUMO of RIVA calculated at B3LYP/6-31+G (d,p) level.

Figure 4 shows the MEP for all drugs studied. The maps indicate that all AChE have well-defined regions of high electron density (red) and low density (dark blue). These regions are likely points of interaction with AChE through electrostatic interactions like hydrogen bonding. Other electron density regions of AChE present mostly negative, indicating that these regions may have electrostatic interactions such as charge transfer, interactions with $\pi-\pi$ systems of aromatic residues in the active site of AChE. According to the literature (2003; de Paula et al., 2009; Nascimento, et al., 2011) RIVA interacts with the Ser200 residue of the catalytic triad through the carbamoyl moiety.

The regions of highest electronic density are located under the oxygen atoms in DDVP, METRI, HUPE, GALA, E2020, PHYSO, RIVA suggesting that these oxygens are receptors of hydrogens that can share in interactions, such as hydrogen bonding with the residues of the AChE catalytic triad or promoting covalent bonds, with the hydroxyl group of Ser200 (Camps, 2002). The rivastigmine (Figure 4h) and donepezil (Figure 4f) have more regions with low electronic density than other drugs, which is important for the interaction with the receptor site. The molecular volume is an important descriptor for drug design, and it is shown that some inhibitors are more linear which should improve the binding to two or more parts of the enzyme.

### 2.4 Mapping of the pharmacophoric profile of AChEIs via PCA

The mapping of pharmacophoric profile of AChEIs was conducted by means of PCA analysis, using of the data set obtained from calculations of the properties obtained for each molecule. For the analysis of principal components the geometry optimizations were performed at the level B3LYP/6-31 + G (d, p). The properties amount is shown in Table 1.

PCA was used to correlate 18 properties of 10 AChEI molecules to reduce the initial parameter set and determine the most relevant data for the acetylcholinesterase inhibition. PCA was conducted using the autoscaling method because the structural and electronic properties have different dimensions.

Since the structural and electronic properties have very different dimensions, autoscaling method is required. This normalization method consists in moving and stretching axes, centering the data on average and divides them by their standard deviation. Thus, each variable present zero mean and variance equal 1, giving the same significance for all variables (Jolliffe, 2002).

The first PCA showed that 83.3% on average of the information set formed by 10 AChEIs and 18 properties, obtained through the calculation performed in this work can be represented by four principals components (38.3% variance to PC1; 59.2% cumulative variance to PC2, 73.2% cumulative variance to PC3, 83.3% cumulative variance to PC4).

In studies of systems of many variables with PCA, it is desirable the lowest possible number of principal components that properly represent over 90% of cumulative variance. Thus, to increase the accuracy and determine the principal components we have used all combinations of 18 properties for the 10 AChEIs studied. Then the PCA was taken from the most relevant properties in relation to the total variance with the lower number of principal components.

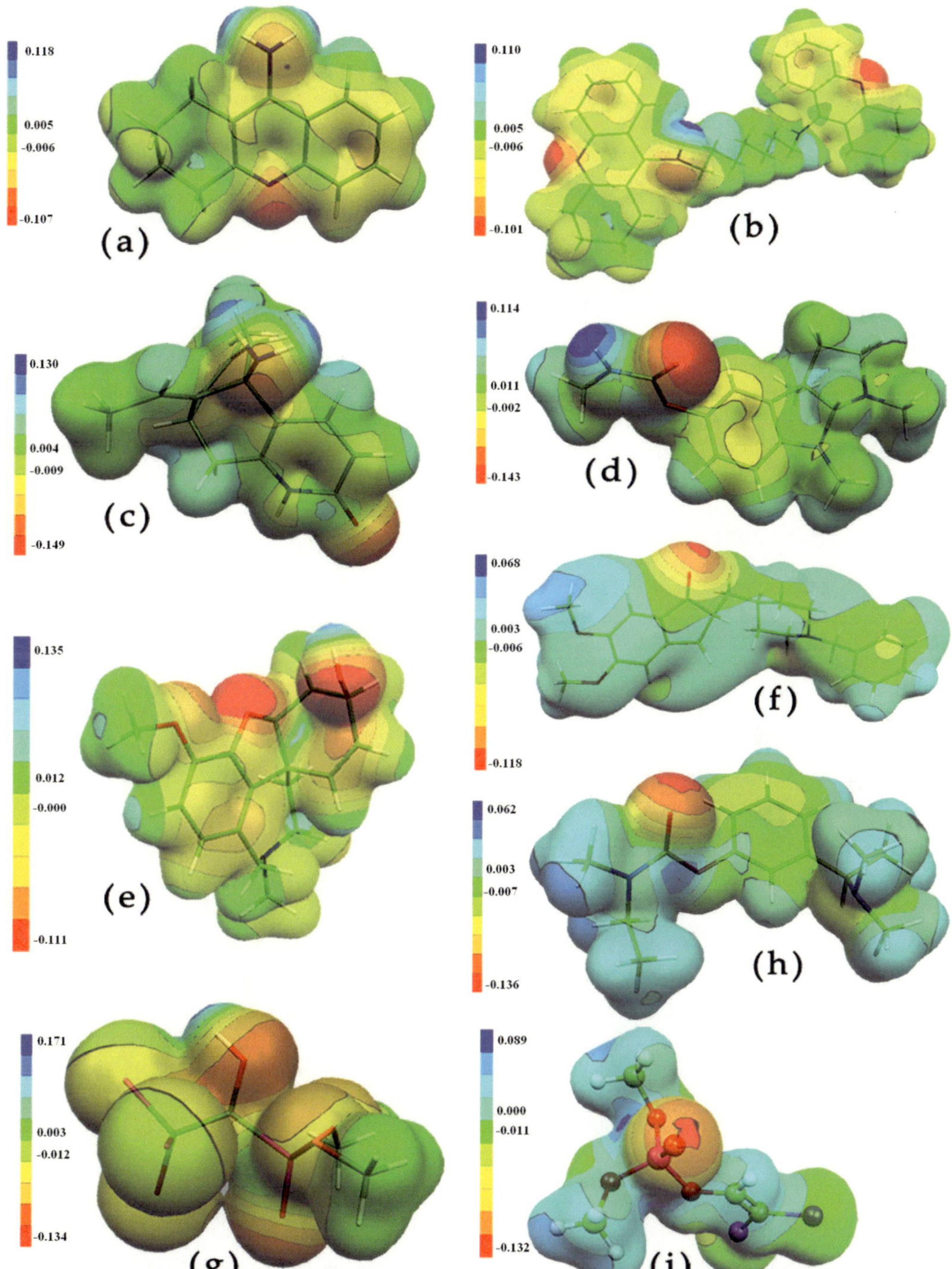


Fig. 4. Map of electrostatic potential of AChEIs calculated at B3LYP/6-31+G(d,p) level. (a) tacrine, (b) tacrine dimer, (c) huperzine A, (d) phenserine, (e) galantamine, (f) donepezil, (g) metrifonate, (h) rivastigmine, (i) dichlorvos.

In a second part, only 6 variables were used (shown below) with the 10 objects (AChEIs), these results in a 92% of total variance with information that characterizes these drugs in three principal components. Therefore, PCA (Figure 5) suggests that following properties are responsible for 92% of variance (63.5% for PC1 and 19.1% for PC2): volume, size of the molecule, distance between the two most acid hydrogens, energy of HOMO-1 frontier orbital, logP partition coefficient, and the number of aromatic rings (Table 2). In other words, these properties describes well the whole set of molecules data.

| | PC1 | PC2 | PC3 |
|---|---|---|---|
| B3LYP/6-31+G(d,p) | 63.5 % | 82.6% | 91.7% |

Table 2. Total cumulative variance (%) using 10 AChEIs and 6 properties

Equations 5, 6 and 7 show the scores of the PC1, PC2 and PC3. The PC1 is essentially the volume, size, partition coefficient- log P and number of aromatic rings of the drug (structural and electronic parameters), the PC2 represents the distance H-H (structural parameter) while the PC3 mainly represents the energy of the HOMO-1 orbital (electronic parameter).

$$PC1= 0.48_{volume} + 0.46_{Size} + 0.13_{H\text{-}H} + 0.37_{HOMO\text{-}1} + 0.42_{logP} + 0.44_{Arom.} \quad (5)$$

$$PC2=-0.01_{volume} + 0.02_{Size} + 0.88_{H\text{-}H} + 0.28_{HOMO\text{-}1} - 0.33_{logP} - 0.19_{Arom.} \quad (6)$$

$$PC3=-0.38_{volume} - 0.40_{Size} - 0.16_{H\text{-}H} + 0.79_{HOMO\text{-}1} + 0.18_{logP} + 0.04_{Arom.} \quad (7)$$

Figure 5 shows that the molecules are grouped into some main groups in regard to PC1, one of these are the GALA, RIVA, PHEN and PHYSO which forms a cluster and these are the FDA approved drugs. HUPE, THA form other group with two potent inhibitors acetylcholinesterase. Therefore, volume, $\pi-\pi$ systems and drug size are important parameters that correlate the AChEIs.

On other hand PC2 is dominated by the HH distance (score: +0.88), which separates compounds according to the distance between the two hydrogens more acidic in two groups: one that found them have H-H values smaller than 2.0 Å, and other group of objects that have values greater than those cited.

Figure 6 depicts the scores of PC1 versus PC3. In this case there are two standards in PC3 DDVP / METRO and GALA / PHYSO / RIVA / PHEN, this is reasonable, since PC3 represent the orbital energy of HOMO-1 (+0.79) (Equation 7), these AChEIs have values close to the HOMO-1 and the volume, as shown in Table 1. Figures 5 and 6, Equations (5, 6 and 7) generated in the PC indicates that the electronic parameters - orbital energy of HOMO-1 , log P and number of aromatic systems - and the structural parameters - volume, size of the drug and H-H- are the most significant properties of AChE in this multivariate analysis.

## 3. Applications of PCA from pharmacophoric profile of the AChEIs

From the pharmacophore profile determined by means of PCA and electronic structure calculations, other classes of molecules can be considered to be classified as AChEI. Therefore, it is important to note that, when added other objects, the new plotted PCs should have accumulative variance equal or larger than the original PCA (Table 3). That is, the original data set is the reference.

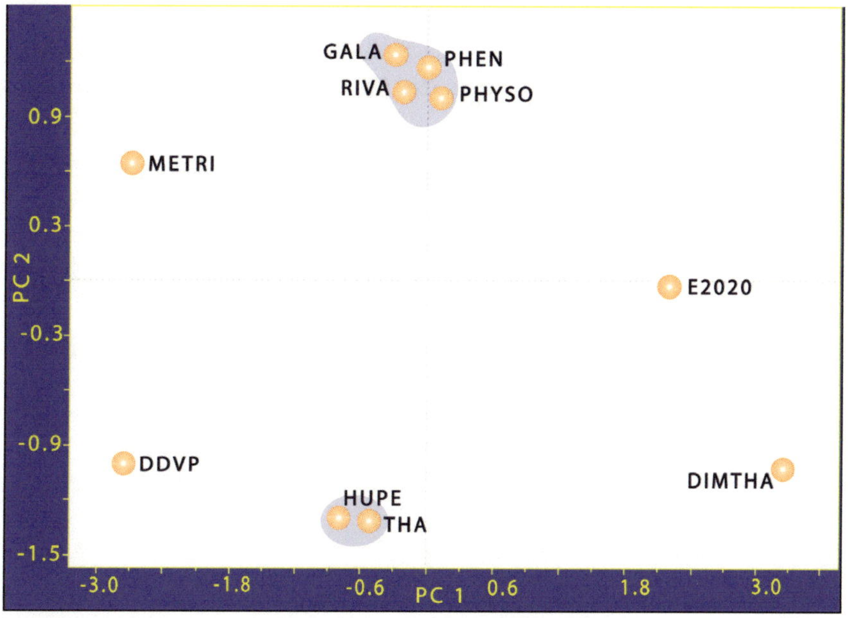


Fig. 5. PC1 versus PC2 scores from B3LYP/6-31+G(d,p) data.

| | PC1 | PC2 | PC3 |
|---|---|---|---|
| B3LYP/6-31+G(d,p) | 71 % | 86% | 94% |

Table 3. Total cumulative variance (%) using 10 AChEIs, 2 new molecules, 1 substrate and 6 properties.

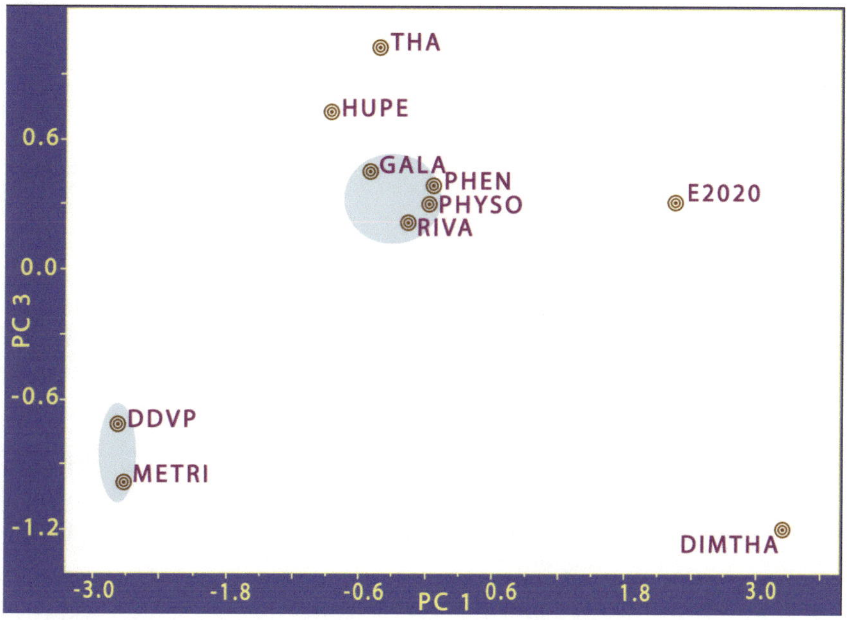


Fig. 6. PC1 versus PC3 scores from B3LYP/6-31+G(d,p) data.

The following are some applications explored. One possible application is the identification of dual biological activity of molecules. Another is the identification of activity against the function of acetylcholinesterase inhibitor, i.e., how behaves the natural substrate of the enzyme.

MOL1 and MOL2 with different biological activity from AChEIs were studied. We have also added another compound on the data set, acetylcholine (acetylcholinesterase substrate). Within these data from the properties obtained through electronic structure calculations, the new PCA was generated. These three objects were added to the 10 other objects of the first PCA which determined the pharmacophoric profile of AChEIs.

As shown in Figure 7, the natural substrate reorients the axis of the PCs and is completely opposed to the other (active) drugs. The acetylcholine is close to some of AChEIS by PC1, which have the properties of volume, size and number of aromatic rings with more high scores. Although acetylcholine is not an inhibitor it must have some properties correlated with the AChEIs. All molecules that are recognized by a protein have a certain set of properties that allows this recognition.

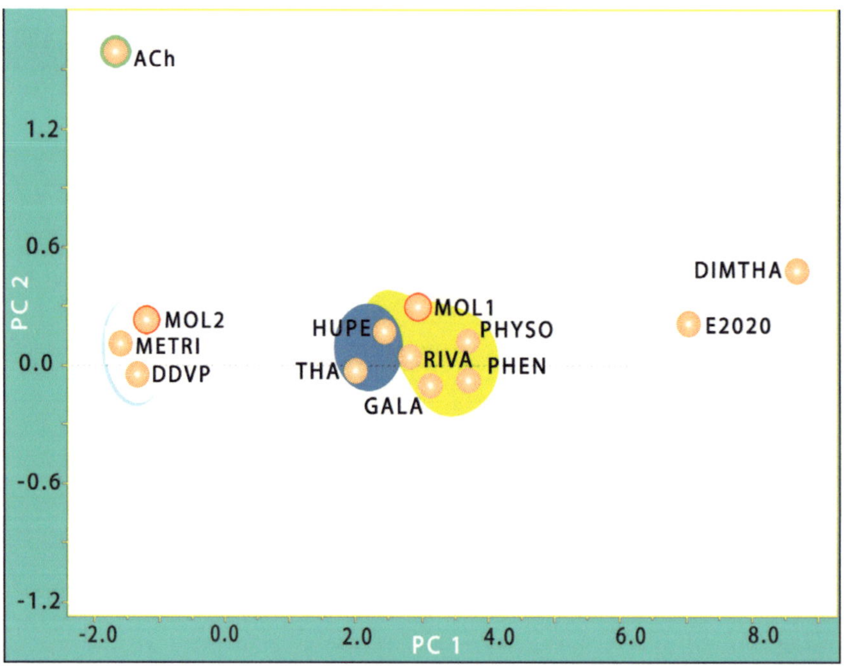


Fig. 7. PC1 versus PC2 scores from 10 AChEIs, 2 new molecules and ACh at B3LYP/6-31+G(d,p) level.

According to the PCA the two new drugs can be classified as AChEIs, although it is not possible to determine whether they have efficacy in the treatment of Alzheimer's disease. It is important to note that the effectiveness of a drug depends also on other factors that are difficult or much complex for modeling.

As shown in Figure 7, the natural substrate reorients the axis of the PCs and is completely opposed to the other (active) drugs. The acetylcholine is close to some of AChEIs by PC1, which have the properties of volume, size and number of aromatic rings with large scores. Although acetylcholine is not an inhibitor it must have some properties correlated with the AChEIs.

All molecules that are recognized by a protein have a certain set of properties that allows this recognition. As shown in Figure 8, in the case of acetylcholinesterase specifically, molecular modeling studies showed that during the molecular recognition of acetylcholine, the quaternary nitrogen of the substrate has a strong ionic interaction with the carboxyl group of the residue Asp72. In addition, there is formation of hydrogen bonds between the amine and ester group of the residue Asn65 and Tyr130 (Patrick, 2005).

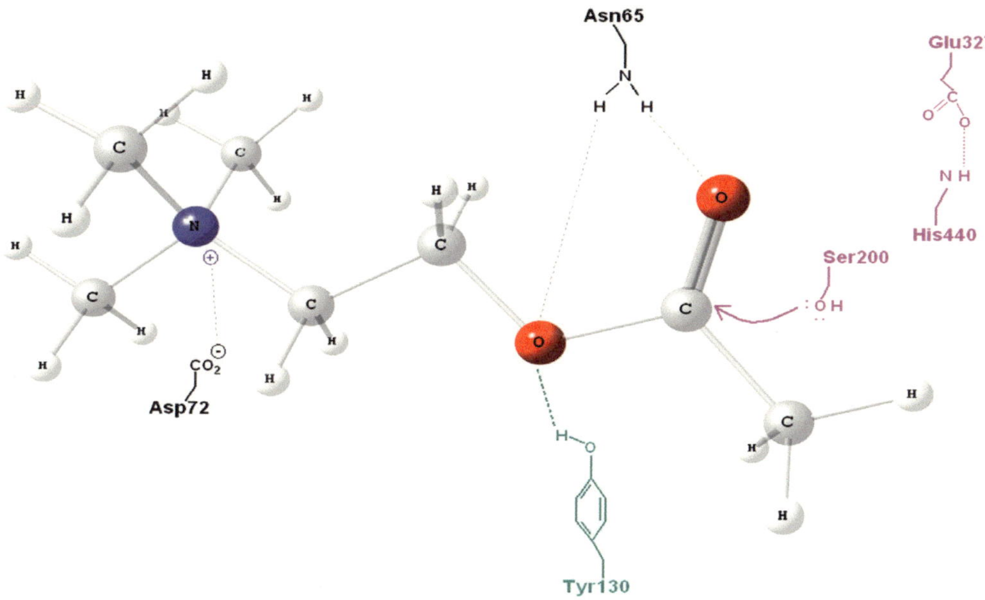


Fig. 8. PC1 Optimized structure of acetylcholine and main points of molecular recognition in acetylcholinesterase at B3LYP/6-31+G(d,p) level.

## 4. Conclusion

All studied AChE inhibitors are recognized as drugs that act effectively in the treatment of AD, even if some are not recommended due to unwanted side effects. From the computer modeling, applied by electronic structure calculations and PCA of known AChEIs with potential for the treatment of AD, it is possible to conclude that the properties relevant to the inhibitory activity are:

1. The HOMO-1 orbital is the most important property, characteristic of all drugs.
2. The study of the electrostatic potential map indicates that all AChEIs have well-defined regions of high and low electronic density, and these regions are the likely points of interaction with acetylcholinesterase, through electrostatic interactions such as hydrogen bonding. In some cases, such as dichlorvos, rivastigmine and physostigmine, it is possible to predict a bond with covalent character to the catalytic triad of the enzyme.

3. The H-H distance between 1.625-2.460Å, volume, and logP are also relevant for the pharmacophoric profile.
4. The electronic properties - orbital energy of the HOMO-1, log P and number of aromatic systems - and the structural parameters volume, size of the drug and H-H distance are the most significant properties in this study, the main components of the pharmacophore profile of AChEIs.
5. From the pharmacophore profile determined using PCA and electronic structure calculations, other classes of molecules can be considered to be classified as AChEI.

In addition to the structural and electronic properties mentioned above, an efficient inhibitor should have good bioavailability, cross the blood brain barrier with relative ease, to have high selectivity for acetylcholinesterase compared to butilcholinesterase; be able to form a complex reversible or pseudo-reversibly with AChE, be non-competitive compared to natural substrate. All these properties are jointly sharing in the profile of AChE pharmacophore studied in this work.

## 5. Acknowledgments

The authors are indebted to CNPq, INCTMN, FINEP, and CAPEs for financial support and Laboratório de Química Computacional/Universidade de Brasília (LQC/UnB) for computation resource.

## 6. References

Barak, D.; Ordentlich, A.; Kaplan, D.; Kronman, C.; Velan, B. & Shafferman, A. (2005). Lessons from functional analysis of AChE covalent and noncovalent inhibitors for design of AD therapeutic agents. *Chemico-Biological Interactions*, Vol.157, No.SI, (December 2005), PP 219-226, ISSN 0009-2797

Bartolucci, C.; Siotto, M.; Ghidini, E.; Amari, G.; Bolzoni, P. N.; Racchi, M.; Villetti, G.; Delcanale, M. & Lamba, D. (2006). Structural Determinants of Torpedo californica Acetylcholinesterase Inhibition by the Novel and Orally Active Carbamate Based Anti-Alzheimer Drug Ganstigmine (CHF-2819). *Journal of Medicinal Chemistry*, Vol.49, No.17, (August 2006), pp.5051-5058. ISSN 0022-2623

Camps, P. & Munoz-Torrero, D. (2002). Cholinergic Drugs in Pharmacotherapy of Alzheimer´s Disease. *Mini Reviews in Medicinal Chemistry*, Vol.2, No.1, (February 2002), pp.11-25, ISSN 1389-5575

Camps, P.; Formosa, X.; Munoz-Torrero, D.; Petrignet, J.; Badia, A. & Clos, M. V. (2005). Synthesis and pharmacological evaluation of huprine-tacrine heterodimers: Subnanomolar dual binding site acetylcholinesterase inhibitors. *Journal of Medicinal Chemistry*, Vol.48, No.6, (March 2005), pp. 1701-1704, ISSN 0022-2623

Carotti, A.; Altornare, C.; Savini, L.; Chlasserini, L.; Pellerano, C.; Mascia, M. P.; Maciocco, E.; Busonero, F.; Mameli, M.; Biggio, G. & Sanna, E. (2003). High affinity central benzodiazepine receptor ligands. Part 3: Insights into the pharmacophore and pattern recognition study of intrinsic activities of pyrazolo[4,3-c]quinolin-3-ones. Bioorganic & Medicinal Chemistry, Vol.11, No.23, (November 2003), pp. 5259-5272, ISSN 0968-0896

Costantino, H. R.; Leonard, A. K.; Brandt, G.; Johnson, P. H. & Quay, S. C. (2008). Intranasal administration of acetylcholinesterase inhibitors. *Bmc Neuroscience*, Vol.9, No.3, (December 2008), pp. 1-3, ISSN 1471-2202

Cramer, C. J. (November 22, 2004 ). *Essentials of Computational Chemistry* (2nd), Wiley, ISBN 047-0091-82-7, West Sussex

de Paula, A. A. N.; Martins, J. B. L.; dos Santos, M. L.; Nascente, L. D.; Romeiro, L. A. S.; Areas, T.; Vieira, K. S. T.; Gamboa, N. F.; Castro, N. G. & Gargano, R. (2009). New potential AChE inhibitor candidates. *European Journal of Medicinal Chemistry*, Vol.44, No.9, (September 2009), pp. 3754-3759, ISSN 0223-5234

de Paula, A. A. N.; Martins, J. B. L.; Gargano, R.; dos Santos, M. L. & Romeiro, L. A. S. (2007). Electronic structure calculations toward new potentially AChE inhibitors. *Chemical Physics Letters*, Vol.446,No.4-6, (October 2007), pp. 304-308, ISSN 0009-2614

Frisch, M. J., G.W. Trucks, H.B.S., G.E. Scuseria, M.A. Robb, J.R.C., Zakrzewski, V.G., Montgomery, Jr. J.A., Stratmann, R.E., Dapprich, J.C.B.S., Millam, J.M., Daniels, A.D., Strain, K.N.K.M.C., Farkas, O., Tomasi, J., Barone, V., Cammi, M.C.R., Mennucci, B., Pomelli, C., Adamo, C., Clifford, S., Petersson, J.O.G.A., Ayala, P.Y., Cui, Q., Morokuma, K., Rabuck, D.K.M.A.D., Raghavachari, K., Foresman, J.B., Ortiz, J.C.J.V., Stefanov, B.B., Liu, G., Liashenko, A., Komaromi, P.P.I., Gomperts, R., Martin, R.L., Fox D.J., Al-Laham, T.K.M.A., Peng, C.Y., Nanayakkara, A., Gonzalez, C., Challacombe, M., Gill, P.M.W., Johnson, B., Chen, W., Wong, M.W., Andres, J.L., Gonzalez, C., Head-Gordon, M., Replogle E.S. & Pople, J.A. (2003), Gaussian03, Gaussian Inc.,Pittsburgh, PA

Jolliffe, I. T. (October 1, 2002). *Principal Component Analysis* (2nd), Springers, ISBN 038-7954-42-2, New York

Lima, P.; Bonarini, A. & Mataric, M. (2004). *Application of Machine Learning*, InTech, ISBN 978-953-7619-34-3, Vienna, Austria

Nascimento, E. C. M. & Martins, J. B. L. (2011). Electronic structure and PCA analysis of covalent and non-covalent acetylcholinesterase inhibitors. *Journal of Molecular Modeling*, Vol.17, No.6, (June 2011), pp. 1371-1379, ISSN 1610-2940

Nascimento, E. C. M.; Martins, J. B. L.; dos Santos, M. L. & Gargano, R. (2008). Theoretical study of classical acetylcholinesterase inhibitors. *Chemical Physics Letters*, Vol.458,No.4-6, (June 2008), pp. 285-289, ISSN 0009-2614

Patrick, G. L. (2005) *An Introduction to Medicinal Chemistry* (3rd), Oxford University Press, ISBN 019-9275-00-9, New York

Rocha, J. R.; Freitas, R. F. & Montanari, C. A. (2011). The GRID/CPCA approach in drug Discovery. *Expert Opinion on Drug Discovery*, Vol.5, No.4, (April 2011), pp. 333-346, ISSN 1746-0441

Sippl, W.; Contreras, J. M.; Parrot, I.; Rival, Y. M. & Wermuth, C. G. (2001). Structure-based 3D QSAR and design of novel acetylcholinesterase inhibitors. *Journal of Computer-Aided Molecular Design*, Vol.15, No.5, (May 2001) pp.395-410, ISSN 0920-654X

Steindl, T. M.; Crump, C. E.; Hayden, F. G. & Langer, T. J. (2005). Pharmacophore modeling, docking, and principal component analysis based clustering: Combined computer-assisted approaches to identify new inhibitors of the human rhinovirus coat protein. *Journal of Medicinal Chemistry* , Vol. 48, No.20, (October 2005), pp. 6250-6260, ISSN 0022-2623

Steindl, T. M.; Schuster, D.; Wolber, G.; Laggner, C. & Langer, T. J. (2006). High-throughput structure-based pharmacophore modelling as a basis for successful parallel virtual screening. *Journal of Computer-Aided Molecular Design*, Vol.20, No.12, (December 2006), pp. 703-715, ISSN 0920-654X

Sugimoto, H.; Ogura, H.; Arai, Y.; Iimura, Y. & Yamanishi, Y. (2002). Research and development of donepezil hydrochloride, a new type of acetylcholinesterase inhibitor. *Japanese Journal of Pharmacology*, Vol.89, No.1, (May 2002), pp. 7-20, ISSN 0021-5198

Tasso, S. M.; Bruno-Blanch, L. E.; Moon, S. C. & Estiú, G. L. (2000). Pharmacophore searching and QSAR analysis in the design of anticonvulsant drugs. Journal of Molecular Structure-Theochem, Vol. 504, No. , (June 2000), pp. 229-240, ISSN 0166-1280

Tezer, N. (2005). Ab initio molecular structure study of alkyl substitute analogues of Alzheimer drug phenserine: structure-activity relationships for acetyl- and butyrylcholinesterase inhibitory action. *Journal of Molecular Structure-Theochem*, Vol.714, No.2-3, (February 2005), pp. 133-136, ISSN 0166-1280

Ul-Haq, Z.; Khan, W.; Kalsoom, S. & Ansari, F. L. (2010). In silico modeling of the specific inhibitory potential of thiophene-2,3-dihydro-1,5-benzothiazepine against BChE in the formation of beta-amyloid plaques associated with Alzheimer's disease. *Theoretical Biology and Medical Modelling*, Vol.7, No. 22, (June 2010), pp. 2-26, ISSN 1742-4682

Wermuth, C. G.; Ganellin, C. R.; Lindberg, P. & Mitscher, L. A. (1998). Glossary of terms used in medicinal chemistry (IUPAC Recommendations 1998). *Pure and Applied Chemistry*, Vol.70, No.5, (May 1998), pp. 1129–1143, ISSN: 0033-4545

5

# Empirical Study: Do Fund Managers Herd to Counter Investor Sentiment?

Tsai-Ling Liao[1], Chih-Jen Huang[1] and Chieh-Yuan Wu[2]
[1]*Providence University, Taichung,*
[2]*Department of the Treasury Taichung Bank,*
*Taiwan*

## 1. Introduction

Behavior among investors often influences one another. Investors may forego their own rational analysis but instead adopt behavior that is similar to the group. Recently, several studies note that this herding phenomenon (simultaneously trade the same stocks in the same direction) even exists in the behavior of institutional investors (Choe, Kho, and Stulz, 1999; Grinblatt, Titman, and Wermers, 1995; Kyrolainen and Perttunen, 2003; Lakonishok, Shleifer, and Vishny, 1992; Walter and Weber, 2006; Wermers, 1999; Wylie, 2005). There are three theoretical foundations for explaining institutional investor herding. First, in order to maintain or build a reputation when markets are imperfectly informed, managers may ignore their private information and hide in the herd (Brandenburger and Polak, 1996; Prendergast, 1993; Scharfstein and Stein, 1990; Trueman, 1994; Zwiebel, 1995). Second, managers may infer private information from the prior actions of agents (peer-group effects), and optimally decide to act alike (Bala and Goyal, 1998; Fung and Hsieh, 1999; Pound and Shiller, 1989). Third, institutional investors may receive the correlative private information from analyzing the same indicators (Banerjee, 1992; Bikhchandani, Hirshleifer, and Welch, 1992; Block, French, and Maberly, 2000; Froot, Scharfstein, and Stein, 1992; Hirshleifer, Subrahmanyam, and Titman, 1994) or favoring securities with specific characteristics (Del Guercio, 1996; Falkenstein, 1996; Gompers and Metrick, 2001; Payne, Prather, and Bertin, 1999).

What are the other factors that may explain the behavior of institutional investors herding? Investor sentiment may be one of the significant factors that cause this behavior. For example, Lakonishok et al. (1992, p.26) state in their pioneer research that, "...they (fund managers) might herd if they all counter the same irrational moves in individual investor sentiment." Barberis and Shleifer (2003), De Long et al. (1990), Lee et al. (1991), and Shleifer (2000) also make similar statements. Accordingly, one may intuitively expect that institutional investors will herd in their sell (buy) decisions in the presence of optimistic (pessimistic) sentiment. However, the literature appears to include no research that conducts an empirical investigation on this interesting hypothesis. The purpose of this study is thus to examine whether prior investor sentiment cross-sectionally explains the level of fund manager herding.

This study focuses on the U.S. fund trading sample because herd behavior becomes increasingly important when large institutional investors dominate the market such as in the U.S. To estimate herding by fund managers, the study in this report applies the Lakonishok et al.'s (1992) measure and Wylie's (2005) trinomial-distribution approach which considers buy, hold and sell positions for a given stock in a period. The study calculates conditional herding measures for stock-months that have a higher or lower proportion of fund buyers relative to the expected proportion in each month, which are the buy-herding measure and the sell-herding measure. To estimate investor sentiment for each stock-month, this paper employs the principal component analysis as the means of extracting the composite unobserved sentiment measure, rather than just selects a single indicator to proxy sentiment (Fisher and Statman, 2000; Neal and Wheatley, 1998). Brown and Cliff (2004) indicate that this estimating procedure is able to successfully extract measures of unobserved sentiment from various indicators. According to Baker and Stein (2004), Brown and Cliff (2004) and Baker and Wurgler (2006), the present study examines ten market weather vanes that can be categorized into three groups: individual stock sentiment indicator (individual stock return and individual stock trading volume), overall market sentiment indicator (S&P500 index return, S&P500 index trading volume, S&P500 index option Put/Call ratio, the number of IPOs, the average first-day return on IPOs, NYSE share turnover, and Russell 2000 index return), and fund sentiment indicator (net purchases of mutual funds).

After excluding observations with incomplete data to estimate herding, the final sample contains 770 U.S. mutual funds' holding records on 527 stocks from January 2001 to December 2005. Incorporating both bear (roughly 2001-2002) and bull (roughly 2003-2005) market periods under examination helps control for the effect of market conditions on the empirical results. The findings indicate a significantly positive association between the sentiment measure and subsequent sell-herding, after controlling the fund performance deviation, the capitalization of the stock, the number of funds trading the stock, the net mutual fund redemption of the stock, and the market-to-book ratio of the stock. However, the evidence shows no significant correlation between the composite sentiment measure and subsequent buy-herding. These findings suggest that institutional investors herd on the selling side when they observe high level of investor optimism, consistent with the intuition that rational institutional investors tend to counteract the optimistic sentiment of the investors (Wermers, 1999).

In sum, this paper contributes to the literature in three ways. First, by extending prior research that examines the impacts of fund herding on subsequent stock returns (Grinblatt et al., 1995; Klemkosky, 1977; Wermers, 1999), and the short-run predictability of sentiment in stock returns (Baker and Wurgler, 2006; Brown and Cliff, 2004; Fisher and Statman, 2000), this article substantiates the sentiment countering hypothesis and thereby contributes to the research realm of negative-feedback strategies, i.e. selling stocks that rise too far (Jegadeesh and Titman, 1993; Nofsinger and Sias, 1999; Scharfstein and Stein, 1990; Wermers, 1997, 1999). Second, the empirical finding indicates an explanatory power of investor sentiment for fund manager herding, providing support for the informational cascade theory where fund managers herd as a result of analyzing the sentiment indicator in this case. In other words, fund managers herd sell because, at least in part, they observe and try to counteract the optimistic sentiment. Finally, the significant association between sentiment and fund manager herding suggests potential directions for further research; specifically, the effect of fund herding behavior of countering optimistic market sentiment on fund performance. Also, researchers may employ the Structural Equation approach to examine the mediating role of institutional investor herding in the causal relationship between investor sentiment and stock returns. Such investigations will help corroborate the beneficial role of fund herding in speeding the price adjustment, as Wermers (1999) suggests, in the presence of mispricing resulting from investor sentiment.

This study proceeds in four sections. The next section discusses the estimation processes of fund herding and investor sentiment, the test methods and the variables. Section 3 describes the sample and data collection. Section 4 reports the empirical results. Finally, Section 5 summarizes the main conclusions.

## 2. Methods

### 2.1 Trinomial-distribution herding measure

To measure fund herding in stock *i* in month *t* (HMi,t), the present report follows the Lakonishok et al. (1992) method:

$$HMi,t = |Pi,t - E(Pi,t)| - E|Pi,t - E(Pi,t)|, \quad (1)$$

where,

$$Pi,t = B_{i,t} / (B_{i,t} + S_{i,t}) \quad (2)$$

$$E(P_{i,t}) = \frac{\sum_{i=1}^{n} B_{i,t}}{\sum_{i=1}^{n} (B_{i,t} + S_{i,t})} \quad (3)$$

$B_{i,t}$ ($S_{i,t}$) is the number of fund managers who buy (sell) stock *i* in month *t*. *Pi,t* is the proportion of mutual funds buying stock *i* in month *t* relative to the total number of funds trading stock *i* in month *t*. *E(Pi,t)* is the sample estimate of *Pi,t*, the expected proportion of buyers for stock *i* in month *t*. *E | Pi,t – E(Pi,t) |* is an adjustment which controls for random variation around the expected proportion of buyers under the null hypothesis of no herding. *| Pi,t – E(Pi,t) |* will be large if the trading of managers polarizes in the direction of either buying or selling. Averaging *HM* over all stock-months of interest can measure the extent to which any fund herds in a given stock-month.

To examine mutual funds' reaction to signals conveying upward or downward consensus, the study calculates conditional herding measures for stock-months that have a higher or lower proportion of buyers than the average stock in the same month as follows:

$$BHM = HMi,t|Pi,t > E[Pi,t] \quad (4)$$

$$SHM = HMi,t|Pi,t < E[Pi,t] \quad (5)$$

This approach is useful in analyzing fund herding into stocks separately from herding out of stocks. With Equation (4) as an example, the buy-herding measure (*BHM*) is equal to *HM* conditional on $P_{i,t}>E[P_{i,t}]$. If mutual funds tend to herd in their selling of stocks more frequently than in their buying of stocks, then the sell-herding measure (*SHM*) will be larger than the buy-herding measure (*BHM*).

Previous studies (e.g., Choe et al., 1999; Wermers, 1999) focus mostly on the binomial distribution of herding measure and consider only the buy and sell positions. Wylie (2005) claims that the accuracy of the Lakonishok et al.'s (1992) measure may be in question because the analysis rests on the assumption that all fund managers can short sell all stocks. However, the U.S. law technically neither forbids nor allows mutual funds to undertake short sales. As Wylie (2005) notes, Section 12(a) of the Investment Companies Act 1940 prohibits short sales by registered mutual funds in contravention of SEC rules, while the SEC issues no rules under Section 12(a). As a result, few U.S. mutual funds undertake any short selling. Wylie thereby argues that essentially herding could arise because of this invalid assumption. To avoid obtaining systematic herding estimate, the following tests apply Wylie's (2005) trinomial-distribution approach to measure fund manager herding, including buy, hold and sell positions.

### 2.2 Investor sentiment measure: Factor analysis approach

This study next conducts the principal component analysis to extract indicators explaining investor sentiment. We use an orthogonal transformation to convert 10 possibly correlated indicators of investor sentiment, which are taken from the literature, into a set of values of uncorrelated indicators. We then base on the factor analysis to calculate the corresponding factor score for every sentiment proxy. The composite sentiment measure is the sum of multiplying the transformed factor score (weighting) by each indicator's value. Table 1 summarizes the definition and the corresponding weighting for each proxy. These ten sentiment indicators can be classified into three groups, including individual stock sentiment indicator, overall market sentiment indicator, and fund sentiment indicator.

#### 2.2.1 Individual stock sentiment indicator

a. Individual stock return (*ISR*): De Long et al. (1990) and De Bondt (1993) find that investors predict the trend of stock returns according to past stock returns, so that the continuity of forecastable stock returns arises. For example, when the return continues for a period of time, the bullish psychology will pervade the investors. Notwithstanding the bull quotation occurs for a period of time and investors believe that the overall market would revise the quotation in the short term, the investors' bullish psychology remains.
b. Individual stock trading volume (*ISV*): Baker and Stein (2004) suggest that the market liquidity, like spreads, depth, trading volume and turnover, carrying information about the market can serve as a sentiment indicator. The trading volume is used as one of the sentiment proxy accordingly.

#### 2.2.2 Overall market sentiment indicator

a. S&P500 index return (*SP500R*) and Russell 2000 index return (*RU2000R*): The reason for incorporating these two proxies is the same as the *ISR*.
b. S&P500 index trading volume (*SP500V*): The reason for this proxy is the same as the *ISV*.

c. S&P500 index option put to call ratio (*SP500PC*): Brown and Cliff (2004) note that a higher (lower) put to call ratio indicates the pessimistic (optimistic) atmosphere getting stronger. The S&P500 index option put to call ratio is included as a proxy of the market sentiment.
d. Initial public offering first-day returns (*IPOR*) and the number of offerings (*IPON*): Due to the information asymmetries between IPO managers and outside investors, IPO activity tends to occur during bullish periods. Following Brown and Cliff (2004) and Baker and Wurgler (2006), the report includes both the IPO first-day returns and the number of offerings.
e. NYSE share turnover (*NYSET*): NYSE share turnover is the ratio of reported share volume to average shares listed from the *NYSE Fact Book*. In a market with short-sales constraints, Baker and Stein (2004) indicate that irrational investors will only be active in the market when they are optimistic and high liquidity is, as a result, a symptom of overvaluation.

### 2.2.3 Fund sentiment indicator

Net purchases of mutual funds (*NP*): Brown and Cliff (2004) indicate that net purchases by mutual funds and foreign investors positively reflect the extent to which they are optimistic about the individual stock. *NP* is the difference between dollar purchases and dollar sales by fund managers in a given stock.

| | Indicator | Definition | Weighting |
|---|---|---|---|
| Individual stock sentiment | Individual stock return (ISR) | $Log(P_{i,t})-Log(P_{i,t-1})$ | 0.097 |
| | Individual stock trading volume (ISV) | $Log(V_{i,t})-Log(V_{i,t-1})$ | 0.073 |
| Market sentiment | S&P500 index return (SP500R) | $Log(SP_t)-Log(SP_{t-1})$ | 0.055 |
| | Russell 2000 index return (RU2000R) | $Log(Ru_t)-Log(Ru_{t-1})$ | 0.089 |
| | S&P500 index trading volume (SP500V) | $Log(SPV_t)-Log(SPV_{t-1})$ | 0.093 |
| | S&P500 index option Put/Call ratio variation (SP500PC) | $(Put_t/Call_t)-(Put_{t-1}/Call_{t-1})$ | 0.105 |
| | Initial public offering first-day return (IPOR) | See Jay Ritter website | 0.152 |
| | Number of offerings (IPON) | See Jay Ritter website | 0.146 |
| | NYSE share turnover (NYSET) | Volume/shares listed | 0.116 |
| Mutual fund managers sentiment | Net purchases of mutual funds (NP) | $B_t-S_t$ | 0.074 |

Table 1. Investor Sentiment Indicators

Note. $P_{i,t}$ is the closing price at the end of month *t* for stock *i*. $V_{i,t}$ is the trading volume of stock *i* in month *t*. $SP_t$ is the closing price at the end of month *t* for S&P500 index. $Ru_t$ is the closing price at the end of month *t* for Russell 2000 index. $SPV_t$ is the S&P500 trading volume in month *t*. $Call_t$ ($Put_t$) is the open interest of S&P500 call (put) option in month *t*. *NP* is the difference between dollar purchases and dollar sales by funds in a given stock, where $B_t$ ($S_t$) is the dollar purchases (sales) by funds in stock *i* in month *t*. The weighting for each sentiment indicator is the transformed factor score by principal component analysis.

### 2.3 The effect of sentiment on fund herding

To gauge the effect of investor sentiment on fund herding, the report estimates the following regression equations:

$$HM_{i,t}=\alpha_0+\alpha_1*S_{i,t-1}+\alpha_2*RSTD_{t-1}+\alpha_3*CAP_{i,t}+\alpha_4*TNUMBER_{i,t}+\alpha_5*PR_t+\alpha_6*PB_{i,t-1}+\varepsilon_{i,t} \quad (6)$$

$$BHM_{i,t}=\alpha_0+\alpha_1*S_{i,t-1}+\alpha_2*RSTD_{t-1}+\alpha_3*CAP_{i,t}+\alpha_4*TNUMBER_{i,t}+\alpha_5*PR_t+\alpha_6*PB_{i,t-1}+\varepsilon_{i,t} \quad (7)$$

$$SHM_{i,t}=\alpha_0+\alpha_1*S_{i,t-1}+\alpha_2*RSTD_{t-1}+\alpha_3*CAP_{i,t}+\alpha_4*TNUMBER_{i,t}+\alpha_5*PR_t+\alpha_6*PB_{i,t-1}+\varepsilon_{i,t} \quad (8)$$

Here, *HM* is the aggregate fund herding measure, *BHM* is the buy-herding measure and *SHM* is the sell-herding measure. The main dependent variable, $S_{i,t-1}$, is the composite sentiment index of stock *i* in month *t-1*. To support the sentiment countering argument, $\alpha_1$ is expected to be positive.

As for control variables, *RSTDt-1* is the standard deviation of fund returns in month *t-1*. Bikhchandani et al. (1992) state that herding may result from some fund managers following other leader funds in their trades. A larger return deviation among funds in previous period indicates managers sharing dissimilar information and implies a higher reputation risk for underperformed managers, giving mutual funds a greater incentive to herd. Therefore, $\alpha_2$ is expected to be positive. Lakonishok et al. (1992), Wermers (1999) and Choe et al. (1999) establish that managers herd more on small stocks. $CAP_{i,t}$ is defined as the capitalization of stock *i* in month *t*, and the coefficient ($\alpha_3$) is expected to be negative. Wermers (1999) and Wylie (2005) indicate a positive relation between the level of herding in a given stock and the number of funds trading that stock ($TNUMBER_{i,t}$). The coefficient of *TNUMBER* ($\alpha_4$) is thus expected to be positive.

Wermers (1999) argues that fund may herd in response to sudden increases in cash inflow (mutual fund purchases) or cash outflows (mutual fund redemptions). The disproportional purchase (redemption) waves initiated by fund investors may force managers to increase (decrease) their stock holding simultaneously, and hence result in a buy (sell) herding. The net mutual fund redemption ($PR_t$) is the ratio of mutual fund purchases minus mutual fund redemptions in month *t* to fund assets. The coefficient ($\alpha_5$) is expected to be positive (negative) when *BHM* (*SHM*) is the dependent variable of the regression equation. The ratio of market value to book equity at the beginning of the trading quarter for stock *i* ($PB_{i,t-1}$) is used as a proxy for growth opportunity. The coefficient ($\alpha_6$) is expected to be positive as Wermers (1999) shows that higher levels of herding by growth-oriented funds are consistent with growth fund possessing less precise information about the true value of their stockholdings (mainly growth stocks). Finally, an industry dummy (equal to one for stocks of high-tech firms, zero otherwise) and four yearly dummies (2001 as the base year) are included in Equations (6) through (8) to control the possible industry and macroeconomic effects on fund herding behavior.

## 3. Sample and data collection

The sample contains 770 U.S. mutual funds' trading records on 527 stocks from 2003 to 2007, after excluding observations with incomplete data to estimate fund herding. Taking each monthly change in a stock as a separate trading, the total stock-month observations amount to 31,093, including 17,095 buy-herding and 13,998 sell-herding observations. Instead of using quarterly fund trading data, this study analyzes the fund herding behavior with monthly data to reduce the impact of intra-period round-trip transactions on the results.

Monthly data on portfolio holdings of the U.S. mutual funds are collected from the CRSP Mutual Fund Database. These data include periodic share holdings of equities for each fund at the end of each month. Individual stock price, trading volume and capitalization are obtained from the CRSP as well. S&P500 index return, S&P500 trading volume, S&P500 index option call/put open interest, NYSE share turnover and Russell 2000 index return are compiled from the Datastream. The number of IPOs and the average first-day returns of IPOs are available at the Jay Ritter website (http://bear.cba.ufl.edu/ritter).

## 4. Empirical results

### 4.1 Descriptive statistics and correlations between selected variables

Table 2 provides the descriptive statistics of the three herding measures, ten sentiment-related indicators and selected control variables in the regression analyses. The mean herding measures are 1.779% (*HM*), 2.164% (*BHM*) and 1.308% (*SHM*), respectively. The medians of the three herding measures are even smaller. These statistics indicate that fund managers seem not to herd much, in line with the findings of Lakonishok et al. (1992) and werners (1999).

| Variable | Mean | Std. Dev. | Min | Q1 | Median | Q3 | Max |
|---|---|---|---|---|---|---|---|
| HM (%) | 1.779 | 5.606 | -16.232 | -1.612 | 0.287 | 3.798 | 97.664 |
| BHM (%) | 2.164 | 5.913 | -16.648 | -1.577 | 0.548 | 4.335 | 77.654 |
| SHM (%) | 1.308 | 5.267 | -16.233 | -1.686 | 0.196 | 3.280 | 96.685 |
| ISR | 0.007 | 0.040 | -0.614 | -0.013 | 0.009 | 0.031 | 0.375 |
| ISV | 0.046 | 0.055 | -0.085 | -0.038 | -0.006 | 0.040 | 1.142 |
| SP500R | 0.006 | 0.070 | -0.015 | -0.003 | 0.005 | 0.007 | 0.025 |
| SP500V | 0.041 | 0.049 | -0.079 | -0.041 | 0.018 | 0.058 | 0.101 |
| RU2000R | 0.008 | 0.042 | -0.068 | -0.022 | 0.021 | 0.044 | 0.089 |
| SP500PC | -0.012 | 0.076 | -0.147 | -0.034 | -0.025 | 0.030 | 0.139 |
| IPOR | 0.122 | 0.013 | 0.102 | 0.121 | 0.122 | 0.123 | 0.140 |
| IPON | 142.600 | 45.026 | 63 | 156 | 159 | 161 | 174 |
| NYSET | 0.432 | 0.421 | 0.363 | 0.621 | 0.625 | 0.632 | 0.648 |
| NP | 267,447 | 33,103,284 | -1.079E9 | -1,970,434 | 126,690 | 2,391,904 | 601,787,572 |
| $S_{t-1}$ | 54.961 | 13.164 | 33.969 | 47.003 | 56.552 | 60.168 | 69.347 |
| $RSTD_{t-1}$ | 0.022 | 0.013 | 0.015 | 0.016 | 0.022 | 0.028 | 0.045 |
| $CAP_t$ | 6,550,706 | 21,769,725 | 42,230 | 557,247 | 1,417,585 | 4,107,318 | 317,133,094 |
| $TNUMBER_t$ | 120.166 | 121.391 | 2 | 56 | 87 | 146 | 651 |
| $PR_t$ | 0.004 | 0.004 | -0.015 | 0.002 | 0.004 | 0.009 | 0.017 |
| $PB_{t-1}$ | 7.970 | 213.206 | -136.187 | 1.594 | 2.621 | 3.966 | 13372.830 |

Table 2. Descriptive Statistics of Herding Measures and Selected Variables

Note. The sample contains 770 U.S. mutual funds' holding records on 527 stocks from 2001 to 2005, including 31,093 (17,095 buy-herding and 13,998 sell-herding) observations. HM is the aggregate herding measure. BHM is the buy-herding measure. SHM is the sell-herding measure. ISR is the individual stock return. ISV is the change of the logarithm of individual stock trading volume. SP500R is the S&P500 index return. SP500V is the change of the logarithm of S&P500 index trading volume.

RU2000R is the Russell 2000 index return. SP500PC is the S&P500 index option Put/Call ratio variation. IPOR is the mean first-day return of IPOs. IPON is the number of IPOs. NYSET is the NYSE share turnover. *NP* is the difference between dollar purchases and dollar sales by fund managers in a given stock. S is the composite sentiment measure. RSTD is the standard deviation of fund return. CAP is the capitalization of a given stock. TNUMBER is the number of funds trading a given stock. PR is the ratio of mutual fund purchases minus mutual fund redemptions in a given month to net assets of the fund. PB is the ratio of market value to book equity at the beginning of the trading quarter.

Table 3 presents Spearman (upper-triangle) and Pearson (lower-triangle) correlation coefficients between selected variables. In the Spearman measure, the correlation between the composite sentiment measure ($S_{t-1}$) and the aggregate herding ($HM_t$) is significantly positive (0.032). Likewise, the correlation between $S_{t-1}$ and the sell-herding ($SHM_t$) is also significantly positive (0.039). However, the correlation between $S_{t-1}$ and the buy-herding ($BHM_t$) is not significantly different from zero (0.013). The bivariate correlation evidence shows that investor sentiment does affect fund herding, especially on the sell-side. Although some of the figures are insignificant, the correlation coefficients between control variables and the herding measures correspond mostly with the predicted signs. The results in the Pearson measure are qualitatively similar to the findings in the Spearman measure.

| Panel A: Herding measure (HM) | | | | | | | |
|---|---|---|---|---|---|---|---|
| | $HM_t$ | $S_{t-1}$ | $RSTD_{t-1}$ | $CAP_t$ | $TNUMBER_t$ | $PR_t$ | $PB_{t-1}$ |
| $HM_t$ | | 0.032*** | 0.066*** | 0.005 | 0.012 | 0.010 | 0.019 |
| $S_{t-1}$ | 0.031** | | -0.304*** | 0.035*** | 0.043*** | 0.160*** | 0.019 |
| $RSTD_{t-1}$ | 0.066*** | -0.414*** | | -0.007 | -0.077*** | 0.092*** | -0.013 |
| $CAP_t$ | -0.014 | 0.028** | -0.001 | | 0.551*** | 0.030*** | 0.014 |
| $TNUMBER_t$ | 0.026** | 0.046*** | -0.058*** | 0.433*** | | -0.008 | 0.007 |
| $PR_t$ | 0.016 | 0.170*** | 0.134*** | 0.022* | -0.027** | | 0.006 |
| $PB_{t-1}$ | 0.005 | 0.009 | -0.016 | -0.005 | -0.014 | -0.008 | |
| Panel B: Buy-herding measure (BHM) | | | | | | | |
| | $BHM_t$ | $S_{t-1}$ | $RSTD_{t-1}$ | $CAP_t$ | $TNUMBER_t$ | $PR_t$ | $PB_{t-1}$ |
| $BHM_t$ | | 0.013 | 0.032*** | -0.065*** | 0.046*** | 0.034*** | -0.007 |
| $S_{t-1}$ | -0.013 | | -0.285*** | 0.028** | 0.014 | 0.226*** | 0.019 |
| $RSTD_{t-1}$ | 0.027** | -0.411*** | | 0.037*** | 0.019 | -0.188*** | 0.011 |
| $CAP_t$ | -0.034** | 0.011 | 0.086*** | | 0.474*** | 0.110*** | 0.005 |
| $TNUMBER_t$ | 0.073*** | 0.016 | 0.052*** | 0.412*** | | 0.061*** | -0.004 |
| $PR_t$ | 0.050*** | 0.198*** | -0.021* | 0.094*** | 0.081*** | | 0.008 |
| $PB_{t-1}$ | 0.005 | 0.013 | -0.024* | -0.005 | -0.017 | -0.008 | |
| Panel C: Sell-herding measure (SHM) | | | | | | | |
| | $SHM_t$ | $S_{t-1}$ | $RSTD_{t-1}$ | $CAP_t$ | $TNUMBER_t$ | $PR_t$ | $PB_{t-1}$ |
| $SHM_t$ | | 0.039*** | 0.123*** | 0.113*** | 0.121*** | -0.066*** | 0.026* |
| $S_{t-1}$ | 0.026** | | -0.300*** | 0.067*** | 0.075*** | 0.117*** | 0.008 |
| $RSTD_{t-1}$ | 0.101*** | -0.443*** | | -0.083*** | -0.170*** | 0.156*** | -0.022 |
| $CAP_t$ | 0.049*** | 0.037*** | -0.073*** | | 0.522*** | -0.078*** | 0.017 |
| $TNUMBER_t$ | 0.039*** | 0.086*** | -0.150*** | 0.455*** | | -0.114*** | 0.010 |
| $PR_t$ | -0.098*** | 0.126*** | 0.242*** | -0.060*** | -0.104*** | | 0.007 |
| $PB_{t-1}$ | 0.016 | -0.001 | 0.003 | -0.002 | -0.007 | -0.016 | |

Table 3. Correlation Coefficients between Regression Variables

Note. The Spearman (Pearson) correlation coefficients are reported at the upper-triangle (lower-triangle). The sample contains 770 U.S. mutual funds' holding records on 527 stocks from 2001 to 2005, including 31,093 (17,095 buy-herding and 13,998 sell-herding) stock-month observations. HM is the aggregate herding measure. BHM is the buy-herding measure. SHM is the sell-herding measure. S is the composite sentiment measure. RSTD is the standard deviation of fund return. CAP is the capitalization of a given stock. TNUMBER is the number of funds trading a given stock. PR is the ratio of mutual fund purchases minus mutual fund redemption in a given month to net assets of the fund. PB is the ratio of market value to book equity at the beginning of the trading quarter. ***, ** and * indicate significant at the 0.01, 0.05 and 0.10 level, respectively.

### 4.2 Herding comparisons between high and low sentiment sub-samples

The bivariate correlation results in Table 3 indicate that $S_{t-1}$ relates positively with *HM* and *SHM*. Fund managers show a tendency to herd in trading stocks with high investor sentiment, especially on the sell-side. To further support this view, Table 4 presents univariate comparisons of herding between high and low $S_{t-1}$ sub-samples. High (low) $S_{t-1}$ sub-sample refers to observations with $S_{t-1}$ greater (less) than the median $S_{t-1}$ of the full sample. Relative to low $S_{t-1}$ stocks, stocks with high $S_{t-1}$ also have higher mean and median *HM* ($t$=5.086 and $Z$=3.475 respectively). Similarly, both the mean and median *SHM* for high $S_{t-1}$ subsample are significantly higher than the figures for low $S_{t-1}$ subsample ($t$=5.755 and $Z$=3.969 respectively). However, neither the mean nor median of *BHM* for high $S_{t-1}$ subgroup is significantly different from that of low $S_{t-1}$ subgroup ($t$=1.419 and $Z$=1.160 respectively). By and large, the observed fund herding behavior in high sentiment stocks mostly comes from selling herding rather than from buying herding. Consequently, Table 4 provides evidence consistent with the correlation coefficient findings in Table 3, suggesting that fund herds form more often on the sell-side than on the buy-side in stocks with high past investor sentiment.

| | | HM | BHM | SHM |
|---|---|---|---|---|
| High St-1 | Mean | 0.017 | 0.018 | 0.013 |
| | Std Dev. | 0.053 | 0.066 | 0.052 |
| | Median | 0.004 | 0.004 | 0.003 |
| Low St-1 | Mean | 0.014 | 0.017 | 0.010 |
| | Std Dev. | 0.051 | 0.058 | 0.039 |
| | Median | 0.002 | 0.003 | 0.001 |
| t value for the difference | | 5.086 (<.001)*** | 1.419 (0.156) | 5.755 (<.001)*** |
| Wilcoxon Z the difference | | 3.475 (<.001)*** | 1.160 (0.246) | 3.969 (<.001)*** |

Table 4. Herding Comparisons between High and Low Sentiment Sub-samples

Note. The sample contains 770 U.S. mutual funds' holding records on 527 stocks from 2001 to 2005, including 31,093 (17,095 buy-herding and 13,998 sell-herding) stock-month observations. $S_{t-1}$ is the sentiment index in month *t-1*. High (low) $S_{t-1}$ sub-sample refers to observations with $S_{t-1}$ greater (less) than the median $S_{t-1}$ of the full sample. HM is the aggregate herding measure. BHM is the buy-herding measure. SHM is the sell-herding measure. The test statistics are heteroskedastic $t$-tests of equal means and non-parametric Wilcoxon $Z$-values of equal medians comparing high sentiment sub-sample with low sentiment sub-sample. Figures in parentheses are p values. *** indicates significant at the 0.01 level.

### 4.3 The effect of prior sentiment on fund herding

Table 5 reports the regression results for different fund herding measures: aggregate herding (*HM*), buy-herding (*BHM*) and sell-herding (*SHM*). The $t$-statistics in Table 5 are based on White's heteroskedasticity consistent estimator for standard errors. The VIFs for all independent variables are smaller than 3, indicating that collinearity does not pose an issue in model estimation (Kennedy, 2003).

| Expected Sign | | Dependent Variable | | |
|---|---|---|---|---|
| | | HM | BHM | SHM |
| | | Estimated Coefficient | Estimated Coefficient | Estimated Coefficient |
| Intercept | | 0.0865 | 0.1514 | 0.1025 |
| | | (0.9307) | (1.1844) | (1.0626) |
| St-1 | + | 0.0003 | 0.0002 | 0.0003 |
| | | (2.2866)** | (1.1715) | (1.9429)* |
| | | [1.4640] | [1.3662] | [1.3817] |
| RSTDt-1 | + | 0.2461 | 0.1707 | 0.4538 |
| | | (6.5192)*** | (1.8334)* | (7.2789)*** |
| | | [1.2206] | [1.2188] | [1.4552] |
| CAPt | - | | | |
| | | -2.4660E-10 | -3.6311E-10 | -7.2224E-11 |
| | | (-5.8024)*** | (-6.6908)*** | (-4.4602)*** |
| | | [2.5679] | [2.4828] | [2.5714] |
| TNUMBERt | + | | | |
| | | 6.7002E-5 | 8.7430E-5 | 3.2909E-5 |
| | | (6.8879)*** | (3.3417)*** | (2.8771)*** |
| | | [2.3489] | [2.1701] | [2.4336] |
| PRt | +/+/- | 0.0322 | 0.0445 | -0.0103 |
| | | (0.9796) | (3.8710)*** | (-2.0621)** |
| | | [1.1518] | [1.1440] | [1.2581] |
| PBt-1 | + | 1.6133E-6 | 8.8284E-7 | 2.7620E-5 |
| | | (0.3765) | (0.1917) | (0.9920) |
| | | [1.0104] | [1.0399] | [1.0530] |
| N | | 31093 | 17095 | 13998 |
| Adj. R-sq | | 0.0240 | 0.0129 | 0.0218 |
| Prob. (F > F*) | | <0.0001 | <0.0001 | <0.0001 |

Table 5. The Effect of Investor Sentiment on Fund Herding

Note. The sample is the 770 U.S. mutual funds' holding records on 527 stocks from 2001 to 2005, including 31,093 (17,095 buy-herding and 13,998 sell-herding) observations. HM is the aggregate herding measure. BHM is the buy-herding measure. SHM is the sell-herding measure. S is the composite sentiment measure. RSTD is the standard deviation of fund return. CAP is the capitalization of a given stock. TNUMBER is the number of funds trade in a given stock. PR is the dollar amount of purchase minus the dollar amount of redemption by mutual funds in a given month, scaled by net assets of the fund. PB is the ratio of market value to book equity at the beginning of the trading quarter. An industry dummy (equal to one for stocks of high-tech firms, zero otherwise) and four yearly dummies (2001 as the base year) are included in each model to control the possible industry and macroeconomic effects on herding behavior (not shown in the table). The *t*-statistics are based on White's heteroskedasticity consistent estimator for standard errors. t values are in ( ); VIFs are in [ ]; ***, ** and * indicate significant at the 0.01, 0.05 and 0.10 level, respectively.

For factors affecting *HM*, the concern of this study is the composite sentiment measure ($S_{t-1}$). $S_{t-1}$ has a positive and significant coefficient ($t$=2.2866), which is consistent with the notion that funds trade in herd more frequently on stocks with prior high sentiment. $S_{t-1}$ positively relates to *SHM* ($t$=1.9429) but insignificantly relates to *BHM* ($t$=1.1715), revealing that the positive effect of $S_{t-1}$ on *HM* mostly comes from the sell-side herding rather than the buy-side herding. These findings indicate that fund managers share an aversion to stocks with high past optimistic sentiment. Overall, the evidence again corroborates the results in Table 3 and Table 4 and provides stronger support for the sentiment countering argument that institutional investors have an inclination to counter optimistic investor sentiment. Also, the evidence implicates the use of negative-feedback strategies by mutual funds as an important source of herding.

Of the control variables, the deviation of fund returns in previous month (*RSTDt-1*) has a positive effect on all three herding measures ($t$=6.5192, 1.8334 and 7.2789 respectively), consistent with Bikhchandani et al.'s (1992) reputational risk argument. The stock capitalization (*CAPt*) negatively relates with all three herding measures ($t$=-5.8024, -6.6908 and -4.4602 respectively), showing that herding is more pronounced in small stocks. This result is consistent with the various herding theories where fund managers may receive lower precision information from these firms and are more likely to ignore this information if consensus opinion is different, or fund managers may share an aversion to holding small stocks because these stocks are less liquid (Lakonishok et al., 1992; Wermers, 1999; Choe et al., 1999). The positive effect of the number of funds trading a stock ($TNUMBER_t$) on all three herding measures ($t$=6.8879, 3.3417 and 2.8771 respectively) is consistent with the findings of Wermers (1999) and Wylie (2005). The net mutual fund redemption ($PR_t$) positively correlates with *BHM* ($t$=3.8710) and negatively correlates with *SHM* ($t$=-2.0621), similar to the findings in Edelen (1999) and Gallagher and Jarnecic (2004). Finally, the coefficient of market-to-book ratio is not significant in all three models, suggesting that the growth orientation of stocks does not play a role in explaining the level of fund managers herding.

### 4.4 Sensitivity tests

To assess the robustness of the results, the study includes four sensitivity tests : (1) re-estimating Table 4 and Table 5 for sub-periods "2001-2002" (a bear market) and "2003-2005" (a bull market) separately; (2) excluding 5% of upper outliers to alleviate the effect of extreme *BHM* and *SHM* and this process reduces the sample size to 16240 buy-herding and 13298 sell-herding observations; (3) replacing the trinomial-distribution herding measure with the popular binomial- distribution herding measure (Lakonishok et al., 1992); and (4) replacing the investor sentiment measure (*St-1*) in Table 4 and Table 5 with a two-period lag measure (*St-2*). The results (not reported) of these tests are qualitatively similar to those reported.

## 5. Conclusions

This paper assesses the relation between investor sentiment and the extent to which fund managers herd in their trades of stocks. The study applies the Lakonishok et al.'s (1992) measure and Wylie's (2005) trinomial-distribution approach to gauge fund manager herding, and uses the principal component analysis as the means of extracting the composite unobserved sentiment measure from ten market weather indicators that can be categorized into three groups: individual stock sentiment indicator, overall market sentiment indicator, and fund sentiment indicator.

The empirical results suggest that investor sentiment plays a significant role in explaining mutual fund herding cross-sectionally, especially on the sell-side. Fund managers show a stronger tendency to herd out of stocks with high prior investor sentiment than to herd into stocks with high prior sentiment. In other words, managers herd sell because, at least in part, they observe and counteract the optimistic sentiment. This finding is consistent with the funds sharing an aversion to stocks that have previously demonstrated high optimistic sentiment, supporting the sentiment countering hypothesis. The results also provide support for the informational cascade theory where managers herd because of analyzing the same sentiment-related indicators.

## 6. References

Baker, M., Stein, J.C., 2004. Market liquidity as a sentiment indicator. Journal of Financial Markets 7(3), 271-299.

Baker, M., Wurgler, J., 2006. Investor sentiment and the cross-section of stock returns. Journal of Finance 61(4), 1645-1680.

Bala, V., Goyal, S., 1998. Learning from neighbors. Review of Economic Studies 65(3), 595-621.

Banerjee, A., 1992. A simple model of herd behavior. Quarterly Journal of Economics 107(3), 797-817.

Barberis, N.C., Shleifer, A., 2003. Style investing. Journal of Financial Economics 68(2), 161-199.

Bikhchandani, S., Hirshleifer, D., Welch, I., 1992. A theory of fads, fashion, custom, and cultural change as informational cascades. Journal of Political Economy 100(5), 992-1026.

Block, S.B., French, D.W., Maberly, E.D., 2000. The pattern of intraday portfolio management decisions: a case study of intraday security return patterns. Journal of Business Research 50(3), 321-326.

Brandenburger, A., Polak, B., 1996. When managers cover their posteriors: Making the decisions the market wants to see. Rand Journal of Economics 27(3), 523-541.

Brown, G.W., Cliff, M.T., 2004. Investor sentiment and the near-term stock market. Journal of Empirical Finance 11(1), 1-27.

Choe, H., Kho, B.C., Stulz, R.M., 1999. Do foreign investor destabilize stock markets? The Korean experience in 1997. Journal of Financial Economics 54(2), 227-264.

De Bondt, W.F., 1993. Betting on trends: Intuitive forecasts of financial risk and return. International Journal of Forecasting 9(3), 355-371.

De Long, J.B., Shlerter, A., Summers, L.H., Waldmann, R.J., 1990. Noise trader risk in financial markets. Journal of Political Economy 98(4), 703-738.

Del Guercio, D., 1996. The distorting effect of the prudent-man laws on institutional equity investments. Journal of Financial Economics 40(1), 31-62.

Edelen, R.M., 1999. Investor flows and the assessed performance of open end fund managers. Journal of Financial Economics 53(3), 439-466.

Falkenstein, E.G., 1996. Preferences for stock characteristics as revealed by mutual fund portfolio holdings. Journal of Finance 51(1), 111-135.

Fisher, K.L., Statman, M., 2000. Investor sentiment and stock returns. Financial Analysts Journal 56(2), 16-23.

Froot, K.A., Scharfstein, D.S., Stein, J.C., 1992. Herd on the street: Informational inefficiencies in a market with short-term speculation. Journal of Finance 47(4), 1461-1484.

Fung, W., Hsieh, D.A., 1999. A primer on hedge funds. Journal of Empirical Finance 6(3), 309-331.

Gallagher, D.R., Jarnecic, E., 2004. International equity funds, performance, and investor flows: Australian evidence. Journal of Multinational Financial Management 14(1), 81-95.

Gompers, P., Metrick, A., 2001. How are large institutions different from other investors? Quarterly Journal of Economics 16(1), 229-259.

Grinblatt, M., Titman, S., Wermers, R., 1995. Momentum investment strategies, portfolio performance, and herding: A study of mutual fund behavior. American Economic Review 85(5), 1088-1105.

Hirschleifer, D., Subrahmanyam, A., Titman, S., 1994. Security analysis and trading patterns: When some investors receive information before others. Journal of Finance 49(5), 1665-1698.

Jegadeesh, N., Titman, S., 1993. Returns to buying winners and selling losers: Implications for stock market efficiency. Journal of Finance 48(1), 65-91.

Kennedy, P., 2003. A guide to econometrics. Cambridge, MA: MIT Press.

Klemkosky, R.C., 1977. The impact and efficiency of institutional net trading imbalances. Journal of Finance 32(1), 79-86.

Kyrolainen, P., Perttunen, J., 2003. Investors' activity and trading behavior. Working paper, University of Oulu, Finland.

Lakonishok, J., Shleifer, A., Vishny, R.W., 1992. The impact of institutional trading on stock prices. Journal of Financial Economics 32(1), 23-43.

Lee, C., Shleifer, A., Thaler, R., 1991. Investor sentiment and the closed-end fund puzzle. Journal of Finance 46(1), 75-109.

Neal, R., Wheatley, S.M., 1998. Do measures of investor sentiment predict returns? Journal of Financial and Quantitative Analysis 33(4), 523-548.

Nofsinger, J.R., Sias, R.W., 1999. Herding and feedback trading by institutional and individual investors. Journal of Finance 54(6), 2263-2295.

Payne, T.H., Prather, L., Bertin, W., 1999. Value creation and determinants of equity fund performance. Journal of Business Research 45(1), 69-74.

Pound, J., Shiller, R.J., 1989. Survey evidence on diffusion of interest and information among investors. Journal of Economic Behavior and Organization 12(1), 47.67.

Prendergast, C., 1993. A theory of yes men. American Economic Review 83(4), 757-770.

Scharfstein, D., Stein, J., 1990. Herd behavior and investment. American Economic Review 80(3), 465-479.

Shleifer, A., 2000. Inefficient markets: An introduction to behavioral finance. Oxford University Press, Oxford.

Trueman, B., 1994. Analyst forecasts and herding behavior. Review of Financial Studies 7(1), 97-124.

Walter, A., Weber, F.M., 2006. Herding in the German mutual fund industry. European Financial Management 12(3), 375-406.

Wermers, R., 1997. Momentum investment strategies of mutual funds, performance persistence, and survivorship bias. Working paper, University of Colorado.

Wermers, R., 1999. Mutual fund herding and the impact on stock price. Journal of Finance 54(2), 581-622.

White, H., 1980. A heteroskedasticity-consistent covariance matrix estimator and a direct test for heteroskedasticity. Econometrica 48(1), 817-838.

Wylie, S., 2005. Fund manager herding: a test of the accuracy of empirical results using U.K. data. Journal of Business 78(1), 381-403.

Zwiebel, J., 1995. Corporate conservatism and relative compensation. Journal of Political Economy 103(1), 1-25.

# 6

# Application of the Principal Component Analysis to Disclose Factors Influencing on the Composition of Fungal Consortia Deteriorating Remained Fruit Stalks on Sour Cherry Trees

Donát Magyar[1] and Gyula Oros[2]
[1]*National Institute of Environmental Health, Budapest,*
[2]*Plant Protection Institute of the Hungariain Academy of Sciences, Budapest, Hungary*

## 1. Introduction

The sour cherry production is concentrated in eastern region of Hungary. During last two decades the commercialized yield varied between 42000-90000 metric tonnes. One of the main constraints is the fruit decay (anthracnose) caused by *Colletotrichum gloeosporioides* (Penz.) Penz. & Sacc. in Penz. (teleomorph: *Glomerella cingulata* [Stoneman] Spauld. & H. Schrenk). This disease has been known for centuries, and in certain orchards yield losses greater than 90% periodically occurred under epidemic conditions. In last decade, however, the grave infection has evolved in each season that resulted in disastrous yield losses and led to decrease of harvested area at about 25 %. Series of control measures were tried out, many of them with success with some degree (Børve & Stensvand, 2006; Børve et al., 2010), among them a newly developed biocontrol preparation was applied (Oros & Naár, 2008; Oros et al., 2011). Detailed analysis of weak efficiency of disease control revealed that beside *C. gloeosporioides* four other anamorphs of *Glomerella* were present (Table 1), among them *C. acutatum* (J.H. Simmonds, 1968), - a new pathogen for the region, - became recently dominant. Its strains, tolerant to recently applied fungicides to control the anthracnose, could be isolated of sour cherry orchards that might be the cause of ineffectiveness of control measures in 2006-2010 (Oros et al., 2010). At the increasing costs of cultivation only high yields may secure a profitable production of sour cherry. From this point of view, the anthracnose caused by *Glomerella* anamorph with special regards to *C. acutatum* is an important element of uncertainty that means elaboration of efficient control measures is the critical challenge to unconquerable sour cherry production. Efficient control measures reduce the yield loss in the orchard by suppressing the pathogen (i.e. by the reduction of the number of viable spores and infected fruits). On our major surprise, the pathogen can not be isolated of overwintering fruit mummies in industrial plantations of East Hungary contrarily to other sour cherry producing regions. A survey was carried out to disclose the cause of this phenomenon collecting samples of remained fruit stalks of varieties in sour cherry gene bank of East Hungary. The relationships between host (sour cherry variety) and fungi associated to stalks have been analyzed by multivariate methods: Non-linear Mapping and Principal Component Analysis (PCA) combined with Potency Mapping (PM) and Spectral Mapping (SPM) techniques.

| | Location | Cglo | Cacu | Ctru | Ctri | n.d. |
|---|---|---|---|---|---|---|
| Újfehértó | 47°46′46.7″N, 21°39′10.2″E | + | + | + | - | + |
| Újfehértó | 47°40′22.2″N, 21°41′24.7″E | + | + | - | - | - |
| Pilis | 47°15′45.9″N, 19°34′13.1″E | + | + | + | + | + |
| Szentendre | 47°44′94.3″N, 19°03′10.7″E | + | - | - | - | - |
| Budapest | 47°31′34.3″N, 19°01′44.7″E | + | + | - | - | - |
| Budapest[b] | 47°30′50.8″N, 19°00′38.5″E | - | - | - | + | - |
| Érd | 47°24′07.3″N, 18°55′26.1″E | - | + | - | - | + |
| Agárd | 47°11′03.8″N, 18°35′15.1″E | + | + | - | - | + |
| Szombathely[c] | 47°13′51.3″N, 16°33′33.8″E | - | + | - | - | + |
| Jakabszállás | 47°45′16.3″N, 14°37′24.4″E | + | + | - | - | - |
| Kecskemét | 46°53′45.9″N, 19°42′24.8″E | + | + | - | - | - |
| Jakabszállás | 46°29′28.7″N, 19°16′22.7″E | + | + | - | - | - |

Table 1. Presence of *Colletotrichum* (Corda) species related to sour cherry in various regions of Hungary[a].

[a]=Isolates of each species gangrene the fruits of sour cherry (*Prunus cerasus* L. cv. Kántorjánosi), [b]=collected of *Hedera helix* L., [c]= herbarium specimen of *Fallopia sachalinensis* (F. Schmidt) Ronse Decr.; Cglo=*C. gloeosporioides* (Penz.) Penz. & Sacc. (*Glomerella cingulata* (Stoneman) Spauld. & H. Schrenk); Cac=*C. acutatum* J.H. Simmonds (*G. acutata* Guerber & J.C. Correll), Ctrun=*C. truncatum* (Schwein.) Andrus & Moore (*G. truncata* C.L. Armstr. & Banniza), Ctri=*C. trichellum* (Fr.) Duke; n.d.=unidentified.

Various multivariate mathematical-statistical methods have been widely applied to elucidate the relationship between the columns and rows of any data matrix. In general, the choice of a particular multivariate method for data analysis is dictated by the properties of the data and by the objectives. The PCA, introduced by Pearson (1901), is a useful and widely applied technique to reduce the number of multidimensional contrasts and to reproduce the original data in low dimensional space. In this perspective PCA is a versatile and an easily applicable multivariate method, which forms clusters of variables and observations taking into consideration simultaneously the overall effect of factors under investigation without being one of dependent variable (Mardia et al., 1979). In last four decades over 43000 scientific papers were published that applied PCA for data analysis among them more than 800 intensively reviewed them (Fig 1). There are large literatures in diverse areas such as psychometrics, chemometrics, computer science and ecology applying PCA. Biological and environmental sciences are among the top users. Such statistic is widely used in these works to determine the source of a variety of soil, plant and atmospheric contaminants, e.g. heavy metals (Titseesang et al., 2008; Schwarz et al., 1999; Tokalioglu et al ., 2010), trace metals (Rungratanaubon et al., 2008; Shaheen et al., 2005; Shah et al., 2006; Vasconcellos et al., 2007), PM 2.5 and PM10 particles (Jacquemin et al., 2009; Quraishi et al., 2009) and volatile organic compounds (You et al., 2008).

Especially in agricultural sciences, where complex datasets with biotical and abiotic variables are often used, the application of multivariate statistics is fundamental. In these studies, PCA is mostly applied to reduce the number of input variables (Uno et al., 2005). Also, these calculations are used to develop precision agriculture applications using crop monitoring (Yang et al., 2004), to uncover gradients of canopy structure (Frazer et al., 2005), to determine the time of harvest (Garcia-Mozo et al., 2007) or to evaluate sampling methods (Magyar et al., 2011). Although, in particular, atmospheric science is a rich source of both applications and methodological developments that influenced applications of PCA. In the field of aerobiology relatively few papers have been published. PCA was applied to analyze the variability of fruit production based on aerobiological pollen data (Ribeiro et al., 2007, 2008), as well as to connect meteorological parameters with the periods of high pollen concentrations (Parrado et al., 2009). This method proved to be reliable to identify the sources and dispersal patterns of airborne bacteria (Lighthart et al., 2009) and spores of plant pathogenic fungi (Frank et al., 2008; Magyar, 1998) that resulted the stressing of its possible use for both signalization of occurrence and identification of sources of plant pathogens. The latter is important in relation to successful control measures in the agricultural areas as the knowledge on the type of dispersal of the pathogen (and biocontrol agent if applied) and the composition of fungal consortia on the target area is crucial for an effective pest control (Magyar 2007).

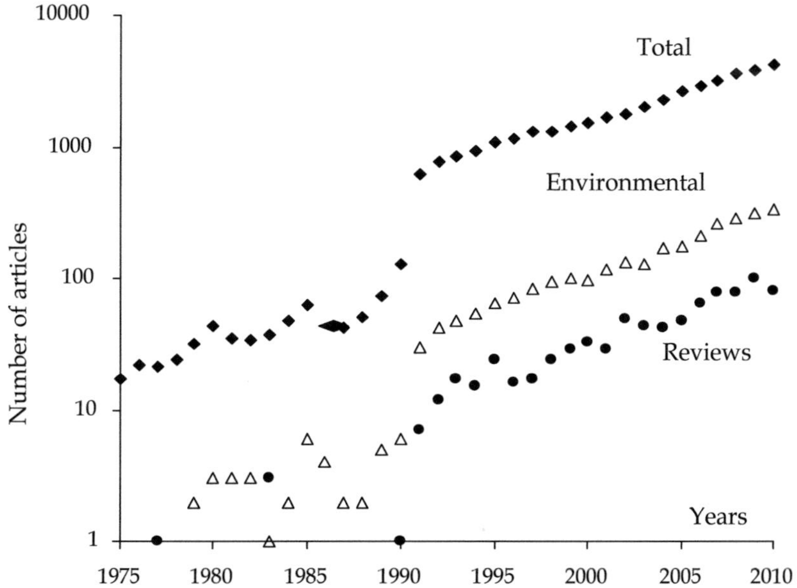


Fig. 1. Changes in abundance of scientific publications that have applied Princpal Component Analysis. Source: Thompson Data Base

The PCA is an exploratory tool to investigate patterns in the data. Usually, the first Principal Component (PC) is accepted to comprise the summarized quantitative aspects of multifactorial effects as the arithmetic mean values of observations correlate with first PC, while the remaining PCs are considered to relate to spectrum of effects. However, the quantitative moments of PC variables are distributed within hidden factors, thus PCA cannot be employed when the separation of the strength and selectivity of the effect is required. The spectral mapping technique (SPM) was developed to overcome this difficulty (Lewi, 2005). Using the normalized original data this method divides the information into two matrices. The first one is a vector containing so-called potency values (PV) proportional to the overall effect of factors or the response of organism to them (quantitative measure of the effect). The second matrix (selectivity or spectral map) contains the information related to the variation in manifestation (spectrum) of the effect (qualitative characteristics of the effect). In our case PV relates to the overall abundance of each fungal species noticed on stalks, i.e., it is a quantitative measure of the diversity of fungal consortium, while the spectral map (SM) comprises qualitative characteristics of variety dependent distribution of species in the orchard. The matrices of PC loadings and variables as well as the SM are generally multidimensional. As the human brain is not suitable for the evaluation of the distribution of data in multidimensional space a nonlinear mapping technique (NLMAP) was developed for the reduction of the dimensionality of such matrices (Sammon, 1969).

Our objectives of this study were to evaluate diversity of mycobiota associated to stalks of sour cherry remained on trees, and in this paper we show the use of PCA in conjunction with SPM and NL-Mapping to reveal patterns in distribution of detected fungal species.

## 2. Experimental

### 2.1 Location and method of sampling

**2.1.1** Sour cherry fruits were collected in ten different locations (Table 1) of Hungary in period of fruit ripening during 2006-2010 and searched for *Colletotrichum* acervuli immediately after taking them into the laboratory. The *Colletotrichum* species are anamorphs of some *Glomerella* (Ascomycota) fungi and their presence was proved microscopically by usual manner. Of the same locations twig s with 15-25 buds and mummies with stalks were collected in February of subsequent years.

**2.1.2** The sour cherry gene bank is located in Újfehértó (East Hungary, 47°46′46″N, 21°39′10″E), and it is surrounded by big sized industrial plantations that comprise mainly three varieties (*Prunus cerasus* L. cv. Érdi bőtermő, Kántorjánosi and Újfehértói fürtös). The up to date management practices have usually been carried out in the area. On the nearby two plots (4 ha size of each) *Trichoderma*-based biopreparations (Oros and Naár, 2008) were applied during past three vegetation periods (Oros et al., 2010, 2011). Fruit stalks remained on trees were collected in March 2011 (10 samples of each tree and 10-30 stalks per sample) of 18 varieties of gene bank as well as of trees of the two industrial plots and kept in freezer up to analysis.

### 2.2 Survey of fungal population on stalks

**2.2.1** The twigs collected in diverse areas in first week of February (2.1.1.) were plug in pots filled with wetted perlite and tucked into plastic bag for 48 hours at ambient temperature to promote acervuli formation on surface of tissues infected with *Colletotrichum*. The acervulus is a characteristic reproductive organ of *Glomerella* anamorphs that easy to recognize of dark brown, thorn like, prominent setae examining samples under dissecting microscope.

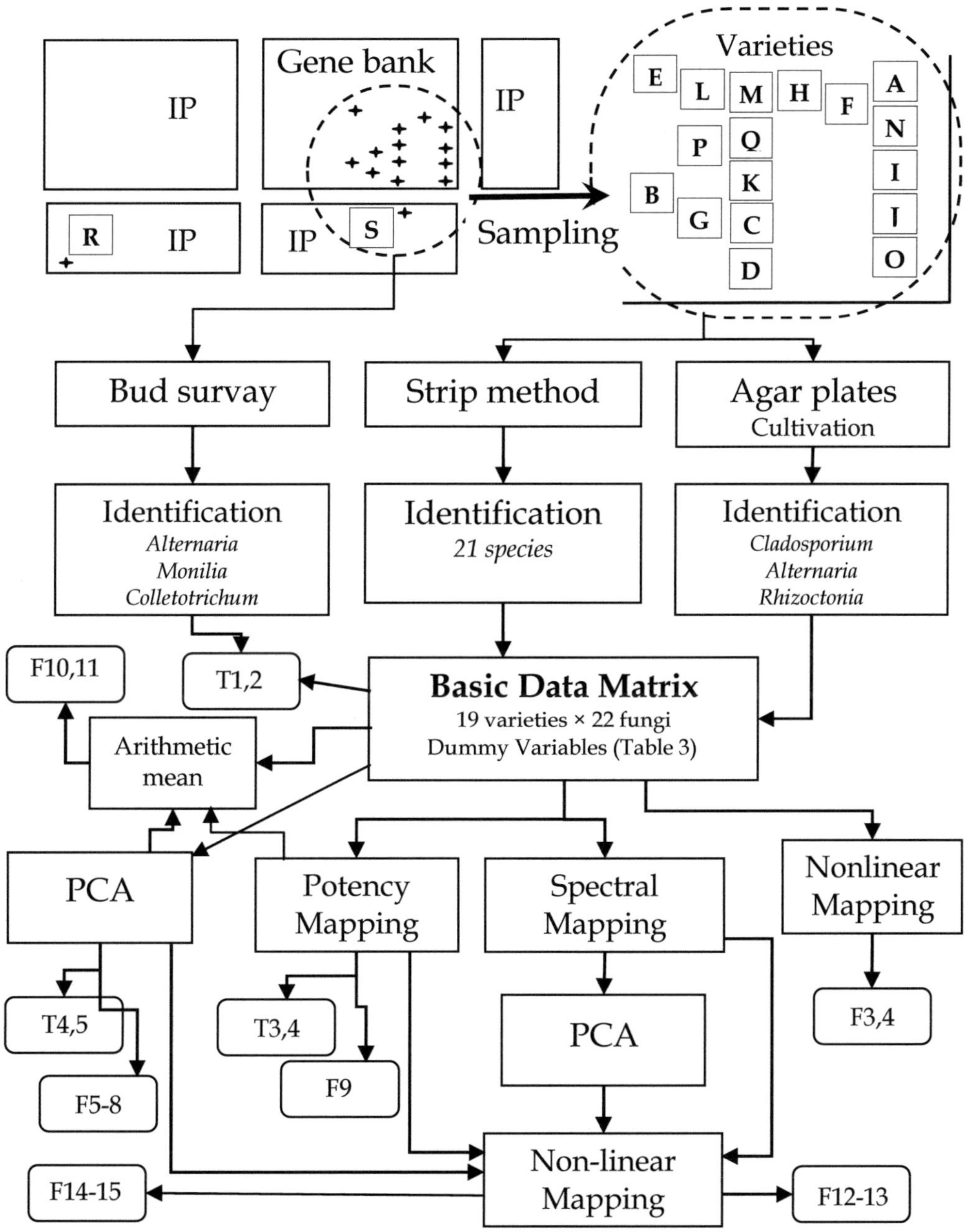


Fig. 2. Flow diagram of the experimental protocol.

IP= industrial plantation. The capital letter codes of varieties are given in Table 2.
The labels in ellipses (T1,2..n) and (F1,2..n) mark number of tables and figures in body text where the results of computation was used for demonstration.

**2.2.2** Stalks collected of trees in winter were placed on the surface of Czapek Agar plates amended with antibiotics (ampicillin, kanamycin and streptomycin) to supress bacteria. The growth of fungi on the surface of stalks as well as on the agar plates had been survayed daily during three weeks.

**2.2.3** To study the fungal spectra of the fruit stalks, a microscopic method was applied which was developed to enable *in situ* investigations by the use of pressure sensitive acrylic strips (MACbond B 1200, MACtac Europe S.A., Brussels). The strip consists of a thin (20 µm) polypropylene film coated on both sides with a rubber based adhesive, which offers very high adhesion property on the stalks. The tape is viewed directly under a microscope to identify the fungi present on the sampled surface under ×800 magnification.

All mycological work was done applying usual manners widely accepted in microbiology. Spores have been grouped into hydrophobic dry spores (xerospores) and readily wettable slime spores (gloiospores) (Mason, 1937, Magyar, 2007).

### 2.3 Data treatment

The basic data matrix consisted of the dummy variables that comprise presence (1) or absence (0) of fungal species (n=22) on the stalks of the sour cherry variety (n=19) concerned, which strongly limits the use of data transformation procedures. The yeast species, which occurred on each stalk examined, had been excluded of further data analyses, because PCA and SPM can be employed for the analysis of data matrices which contain variables showing variance. First the Nonlinear Mapping (NLMAP) was applied for clustering both sour cherry varieties and fungi detected on stalks following Sammon (1969).

PCA was carried out on the correlation matrix and only the components having an eigenvalue greater than one were included in the evaluation of data. To facilitate the evaluation of the multidimensional maps of principal component (PC) loadings, their dimensionality was reduced to two by the NLMAP technique.

Potency and Spectral Mapping techniques (Lewi, 2005) were applied on the matrix used for PCA, and applied in order to calculate the abundance and the variety dependent distribution of fungal species as well as the degree of diversity and the heterogeneity spectrum of settlement on proper cherry variety. Potency and Spectral Mapping were carried out twice: (a) The abundance of 19 fungal species represented the variables and the 19 sour cherry variables the observations (spectral map calculated comprises the heterogeneity spectrum of fungal settlement on varieties). (b) The cherry varieties represented the variables and the abundance of 19 fungal species the observations (spectral map calculated comprises the variety dependent distribution of fungi). The potency values calculated relate to either the abundance of fungal species in sour cherry plantation or the degree of diversity of fungal consortium found on stalks of proper variety. The dimensionality of the maps was reduced to two by the NLMAP technique or hidden variables were extracted by PCA as the evaluation of the multidimensional selectivity (spectral) maps of varieties and associated fungi is difficult.

Statistical functions of Microsoft Office Excel 2003 (Microsoft, Redmondton, USA) and Statistica5 program (StatSoft, Tusla, USA) were used for analysis of data. The graphical presentation of result of data analysis was edited uniformly in MSOffice Power Point 2003.

| Sour cherry varieties | | | Fungal species detected on stalk remained on trees | | |
|---|---|---|---|---|---|
| No. | Code[a] | Gene bank | No. | Code | Taxonomic position[c] and scientific name |
| 1 | A | Aranka | Ascomycota | | |
| 2 | B | Bigarreau | | Pezizomycotina | |
| 3 | C | Bosnyák | 1 | tol | *Trullula olivascens* (Sacc.) Sacc. |
| 4 | D | Csengődi | | Dothideomycetes, Botryosphaeriales | |
| 5 | E | Debreceni b. | 2 | dte | *Diplodia tecta var. cerasii* Berk. & Broome |
| 6 | F | Érdi bőtermő | | Dothideomycetes, Capnodiales | |
| 7 | G | Géczi | 3 | cla | *Cladosporium sp.* |
| 8 | H | Germersdorfi | | Dothideomycetes, Pleosporales | |
| 9 | I | Halka | 4 | agl | *Aglaospora profusa* (Fr.) De Not.[g] |
| 10 | J | Horka | 5 | alt | *Alternaria sp.* |
| 11 | K | Katalin | 6 | ppo | *Phoma pomorum* Thüm. |
| 12 | L | Maliga emléke | 7 | epu | *Epicoccum purpurascens* Ehrenb. |
| 13 | M | Margit | 8 | lep | *Leptosphaeria sp.* |
| 14 | N | Pándy 56 | 9 | phe | *Pleospora herbarum* P. Karst. |
| 15 | O | Sylvana | 10 | ple | *Pleospora sp.*[g] |
| 16 | P | Szomolyai | 11 | sfo | *Splanchnonema foedans* (Fr.) Kuntze[g] |
| 17 | Q | Vanda | 12 | ste | *Stemphylium sp.* |
| | | | | Sordariomycetes[c] | |
| Industrial plantation[b] | | | 13 | sor | unidentified sp.[d] |
| | | | | Sordariomycetes, Glomerellales[e] | |
| 18 | R | KántorjánosiU | 14 | cgl | *Colletotrichum gloeosporioides* (Penz.) Penz. & Sacc. |
| 19 | S | KántorjánosiH | | Sordariomycetes, Xylariales | |
| | | | 15 | dpl | *Dendrophoma pleurospora* (Sacc.) Sacc. |
| | | | | Sordariomycetes, Diaporthales | |
| | | | 16 | cyt | *Cytospora sp.* |
| | | | | Pezizomycotina, Helotiales | |
| | | | 17 | mon | *Monilia sp.* |
| | | | 18 | m-s | Steril mycelium[f] |
| | | | 19 | y-w | yeast 1 - white |
| | | | 20 | y-r | yeast 2 - red |
| | | | 21 | y-b | yeast 3 - black |
| | | | Basidiomycota | | |
| | | | | Agaricomycetes, Cantharellales | |
| | | | 22 | rhi | *Rhizoctonia sp.* |

Table 2. Stalk samples collected in gene bank of Újfehértó and the list of detectable fungi associated to.

[a]=The codes of sour cherry varieties (capital letters) and fungal species colonized them during the winter are used in following tables and figures. [b]=These plantations are adjacent to the gene bank. [c]=Species were assigned according to Index Fungorum. [d]=Detected only on a single stalk, [e]=Immature perithecia, [f]=Mycelium of an ascomycetaceous fungal species, [g]= free spores.

## 3. Results

**3.1** The presence of *Colletotrichum* species was in each location stated on ripening fruits during vegetation periods of 2006-2010, and among them *C. acutatum* proved to be dominant being followed by *C. gloeosporioides* (Table 1). The occurrence of other *Glomerella* anamorphs was sporadic. In several samples acervuli of a *Colletotrichum* species unknown for Hungarian mycoflora were observed. Neither on leaves collected during vegetation period nor on decaying fruits were found *Glomerella* perithecia (sexual reproductive organ) in 2006-2011. The yield losses caused by anthracnose were significant in each location. The role of other pathogenic fungi was insignificant, among them *Monilia* sp. was most frequent, although in few cases *P. pomorum, D. tecta* and *Gibberella avenacea* R.J. Cook associated also to fungal consortium. The sporadic appearance of other than *Colletotrichum* fungal species was seemingly connected to intensive pest management in sour cherry plantations.

**3.2** On twigs collected in February of 2007-2011 the rate of infection varied between 2 and 90 percents (method 2.2.1). *C. acutatum, C. gloeosporioides* and *C. trichellum* could be detected (Table 1) in listed order, but the samples of Újfehértó where *C. gloeosporioides* was only found. In the gene bank (L-1) the manifested infection rate on the twig s was low, moreover, few infected buds could be found only on two varieties (*P. cerasus* L. cv. Érdi bőtermő and Újfehértói fürtös). Also, *C. gloeosporioides* was the single overwintering *Colletotrichum* in Central Hungary (L-4), although in this orchard acervuli of other *Glomerella* anamorphs were not discovered on decayed fruits in vegetation period. *C. trichellum* was detected in one orchard only (L-3) on fruits, and in the same place it was found on basal parts of stalks. *C. truncatum* was found only on fruits in two locations (L1 and L3) suggesting that this species does not overwinter on sour cherry trees. Other fungal species were detected such as *Alternaria spp., Cytospora* sp. and *Phoma pomorum.*

**3.3** Counteracting to special analysis cosmopolite fungi *Alternaria* and *Cladosporium* overgrew rapidly all others on stalks placed on the Czapek Agar plates (method 2.2.2) except *Rhizoctonia solani* J.G. Kühn (teleomorph: *Thanatephorus cucumeris* [A.B. Frank] Donk) that was identified of two varieties (Pándy 56 of gene bank and Kántorjánosi of industrial plantation). The *Trichoderma* strains of biopreparate were not detectable. The acervulus of *C. gloeosporioides* could be detected on a single stalk (variety Aranka of gene bank).

**3.4** Following protocol 2.2.3. we could detect presence of 8 known phytopathogenic and 13 saprobiont fungal species in samples collected in gene bank of Újfehértó (Table 2). Five species occurred only in a single variety (Table 3). Thus, intensively growing sterile mycelium of an ascomycete was observed on stalks of Debreceni bőtermő (N, letter codes of varieties are given in Table 2); however, this did not counteract the detection of other fungal species. Also, immature perithecia could be seen on stalks of Géczi (E). Spores of three phytopathogenic species causing branch dieback, *Leptosphaeria, Splanchonema* and *Cytospora* were detected on stalks of varieties B, P and R, respectively. The most frequent *Alternaria* and *Stemphylium* being facultatively parasitic can take part in fruit decay. The occurrence of *Monilia*, one of the most devastating pathogens of sour cherry, was less pronounced as expected. The other species are known as saprobionts and we have no information about their possible phytopathogenic properties, nevertheless, these can also have importance in postharvesting fruit rot. Presence of none of *Glomerella* anamorph but *C. gloeosporioides* on some stalks of a single variety (A) could be detected.

| Fungi[a] | Sour cherry varieties[a] | | | | | | | | | | | | | | | | | | |
|---|---|---|---|---|---|---|---|---|---|---|---|---|---|---|---|---|---|---|---|
| Code | A | B | C | D | E | F | G | H | I | J | K | L | M | N | O | P | Q | R | S | PV[b] |
| tol | 0 | 0 | 1 | 0 | 0 | 1 | 0 | 1 | 1 | 0 | 0 | 1 | 1 | 1 | 1 | 1 | 0 | 0 | 1 | 52.6 |
| dte | 0 | 0 | 0 | 0 | 0 | 0 | 0 | 0 | 0 | 0 | 0 | 0 | 0 | 1 | 1 | 0 | 0 | 0 | 0 | 10.5 |
| cla | 1 | 0 | 0 | 0 | 1 | 0 | 0 | 0 | 0 | 0 | 0 | 0 | 0 | 0 | 1 | 0 | 1 | 0 | 1 | 26.3 |
| agl | 0 | 0 | 1 | 0 | 0 | 0 | 0 | 1 | 0 | 0 | 0 | 0 | 0 | 0 | 0 | 0 | 0 | 1 | 1 | 21.1 |
| alt | 1 | 1 | 1 | 0 | 1 | 1 | 1 | 1 | 1 | 1 | 1 | 1 | 1 | 1 | 1 | 1 | 1 | 1 | 1 | 94.7 |
| ppo | 0 | 0 | 1 | 0 | 0 | 1 | 0 | 0 | 0 | 0 | 0 | 0 | 0 | 1 | 1 | 0 | 0 | 0 | 0 | 21.1 |
| epu | 0 | 0 | 0 | 0 | 0 | 0 | 0 | 0 | 0 | 0 | 0 | 0 | 0 | 0 | 1 | 1 | 1 | 0 | 0 | 15.8 |
| lep | 0 | 1 | 0 | 0 | 0 | 0 | 0 | 0 | 0 | 0 | 0 | 0 | 0 | 0 | 0 | 0 | 0 | 0 | 0 | 5.3 |
| phe | 1 | 0 | 0 | 1 | 1 | 0 | 0 | 0 | 0 | 0 | 1 | 0 | 0 | 0 | 1 | 0 | 0 | 1 | 0 | 31.6 |
| ple | 0 | 0 | 0 | 0 | 1 | 0 | 0 | 0 | 0 | 0 | 0 | 0 | 0 | 0 | 0 | 1 | 0 | 0 | 0 | 10.5 |
| sfo | 0 | 0 | 0 | 0 | 0 | 0 | 0 | 0 | 0 | 0 | 0 | 0 | 0 | 0 | 0 | 1 | 0 | 0 | 0 | 5.3 |
| ste | 1 | 0 | 0 | 1 | 1 | 0 | 0 | 0 | 1 | 1 | 1 | 0 | 0 | 1 | 1 | 1 | 1 | 0 | 0 | 52.6 |
| sor | 0 | 0 | 0 | 0 | 0 | 0 | 1 | 0 | 0 | 0 | 0 | 0 | 0 | 0 | 0 | 0 | 0 | 0 | 0 | 5.3 |
| cgl | 1 | 0 | 0 | 0 | 0 | 0 | 0 | 0 | 0 | 0 | 0 | 0 | 0 | 0 | 0 | 0 | 0 | 0 | 0 | 5.3 |
| dpl | 0 | 0 | 1 | 0 | 0 | 0 | 0 | 0 | 0 | 0 | 1 | 0 | 0 | 0 | 1 | 0 | 1 | 0 | 0 | 21.1 |
| cyt | 0 | 0 | 0 | 0 | 0 | 0 | 0 | 0 | 0 | 0 | 0 | 0 | 0 | 0 | 0 | 0 | 0 | 1 | 0 | 5.3 |
| mon | 0 | 1 | 0 | 0 | 0 | 0 | 1 | 0 | 0 | 0 | 0 | 0 | 0 | 1 | 0 | 0 | 0 | 0 | 1 | 21.1 |
| ms | 0 | 0 | 0 | 0 | 1 | 0 | 0 | 0 | 0 | 0 | 0 | 0 | 0 | 0 | 0 | 0 | 0 | 0 | 0 | 5.3 |
| y1 | 1 | 1 | 1 | 1 | 1 | 1 | 1 | 1 | 1 | 1 | 1 | 1 | 1 | 1 | 1 | 1 | 1 | 1 | 1 | 100 |
| y2 | 1 | 1 | 1 | 1 | 1 | 1 | 1 | 1 | 1 | 1 | 1 | 1 | 1 | 1 | 1 | 1 | 1 | 1 | 1 | 100 |
| y3 | 1 | 1 | 1 | 1 | 1 | 1 | 1 | 1 | 1 | 1 | 1 | 1 | 1 | 1 | 1 | 1 | 1 | 1 | 1 | 100 |
| rhi | 0 | 0 | 0 | 0 | 0 | 0 | 0 | 0 | 0 | 0 | 0 | 0 | 0 | 1 | 0 | 0 | 0 | 0 | 1 | 10.5 |

Table 3. Mycological analysis of stalks collected in gene bank of Ujfehértó.

[a]=codes of fungi and varieties are shown in Table 2. [b]=Potency Values as percent of the maximum abundance. 0=not found, 1=presented on more than 1 specimen.

## 3.5 Data analysis

### 3.5.1 Non-linear mapping

Both sour cherry varieties and associated fungal species were mapped following method of Sammon (1969). The fungi might be grouped by their relation to host plant, one of clusters comprise pathogens (dotted line) while the other the saprobionts mainly (Fig 3).

In the case of varieties no clusters could be recognized (Fig 4). The four varieties (E, F, S and R) are near relatives, and their position on NLMap is not linked. These were selected for free pollination and improving the mechanic harvest. Unfortunately, data on the genetic background of other varieties is absent; the varieties B, O and Q were introduced of other geographic regions. The varieties F, R, S and E are extremely susceptible to anthracnose caused by *C. acutatum* in decreasing order; moreover their susceptibility to *Monilia* is outstanding too. Nevertheless, due to the high commercial value of their fruits and favorable biological properties these are the most run varieties in Hungary.

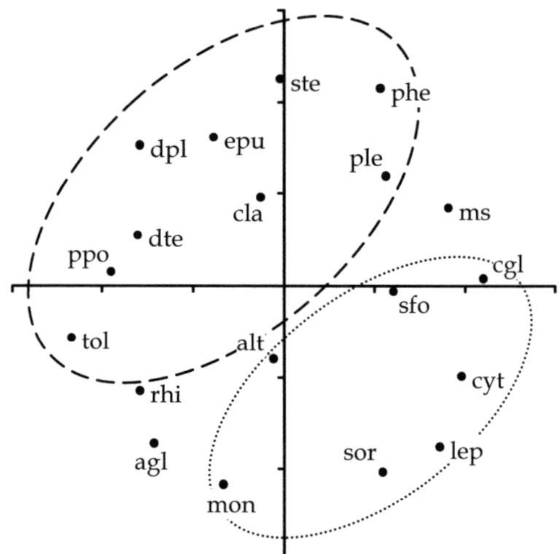


Fig. 3. Non-linear Map of fungal species as variables. The codes of fungi are given in Table 3. Ellipses with dotted and broken lines mark groups of pathogenic and saprotrophic fungi, resp.

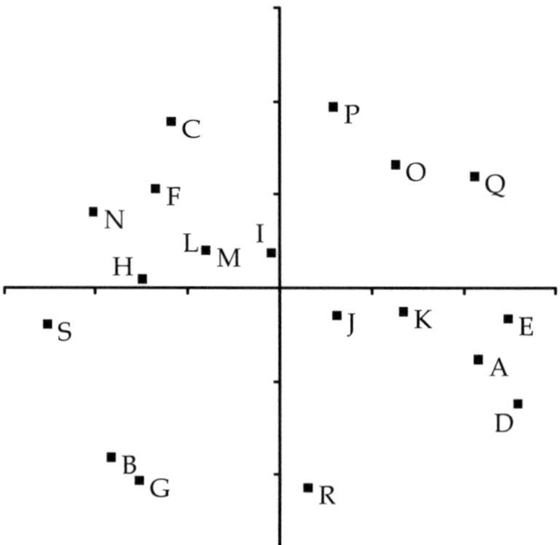


Fig. 4. Non-linear Map of sour cherry varieties as variables. The codes of varieties as given in Table 3.

### 3.5.2 Principal Component Analysis

### 3.5.3 Influence of sour cherry varieties on distribution of associated fungi

The PC loadings (sour cherry varieties as variables) computed by PCA are compiled in Table 4. Five hypothetical (background) variables explained the majority of total variance with 22.57 % loss of information. This fact suggests that the 19 original variables can be reduced to 5 theoretical variables. Most of varieties have high loadings in the first two PCs, and those being in gene bank are positively related, that means these two factors act similarly on distribution of associated fungi when deteriorating stalks of varieties concerned. The two sampling points of industrial plantations (R and S) have not common weight with the former group.

On bivariate plot of varieties as PC variables no connection can be observed similarly to NL-map (Fig 4), though these two plots are different. The distribution fungal species plotted as PC scores does not show connection between them, although the xerospore producing anemophilous species seem to deviate of the main group comprising gloeospore producing ones.

| Variables | | | Principal components | | | | | PV[a] |
|---|---|---|---|---|---|---|---|---|
| No. | Sour cherry varieties | Code | PC1 | PC2 | PC3 | PC4 | PC5 | |
| 1 | Aranka | A | 0.44 | **0.65** | -0.22 | 0.14 | 0.01 | 36.4 |
| 2 | Bigarreau | B | 0.34 | -0.23 | **-0.69** | -0.25 | 0.15 | 27.3 |
| 3 | Bosnyák | C | **0.60** | -0.40 | 0.32 | 0.36 | 0.31 | 36.4 |
| 4 | Csengődi | D | 0.20 | **0.80** | -0.02 | 0.12 | 0.02 | 22.7 |
| 5 | Debreceni bőtermő | E | 0.35 | **0.64** | -0.18 | 0.07 | -0.29 | 40.9 |
| 6 | Érdi bőtermő | F | **0.74** | -0.40 | 0.26 | -0.01 | 0.12 | 27.3 |
| 7 | Géczi | G | 0.34 | -0.23 | **-0.69** | -0.25 | 0.15 | 27.3 |
| 8 | Germersdorfi | H | **0.73** | -0.44 | 0.01 | 0.39 | -0.18 | 27.3 |
| 9 | Halka | I | **0.90** | 0.09 | 0.09 | -0.19 | -0.18 | 27.3 |
| 10 | Horka | J | **0.75** | 0.43 | -0.21 | -0.21 | -0.05 | 22.7 |
| 11 | Katalin | K | **0.56** | **0.64** | -0.07 | 0.18 | 0.26 | 31.8 |
| 12 | Maliga emléke | L | **0.88** | -0.35 | 0.06 | 0.01 | -0.19 | 22.7 |
| 13 | Margit | M | **0.88** | -0.35 | 0.06 | 0.01 | -0.19 | 22.7 |
| 14 | Pándy 56 | N | **0.56** | -0.20 | 0.04 | -0.44 | 0.35 | 45.5 |
| 15 | Sylvana | O | **0.56** | 0.35 | 0.49 | -0.02 | 0.42 | 54.5 |
| 16 | Szomolyai | P | **0.54** | 0.12 | 0.28 | -0.35 | **-0.58** | 40.9 |
| 17 | Vanda | Q | 0.47 | 0.49 | 0.12 | -0.16 | 0.21 | 36.4 |
| 18 | Kántorjánosi-U | R | 0.33 | 0.07 | -0.33 | **0.78** | -0.06 | 31.8 |
| 19 | Kántorjánosi-H | S | 0.49 | -0.44 | -0.28 | 0.10 | 0.08 | 40.9 |
| Eigenvalues (Explained variation) | | | 6.74 | 3.54 | 1.79 | 1.51 | 1.13 | |
| Proportion total | | | 35.5 | 18.6 | 9.4 | 8.0 | 6.06 | |

Table 4. Principal Component Loadings

[a]=Potency Values as percent of the maximum abundance.

### 3.5.4 Similarities in distribution of fungi on stalks of sour cherry

Nine hidden (background) variables explained the 87.82 % of total variance, however, altering of the variations in susceptibility of sour cherry none of eigenvalues exceeded 20 %. Six variables (fungal species) dominated two PCs, while only three PCs were dominated by more then one variable on bivariate blots by two major PCs neither PC loadings (Fig 7, fungi) nor PC scores (Fig 8, varieties) are designated to explainable clusters, because neither taxonomic position nor mode of distribution of fungi could be linked to groups as well as relations of sour cherry varieties with similar genetic background (E, F, R and S) or their spatial position in gene bank could be linked on plot (Fig 8).

On our view the above facts indicate the multifactorial and highly specific character of interaction between sour cherry and associated mycobiota. This aspect seems to be support by formation of two big clusters on Figure 5, since genetic background of the mode of life is very complex and the strategy of infection of these pathogens is also different.

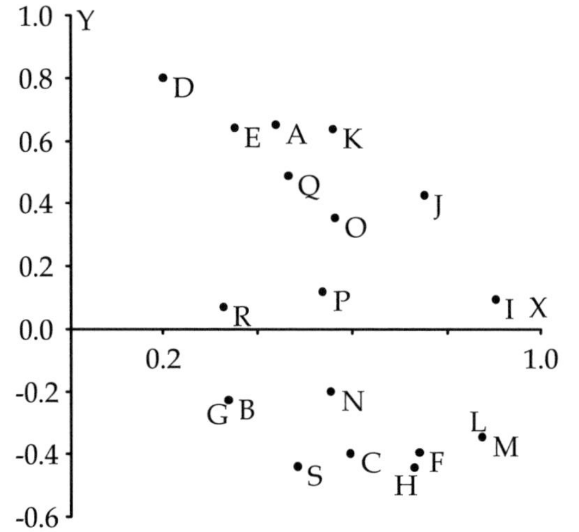


Fig. 5. Plotting sour cherry varieties as PC loadings by two major PCs. The capital letters mark varieties as given in Table 3. The distribution of variables is determined by 54 % of total variation.

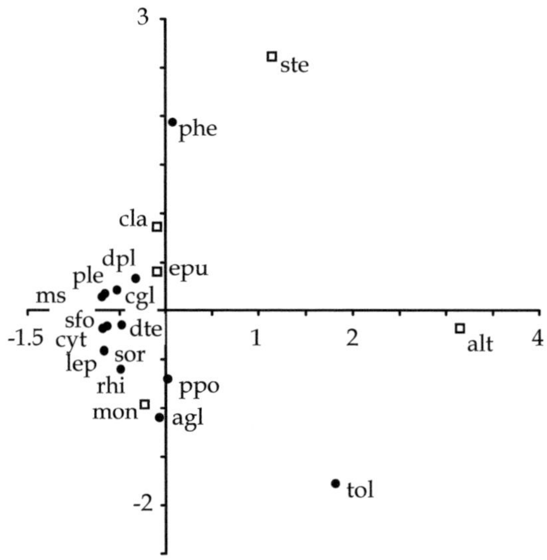


Fig. 6. Plotting fungal species as PC scores by two major PCs. The codes of fungi are given in Table 3. Full circles and opened squares mark species producing xero- or gloeospores, resp.

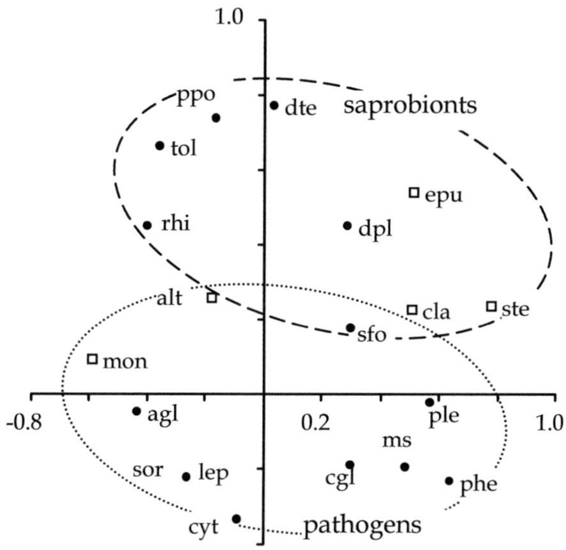


Fig. 7. Plotting fungal species as PC loadings by two major PCs components. The codes of fungi are given in Table 3. Full circles and opened squares mark species producing xero- or gloeospores, resp.

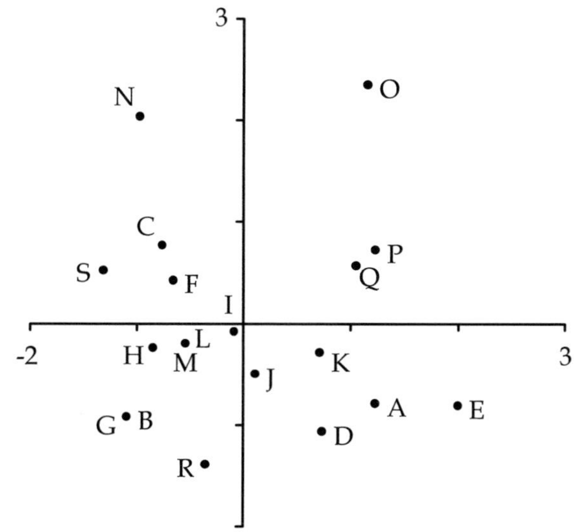


Fig. 8. Plotting sour cherry varieties as PC scores by two major principal components. The capital letters mark varieties as given in Table 3. The distribution of variables is determined by 33 % of total variation.

### 3.5.5 Potency mapping

The potency values (PV-SC) of individual sour cherry varieties and associated fungi calculated by potency mapping technique are included into Tables 4 and 3; all PVs are transformed into percents of the possible maximum for correlation analyses.

The potential diversity of mycobiota varied between 10-50 percents in gene bank (Fig 9A) that means no one of varieties was colonized with full spectrum of fungi detected in the area. The genetic background has seemingly minor importance in this respect because the PVs of related varieties are different. However, the highest PVs were measured one trees located near the edge of the orchard suggesting the importance of spatial position of trees influences in major extent the diversity of associated mycobiota than the genetic background. The potential incidence of fungal species (PV-FI) varied in higher extent (5-100 %) than PV-SC (Fig 9 B); being the yeasts the most dispersed fungi. The saprobionts had higher values than phytopathogens in general, although the facultatively parasitic *Alternaria* was dominant in majority of samples. The xerospore producing fungi were more frequent than the gloeospore-forming ones that can be explained by the season of sampling. Gloeospores, tend to be dispersed by water, e.g. rain splash (rain drops hit spore-filled water film on branches and leaves of trees, aerosolizing spores.) Occasionally, gloeospores are also dispersed by insects. In winter time both treetop and most of insect were absent.

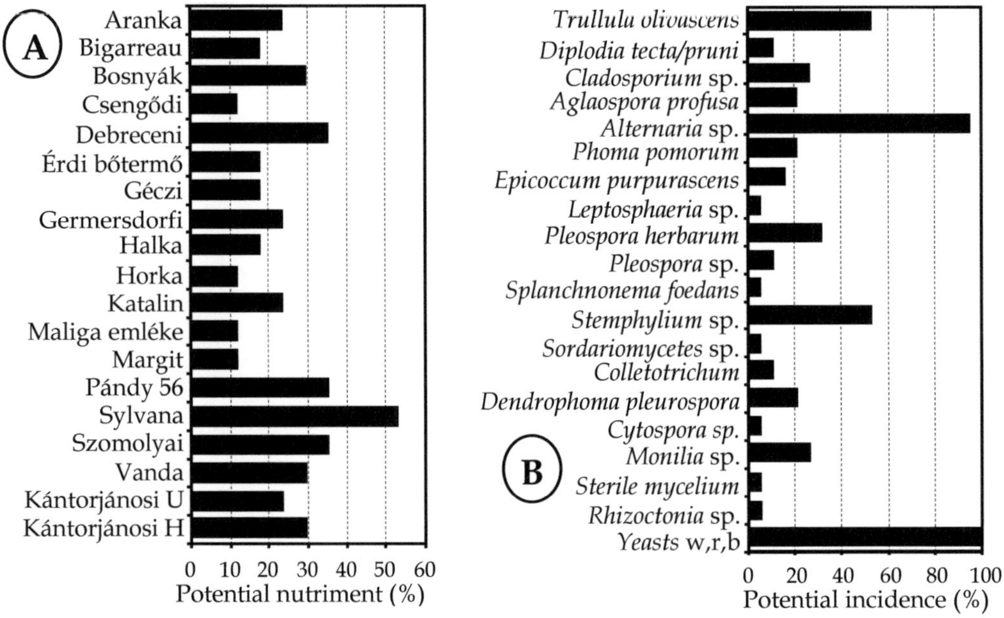


Fig. 9. Potency maps of sour cherry varieties and associated fungal species.

A - diversity of mycobiota associated to stalks takes maximum value in the case when all species detected in gene bank are represented in the sample
B - maximum incidence takes places when the species concerned is presented in all samples.

### 3.5.6 Regression analysis

The arithmetic means of columns and rows (host plant×fungi), potency values of varieties and fungi, the principal components scores of Tables 4 and 5 were correlated by linear regression analysis (Fig 10 and 11).

| Variables[a] | | Principal Components | | | | | | | | | Type[b] |
|---|---|---|---|---|---|---|---|---|---|---|---|
| No. | Code | PC1 | PC2 | PC3 | PC4 | PC5 | PC6 | PC7 | PC8 | PC9 | |
| 1 | tol | -0.36 | **0.66** | -0.29 | 0.31 | 0.17 | -0.22 | 0.26 | 0.13 | 0.00 | G |
| 2 | dte | 0.03 | **0.77** | 0.30 | -0.26 | 0.00 | -0.15 | -0.27 | -0.12 | 0.00 | G |
| 3 | cla | **0.51** | 0.22 | 0.32 | -0.32 | 0.38 | 0.36 | 0.24 | -0.02 | 0.00 | X |
| 4 | agl | -0.43 | -0.05 | 0.19 | 0.46 | **0.57** | 0.14 | -0.03 | -0.14 | 0.00 | G |
| 5 | alt | -0.18 | 0.26 | -0.18 | 0.01 | 0.31 | **0.64** | 0.23 | 0.13 | 0.00 | X |
| 6 | ppo | -0.16 | **0.74** | 0.26 | 0.10 | -0.02 | -0.10 | -0.15 | 0.22 | 0.00 | G |
| 7 | epu | **0.51** | **0.54** | -0.26 | 0.13 | -0.18 | 0.37 | -0.14 | -0.30 | 0.00 | X |
| 8 | lep | -0.26 | -0.22 | -0.11 | -0.35 | -0.22 | 0.29 | -0.20 | -0.02 | **-0.73** | G |
| 9 | phe | **0.64** | -0.23 | **0.55** | 0.02 | 0.16 | -0.12 | -0.24 | -0.13 | 0.00 | G |
| 10 | ple | **0.57** | -0.02 | **-0.66** | -0.11 | 0.41 | -0.03 | -0.16 | 0.07 | 0.00 | G |
| 11 | sfo | 0.30 | 0.18 | **-0.79** | 0.14 | 0.02 | -0.02 | -0.02 | -0.43 | 0.00 | G |
| 12 | ste | **0.78** | 0.24 | 0.06 | -0.24 | -0.18 | -0.25 | 0.01 | -0.13 | 0.00 | X |
| 13 | sor | -0.26 | -0.22 | -0.11 | -0.35 | -0.22 | 0.29 | -0.20 | -0.02 | **0.73** | G |
| 14 | cgl | 0.30 | -0.19 | 0.33 | -0.20 | 0.03 | 0.07 | **0.69** | -0.30 | 0.00 | G |
| 15 | dpl | 0.29 | 0.45 | 0.36 | 0.29 | -0.26 | 0.46 | -0.18 | 0.17 | 0.00 | G |
| 16 | cyt | -0.09 | -0.34 | 0.25 | 0.42 | 0.38 | 0.09 | -0.40 | -0.43 | 0.00 | G |
| 17 | mon | **-0.59** | 0.09 | -0.04 | **-0.72** | 0.08 | 0.12 | -0.19 | -0.21 | 0.00 | X |
| 18 | ms | 0.48 | -0.20 | -0.12 | -0.30 | **0.55** | -0.02 | -0.20 | **0.52** | 0.00 | u |
| 19 | rhi | -0.40 | 0.45 | 0.11 | -0.44 | 0.43 | -0.26 | 0.03 | -0.25 | 0.00 | u |
| Eigenvalue | | 3.38 | 2.83 | 2.18 | 1.95 | 1.66 | 1.32 | 1.19 | 1.12 | 1.06 | |
| Prop. total | | 17.8 | 14.9 | 11.5 | 10.3 | 8.73 | 6.97 | 6.28 | 5.89 | 5.56 | |

Table 5. Principal component loadings

[a] = Codes are deciphered in Table 2. Yeast species were presented in each sample thus these three variables did not show variance and were omitted of PCA.
[b] = Xerospores (X) mostly transported by wind, gloeospores (G) may be transported either by water of animals.

The arithmetic mean on diversity of mycobiota of sour cherry varieties in our case correlated only with potency values ($r^2$=1) but no correlation was revealed ($r^2$=0.12) with PC scores of first PC and poor correlation ($r^2$=0.49) with PC scores of PC2 that well demonstrated on Fig 10. Contrarily, the arithmetic mean of dispersal significantly correlated both with potential incidence of fungal species ($r^2$=1) and PC scores of first PC ($r^2$=0.968) and the relation with scores of second PC is evidently non-linear (Fig11). This entire well demonstrates that the first PC does not comprise always the majority of quantitative measures of the observations, and it is necessary to check this relationship before using PC1 in further calculations.

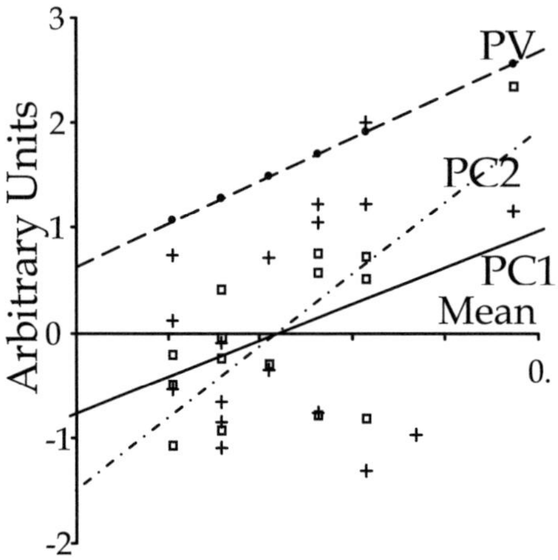


Fig. 10. Relationships between arithmetic mean of diversity of mycobiota and PVs and PC scores of sour cherry varieties.

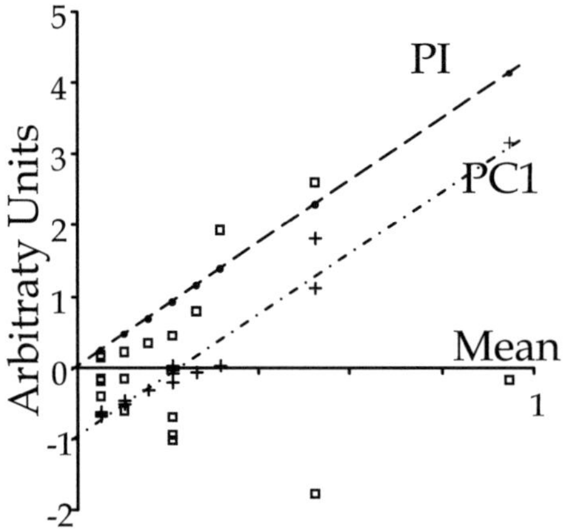


Fig. 11. Relationships between arithmetic mean of incidence of fungal species and PVs and PC scores.

### 3.5.7 Spectral component analysis

The two-dimensional non-linear selectivity map of sour cherry varieties is shown in Figure 12. This map shows similarities to Figure 4. However, in this case there is possible to distinguish a group of varieties where anemophilous fungi dominate the associated fungal consortium deteriorating the remained stalks on trees.

Carrying out the PCA with SPM when varieties serve as variables resulted five significant PC components indicating the elimination of part of total variance of basic data matrix by Potency Mapping. None of them was dominant and on the bivariate plots any patterns could be recognized in grouping of sour cherry varieties (these details are not shown). However, the Non-linear mapping of PC variables revealed three cluster (Fig 13), where varieties seemingly grouped according to composition of the fungal consortia. Moreover, this pattern is better pronounced than on Figure 12 supporting the advantage of combination of spectral mapping with PCA as well as the use of NL mapping for studying matrices comprising variables where the setting of observed phenomena influenced by multidunous complex of factors synchronously (environmental, ecological, genetic, etc).

The varietal distribution of fungal species is shown on two-dimensional non-linear selectivity map in Figure 14. The species frequently found on decaying fruits formed a separated cluster, while those parasiting on leaves and phloeme distribute without clear grouping.

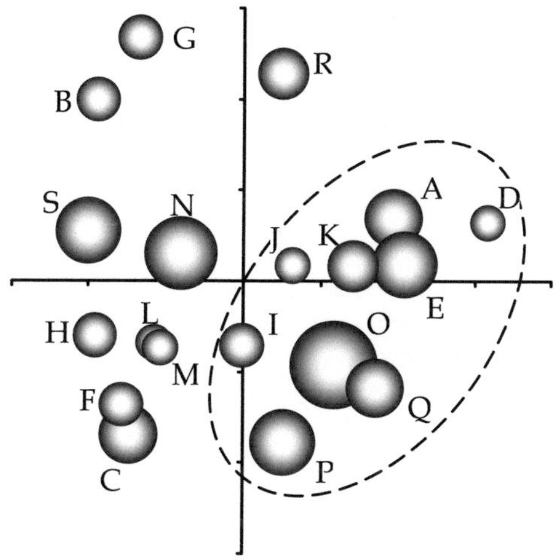


Fig. 12. Non-linear Map of sour cherry varieties as spectral variables. The capital letters mark varieties as given in Table 4. The size of balls is proportional to potential diversity of mycobiota recorded (Table 4). The ellipse marks the varieties colonized predominantly with xerospore producing fungi.

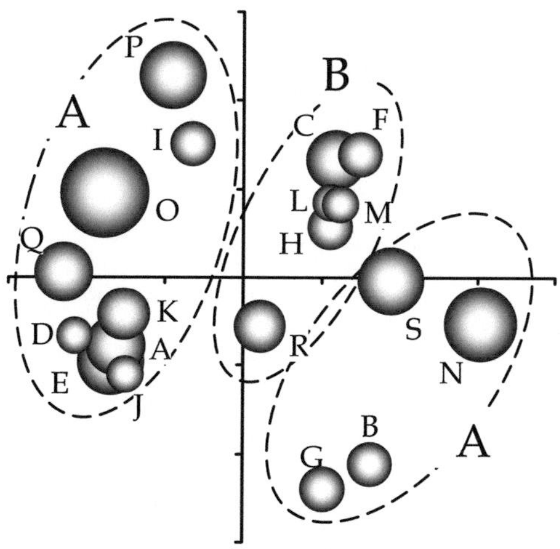


Fig. 13. Similarities in composition of mycobiota on stalk of sour cherry varieties. The capital letters mark varieties as given in Table 4. The size of balls is proportional to potential diversity of mycobiota recorded (Table 4). The ellipses mark the varieties colonized predominantly with xerospore (A) or gloeospore (B) producing fungi, resp.

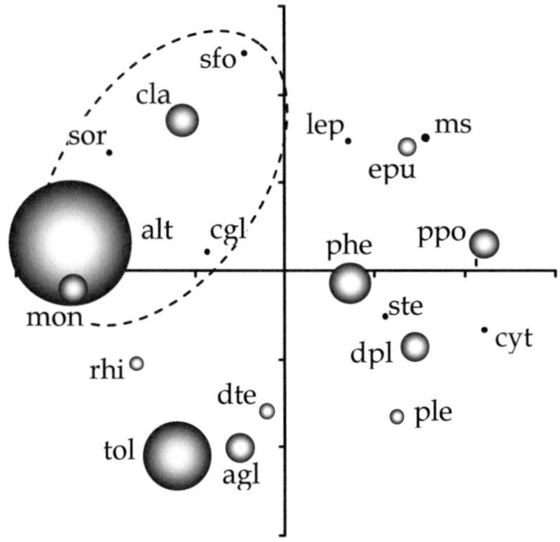


Fig. 14. Non-linear Map of fungal species as spectral variables. The codes of fungi are as given in Table 4. The size of balls is proportional to potential incidence of species recorded (Table 4). The ellipse marks the species causing fruit rot.

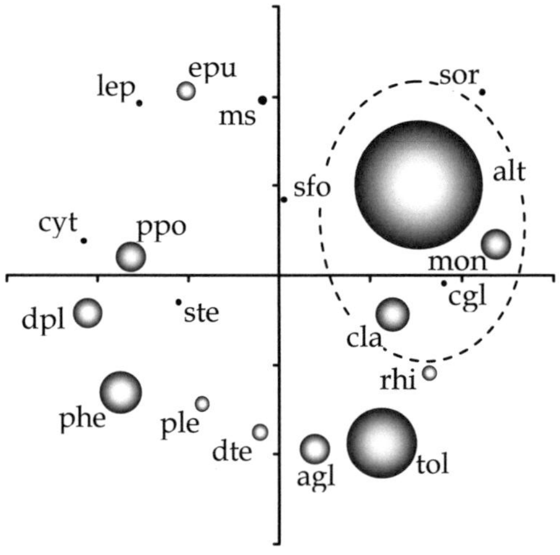


Fig. 15. Similarities in distribution of fungal species on stalk of sour cherry. The codes of fungi are as given in Table 4. The size of balls is proportional to potential incidence of species recorded (Table 4). The ellipse marks the species causing fruit rot.

The PCA of SC variables resulted eight components with few differences of their eigen-values. On our opinion this indicates the similar weight of numerous factors influencing on the incidence of fungal species in the orchard. On the NL map of PC variables of SPM the cluster of fruit rotting fungi became more compact. The grouping of the species remained unclear. All this underlines again the value of combined application of spectral mapping with PCA.

Neither sour cherry varieties nor associated fungi formed separated clusters on bivariate NLMaps (graphs are not shown). Most probably the two dimensional mapping distributed the information explained by PCs (see Tables 4 and 5) moderately thus the alterations shown by PC loadings were not reflected on scattering of PC scores (varieties or fungi) on the plot, i.e., the two dimensional NL-mapping smashed the variance and divided between two axes. One can frequently meet the clustering of variables on scatter plots and have no success in explanation. Nevertheless, this information is serviceable, because it might orient the further search and design of research project.

However, there were clear groups when tridimensional mapping carried out. In the case of sour cherry (Fig 16) neither the diversity nor the genetic background of varieties channelized the clustering.

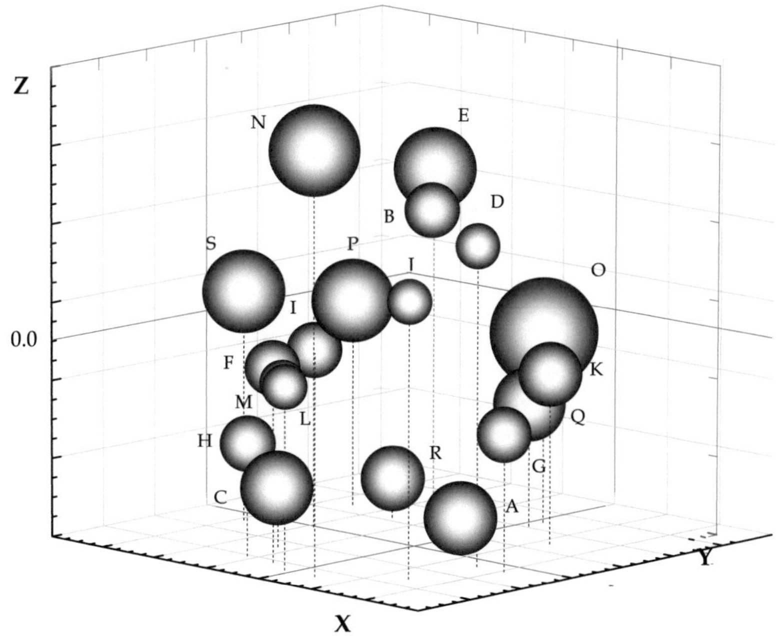


Fig. 16. Non-linear Map of sour cherry varieties as PC scores. The capital letters mark varieties as given in Table 4. The size of balls is proportional to diversity of colonization recorded (Table 4).

In the case of fungi the separation of species into two groups was more evident (Fig 17). Neither potential incidence nor mode of distribution is seemingly connected to clustering in tridimensional NL map. The first group comprises mainly fungi, which are pathogens or take part in fruit rot both on the tree and post harvesting in stores. Most of saprobionts are in the second groups. There was not found any relationships between taxonomic position of fungal species and their varietal dependent incidence in orchards. The high divergence of the composition of consortia on stalks might cause the absence of *Trichoderma* strains of biopreparation.

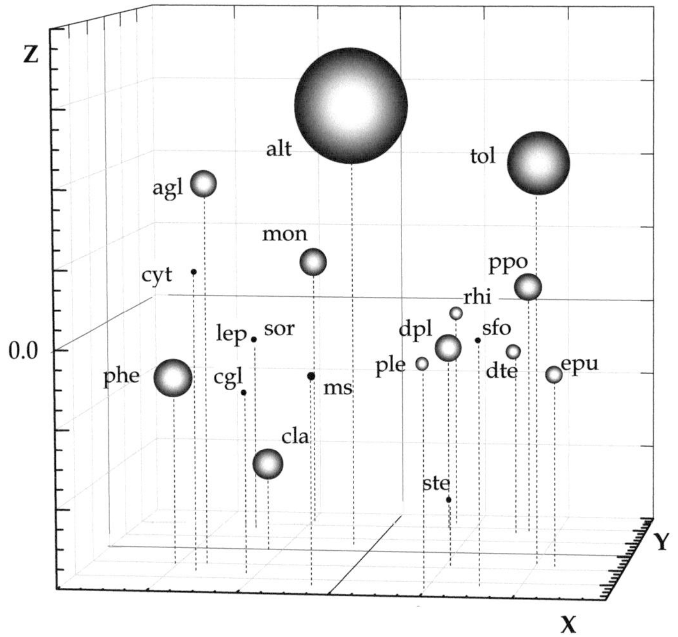


Fig. 17. Non-linear Map of fungal species as PC scores. The codes of fungi are as given in Table 4. The size of balls is proportional to potential incidence of species recorded (Table 3).

## 4. Conclusions

Presence of five *Glomerella* anamorphs has been established in Hungarian sour cherry orchards, among them *C. acutatum*, the new pathogen for Carpathian Basin became dominant over *C. gloeosporioides* in past five year. Using multivariate statistical methods for detailed analysis of mycobiota revealed two different fungal consortia associated to sour cherry in winter period.

PCA proved to be superior to multidimensional scaling and possible as well as valuable amendment of data exploration when combined with Potency and Spectral Mapping.

The correct order of organization of modules in multivariate analysis of data matrices comprising results of observations or experiments undergone by multifactorial effect can reveal unexpected factors thus increase the profitability of approach.

## 5. Acknowledgements

The research work was supported by the Baross Regional Innovation Programme (No. EA07-EA-KOZKFI-2008-0066). We thanks to Á. Szécsi for the critical reading of the manuscript.

## 6. References

Børve, J. & Stensvand, A. (2006). Timing of fungicide applications against anthracnose in sweet and sour cherry production in Norway. *Crop Protection*, 25(8), pp. 781-787. ISSN: 0261-2194

Børve, J., Djønne ,R. & Stensvand A. (2010). *Colletotrichum acutatum* occurs symptomatically on sweet cherry leaves. *European Journal of Plant Pathology,* 127(3), pp. 325-332. ISSN: 0929-1873

Frank, K.L., Kalkstein, L.S., Geils, B.W. & Thistle, H.W. (2008). Synoptic climatology of the long-distance dispersal of white pine blister rust. I. Development of an upper level synoptic classification. *International Journal of Biometeorology,* 52(7), pp.641-652. ISSN: 0020-7128

Frazer, G.W., Wulder, M.A. & Niemann, K.O. (2005). Simulation and quantification of the fine-scale spatial pattern and heterogeneity of forest canopy structure: A lacunarity-based method designed for analysis of continuous canopy heights. *Forest Ecology and Management,* 214(1-3), pp. 65-90. ISSN: 0378-1127

Garcia-Mozo, H., Gomez-Casero, M.T., Dominguez, E. & Galan, C. (2007). Influence of pollen emission and weather-related factors on variations in holm-oak (*Quercus ilex* subsp. *ballota*) acorn production. *Environmental and Experimental Botany,* 61(1), pp. 35-40. ISSN: 0098-8472

Jacquemin, B., Lanki, T., Yli-Tuomi, T., Vallius, M., Hoek, G., Heinrich, J., Timonen, K. & Pekkanen, J., (2009). Source category-specific PM(2.5) and urinary levels of Clara cell protein CC16. The ULTRA study. *Inhalation Toxicology,* 21(13), pp. 1068-1076. ISSN: 0895-8378

Lewi, PJ (2005). Spectral mapping, a personal and historical account of an adventure in multivariate data analysis. *Chemometrics and Intelligent Laboratory Systems,* 77(1-2), pp. 215-223. ISSN: 0169-7439

Lighthart, B., Shaffer, B.T., Frisch, A.S. & Paterno, D. (2009). Atmospheric culturable bacteria associated with meteorological conditions at a summer-time site in the mid-Willamette Valley, Oregon. *Aerobiologia,* 25(4), pp. 285-295. ISSN: 0393-5965

Magyar D (1998) Adatok a Budakeszi-erdQĠmikroszkópikus gombáinak ismeretéhez. MSc, St. Steven University [Data to the Knowledge of the Microscopic Fungi in the Forests around Budakeszi]

Magyar D (2007) Aeromycological aspects of mycotechnology. In: *Mycotechnology: Current Trends and future Prospects.* (ed. M.K. Rai) I.K. International Publ. House, New Delhi: 226-263. ISBN: 9788189866082

Magyar, D., Eszeki, E.R., Oros, G., Szecsi, A., Kredics, L., Hatvani, L. & Körmöczi, P. (2011). The air spora of an orchid greenhouse. *Aerobiologia,* 27(2), pp. 121-134. ISSN: 0393-5965

Mardia, K.V., Kent, J.T. & Bibby, J.M. (1979). *Multivariate Analysis,* Academic Press, London, 213–254.

Mason E.W. (1937). *Annotated account of fungi received at the Imperial Mycological Institute,* List II, Fasc. 3., Kew, Surrey, IMI, pp . 69-99. ISSN 0027-5522 ; 3

Oros, G. & Naár, Z. (2008). Környezetkímélő biocidek és alkalmazásuk. HPO 0800405

Oros, G., Vajna, L., Balázs, K., Fekete, Z., Naár, Z. & Eszéki, E. (2010). Anthracnose and possibilities of the control with special regard to resident *Glomerella* population in sour cherry plantations of East Hungary. *Agricultural Research,* 39, pp. 12-17. ISSN 1587-1282

Oros, G., Naár, Z. & Cserháti, T. (2011). Growth response of *Trichoderma* species to organic solvents. *Molecular Informatics,* 30(2-3):276-285. ISSN: 1868-1743

Parrado, Z.G., Barrera, R.M.V., Rodriguez, C.R.F., Maray, A.M.V., Romero, R.P., Fraile, R. & Gonzalez, D.F. (2009). Alternative statistical methods for interpreting airborne Alder (*Alnus glutimosa* [L.] Gaertner) pollen concentrations. *International Journal of Biometeorology,* 53(1), pp. 1-9. ISSN: 0020-7128

Quraishi, T.A., Schauer, J.J. & Zhang, Y.X. (2009). Understanding sources of airborne water soluble metals in Lahore, Pakistan. *Kuwait Journal of Science & Engineering*, 36(1A), pp. 43-62. ISSN: 1024-8684

Pearson, K. (1901). On lines and planes of closest fit to systems of points in space. *Philosophical Magazine*, 6(2), pp. 559–572. ISSN: 1478-6435

Ribeiro, H., Cunha, M. & Abreu, I. (2007). Improving early-season estimates of olive production using airborne pollen multi-sampling sites RID A-6708-2010 RID A-6711-2010. *Aerobiologia*, 23(1), pp. 71-78. ISSN: 0393-5965

Ribeiro, H., Cunha, M. & Abreu, I. (2008). Quantitative forecasting of olive yield in Northern Portugal using a bioclimatic model RID A-6708-2010 RID A-6711-2010. *Aerobiologia*, 24(3), pp. 141-150. ISSN: 0393-5965

Rungratanaubon T., Panich , N. & Wangwongwattana, S. (2008). Characterization and source identification of trace metals inairborne particulates of Bangkok, Thailand. *Annals of the New York Academy of Sciences*, 1140, pp. 297-307. ISSN: 0077-8923

Sammon, J.W. (1969). A nonlinear mapping for data structure analysis, *IEEE Transactions on Computers*, 18, pp. 401–407. ISSN: 0018-9340

Schwarz, A., Wilcke, W., Kobza, J. & Zech, W. (1999). Spatial distribution of soil heavy metal concentrations as indicator of pollution sources at Mount Krizna (Great Fatra, central Slovakia) RID A-1573-2008. *Journal of Plant Nutrition and Soil Science*, 162(4), pp. 421-428. ISSN: 1436-8730

Shah, M.H., Shaheen, N., Jaffar, M., Khalique, A, Tariq, S.R. & Manzoor, S. (2006). Spatial variations in selected metal contents and particle size distribution in an urban and rural atmosphere of Islamabad, Pakistan. *Journal of Environmental Management*, 78(2), pp. 128-137. ISSN: 0301-4797

Shaheen, N., Shah, M.H. & Jaffar, M. (2005). A study of airborne selected metals and particle size distribution in relation to climatic variables and their source identification. *Water Air and Soil Pollution*, 164(1-4), pp.275-294. ISSN: 0049-6979

Titseesang, T., Wood, T. & Panich, N. (2008). Leaves of Orange Jasmine (*Murraya paniculata*) as Indicators of Airborne Heavy Metal in Bangkok, Thailand. *Annals of The New York Academy of Sciences*, 1140, pp. 282-289. ISSN: 0077-8923

Tokalioglu, S., Yilmaz, V. & Kartal, S. (2010). An Assessment on Metal Sources by Multivariate Analysis and Speciation of Metals in Soil Samples Using the BCR Sequential Extraction Procedure. Clean-Soil Air Water, 38(8), pp. 713-718. ISSN: 1863-0650

Uno, Y., Prasher, S.O., Lacroix, R., Goel, P.K., Karimi, Y., Viau, A. & Patel, R.M. (2005). Artificial neural networks to predict corn yield from Compact Airborne Spectrographic Imager data. *Computers and Electronics in Agriculture*, 47(2), pp. 149-161. ISSN: 0168-1699

Yang, C., Everitt, J.H. & Bradford, J.M. (2004). Airborne hyperspectral imagery and yield monitor data for estimating grain sorghum yield variability. *Transactions of The Asae*, 47(3), pp. 915-924. ISSN: 0001-2351

You, X.Q., Senthilselvan, A., Cherry, N.M., Kim, H.M. & Burstyn, I. (2008). Determinants of airborne concentrations of volatile organic compounds in rural areas of Western Canada. *Journal of Exposure Science and Environmental Epidemiology*, 18(2), pp. 117-128. ISSN: 1559-0631

Vasconcellos, P.C., Balasubramanian, R., Bruns, R.E., Sanchez-Ccoyllo, O. Andrade, M.F., Flues, M. (2007). Water-soluble ions and trace metals in airborne particles over urban areas of the state of Sauo Paulo, Brazil: Influences of local sources and long range transport RID C-2243-2011. *Water Air and Soil Pollution*, 186(1-4), pp. 63-73. ISSN: 0049-6979

# 7

# Application of PCA in Taxonomy Research – Thrips (Insecta, Thysanoptera) as a Model Group

Halina Kucharczyk[1], Marek Kucharczyk[2],
Kinga Stanisławek[1] and Peter Fedor[3]
[1]*Department of Zoology, Maria Curie-Skłodowska University, Lublin,*
[2]*Department of Nature Conservation, Maria Curie-Skłodowska University, Lublin,*
[3]*Department of Ecosozology, Faculty of Natural Sciences, Comenius University, Bratislava,*
[1,2]*Poland*
[3]*Slovakia*

## 1. Introduction

Phenetic taxonomy is based on analysis of many unweighted characters. The number of variables that can be analyzed for a plant or animal species is so high that it is necessary to use a mathematical tool for grouping them into units corresponding to taxa.

Principal Component Analysis enables researchers to reduce the number of possible groupings. It is significant due to the occurrence of some redundancy in variables. In this case, redundancy means that some of the variables are correlated with one another, because they are measuring the same construct.

Principal Component Analysis replaces many original characters with only a few most significant principal components (PCs) which represent combinations of closely correlated original characters.

Principal Component Analysis was first described by Pearson (1901); in the 1930's Hotteling (1933) prepared a fully functional method that generates a set of orthogonal axes, placed in decreasing order and determining the main directions of variability of samples.

In analysis of principal components, eigenvalues represent the relative participation of each principal component in presenting the general variability of sampled material. The numerical value of a given eigenvalue is a direct indicator of the weight of a particular component in the general characteristics of the variability of a set of data. In practice, the distribution of the elements of the analyzed set in the space of the first three or four components allows one to present almost the complete diversity of the set.

Eigenvectors are sets of numbers which show the weights of individual characters for each principal component. Like the correlation coefficient, eigenvalues are scaled from -1 to +1. The higher the value, the more closely a given trait is connected with a component. On the basis of eigenvectors it is then possible to interpret principle components, e.g. determine which character (or characters) are the most representative.

Before principal components are determined, the data can be processed in many ways. It is necessary if variables (features) are expressed in different units or the range of their variability is different. The two methods used are centering and standardization. In the former, the variable axes are moved so that the beginning of one axis is in the centre of inertia of the axis. Standardization of data involves changing the values of characters in such a way that their mean is 0 and standard deviation is 1.

If analysis covers calculating the covariance matrix, the variables are centred; if, on the other hand, the correlation coefficient matrix needs to be calculated, the variables are centred and standardized at the same time.

The Principal Component Analysis (PCA) is widely used in taxonomic research of plants and animals ( Apuan et al., 2010; Chiapella, 2000; Kucharczyk & Kucharczyk 2009; Lilburn & Garrity, 2004; Sahuquillo et al., 1997; Wolff et al., 1997; Véla et al., 2007). Below we present application of PCA in taxonomy of the order Thysanoptera (Insecta). The numerical calculations were made with the Multi-Variate Statistical Package (MVSP) (Kovach, 2005) and Statistica PL, version 6 (StatSoft Inc. 1984-2001).

### 1.1 Characteristic of the order Thysanoptera

Up to now almost 6000 species of the order Thysanoptera have been described worldwide but many other are added to this list every year. The asymmetric, with only one left mandibule, mouth cone is a synapomorphic character state which differs thrips from the other insect orders (Mound et al., 1980; Mound & Morris, 2004, 2007). In currently accepted systematic position the order Thysanoptera is divided into two suborders: Terebrantia and Tubulifera. The former consists of eight families which comprise very small and tiny insects, most of them reaching 1-3 mm in length. They are mostly herbivorous insects feeding both on dicotyledonous and monocotyledonous plants, only a small part of them are facultative or obligate predators (fam. Aeolothripidae). The family Thripidae with nearly 2,500 species is the largest within Terebrantia but the relationships within this group is not clear (Mound & Morris, 2007). Only one tubuliferan family - Phlaeothripidae with almost 3,500 known species includes larger thrips, the biggest, mostly tropical taxa, reach up to 15 mm. Excepting herbivores most of them live on dead wood or in leaf litter and feed on fungi. Some phytophagous species are regarded as pests, feeding and breeding on different parts of plants they cause deformations of leaves, flowers and fruits, and in the final result stop their development. A limited number of terebrantians may transmit fungi, bacteria and viruses, which may infect the host plants reducing the quality of yields and their market value (Lewis, 1997; Tommasini & Maini, 1995).

Originally taxonomy referred to the description and naming of the organisms (alpha taxonomy). Currently it is the science based on different fields of knowledge and that uses various tools to classify the organisms and determine the relationships amongst them. There are two systems of the Thysanoptera order classification: phenetic based mainly on the morphological characteristics of adults' specimens and phylogenetic one based on the evolutionary relationships (Mound, 2010). The former one is more practical and widely used in construction of the identification keys. The latter based on the molecular and genetic data is not satisfactory enough because of insufficient data. Therefore in practice, morphology and other biological aspects, e.g. observations of developmental stages and relations with host plants provide more data and may be useful in comprehension of the relationships amongst taxa (Crespi et al., 2004; Mound & Morris, 2007).

Some disagreements exist concerning classification system of Thysanoptera on different taxonomic levels. Because of the great differences in body structures between Terebrantia and Tubulifera Bhatti (1988) proposed to raise them to the order rank in new superorder Thysanopteroidea composed of 40 families. In the next works this author divided Terebrantia into 28 families (Bhatti, 2005, 2006). However, this classification is not accepted by most of thysanopterologists now. On the other hand the current state of Terebrantia with eight families has not been taken under consideration by zur Strassen (2003) in his latest key. In contrast to the currently accepted division zur Strassen classified the species of the genera *Melanthrips* and *Ankothrips* into the family Aeolothripidae. Many revisions at the genus level took place in the past, e.g. changes within the genera: *Thrips* Linnaeus 1758 and *Taeniothrips* Amyot & Serville 1843, *Mycterothrips* Trybom 1910 and *Taeniothrips*, *Anaphothrips* Uzel 1895 and *Rubiothrips* Schliephake 1975 etc. (Strassen zur 2003).

The correct identification of specimens to the species level is a basis for further taxonomic study. Often there are many problems with recognition of adults, as well as immature stages because of diversity of variation within and between species, particularly in the species rich genera, e.g. *Thrips* (Terebrantia) and *Haplothrips* Amyot & Serville 1843 (Tubulifera). These genera include mainly the Holarctic species feeding and breeding on dicotyledonous plants, though occasionally they are graminicolous. Very numerous species representing both of them may suggest that these genera are relatively young in evolutionary history. Morphologically, many species are very similar and are treated by some researchers as the same species in two forms, e.g. *Thrips atratus* Haliday 1836 and *Thrips montanus* Priesner 1921, *Thrips sambuci* Heeger 1854 and *Thrips fuscipennis* Haliday 1836, *Thrips fuscipennis* and *Thrips menyanthidis* Bagnall 1923, *Haplothrips leucanthemi* (Schrank 1781) and *Haplothips niger* (Osborn 1883), *Haplothrips tritici* (Kurdjumov 1912) and *Haplothrips cerealis* Priesner 1939 (Mound & Minaei, 2007, 2010; Strassen zur, 2003).

There are even more problems with recognizing larval stages. The larvae are less mobile therefore have a stronger relationship with their host plants. Morphological dissimilarity amongst larvae of different species are often larger than amongst their adults (Kucharczyk, 2010). So, the detailed study both on adults and immature stages with application of the statistical tools, within PCA method, allows to solve several problems in the taxonomy of thrips and at least partially explain their phylogenetic relationships.

## 2. Application of Principal Component Analysis method in taxa recognizing

### 2.1 Identification of the second larval instar of the *Haplothrips* genus species

The Principal Component Analysis method (PCA) may be useful in selecting from among the great number of morphometric characters, especially those that have some taxonomical value. Such a necessity is occuring within genera which species are very uniform in morphological structure and there are weak qualitative characters differentiating them. The only potential differences are related to the measurements of some body parts. The genus

*Haplothrips* (Thysanoptera, Phlaeothripidae) is a good example for the discussed case. To this genus belong species that are very similar in body structure and therefore they are difficult to identify.

The *Haplothrips* genus is one of the most numerous in species of the Phlaeothripidae family (about 230 species). It is distributed worldwide but mainly in northern hemisphere, almost 70% known hitherto species of this genus have been noted in Holarctic. Most of them are phytophagous, feeding and breeding in flowers of dicotyledons, mostly of *Asteraceae* family, only a few taxa are connected with monocotyledons (Mound & Minaei, 2007; Pitkin, 1976; Zawirska & Wałkowski, 2000). Because of the fact that the research on larval morphology is very scarce (especially on species belonging to Phlaeothripidae) the morphological analysis of the second instar larvae of *Haplothrips* species have been undertaken.

The measurements were made for 165 specimens of larvae belonging to 11 species (each species was represented by 15 individuals) collected from various habitats and regions of Poland. Among examined species were: *Haplothrips aculeatus* (Fabricius 1803) (inhabitant of various *Poaceae* plants, the ubiquistic species); *H. tritici* (Kurdjumov 1912) (also *Poacae*-related species, but mainly with *Triticum vulgare;* occurring in south-eastern part of Europe and Asia Minor); *H. setiger* Priesner 1921 (the polyphagous species living on flowers of *Asteraceae* for example *Senecio fuchsii, Crepis* spp., *Matricaria* spp., *Achillea* spp., *Anthemis arvensis;* having Palaearctic range of occurrence); *H. subtilissimus* (Haliday 1852) (present on leaves of bushes and trees such as *Quercus* spp. and *Fagus* spp.; considered as predatory species on mites, lepidopteran eggs and larval stages of *Coccidae* and *Aleyrodidae*; noted from Europe). The rest of species are regarded as mono- or oligophags: *H. arenarius* Priesner 1920 (especially connected with *Helichrysum arenarium;* with Palaearctic range); *H. dianthinus* Priesner 1924 (on *Dianthus carthusianorum;* noted from central and south Europe); *H. jasionis* Priesner 1950 (on *Jasione montana;* known from Europe only); *H. leucanthemi* (Schrank 1781) (on *Leucanthemum* spp.; reaching Holarctic, Oriental and Australian regions); *H. setigeriformis* Fabian 1938 (on *Potentilla argentea,* known from Europe and Near East); *H. statices* (Haliday 1836) (on *Armeria elongata;* occurring in Europe) and *H. angusticornis* Priesner 1921) (connected with *Asteraceae* but mainly with *Achillea millefolium;* noted from Palaearctic).

Each individual of larva have been measured in respect of 72 potentially important features, most of them concerned lengths of selected both dorsal (d) and ventral (v) setae on all parts of body (h - head, pro - pronotum, mes - mesonotum, meta - metanotum), distances between setae, measurements of apical abdominal segments (8-11) and antennal segments (ant, III-VII). Finally, the data matrix consisted of 11880 measurements was constructed (comprised of 72 character states of 165 individuals belonging to 11 species) (Fig. 1).

The specimens were ordinated along first two PCA axes (transformed data). The results of PCA showed that the cumulative variance of the two principal components reached 62.4%: Axis 1 - 38.5%, Axis 2 – 23.9%. Figure 2 shows two groups of specimens belonging to different species which are clearly isolated (*H. aculeatus* and *H. subtilissimus*). The most discriminative features with the highest eigenvalues are the lengths of abdominal setae (9-d2, 9-v2, 11-d1, 11-v2). On the mentioned PCA graph the most numerous group of specimens, belonging to the nine other and listed above species, is creating compact cloud. To choose more selective character states discriminating the rest of examined species we can remove the measurements of the two first separated species (*H. subtilissimus* and *H. aculeatus*) from the primary data matrix.

| | A | B | C | D | E | F | G | H | I | J | K | L | M | N | O | P | Q | R | S | T | U | V | W | X | Y | Z | AA | AB | AC |
|---|---|---|---|---|---|---|---|---|---|---|---|---|---|---|---|---|---|---|---|---|---|---|---|---|---|---|---|---|---|
| 1 | | ant-l | ant-dist | I-l | I-w | I-l/w | II-l | II-w | II-l/w | III-l | III-w | III-l/w | IV-l | IV-w | IV-l/w | V-l | V-w | V-l/w | VI-l | VI-w | VI-l/w | VII-l | VII-w | VII-l/w | t-3 | t-4-int | t-4-ex | t-5 | t-6 |
| 2 | acul1 | 230,0 | 25,0 | 20,0 | 32,5 | 0,6 | 30,0 | 21,3 | 1,4 | 50,0 | 22,5 | 2,2 | 47,5 | 25,0 | 1,9 | 32,5 | 22,5 | 1,4 | 25,0 | 17,5 | 1,4 | 22,5 | 8,8 | 2,6 | 10,0 | 12,5 | 5,0 | 10,0 | 13,8 |
| 3 | acul2 | 220,0 | 22,5 | 20,0 | 30,0 | 0,7 | 30,0 | 21,3 | 1,4 | 47,5 | 22,5 | 2,1 | 47,5 | 23,8 | 2,0 | 32,5 | 22,5 | 1,4 | 22,5 | 15,0 | 1,5 | 22,5 | 8,8 | 2,6 | 10,0 | 13,8 | 5,0 | 11,3 | 17,5 |
| 4 | acul3 | 230,0 | 22,5 | 20,0 | 32,5 | 0,6 | 30,0 | 20,0 | 1,5 | 50,0 | 22,5 | 2,2 | 47,5 | 25,0 | 1,9 | 35,0 | 22,5 | 1,6 | 25,0 | 16,3 | 1,5 | 22,5 | 7,5 | 3,0 | 10,0 | 12,5 | 5,0 | 12,5 | 15,0 |
| 5 | acul4 | 205,0 | 22,5 | 17,5 | 27,5 | 0,6 | 27,5 | 18,8 | 1,5 | 45,0 | 21,3 | 2,1 | 42,5 | 22,5 | 1,9 | 30,0 | 21,3 | 1,4 | 22,5 | 15,0 | 1,5 | 20,0 | 7,5 | 2,7 | 7,5 | 12,5 | 5,0 | 10,0 | 12,5 |
| 6 | acul5 | 225,0 | 22,5 | 22,5 | 28,8 | 0,8 | 30,0 | 21,3 | 1,4 | 50,0 | 21,3 | 2,4 | 45,0 | 22,5 | 2,0 | 32,5 | 20,0 | 1,6 | 25,0 | 15,0 | 1,7 | 22,5 | 7,5 | 3,0 | 10,0 | 12,5 | 7,5 | 10,0 | 15,0 |
| 7 | acul6 | 205,0 | 22,5 | 20,0 | 30,0 | 0,7 | 27,5 | 20,0 | 1,4 | 45,0 | 20,0 | 2,3 | 40,0 | 22,5 | 1,8 | 27,5 | 20,0 | 1,4 | 20,0 | 16,3 | 1,2 | 20,0 | 8,8 | 2,3 | 10,0 | 12,5 | 6,3 | 10,0 | 12,5 |
| 8 | acul7 | 225,0 | 22,5 | 25,0 | 32,5 | 0,8 | 30,0 | 21,3 | 1,4 | 50,0 | 22,5 | 2,2 | 45,0 | 25,0 | 1,8 | 32,5 | 22,5 | 1,4 | 25,0 | 16,3 | 1,5 | 22,5 | 8,8 | 2,6 | 10,0 | 12,5 | 5,0 | 10,0 | 15,0 |
| 9 | acul8 | 235,0 | 25,0 | 30,0 | 30,0 | 1,0 | 30,0 | 22,5 | 1,3 | 50,0 | 22,5 | 2,2 | 47,5 | 23,8 | 2,0 | 35,0 | 22,5 | 1,6 | 25,0 | 16,3 | 1,5 | 25,0 | 8,8 | 2,9 | 8,8 | 12,5 | 5,0 | 10,0 | 12,5 |
| 10 | acul9 | 225,0 | 22,5 | 25,0 | 30,0 | 0,8 | 30,0 | 20,0 | 1,5 | 52,5 | 22,5 | 2,3 | 45,0 | 25,0 | 1,8 | 35,0 | 22,5 | 1,6 | 25,0 | 15,0 | 1,7 | 22,5 | 7,5 | 3,0 | 7,5 | 12,5 | 5,0 | 8,8 | 13,8 |
| 11 | acul10 | 200,0 | 22,5 | 22,5 | 27,5 | 0,8 | 27,5 | 20,0 | 1,4 | 42,5 | 20,0 | 2,1 | 42,5 | 21,3 | 2,0 | 30,0 | 18,8 | 1,6 | 22,5 | 15,0 | 1,5 | 17,5 | 7,5 | 2,3 | 8,8 | 12,5 | 3,8 | 10,0 | 13,8 |
| 12 | acul11 | 220,0 | 22,5 | 25,0 | 30,0 | 0,8 | 30,0 | 20,0 | 1,5 | 42,5 | 22,5 | 1,9 | 47,5 | 25,0 | 1,9 | 37,5 | 22,5 | 1,7 | 25,0 | 15,0 | 1,7 | 22,5 | 7,5 | 3,0 | 10,0 | 12,5 | 5,0 | 11,3 | 15,0 |
| 13 | acul12 | 215,0 | 22,5 | 25,0 | 30,0 | 0,8 | 30,0 | 20,0 | 1,5 | 47,5 | 22,5 | 2,1 | 42,5 | 25,0 | 1,7 | 32,5 | 22,5 | 1,4 | 25,0 | 15,0 | 1,7 | 20,0 | 7,5 | 2,7 | 7,5 | 12,5 | 5,0 | 11,3 | 13,8 |
| 14 | acul13 | 215,0 | 22,5 | 20,0 | 30,0 | 0,7 | 30,0 | 21,3 | 1,4 | 47,5 | 22,5 | 2,1 | 42,5 | 25,0 | 1,7 | 30,0 | 22,5 | 1,3 | 22,5 | 15,0 | 1,5 | 20,0 | 8,8 | 2,3 | 7,5 | 12,5 | 5,0 | 10,0 | 15,0 |
| 15 | acul14 | 215,0 | 22,5 | 20,0 | 30,0 | 0,7 | 27,5 | 20,0 | 1,4 | 47,5 | 21,3 | 2,2 | 42,5 | 22,5 | 1,9 | 32,5 | 20,0 | 1,6 | 22,5 | 15,0 | 1,5 | 20,0 | 8,8 | 2,3 | 10,0 | 12,5 | 5,0 | 11,3 | 12,5 |
| 16 | acul15 | 210,0 | 25,0 | 17,5 | 30,0 | 0,6 | 27,5 | 20,0 | 1,4 | 45,0 | 20,0 | 2,3 | 42,5 | 22,5 | 1,9 | 32,5 | 20,0 | 1,6 | 22,5 | 15,0 | 1,5 | 20,0 | 8,8 | 2,3 | 7,5 | 12,5 | 5,0 | 11,3 | 12,5 |
| 17 | aren1 | 205,0 | 10,0 | 20,0 | 27,5 | 0,7 | 27,5 | 21,3 | 1,3 | 35,0 | 21,3 | 1,6 | 37,5 | 22,5 | 1,7 | 35,0 | 20,0 | 1,8 | 27,5 | 15,0 | 1,8 | 20,0 | 7,5 | 2,7 | 5,0 | 8,8 | 3,8 | 6,3 | 6,3 |
| 18 | aren2 | 205,0 | 15,0 | 20,0 | 32,5 | 0,6 | 30,0 | 22,5 | 1,3 | 37,5 | 22,5 | 1,7 | 37,5 | 22,5 | 1,7 | 35,0 | 20,0 | 1,8 | 27,5 | 15,0 | 1,8 | 22,5 | 7,5 | 3,0 | 6,3 | 10,0 | 2,5 | 7,5 | 7,5 |
| 19 | aren3 | 210,0 | 15,0 | 20,0 | 32,5 | 0,6 | 32,5 | 22,5 | 1,4 | 37,5 | 22,5 | 1,7 | 40,0 | 22,5 | 1,8 | 35,0 | 20,0 | 1,8 | 27,5 | 15,0 | 1,8 | 25,0 | 7,5 | 3,3 | 6,3 | 10,0 | 5,0 | 8,8 | 10,0 |
| 20 | aren4 | 205,0 | 10,0 | 20,0 | 27,5 | 0,7 | 30,0 | 21,3 | 1,4 | 40,0 | 21,3 | 1,9 | 37,5 | 22,5 | 1,7 | 32,5 | 20,0 | 1,6 | 27,5 | 15,0 | 1,8 | 22,5 | 7,5 | 3,0 | 5,0 | 10,0 | 3,8 | 7,5 | 7,5 |
| 21 | aren5 | 200,0 | 10,0 | 20,0 | 27,5 | 0,7 | 30,0 | 20,0 | 1,5 | 35,0 | 22,5 | 1,6 | 37,5 | 22,5 | 1,7 | 32,5 | 22,5 | 1,4 | 27,5 | 15,0 | 1,8 | 22,5 | 8,8 | 2,6 | 6,3 | 8,8 | 3,8 | 7,5 | 8,8 |
| 22 | aren6 | 211,0 | 12,4 | 21,9 | 32,3 | 0,7 | 30,6 | 22,1 | 1,4 | 36,6 | 23,9 | 1,5 | 38,3 | 24,5 | 1,6 | 33,8 | 21,5 | 1,6 | 28,0 | 16,7 | 1,7 | 23,9 | 9,8 | 2,4 | 5,0 | 8,7 | 4,0 | 7,0 | 8,3 |
| 23 | aren7 | 227,7 | 15,0 | 23,9 | 32,2 | 0,7 | 33,0 | 23,5 | 1,4 | 42,7 | 24,0 | 1,8 | 43,2 | 24,6 | 1,8 | 37,5 | 22,4 | 1,7 | 29,8 | 16,7 | 1,8 | 24,9 | 9,5 | 2,6 | 5,9 | 10,7 | 5,3 | 9,0 | 9,2 |
| 24 | aren8 | 213,8 | 14,0 | 20,4 | 29,3 | 0,7 | 30,2 | 21,8 | 1,4 | 36,9 | 22,7 | 1,6 | 40,0 | 24,5 | 1,6 | 36,2 | 21,1 | 1,7 | 29,9 | 15,8 | 1,9 | 23,3 | 8,5 | 2,7 | 5,8 | 8,1 | 3,3 | 6,5 | 7,3 |
| 25 | aren9 | 199,5 | 9,8 | 21,1 | 28,6 | 0,7 | 29,5 | 20,6 | 1,4 | 34,9 | 21,5 | 1,6 | 35,3 | 22,2 | 1,6 | 32,6 | 18,6 | 1,8 | 27,3 | 14,7 | 1,9 | 22,7 | 9,0 | 2,5 | 5,1 | 7,8 | 3,7 | 6,6 | 6,9 |
| 26 | aren10 | 207,9 | 15,0 | 20,1 | 30,9 | 0,7 | 31,0 | 21,9 | 1,4 | 32,9 | 24,0 | 1,4 | 40,6 | 25,6 | 1,6 | 35,6 | 22,5 | 1,6 | 28,9 | 15,7 | 1,8 | 23,9 | 8,9 | 2,7 | 5,7 | 9,9 | 4,5 | 7,4 | 8,7 |
| 27 | aren11 | 197,8 | 9,8 | 19,9 | 29,7 | 0,7 | 26,9 | 20,6 | 1,3 | 35,0 | 21,9 | 1,6 | 36,2 | 22,4 | 1,6 | 31,1 | 19,9 | 1,6 | 28,3 | 15,8 | 1,8 | 23,2 | 9,2 | 2,5 | 5,3 | 7,4 | 4,8 | 6,0 | 7,5 |
| 28 | aren12 | 205,8 | 13,5 | 19,3 | 28,7 | 0,7 | 30,0 | 21,6 | 1,4 | 34,8 | 24,0 | 1,5 | 36,6 | 24,3 | 1,5 | 33,2 | 20,6 | 1,6 | 29,5 | 16,0 | 1,8 | 23,0 | 8,8 | 2,6 | 6,0 | 8,9 | 4,8 | 8,2 | 8,9 |
| 29 | aren13 | 204,6 | 12,0 | 18,6 | 28,9 | 0,6 | 29,2 | 21,0 | 1,4 | 36,6 | 21,1 | 1,7 | 37,0 | 22,9 | 1,6 | 31,9 | 20,3 | 1,6 | 28,0 | 15,5 | 1,8 | 25,3 | 8,8 | 2,9 | 5,0 | 8,5 | 4,6 | 7,0 | 6,9 |
| 30 | aren14 | 201,5 | 9,8 | 19,0 | 28,4 | 0,7 | 29,2 | 19,6 | 1,5 | 36,1 | 20,2 | 1,8 | 38,7 | 21,8 | 1,8 | 32,4 | 19,2 | 1,7 | 27,0 | 14,0 | 1,9 | 21,9 | 8,9 | 2,5 | 6,3 | 9,2 | 4,0 | 7,4 | 8,2 |
| 31 | aren15 | 214,2 | 13,3 | 21,3 | 30,9 | 0,7 | 29,8 | 22,3 | 1,3 | 37,0 | 22,9 | 1,6 | 38,7 | 24,7 | 1,6 | 34,6 | 20,8 | 1,7 | 28,5 | 16,2 | 1,8 | 26,0 | 8,3 | 3,1 | 4,6 | 8,3 | 3,6 | 6,2 | 8,0 |
| 32 | dianth1 | 265,0 | 17,5 | 22,5 | 35,0 | 0,6 | 35,0 | 23,8 | 1,5 | 57,5 | 25,0 | 2,3 | 50,0 | 25,0 | 2,0 | 45,0 | 22,5 | 2,0 | 32,5 | 17,5 | 1,9 | 22,5 | 10,0 | 2,3 | 7,5 | 12,5 | 2,5 | 10,0 | 10,0 |
| 33 | dianth2 | 265,0 | 25,0 | 22,5 | 35,0 | 0,6 | 37,5 | 22,5 | 1,7 | 57,5 | 25,0 | 2,3 | 50,0 | 27,5 | 1,8 | 45,0 | 25,0 | 1,8 | 32,5 | 17,5 | 1,9 | 22,5 | 10,0 | 2,3 | 7,5 | 12,5 | 2,5 | 10,0 | 10,0 |
| 34 | dianth3 | 272,5 | 17,5 | 25,0 | 35,0 | 0,7 | 37,5 | 22,5 | 1,7 | 60,0 | 25,0 | 2,4 | 52,5 | 25,0 | 2,1 | 42,5 | 22,5 | 1,9 | 35,0 | 17,5 | 2,0 | 22,5 | 8,8 | 2,6 | 5,0 | 12,5 | 2,5 | 7,5 | 10,0 |
| 35 | dianth4 | 270,0 | 17,5 | 25,0 | 37,5 | 0,7 | 37,5 | 22,5 | 1,7 | 60,0 | 25,0 | 2,4 | 52,5 | 25,0 | 2,1 | 45,0 | 22,5 | 2,0 | 35,0 | 17,5 | 2,0 | 22,5 | 8,8 | 2,6 | 7,5 | 12,5 | 2,5 | 10,0 | 10,0 |
| 36 | dianth5 | 275,0 | 20,0 | 25,0 | 35,0 | 0,7 | 35,0 | 23,8 | 1,5 | 57,5 | 25,0 | 2,3 | 55,0 | 25,0 | 2,2 | 45,0 | 23,8 | 1,9 | 32,5 | 17,5 | 1,9 | 22,5 | 10,0 | 2,3 | 7,5 | 12,5 | 3,8 | 8,8 | 10,0 |
| 37 | dianth6 | 260,0 | 22,5 | 25,0 | 33,8 | 0,7 | 35,0 | 23,8 | 1,5 | 62,5 | 26,3 | 2,4 | 52,5 | 27,5 | 1,9 | 42,5 | 23,8 | 1,8 | 32,5 | 17,5 | 1,9 | 20,0 | 10,0 | 2,0 | 7,5 | 12,5 | 3,8 | 7,5 | 12,5 |
| 38 | dianth7 | 270,0 | 20,0 | 22,5 | 35,0 | 0,6 | 35,0 | 25,0 | 1,4 | 60,0 | 25,0 | 2,4 | 52,5 | 26,3 | 2,0 | 42,5 | 23,8 | 1,8 | 35,0 | 17,5 | 2,0 | 22,5 | 10,0 | 2,3 | 7,5 | 12,5 | 2,5 | 7,5 | 10,0 |
| 39 | dianth8 | 270,0 | 20,0 | 22,5 | 40,0 | 0,6 | 37,5 | 23,8 | 1,6 | 62,5 | 25,0 | 2,5 | 52,5 | 25,0 | 2,1 | 43,8 | 22,5 | 1,9 | 33,8 | 17,5 | 1,9 | 22,5 | 10,0 | 2,3 | 7,5 | 12,5 | 5,0 | 10,0 | 10,0 |
| 40 | dianth9 | 265,0 | 22,5 | 22,5 | 37,5 | 0,6 | 37,5 | 22,5 | 1,7 | 60,0 | 25,0 | 2,4 | 50,0 | 25,0 | 2,0 | 42,5 | 22,5 | 1,9 | 32,5 | 17,5 | 1,9 | 22,5 | 10,0 | 2,3 | 7,5 | 12,5 | 3,8 | 10,0 | 7,5 |

Fig. 1. A piece of data matrix comprising (in the whole) 72 metric characters of the second larval instar of eleven *Haplothrips* species

The result of the next PCA is separation of the next group of individuals belonging to *H. dianthinus* species (Axis 1 - 40.5%, Axis 2 - 18.1%; Fig. 3). The most discriminative features are short setae 9-v1 and 11-d1. There are some additional selective features (e.g. 8-v2, h-s2) but they are less significant. After elimination of *H. dianthinus* data from the matrix next two species are emerging: *H. arenarius* and *H. angusticornis* (Fig. 4) and three main characters are discriminated (setae: 8-v2, 9-d2, 9-v2). Now we can establish the value-range of selective features for examined species both with the help of data matrix and PCA graph.

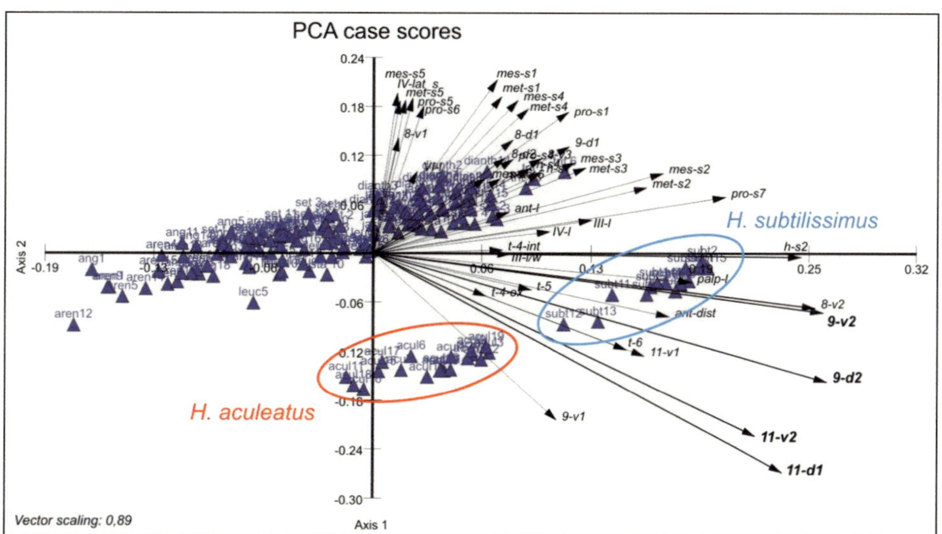


Fig. 2. PCA - scatter diagram of studied *Haplothrips* spp. specimens as OTUs along PC1 and PC2 based on 72 quantitative features (names of species and characters abbreviated)

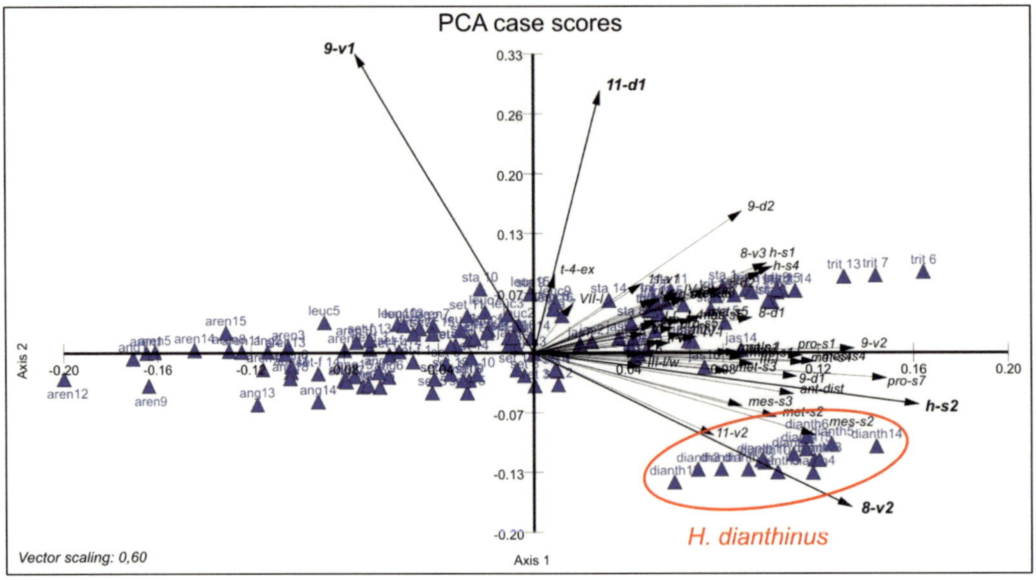


Fig. 3. PCA - scatter diagram of studied *Haplothrips* spp. specimens as OTUs along PC1 and PC2 based on 72 quantitative features (without specimens of *H. aculeatus* and *H. subtilissimus*)

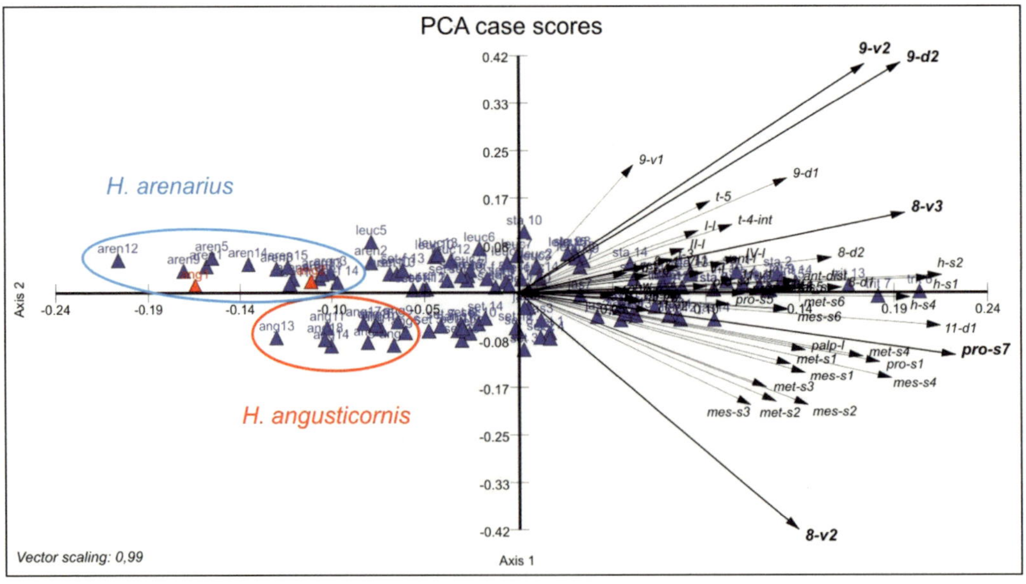


Fig. 4. PCA - scatter diagram of studied *Haplothrips* spp. specimens as OTUs along PC1 and PC2 based on 72 quantitative features (without specimens of *H. aculeatus*, *H. subtilissimus* and *H. dianthinus*)

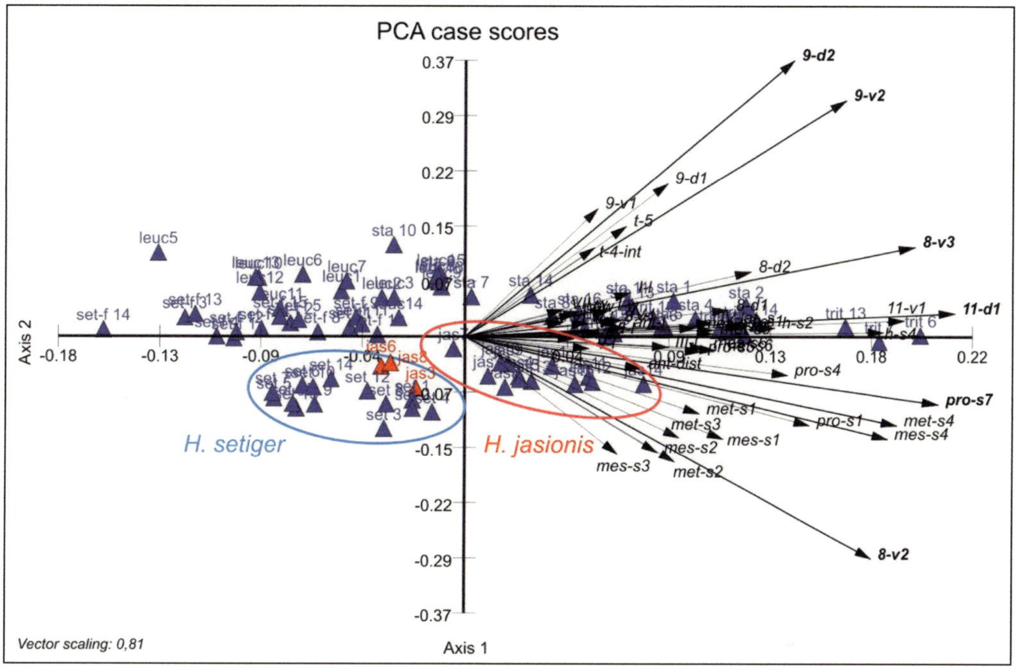


Fig. 5. PCA - scatter diagram of studied *Haplothrips* spp. specimens as OTUs along PC1 and PC2 based on 72 quantitative features (without specimens of *H. aculeatus, H. subtilissimus, H. dianthinus, H. arenarius* and *H. angusticornis*)

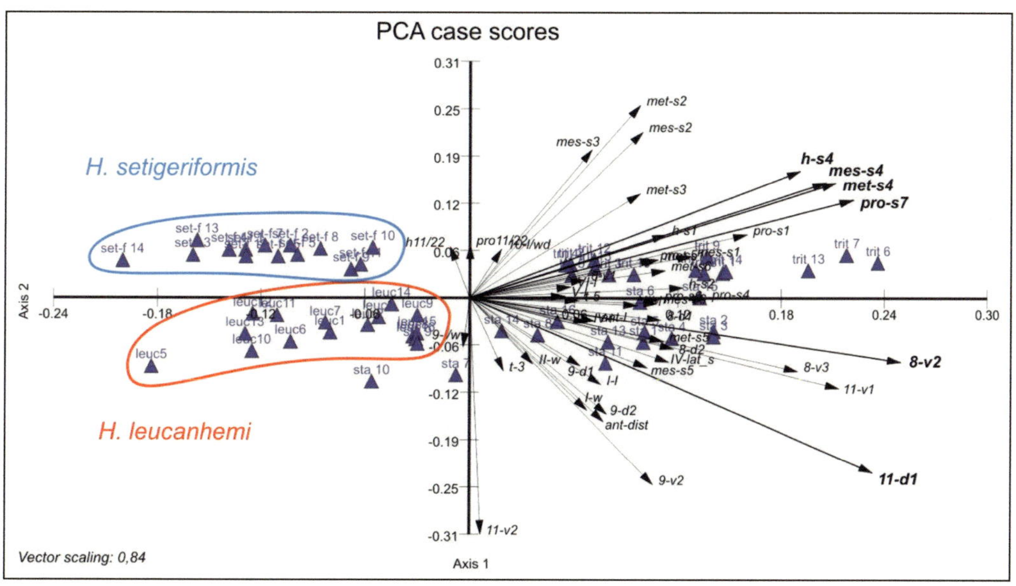


Fig. 6. PCA - scatter diagram of studied *H. setigeriformis, H. leucanthemi, H. tritici* and *H. statices* specimens as OTUs along PC1 and PC2 based on 72 quantitative features

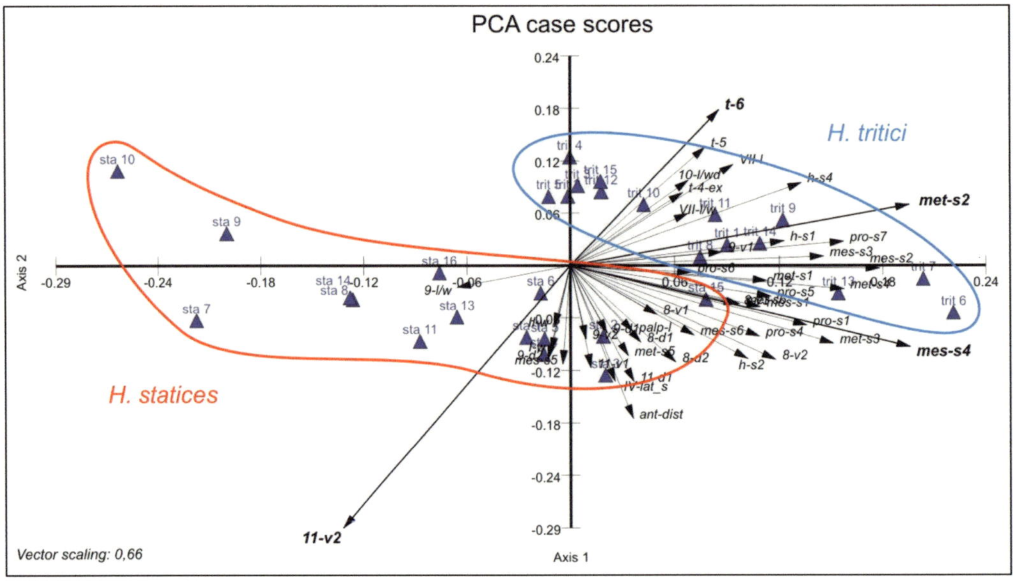


Fig. 7. PCA - scatter diagram of *H. tritici* and *H.statices* specimens as OTUs along PC1 and PC2 based on 72 quantitative features

Next steps of the analysis are similar to the previous ones (Figs. 5-7). During the analysis, it may turn out, that some specimens have ambiguous position (like "jas3, jas6, jas8" Fig. 5) and this may be a hint to check the data of the doubtful ones. At the end of the elimination, two species which are the most resembling (*H. tritici* and *H. statices*) remain. The selected distinguishing features can be used to construct an identification key. Additional benefit of presented method is elimination of characters which have no taxonomical value (in this case, out of 72 tested features there are 22 unimportant states). That would speed up further measurements of more specimens and improve the precision of determined features range.

## 2.2 Identification of the second larval instar of the *Thrips* genus species

The study on the morphology of the second larval instar of Central European *Thrips* species is another example of using PCA method in taxonomic research (Kucharczyk, 2010). In contrast to the previously discussed larvae of the genus *Haplothrips*, larvae of 34 researched species of the genus *Thrips* may be recognized mainly on the basis of qualitative characters. The data matrix covered 26 multistate discontinuous characters, amongst them the most important related to the sclerotisation and sculpture of integument, the structure of spiracles and antennae (Tab. 1). This analysis was conducted in two steps. At the beginning all researched specimens were treated as operational taxonomic units (OTUs) which have been characterized by 26 variables. On the graph OTUs form the clouds which are corresponding to the studied species. During the second step not specimens but species were treated as OTUs. In these cases the PCA was applied as a method for ordination and reducing the number of variables, the characters which had the highest loadings to PC1 and PC2 were extracted and they were regarded as the most important and useful in distinguishing the studied taxa (Fig. 8). Finally, these selected features have been used in constructing the identification key to second larval instar of studied *Thrips* species.

The characteristics tested in PCA method have been also used as variables in Claster Analysis (CA). The results in the two-dimension ordination of PCA were consistent with the results in the hierarchical clustering analysis (Fig. 9). The results of numerical analysis sheds some new light on the relationships amongst studied species. Moreover, the received dendrogram showed the similarity within studied *Thrips* species and allowed to propose the ancestral (plesiomorphic) and advanced (apomorphic) characteristics of immature stage which have not been studied hitherto.

| Feature | Abbrev. | Feature | Abbrev. |
|---|---|---|---|
| head sclerotization | HPLATE | sculpture of tergite VIII | T8SCP |
| pronotum sclerotization | PRPLATE | tergite IX - sculpture anterior to and between campaniform sensilla (Cs) | T9SCP |
| mesonotum sclerotization | MSPLATE | tergite IX - sculpture between setae level and Cs | T9DSCP |
| metanotum sclerotization | MTPLATE | tergite IX - sculpture, teeth between D1 setae | T9D1TH |
| tergite IX sclerotization | T9SCL | sculpture of tergite X | T10SCP |
| tergite X sclerotization | T10SCL | microtrichial comb at base of hind legs | COXAT |
| sternite XI sclerotization | S11SCL | spiracles on abdominal segment II | AT2SP |
| posteromarginal comb on tergite IX | T9COMB | number of facets in spiracle on abdominal segment II | NOOFF |
| length of teeth on posteromarginal comb | LCOMB | width of spiracle on abdominal segment II | WOFSP |
| sculpture of mesonotum | MSSCP | microtrichia on antennal segment III | ANS3MT |
| sculpture of metanotum | MTSCP | length of trichome on antennal segment VI | ANS6TR |
| sculpture of tergites III-VII | T3-7SC | shape of abdominal tergal setae D1, D2 | TSETA |
| sculpture of sternites III-VII | S3-7SC | furca on metasternum | FURCA |

Table 1. Features of second larval instar of *Thrips* species used in numerical analyses (after Kucharczyk, 2010)

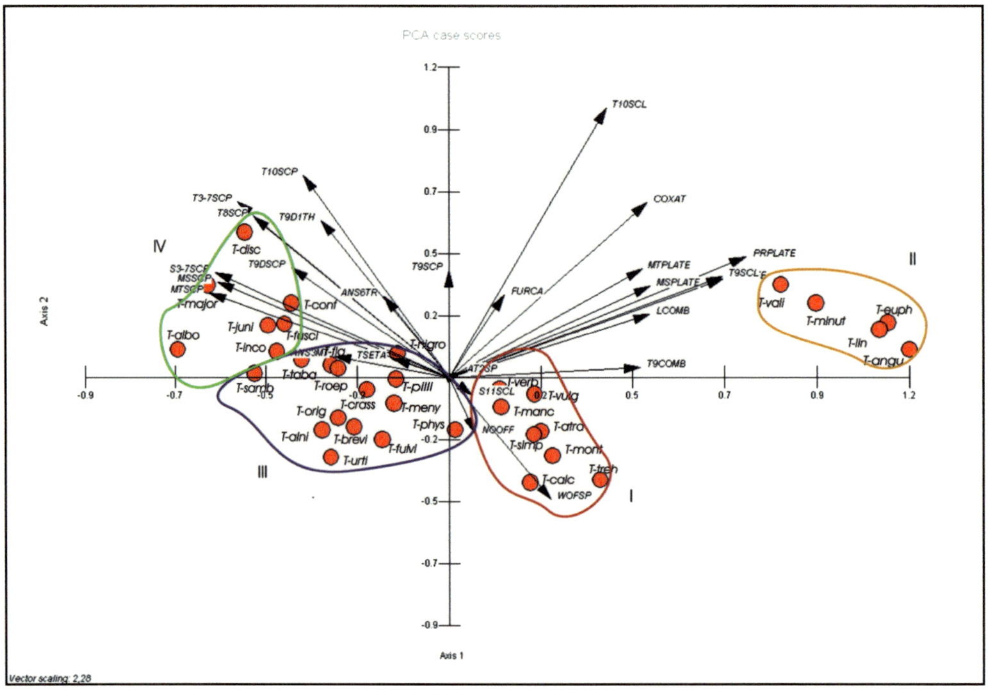


Fig. 8. PCA – scatter diagram of *Thrips* genus species as OTUs along PC1 and PC2 based on 26 qualitative features (names of species abbreviated, abbreviations of characters as in Table 1)

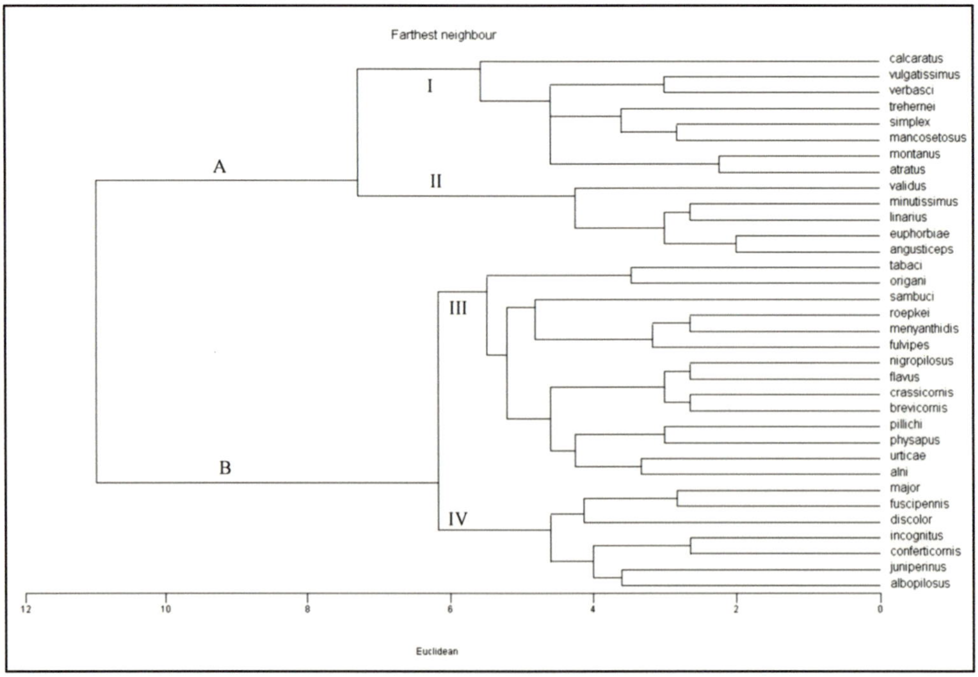


Fig. 9. Dendrogram (CA) of the similarity amongst analyzed *Thrips* species based on the morphological characters of the second larval instar (after Kucharczyk, 2010)

### 2.3 Identification of *T. atratus* Haliday 1836 and *T. montanus* Priesner 1920 species

For the first time Kucharczyk & Kucharczyk (2009) have used PCA method in the study on two similar *Thrips* species: *T. atratus* and *T. montanus*. Since the description, both of them were re-classified to *Thrips* Linnaeus 1758, *Taeniothrips* Amayot et Serville 1843 or *Similothrips* Schliephake 1972 genera (Priesner, 1964; Schliephake, Klimt, 1979; Schliephake, 2001). Zur Strassen ( 2003) replaced the latter species by *T. atratus* as its mountainous form.

During the study on the *Thrips* genus larvae the morphological differences between second larval instar of *T. atratus* and *T. montanus* were observed (Kucharczyk, 2010). These species also showed different food preferences. PCA method was applied for distinguishing the most important morphological, measurable features (8 for females and 12 for males) which may be useful in recognizing these species.

On the graphs prepared for females and males separately the specimens of both species were ordered along the first two principal components, the lengths of vectors were correlated with the features significant in recognizing the studied taxa (Figs. 10, 11, Tab. 2). Two principal components sequentially accounted for cumulatively 76.0% for females (62.7% and 13.3% for PC1 and PC2 respectively) and 63.6% for males (41.4% and 22.0% for PC1 and PC2 respectively). The quantitative characters together with the qualitative ones both of adults and larvae and the host preferences: polyphagy for *T. atratus* and monophagy (on *Rhinanthus* spp.) for *T. montanus* allowed to consider them as valid taxa.

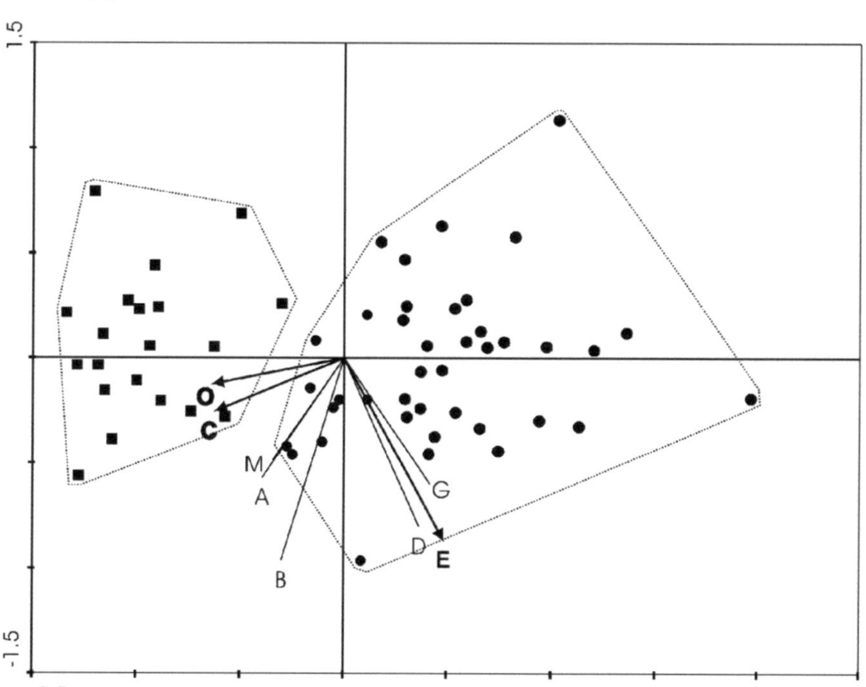


Fig. 10. PCA scatter diagram of female specimens of *Thrips atratus* (squares) and *Thrips montanus* (circles) as OTUs along PC1 and PC2 based on 8 quantitative characters (symbols as in Table 2) (after Kucharczyk & Kucharczyk, 2009)

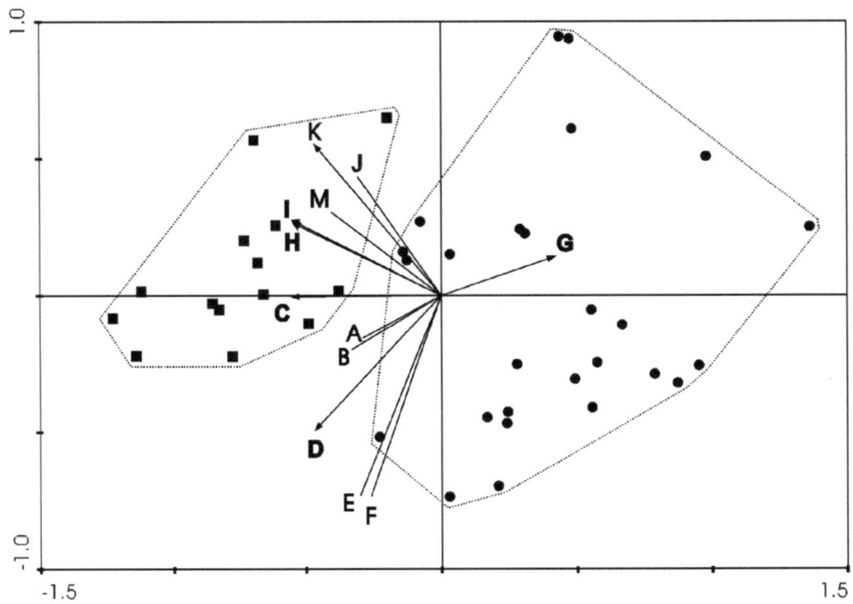


Fig. 11. PCA scatter diagram of male specimens of *Thrips atratus* (squares) and *Thrips montanus* (circles) as OTUs along PC1 and PC2 based on 12 quantitative characters (symbols as in Table 2) (after Kucharczyk & Kucharczyk, 2009)

| Features | Female | Male |
|---|---|---|
| length of pronotal posteroangular seta interna | A | A |
| length of pronotal posteroangular seta externa | B | B |
| length of pronotal posteromarginal median seta | C | C |
| number of discal setae on sternum V | D | D |
| number of discal setae on sternum VII | E | E |
| number of discal setae on sternum VIII | - | F |
| max. number of distal setae on first vine of forewing | G | G |
| width of pore plate on sternum V | - | H |
| width of pore plate on sternum VII | - | I |
| length of antennal segment VII | - | J |
| length of antennal segment VIII | - | K |
| length of antennal segment III | M | M |
| max. length of median microtrichia on tergum VIII | O | - |

Table 2. Features of *Thrips atratus* and *Thrips montanus* adults used in quantitative analyses (after Kucharczyk & Kucharczyk 2009)

### 2.4 Identification of *Thrips fuscipennis* Haliday 1836 and *T. sambuci* Heeger 1854 species

Similar problem exist in distinguishing *Thrips fuscipennis* Haliday 1836 and *T. sambuci* Heeger 1854. The former is a polyphagous species most often feeding in flowers while the latter is a monophagous insect feeding, breeding and developing on the lower side of *Sambucus* spp. leaves. In spite of both of the species being recognized as valid, their adults may be distinguished mainly by differences in color of antennal segments (Schliephake & Klimt, 1979; zur Strassen, 2003). Due to the fact that the color characters are very variable in specimens originated from different populations and stations, it is not possible to accurately identify species by using only their color features (Strassen zur, 1997; Mound & Minaei, 2010).

The aim of this task was to find new characters and test their usefulness in distinguishing these taxa. Fedor et al. (2008) proposed to use an artificial neural network method (ANN) for identifying species. This method was successfully applied according to 18 species of four genera: *Aeolothrips, Chirothrips, Dendrothrips* and *Limothrips*. Finally the authors selected 19 morphometric features which have been used in ANN analysis. In the current study seven of them were selected to distinguish *T. fuscipennis* and *T. sambuci*. Additionally five new quantitative features present in both sexes and three typical for males, and one qualitative feature were proposed for using in comparative study (Tab. 3).

| Features | Female | Male |
|---|---|---|
| head width | H-w | H-w |
| head length (dorsal) | H-l-d | H-l-d |
| head length (ventral) | H-l-v | H-l-v |
| eye length | E-l | E-l |
| ovipositor length | Ov-l | - |
| antenal s. V length | A-V-l | A-V-l |
| antenal s. VI length | A-VI-l | A-VI-l |
| distance between anterior and posterior ocelli | D-oc | D-oc |
| distance between CS - metanotum | D-Cs-mt | D-Cs-mt |
| distance between D1 - metanotum | D-D1-mt | D-D1-mt |
| length of posteroangular seta interna | L-p-s-int | L-p-s-int |
| length of posteroangular seta externa | L-p-s-ext | L-p-s-ext |
| number of campaniform sensillae - mesonotum | N-cs-ms | N-cs-ms |
| distance between setae D1 and fore edge of metanotum | D-D1-e-mt | D-D1-e-mt |
| width of area porosae on sternum V | - | W-ap-sV |
| width of area porosae on sternum VI | - | W-ap-sVI |
| width of area porosae on sternum VII | - | W-ap-sVII |

Table 3. Features of *Thrips fuscipennis* and *Thrips sambuci* adults used in quantitative analyses

Similarly as in *T. atratus* and *T. montanus* characters of females and males were analyzed separately. On the graphs being the result of PCA method the specimens of both species are well separated and are located on opposite sides of Axis 2 (Figs 12, 13). The number of campaniform sensilla on the mesonotum – two (sporadically one) in *T. fuscipennis* and lack of them in *T. sambuci* is the characteristic which in the highest degree differentiate these species. Additionally the measured setae are shorter in specimens of the former species. Males of the latter one are characterized by narrower area porosae on abdominal sternites V and VI and very often lack of them on sternite VII. Moreover, this analysis shows low significance such characteristics in recognizing these species as: eye length, distance between ocelli and length of antennal segments. The characteristics mentioned above were used in ANN analysis, the results obtained with PCA method tend to reflect on their usefulness in further similar studies, particularly on *Thrips* species identification.

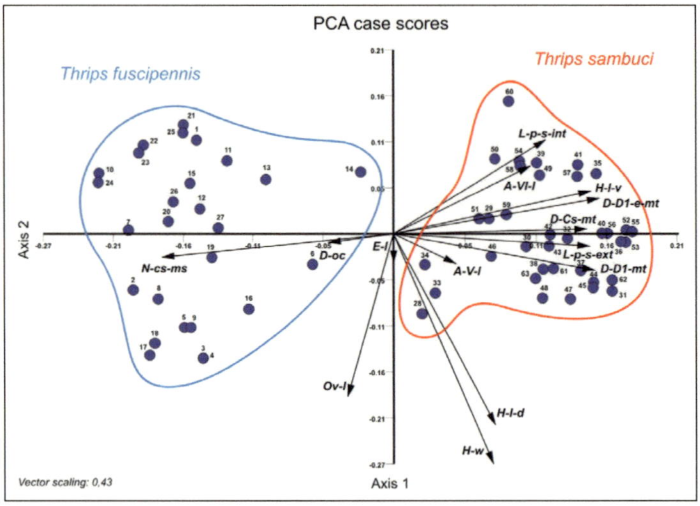


Fig. 12. PCA scatter diagram of female specimens of *Thrips fuscipennis* and *Thrips sambuci* as OTUs along PC1 and PC2 based on 14 quantitative characters (abbreviations as in Table 3)

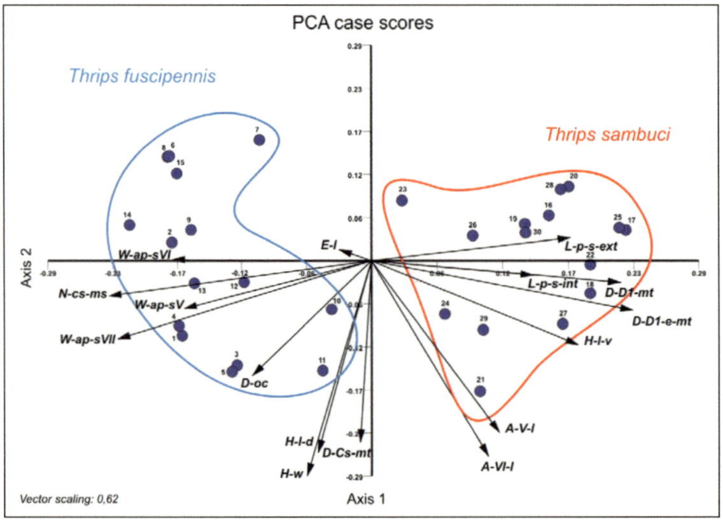


Fig. 13. PCA scatter diagram of male specimens of *Thrips fuscipennis* and *Thrips sambuci* as OTUs along PC1 and PC2 based on 16 quantitative characters (abbreviations as in Table 3)

## 3. Conclusion

The above mentioned examples show that PCA method is a valuable tool in identifying species. Its' application allows to select morphometric and qualitative characteristics which discriminate taxa and may be useful in construction of identification keys. This method also allows to verify the significance of some characteristics in taxonomic study and select these most relevant from large set of data.

The results obtained using the numerical taxonomy methods could objectively reflect the taxonomic position of studied taxa.

## 4. Acknowledgements

Financial support for this project was partially provided by the Slovak Science Foundation VEGA 1/0137/11.

## 5. References

Apuan D.A.; Torres M.A.J. & Demayo C. G. (2010). Describing variations and taxonomic status of earthworms collected from selected areas in Misamis Oriental, Philippines using principal component and parsimony analysis. *Egyptian Academic Journal of Biological Sciences*, B Zoology, Vol. 2, No. 1, pp. 27-36, ISSN 2090-0759

Bhatti J.S. (1988). The orders Terebrantia and Tubulifera of the superorder Thysanoptera (Insecta). A Critica Appraisal, *Zoology (Journal of Pure and Applied Zoology)*, Vol. 1: pp. 167-240, ISSN 0970-1516

Bhatti J.S. (2005). Fifteen new families in the Order Terebrantia (Insecta). *Thysanoptera*, Vol. 1, pp. 27-49.

Bhatti J.S. (2006). The classification of Terebrantia (Insecta) into families. *Oriental Insects*, Vol. 40, pp. 339-375, ISSN 0030-5316

Chiapella J. (2000). The Deschampsia cespitosa complex in central and northern Europe: morphological analysis. *Botanical Journal of the Linnean Society*, Vol.134. No. 4, (August 2006), pp. 495-512, ISSN 1095-8339

Crespi, B.J.; Morris, D.C. & Mound, L.A. (2004). *Evolution of Ecological and Behavioural Diversity: Australian Acacia Thrips as Model Organisms*. Australian Biological Resources Study & Australian National Insect Collection. ISBN: 9780975020616, Canberra, Australia

Fedor P.; Malenovský J.; Vaňhara J.; Sierka W. & Havel J. (2008). Thrips (Thysanoptera) identification using artificial neural networks. *Bulletin of Entomological Research*, Vol. 98, No.5, (April 2008), pp. 437-447, ISSN 0007-4853

Hotelling, H. (1933). Analysis of a complex of statistical variable into principal components. *Journal of Educational Psychology*, Vol. 24, No. 6, (September 1933), pp. 417-441, ISSN 0022-0663

Kovach Computing Services (2005). *Multi-Variate Statistical Package Plus, Version 3.1*. Kovach Computing Services, Pentraeth, Wales, U.K.: 137 pp.

Kucharczyk H. (2004). Larvae of the Genus Thrips - Morphological Features in Taxonomy. *Acta Phytopathologica et Entomologica Hungarica*, Vol. 39 No. 1-3, (May 2004), pp. 211-219, ISSN 0238-1249

Kucharczyk H. (2010). *Comparative morphology of the second larval instar of the Thrips genus species (Thysanoptera: Thripidae) occurring in Poland*. Mantis Publishing Company, ISBN 978-83-929997-7-5, Olsztyn, Poland

Kucharczyk H. & Kucharczyk M. (2009). *Thrips atratus* HALIDAY, 1836 and *Thrips montanus* PRIESNER, 1920 (Thysanoptera: Thripidae) - one or two species? Comparative morphological studies. *Acta Zoologica Academiae Scientiarum Hungaricae*, Vol. 55, No. 4, (November 2009), pp. 349-364, ISSN 1217-8837

Lilburn T.G. & Garrity G.M. (2004). Exploring prokaryotic taxonomy. *International Journal of Systematic and Evolutionary Microbiology*, Vol. 54, No. 1, (January 2004), pp. 7-13, ISSN 1466-5026

Lewis T. (Ed.). (1997). *Thrips as Crop Pests*. CAB International, ISBN 0 85199 178 5, Wallingford, UK

Mound L.,A.; Heming C., S. & Palmer J., M., (1980). Phylogenetic relationships between the families of recent Thysanoptera (Insecta). *Zoological Journal of the Linnean Society*, Vol. 69, No. 2, (June 1980), pp. 111-141, ISSN 1096-3642

Mound L.A. & Morris D.C. (2004). Thysanoptera Phylogeny - the Morphological Background. *Acta Phytopathologica et Entomologica Hungarica*, Vol. 39, No. 1-3, (May 2004), pp. 101-113, ISSN 0238-1249

Mound L.A. & Morris D.C. (2007). The insect order Thysanoptera: Classification versus Systematics. *Zootaxa*, Vol. 1668, (December 2007), pp. 395-411, ISSN 1175-5326

Mound L.A. (2010). Classifying thrips into families. http://anic.ento.csiro.au/thrips/identifying_thrips/classification.html [accessed October 2011].

Mound, L. A., Minaei, K. (2007). Australian insects of the Haplothrips lineage (Thysanoptera - Phlaeothripinae). *Journal of Natural History*, Vol. 41, pp. 2919-2978, ISSN 0022-2933

Mound, L.A. & Minaei, K. (2010). Taxonomic problems in character state interpretation: variation in the wheat thrips *Haplothrips tritici* (Kurdjumov) (Thysanoptera, Phlaeothripidae) in Iran. *Deutsche Entomologische Zeitschrift*, Vol. 57, No. 2, (November 2010), pp. 233-241, ISSN 1435-1951

Pearson, K. (1901). On lines and planes of closest fit to systems of points in space. *Philosophical Magazine*, Vol. 2, pp. 559-572, ISSN 1478-6443 http://pbil.univ-lyon1.fr/R/pearson1901.pdf [accessed October 2011].

Pitkin, B. R. (1976). A revision of the Indian species of *Haplothrips* and related genera (Thysanoptera: Phlaeothripidae). *Bulletin of the British Museum (Natural History), Entomology series*, Vol. 34, pp. 223-280, ISSN 0524-6431

Sahuquillo E.; Fraga M.I. & Martinez Cortizas A. (1997). Comparative study of classical and numerical taxonomic methods for infraspecific taxa of *Triticum aestivum* L. traditionally cultivated in Galicia (NW Spain), pp. 919-926. Valdes B. & Pastor J. (Eds.), *Proceedings of the VIII optima meeting*. Sevilla 25 Sep.-1 Oct. 1995.

Schliephake G. (2001). Verzeichnis der Thysanoptera (Fransenflügler) - Physopoda (Blasenfüße). [In] *Entomofauna Germanica* Klausnitzer B. (Ed.).Vol. 5, pp. 91-106, ISSN 0232-5535

Schliephake G. & Klimt K. (1979). *Thysanoptera, Fransenflügler*. Die Tierwelt Deutschlands, Vol. 66. Veb Gustav Fischer Verlag Jena, Germany

Strassen zur R. (1997). How to classify the species of the genus *Thrips* (Thysanoptera)? *Folia Entomologica Hungarica*, Vol. 58, pp. 227–235, ISSN 0373-9465

Strassen zur R. (2003). *Die terebranten Thysanopteren Europas und des Mittelmeer-Gebietes*. [In:] Die Tierwelt Deutschlands Vol. 74 (Goecke & Evers eds). ISBN 3-931374-58-0, Keltern, Germany

Tommasini M. G. & Maini S. (1995). Frankliniella occidentalis and other thrips harmful to vegetable and ornamental crops in Europe, In: *Biological control of thrips pests* Loomans A., J.; M., van Lenteren J.,C.; Tomasini M., G.; Maini S. & Riudavets J. (Eds.), pp. 1-42, Wageningen Agricultural University Papers, 95.1, ISBN 90-6754-395-0, Wageningen, The Netherlands

Wolff R. L.; Comps B.; Marpeau A. M. & Deluc L.G. (1997). Taxonomy of *Pinus* species based on the seed oil fatty acid compositions. *Trees – Structure and Function*, Vol. 12, No. 2, (December 1977), pp. 113-118, ISSN 0931-1890

Véla E.; Tirard A.; Rinucci M.; Suesh C.M. & Provost E. (2007). Floral Chemical Signatures in the Genus *Ophrys* L. (Orchidaceae): A Preliminary Test of a New Tool for Taxonomy and Evolution. *Plant Molecular Biology Reporter*, Vol. 25, No. 3-4, (December 2007), pp. 83-97, ISSN 0735-9640

Zawirska I. & Wałkowski W. (2000). Fauna and importance of Thrips (Thysanoptera) for rye and winter wheat in Poland. *Journal of Plant Protection Research*, Vol. 40, No. 1, pp. 35–55, ISSN 1427-4345

# 8

# PCA – A Powerful Method for Analyze Ecological Niches

Franc Janžekovič and Tone Novak
*University of Maribor, Faculty of Natural Sciences and Mathematics, Department of Biology, Maribor Slovenia*

## 1. Introduction

Principal Component Analysis, PCA, is a multivariate statistical technique that uses orthogonal transformation to convert a set of correlated variables into a set of orthogonal, uncorrelated axes called principal components (James & McCulloch 1990; Robertson et al., 2001; Legendre & Legendre 1998; Gotelli & Ellison 2004). Ecologists are most frequently dealing with multivariate datasets. This is especially true in field ecology, and this is why PCA is an attractive and frequently used method of data ordination in ecology. PCA enables condensation of data on a multivariate phenomenon into its main, representative features by projection of the data into a two-dimensional presentation. The two created resource axes are independent, and although they reduce the number of dimensions–i.e. the original data complexity–they maintain much of the original relationship between the variables: i.e., information or explained variance (Litvak & Hansell 1990). This is helpful in focusing attention on the main characteristics of the phenomenon under study. It is convenient that, if the first few principal components (PCs) explain a high percentage of variance, environmental variables that are not correlated with the first few PCs can be disregarded in the analysis (Toepfer et al., 1998). In addition, applying PCA has become relatively user-friendly because of the numerous programs that assist in carrying out the computational procedure with ease (Dolédec et al., 2000; Guisan & Zimmerman 2000; Robertson et al., 2001; Rissler & Apodaca 2007; Marmion et al., 2009).

PCA has been widely used in various fields of investigation and for different tasks. Many authors have used PCA for its main purpose: i.e., to reduce strongly correlated data groups or layers. These studies concern either environmental variation (e.g., Kelt et al., 1999; Johnson et al., 2006; Rissler & Apodaca 2007; Glor & Warren 2010; Novak et al., 2010a; Faucon et al., 2011; Grenouillet et al., 2011), the investigated species or community characteristics (e.g., Kingston et al., 2000; Pearman 2002; Youlatos 2004; Kitahara & Fujii 2005), or both, sometimes in combination with detrended correspondence analysis, DCA, canonical correspondence analysis, CCA, and other ordination methods (e.g., Warner et al., 2007; González-Cabello & Bellwood 2009; Marmion et al., 2009; Mezger & Pfeiffer 2011). The application of PCA has helped in various fields of ecological research, e.g., in determination of enterotypes of the human gut microbiome on the basis of specialization of their trophic niches (Arumugam et al. 2011). In aquatic habitat studies, it has been applied for evaluation of aquatic habitat suitability, its regionalization, analysis of fish abundance, their seasonal and spatial variation, lake ecosystem organization change etc. (Ahmadi-Nedushan et al., 2006; Blanck et al., 2007; Catalan et al., 2009). However, it has been often applied in analyzing farming system changes (Amanor & Pabi 2007).

In many cases, PCA has been used as a source or supporting analysis in the performance of more complex analysis, such as the study of adaptive fish radiation, strongly influenced by trophic niches and water depth (Clabaut et al., 2007), predicting the potential spatial extent of species invasion (Broennimann et al., 2007) and multi-trait analysis of intra- and interspecific variability of plant traits (Albert et al., 2010). Chamaillé et al. (2010) performed PCA and Hierarchical Ascendant Classification to evaluate environmental data, on the one hand, and human and dog population density data, on the other, in order to detect possible ranking of regions differently threatened by leishmaniasis.

Niche differentiation and partitioning is an ecological issue where PCA is frequently used. It enables efficient differentiation among related parapatric species (Dennis & Hellberg 2010). To access the problem authors use various available input data, which may be other than direct measurements of the niche. Since body shape and composition can readily be related to adaptation to the environment, morphometry figures as an adequate surrogate approach for studying the niche. Morphometric characteristics represent a data set vitable for evaluating the organism−environment relationship; besides PCA, Lecomte & Dodson (2005) additionally used discriminant analysis for this purpose. Inward et al. (2011) applied PCA to determine the morphological space of dung beetles representing regional faunas. Claude et al. (2003) demonstrated, using the case of turtles, that geometric morphometry, evaluated with PCA, can help to analyze the evolution of convergence. Morphometric differences between related species can easily refer to niche partitioning, reflecting differences in spatial or trophic level (Catalan et al. 2009; Niet-Castañda & Jiménez-Jiménez 2009; Novak et al., 2010b).

Hypogean habitats such as caves and artificial tunnels are relatively simple habitats owing to their low diversity, low production, and the constancy of their environmental factors (Culver 2005); they are thus suitable for investigating an environmental niche in situ. In this contribution we demonstrate the use of PCA in exploring an ecological niche in two case studies from caves in Slovenia. In the study of the three most abundant hymenopteran species that settled in the caves for rest, we only could evaluate their spatial niches on the basis of the usually measured environmental parameters. PCA was applied in two levels: 1. In the exploratory data analysis it was used as an efficacious tool to reduce the parameters into two principal components, PCs. 2. In the test hypothesis, the PCs of all three species were subjected to variance analysis to detect differences between the spatial niches of the three species.

### 1.1 Ecological niche concept

Ecological or environmental niche is one of the most useful concepts for exploring how and where organisms live and how are they related to their environment. After its introduction (Grinnel 1917), this concept changed considerably as new knowledge about the habitat and functioning of the organisms within was acquired. There are three main views in the evolution of the concept. The first view is that niche equals habitat, which is a multidimensional presentation of conditions in the local physical place where an organism lives. The second and most frequently applied understanding is the functioning of an organism within its concrete environment, which concerns acting on and responding to the organism's physical environment as well as to other organisms, originally, within a community. Since the community concept is in the course of radical change (Ricklefs 2008), it is convenient to replace the term community with a more general one, an assemblage. In practice, habitat and the function of an organism are often discussed as spatial, temporal and trophic niches. The third view is that the niche refers to variables within the whole range of the distribution area of an organism, which provides much information about organism–environment relationships on the global level (Soberón 2007).

The century-old niche concept has had many peaks and falls in the history. Since the initial idea was generated, it has evolved much over the years, being forgotten or neglected and/or misinterpreted, and recovering in different ways (Collwell & Rangel 2009). In the last decade, an intensive debate has taken place about redefining the meaning, importance and suitability of different aspects of the niche, including the measuring methods. This development accurately reflects the importance of its resurrection for the progress of ecological, evolutionary and related investigations. Today there are numerous niche concepts (Chase & Leibold 2003). Grinnell (1917) conceptualized the idea on a case study of a bird, the California Thrasher. He wrote that the niche comprehends the various circumstances to which a species is adapted by its constitution and way of living. He also wrote that no two species in a single fauna have precisely the same niche relationship, a fact which indicates the different roles of species within a community. What Grinnell called a niche was later understood for a long time as habitat, i.e., the sort of place where an organism lives. By a niche, Elton (1927) meant the place of an (animal) organism in its community, its relation to food and enemies and to some extant to other factors. He stressed the importance of the food (trophic) dimension of the ecological niche. Later (Elton 1933) he denoted a niche as a species mode of life, as for instance, professions in a human community. Instead of environments, Hutchinson (1957) attributed niches to species (Collwell & Rangel 2009), and explained the niche as part of an abstract multidimensional space, the ecospace, representing the whole range of intracommunity variables and interactions. Within this space, the way of life of a species is balanced by its tolerances and requirements. He called the overall potential niche the fundamental niche, in contrast to the realized niche, which is narrower for the negative impact of competitors and predators. The niche concept evolved additionally through discussion of several points of the competitive exclusion principle: i.e., the assumption that two species with identical environmental requirements cannot coexist indefinitely in the same location. According to Hardin (1960), the niche dimensions are represented by different abiotic and biotic variables concerning a species, such as its life history, habitat, position in the food chain, and its geographic range. Whittaker & Levin (1975) understood the niche as a species' requirements and its position in relation to other species in a given community. Recently, the importance of studying the niche to improve our understanding of the functioning of species and whole ecosystems has again become a widely discussed topic in global ecology. Pullian (2000) showed that, besides competition, other factors, such as niche width, habitat availability, dispersal, etc. influence the observed relationship between species distribution and the availability of suitable habitat, and should thus be incorporated into Hutchinson's niche concept. Soberón (2007) justified the separation of niches into two–the Grinnellian and the Eltonian class–on the basis of their focuses. Hutchinson's (1957) consideration can be applied to both groups. The Grinnellian class of niches is based on consideration of their non-interactive variables, such as average temperature, precipitation and solar radiation, and environmental conditions on a broad scale. These variables are relevant to understanding coarse-scale ecological and geographic properties of species. The Eltonian class niches, in contrast, focus on bionomic variables, such as biotic interactions and resource–consumer dynamics, which can be measured principally on a local scale. Whereas datasets of variables of the Grinnellian niche group have been rapidly compiled in the World, very little theory has been developed explicitly about this. On the other hand, variables for considering much more dynamic and complex Eltonian niches have never been available (Soberón 2007). Both classes of niches are relevant to understanding the distribution of individuals of a species, but the Eltonian class is easier to measure at the high spatial resolutions characteristic of most ecological studies, whereas the Grinnellian class is suited to the low spatial resolution at which distributions are typically defined (Soberón 2007). Applying the modelling of species distribution to the distribution constraints is strongly encouraged to provide better insight in species distributions (Kearney & Porter 2009; Bellier et al., 2010). It is important to understand that a niche is not a conservative concept, but a consequence of the complexity of the subject, which may refer to very different features of the fundamental niche, with different ecological and evolutionary properties (Soberón & Nakamura 2009). It has been demonstrated that, on the one hand, inconsistent adaptive pressures may give rise to a whole palette of niche diversification (e.g., Romero 2011), while, on the other hand, convergent evolution in various combinations takes place within the multidimensional niche space (e.g., Hormon et al., 2005).

### 1.2 Ordination and the PCA concept

Ordination is a method in multivariate analysis used in exploratory data analysis. Exploratory data analysis is an approach to analyzing data sets to summarize their main characteristics in an easy-to-understand form, often in graphs. In this procedure no statistical modelling is used. The order of objects in ordinations is characterized by values of multiple variables. Similar objects are ordinated near each other and vice versa. Many ordination techniques exist, including principal components analysis (PCA), non-metric multidimensional scaling (NMDS), correspondence analysis (CA) and its derivatives, like detrended CA (DCA), canonical CA (CCA), Bray–Curtis ordination, and redundancy analysis (RDA), among others (Legendre & Legendre 1998; Gotelli & Ellison 2004).

PCA is widely useful in considering species; it is appropriate for the analysis of community composition data or as gradient analysis. Gradient analysis is an analytical method used in plant community ecology to relate the abundance of various species within a plant community to various environmental gradients by ordination or by weighted averaging. These gradients are usually important in plant species distribution, and include temperature, water availability, light, and soil nutrients, or their closely correlated surrogates (Lepš & Šmilauer 2003).

## 2.1 Environmental niche of three hymenopteran and two spider species

Between 1977 and 2004, 63 caves and artificial tunnels were ecologically investigated in Slovenia; the three most abundant Hymenoptera species found in these studies have been ecologically evaluated (details in Novak et al. 2010a). In the caves, many environmental data were collected, as follows. The following abbreviations of the environmental variables are used: Dist-E = distance from entrance; Dist-S = distance from surface; Illum = illumination; PCS = passage cross-section; Tair =air temperature; RH = relative air humidity; Tgr = ground temperature; HY = substrate moisture. The hymenopteran spatial niche breadth was originally represented by nine variables. The variation was subjected to PCA, and differences in niche overlap were tested using One-way ANOVA. In the following, we demonstrate the analysis of occupied physical space in the three species: *Amblyteles armatorius*, n=16, *Diphyus quadripunctorius*, n=42, and *Exallonyx longicornis*, n=44. These variables refer to the environmental conditions for the individual placements within the caves.

PCA requires normal data distribution. This is often not the case with the environmental data provided by field investigations, as in our case. In variables presented as proportions or ratios, e.g., humidity, this problem can be overcome with the arc-sin transformation. In those variables stretched over a large scale of values, e.g., illumination and passage cross section, this can be achieved by transformation in the logarithmic scale. In our study, we used the Kolmogorov-Smirnov test, K-S, to check the data for normality. To normalize distribution, we transformed air humidity and substrate moisture data (*arcsin*) (Fig. 1), and passage cross section and illumination data (*log*) (Fig. 2). PCA is sensitive to the relative scaling of the original variables. We therefore *z*-standardized the data. Here we demonstrate relations between nine environmental variables with Pearson correlation coefficients (Table 1).

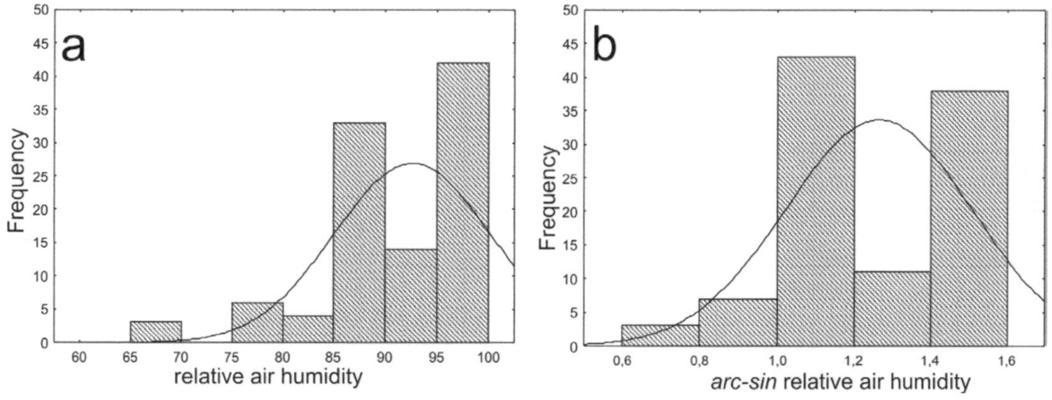


Fig. 1. Distribution of row relative air humidity data (a) and the data after *arc-sin* transformation (b) with normal distribution curve.

To obtain detailed information on the pattern of variation, the sets of nine environmental variables were subjected to PCA. In this way, we obtained nine PCs. These new values are called principal component scores. The Eigenvalue and ratios of explained variances are presented in Table 2, where PC variance is in progressive decline. The last four components represent such a small ratio of the total variance that it is reasonable to ask whether they describe any biotic response or not. A common rule is to interpret only those components that contribute more than 5% of the total variance. In this study case on Hymenoptera, PCs1 to PCs5 meet this criterion in the total account of 92.5% of the variance explained, while 7.5% of the variance remains unexplained. The explained contribution of variances to the total variance is shown in a scree plot (Fig. 3). The large differences between the variances of the first three PCs and much smaller ones of the other scores are clearly evident.

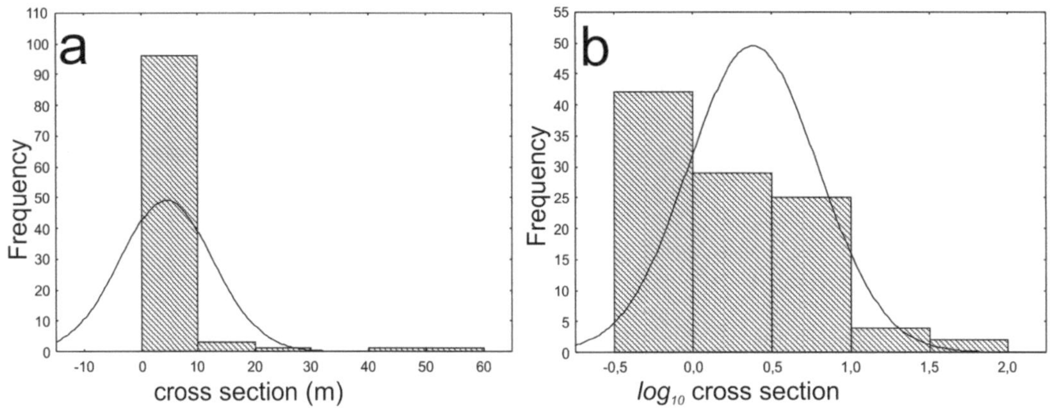


Fig. 2. Distribution of row cross section data (a) and the data after logarithmic transformation (b) with normal distribution curve.

| | 1 | 2 | 3 | 4 | 5 | 6 | 7 | 8 | 9 |
|---|---|---|---|---|---|---|---|---|---|
| 1 Air temperature | 1.00<br>--- | | | | | | | | |
| 2 *arc-sin* relative air humidity | 0.15<br>0.133 | 1.00<br>--- | | | | | | | |
| 3 Ground temperature | **0.94<br><0.001** | 0.18<br>0.079 | 1.00<br>--- | | | | | | |
| 4 *arc-sin* substrate moisture | **0.388<br><0.001** | **0.59<br><0.001** | **0.37<br><0.001** | 1.00<br>--- | | | | | |
| 5 Airflow | **-0.48<br><0.001** | **-0.36<br><0.001** | **-0.43<br><0.001** | **-0.55<br><0.001** | 1.00<br>--- | | | | |
| 6 Distance from entrance | **-0.34<br><0.001** | 0.14<br>0.153 | **-0.41<br><0.001** | 0.10<br>0.312 | 0.04<br>0.712 | 1.00<br>--- | | | |
| 7 Distance from surface | -0.02<br>0.837 | **0.24<br>0.017** | -0.04<br>0.683 | **0.46<br><0.001** | -0.11<br>0.275 | **0.67<br><0.001** | 1.00<br>--- | | |
| 8 Passage cross-section | **0.35<br><0.001** | 0.17<br>0.089 | **0.23<br>0.025** | **0.39<br><0.001** | **-0.40<br><0.001** | -0.11<br>0.274 | 0.05<br>0.656 | 1.00<br>--- | |
| 9 *log* illumination | **0.45<br><0.001** | -0.18<br>0.077 | **0.46<br><0.001** | -0.04<br>0.690 | -0.07<br>0.494 | **-0.821<br><0.001** | **-0.679<br><0.001** | **0.37<br><0.001** | 1.00<br>--- |

Table 1. Pearson correlations coefficient among nine environmental variables. Significant correlations in bold. (Upper row r, lower row p).

| PC | Eigenvalue | % Total | Cumulative Eigenvalue | Cumulative % |
|---|---|---|---|---|
| 1 | 3.38 | 37.61 | 3.38 | 37.61 |
| 2 | 2.65 | 29.48 | 6.04 | 67.09 |
| 3 | 0.94 | 10.44 | 6.98 | 77.52 |
| 4 | 0.80 | 8.94 | 7.78 | 86.47 |
| 5 | 0.54 | 6.05 | 8.33 | 92.52 |
| 6 | 0.37 | 4.06 | 8.69 | 96.58 |
| 7 | 0.18 | 1.95 | 8.87 | 98.53 |
| 8 | 0.09 | 1.03 | 8.96 | 99.55 |
| 9 | 0.04 | 0.45 | 9.00 | 100.00 |

Table 2. Eigenvalues and percentages of explained variability.

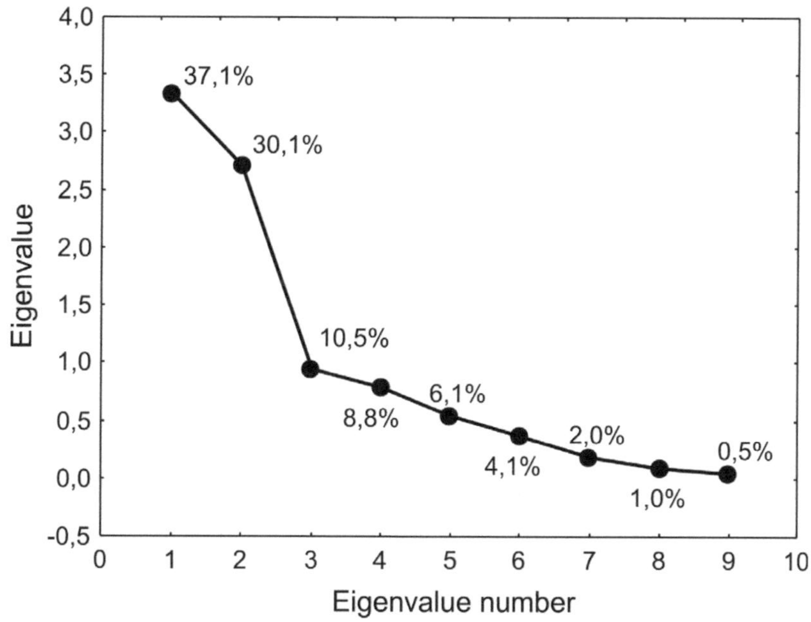


Fig. 3. Scree plot of the eigenvalue and the percent of variance explained by each component is shown in decreasing order.

Projection of the variables on the factor plane revealed that the 1st and the 2nd axes of the PCs explained 37.6% and 29.5% of the total variance. The Pearson correlation coefficients and elementary graphics associated with the relations between PCs and environmental variables are presented in Table 3 and Fig 4, respectively. In this graphic presentation, they are placed on a circle, called the correlation circle, with the pair of factor axes as its axes. The stronger the correlation between a variable and the factor, the greater the correlation of the corresponding variable with the factor axes. The variables that are correlated with a particular factor can thus be identified, thereby providing information as to which variables can explain the given factor. This is demonstrated in Fig.5. PC1 best explains the variability of air and ground temperature, and illumination: these values increase with the decreasing PC1, while the values of airflow and distance from the entrance increase with the increasing

PC1. PC2 best explains the variability of air humidity, substrate moisture and both distance from the entrance and from the surface: these values increase with the increasing PC2. In the case presented the explanatory power of PCA with respect to variable importance is evident. PCs thus fully represent adequate surrogates to explain the spatial component of the niche. For the interpretation of these outputs, one needs good biological and ecological knowledge about the organisms under study. The projections of the environmental dimensions of the three species are represented by polygons in Fig. 5. A more elaborate figure has been published elsewhere (Novak et al. 2010a).

| Parameter \ PC | PC1 | PC2 |
|---|---|---|
| Air temperature | -0.87 | -0.10 |
| *arc-sin* relative air humidity | -0.29 | -0.60 |
| Ground temperature | -0.85 | -0.06 |
| *arc-sin* substrate moisture | -0.49 | -0.72 |
| Airflow | 0.62 | 0.45 |
| Distance from entrance | 0.60 | -0.67 |
| Distance from surface | 0.26 | -0.82 |
| *log* passage cross-section | -0.55 | -0.20 |
| *log* illumination | -0.69 | 0.63 |
| $F_{2,99}$ ; p | 19.85; <0.001 | 9.19; <0.001 |

Table 3. Pearson correlation coefficient between environmental variables and the first two Principal Components (PCs) F and p values of one-way ANOVA in testing the first two PCs according to the three hymenopteran species.

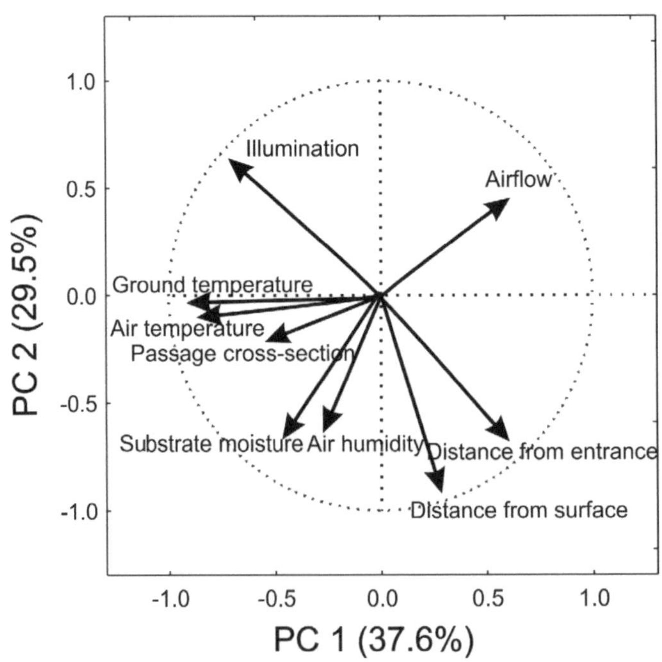


Fig. 4. Projection of the nine ecological variables on the 1st and 2nd factor planes. Graphical associated fall of the variables (arrows) in the correlation circle.

Moreover, PCs enable the testing of differences between environmental niches. For this purpose, in the test hypothesis, the PCs defining niches were subjected to variance analysis for differences between the three species. One-way ANOVA was used to test differences between species in the 1st and 2nd principal components (F and p values in Table 3). In this way, PCA allows testing of differences between niches.

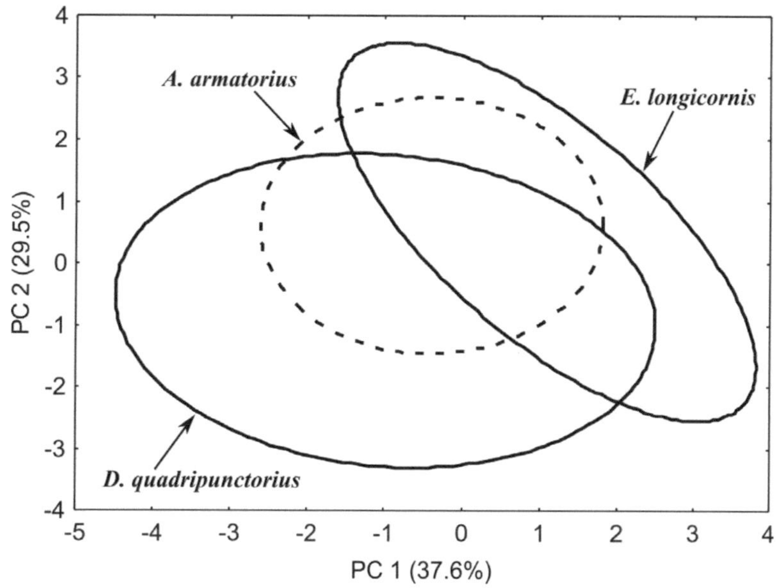


Fig. 5. Ordination of the nine environmental variables in 1st and 2nd PC axes. Ellipses (95% confidence) represent spatial niches in the three hymenopteran species.

The same analyses of the spatial niches were carried out on two co-existing spider species, *Meta menardi* and *Metellina meriannae* (Novak et al. 2010b). In this case, the variations in temperature, humidity, airflow and illumination were subjected to PCA. The 1st and the 2nd PCs together explained 70.4% of variation (Figs. 6 and 7). In this way, we presented the course of temporal changes in the spatial niches of the two spiders.

## 3. Discussion

Since computer techniques and technologies have enabled efficacious computation of PCA, it has become one of the most useful tools in ecology in various fields of use. Still, one can readily notice that many problems appear when its applicability for different purposes is to be estimated. On the one hand, reservations occur because of the credibility or interpretability of the data. Yet Austin (1985), e.g., stated that animal ecologists often use PCA without discussion of the ecological implications of its linear model, although the PCA axes are not necessarily ecologically independent, and there is no necessary ecological interpretation of components. Besides, it is particularly notable that two- and three-dimensional data using Gauss species response curves can produce complex flask-like distortions in which the underlying gradient structure is impossible to recognize without prior knowledge. In this sense, some authors (e.g., Hendrickx et al., 2007), in a specific context, decided not to rely on the obtained PCA axes, since they obscured additive and

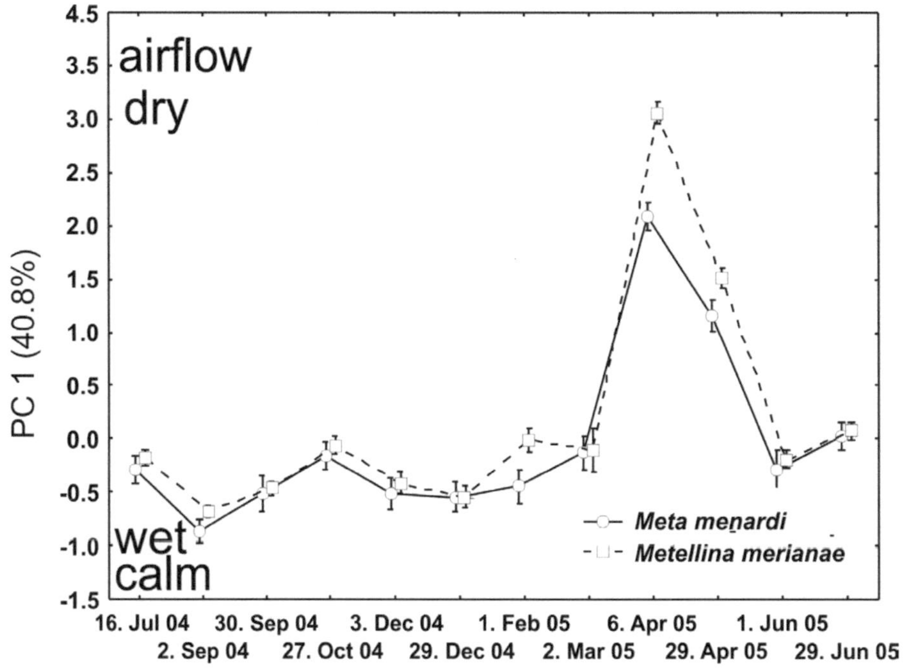


Fig. 6. Comparison between temporal spatial niches of *Meta menardi* and *Metellina merianae* in their 1st principal component (mean ± SE).

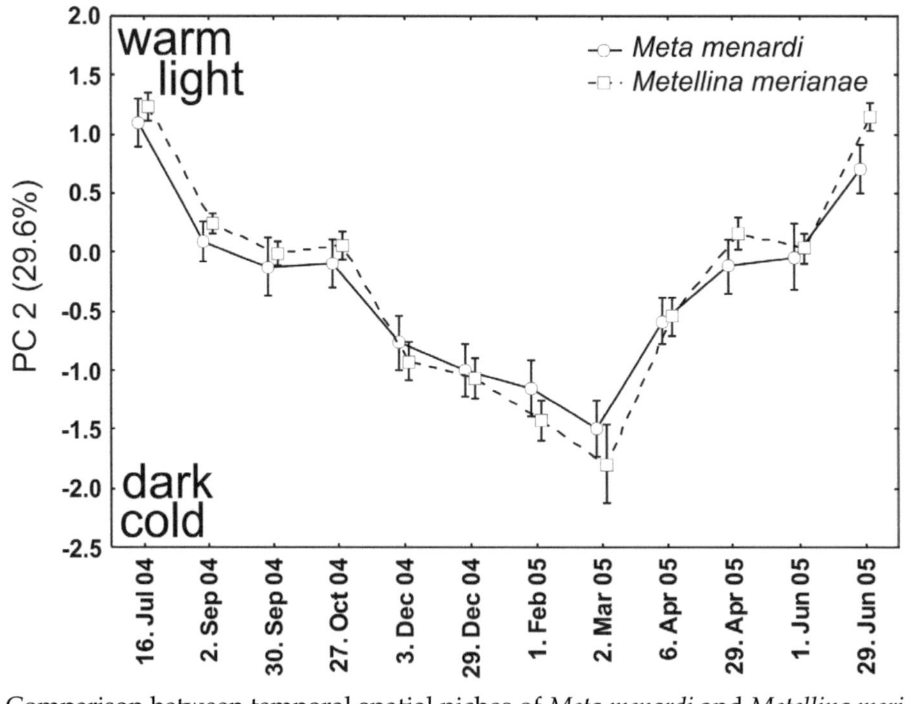


Fig. 7. Comparison between temporal spatial niches of *Meta menardi* and *Metellina merianae* in their 2nd principal component (mean ± SE).

interactive effects among variables that were partially correlated. The others highlighted that the use of PCA on spatially autocorrelated data is not appropriate in a spatial context (Novembre & Stephens 2008). Indirectly, this is the identical aspect of the problem that Nakagawa & Cuthill (2007) explain on a general level; they direct attention to the mostly neglected fact that we are still dealing with and discussing statistical rather than biotic importance. Null hypothesis significance testing, e.g., does not provide us with the magnitude of an effect of interest, nor with the precision of the estimate of the magnitude of that effect. They thus advocate (ibid.) presentation of biotic relevance: i.e., measurement of the magnitude of an effect on an organism and its confidence interval. These are the real goals in biology irrespective of the statistical values measured. In this sense, biologists and ecologists are often "trapped in statistics", into "statitraps", rather than dealing with biotic phenomena themselves. In addition, in comparison with other methods, PCA sometimes proves not to be as efficacious. In their study on ecological niches of two *Ceratitis* flies, De Meyer et al. (2008) aver that the PCA model is apparently not as good as the GARP (genetic algorithm for rule-set prediction) model at capturing the species–environment relationship. This is probably because the PCA model cannot account for nonlinear species–environment relationships in the way that GARP can.

Additionally, Ricklefs (2008) established that a local ecological community, which consists of those species whose distribution include a particular point in space and time, is an epiphenomenon with relatively little explanatory power in ecology and evolutionary biology. To understand the coexistence of species locally, one must understand what shapes species distribution within regions, but factors that constrain distribution within regions are poorly understood. This evidence of the disintegrating concept of "community" requires reconsideration of our prior explanations of coexisting species in the subterranean environment. Although many authors use the term community when dealing with its biota, their methods of discussion reveal that they mostly use the term in its wide, much looser meaning rather than referring to specific species composition and their inter- and intraspecific relations. This is especially evident in their continual references to the fact that the biology and/or ecology of many species remain unknown.

In this contribution we had presented common uses of PCA and its efficacy in accessing multivariate data, such as provided in the most usual and convenient field investigations in caves. With respect to the dynamic niche conception (Soberón 2007; Soberón & Nakamura 2009), our case studies deserve additional comment. On the one hand, the essence of the "disintegration community concept" sensu Ricklefs (2008) can well be perceived in the widely distributed species that we encounter in caves. These include the animal species appertaining to the parietal association, i.e., those species found especially on the walls and ceiling near cave entrances (Jennings 1997). The three hymenopteran and two spider species discussed in this study belong among them. The parietal association is a heterogeneous assemblage of species with respect to their geographical distribution and activity in the subterranean environment, thus directly raising the need for more complete knowledge of their "biogeography of the species" sensu Ricklefs (2008). On the other hand, in highly endemic troglomorphic species–i.e., those well adapted to the hypogean habitat (e.g., Gibert and Deharveng 2002, Christman et al., 2005, Culver and Pipan 2009)–the community disconception might have little or no significant impact on the findings under discussion.

## 4. Conclusions

PCA is a useful tool enabling ordination of environmental variables in ecology. An environmental niche comprises multidimensional data set, while PCA is an appropriate statistical tool to handle such data sets. In this way, PCA readily provides the means to explain the variance magnitudes related to environmental variables, which represent the environmental niche. Despite that, one must be aware that PCA output depends on input data, which can never cover all dimensions of an environmental niche. Besides, PCA may obscure many effects among partially correlated variables. As with other statistical approaches, it is necessary to consider the results carefully, implementing a broad knowledge of the biology and ecology of the organisms under study in order to avoid statistical artifacts.

## 5. Acknowledgement

We are indebted to Michelle Gadpaille for valuable improvement of the language. This study was partly supported by the Slovene Ministry of Higher Education, Science and Technology within the research program Biodiversity (grant P1-0078).

## 6. References

Ahmadi-Nedushan, B., St-Hilaire, A., Bérubé, M., Robichaud, E., Thiémonge, N. & Bobée, B. (2006). A review of statistical methods for the evaluation of aquatic habitat suitability for instream flow assessment. *River Research and Applications*, Vol. 22, pp. 503–523.

Albert, C.H., Thuiller, W., Yoccoz, N.G., Douzet, R., Aubert, S. & Lavorel, S. (2010). A multi-trait approach reveals the structure and the relative importance of intra- vs. interspecific variability in plant traits. *Functional Ecology*, Vol. 24, pp. 1192–1201.

Amanor. K.S. & Pabi, O. (2007). Space, time, rhetoric and agricultural change in the transition zone of Ghana. *Human Ecology*, Vol. 35, pp. 51−67.

Arumugam, M., Raes, J., Pelletier, E. *et al.* (2011). Enterotypes of the human gut microbiome. *Nature*, Vol. 473, pp. 174–180.

Austin, M.P. (1985). Continuum concept, ordination methods, and niche theory. *Annual Review of Ecology and Systematics*, Vol. 16, pp. 39−61.

Bellier E., Certain G., Planque B., Monestiez P. & Bretagnolle, V. (2010). Modelling habitat selection at multiple scales with multivariate geostatistics: an application to seabirds in open sea. *Oikos*, Vol. 119, pp. 988–999.

Blanck, A., Tedesco, P.A. & Lamouroux, N. (2007). Relationships between life-history strategies of European freshwater fish species and their habitat preferences. *Freshwater Biology*, Vol. 52, pp. 843–859.

Broennimann, O., Treier, U.A., Müller-Schärer, H., Thiuller, W., Peterson, A.T. & Giusan, A. (2007). Evidence of climatic niche shift during biological invasion. *Ecology Letters*, Vol. 10, pp. 701–709.

Catalan, J., Barbieri, M.G., Bartumeus, F., Bitušík, P., Botev., I., Brancelj, A., Cogălcineanu, D. et al. (2009). Ecological thresholds in European alpine lakes. *Freshwater Biology*, Vol. 54, pp. 2494–2517.

Chamaillé. L., Tran, A., Meunier, A., Bourdoiseau, G., Ready, P. & Dedet, J.-P. (2010). Environmental risk mapping of canine leishmaniasis in France. *Parasites & Vectors*, Vol. 3, No.31, pp. 8.

Chase, J.M. & Leibold, M. (2003). *Ecological Niches: Linking Classical and Contemporary Approaches*. University of Chicago Press.

Christman, M.C., Culver, D.C., Madden, M. & White, D. (2005). Patterns of endemism of the eastern North American cave fauna. *Journal of Biogeography*, Vol. 32, pp. 1441−1452.

Clabaut, C., Bunje, P.M.E., Salzburger, W. & Meyer, A. (2006). Geometric morphometrical analyses provide evidence for the adaptive character of the Tanganyikan cichlid fish radiation. *Evolution*, Vol. 61, pp. 560−578.

Claude, J., Paradis, E., Tong, H., Auffray, J.-C. (2003). A geometric morphometric assessment of the effects of environment and cladogenesis on the evolution of the turtle shell. *Biological Journal of the Linnean Society*, Vol. 79, pp. 485–501.

Collwell, R.K. & Rangel, T.F. (2009). Hutchinson's duality: The once and future niche. *Proceedings of the National Academy of Sciences of the United States of America*, Vol. 106, Suppl. 2, pp. 19651–19658.

Culver, D.C. (2005). Ecotones. In: Culver, D., White, W.B. (Eds.), *Encyclopedia of Caves*. Elsevier, Amsterdam, pp. 206−208.

Culver, D.C. & Pipan, T. (2009). *The Biology of Caves and Other Subterranean Habitats*. Oxford University Press.

De Meyer, M., Robertson, M.P., Peterson, A.T. & Mansell, M.W. (2008). Ecological niches and potential geographical distributions of Mediterranean fruit fly (*Ceratitis capitata*) and Natal fruit fly (*Ceratitis rosa*). *Journal of Biogeography*, Vol. 35, pp. 270–281.

Dennis, A.B. & Hellberg, M.E. (2010). Ecological partitioning among parapatric cryptic species. *Molecular Ecology*, Vol. 19, pp. 3206–3225.

Dolédec, S., Chessel, D. & Gimaret-Carpentier, C. (2000). Niche Separation in Community Analysis: A New Method. *Ecology*, Vol. 81, pp. 2914-2927.

Eaton, M.D., Soberón, J. & Townsend Petersen, A. (2008). Phylogenetic perspective on ecological niche evolution in American blackbirds (Family Icteridae). *Biological Journal of the Linnean Society*, Vol. 94, pp. 869–878.

Ecke, F., Löfgren, O., Hörnfeldt, B., Eklund, U, Ericsson, P. & Sörlin, D. (2001). Abundance and diversity of small mammals in relation to structural habitat factors. *Ecological Bulletins*, Vol. 49, pp. 165−171.

Elton C. (1927). *Animal Ecology*. Sidgwick & Jackson. London.

Elton, C. (1933) *The Ecology of Animals*. Methuen, London.

Faucon, M.P., Parmentier, I., Colinet, G., Mahy, G., Luhembwe, M.N. & Meerts, P. (2011). May rare metallophytes benefit from disturbed soils following mining activity? The Case of the *Crepidorhopalon tenuis* in Katanga (D. R. Congo). *Restoration Ecology*, Vol. 19, pp. 333–343.

Gibert, J. & Deharveng, L. (2002). Subterranean ecosystems: a truncated functional biodiversity. *Bioscience*, Vol. 52, pp. 473−481.

Glor, R.E. & Warren, D. (2010). Testing ecological explanations for biogeographic boundaries. *Evolution*, Vol. 65, pp. 673–683.

González-Cabello, A. & Bellwood, D.R. (2009). Local ecological impacts of regional biodiversity on reef fish assemblages. *Journal of Biogeography*, Vol. 36, pp. 1129–1137.

Gordon, W., Frazer, G.W., Wulder, M.A. & Niemann, K.O. (2005). Simulation and quantification of the fine-scale spatial pattern and heterogeneity of forest canopy structure: A lacunarity-based method designed for analysis of continuous canopy heights. *Forest Ecology and Management*, Vol. 3, pp. 65-90.

Gotelli, N.J. & Ellison, A.M. (2004). *A Primer of Ecological Statistics*. Sinauer Associates, Inc. Sunderland. 510 pp.

Grenouillet, G., Buisson, L., Casajus, N. & Lek, S. (2011). Ensemble modelling of species distribution: the effects of geographical and environmental ranges. *Ecography*, Vol. 34, pp. 9−17.

Grinnell, J. (1917). The niche-relationships of the California Thrasher. *Auk*, Vol. 34, pp. 427–433.

Guisan, A. & Zimmerman, N.E. (2000). Predictive habitat distribution models in ecology. *Ecological Modelling*, Vol. 135, pp. 147−186.

Hardin, G. (1960). The competitive exclusion principle. *Science*, Vol. 131, pp. 1292−1297.

Hendrickx, F., Maelfait, J.-P., Van Wingerden, W., Schweiger, O., Speelmans, M., Aviron, S., Augenstein, I., et al. (2007). How landscape structure, land-use intensity and habitat diversity affect components of total arthropod diversity in agricultural landscapes. *Journal of Applied Ecology*, Vol. 44, pp. 340–351.

Hormon, L.J., Kolbe, J.J., Cheverud, J.M. & Losos, J.B. (2005). Convergence and the multidimensional niche. *Evolution*, Vol. 59, pp. 409−421.

Hutchinson, G.E. (1957). *Concluding remarks*. In: *Cold Spring Harbor Symposia on Quantitative Biology*, Vol. 22, pp. 415–427.

Inward, D.J.G., Davies, R.G., Pergande, C., Denham, A.J. & Vogler, A.P. (2011). Local and regional ecological morphology of dung beetle assemblages across four biogeographic regions. *Journal of Biogeography*, Vol. 38, pp. 15.

James, F.C. & McCulloch, C.E. (1990). Multivariate Analysis in Ecology and Systematics: Panacea or Pandora's Box? *Annual Review of Ecology and Systematics*, Vol. 21, pp. 129−166.

Jennings, J.N. (1997). *Cave and karst terminology*. Coggan M, Nicholson P, eds. Australian Speleological Federation Incorporated Administrative Handbook. http://home.mira.net/~gnb/caving/papers/jj-cakt.html

Johnson, R.K., Hering, D., Furse, M.T. & Clarke, R.T. (2006). Detection of ecological change using multiple organism groups: metrics and uncertainty. In: Furse, M.T., Hering, D., Brabec, K., Buffagni, A., Sandin L. & Verdonschot P.F.M. (eds), The Ecological Status of European Rivers: Evaluation and Intercalibration of Assessment Methods. *Hydrobiologia*, Vol. 566, pp. 115–137.

Kearney, M. & Porter, W. (2009). Mechanistic niche modelling: combining physiological and spatial data to predict species' ranges. *Ecology Letters*, Vol. 12, pp. 334–350.

Kelt, D.A., Meserve, P.L., Bruce, Patterson, D. & Lang, B.K. (1999). Scale Dependence and Scale Independence in Habitat Associations of Small Mammals in Southern Temperate Rainforest. *Oikos*, Vol. 85, pp. 320−334.

Kingston, T., Jones, G., Zubaid, A. & Kunz, T.H. (2000). Resource partitioning in Rhinolophoid bats revisited. *Oecologia*, Vol. 124, pp. 332−342.

Kitahara, M. & Fujii, K. (2005). Analysis and understanding of butterfly community composition based on multivariate approaches and the concept of generalist/specialist strategies. *Entomological Science*, Vol. 8., pp. 137-149.

Lecomte, F. & Dodson, J.J. (2005). Distinguishing trophic and habitat partitioning among sympatric populations of the estuarine fish *Osmerus mordax* Mitchill. *Journal of Fish Biology*, Vol. 66, 1601-1623.

Legendre, P., & Legendre, L., (1998). Numerical Ecology. Elsevier: Amsterdam, 853 pp.

Lepš, J. & Šmilauer, P. (2003). *Multivariate analysis of ecological data using CANOCO*. Cambridge University Press. Cambridge.

Litvak, M.K. & Hansell, R.I.C. (1990). A community perspective on the multidimensional niche. *Journal of Animal Ecology*, Vol. 59, pp. 931–940.

Marmion, M., Parviainen, M., Luoto, M., Heikkinen, R.K. & Thuiller, W. (2009). Evaluation of consensus methods in predictive species distribution modelling. *Diversity and Distributions*, Vol. 15, pp. 59-69.

Mezger, D. & Pfeiffer, M. (2011). Partitioning the impact of abiotic factors and spatial patterns on species richness and community structure of ground ant assemblages in four Bornean rainforests. *Ecography*, Vol. 34, pp. 39–48.

Nieto-Castañeda, I.G. & Jiménez-Jiménez, M.L. (2009). Possible niche differentiation of two desert wandering spiders of the genus *Syspira* (Araneae: Miturgidae). *The Journal of Arachnology*, Vol. 37, pp. 299–305.

Nakagawa, S. & Cuthill, I.C. (2007). Effect size, confidence interval and statistical significance: a practical guide for biologists. *Biological Reviews*, Vol. 82, pp. 591-605.

Novak, T. (2005). Terrestrial fauna from cavities in Northern and Central Slovenia, and a review of systematically ecologically investigated cavities. *Acta Carsologica*, Vol. 34, 169–210.

Novak, T., Thirion, C., Janžekovič, F. (2010a). Hypogean ecophase of three hymenopteran species in Central European caves. *Italian Journal of Zoology*, Vol. 77, pp. 469–475.

Novak, T., Tkavc, T., Kuntner, M., Arnett, A.E., Lipovšek Delakorda, S., Perc, M. & Janžekovič, F. (2010b). Niche partitioning in orbweaving spiders *Meta menardi* and *Metellina merianae* (Tetragnathidae). *Acta Oecologica*, Vol. 36, pp. 522–529.

Novembre, J. & Stephens, M. (2008). Interpreting principal component analysis of spatial population genetic variation. *Nature Genetics*, Vol. 40, pp. 646-649.

Pearman, P.B. (2002). The scale of community structure: Habitat variation and avian guilds in tropical forest understory. *Ecological Monographs*, Vol. 72, pp. 19–39.

Pulliam, H.R. (2000). On the relationship between niche and distribution. *Ecology Letters*, Vol. 3, pp. 349-361.

Ricklefs, R. (2008). Disintegration of the ecological community. *The American Naturalist*, Vol. 172, pp. 741–750.

Rissler, L.J. & Apodaca, J.J. (2007). Adding more ecology into species delimitation: Ecological niche models and phylogeography help define cryptic species in the black salamander (*Aneides flavipunctatus*). *Systematic Biology*, Vol. 56, pp. 924–942.

Robertson, M.P., Caithness, N. & Villet, M.H. (2001). A PCA-based modelling technique for predicting environmental suitability for organisms from presence records. *Diversity and Distributions*, Vol. 7, pp. 15–27.

Romero, A. (2011). The evolution of cave life. *American Scientist*, Vol. 99, pp. 144–151.

Soberón, J. (2007). Grinnelian and Eltonian niches and geographic distributions of species. *Ecology Letters*, Vol. 10, pp. 1115-1123.

Soberón, J. & Nakamura, M. (2009). Niches and distributional areas: Concepts, methods, and assumptions. *Proceedings of the National Academy of Sciences of the United States of America*, Vol. 106, Suppl. 2, pp. 19644-19650.

Toepfer, C.S., Williams, L.R., Martinez, A.D. & Fisher, W.L. (1998). Fish and habitat heterogeneity in four streams in the central Oklahoma/Texas plains ecoregion. *Proceedings of the Oklahoma Academy of Science* 78: 41–48.

Warner, B.G., Asada, T. & Quinn, N.P. (2007). Seasonal influences on the ecology of testate amoebae (Protozoa) in a small *Sphagnum* peatland in Southern Ontario, Canada. *Microbial Ecology*, Vol. 54, pp. 91−100.

Whittaker, R.H. & Levin S.A., eds. (1975). *Niche*. Benchmark Papers in Ecology. Dowden, Hutchinson & Ross Inc., Stroudsburg.

Youlatos, D. (2004). Multivariate analysis of organismal and habitat parameters in two neotropical primate communities. *American Journal of Physical Anthropology*, Vol. 123, pp. 181-194.

# 9

# The Health Care Access Index as a Determinant of Delayed Cancer Detection Through Principal Component Analysis

Eric Belasco[1], Billy U. Philips, Jr.[2] and Gordon Gong[2]
[1]*Montana State University, Department of Agricultural Economics and Economics,*
[2]*Texas Tech University Health Sciences Center, F. Marie Hall Institute of Rural Community Health,*
*USA*

## 1. Introduction

In the past two decades, cancer mortality declined significantly in the United States (Byers, 2010). Although the reasons for the decline have not been well-established, many factors such as the reduction in the number of smokers, increased cancer screening, and better treatment may have played an important role (Byers, 2010, Richardson et al. 2010). However, disparities in cancer mortality persisted among different ethnic groups and social classes (Byers, 2010). Health status and health disparities among different social and ethnic groups are to a large degree determined by socioeconomic status and living conditions in general (Pamies and Nsiah-Kumi, 2008; World Health Organization [WHO], 2008). For example, life expectancy worldwide increased from 48 years in 1955 to 66 years in 2000 mainly as a result of improvement of overall living conditions in addition to advancement in medical science and large-scale preventive interventions (Centers for Disease Control and Prevention [CDC], 2011). Large health disparities exist between poor and rich countries or within any given rich or poor country (WHO 2008). In the case of cancer mortality due to delayed detection, socioeconomic status may determine health insurance coverage status, which in turn affects health behaviour including regular check-ups and participation in cancer surveillance among high risk groups. Regular cancer surveillance is critical for cancer control (Byers, 2010, Richardson et al. 2010). Lack of health insurance due to economic hardship may result in the delay in cancer detection.

A vexing question is how to determine socioeconomic status. Early studies associated cancer mortality with single socioeconomic indicators such as individual income, education level, below or above poverty level among others. For example, Ward et al. (2004) used the percentage of the population below the poverty line as a socioeconomic indicator and found that cancer mortality rate was 13% and 3% higher in men and women, respectively, among U.S. counties with ≥ 20% of the population below the poverty line as compared with those with < 10% below the poverty line from 1996 to 2000. On the other hand, Clegg et al. (2009) used education level as an indicator, and reported that lung cancer incidence was significantly higher among Americans with less than a high school education than those with a college education. Clegg et al (2009) also used family annual income as an indicator and found that lung cancer incidence rate was 70% higher in those with less than $12,500 annual income compared with those with incomes $50,000 or higher.

An alternative approach to assessing socioeconomic status is to build a composite index based on many aspects of socioeconomic status using readily available census data, which is then employed to predict health status using information from disease (e.g., cancer) registries. For example, Singh (2003) used 17 socioeconomic indicators (such as household income, median home value) derived from US census data to build a composite index for socioeconomic deprivation by factor analysis and principal component analysis (PCA). Such a composite index is believed to reflect socioeconomic status more thoroughly with multiple indictors, while PCA address the issue of inter-correlation among factors. Such a composite index tends to have a high reliability coefficient ($\alpha$ equal to 0.95) (Singh, 2002). The composite index is then used to predict health status. For example, Singh (Singh, 2003) reported that US mortality of all-causes was significantly and positively correlated with a deprivation index derived from US census data. Crampton et al. (1997) developed a relatively simple socioeconomic status index termed the New Zealand Index of Relative Deprivation (NZDep91) which was constructed based on nine socioeconomic variables from New Zealand census data. The NZDep91 is subsequently used to predict hospital discharge rate and all-cause mortality (Salmond et al. 1998). Albrecht and Ramasubramanian (2004) (Henceforth, A&R) modified the NZDep91 and developed a Wellbeing Index (WI) by principal component analysis using ten socioeconomic variables from US Census 2000 data. The WI is recently shown to be highly correlated with delayed cancer detection (as assessed by the ratio of late- to early-stage cases) of female genital system (FGS), lung-bronchial and all-type cancers at diagnosis among Texas counties (Philips et al., 2011).

One of the main purposes of the current study is to determine whether the percentage of late-stage cancer incidence is correlated with a newly developed index of health accessibility, which is an extension of the previously mentioned WI. We term the new index the Health Care Accessibility Index (HCAI), which is derived from principal component analysis of ten socioeconomic indicators derived from US Census Bureau's 2005-2009 American Community Survey plus two additional factors that are more closely related to health, i.e., health insurance coverage and physician supply. By examining the relationship between HCAI and late-stage cancer detection, we are able to establish whether health inequities exist in certain communities that can be related to access to health care. A high percentage of late-stage cancer cases is problematic for communities due to the often relatively low survival rates of costly procedures. The derived HCAI is compared with WI in their association with delayed cancer detection in Texas counties to determine the optimal model by the Akaike's Information Criteria (AIC) (Akaike, 1974) and Schwartz Information Criteria (SIC) (Schwartz, 1978). Another difference between WI and HCAI is that the ten socioeconomic variables for computing the HCAI includes a new variable, the median income of a county and excludes the percentage of people with disability (because of its absence in the US Census Bureau's 2005-2009 American Community Survey database).

This study also addresses several practical statistical issues regarding the choice of socioeconomic variables in association with delayed cancer detection. Firstly, A&R arbitrarily classify the WI rankings as the deciles of the first component scores retrieved from PCA. It is quite frequent for groupings to be assigned based on terciles, quartiles, or deciles, which may or may not produce a proper classification. If ad hoc metrics are used to denote an optimal number of groups to be used in any particular index, then the inference that derive from the index might not have meaning and might even add uncertainty to the representativeness of the index. For example, the grouping used for Texas might be substantially different from Delaware, given the different sizes and regional dynamics. We propose to use AIC and SIC to find the optimal number of groups and compare the goodness-of-fit between models. Secondly, it has not been statistically demonstrated that using the composite WI is superior to using each of the 10 individual socioeconomic variables in their correlation with health status such as delayed cancer detection. The optimal model of the two (a composite vs. multiple variables) will also be determined by AIC and SIC. Thirdly, we also propose the use of raw principal component scores in a regression in order to improve goodness-of-fit and compare these results to the previously mentioned models based on AIC and SIC. Using these proposed methods, we look to characterize the relationship between late-stage cancer detection with the HCAI in order to identify the existence or lack of existence of economically-rooted health inequities.

## 2. Review of PCA in regression analysis

### 2.1 Review of creating indices using PCA

Using PCA in economics is particularly appealing for applications where comparisons are warranted that comprise over a collection of variables. This method is particularly convenient and informative when the researcher is interested in the "ability" of a model to characterize a collection of variables rather than the marginal impacts between one variable and another (Greene, 2012, pg. 93). Further, the assumption of an exogenous shock is often used when evaluating marginal impacts, which are unrealistic with covariates that are highly correlated. For example, in evaluating the relative wealth or poverty in developing counties, asset indices are often built to reflect the relative wealth in order to make cross-country comparisons. For example, Booysen et al (2008) conducted a transregional survey in sub-Saharan Africa to evaluate the movement of poverty across regions over a particular time span. While "poverty" is a loosely defined term, this line of research commonly utilizes an asset index in order to evaluate the ability of citizens in each country to consume durable goods. Another example is Gwatkin et al. (2000) who use data from the Demographic and Health Survey (DHS) program to evaluate socio-economic differences between developing countries. They create a socio-economic status (SES) index, which was also used in studies such as Vyas and Kumaranayake (2006).

SES indices have also been developed in order to evaluate health outcomes in developing counties in studies such as Deaton (2003), where it is argued that health outcomes and the utilization of health services are largely different across different socioeconomic classes. Ruel and Menon (2002) create a child feeding index using responses in the DHS survey to assess the influences on child feeding practices. As in many related studies, the creation of an index is used to identify problem areas that need improved policy design or interventions.

Ewing et al. (2008) develop a sprawl index in order to evaluate the relationship between urban sprawl and health-related behaviours. As in the case of other index variables, urban sprawl can be defined only when many factors are combined. The main four factors that are include are residential density, mixture between homes, jobs, and services, strength of business centers, and accessibility of the street network. Data were from the Behavioral Risk Factor Surveillance Systems (BRFSS) from 1998 to 2000 to evaluate health activity and lifestyle. On a similar topic is a study by Pomeroy et al (1997) who evaluated perceptions and the factors that led to the appearance of a successful resource management program.

While many of these indices include continuous variables, Koenikov and Angeles (2009) and Filmer and Pritchett (2001) explore ways to incorporate discrete data into PCA. Savitry (2005) provides a comprehensive overview of all multivariate methods used in developing indices.

### 2.2 Review of using PC scores in regression analysis

While the developments of a reliable and robust index are quite extensive in the literature, little research has been done to evaluate the impact of incorporating an index into a regression. This is precisely because usually the index itself is the primary interest. For example, in the identification of poverty traps, one needs only an index for identification. However, if one were interested in the impact of a poverty trap on say personal liberties, then a regression would likely be needed.

There are many studies that conduct PCA to create an index that is used in a secondary regressions (Everitt and Hothorn, 2011; Everitt, 2011; Vyas and Kumaranayake, 2006). One method that is used is to rescale the principal component values. For example, Ewing et al. (2008) scale the raw principal component scores to have a mean of 100 and a standard deviation of 25. This is similar to normalizing the principal components. However, the distribution of scores might not be normal. Vyas and Kumaranayake (2006) report finding scores that are different distributions by county that include counties with normal and uniform distributions along with distributions that possess negative or positive skewness. Their results suggest that the characterization groups might be different across cases, which leads one to consider a data-driven approach. Pomeroy et al (1997) use the raw and unscaled principal component scores as the dependent variable. While marginal impacts are not easily interpreted, they can loosely be discussed directionally.

Another common approach is to use cut-off points. For example, Filmer and Pritchett (2001) split their sample into three populations based on cut-off points at the 60$^{th}$ and 20$^{th}$ percentile of the principal component scores. They define these groups to be 'low', 'medium', and 'high' socioeconomic groups. Cut-off points can also be defined a bit more arbitrarily. For example, A&R use deciles (ten groups) to define different ranges of socio-economic status. The implicit assumption made is that each group is distinctly different from the other. However, just as Vyas and Kumaranayake (2006) found differences by country it might also be possible to find that some populations need more grouping than others. Others, such as Booyens et al (2008) use quintiles in breaking up the population.

Other approaches that are more data-driven include the use of cluster analysis, which is described in some detail in Everitt et al. (2011). However, one limitation of cluster analysis as well as the previously discussed methods is that the existence of an observation in a particular group is mutually exclusive of other groups. This means that each observation must with certainty fit into a single category. However, an evolving area of research includes latent class analysis where the residency within each group is treated as an unobserved factor that is estimated with some probability, rather than absolute certainty. However, these models are also with their own set of limitations. For example, the number of classes must be defined *a priori*. Additionally, the models have been shown to be highly non-linear and can often have difficulty finding optimized values when many classes are added.

## 3. Data and application

### 3.1 Data summary and description

This study was approved by Texas Tech University Health Sciences Center Institutional Review Board with exemption for review because of its use of published data.

### 3.2 Cancer data

Cancer stage data from 2004 to 2008 were provided by the Texas Cancer Registry, Cancer Epidemiology and Surveillance Branch, Texas Department of State Health Services (note that cancer data from 2008 are the latest data available). This database provides cancer data by year, age, county, Hispanic origin (Hispanic vs. Non-Hispanic) as well as population size for each county. We used cancer data between 2004 and 2008 cancer data (the latest available data is the 2008 data) to match the five years of American Community Survey data although there is a one-year lag. The five most common categories of cancer are studied including breast, colorectal, FGS, lung-bronchial, and prostate cancers. Female genital system includes cervix uteri, ovary, corpus and uterus, vagina, vulva and others. We pool the five-year (2004-2008) data to calculate the numbers of age-adjusted late- and early-stage cancer cases per unit (100,000) population using 2000 USA standard population (National Cancer Institute, NCI, n.d.). We use the percentage of late-stage cases among all staged cancer cases in our analysis. The number of unstaged cancer cases is not included in the denominator because the percentage of unstaged cases varies significantly by cancer type as well as by county, and inclusion of such cases in the denominator would result in uncertainty in estimating the percentage of late-stage cancer cases. Carcinoma in situ and localized cancers are considered as early-stage while cancers defined as "regional, direct extension only", "regional, regional lymph nodes only", "regional, direct extension and regional lymph nodes", "regional, NOS" and "distant" are considered as late-stage (Philips et al. 2011).

### 3.3 Socioeconomic status data

Socioeconomic status data are derived from the U.S. Census Bureau's (n.d.) 2005-2009 American Community Survey. Since this survey does not provide percentage of people with disability, the present study uses the remaining nine of the 10 socioeconomic variables originally used to build the Wellbeing Index (WI) developed by A&R and are listed in Table 1. We add median income (from the 2005-2009 American Community Survey) so that the total number of socioeconomic variables is still ten in the current study.

### 3.4 Data of factors more closely related to health

Data for the percentage of uninsured and percentage of obese individuals are obtained from Texas State Data Center (n.d.). The number of physicians and estimated population size in each county from 2004 to 2008 are derived from Texas Department of State Human Services (DSHS, n.d.). Physician supply is the number of physicians per 1,000 residents in each county. Physicians considered are those with medical doctor (MD) and/or doctor of osteopathy (DO) degrees who worked directly with patients. Residents and fellows; teachers; administrators; researchers; and those who were working for the federal government, military, retired, or not in practice were excluded from the total of physicians by DSHS (n.d.).

| | Description | Weighted Mean | Std Error | Min | Max |
|---|---|---|---|---|---|
| % Single Parent | % people in single parent households | 18.89 | 0.54 | 0.00 | 27.55 |
| % No High School | % people over 18 without high school | 21.38 | 1.17 | 2.60 | 53.59 |
| % Unemployed | % people unemployed | 6.87 | 0.15 | 0.00 | 21.58 |
| % Income Support | % people with income support | 28.20 | 1.47 | 0.00 | 66.64 |
| % Below Poverty | % people in households below poverty level | 15.90 | 0.88 | 0.00 | 46.81 |
| % No House | % people not living in own home | 34.19 | 1.74 | 11.88 | 70.83 |
| % Few Room | % people living in homes with too few bedrooms | 4.97 | 0.47 | 0.00 | 15.79 |
| % No Phone | % people in households without phone | 5.33 | 0.20 | 0.00 | 18.23 |
| %No Car | % people in households without car | 6.55 | 0.40 | 0.00 | 17.55 |
| Median Income | Median Income (in $1,000s) | 49.63 | 1.73 | 20.38 | 82.55 |
| %Uninsured | % people without health insurance | 24.47 | 1.07 | 14.20 | 38.10 |
| Physician supply | Number of physicians per 1,000 residents | 0.16 | 0.01 | 0.00 | 0.33 |
| % Obese | % people with BMI above 30 | 28.90 | 0.33 | 23.80 | 32.80 |
| % Hispanic | % Hispanic | 35.60 | 3.19 | 2.44 | 97.15 |
| Population | Population (in 1,000s) | 636.02 | 218.75 | 0.02 | 1,912 |

Table 1. Description and weighted summary statistics of relevant covariates

## 4. Statistical analysis and results

### 4.1 Computing the health care accessibility index

While access to health care may be determined by socioeconomic status in general, health insurance coverage and health care services (number of physicians and or hospitals relative to local population) may more directly impact on health as discussed above. In this study we add the latter two variables to the 10 socioeconomic indicators for a principal component analysis to build Health Care Accessibility Index (HCAI).

There are two well-known benefits to creating an index rather than including all of the components in the index into a regression. First, the index itself can serve as a variable of interest to identify (in our study) areas of extreme health care obstacles. Second, the index can significantly reduce the degrees of freedom in a regression while preserving information from the variables. In this study we also find a third benefit, which is that if we are careful about how we use the index in regression analysis, we can also improve goodness-of-fit.

While the A&R study presents an appealing start to our research, it lacks two important components. First, it only includes socioeconomic variables that are intended to provide an index for wellbeing or economic deprivation. While their index provides a good basis to evaluate the ability to pay for health services, it lacks health-specific variables. Second, the study presents a less than appealing method for incorporating the index into regression by ranking the first principal component scores into deciles. This essentially groups the observations into 10 different bins which, may or may not be the optimal number of bins to use in grouping regions. To fix the second issue of rank ordering, we use AIC and SIC to determine the optimal number of bins (more discussion on this follows in the next subsection).

In order to fix the first issue, we add three components which are essential to access to health care services which include median income, the percentage of the uninsured, and the number of physicians per 1,000 residents. All variables are computed at the county-level. These variables are intended to account for the access residents within a county have to health care services. Lack of health care insurance in clearly a hurdle in obtaining affordable health care services and is another factor included in this analysis. Finally, the number of physicians allows for an evaluation into geographic access to health care services. Does the county in question have an adequate medical infrastructure to prevent and detect illness when it arises? Some Counties in Texas (particularly in the western region) are geographically isolated. If adequate care is not geographically close to residents, the economic cost to receiving care increases in terms of time off work and travel costs.

In compiling the twelve variables, we are able to derive an index for access to health care services through the use of PCA. While the goal is to consolidate a group of many variables into a smaller set of linearly related variables (principal components), it is often the case that multiple principal components are needed to explain a substantial proportion of the variation in the independent variables. Results from the initial PCA is shown below in table 2.

The first principal component explains 41% of all variation in the variables included in the index. The influence is also spread across relatively evenly across all included variables, with the exception of direct patient physician supply (*DPC*). The principal component scores from this regression are used as the HCAI. Notice the negative score associated with median income, which is consistent with a negative relationship between income and obstacles to receiving health care services. The second principal component explains an additional 16% of the variation in the variables and is largely influenced by physician supply, percent uninsured, percent without their own house, and percent of single parents.

These results provide us with a couple of items. First, we have the first five principal components, which can be used rather than the original 13 variables in order to shrink the necessary variables while still preserving almost all of the variation in the variables. One notable and helpful point in this analysis is that each component is orthogonal to the others,

| | Principal Component Score | | | | |
|---|---|---|---|---|---|
| Variable | 1 | 2 | 3 | 4 | 5 |
| % Single Parent | 0.31 | 0.22 | 0.22 | 0.01 | 0.21 |
| % No High School | 0.36 | -0.13 | -0.36 | -0.03 | 0.15 |
| % Unemployed | 0.24 | 0.05 | 0.31 | 0.71 | -0.43 |
| % Income Support | 0.18 | -0.56 | 0.23 | -0.05 | 0.28 |
| % Below Poverty | 0.39 | -0.08 | 0.09 | 0.00 | -0.02 |
| % No House | 0.18 | 0.45 | 0.21 | -0.30 | -0.24 |
| % Few Room | 0.27 | 0.17 | -0.48 | 0.24 | 0.17 |
| % No Phone | 0.28 | 0.09 | -0.04 | -0.50 | -0.48 |
| %No Car | 0.36 | -0.03 | 0.22 | 0.12 | 0.05 |
| Median Income | -0.35 | 0.31 | -0.18 | 0.26 | -0.08 |
| %Uninsured | 0.31 | 0.24 | -0.41 | 0.07 | 0.11 |
| DPC | -0.01 | 0.47 | 0.37 | -0.04 | 0.57 |
| Eigenvalue | 4.89 | 1.94 | 1.36 | 0.76 | 0.74 |
| Difference | 2.95 | 0.58 | 0.60 | 0.02 | 0.11 |
| Proportion | 0.41 | 0.16 | 0.11 | 0.06 | 0.06 |
| Cumulative | 0.41 | 0.57 | 0.68 | 0.75 | 0.81 |

Table 2. PCA results from variables in health access index

meaning the independent variables will be uncorrelated and avoids the issue of multicollinearlity which arises from including all 13 variables into the regression. A notable problem when multicollinearity is particularly acute is that it is difficult to isolate marginal relationships between competing variables, which often leads to high standard errors. Second, the first principal component scores can be used to develop the index of interest in the following way. The index is created by recognizing that the first principal component can be written as a linear combination of the original variables such that

$$Prin1 = a_{1,1}x_1 + a_{1,2}x_2 + \cdots + a_{1,13}x_{13} \tag{1}$$

where $a_1 = (a_{1,1}, a_{1,2}, \ldots, a_{1,13})$ include the parameter estimates in table 2 and $x = (x_1, x_2, \ldots, x_{13})$ include the 13 parameters used in the index from table 1. $Prin1$ is then computed for each observation and can be ranked to present a ordering of health accessibility. The second component is then derived based on the remaining variability and results in $a_2$ which leads to $Prin2$. The second component is derived based on the restriction that $a_2'a_1 = 0$ so that $Corr(Prin1, Prin2) = 0$. Additional components can be derived in the same fashion so that they are uncorrelated with all prior components.

The rankings associated with the first principal component, *Prin*1, is used as the basis of our HCAI. To illustrate the regional dynamics of this new index, we present figure 1 below.

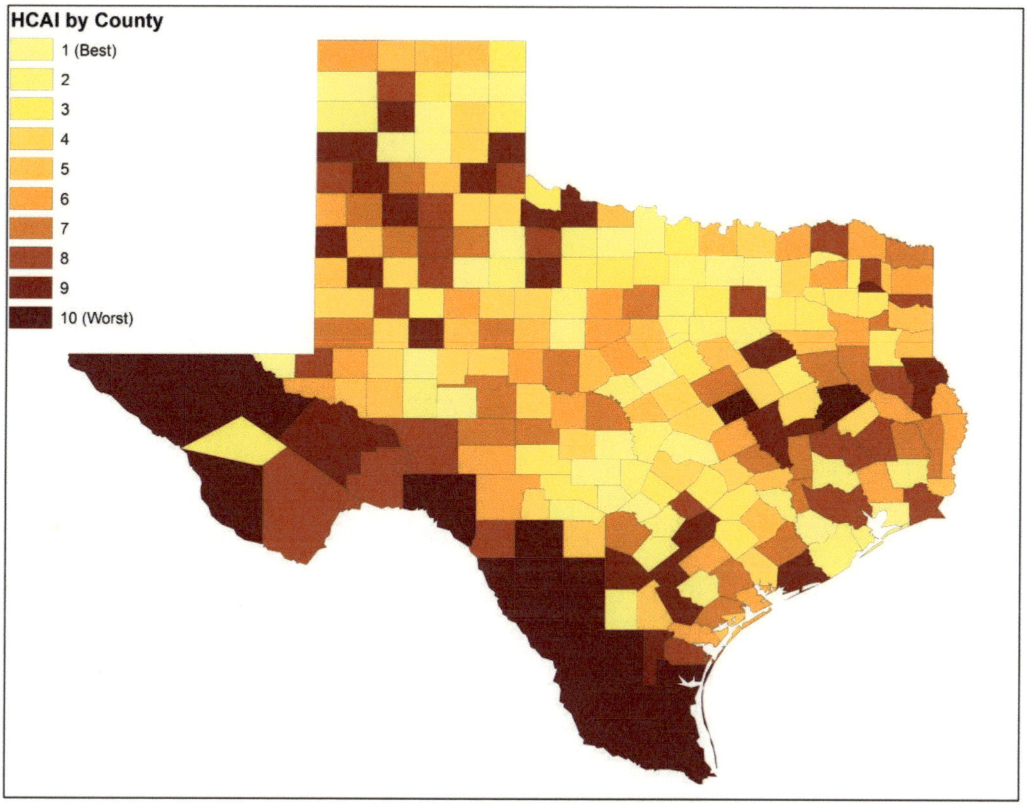


Fig. 1. Map of ordinal ranking of Texas counties by HCAI

Figure 1 shows that the counties that have poor health care access are focused along the Mexican border. These counties tend to have a large percentage of Hispanics in the population. Other counties further north are also found to have high HCAI rankings. Some of the better HCAI rankings are in the more urban areas of Texas that are near the largest metropolitan areas of Dallas, Fort Worth, Houston, San Antonio, and Austin. The large size of Texas, as well as its diversity in the variables used in this study, provides an opportunity to evaluate the how these factors influence late-stage cancer detection.

In the present study, the percentage of late stage cancer cases is hypothesized to correlate with the developed HCAI. Figure 2 below shows a scatter plot of the dependent variables along with the first component scores of HCAI for each cancer type that is evaluated in this study. Given the principal component scores shown in table 2, it is clear that a county that has a low degree of health access will possess a high component score. These same areas with a low degree of health access correspond to high rankings. Given this, it is not surprising that some of the slopes shown in figure 2 are significantly positive. Lung cancer is one example where the positive slope is particularly striking. On the other hand, the relationship with the HCAI is shown to be positive but relatively flat for colon and prostate cancers. Given the different speeds of progression in different cancers, it should not be surprising that the relationships differ across cancer types.

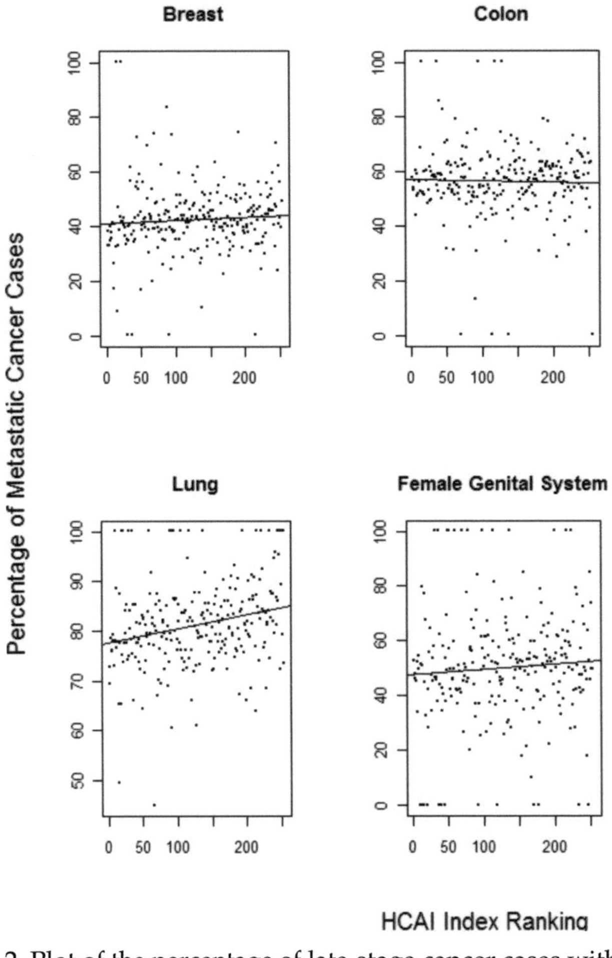


Fig. 2. Plot of the percentage of late-stage cancer cases with HCAI first principal component score, by cancer type

### 4.2 Weighted tobit regression model

The dependent variable of interest used in this analysis is the log of the sum between one and the percentage of late-stage cancer cases among all staged cancer cases. Some counties experienced no late-stage cancer cases and other experiences all late-stage cancer cases. These are all from relatively small counties. However, this does mean that censoring is an issue that will need to be dealt with in this study. As shown in table 3, there is some degree of censoring for all regressions in our analysis. For example, the percentage of late stage breast cancer cases was lower censored at 0 in four cases and upper censored in two cases. In the lower censoring case, the county experienced no late-stage cancer cases but experienced only early-stage cases. Alternatively, some counties experienced only late-stage cases resulting in an upper bound of 100%. This double-bounded censoring occurred for all cancer types to varying degrees.

If censoring is not accounted for and usual least squares methods are used, we are assured of biased estimates (Amemiya, 1973). So, in order to obtain unbiased parameter estimates, we use the Tobit model (Tobin, 1958) to correct for censoring. The Tobit model assumes the dependent variable, $Y$, is a partially observed derivative of a latent variable, $Y^*$. When censoring occurs at zero, the two are related with the following relationship, $Y = \max(0, Y^*)$. In this particular application of a double-bounded Tobit model, we assume a latent variable, $Y^*$, with the following relationship to the observed variable, $Y$.

$$Y = \begin{cases} 0 & if\ Y^* < 0 \\ Y^* & if\ 0 \leq Y^* \leq 100 \\ 100 & if\ Y^* > 100 \end{cases} \tag{2}$$

A more comprehensive discussion on the Tobit model can be found in Amemiya (1974). Residuals are weighted by population size in order to account for the wide variety in the size of counties. Weighted statistics provide a more accurate representation of the population of interest, which is the state of Texas in this application.

| Type | Weighted Mean | Std Error | Min | Max | Limit obs. | |
|---|---|---|---|---|---|---|
| | | | | | Lower | Upper |
| Breast | 40.46 | 0.48 | 0.00 | 100.00 | 4 | 2 |
| Colon | 57.92 | 0.81 | 0.00 | 100.00 | 4 | 5 |
| Prostate | 14.45 | 0.41 | 0.00 | 48.87 | 18 | 0 |
| Lung | 79.28 | 0.59 | 44.98 | 100.00 | 0 | 20 |
| FGS | 50.69 | 0.55 | 0.00 | 100.00 | 13 | 13 |

Table 3. Percentage of late-stage cancer cases, by cancer type

### 4.3 Incorporating the index into regression analysis

While the selection of principal components is well established and can be found in most intermediate econometrics text books, the usage of that information in a regression is less understood. For this reason, we use five separate regressions for each cancer type to compare different methodologies that could be used to integrate an index or information in an index within a regression context. We use AIC and SIC to compare model fit and include two additional variables that are hypothesized to influence the percentage of late stage cancer cases. These variables include the log of the percentage of obese individuals and the log of percentage of Hispanic individuals. The five models of interest include the following specifications:

Model 1. The composite ordinal WI (from A&R) based on deciles according to the first principal component scores

Model 2. All 12 variables in the composite HCAI

Model 3. The raw scores of the first two principal components using variables in HCAI

Model 4. The composite ordinal HCAI based on grouping of the first principal component scores that minimizes AIC

Model 5. The composite ordinal HCAI based on grouping of the first principal component scores that minimizes SIC

The first method uses the method proposed by A&R, which assumes the use of deciles to group the ordered WI. This model provides a prior baseline under which to evaluate all of the proposed models. As A&R point out, once the index is used to define groups, those groups are compared in an ordinal manner, implying the se of ordinal and not continuous variables in a regression. Model 2 does not make use of PCA and places all the variables used in the HCAI directly into the regression. This model also provides another baseline model in that we can compare the methods used in PCA to that of this model which simply uses the variables without any PCA. In order for PCA to be effective in our analysis, it will need to demonstrate an improvement over model 2. In our original analysis we included two additional regressions that were omitted from our final analysis because they were not found to improve upon the utilized models. These models included (1) the raw WI variables and (2) decile grouping from the HCAI. Using raw WI variables was consistently outperformed by model 2, while the decile grouping of HCAI was consistently outperformed by models 4 and/or 5 unless they selected 10 groups (as occurred in a couple of instances).

The final three models make use of the PCA results based on the HCAI. These are the models of interest in the sense that we hypothesize that they will improve upon Models 1 and 2. Model 3 uses the raw scores of first two principal components in the regression. The scores of first two components are used since they minimize both AIC and SIC in all of the used regressions. Component scores are fitted values using the parameter values listed in table 2 and the associated variables for each observation. Therefore, each observation will contain a unique component score. The first two principal component scores explain 57% of the total variation in the used variables based on the results in table 2. While it might seem more appropriate to include say the first five principal components in order to explain 81% of the total variation, the final three components tend to be significant are the model does not fit the data as well (in terms of AIC and SIC) as when only the first two scores are included. This model provides a set of continuous variables that can be used to determine late-stage cancer detection. Models 4 and 5 use AIC and SIC to find the optimal number of groups to be used for HCAI. This search method is conducted by running all models up to 40 groups in order to find the single model that minimizes AIC or SIC. Observations are evenly divided across the selected number of groups which are used as categorical variables in our analysis.

These five models are then compared by using AIC and SIC metrics to assess goodness-of-fit. In order to declare one of the proposed models as an improvement, it will need to have values of both AIC and SIC that are lower than models 1 and 2. Since maximum likelihood methods are commonly used to estimate Tobit estimates, an appealing goodness-of-fit measure is that of AIC and/or SIC since they are both easily derived from maximum likelihood outputs. The two can be expressed as follows

$$AIC = 2k - 2LL \tag{3}$$

$$SIC = klog(k) - 2LL \tag{4}$$

where $k$ is the number of parameters and $LL$ is the maximized log-likelihood value that corresponds with the optimal parameter estimates. Both information criteria measure are composed of two parts, which include two times LL and the penalty for adding new parameters. For models that contain more than 8 parameters ($k > 8$), SIC provides a heavier penalty for adding new variables. For this reason, AIC tends to support more overfit models while SIC tends to support more underfit models. Both AIC and SIC provide a basis upon which to select from nested models by minimizing the associated criteria. However, Koehler and Murphree (1988) point out that in their analysis the two criteria indicate different models 27% of the time. Their research suggests that SIC is often preferred due to the improved predication accuracy from SIC-preferred models. However, as pointed out by Kuha (2004), the preference for AIC or SIC can often rely on the data generation process and when AIC and SIC both select the same model, it is shown to be a more robust result. With these points in mind, we proceed by using AIC and SIC criteria separately to determine the appropriate model and determine whether our proposed models provide an improvement on past models.

A&R suggest that when using their index of socioeconomic deprivation, one should rank the first principal component then group this ranking into deciles, so that groupings will occur at 10% intervals. For example, if a county receives a grouping of 1, it is found to be in the lower 10% with regard to the index. We find this ranking method to be arbitrary and therefore suggest the use of AIC/SIC to determine the optimal number of deciles. This is achieved by re-running each model with the number of groups changing every time starting with 1 group (or rather no grouping at all) and up to 40 groups. Forty groups were selected because AIC and SIC did not show significant improvement after that point. With such a large and diverse state as Texas, it seems that the same rank grouping would be substantially different than for some smaller state. Of importance for this study is the generalizeability of our study to other areas. For example, does the method that we propose provide a general enough method on which to evaluate any region? While the selection of a specific grouping may work best in one situation, a more data-driven approach (such as the one we propose) will provide more flexibility in evaluating different situations.

Also included in each model will be two additional variables in order to evaluate the impact of different specifications on parameter estimates, since typically marginal inference is an important component of any analysis (particularly in economics). These variables include the log of percentage of Hispanic and the log of percentage classified as obese. Both variables are likely to have some degree of correlation with the index ranking but to varying degrees. This provides a look at two variables that have varying relationships with the index in question.

### 4.4 Empirical results

#### 4.4.1 Model selection

The first tasks in the empirical section of this analysis is to determine grouping based on the candidate model that minimizes AIC and SIC (models 4 and 5) for all 5 cancer types models (as outlined in section 3.2). The selected numbers of groups are shown in table 4.

| | AIC | SIC |
|---|---|---|
| Breast | 11 | 5 |
| Colon | 10 | 1 |
| Prostate | 9 | 1 |
| Lung | 34 | 8 |
| FGS | 1 | 1 |

Table 4. Number of groups determined based on minimum AIC and SIC criteria for models 4 and 5

The variability in the selected number of groups, suggests that arbitrarily extracting a grouping number might not be optimal. For example, in our analysis, lung cancer makes use of many groups while FGS cancer suggests only 1 group (which means no group distinctions at all). To illustrate the meaning of a group, the use of 34 groups implies there are 34 groups (each with 7-8 observations). In the regression, this implies that 33 new classification variables are added which are evaluated relative to the omitted group. Each model is run using the weighted Tobit model to assess the goodness-of-fit as well as parameter estimates of variables outside of the proposed index for all 5 cancer-types as well as with all cancer types. The regression results are below in table 5.

Table 5 shows the five models used in this study along the column headers, while row headers provide space for AIC and SIC measures for each cancer type. The purpose of this analysis is to evaluate the newly developed models (3-5) and compare them to two baseline models (1-2). Recall that in order for models 3-5 to unambiguously improve the baseline models 1-2, they must improve both AIC and SIC measures. Each model that improves upon the baseline models is shaded in grey in table 5.

For each cancer type, at least one of the three proposed models improved over the baseline models. Additionally, the use of a data-driven method such as AIC or SIC provides a clear improvement over the baseline models in each cancer-type. It is interesting to note that the best fitting model is not unambiguously from using AIC or SIC. This is a common problem that has been discussed in other studies such as Koehler and Murphree (1988). The only instance when AIC does not improve upon the base models is in the case of lung cancer, which suggests the use of 34 groups, which is unusually high based on the other selections. Alternatively, SIC provides a heavier penalty for new variables and tends to support lower parameterized models. This is clearly the case when SIC suggests the use of 1 group 60% of the time, where no group distinctions are provided. In two out of those three instances, SIC provided a clear improvement over baseline models. The use of raw principal component scores are also considered in model 3 and are shown to clearly improve on the baseline models in breast and FGS cancer types. In the case of breast cancer, it is interesting to note that this model provided the lowest AIC and SIC, making it a robust selection for best model fit.

The results from regressions on colon cancer provide some interesting insights. First, AIC suggests the use of deciles, which is in line with A&R. While this research makes the argument that deciles will not always be satisfactory, it can be optimal in some situations. It is also worth noting that model 4 improves in both AIC and SIC over model 1. These two models are identical with the exception of the new index that is used in model 4, which indicates its improved predictive power over the WI. The model suggested by SIC is to include only 1 group, which are really no group distinctions at all. This also presents the possibility that when an index might not always provide valuable information for a model, which can only be detected through the use of a data-driven method, such as information criteria. However, model 5 is not the favored model in this scenario as it does not improve upon the earlier models. For this reason, model 4 is the preferred model for colon cancer.

| | | WI | HCAI | | | |
|---|---|---|---|---|---|---|
| | | Deciles | All Vars | Prin Scores | Optimal AIC | Optimal SIC |
| Type | Model | 1 | 2 | 3 | 4 | 5 |
| Breast | AIC | -384.601 | -389.795 | -404.036 | -392.419 | -383.926 |
| | SIC | -338.822 | -333.451 | -382.907 | -343.118 | -355.755 |
| Colon | AIC | -397.097 | -405.139 | -395.164 | -408.757 | -392.295 |
| | SIC | -351.266 | -348.732 | -374.011 | -362.926 | -378.193 |
| Prostate | AIC | -24.605 | -23.846 | -25.446 | -34.810 | -28.660 |
| | SIC | 21.174 | 32.498 | -4.318 | 7.447 | -14.574 |
| Lung | AIC | -859.488 | -855.730 | -851.945 | -878.167 | -869.566 |
| | SIC | -813.554 | -799.195 | -830.744 | -747.431 | -830.699 |
| FGS | AIC | -132.891 | -131.983 | -144.948 | -143.995 | -143.995 |
| | SIC | -87.269 | -75.833 | -123.891 | -129.958 | -129.958 |

Table 5. Weighted Tobit goodness-of-fit regression results from 5 alternative models, by cancer type (Dependent variable: logged sum of one plus percentage late-stage cancer cases)

Based on figure 3, lung cancer is one particular type of cancer that is expected to be highly correlated with the HCAI. Because of the high degree of positive correlation, it is not surprising that AIC and SIC both suggest relatively high grouping values. For example, AIC suggests using 34 groups, while SIC suggests the use of 8 groups. However, while the AIC selected model has a very low AIC value (-878.167), SIC is not as impressive given the relatively large penalty factor for each of the 34 groups. Thus, the SIC selected group is able to improve over the base models in terms of AIC and SIC and is therefore the preferred model for lung cancer.

Many variables under study are inter-correlated. For example, figure 3 shows that percentage of the uninsured in Hispanic is significantly higher than that in non-Hispanic populations in Texas. Figure 4 shows that the percentage of late-stage cancer cases in Hispanics is higher than non -Hispanics in Texas. Hispanics also tend to have higher percentage obese individuals and socioeconomically deprived individuals. These facts suggest the necessity to adjust for covariates in assessing the association between the HCAI and delayed cancer detection. This will also allow determination of the role of ethnicity in the delayed cancer diagnosis.

We acknowledge in this research that not all types of cancer will have a significant relationship with the HCAI index. While the types of cancers that do have a correlation show health inequities, not all cancer types are thought to have inequities. In particular, we expect for cancers where detection is at a high cost (such as breast, colon, and lung cancer) to be particularly susceptible to health inequities. For example, in order to detect lung colon cancer, a costly colonoscopy is necessary which will have a lower compliance rate in individuals with poor access to health care services.

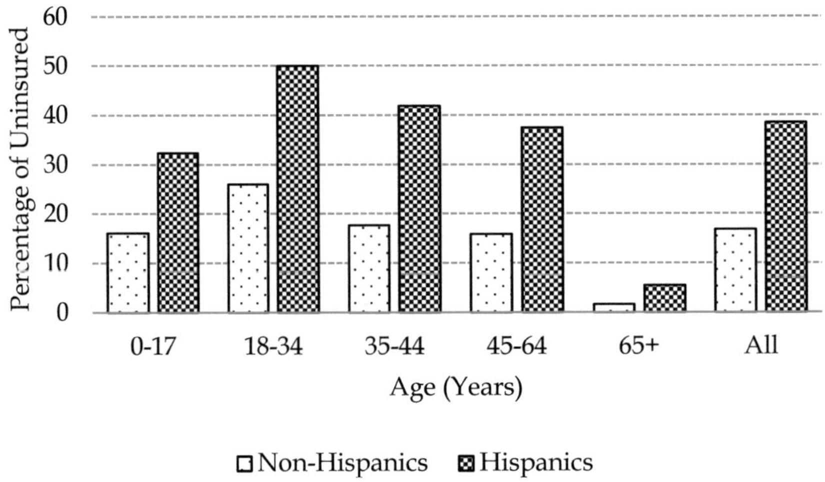


Fig. 3. Percentage of the Uninsured in Hispanic and Non-Hispanic populations in Texas.

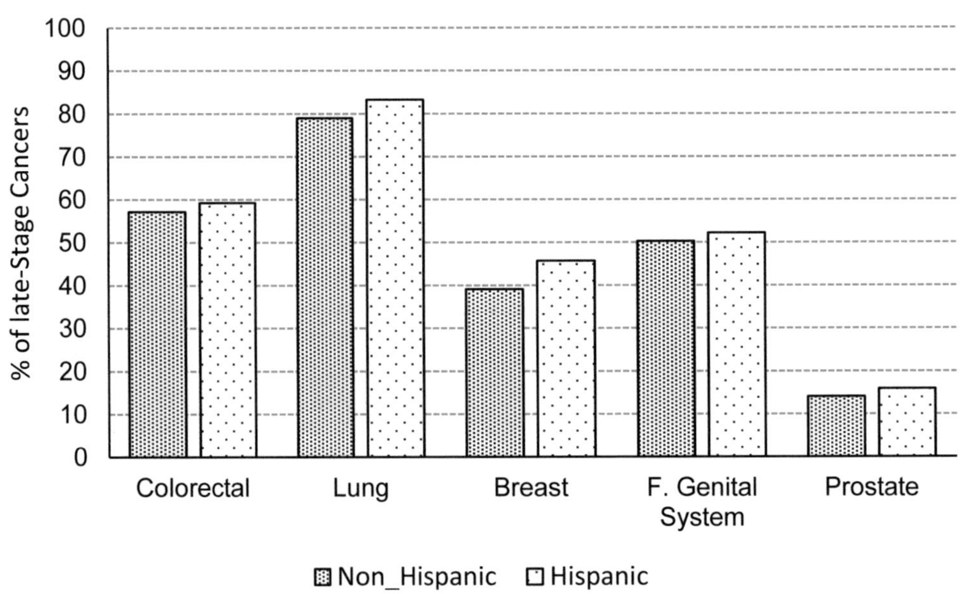


Fig. 4. Percentage of late-stage cancer cases in Hispanics and non -Hispanics in Texas.

#### 4.4.2 Regression results

Below are included the parameter estimates resulting from the optimally selected model based on the criteria stated above. In all four cases, at least one of the newly proposed models outperformed the baseline models in terms of AIC and SIC. In the case of breast cancer, late-stage cancer detection was best determined using the raw principal component scores. Parameter estimate results are listed in table 6.

| Variables | Estimate | SE | p-value |
|---|---|---|---|
| Intercept | 2.800 | 0.691 | <.0001 |
| log(% Hisp) | -0.039 | 0.019 | 0.040 |
| log(% Obese) | 0.325 | 0.218 | 0.136 |
| Prin1 | 0.024 | 0.005 | <.0001 |
| Prin2 | -0.021 | 0.006 | 0.001 |
| Sigma | 0.106 | 0.005 | <.0001 |

Table 6. Weighted Tobit regression results of parameter estimates of the % late-stage cases for breast cancer, using raw principal component scores from composite HCAI (model 3).

Based on the parameter estimates in table 2, it is clear that a high principal component score is associated with lower health care access. For this reason, it's not surprising to see a positive and significant parameter estimate of the %late-stage breast cancer cases on *Prin*1. However, *Prin*2 is negatively associated with % late-stage breast cancer cases. Percent Hispanic appears to have a negative impact on late-stage cancer detection. As shown in Figure 3, Hispanics, as a group, actually have a higher percentage of late-stage cancer cases relative to non-Hispanics. However, once we control for obesity and access to health care, we see that a 10% increase in the % Hispanics in a population is associated with a 0.39% decrease in late-stage breast cancer cases. As shown in figure 3, the percentage of uninsured (which is included in HCAI) is much higher in Hispanic populations than non-Hispanic populations. This speaks to the high degree of correlation between HCAI and % Hispanic variables. Obesity appears to have an insignificant impact, although it is worth noting that obesity is highly correlated with % Hispanic as well as the HCAI.

Table 7 shows the results for late-stage colon cancer incidents, where the use of deciles based on model 4 was selected based on its low AIC and SIC. Paradoxically, counties with the worst degree of access to health care (rank9) have a lower percentage of late-stage cancer cases. After detection of adenomatous polyps (not carcinoma yet) by screening, polypectomy is generally performed before the polyps evolve to early-stage cancer (Philips et al., 2011). Since counties with better health care accessibility tend to have higher CRC screening rate, many would-be early stage CRC cases are eliminated by polypectomy, resulting in a reduction in early-stage CRC cases. This may explain the paradoxical negative association between the percentage of late-stage CRC cases and HCAI. In order for the HCAI to be more useful in finding areas that are problematic in late-stage colon cancer determination, the data necessary needs to include the finding of colon cancer precursors, which is not present in our data.

| Variables | Estimate | SE | p-value |
|---|---|---|---|
| Intercept | 2.592 | 0.710 | 0.001 |
| log(% Hisp) | 0.026 | 0.020 | 0.182 |
| log(% Obese) | 0.390 | 0.220 | 0.076 |
| rank1 | 0.106 | 0.041 | 0.009 |
| rank2 | 0.119 | 0.043 | 0.006 |
| rank3 | 0.104 | 0.038 | 0.006 |
| rank4 | 0.021 | 0.039 | 0.582 |
| rank5 | 0.079 | 0.051 | 0.120 |
| rank6 | 0.076 | 0.047 | 0.106 |
| rank7 | 0.053 | 0.032 | 0.092 |
| rank8 | 0.121 | 0.027 | <.0001 |
| rank9 | 0.062 | 0.047 | 0.189 |
| Sigma | 0.102 | 0.005 | <.0001 |

Table 7. Weighted Tobit regression results of % late-stage colon cancer cases, composite HCAI grouping based on 10 groups (model 4).

The parameter estimate results for table 8 are shown below for prostate cancer. Almost all of the included parameter estimates are insignificant, which is consistent with previous results likely for the same reason (Philips et al., 2011).

| Variables | Estimate | SE | p-value |
|---|---|---|---|
| Intercept | 2.395 | 1.420 | 0.092 |
| log(% Hisp) | 0.014 | 0.041 | 0.731 |
| log(% Obese) | 0.077 | 0.443 | 0.862 |
| rank1 | -0.009 | 0.080 | 0.916 |
| rank2 | -0.005 | 0.089 | 0.964 |
| rank3 | -0.046 | 0.074 | 0.538 |
| rank4 | 0.054 | 0.084 | 0.522 |
| rank5 | 0.095 | 0.107 | 0.374 |
| rank6 | 0.044 | 0.064 | 0.495 |
| rank7 | 0.139 | 0.062 | 0.024 |
| rank8 | -0.056 | 0.058 | 0.338 |
| Sigma | 0.214 | 0.010 | <.0001 |

Table 8. Weighted Tobit regression results of % late-stage cases for prostate cancer, composite HCAI grouping based on 9 groups (model 4).

Table 9 shows the results for late-stage lung cancer detection. Based on figure 2, the relationship between lung cancer and HCAI has the strongest correlation. Based on this, it is not surprising to see so many significant variables in the results. For example, groups 1, 2, and 6 are statistically lower than the reference group 8 (omitted). Relative to the other groups, the omitted group (8) has the worst access to health care. It is worth mentioning that as the ranking group increases, the estimate decreases in general. Given that the HCAI also captures socioeconomic factors, it is also likely that counties with high HCAI have higher smoking rates than other counties given the relationship between income and the prevalence of smoking (Laaksonen et al., 2005).

| Variables | Estimate | SE | p-value |
|---|---|---|---|
| Intercept | 4.469 | 0.275 | <.0001 |
| log(% Hisp) | 0.002 | 0.008 | 0.894 |
| log(% Obese) | -0.018 | 0.085 | 0.836 |
| rank1 | -0.073 | 0.016 | <.0001 |
| rank2 | -0.019 | 0.017 | 0.266 |
| rank3 | -0.048 | 0.015 | 0.001 |
| rank4 | -0.025 | 0.016 | 0.115 |
| rank5 | -0.020 | 0.017 | 0.226 |
| rank6 | -0.040 | 0.011 | 0.001 |
| rank7 | 0.004 | 0.011 | 0.777 |
| Sigma | 0.042 | 0.002 | <.0001 |

Table 9. Weighted Tobit regression results of estimate of % late-stage cases for lung cancer, composite HCAI grouping based on 8 groups (model 5).

Finally, table 10 shows the parameter estimates associated with late-stage detection for FGS cancer. While the optimal model does not contain any information for the HCAI, the results show significant relationships with the percentage of Hispanics (a negative association) and obese (a positive association) individuals. The latter finding is consistent with previous report that obesity is a risk factor for endometrial cancer (a type of FGS cancer), while the former again suggests that the higher percentage of late-stage FGS cancer in Hispanics (Fig. 3) is due to their lack of access to health care and or socioeconomic deprivation.

| Variables | Estimate | SE | p-value |
|---|---|---|---|
| Intercept | 0.996 | 0.910 | 0.274 |
| log(% Hisp) | -0.080 | 0.031 | 0.010 |
| log(% Obese) | 0.955 | 0.296 | 0.001 |
| Sigma | 0.177 | 0.008 | <.0001 |

Table 10. Weighted Tobit regression results for female genital system (FGS) cancer with no HCAI grouping (models 4 and 5).

## 5. Final discussion

These results demonstrated in the previous section suggest that data-driven methods such as AIC and SIC as well as the use of raw principal component scores show significant improvement in terms of model-fit when explaining the percentage of late stage cancer cases. Additionally, the newly developed HCAI improves upon the index derived by A&R when evaluating late stage cancer detection. These findings add insight into future studies that will hopefully more effectively utilize PCA in different applications.

The use of data-driven techniques clearly has the ability to provide more flexibility in using PCA in a variety of applications without using rigid rules for grouping ordered indices. While some indices may be appropriately developed, the usage of this information in a regression or grouping context still desires some advances. Using this methodology, we found that the HCAI is positively correlated with the percentage of late-stage cases for breast and lung cancers. This research identifies that in order to promote early cancer detection policy makers should increase the rate of health access as defined in the HCAI. Additionally, the relative lack of explanatory power for the other cancers in this study points to some necessary areas for future research. First, in the case of colon cancer, more detailed data are needed. In order to properly assess colon cancer, data are needed that take account for pre-cancer detection of polyps and adenomas. Without that information, late- and early-stage cancer cannot be properly identified because early detection in the case of colon cancer is detection before it is cancerous. For prostate and FGS cancers, the index was not very explanatory which indicates a lack of health inequality in these areas that can be traced back to health access. Without a doubt, each cancer type is different and in this research we have identified a few types of cancers (breast and lung primarily) where health inequalities exist. These cancers tend to have relatively high costs associated with detection.

The issue of creating an index of health accessibility to explain disparities in health outcomes is an area of much importance in developing and developed counties. The present study shows that health care accessibility as measured by the HCAI impacts delayed detection of several cancers consistent with the results based on data in 2000 (Philips et al., 2011). The positive correlations between the two variables are statistically significant or marginal for all these cancers studied after controlling for several other potential determinants. This finding suggests that socioeconomic deprivation, health insurance coverage and health care service significantly impacts delayed detection of these cancers independent of percentage of Hispanics or percentage of obese individuals in counties of Texas which is one the states with most severe physician shortage and lowest health insurance coverage. Thus, we not only have for the first time proposed the HCAI but also for the first time validated its utility in its correlation with delayed cancer detection.

Previous study showed that WI was significantly associated with delayed detection (assessed by the ratio of late- to early-stage cancer cases similar to the percentage of late-stage cases) of FGS and lung-bronchial cancers but not breast cancer (Philips et al., 2011) in contrast to the results of the current study. The difference in the results is due to difference in study design with which several covariates are entered in the regression in the current study. Particularly, the HCAI covers not only socioeconomic variables but health insurance and health service (physician supply) as well.

It is of interest to note that Hispanics have higher percentage of late-stage cancer cases and also have higher percentages of obese individuals, higher percentages of uninsured individuals. In our multiple regression analysis, we find that delayed cancer detection is actually negatively associated with Hispanic after adjusting for socioeconomic status, physician supply, and percentage of uninsured, which are all included in the HCAI, and percentage of obese individuals. This suggests that their delay in cancer diagnosis is likely due to these factors rather than Hispanic culture per se or genetic predisposition.

These findings provide the evidence-base critical for decision makers to establish policies to promote early detection for effective cancer control targeting specific berries such as physician shortage, lack of health insurance and to improve socioeconomic conditions in general.

One promising avenue that was not undertaken in this study is that of latent class models in grouping. Latent class models provide a couple of advantages over the methods used here: (1) class identification is treated as an unknown variable, which is different from the mutually exclusive technique use in most studies; and (2) classes need not be similar sizes. One major limitation of finite mixture models is that they have been shown to be highly non-linear when more than a few groups are used. In our particular application, this would be a major obstacle given the high number of classes suggested by some models.

## 6. Acknowledgment

The authors wish to acknowledge the assistance from Kris Hargrave. Cancer incidence data have been provided by the Texas Cancer Registry, Cancer Epidemiology and Surveillance Branch, Texas Department of State Health Services, 1100 W. 49th Street, Austin, Texas, 78756, http://www.dshs. state.tx.us/tcr/default.shtm, or (512) 458-7523.

## 7. References

Abseyasekera, S. (2005). Multivariate Models for Index Construction, In: *Household Surveys in Developing and Transition Counties: Design, Implementation, and Analysis*, pp. 367-388, United Nations, 92-1-161481-3, New York, NY, USA.

Albrecht, J. and Ramasbramanian, L. (2004). The Moving Target: A Geographic Index of Relative Wellbeing, *Journal of Medical Systems*, Vol.28, No.4, pp. 371-384, 1573-589X

Amemiya, T. (1973). Regression Analysis when the Dependent Variable is Truncated Normal, *Econometrica*, Vol.41, No.6, pp. 997-1016, 1468-0262

Amemiya, T. (1984). Tobit Models: A Survey, *Journal of Econometrics*, Vol.24, No.1-2, pp. 3-61, 0304-4076

Akaike, H. (1974). A New Look at the Statistical Identification Model, *IEEEE Transactions on Automatic Control*, Vol.19, No.6, pp. 716-723, 0018-9286

Berenger, V. and Verdier-Chouchane, A. (2007). Multidimensional Measures of Well-Being: Standard of Living and Quality of Life Across Counties, *World Development*, Vol, 35, No. 7, 1259-1276.

Booysen, F., Van Der Berg, S., Burger, R., Von Maltitz, M. (2008). Using an Asset Index to Assess Trends in Poverty in Seven Sub-Saharan African Countries, *World Development*, Vol. 36, No. 6, 1113-1130.

Byers T. Two decades of declining cancer mortality: progress with disparity. Annu Rev Public Health 2010, 31:121-32.

Centers for Disease Control and Prevention (CDC). Ten great public health achievements--worldwide, 2001-2010. MMWR Morb Mortal Wkly Rep. 2011;60:814-818

Clegg LX, Reichman ME, Miller BA, Hankey BF, Singh GK, Lin YD, Goodman MT, Lynch CF, Schwartz SM, Chen VW, Bernstein L, Gomez SL, Graff JJ, Lin CC, Johnson NJ, Edwards BK. Impact of socioeconomic status on cancer incidence and stage at diagnosis: selected findings from the surveillance, epidemiology, and end results: National Longitudinal Mortality Study. Cancer Causes Control. 2009;20:417-435.

Cortinovis, I., Vella, V., and Nkiku, J. (1993). Construction of a Socio-Economic Index to Facilitate Analysis of Health Data in Developing Countries, *Social Science & Medicine*, Vol, 36, No. 8, 1087-1097, 0277-9536.

Crampton P, Salmond C, Sutton F. NZDep91: a new index of deprivation. Soc Policy J New Zealand. 1997;9:186-193

Davis, B. (2003). *Choosing a Method for Poverty Mapping, Food and Agriculture Organization of the United Nations*, 92-5-104920-3, Rome, Italy. Accessed at http://www.fao.org/DOCREP/005/y4597e/y4597e00.HT on Oct. 7, 2011.

Deaton, A. (2003). Health, Inequality, and Economic Develoment. *Journal of Economic Literature*, Vol 41, 113-158,

Everitt, B. and Hothorn, T. (2011). *An Introduction to Applied Multivariate Analysis with R*, Springer Science, 978-1-4419-9649-7, New York, NY, USA.

Everitt, B., Landau, S., Leese, M. and Stahl, D. (2011). *Cluster Analysis* (5th ed.), John Wiley & Sons, 978-0470220436, Chichester, UK.

Ewing, R., Schmid, T., Killingsworth, R., Zlot, A., Raudenbush, S. (2003). Relationship Between Urban Sprawl and Physical Activity, Obesity, and Morbidity. *American Journal of Health Promotion*, Vol. 18, No. 1, 47-57.

Filmer, D. & Pritchett, L.H. (2001). Estimating Wealth Effects without Expenditure Data – or Tears: An Application to Educational Enrollments in States of India, *Demography*, Vol, 38, No. 1, 115-132.

Greene, W. (2012). *Econometric Analysis* (7th edition), Pearson Education, Inc., 0-13-139538-6, Upper Saddle River, NJ, USA.

Gwatkin, D.R., Rutstein, S., Johnson, K., Pande, R., and Wagstaff, A. (2000). Socio-Economic Differences in Health, Nutrition, and Population. HNP/Poverty Thematic Group, World Bank, Washington, DC.

Koehler, A. & Murphree, E. (1988). A Comparison of the Akaike and Schwarz Criteria for Selecting Model Order, *Journal of the royal Statistical Society. Series C (Applied Statistics)*, Vol.37, No.2, pp. 187-195, 1467-9876

Kolenikov, S. and Angeles, G. (2009). Socioeconomic Status Measurement with Discrete Proxy Variables: Is Principal Component Analysis a Reliable Answer? *Review of Income and Wealth*, Vol. 55, No. 1, 128-168.

Kula, J. (2004). AIC and BIC: Comparisons of Assumptions and Performance, *Sociological Methods & Research,* Vol.33, No.2, pp. 188-229, 1552-8294.

Laaksonen, M., Rahkonen, O., Karvonen, S. & Lahelma, E. (2005). Socioeconomic Status and Smoking: Analysing Inequalities with Multiple Indicators, *European Journal of Public Health,* Vol.15, No.3, pp. 262-269, 1464-360X.

National Cancer Institute . Surveillance Epidemiology and End Results. Standard populations - 19 age groups. http://seer.cancer.gov/stdpopulations/stdpop.19ages.html (accessed 09 Mar 2011

Philips BU, Jr., Gong G, Hargrave KA, et al. Correlation of the Ratio of Late-stage to Non-Late-stage Cancer Cases with the Degree of Socioeconomic Deprivation among Texas Counties. International Journal of Health Geographics 2011;10: 12.

Pomeroy, R.S., Pollnac, R.B., Katon, B.M., & Predo, C.D. (1997). Evaluating Factors Contributing to the Success of Community-Based Coastal Resource Management: The Central Visayas Regional Project-1, Philippines. *Ocean and Coastal Management,* Vol. 36, No. 1-3, 97-120.

Richardson LC, Royalty J, Howe W, et al. Timeliness of breast cancer diagnosis and initiation of treatment in the National Breast and Cervical Cancer Early Detection Program, 1996-2005. Am J Public Health 2010, 100:1769-1776.

Ruel, M.T. and Menon, P. (2002). Child Feeding Practices Are Associated with Child Nutritional Status in Latin America: Innovative Uses of the Demographic and Health Surveys. *Journal of Nutrition,* Vol. 132, No. 6, 1180-1187, 1541-6100.

Salmond C, Crampton P, Sutton F. NZDep91: A New Zealand index of deprivation. Aust N Z J Public Health. 1998;22:835-837

Schwartz, G. (1978). Estimating the Dimension of a Model, *Annals of Statistics,* Vol.6, No.2, pp. 461-464, 0090-5364

Singh GK. Area deprivation and widening inequalities in US mortality, 1969-1998. Am J Public Health. 2003;93:1137-1143.

Singh GK, Siahpush M. Increasing inequalities in all-cause and cardiovascular mortality among US adults aged 25-64 years by area socioeconomic status, 1969-1998. Int J Epidemiol. 2002;31:600-613.

Texas State Data Center. Published Reports. http://txsdc.utsa.edu/Reports. accessed 13 Oct. 2011.

Texas Department of State Health Services . County supply and distribution Tables. http://www.dshs.state.tx.us/chs/hprc/PHYS-lnk.shtm (accessed 17 Mar 2011).

Tobin, J. (1958). Estimation of Relastionships for Limited Dependent Variables, *Econometrica,* Vol.26, No.1, pp. 24-36, 1468-0262

U.S. Census Bureau. 2005-2009 American Community Survey 5-Year Estimates. http://factfinder.census.gov/servlet/DCGeoSelectServlet?ds_name=ACS_2009_5YR_G00_

Vyas, S. and Kumaranayake, L. (2006). Constructing Socio-Economic Status Indices: How to Use Principal Component Analysis. *Health Policy and Planning,* Vol. 21, No. 6, 459-468, 1460-2237.

Ward E, Jemal A, Cokkinides V, Singh GK, Cardinez C, Ghafoor A, Thun M. Cancer disparities by race/ethnicity and socioeconomic status. CA Cancer J Clin. 2004 Mar-Apr;54(2):78-93

World Health Organization. Commission on Social Determinants of Health. Closing the gap in a generation: Health equity through action on the social determinants of health. http://www.who.int/social_determinants/thecommission/finalreport/en/index.html (accessed 25 August 2011)

# 10

# Principal Component Analysis Applied to SPECT and PET Data of Dementia Patients – A Review

Elisabeth Stühler and Dorit Merhof
*University of Konstanz*
*Germany*

## 1. Introduction

Alzheimer's disease (AD) is the most common cause of dementia, followed by vascular and frontotemporal dementia. Approximatly 8% of the population in developed countries is impaired by AD at the age of 65, with the risk expanding to 30% for individuals over the age of 85 years (Petrella et al. (2003)). Due to the increasing life expectancy, the spread of AD is estimated to triple over the next 50 years (Petrella et al. (2003)). If AD remained untreated, the economic impact on society would increase dramatically (Carr et al. (1997); Mueller et al. (2005)), but it is even more important to alleviate the psychological strain on patients and their relatives. Normally, a patient affected by AD has an anticipated average life expectancy of 8-10 more years, divided into several stages of the disease. The neuropathological stages of AD are described in detail in Braak & Braak (1991), where the development of amyloid deposition and neurofibrillary changes within the brain are explained. These changes can already be observed in the preclinical phase, i.e., before clinical symptoms occur. Clinical symptoms usually begin (in early stages) with memory and learning impairment, followed by alterations in judgement, display of social behavioral problems and reduced faculty of speech. In late stages of AD, motoric and sensory functions are affected as well (Selkoe (2001)).

First pharmaceuticals for treatment of AD symptoms were recently developed, and there are several more under clinical trials at the moment, which in turn require the early detection of AD (Petrella et al. (2003)). Cases with early-onset AD are usually diagnosed with mild cognitive impairment (MCI). According to Tabert et al. (2006), about 10% of cases with amnestic MCI (i.e., patients with memory deficits) and about 50% of MCI cases with further cognitive domain deficits will convert to AD within three years.

In early stages of AD, structural changes within the brain are difficult to detect, as they are restrained to very specific areas (e.g., hippocampal atrophy) until AD is advanced to a middle or later stage. Petrella et al. (2003) advise therefore to resort to nuclear medicine imaging which captures more subtle pathological changes, rather than to magnetic resonance imaging (MRI) or X-ray computed tomography (CT) as they are less capable for early detection of dementia. Prevalent in clinical assessment of AD are positron emission tomography (PET) and single-photon emission computed tomography (SPECT), where PET is observed to perform superior to SPECT for distinguishing between AD and a control group (CTR), e.g., in Herholz et al. (2002b).

In nuclear medicine, the biomarkers used for detection of AD include increased $\beta$-amyloid deposition, decreased glucose metabolism and reduced blood flow in the brain, which are among many indicators for AD. Furthermore, AD can be correlated to several risk factors, such as the genetic inheritance of the $\epsilon4$ allele of the apolipoprotein E (APOE) or the increased accumulation of tau proteins in the cerebrospinal fluid (CSF).

SPECT or PET images are typically evaluated by clinical reading, but this procedure requires expert knowledge, is time-consuming and rater-dependent. Therefore, statistical analyses for automated detection or prediction of AD progression in MCI have been subject to recent research.

As SPECT or PET datasets contain a large amount of information, i.e., more than $10^5$ voxel-values within the whole-brain region, and as usually up to 100 subjects are considered in a study, statistical analysis of the 3D-images is very challenging. It includes univariate analysis where a voxel-wise comparison is performed to differentiate between AD and normal controls (CTR), e.g., in Dukart et al. (2010) and Habeck et al. (2008). More recently also multivariate analysis, such as principal component analysis (PCA), has been applied to enable statistical evaluation of all voxel-values at the same time, thereby accounting not only for differences in single intensity values but also correlations between regions. This usually outperforms univariate analysis in the early identification of AD (Habeck et al. (2008)).

In many studies, PCA is therefore employed to either reduce the high dimensionality of the data (Markiewicz et al. (2009; 2011a;b); Merhof et al. (2009; 2011)), to discriminate between dementia of Alzheimer type and asymptomatic controls (Fripp et al. (2008a); Habeck et al. (2008); Scarmeas et al. (2004)) or to assess the amount of variability of the data (Fripp et al. (2008a;b)).

The objective of this review is to present and discuss these applications of PCA, and also to give an insight into adequate preprocessing of the data and implementation of PCA:

Basically, any analysis of PET or SPECT data requires preprocessing of the data in a first step, comprising registration of each subject to a brain atlas (a.k.a. spatial normalization), smoothing of all voxel-values and normalization of intensities as briefly described in Section 2.2. This enables voxel-wise comparisons between images in univariate analysis (see Section 6.2.1) and the correlation (or interpretation of covariance) of all voxels within the whole-brain region in multivariate analysis.

After preprocessing, neuroimaging data is commonly reduced to a lower-dimensional subspace in the studies reviewed in this work. In most cases, this is achieved by PCA implemented as in Section 3.1, but also by the scaled subprofile model (SSM), which is a modification of PCA described in Section 3.2. Partial least squares correlation or regression (PLSC/ PLSR) is also related to PCA as it is based on the same decomposition procedure (Section 6.2.2).

The method to be used for dimensionality reduction and further analysis depends on the purpose of the study, and also on different criteria regarding stability of the dimensionality reduction. In Section 5, some criteria for the validation of PCA regarding stability and robustness are presented.

After PCA is accomplished on the neuroimaging data of AD patients and a CTR group, the resulting projections of all subjects can be used to train discrimination as described in Section 4.3. Employing MCI cases where AD is prognosed or was already confirmed, the disrimination can then be tested regarding its potential to detect AD in early stages.

A detailed outline of all methods presented in this review and a workflow for the analysis of PET and SPECT data is depicted in Figure 1.

## 2. Constitution of the data matrix

In all studies reviewed in this work, PET or SPECT images of asymptomatic controls and patients with AD are considered.

Both techniques generate three-dimensional images of the brain, depicting the aggregation of a radioactive tracer and therefore providing metabolic information (e.g., glucose metabolism, brain perfusion or plaque deposition) within distinct brain areas. Although PET produces images with higher resolution, SPECT is considered to be adequate to detect abnormalities of perfusion which are specific for AD (e.g., Caroli et al. (2007); Herholz et al. (2002b); Ishii et al. (1999); Matsuda (2007)). As SPECT is – in comparison to PET – more prevalent and economical, it is commonly the preferred imaging method according to Minati et al. (2009).

Overall, three tracers were used for the SPECT and PET data examined in this review:

SPECT-scans based on the tracer technetium-99m-ethyl cysteinate dimer ($^{99m}$Tc-ECD) show perfusion patterns of the brain. In Herholz et al. (2002b) it is observed, that superior results regarding the detection of AD and the assessment of affected brain regions can be achieved by $^{18}$F-2-fluoro-2-deoxy-D-glucose (FDG) PET-imaging, which measures the changes in glucose metabolism (Ishii et al. (1999)). The tracer $^{11}$Carbon-Pittsburgh compound B ($^{11}$C PiB) is able to quantify $\beta$-amyloid deposition in the diseased brain as pointed out by Klunk et al. (2004).

### 2.1 Sample selection

If a groupwise comparison of subjects with AD and CTR is intended by statistical analysis of PET or SPECT images, not all datasets are apt to be included in the sample. Especially the CTR group should be gender- and age-matched to account for age-related atrophy within the brain. The effect of age-related changes of the brain on multivariate analysis such as PCA is discussed in Zuendorf et al. (2003), where at least two principal components, i.e., two independent directions of variance, could be correlated with age.

Subjects representing the AD group should not be affected by other neuro-degenerative diseases, and are also recommended to be in a stage of mild to moderate AD. Cases of late AD, where almost the whole brain is affected, would put too much emphasis on regions still unaffected by early-onset AD.

### 2.2 Preprocessing of the images

In each study, the PET or SPECT images selected for statistical analysis are registered to an atlas of the brain (a.k.a. spatially normalized), smoothed and intensity normalized. An optimized preprocessing method for SPECT images is presented in Merhof et al. (2011), where a dataset containing AD and CTR subjects is preprocessed by various methods, and subsequently tested by PCA and discrimination analysis. Best results regarding robustness and classification accuracy are achieved by affine registration (Bradley et al. (2002); Herholz et al. (2002a)), smoothing of voxel intensity values based on the standard isotropic Gaussian filter with full width half maximum (FWHM) of 12mm (Herholz et al. (2002a); Ishii et al.

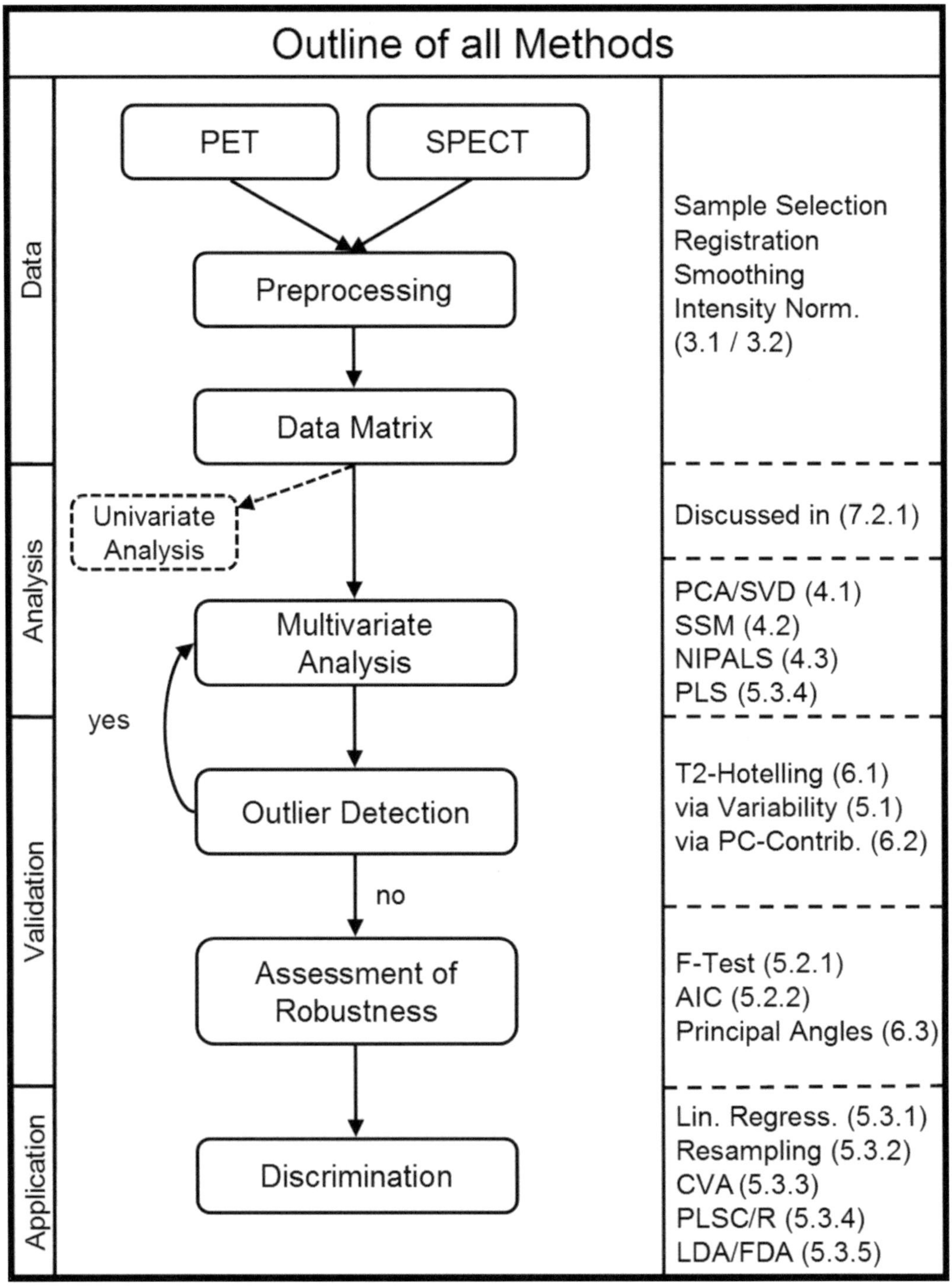


Fig. 1. Application flow and methods presented in this review

(2001); Matsuda et al. (2002)) and normalization according to the 25% brightest voxels within the whole-brain region.

To our knowledge, a detailed review of preprocessing methods and their impact on PCA applied to PET images (and subsequent analysis, with regard to discrimination of AD and CTR) has not yet been published. However, Herholz et al. (2004) present a detailed and effectual survey of the general handling of PET images in neuroscience.

After sample selection and preprocessing, the voxel-values of each scan are converted into a vector and all datasets are stored column-wise in a data matrix $X$ as depicted in Figure 2. This enables univariate (i.e., voxel-wise) comparison, or multivariate analysis (e.g., PCA and in some cases subsequent discriminant analysis), as the observations for each voxel or brain region are now represented row-wise in $X$.

## 3. Principal component analysis

Two main implementations of PCA are considered in this review:

- The first and widely used approach is based on variance, where principal components (PCs) are determined by singular value decomposition (SVD) of the $m \times n$ data matrix $X$ (e.g., Markiewicz et al. (2009); Merhof et al. (2011)). In this way, it is not necessary to compute the $m \times m$ covariance matrix $XX^T$ which is time-consuming due to the very high dimensionality $m$ of the data (in SPECT and PET images, the whole-brain region contains more than $10^5$ voxels) and might even lead to a loss of precision.
- In a second implementation, PCA is modified to a scaled subprofile model (SSM) (e.g., in Habeck et al. (2008); Scarmeas et al. (2004)). SSM is also covariance-based, but also captures the regional patterns of brain function and thereby advances subsequent discriminant analysis. PCA is performed, and afterwards subject scaling factors are calculated to convey each subject's contribution to a fixed PC as described in Alexander & Moeller (1994) and Moeller et al. (1987).

Another framework is presented in Miranda et al. (2008) and Duda et al. (2001), where an approximation of PCA is achieved by minimizing the mean square error of approximation, also characterized as a total least squares regression problem (Van Huffel (1997)). However, to our knowledge this method has not been applied to SPECT or PET data of patients affected by AD and a CTR group so far and is therefore not considered further in this review.

As PCA is sensitive to outliers within the data, methods to perform a more robust PCA are also considered, e.g., in Serneels & Verdonck (2008). However, for analysis of SPECT or PET images the underlying data usually contains a manageable amount of subjects and can therefore be sorted manually or by applying tests as presented in Section 5. It is also advisable to visualize those PCs intended to remain in the subsequent analysis as explained in Section 3.4. Thereby, it can be assured that only those regions of the brain which explain the difference to CTR in mild to moderate AD are considered, and that there are no abnormally prominent regions identified by the PCA.

In this review, PCA via SVD and SSM are presented in Sections 3.1 and 3.2. During resampling, both of these methods may become unstable; therefore, an alternative implementation is indicated in Section 3.3.

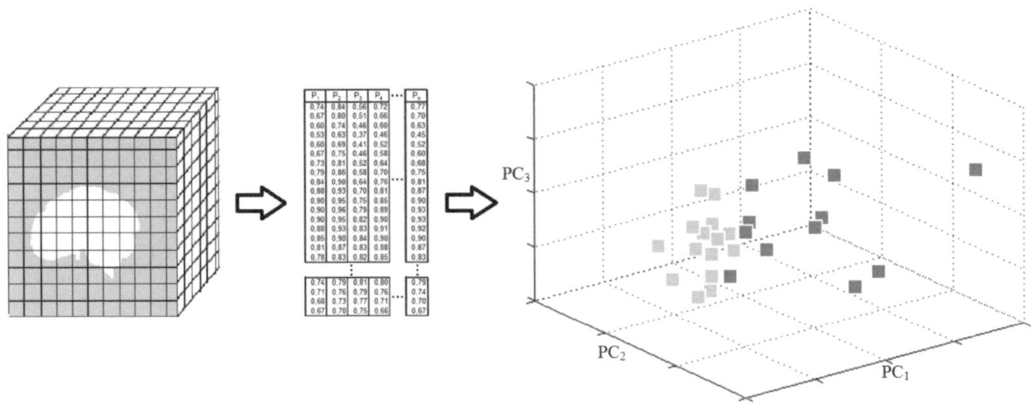


Fig. 2. Exemplary development of PCA on neuroimaging data. *Left*: Volume dataset. *Middle*: Data matrix $X$ containing one volume dataset per column. *Right*: Projection into subspace spanned by three PCs.

A general outline of the PCA on neuroimaging data is depicted in Figure 2, where each image contained column-wise in the data matrix $X$ is projected into a subspace spanned by the first three PCs.

### 3.1 PCA via singular value decomposition

Prior to multivariate analysis, the overall mean of the data matrix $X$ is usually set to zero by subtracting the mean vector from each column. This is not compulsive but considerably simplifies further analysis (Habeck et al. (2010); Miranda et al. (2008)).

Singular value decompositon (SVD) of the data matrix is applied by $X = VSU^T$ (as in Markiewicz et al. (2009; 2011b)). As $X$ is of size $m \times n$ with $m >> n$, it is sufficient to compute only the first $n$ columns of $V$, i.e., the first $n$ PCs. If the datasets contained in $X$ were mean-centered beforehand, $X$ is of rank $n - 1$ at most, so the number of PCs to be computed is furthermore reduced to $n - 1$ (this follows from the properties of the associated centering matrix, i.e., it is idempotent and therefore of rank $n - 1$).

The columns of $V$ are sorted according to the magnitude of their associated singular values, i.e., the diagonal elements of $S$. PC scores for all subjects are computed by $V^T X$, i.e., each subject is projected into a PC-subspace as depicted in Figure 2. If all PCs were used, all variance of the data would be maintained, but a subset of only a few PCs is sufficient to represent more than 60% of the variance (see Section 4.2).

### 3.2 PCA modified to scaled subprofile model analysis

Scaled subprofile model (SSM) analysis enhances the discriminative powers of the PCA as it not only extracts the covariance structure within groups but also assesses the contribution of each subject to the covariance pattern. As explained in detail in Alexander & Moeller (1994), the data matrix $X$ is natural log-transformed, and subsequently mean-centered over brain regions and subjects. Then PCA is performed on $X$ as in Section 3.1 by $X = VSU^T$ and $n$ PCs are contained in $V$. Furthermore, PCA via eigenvalue decomposition of the $n \times n$

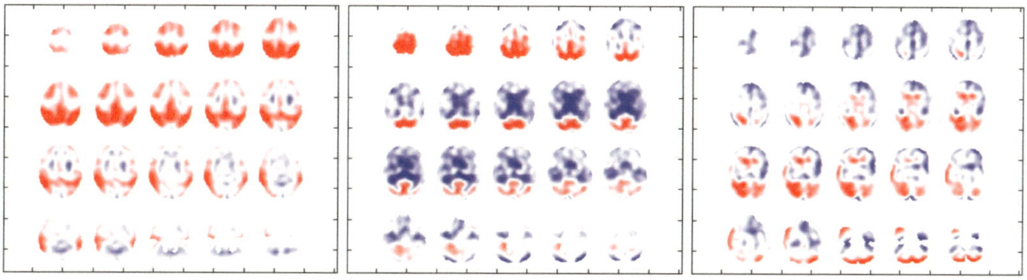

(a) Principal Component 1 (b) Principal Component 2 (c) Principal Component 3

Fig. 3. Examples for the first three principal components of a dataset containing SPECT images of 23 asymptomatic controls and 23 patients with Alzheimer's disease.

covariance matrix $X^T X$ is applied, resulting in $n$ eigenvectors which represent sets of subject scaling factors (SSFs). The associated eigenvalues to the PCs, SSFs respectively, of both decompositions are equal. Whereas the PCs describe the main directions of variance in the data, the SSFs describe the degree of subjects' expression of the fixed PC (Habeck et al. (2008)). The expression of the PC-scores $V^T X$ for each subject is quantified by the associated SSFs in accordance with the procedure described in Alexander & Moeller (1994) and Habeck et al. (2008). As above in Section 3.1, only a few PCs and associated SSFs are sufficient to reflect pathological differences within the data.

### 3.3 PCA via non-linear iterative partial least squares

During bootstrap resampling (e.g., to assess robustness of the PCA as described in Section 5.3), individual subjects may be repeatedly present within the resampled data matrix, thereby rendering the SVD unstable.

In this case, Markiewicz et al. (2009) propose the application of the non-linear iterative partial least squares (NIPALS) algorithm. The (resampled) data matrix $X$ is decomposed by $X = v_1 t_1^T + R$, where $v_1$ denotes an estimate of the first PC of $X$, $t_1$ represents the appendent PC scores of each subject and $R$ is the remaining residual. As an estimate for $v_1$, Wold et al. (1987) propose the (normalized) column of $X$ with the largest variance, but the employment of other start vectors is possible as well (Miyashita et al. (1990)). The NIPALS algorithm is iterated with $R$ acting as new start matrix until all PCs required for further analysis are computed. The NIPALS method is related to canonical correlation analysis (Höskuldsson (1988)), and thereby also to canonical variate analysis as presented in Section 4.3.3.

### 3.4 Visualization of PCs

Axial slices of PCs can be visualized as illustrated in Figure 3, where $^{99m}$Tc-ECD SPECT images of 23 subjects with Alzheimer's disease and 23 asymptomatic controls were decomposed by PCA via SVD. As PCA seeks directions for representation (rather than discrimination), the displayed patterns are not to be mistaken with discriminant images.

The voxel-values of each PC are converted back into a three-dimensional image (reverse to the procedure in Section 2.1), and every third slice of the PC-image between slice 15 and 72 is depicted. The intensities of the voxel-values are mapped to a colormap ranging from blue negative values to red positive values. Neutral voxel-values (= 0, as the data was mean-centered) correspond to white.

The main variance observed in the temporal lobes is captured in the first PC, whereas the second PC expresses changes in the area of the ventricles, which could be attributed to the expansion of ventricles in AD patients. A first intuitive conclusion might be to maintain only the first two PCs for further analysis, as those describe the regions usually affected by AD within the brain most distinctly. However, there are more reliable methods to decide which PCs to keep (see Section 4.2).

## 4. Applications

In the statistical evaluation of neuroimages, the main purpose of PCA is primarily an efficient reduction of the very high dimensionality and the removal of noise and redundant information within the data. The PCs produced during PCA represent the axes of the new subspace, into which the original datasets containing the voxel-values are transformed. The decision which PCs are suited to represent the data sufficiently has a great impact on all further analysis. Therefore, the contribution of each PC should be thoroughly evaluated. Different criteria for choosing a well-fitting subset of PCs are presented in Section 4.2. Also, the measurement of the amount of variability maintained within each PC is closely connected with the question of its significance (see also Section 4.1).

If the dataset at hand contains two (or more) groups of subjects, the PCs established to be relevant for further analysis are found to notably describe those regions within the brain, which differ significantly across groups. PCA can therefore be useful to train a discrimination or to provide the basis for subsequent discriminant analysis as presented in Section 4.3.

### 4.1 Explanation of the variability

Under the condition that the variables (voxels) of all subjects are on the same scale (this has to be ensured during preprocessing of the images), the variance of the $i$th PC equals its associated eigenvalue $e_i$ (Massy (1965)). The percentage of the accumulated variance represented by any number $n$ of all $N$ PCs is then calculated by

$$var(n) = \frac{\sum_{i=1}^{n} e_i}{\sum_{i=1}^{N} e_i}. \tag{1}$$

In several studies it is observed that the first few PCs generally account for more than 60% of the variability (e.g., in Habeck et al. (2008); Markiewicz et al. (2009)). The percentage of the cumulative variance explained is used by Fripp et al. (2008a) to compare different methods for preprocessing of the data, e.g., spatial registration to different brain atlases.

### 4.2 Dimensionality reduction

In neuroimaging, the number of variables $m$ (i.e., voxels of the whole-brain region) greatly outnumbers the number of observations $n$ (i.e., subjects included in the study). For this reason, a dimensionality reduction of the data before further analysis, such as discrimination or correlation (as in Pagani et al. (2009)), is commonly applied. PCA is well suited for this purpose, as it reduces the variable space to a few dimensions only. It also helps to focus on the main directions of variance within the data (i.e., the first few PCs) and treats unused PCs corresponding to lower eigenvalues as noise in the data.

In each of the reviewed studies, only the first few principal components (PCs) are used to represent the main variance of the data. In some cases, this is justified by execution of the partial F-test as presented in Section 4.2.1(Markiewicz et al. (2009)), by calculation of the cumulative variance explained by the PCs (e.g., Fripp et al. (2008a); Zuendorf et al. (2003), see also Section 4.1) or by application of Akaike's information criterion (Habeck et al. (2008); Scarmeas et al. (2004), see also Section 4.2.2).

### 4.2.1 Partial F-test

The partial F-test measures which PCs add significant variance to the data (Markiewicz et al. (2009)). In the beginning, the mean-centered data matrix $X = X_{start}$ is entered into a regression model, and its residual sum of squares $rss(1)$ is computed. In a first iteration, the first PC $v_1$ is added to the model and prediction values for the original data matrix are calculated by $v_1 v_1^T X_{start}$. Then the residual sum of squares of the deviation matrix $D = X_{start} - v_1 v_1^T X_{start}$ is calculated. In each of the following $N-1$ iterations, $D$ and the next PC are entered into the model.
F-values and p-values for each iteration are calculated by

$$F_n = \frac{(rss(n) - rss(n+1))(N-n)}{rss(n+1)} \tag{2}$$

and

$$p_n = 1 - fcdf(F(n)), \tag{3}$$

where $fcdf$ denotes the F cumulative distribution function with numerator and denominator degrees of freedom 1 and $N-n-1$.
As the limiting factor for number of PCs, Markiewicz et al. (2009) propose $p$ to be lower than 0.05, which is a standard level of significance.

### 4.2.2 Akaike's information criterion

Similar to partial F-test, Akaike's information criterion (AIC) determines the subset of PCs which represents the best fitting model (Akaike (1974)).

AIC-values are calculated by

$$A = -2log(L) + 2K, \tag{4}$$

where $L$ denotes the maximum value of the log-likelihood function of the model and $K$ the number of estimable parameters (Burnham & Anderson (2002)). The model which scores the smallest AIC-value $A$ is considered to be the best fitting one. As AIC may be biased if the ratio of sample size and number of parameters is too small (e.g., $\frac{n}{K} < 40$), Sugiura (1978) proposes a correction factor (AIC$_c$):

$$A_c = A + \frac{2K(K+1)}{n-K-1}. \tag{5}$$

Burnham & Anderson (2002) recommend the usage of AIC$_c$ in any case, as AIC and AIC$_c$ are similar for a sufficiently large ratio $\frac{n}{K}$.

In Habeck et al. (2008), AIC-values $A$ are calculated only for models generated by the first six PCs (explaining more than 75% of all variance), and the best-fitting model with the lowest

AIC-value is chosen for subsequent analysis. However, it should be noted that the AIC does not recognize if none of the models is suited to represent the population, i.e., the PCs entered into the AIC need to be chosen carefully.

### 4.3 Discrimination methods

With regards to the early detection of AD, the discriminative power of PCA can be very valuable. Discrimination should be trained on subjects with mild to moderate AD and asymptomatic CTR, and afterwards be tested on MCI cases, thereby assessing the capability to detect early AD cases among the data collected for the study.

Due to the orthogonality of all eigenvectors, each PC is uncorrelated with all preceding PCs and therefore captures an independent feature of the dataset. As the main variance resides in the first PCs, they depict prominent features of the data (provided that there are no outliers). Hence, the PCs can be employed for the differentiation of groups within the dataset. Those PCs which best discriminate the subjects can either be determined in a linear regression model as presented in Section 4.3.1 (Habeck et al. (2008); Scarmeas et al. (2004)) or as in Section 4.3.2 by a leave-one-out approach (Fripp et al. (2008a)). If necessary, discrimination can be refined further, e.g., by Canonical variate analysis (Section 4.3.3) or Fisher's discriminant analysis (Section 4.3.5).

#### 4.3.1 Linear regression

Linear regression is a subtype of general regression analysis and is widely used for the identification of those independent variables, which relate strongly to the dependent variable (e.g., group membership). After the successful completion of the regression, it can furthermore be applied to predict the group membership of a newly added value.

The achieved PC-scores $\tilde{X}$ of each subject are entered as independent variables into the linear regression model $y = \tilde{X}b + \epsilon$. The vector $y$ of the subjects' group memberships, in this case AD and CTR, contains the dependent variables.

It is common to use only a subset of all PCs (determined by significance tests or the amount of variance they represent), but it should be noted that even a PC which captures little variance might still be related to a dependent variable (Jolliffe (1982)).

The regression results in a linear combination of those PCs which achieve the best differentiation of the two classes (e.g., Habeck et al. (2008); Scarmeas et al. (2004)).

If the dependent variables include more information than group membership (e.g., age or existence of genetic risk factors), partial least squares (PLS) regression can be applied (see also Section 4.3.4). This method generalizes PCA and multiple linear regression.

#### 4.3.2 Leave-one-out resampling

In leave-one-out resampling, one subject is drawn from the underlying data sample per iteration and subsequent analysis is applied. This measures the individual contribution of each subject and can therefore be applied to sort out abnormal interference of particular subjects where necessary.

In Fripp et al. (2008a), $n - 1$ out of $n$ images are decomposed by PCA in each iteration. Then, PC-scores of the subjects contained in the sample are plotted pairwise against each other.

Those PCs which generally provide the best cluster and separation of the groups within iterations are considered for further analysis.

#### 4.3.3 Outline of canonical variate analysis

Canonical variate analysis (CVA) is another regression model considered to enhance the discriminative strength of PCA in neuroimaging. Similarly to linear regression, it identifies the best separation of groups depending on PC-scores. The first canonical variable (CV) is the best of all possible linear combinations of PC-scores for differentiation of the groups and – analogous to PCs – the following CVs are computed under the condition to be orthogonal to all precedent CVs.

PCA is applied for dimensionality reduction and removal of noise (i.e., discarded PCs). The within- and between-group sum-of-squares and crossproduct matrices $W$ and $B$ are computed for the PC-scores of all subjects. Then the CVs, i.e., the eigenvectors of $W^{-1}B$, are linear combinations of PC-scores and are sorted by their discriminative power (Borroni et al. (2006); Kerrouche et al. (2006)). CV-scores of all subjects are calculated analogous to PC-scores.

#### 4.3.4 Outline of partial least squares correlation and regression

As in PCA, the main element of partial least squares (PLS) methods is the SVD, which is applied to the correlation matrix $YX^T$ (rather than the data matrix $X$ containing the mean-centered data, as in PCA). The independent variables are the mean-centered and normalized voxel-values of all brain images stored in $X$, and the $n$ vectors of dependent variables for all subjects (also mean-centered and normalized) form the $k \times n$ matrix $Y$. SVD of $YX^T$ produces $VSU^T$, where $S$ is a diagonal matrix containing singular values and $U$ and $V$ column-wise contain the left respectively right singular vectors. Analogous to PCA, it is sufficient to compute only the first few columns of $V$. Then, the high-dimensional data of $X$ is reduced by $T = X^TU$ (a.k.a. brain scores), and $Y$ is reduced to $Y^TV$ (a.k.a. behavior scores).

It depends on the intention of the study, in which way these results are further analysed and applied. Krishnan et al. (2011) give an elaborate survey of the main PLS methods used in neuroimaging as well as of practical implementations. Generally, they present two basic approaches, i.e., PLS regression and PLS correlation. PLS regression is a generalization of multiple linear regression and PCA (Abdi (2010)), and is used to predict behavior on the basis of neuroimages, in this case PET or SPECT data. PLS correlation focuses on the analysis of the relation between two groups within the dataset and can be subdivided into more specific applications according to the design of the research.

#### 4.3.5 Outline of linear and Fisher's discriminant analysis

Similar to CVA, linear discriminant analysis (LDA) seeks discriminative directions of the data rather than representative directions (as does PCA). It can be applied both to the original mean-centered voxel-values contained in the data matrix $X$ or in a second step after performance of PCA to the PC-scores of all subjects. The latter approach is preferable when dealing with high-dimensional data, as either the inverse of an $m \times m$ scatter matrix has to be computed or a generalized eigenvalue decomposition of $m \times m$ matrices is required.

Fisher's discriminant analysis (FDA) is a special application of LDA, without the constraints of normal distributed groups and equal group covariance. It has lately been applied several times to diffentiate between subjects with AD and normal controls, e.g., in Markiewicz et al. (2009; 2011a); Merhof et al. (2009; 2011).

The purpose of FDA is to maximize the ratio of the between- and the within-group scatter $S_B$ and $S_W$, thereby projecting the data into a one-dimensional subspace. This is achieved by the projection vector $w$, i.e., the solution of the generalized eigenvalue problem $S_W^{-1} S_B w = \lambda w$ (Duda et al. (2001)). Subsequent classification can be computed with very limited effort by a threshold or nearest-neighbor approach.

## 5. Derivation of robustness of the PCA

So far, PCA and its applications in neuroimaging were introduced, but not yet validated and discussed. It is very important to assess the robustness of the PCA (and, where necessary, subsequent procedures) before interpretation of the results, as instability and overtraining may occur for various reasons. PCA is sensitive to conspicuous cases and it is therefore recommended to inspect the resulting PCs before further analysis. In order to ensure that no pathologically abnormal cases (outliers) remain in the training set, the $T^2$-Hotelling test is executed, e.g., by Pagani et al. (2009); Zuendorf et al. (2003) (see Section 5.1). Kerrouche et al. (2006) also propose further measurement of the individual contribution of one observation to each PC (see Section 5.2), to assess if the removal of one observation changes the outcome of PCA significantly. Habeck et al. (2010) also observe that if the first PC contains more than 90% of the variance to the data, it is very probable that the dataset $X$ includes one or more outliers (see Section 4.1).

By bootstrap resampling of the dataset and subsequent PCA the instability caused by removal of a subset of subjects is measured (Markiewicz et al. (2009; 2011a); Merhof et al. (2011)) via principal angles between PC-subspaces.

### 5.1 Hotelling's T-square test

Hotelling's T-square test is an adaption of the Student's T-test to the multivariate case (Hotelling (1931)). As the F-distribution is more prevalent, the $T^2$-distribution is usually transformed to

$$T^2 \sim \frac{p(n-1)}{n-p} F_{p,n-p}, \tag{6}$$

where $n$ denotes the number of subjects and $p$ the number of PCs retained in the model. Let $y_i$ denote the column vector of PC-scores of the $i$th subject, then its $T^2$-value is obtained by $T^2 = y_i^T y_i$. Zuendorf et al. (2003) propose a threshold of $p < 0.01$, and further assess the validity of the $T^2$-test by adding an abnormal case to a set of normal controls in 15 iterations. However, the $T^2$-test can also be applied to a dataset containing two or more groups (Kerrouche et al. (2006); Pagani et al. (2009)).

### 5.2 Contribution of subjects to PCs

The amount of the contribution $c_{i,j}$ of the $i$th subject to the $j$th PC is measured by

$$c_{i,j} = \frac{1}{n-1} \cdot \frac{y_i(j)}{e_j}, \tag{7}$$

where $n$ denotes the number of all subjects, $y_i$ the column vector of PC-scores of the $i$th subject and $e_j$ the eigenvalue corresponding to the $j$th PC. An abnormally large value of $c_{i,j}$ indicates that the removal of the $i$th subject might significantly change the results of PCA (Kerrouche et al. (2006)).

### 5.3 Principal angles of PC-subspaces

In order to compare sets of PCs during resampling iterations, the use of principal angles between PC-subspaces of a fixed dimension is proposed by Markiewicz et al. (2009). If the largest principal angle between an original and resampled subspace is very small, the PCA can be considered to be sufficiently independent of the underlying training set. Otherwise, abnormal large principal angles can indicate that too many PCs (i.e., too much noise) are included in the analysis, or that the sample was not selected carefully enough with respect to outliers.

In bootstrap resampling, $n$ subjects are drawn with replacement from the original training set (Efron & Tibshirani (1993)). For better replication of the original set, AD and CTR cases are stratified in the bootstrap sample (Markiewicz et al. (2009)).

For every sample, PCA is performed and the subspace spanned by the first $i$ PCs is compared to the $i$-dimensional PC-subspace of the original set. This is achieved by calculating the largest principal angle between the two subspaces (Golub & van Van Loan (1996); Knyazev et al. (2002)). For any number of PCs, the mean angle across all iterations is computed. Increased angles indicate the destabilization of the PCA.

The same method can be applied with leave-one-out resampling, i.e., one subject of each group is dropped in every iteration (Markiewicz et al. (2009)).

The computation of prinipal angles should be treated with caution, as round-off errors might cause inaccurate estimates for small angles. A solution to this problem is proposed by Knyazev et al. (2002), where a combined sine and cosine based approach is presented and generalized.

## 6. Discussion

For analysis of SPECT and PET data, PCA is widely applied and commonly reported to deliver stable and efficient results when used correctly. However, some limitations of PCA outlined in Section 6.1 remain, which might interfere strongly with the outcome of the statistical analyses. In some cases it might even be advisable to apply alternative methods to obtain more reliable results. In Section 6.2, examples are given where the performance of PCA on neuroimaging data was investigated and compared to other approaches.

### 6.1 Limitations of the PCA in neuroimaging

As PCA is based solely on the decomposition of the covariance matrix, the underlying data must be dealt with carefully. The preprocessing of the images has a crucial impact on the outcome of the analysis as pointed out by Fripp et al. (2008a), where PCA on $^{11}C$ PiB PET data proved to be sensitive to inaccuracies originating from non-rigid registration and intensity normalization. On $^{99m}Tc$-ECD SPECT, the classification accuracy of AD and CTR subjects via PCA and subsequent FDA depends significantly on the data preprocessing method (Merhof et al. (2011)). Classification accuracy relies also on scanner type and reconstruction method of FDG-PET, if the data is aquired from a more heterogeneous dataset, e.g., from the ADNI database as in Markiewicz et al. (2011b).

Not only the preprocessing of the training set but also its composition is essential. This includes the stratification of groups, the sample sizes and the absence of outliers as described in Sections 6.1.1 and 6.1.2. Moreover, the number of PCs retained in the analysis is important and depends on the purpose of the study (see Section 6.1.3).

### 6.1.1 Sample size

The selection of subjects suited for training is constrained by many premisses, such as age- and gender-matched CTR cases, stage of AD, the absence of other neurodegenerative disorders and the quality of the scan. It is also preferable for all images to be aquired by the same scanner, as this improves comparability of the data. Therefore, most studies only include less than 30 subjects of each group, except for Markiewicz et al. (2011b) where previous results on a smaller and more homogeneously selected training set were validated on more than 160 AD and CTR cases obtained from the ADNI database. ADNI provides generally accessible data of patients diagnosed with AD or MCI and of normal controls collected from various clinical sites (Mueller et al. (2005)).

An under-sized training set might cause the extraction of instable features (Markiewicz et al. (2009)) resulting in overly optimistic results of subsequent analysis (Markiewicz et al. (2011a)). This might be remedied by bootstrap resampling of the training set but must rely on the assumption that the sample is representative of the population (Markiewicz et al. (2009)).

### 6.1.2 Sensitivity to outliers

As the covariance matrix is calculated empirically, the estimates of eigenvectors (PCs) are heavily influenced by outliers, i.e., pathologically abnormal cases within the training dataset. The variance caused by only one outlier may be captured within the first PC, which will thereby not regard the variance within regular cases and dramatically change further results. Approaches which substitute the original covariance matrix by a more robust estimate (e.g., in Debruyne & Hubert (2006)) exist, but these methods are not practical for datasets of high dimensionality. For this reason, it is highly recommended to determine outliers by additional testing when applying PCA to neuroimaging data.

### 6.1.3 Number of PCs

Although several approaches are presented in Section 4.2, the determination of how many and which PCs are best suited to represent the original images remains subject to interpretation. So far, much relies on the purpose of PCA and the further analysis of the data. An elaborate overview over criteria for estimating the number of significant PCs and their application is presented in Peres-Neto et al. (2005), and some of these apply to covariance matrices as well as correlation matrices. The application of such methods can be ambivalent, as reviewed by Franklin et al. (1995). In most studies, PCs are chosen according to their potential to explain data and their impact on robustness. It is therefore advisable to determine the number of retained PCs not only based on one criterion but also on the best possible trade-off between the resulting accuracy and robustness.

### 6.2 Comparison to similar methods

In Section 4, different extensions to PCA such as linear regression or canonical variate analysis are presented, and also an outline of methods with similar properties or intentions as PCA is given. The decision which of these methods is best to employ always depends on the underlying research question, on the data available and the selected sample.

Sections 6.2.1 to 6.2.3 provide a review of the most important methods frequently applied to neuroimaging data and compare them to results obtained from PCA.

#### 6.2.1 Univariate analysis of neuroimaging data

Univariate analysis measures the voxel-by-voxel correlation between groups (e.g., by a voxel-wise T-test in Habeck et al. (2008)) and thereby merely focuses on the identification of significant shifts between voxel-values. In neuroimaging, univariate methods are commonly used during image preprocessing, e.g., in Dukart et al. (2010) or Scarmeas et al. (2004) for intensity normalization. Voxel-wise analysis can also be used for differentiation between groups, but it is unanimously reported that multivariate approaches outperfom univariate analysis in this matter, especially in the detection of early-onset cases of dementia (Habeck et al. (2008); Scarmeas et al. (2004)). Another drawback of voxel-wise analysis is the sensitivity regarding the preprocessing of the data, and even under the assumption that an optimized normalization factor was applied, the interpretion of the results must be addressed in a multivariate fashion (McCrory & Ford (1991)). In another approach, Higdon et al. (2004) tried to apply a between-group T-test for dimensionality reduction, but this proved to be ineffective and even deteriorated accuracy results.

On the other hand, multivariate analysis is found to be more robust as it considers the entire covariance structure of the data (accounting for relations among regions) and withstands the deviation of individual voxel-values (Borroni et al. (2006); Habeck et al. (2008)). It thereby detects correlated alterations in a diseased brain, whereas univariate analysis might not be able to recognize these differences.

#### 6.2.2 Partial least squares

When examining very high-dimensional data, and especially for the discrimination into groups within the dataset, PLS has been reported to perform better than PCA regarding the classification accuracy (Higdon et al. (2004); Kemsley (1996)). This is rather self-evident, as PCA does not take into account further behavioral data (e.g., neuropsychological data such as Mini-Mental State Examination (MMSE) scores, age, years of education). If PLS is applied for dimensionality reduction, Kemsley (1996) reports that fewer PLS dimensions than PCs were required for a successful subsequent differentiation of the groups. This implies that PLS will capture the most discriminative attributes of the subjects within the first dimensions, rather than the representative directions generated by PCA.

Nevertheless, there are certain drawbacks in the application of PLS methods. PLS tends to overfit the data, so the determination of the number of PLS dimensions kept in the analysis is of decisive importance (Abdi (2010)). In addition, PLS may detect differences which are not characteristic of the examined groups but were produced randomly by noise within the underlying dataset (Kemsley (1996)). Furthermore, PLS only works under the assumption that behavior relates linearly to neuroimaging data (McIntosh & Lobaugh (2004)).

Overall, if allowances are made for these effects and significant behavioral data is available, PLS can still be a favourable alternative to PCA.

### 6.2.3 Linear discriminant analysis

As explained above, performance of PCA (or any other dimensionality reduction method) prior to LDA is preferable in neuroimaging due to the high dimensionality of the data and the resulting expensive computation. To our knowlegde, LDA as described above in Section 4.3.5 has not yet been applied to discriminate AD from CTR using voxel-values of PET or SPECT images of the whole-brain region, although McEvoy et al. (2009) utilize a stepwise approach of LDA to identify brain regions significant for differentiation.

In other areas also dealing with high-dimensional data, such as object recognition in images, LDA is usually considered to perform superior to PCA. But this is not necessarily the case for small training sets, as pointed out by Martínez & Kak (2001). In the same study they also observe PCA to be less biased than LDA, i.e., less constrained to the training set.

The overall good results regarding accuracy and robustness of the PCA-LDA or PCA-FDA approach (e.g., as presented in Markiewicz et al. (2009)) also indicate, that a preceding PCA does not impair the discriminant analysis.

## 7. Conclusion

PCA applied to SPECT or PET data is well suited to reduce the high dimensionality of the original dataset containing voxel-values of the whole-brain region. It achieves best results when data is transformed into a subspace spanned by a well-chosen subset of PCs that represents the variability within all datasets and at the same time reduces noise and redundant information. PCA can also be used successfully to train discrimination between AD and a set of asymptomatic CTRs with the intention to enable an early detection of AD, or to provide a stable and effective basis for the subsequent application of discriminant analysis.

## 8. References

Abdi, H. (2010). Partial least squares regression and projection on latent structure regression (PLS Regression), *Wiley Interdisciplinary Reviews: Computational Statistics* 2(1): 97–106.

Akaike, H. (1974). A new look at the statistical model identification, *IEEE Transactions on Automatic Control* 19(6): 716–723.

Alexander, G. E. & Moeller, J. R. (1994). Application of the scaled subprofile model to functional imaging in neuropsychiatric disorders: A principal component approach to modeling brain function in disease, *Human Brain Mapping* 2(1-2): 79–94.

Borroni, B., Anchisi, D., Paghera, B., Vicini, B., Kerrouche, N., Garibotto, V., Terzi, A., Vignolo, L., Luca, M. D., Giubbini, R., Padovani, A. & Perani, D. (2006). Combined 99mTc-ECD SPECT and neuropsychological studies in MCI for the assessment of conversion to AD, *Neurobiology of Aging* 27(1): 24 – 31.

Braak, H. & Braak, E. (1991). Neuropathological stageing of Alzheimer-related changes, *Acta Neuropathologica* 82: 239–259.

Bradley, K. M., O'Sullivan, V. T., Soper, N. D. W., Nagy, Z., King, E. M., Smith, A. D. & Shepstone, B. J. (2002). Cerebral perfusion SPET correlated with Braak pathological stage in Alzheimer's disease, *Brain* 125(8): 1772–1781.

Burnham, K. P. & Anderson, D. R. (2002 2nd edn). *Model Selection and Multi-Model Inference. A Practical Information-Theoretic Approach*, New York Springer-Verlag.

Caroli, A., Testa, C., Geroldi, C., Nobili, F., Barnden, L., Guerra, U., Bonetti, M. & Frisoni, G. (2007). Cerebral perfusion correlates of conversion to Alzheimer's disease in amnestic mild cognitive impairment, *Journal of Neurology* 254: 1698–1707.

Carr, D., Goate, A., Phil, D. & Morris, J. (1997). Current Concepts in the Pathogenesis of Alzheimer's Disease, *The American Journal of Medicine* 103(3, Supplement 1): 3S – 10S.

Debruyne, M. & Hubert, M. (2006). The Influence Function of Stahel-Donoho Type Methods for Robust Covariance Estimation and PCA.

Duda, R. O., Hart, P. E. & Stork, D. G. (2001). *Pattern Classification*, Wiley-Interscience.

Dukart, J., Mueller, K., Horstmann, A., Vogt, B., Frisch, S., Barthel, H., Becker, G., Möller, H. E., Villringer, A., Sabri, O. & Schroeter, M. L. (2010). Differential effects of global and cerebellar normalization on detection and differentiation of dementia in FDG-PET studies, *NeuroImage* 49(2): 1490 – 1495.

Efron, B. & Tibshirani, R. J. (1993). *An Introduction to the Bootstrap*, Chapman & Hall/CRC.

Franklin, S. B., Gibson, D. J., Robertson, P. A., Pohlmann, J. T. & Fralish, J. S. (1995). Parallel analysis: a method for determining significant principal components, *Journal of Vegetation Science* 6(1): 99–106.

Fripp, J., Bourgeat, P., Acosta, O., Raniga, P., Modat, M., Pike, K. E., Jones, G., O'Keefe, G., Masters, C. L., Ames, D., Ellis, K. A., Maruff, P., Currie, J., Villemagne, V. L., Rowe, C. C., Salvado, O. & Ourselin, S. (2008a). Appearance modeling of 11C PiB PET images: Characterizing amyloid deposition in Alzheimer's disease, mild cognitive impairment and healthy aging, *NeuroImage* 43(3): 430 – 439.

Fripp, J., Bourgeat, P., Raniga, P., Acosta, O., Villemagne, V. L., Jones, G., O'Keefe, G., Rowe, C. C., Ourselin, S. & Salvado, O. (2008b). MR-less high dimensional spatial normalization of 11C PiB PET images on a population of elderly, mild cognitive impaired and Alzheimer disease patients., *Medical Image Computing and Computer-Assisted Intervention* 11(Pt 1): 442–449.

Golub, G. H. & van Van Loan, C. F. (1996). *Matrix Computations (Johns Hopkins Studies in Mathematical Sciences)*, 3rd edn, The Johns Hopkins University Press.

Habeck, C., Foster, N. L., Perneczky, R., Kurz, A., Alexopoulos, P., Koeppe, R. A., Drzezga, A. & Stern, Y. (2008). Multivariate and univariate neuroimaging biomarkers of Alzheimer's disease, *NeuroImage* 40(4): 1503–1515.

Habeck, C., Stern, Y. & (2010). Multivariate Data Analysis for Neuroimaging Data: Overview and Application to Alzheimer's Disease, *Cell Biochemistry and Biophysics* 58: 53–67. 10.1007/s12013-010-9093-0.

Herholz, K., Herscovitch, P. & Heiss, W. (2004). *NeuroPET: PET in Neuroscience and Clinical Neurology*, Springer, Berlin.

Herholz, K., Salmon, E., Perani, D., Baron, J., Holthoff, V., Frölich, L., Schönknecht, P., Ito, K., Mielke, R., Kalbe, E., Zündorf, G., Delbeuck, X., Pelati, O., Anchisi, D., Fazio, F., Kerrouche, N., Desgranges, B., Eustache, F., Beuthien-Baumann, B., Menzel, C., Schröder, J., Kato, T., Arahata, Y., Henze, M. & Heiss, W. (2002a). Discrimination between Alzheimer dementia and controls by automated analysis of multicenter FDG PET, *Neuroimage* 17(1): 302–316.

Herholz, K., Schopphoff, H., Schmidt, M., Mielke, R., Eschner, W., Scheidhauer, K., Schicha, H., Heiss, W. & Ebmeier, K. (2002b). Direct Comparison of Spatially Normalized PET and SPECT Scans in Alzheimer's Disease, *Journal of Nuclear Medicine* 43(1): 21–26.

Higdon, R., Foster, N. L., Koeppe, R. A., DeCarli, C. S., Jagust, W. J., Clark, C. M., Barbas, N. R., Arnold, S. E., Turner, R. S., Heidebrink, J. L. & Minoshima, S. (2004). A comparison of classification methods for differentiating fronto-temporal dementia from Alzheimer's disease using FDG-PET imaging, *Statistics in Medicine* 23(2): 315–326.

Höskuldsson, A. (1988). PLS regression methods, *Journal of Chemometrics* 2(3): 211–228.

Hotelling, H. (1931). The generalization of student's ratio, *The Annals of Mathematical Statistics* 2(3): pp. 360–378.

Ishii, K., Sasaki, M., Sakamoto, S., Yamaji, S., Kitagaki, H. & Mori, E. (1999). Tc-99m Ethyl Cysteinate Dimer SPECT and 2-[F-18]fluoro-2-deoxy-D-glucose PET in Alzheimer's Disease: Comparison of Perfusion and Metabolic Patterns, *Clinical Nuclear Medicine* 24(8): 572–575.

Ishii, K., Willoch, F., Minoshima, S., Drzezga, A., Ficaro, E. P., Cross, D. J., Kuhl, D. E. & Schwaiger, M. (2001). Statistical Brain Mapping of 18F-FDG PET in Alzheimer's Disease: Validation of Anatomic Standardization for Atrophied Brains, *Journal of Nuclear Medicine* 42(4): 548–557.

Jolliffe, I. T. (1982). A Note on the Use of Principal Components in Regression, *Journal of the Royal Statistical Society. Series C (Applied Statistics)* 31(3): pp. 300–303.

Kemsley, E. (1996). Discriminant analysis of high-dimensional data: a comparison of principal components analysis and partial least squares data reduction methods, *Chemometrics and Intelligent Laboratory Systems* 33(1): 47 – 61.

Kerrouche, N., Herholz, K., Mielke, R., Holthoff, V. & Baron, J.-C. (2006). 18FDG PET in vascular dementia: differentiation from Alzheimer's disease using voxel-based multivariate analysis, *J Cereb Blood Flow Metab* 26: 1213–1221.

Klunk, W. E., Engler, H., Nordberg, A., Wang, Y., Blomqvist, G., Holt, D. P., Bergström, M., Savitcheva, I., Huang, G.-F., Estrada, S., Ausén, B., Debnath, M. L., Barletta, J., Price, J. C., Sandell, J., Lopresti, B. J., Wall, A., Koivisto, P., Antoni, G., Mathis, C. A. & Långström, B. (2004). Imaging brain amyloid in Alzheimer's disease with Pittsburgh Compound-B, *Annals of Neurology* 55(3): 306–319.

Knyazev, A. V., Merico & Argentati, E. (2002). Principal angles between subspaces in an a-based scalar product: Algorithms and perturbation estimates, *SIAM J. Sci. Comput* 23: 2009–2041.

Krishnan, A., Williams, L. J., McIntosh, A. R. & Abdi, H. (2011). Partial Least Squares (PLS) methods for neuroimaging: A tutorial and review, *NeuroImage* 56(2): 455 – 475.

Markiewicz, P. J., Matthews, J. C., Declerck, J. & Herholz, K. (2009). Robustness of multivariate image analysis assessed by resampling techniques and applied to FDG-PET scans of patients with Alzheimer's disease, *NeuroImage* 46(2): 472 – 485.

Markiewicz, P., Matthews, J., Declerck, J. & Herholz, K. (2011a). Robustness of correlations between PCA of FDG-PET scans and biological variables in healthy and demented subjects, *NeuroImage* 56(2): 782 – 787. Multivariate Decoding and Brain Reading.

Markiewicz, P., Matthews, J., Declerck, J. & Herholz, K. (2011b). Verification of predicted robustness and accuracy of multivariate analysis, *NeuroImage* 56(3): 1382 – 1385.

Martínez, A. M. & Kak, A. C. (2001). PCA versus LDA, *IEEE Trans. Pattern Anal. Mach. Intell.* 23: 228–233.

Massy, W. F. (1965). Principal Components Regression in Exploratory Statistical Research, *Journal of the American Statistical Association* 60(309): pp. 234–256.

Matsuda, H. (2007). Role of Neuroimaging in Alzheimer's Disease, with Emphasis on Brain Perfusion SPECT, *Journal of Nuclear Medicine* 48(8): 1289–1300.

Matsuda, H., Kitayama, N., Ohnishi, T., Asada, T., Nakano, S., Sakamoto, S., Imabayashi, E. & Katoh, A. (2002). Longitudinal Evaluation of Both Morphologic and Functional Changes in the Same Individuals with Alzheimer's Disease, *Journal of Nuclear Medicine* 43(3): 304–311.

McCrory, S. J. & Ford, I. (1991). Multivariate analysis of spect images with illustrations in Alzheimer's disease, *Statistics in Medicine* 10(11): 1711–1718.

McEvoy, L. K., Fennema-Notestine, C., Roddey, J. C., Hagler, D. J., Holland, D., Karow, D. S., Pung, C. J., Brewer, J. B. & Dale, A. M. (2009). Alzheimer Disease: Quantitative Structural Neuroimaging for Detection and Prediction of Clinical and Structural Changes in Mild Cognitive Impairment1, *Radiology* 251(1): 195–205.

McIntosh, A. R. & Lobaugh, N. J. (2004). Partial least squares analysis of neuroimaging data: applications and advances, *NeuroImage* 23, Supplement 1(0): S250 – S263.

Merhof, D., Markiewicz, P. J., Declerck, J., Platsch, G., Matthews, J. C. & Herholz, K. (2009). Classification Accuracy of Multivariate Analysis Applied to 99mTc-ECD SPECT Data in Alzheimer's Disease Patients and Asymptomatic Controls, *Nuclear Science Symposium Conference Record (NSS/MIC)* pp. 3721–3725.

Merhof, D., Markiewicz, P. J., Platsch, G., Declerck, J., Weih, M., Kornhuber, J., Kuwert, T., Matthews, J. C. & Herholz, K. (2011). Optimized data preprocessing for multivariate analysis applied to 99mTc-ECD SPECT data sets of Alzheimer's patients and asymptomatic controls, *J Cereb Blood Flow Metab* 31(1): 371–383.

Minati, L., Edginton, T., Grazia Bruzzone, M. & Giaccone, G. (2009). Reviews: Current Concepts in Alzheimer's Disease: A Multidisciplinary Review, *American Journal of Alzheimer's Disease and Other Dementias* 24(2): 95–121.

Miranda, A. A., Borgne, Y.-A. & Bontempi, G. (2008). New Routes from Minimal Approximation Error to Principal Components, *Neural Process. Lett.* 27: 197–207.

Miyashita, Y., Itozawa, T., Katsumi, H. & Sasaki, S.-I. (1990). Comments on the NIPALS algorithm, *Journal of Chemometrics* 4(1): 97–100.

Moeller, J. R., Strother, S. C., Sidtis, J. J. & Rottenberg, D. A. (1987). Scaled Subprofile Model: A Statistical Approach to the Analysis of Functional Patterns in Positron Emission Tomographic Data, *Jornal of Cerebral Blood Flow & Metabolism* 7(5): 649–658.

Mueller, S. G., Weiner, M. W., Thal, L. J., Petersen, R. C., Jack, C., Jagust, W., Trojanowski, J. Q., Toga, A. W. & Beckett, L. (2005). The Alzheimer's Disease Neuroimaging Initiative, *Neuroimaging Clinics of North America* 15(4): 869 – 877.

Pagani, M., Salmaso, D., Rodriguez, G., Nardo, D. & Nobili, F. (2009). Principal component analysis in mild and moderate Alzheimer's disease – A novel approach to clinical diagnosis, *Psychiatry Research: Neuroimaging* 173(1): 8 – 14.

Peres-Neto, P. R., Jackson, D. A. & Somers, K. M. (2005). How many principal components? stopping rules for determining the number of non-trivial axes revisited, *Computational Statistics & Data Analysis* 49(4): 974 – 997.

Petrella, J. R., Coleman, R. E. & Doraiswamy, P. M. (2003). Neuroimaging and Early Diagnosis of Alzheimer Disease: A Look to the Future, *Radiology* 226(2): 315–336.

Scarmeas, N., Habeck, C. G., Zarahn, E., Anderson, K. E., Park, A., Hilton, J., Pelton, G. H., Tabert, M. H., Honig, L. S., Moeller, J. R., Devanand, D. P. & Stern, Y. (2004).

Covariance PET patterns in early Alzheimer's disease and subjects with cognitive impairment but no dementia: utility in group discrimination and correlations with functional performance, *NeuroImage* 23(1): 35 – 45.

Selkoe, D. J. (2001). Alzheimer's disease: Genes, proteins, and therapy, *Physiological Reviews* 81(2): 741–766.

Serneels, S. & Verdonck, T. (2008). Principal component analysis for data containing outliers and missing elements, *Computational Statistics & Data Analysis* 52(3): 1712 – 1727.

Sugiura, N. (1978). Further analysis of the data by akaike' s information criterion and the finite corrections, *Communications in Statistics - Theory and Methods* 7(1): 13–26.

Tabert, M. H., Manly, J. J., Liu, X., Pelton, G. H., Rosenblum, S., Jacobs, M., Zamora, D., Goodkind, M., Bell, K., Stern, Y. & Devanand, D. P. (2006). Neuropsychological Prediction of Conversion to Alzheimer Disease in Patients With Mild Cognitive Impairment, *Arch Gen Psychiatry* 63(8): 916–924.

Van Huffel, S. (1997). *Recent Advances in Total Least Squares Techniques and Errors-in-Variables Modeling*, SIAM, Philadelphia, PA.

Wold, S., Geladi, P., Esbensen, K. & Öhman, J. (1987). Multi-way principal components-and PLS-analysis, *Journal of Chemometrics* 1(1): 41–56.

Zuendorf, G., Kerrouche, N., Herholz, K. & Baron, J.-C. (2003). Efficient principal component analysis for multivariate 3D voxel-based mapping of brain functional imaging data sets as applied to FDG-PET and normal aging, *Human Brain Mapping* 18(1): 13–21.

# 11

# Public Parks Aesthetic Value Index

M. K. Mohamad Roslan and M. I. Nurashikin
*Universiti Putra Malaysia*
*Malaysia*

## 1. Introduction

### 1.1 Aesthetic

Aesthetics is a branch of philosophy dealing with the nature of beauty, art, and taste with the creation and appreciation of beauty. It is more scientifically defined as the study of sensory or sensory-emotional values. Broadly, scholars in the field define aesthetics as critical reflection on art, culture and nature.

Traditionally, aesthetics is specific to the philosophy of art on level of beauty of any object but later, it was expanded to include the sublime with careful analyses on the characteristics. Studies on the aesthetics level experience and its attitude being conducted had lead to a survey of the scope of aesthetics together with an account of redefinition of the term.

According to Kant (1964), the theory of beauty has four clear aspects which are freedom from concepts, objectivity in the judgment, the disinterest of the viewer, and evenness in obligation. Freedom from concepts refers the cognitive powers of human understanding and imagination judge in identifying an object, such as stone. In this situation, the replacement of a certain individual object by a group of object would bring the experience of pure beauty when the cognitive powers are held to be in harmonious. Consequently, this results the second aspect that is the objectivity in the judgment since the cognitive powers are common to all who can classify that group of object. However, this was not the basis on which the apprehension of pure beauty was obligatory. According to Kant (1964), it derived from its disinterest because pure beauty not necessarily brings the fully agreement of the visitor to its aesthetic level nor does it raise any desire to possess the object but it just make satisfaction in a distinctive intellectual way.

#### 1.1.1 Aesthetic value

Aesthetics value can be defined as theory of the level of beauty of a certain natural resources (Slater, 2006). This term appear when there was some interest among researchers to assess the level of aesthetic quantitatively instead of just qualitatively. It is about the objectivity and universality of judgments of pure beauty. It can also be defined as the individual judgment of the quality of beauty.

The main debate over aesthetic value concerns on social and political matters, in many different points of view. The central question concerns whether there are existing aesthetics expertises that have aesthetic interests and whether their view represents a fair view since, from a sociological perspective, it only just a portion of the whole population.

### 1.1.2 Environmental aesthetics

This paper would explain on public parks which are related to environment. In this case, according to Fisher (2003), environmental aesthetics applies naturally in the study of the aesthetics of nature conducted under the influence of environmental aspects concerns. The rapid growth of concern for the natural environment over the last third of the 20th century has brought the welcome reintroduction of nature as a significant topic in aesthetics such as the aesthetic interactions with nature, the aesthetic value of nature, and the status of art about nature. Nature began to be seen as comprising landscapes compelling in their own wild beauty and objects valuable in their smallest natural detail.

Environmental aesthetics consists of two terms which are environmental roots that emphasize on the aesthetics of nature, and aesthetic appreciation that focusing on the notion of environments of all sorts as objects of appreciation (Fisher, 2003). Based on these, nature is regarded not as an object to be exploited, but as something with an autonomous and worthy existence in itself. A thing is right when it tends to preserve the integrity, stability, and beauty of the biotic community but is wrong if otherwise.

According to Berleant (1998), environmental aesthetics applies to the researches on the aesthetics of nature of all sorts of environments either human made or natural conducted under the influence of environmental concerns. This environmental aesthetics also incorporates city planning, landscape architecture and environmental design, and it is significant because, whether applied to nature or built environments because it directly associating the modern structure or design to the standard theories in aesthetics. In short, Thompson (1995) found that although the authorities have an obligation to preserve beautiful high aesthetic value parks, they also have an obligation to preserve aesthetically valuable nature areas.

### 1.1.3 Aesthetic value judgment

Judgments of aesthetic value rely on our ability to discriminate at a sensory level. Aesthetics examines our affective domain response to an object or phenomenon. According to Slote (1971), a personal view on the level of aesthetic cannot represent the view of the community because each individual in the community also have their own view unless all of them have the same perception.

Since beauty is not a tangible object, judgment of aesthetic value is subjective to any individual perception. For example, it can be related to desirability or expectation on certain aspects of the nature. This assessment would also be based on many other kinds of issues such as senses, emotions, intellectual opinions, culture, preferences, values and behavior (Slater, 2006).

Aesthetic judgment is not only focusing on sensory discrimination but also linked to capacity for pleasure (Kant, 1964). The sensory discrimination that raising the pleasure would results the enjoyment. However, in order to judge something as having high aesthetic value, it needs a third requirement, that is it must involve the viewer's capacities of reflective contemplation. In short, judgments of aesthetic involved sensory, emotional and intellectual all at once.

### 1.1.4 Factors of aesthetic value judgment

There are some factors that could contribute in viewer view or perception on aesthetic value level. Negative reaction such as disgust shows that sensory detection is linked to body language, facial expressions, and behaviors (Slote, 1971). Aesthetic judgments also linked to emotions that partially embodied in physical reactions. For example, watching a well-design landscape parks may give the viewer a reaction of awe and consequently, physical reactions.

In the other hand, aesthetic judgments may be affected by culture or norm of the community. Other factors that could affect the judgments are desirability, economic, political, or moral value.

Aesthetic judgments can be highly parallel or highly contradict between viewers (Mothersill, 2004). In the other hand, aesthetic judgments were partly intellectual and interpretative. Therefore aesthetic judgments were usually based on the senses, emotions, intellectual opinions, will, desires, culture, preferences, values, subconscious behavior, conscious decision, training, instinct, sociological institutions, or combinations.

## 1.2 Index

Index is a numerical standardized value of evaluation of a variable that is in composite form, meaning that this variable was formed through the integration or combination of several other variables.. The evaluation process of this composite variable is not an easy process since there is no standard value used as a base of comparison of the evaluation. Therefore the indices are the best way to be introduced to determine that particular standard value.

A Public Parks Aesthetic Value Index, proposed in this paper is a measurement that aggregates numerous aesthetic value indicators into one consolidates and objective value representing the status of aesthetic value in public parks in relation to some specific use. The mode of assessment was carried out with a full consideration on certain criteria as suggested by Haslina (2006).

## 1.3 Public parks

Public parks is a bounded area of land, either in the form of natural or man-made state, dedicated to recreation use and generally characterized by its natural historic or landscape features. It is used for both passive and active forms of recreation and may be designated to serve the residents of a neighbourhood, community, region or province. This area is maintained by a local or private authority for recreation which can stimulate the morale and transform the cities to become more attractive and entertaining to visitors (Cranz, 1982).

Parks commonly resembles a certain formation of landscaping which any individuals who visit it would find more relaxing and feeling enjoyment of any recreation activities, for example picnics and sporting activities. Many smaller neighbourhood parks are receiving increased attention and valuation as significant community assets and places in heavily populated urban areas.

Traditionally, public parks are classified by their horticulture design, recreational value and public friendliness. It often provide leisure and sports areas for children and teenagers. It may also equip with semi natural or man-made habitats such as woodlands, heaths and wetlands. The formal garden designs and high composition of flora species characterized the horticultural history of public parks.

Rapid population growth rate and prolonged economic expansion consequently results a high degree of urbanization which could led to serious degradation of living environment and public amenities including air and water pollution, expanding temperature diffusion and decrement of public open space areas.

In order to have the functioning public parks in giving relaxation, enjoyment and pleasuring to the community, besides a good management and proper location, public parks should have a good aesthetic value. Aesthetic value for public parks can be defined as the level of beauties of natural beauty of public parks. However, the assessment of aesthetic value cannot be done in a straight forward approach because as mention earlier, this term is subjective to qualitative variables which are in composite form that means it consists of several aspects that need to combine together. Therefore the index of aesthetics value needs to be applied and measure to assess the aesthetic value of the parks. Index is a numerical standardized value of the composite form.

### 1.3.1 Characteristics of aesthetic values in public parks

Aesthetic values in Public Parks are normally characterized with five scopes; tree, fauna, lake, flower and building (Haslina, 2006).

#### 1.3.1.1 Tree

The basic value of the tree in public parks could be based on emotion, aesthetic or strictly utilitarian. However, visitors seldom perceive the value as strictly aesthetic or monetary gain. In some areas, public used to spend on tree care and management which reflects the approximate value of trees. Trees may be selected to be planted in the public parks based on texture, fragrance, size, shape and colour. The replanting of best and cost effective tree species could create beauty environment and pull the visitors to come.

#### 1.3.1.2 Fauna

Fauna is one of the aesthetic aspects in public parks. The existing of fauna species such as birds and fish would make the parks more harmonious, relaxing and enjoyable. Fauna species could revitalized flora survival needs and hence support each other. In the other hand, the abundance of greens in public parks could cause a significant increment in wildlife population.

#### 1.3.1.3 Lake

Public parks, especially in urban areas are normally provides the visitors with lake as a choice for recreation. The physical attractiveness of a particular area contributes to its aesthetic value. The well-managed lake would provide the visitors not only the natural visual element but also a serene sound and smell. These aesthetic characteristics of the lake make up the reason that visitors might find lake as beautiful and attractive.

#### 1.3.1.4 Flower

The function of flowers in public parks is well-documented. The evolving, flourishing process of civilization has increased the usage of flower in the form of civility, art and religion. For example, the Japanese people regarded flowers as materials for spiritual expressions (Haslina, 2006).

#### 1.3.1.5 Building

Besides location, the appropriate design for the intended function of the building is one of the important elements of value The style and functional utility of a building are necessarily interrelated to create a success. In the valuation process, the aspects that are critical are functionality of the layout, design attractiveness, the appropriateness of the material and its quality and workmanship applied.

### 1.4 Elements of design aesthetic value

According to Janick (2010), there are four main design elements which become the visible features of all objects. They are as follows.

a. Colour - Refer to visual sensation produced by different light wavelength that may be described in term of its hue, intensity and chroma.
b. Texture - Refer to the visual effect of tactile surface qualities.
c. Form - Refer to the shape and structure of a three dimensional object
d. Line - Refer to one dimensional interpretation of form that delimits the shape and structure

### 1.5 Aim of study

Non-existence of this public parks aesthetic value index in Malaysia brings the issue of the uncertainty whether the existing public parks are functioning well in giving relaxation, enjoyment and pleasuring to the community and visitors. This study aims to introduce an index to assess the level of aesthetics value of public parks and applied it to assess the aesthetic value level of a certain public parks in Malaysia.

## 2. Methodology

### 2.1 Study area

This research had been conducted in Shah Alam Lake Garden, Selangor, Malaysia which is one of the popular recreational public parks among domestic tourists in Selangor (Fig. 1). This garden, consists of three man-made lakes is located in the city centre and provides the community and visitors with water-sports besides some basic leisure activities. This urban public parks, managed by government has a beautiful, green and unique landscape with the presence of many different bird species.

### 2.2 Data collection

Due to the assessment of aesthetics value is subjective to individual's perception as mention earlier, the first stage of this research is a qualitative study with questionnaire as a tool of data collection.

Fig. 1. The map of Shah Alam Lake Garden

The field survey was conducted on weekdays, weekend and public holidays. The respondents involved those who were present at the lakes at the time when questionnaires were given. Besides that, the respondent age of above 16 years old was also been point out in order to ensure that the respondents could give the grading to the aspects fairly. The questionnaires were designed based on the information obtain through discussions with experts. The questions provided in the questionnaires were straight forward and highly focusing on the aim that was to assess the respondent perception on the aesthetic level of the parks.

This questionnaire has two parts, Part 1 and Part 2. Part 1 involved the ranking of importance of five aspects in study while Part 2 involved the ranking of aesthetic value of each aspects attribute. Likert scale was used as aesthetic value grading scale for each sub-aspect from 1 to 10 where 10 means very high value. Moreover, Likert scale was also used as importance scale for each aspect from 1 to 5 where 5 means not important at all. Five aspects in study were as follows.

a. Aesthetic value of Tree

After the final round of discussion with experts, this section focused on the form, colour, shape, scent, sound, appearance, function, site selection, and quantities and diversity of species that could influencing the visitors to visit the park.

b. Aesthetic value of Animal or Fauna

After the final round of discussion with experts, this section focused on the quantities and diversity of species, sound, disturbance, smell, colour, shape, habitat and its function in the parks.

c. Aesthetic value of Lake

After the final round of discussion with experts, this section focused on water quality in term of presence of fish, colour, lake design, sound, disturbance, odor, site selection and its function to visitors.

d. Aesthetic value of Flower

After the final round of discussion with experts, this section focused on type of species, shape, form, colour, sound, aroma, site selection, design, quantities and its function.

e. Aesthetic value of Buildings

After the final round of discussion with experts, this section focused on design, level of cleanliness, material quality, number of buildings, suitability of the location and its functionality.

### 2.3 Sampling technique

Sampling was conducted using convenience sampling techniques. In convenience sampling, the sample selection is based on availability and accessibility. This technique is quite suitable in this study due to the high mobility of visitors and also unknown activities among visitors.

After calculations of sample size based on the population size obtained, it was found that as many as 407 respondents need to be taken as a sample size from the whole population. Although convenience sampling technique was applied, the sample of respondents was still taken from the whole area of the lake to ensure that the results would represent the whole visitors of the lake from all areas. Although the likelihood of the sample being unrepresentative of the visitor population would be quite high, the information could still provide a fairly significant insight and become a good source of data in exploratory research.

### 2.4 Data analysis

Second stage of this research involved quantitative study where all the qualitative data being analyzed and quantify to give a certain value to indicate the aesthetic level. Descriptive analysis was first used to determine the importance of each aspect in determining the aesthetic level. Table 1 shows the results of the analysis. It was found that the most important aspect for the respondents to visit the parks was tree with mean value of 2.00 followed by lake, flower and fauna with mean value of 2.34, 3.06 and 3.33 respectively. The aspect of buildings was considered as not important to be including in the assessment with the mean of 4.25. From the observations and feedbacks from visitors, the buildings in the parks doesn't have any uniqueness that can be consider as important enough to attract them to visit. Therefore, the aspect of building would not be analyzed in the second stage.

| Aspects | Mean | Median |
|---|---|---|
| Tree | 2.00 | 2.00 |
| Fauna | 3.33 | 3.00 |
| Lake | 2.34 | 2.00 |
| Buildings | 4.25 | 5.00 |
| Flowers | 3.06 | 3.00 |

Table 1. Descriptive analysis of the aspects in study

### 2.4.1 Principal Component Analysis (PCA)

PCA is a technique for simplifying a data set by reducing multidimensional data sets to lower dimensions for analysis. It is an orthogonal linear transformation that transforms the variance covariance matrices to a new coordinate system such that the greatest variance would lie on the first principal component and so on while at the same time still retaining the characteristics of the data set.

The correlation between variables or aspects and component, indicated by factor loading was used as a basic for classifying the dominant variables in each component. If the factor loading value is more than 0.7, the attribute can be considered as dominant and plays a main role in that component (Nunnally, 1978; Hair, 2009) because it would account for more than 50% of the variance in all variables.

The next step of analysis was the usage of Principal Component Analysis (PCA). The reason of applying this analysis was to extract the main attributes from all attributes of each aspect. This is important as in index development, the number of attributes must be optimum in order to represent a good indicator for assessment tools. The index with a lot of attributes would seldom represent a good status indicator. Moreover, another important step in developing or creating an index is the selection of only the most important or dominant attributes as these attributes play a big role in determining the status of the variable such as the status of aesthetic value of the public parks in this study. Since in PCA, the first component is always dominant compare to the other component, the selection of the main attributes was done only in first component.

However, this dominant component cannot be chose straight away as the indicator used in index as there is still some effect of correlation between components. According to Hair (2009), factor loading is the coefficient used to interpret the role of each variable and factor and the unrotated solution may not provide a true pattern of variable loadings due to the presence of this effect. In order to solve this problem, the whole extracted component need to be rotated. According to Rencher (2002), the most popular rotation technique is the varimax technique which seeks rotated loadings that maximize the variance of the squared loadings in each component. Indirectly, this varimax rotation could eliminate the effect of correlation between components.

As mention earlier, only four aspects were being considered after the aspect of building had been omitted. In the preliminary round, PCA was run separately for all five aspects to determine the main or dominant attributes in each aspect. As mention earlier, the index is the composite variable that was formed through the combination of several other variables. Mohamad Roslan et. al (2007) found that the index that can best explaining the real condition or situation is the index that focused on dominant variables in each scope or classification. In this study, it refers to dominant variables in the scope of tree, fauna, lake and flower.

Table 2 shows the component structure of trees after the varimax rotation process of PCA. The Table reports the factor loadings for each variable on the components between the item and the rotated factor. In this case, the factor loading value of 0.7 was chosen as as threshold value explained earlier. It was found that the attributes of maintenance, species selection and form has the factor loading value of 0.773, 0.766 and 0.741 respectively which are higher than the threshold value.

| Attributes | Component | |
|---|---|---|
| | Component 1 | Component 2 |
| Form | 0.741 | -0.027 |
| Colour | 0.425 | 0.419 |
| Maintenance | 0.773 | 0.101 |
| Quality | 0.202 | 0.699 |
| Species selection | 0.766 | 0.031 |
| Function | 0.058 | 0.840 |
| Quantity | 0.499 | 0.376 |
| Location | -0.078 | 0.762 |

Table 2. Results of PCA analysis for the aspect of tree

From the feedback of the visitors, different form of trees can create an attractive view to the park while giving an aesthetic visual to them. At the same time, a proper tree maintenance existed in the park becomes an added value in term of aesthetic value.

Table 3 shows the results of PCA analysis for the aspect of animal or fauna. It was found that higher factor loading recorded for habitat and selection of species with 0.872 and 0.839 respectively. From the feedback of visitors, the varieties of species play a pull factor to attract them to the parks.

| Attributes | Component | |
|---|---|---|
| | Component 1 | Component 2 |
| Species selection | 0.839 | -0.025 |
| Quantity | 0.503 | 0.490 |
| Habitat | 0.872 | 0.094 |
| Sound | 0.067 | 0.779 |
| Cleanliness | 0.663 | 0.260 |
| Function | 0.095 | 0.843 |

Table 3. Results of PCA analysis for the aspect of animal or fauna

Table 4 shows the results of PCA analysis for the aspect of lake. There were two attributes that have the value of factor loading more than threshold, colour and quality with 0.905 and 0.903 respectively. From the feedback of the visitors, both attributes are important in influencing their first impression.

| Attributes | Component | |
|---|---|---|
| | Component 1 | Component 2 |
| Quality | 0.903 | 0.136 |
| Design | 0.309 | 0.712 |
| Colour | 0.905 | 0.162 |
| Cleanliness | 0.575 | 0.524 |
| Sound | 0.261 | 0.650 |
| Function | -0.041 | 0.851 |

Table 4. Results of PCA analysis for the aspect of lake

Table 5 shows the results of PCA analysis for the aspect of flower. It was found that only one component could be form from the analysis and from this component, five attributes were major attributes. They were quality, odour, quantity, colour and species selection. These five attributes were related to each other because if choices of species were good enough, it reflected the qualities, quantities, colour and odour. In short, a good quality species produced a good flower.

| Attributes | Component |
|---|---|
| Species selection | 0.746 |
| Quantity | 0.773 |
| Quality | 0.810 |
| Colour | 0.754 |
| Odour | 0.796 |
| Function | 0.586 |

Table 5. Results of PCA analysis for the aspect of flower

From all these Table 2 to Table 5, it was found that most major attributes that could be a major pull factors for the visitors to visit were in the aspect of flower. In the final analysis of PCA, all main attributes in each aspect were analyzed together. In this stage, the data was separate into two parts, 300 of them in one part, used to determine the main attribute in exploratory while another 107 were used in validating the exploratory results. This process was done because the results of this stage would be use in creating the index and therefore need to be validating to ensure that the index is valid to be use. The data set of 300 respondents still covering the whole public parks areas and the same applied for the data set of 107 respondents.

Table 6 shows the results of PCA analysis for data set in determining the main attributes in exploratory. It was found that odour of flower, colour of water, quality of flower, quantity of flower, water quality and colour of flowers were the main attributes from all selection attributes, having the factor loading value of 0.805, 0.766, 0.756, 0.734, 0.704 and 0.700 respectively.

| Attributes | Component | |
|---|---|---|
| | Component 1 | Component 2 |
| Tree formation | 0.141 | 0.800 |
| Tree maintenance | 0.273 | 0.771 |
| Tree species selection | 0.175 | 0.727 |
| Water quality | 0.704 | 0.255 |
| Colour of water | 0.766 | 0.177 |
| Flower species selection | 0.679 | 0.164 |
| Quantity of flower | 0.734 | 0.186 |
| Quality of flower | 0.756 | 0.175 |
| Colour of flower | 0.700 | 0.172 |
| Odour of flower | 0.805 | 0.172 |

Table 6. Results of PCA analysis for data set to determine the main attrbutes

Table 6 shows that from the exploratory results, it was found that six attributes could be taken as aesthetic value index indicators from 26 attributes. They were odour of flower, colour of water, quality of flowers, quantity of flowers, water quality and colour of flowers following the order. This order or arrangement is important in the next stage of analysis. Consequently, running the same process for the validation data set gave the same arrangement of attributes though the factor loading value is not same.

### 2.4.2 Benchmarking analysis

Benchmarking analysis is the analysis used in comparing the current project, methods or processes with the best practices in order to drive improvement. The objective of this analysis is to set an appropriate reliability and quality matrices for the output based on similar present products.

This method had been used by Mohamad Roslan et. al (2007) in their study on the creation of ground water quality index. They found that this method is suitable and highly sensitive. However, in their study, they used the quantitative data of several prospect water quality indicators.

After the analysis of PCA, the next stage is to develop the index. This study would offer some uniqueness in the index formation; the index is not in statistical equation but in graphical feature (Figure 3). Each vertex represents the attributes, extracted from the PCA and arranged in clock-wise order. This graphical feature which is part of benchmarking analysis was drawn to calculate the index with the help of mean analysis. From this graphics, the index was calculated as the percentage of the polygon area drawn using the data to the whole polygon area as shown in the green area in Figure 3.

Procedure of developing the index using Benchmarking Analysis is as follow.

1. Fix the value of analysis with the scale of 0 to 10 for each main attribute extracted from PCA
2. Built the polygon
3. Calculate each triangle (Fig 2) area inside the polygon drawn based on the value from data using formula (1).

$$A=0.5 \times a \times b \times \sin\theta \tag{1}$$

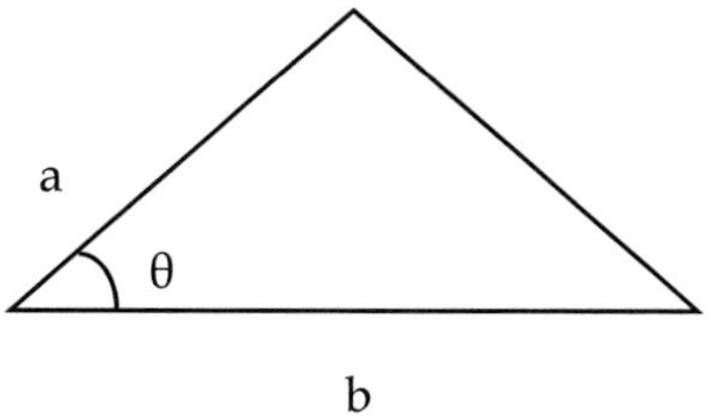


Fig. 2. Triangle in the polygon

4. Determine the ratio of the area using formula (2).

$$IV=A/L \times 100 \tag{2}$$

WhereIV=Index value

A= Total triangle area

L=Total polygon area

5. Evaluate the aesthetic value using the index.

### 2.4.3 Sensitivity analysis

This analysis was also conducted to the model to determine its sensitivity to the change of the value. For example, determine the changing rate of the index when any attribute fall from 10 to 9 in benchmarking scale. In this case, it was found that the index value would fall as much as 3.33% for any single attribute that fall as much as one unit in benchmarking scale.

## 2.5 Creation and calculation of the index in the study

Before the polygon can be drawn, the mean analysis for the main attributes extracted from PCA need to be conduct to determine the level of aesthetic level for each attribute within the Benchmarking scale of 0-10. This analysis involved all respondents without separation. Table 7 shows the mean analysis for six main attributes. It shows that during the study, the attribute of quality of flower has the highest ranking compare to the other five attributes with the value of 5.12.

| Attributes | Mean |
|---|---|
| Water quality | 4.13 |
| Colour of water | 3.99 |
| Quantity of flower | 4.75 |
| Quality of flower | 5.12 |
| Colour of flower | 4.90 |
| Odour of flower | 4.28 |

Table 7. Mean analysis for six main attributes

The values in Table 7 were used in drawing the polygon and consequently the Public Parks Aesthetic Value Index. Fig. 3 shows the polygon used to calculate the index with the scale of 0-10. Using the procedure and formula as mentioned earlier, it was found that the value for Shah Alam Lake Garden Aesthetic Value Index is 20.44.

This study also furnished the index with the grading system; develop using the concept of the polygon area to help in grading the aesthetic value index through benchmarking analysis. Five grades were applied in this product that is excellence, good, satisfactory, poor and very poor (Table 8). These grades were set up using the benchmarking level scale of 9, 7, 5. 3 and 1 for all six attributes.

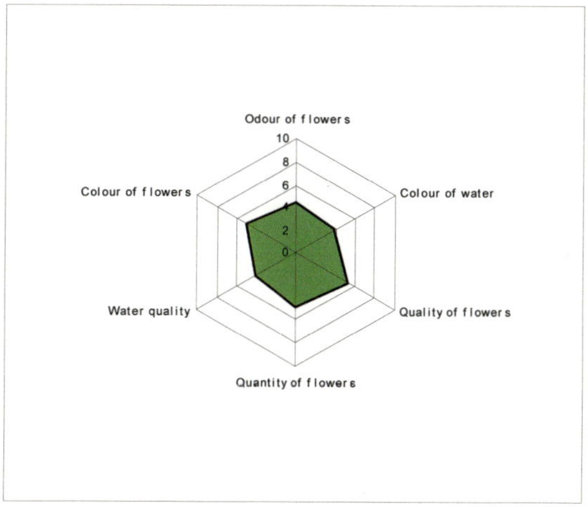


Fig. 3. The index with the mean value for each variables

| Grades | Value range |
|---|---|
| Excellence | 81.00 - 100.00 |
| Good | 49.00 - 80.99 |
| Satisfactory | 25.00 - 48.99 |
| Poor | 9.00 - 24.99 |
| Very poor | 0.00 - 8.99 |

Table 8. Grading system

The final stage of this study is to apply the index to calculate the aesthetic value index of the study area. Based on the grading range (Table 8), the index value for Shah Alam Lake Garden Aesthetic Value Index of 20.44 can be classified as poor. From Figure 3 and Table 7, it was shown that almost all attributes were given grading scales less than 5 except the quality of flower. This contributes the overall grade level.

## 3. Conclusion

Taking into considerations the five criteria related to the assessment of aesthetic value index as mentioned by Haslina (2006) which primarily focuses on the beauties and the 'in' thing, this index would be a valuable tool in the management of aesthetic value in tourist attraction sites such as public parks, lake gardens and recreational forests. The minimal constraint in application made it suitably applied in assessing the aesthetic value under the sun.

This index was created with the data gain during daylight. Therefore, the main indicator variables in the index became dominant during daylight. However, the assessment for the night session may refer to another set of indicator variables. This brings the potential of future study on this aspect.

Another potential scope of future study is to use the quantitative method in measuring the value or grading for each attribute in the index instead of just collecting the data through a large size of respondent perception's. This could be done with the utilization of several high technology equipments.

## 4. References


Brady, E. (2005). *Aesthetics in Practice: Valuing the Natural World.* New York: City University of New York

Berleant, A. (1998). Environmental Aesthetics. In M. Kelly (Ed). *Encyclopedia of Aesthetics.* New York: Oxford University Press, 114-120.

Camp, J.V. (2005). *Freedom of Expression.* New York: The National Endowment for the Arts.

Carlson, A. (2000). *Aesthetics and the Environment: The Appreciation of Nature, Art and Architecture.* New York: Routledge.

Carlson, A. (2010). Environmental aesthetics. In E. Craig (Ed.), *Routledge Encyclopedia of Philosophy.* London: Routledge.

Cranz, G. (1982). *The Politics of Park Design: A History of Urban Parks.* Cambridge Press: Cambridge

Fisher, J. A. (2003). Environmental Aesthetics. In: the *Oxford Handbook of Aesthetics* ed. Levinson, J. Oxford: Oxford University Press

Hair, J. F., Black, W. C., Banin, B. J. & Anderson, R.E. (2009). *Multivariate Data Analysis.* 7th Edition. Prentice Hall: New Jersey.

Haslina, H. (2006). *Visitors Perception on Aesthetic Function of Ornamental Plant in Taman Tasik Titiwangsa, Kuala Lumpur.* UPM Faculty of Forestry Bachelor Thesis.

Hettinger, N. (2005). Allen Carlson's Environmental Aesthetics and the Protection of the Environment. *Environmental Ethics* 27(1): 57-76.

Janick, J. (2010). Plant iconography and art. *Source of information on horticultural technology, 1.*

Kant, I. (1964), *The Critique of Judgement,* trans. J.C.Meredith, Oxford University Press, Oxford.

Kent, A. J. (2005). Aesthetics: A Lost Cause in Cartographic Theory. *The Cartographic Journal.* 42(2). Pg 182-188

Lifestyle Information Network. (2007). Guidelines for Developing Public Recreation Facility Standards. Date of access: 20th October 2011.. http://lin.ca/resource-details/1477.

Mothersill, M. (2004). Beauty and the Critic's Judgment: Remapping Aesthetics. In *The Blackwell Guide to Aesthetics.* New Jersey: Blackwell Publishing Ltd

Mohamad Roslan, M.K., Mohd Kamil, Y., Wan Nor Azmin, S. & Mat Yusoff, A. (2007). Creation of a Ground Water Quality Index for an Open Municipal Landfill Area. *Malaysian Journal of Mathematical Sciences.* 1(2): 181-192

Nunnally, J. C. (1978). *Psychometric theory.* 2nd Edition. New York: McGraw-Hill.

Rencher, A.C. (2002). *Methods of Multivariate Analysis.* 2nd Edition. New York: John Wiley & Sons.

Slater, B.H. (2005). Aesthetics. Internet Encyclopedia of Philosophy. Date of access: 11 June 2011. http://www.iep.utm.edu/aesthetic.

Slote, M. A. (1971). The Rationality of Aesthetic Value Judgments. *The Journal of Philosophy.* 68 (22): 821-839

Thompson, J. (1995). "Aesthetics and the value of nature," *Environmental Ethics* 17, 291-305